Tabulated area = probability

Standard normal (

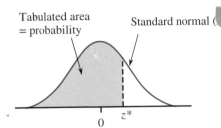

0   z*

z*	.00	.01	.02	.03	.04	.05	.06	.07	.08	.09
0.0	.5000	.5040	.5080	.5120	.5160	.5199	.5239	.5279	.5319	.5359
0.1	.5398	.5438	.5478	.5517	.5557	.5596	.5636	.5675	.5714	.5753
0.2	.5793	.5832	.5871	.5910	.5948	.5987	.6026	.6064	.6103	.6141
0.3	.6179	.6217	.6255	.6293	.6331	.6368	.6406	.6443	.6480	.6517
0.4	.6554	.6591	.6628	.6664	.6700	.6736	.6772	.6808	.6844	.6879
0.5	.6915	.6950	.6985	.7019	.7054	.7088	.7123	.7157	.7190	.7224
0.6	.7257	.7291	.7324	.7357	.7389	.7422	.7454	.7486	.7517	.7549
0.7	.7580	.7611	.7642	.7673	.7704	.7734	.7764	.7794	.7823	.7852
0.8	.7881	.7910	.7939	.7967	.7995	.8023	.8051	.8078	.8106	.8133
0.9	.8159	.8186	.8212	.8238	.8264	.8289	.8315	.8340	.8365	.8389
1.0	.8413	.8438	.8461	.8485	.8508	.8531	.8554	8577	.8599	.8621
1.1	.8643	.8665	.8686	.8708	.8729	.8749	.8770	.8790	.8810	.8830
1.2	.8849	.8869	.8888	.8907	.8925	.8944	.8962	.8980	.8997	.9015
1.3	.9032	.9049	.9066	.9082	.9099	.9115	.9131	.9147	.9162	.9177
1.4	.9192	.9207	.9222	.9236	.9251	.9265	.9279	.9292	.9306	.9319
1.5	.9332	.9345	.9357	.9370	.9382	.9394	.9406	.9418	.9429	.9441
1.6	.9452	.9463	.9474	.9484	.9495	.9505	.9515	.9525	.9535	.9545
1.7	.9554	.9564	.9573	.9582	.9591	.9599	.9608	.9616	.9625	.9633
1.8	.9641	.9649	.9656	.9664	.9671	.9678	.9686	.9693	.9699	.9706
1.9	.9713	.9719	.9726	.9732	.9738	.9744	.9750	.9756	.9761	.9767
2.0	.9772	.9778	.9783	.9788	.9793	.9798	.9803	.9808	.9812	.9817
2.1	.9821	.9826	.9830	.9834	.9838	.9842	.9846	.9850	.9854	.9857
2.2	.9861	.9864	.9868	.9871	.9875	.9878	.9881	.9884	.9887	.9890
2.3	.9893	.9896	.9898	.9901	.9904	.9906	.9909	.9911	.9913	.9916
2.4	.9918	.9920	.9922	.9925	.9927	.9929	.9931	.9932	.9934	.9936
2.5	.9938	.9940	.9941	.9943	.9945	.9946	.9948	.9949	.9951	.9952
2.6	.9953	.9955	.9956	.9957	.9959	.9960	.9961	.9962	.9963	.9964
2.7	.9965	.9966	.9967	.9968	.9969	.9970	.9971	.9972	.9973	.9974
2.8	.9974	.9975	.9976	.9977	.9977	.9978	.9979	.9979	.9980	.9981
2.9	.9981	.9982	.9982	.9983	.9984	.9984	.9985	.9985	.9986	.9986
3.0	.9987	.9987	.9987	.9988	.9988	.9989	.9989	.9989	.9990	.9990
3.1	.9990	.9991	.9991	.9991	.9992	.9992	.9992	.9992	.9993	.9993
3.2	.9993	.9993	.9994	.9994	.9994	.9994	.9994	.9995	.9995	.9995
3.3	.9995	.9995	.9995	.9996	.9996	.9996	.9996	.9996	.9996	.9997
3.4	.9997	.9997	.9997	.9997	.9997	.9997	.9997	.9997	.9997	.9998
3.5	.9998	.9998	.9998	.9998	.9998	.9998	.9998	.9998	.9998	.9998
3.6	.9998	.9998	.9999	.9999	.9999	.9999	.9999	.9999	.9999	.9999
3.7	.9999	.9999	.9999	.9999	.9999	.9999	.9999	.9999	.9999	.9999
3.8	.9999	.9999	.9999	.9999	.9999	.9999	.9999	.9999	.9999	1.0000

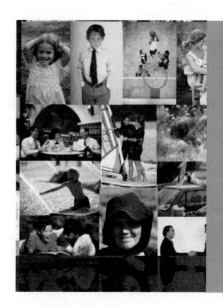

Introduction
to Statistics
and Data Analysis

Roxy Peck
California Polytechnic State University
San Luis Obispo

Chris Olsen
George Washington High School
Cedar Rapids, IA

Jay Devore
California Polytechnic State University
San Luis Obispo

DUXBURY
™
THOMSON LEARNING

Australia • Canada • Mexico • Singapore • Spain • United Kingdom • United States

DUXBURY ™

THOMSON LEARNING

Sponsoring Editor: *Carolyn Crockett*
Assistant Editor: *Seema Atwal*
Marketing Team: *Tom Ziolkowski/ Samantha Cabaluna*
Editorial Assistant: *Ann Day*
Production Editor: *Tessa Avila*
Production Service: *Susan L. Reiland*
Manuscript Editor: *Christine Levesque*

Permissions Editor: *Fiorella Ljunggren*
Interior Design: *Andrew Ogus*
Cover Design: *Laurie Albrecht*
Interior Illustration: *Lori Heckelman*
Print Buyer: *Vena Dyer*
Typesetting: *G & S Typesetters*
Cover Printing: *Phoenix Color Corp.*
Printing and Binding: *R. R. Donnelley/Crawfordsville*

For more information about this or any other Duxbury product, contact:
DUXBURY
511 Forest Lodge Road
Pacific Grove, CA 93950 USA
www.duxbury.com
1-800-423-0563 (Thomson Learning Academic Resource Center)

Library of Congress Cataloging-in-Publication Data
Peck, Roxy.
   Introduction to statistics and data analysis / Roxy Peck, Chris Olsen, Jay Devore.
      p. cm.
   Includes bibliographical references and index.
   ISBN 0-534-37092-6
   1. Statistics.   I. Olsen, Chris, 1948–   II. Devore, Jay L.   III. Title.
QA276.12.P42 2000
519.5 — dc21

00-033734

To my mother, Lucelle.
R.P.

To my wife, Sally, and my daughter, Anna.
C.O.

To Carol, Allie, and Teri.
J.D.

Contents

Preface

Introduction to Statistics and Data Analysis is intended for use as a textbook in introductory statistics courses at two- and four-year colleges and universities, as well as the Advanced Placement Statistics course as presented in the Advanced Placement Course Description. We believe that the following special features of our book distinguish it from other texts.

Features

A Traditional Structure with a Modern Flavor

The topics included in almost all introductory texts are here also. However, we have interwoven some new strands that reflect current and important developments in statistical analysis. These include coverage of sampling and experimental design, the role of graphical displays as an important component of data analysis, transformations, residual analysis, normal probability plots, and simulation. The organization gives instructors considerable flexibility in deciding which of these topics to include in a course.

The Use of Real Data and the Importance of Context

Many students are skeptical of the relevance and importance of statistics. Contrived problem situations and artificial data often reinforce this skepticism. A strategy that we have employed successfully to motivate students is to present examples and exercises that involve data extracted from journal articles, newspapers, and other published sources. Most examples and exercises in the book are of this nature; they cover a very wide range of disciplines and subject areas. These include, but are not limited to, health and fitness, consumer research, psychology and aging, environmental research, law and criminal justice, and entertainment.

Statistics is not about numbers; it is about data — numbers in context. It is the context that makes a problem meaningful and something worth considering. Examples and exercises with overly simple settings do not allow students to practice interpreting results in authentic situations or give students the experience necessary to be able to use statistical methods in real settings. We believe that the exercises and examples are a particular strength of this text, and we invite you to compare the examples and exercises with those in other introductory statistics texts.

Mathematical Level and Notational Simplicity

A good background in second-year algebra constitutes sufficient mathematical preparation for reading and understanding the material presented herein. We want students to focus on concepts without having to grapple unnecessarily with the manipulation of formulas and symbols. To achieve this, we have sometimes used words and phrases in addition to and in place of symbols, as shown here in material from pages 131 and 496. (See page 460 for another example.)

EXAMPLE 4.17   Suppose that two graduating seniors, one marketing major and one accounting major, are comparing job offers. The accounting major has an offer for $35,000 per year, and the marketing student has one for $33,000 per year. Summary information about the distribution of offers follows:

Accounting: mean = 36,000   standard deviation = 1500

Marketing: mean = 32,500   standard deviation = 1000

Then,

$$\text{accounting } z \text{ score} = \frac{35,000 - 36,000}{1500} = -.67$$

(so $35,000 is .67 standard deviation below the mean), whereas

$$\text{marketing } z \text{ score} = \frac{33,000 - 32,500}{1000} = .5$$

Relative to the appropriate data sets, the marketing offer is actually more attractive than the accounting offer (though this may not offer much solace to the marketing major).

Summary of Large-Sample *z* Test for π

Null hypothesis: H_0: π = hypothesized value

Test statistic: $z = \dfrac{p - \text{hypothesized value}}{\sqrt{\dfrac{(\text{hypothesized value})(1 - \text{hypothesized value})}{n}}}$

Alternative hypothesis:

H_a: π > hypothesized value

H_a: π < hypothesized value

H_a: $\pi \neq$ hypothesized value

P-value:

Area under *z* curve to right of calculated *z*

Area under *z* curve to left of calculated *z*

(i) $2 \cdot$ (area to right of *z*) if *z* is positive

(ii) $2 \cdot$ (area to left of *z*) if *z* is negative

Assumptions: 1. *p* is the sample proportion from a *random sample*.
2. The *sample size is large*. This test can be used if *n* satisfies both
$n(\text{hypothesized value}) \geq 10$ and $n(1 - \text{hypothesized value}) \geq 10$.

For those who are apprehensive about their mathematical skills, we trust the verbal descriptions are not only faithful to the statistical concepts but also stepping-stones to the more precise mathematical descriptions.

The Use of Technology

The computer has brought incredible statistical power to the desktop of every investigator. The wide availability of statistical computer packages such as MINITAB, S-Plus, JMP, and SPSS, and the graphical capabilities of the modern microcomputer have transformed both the teaching and learning of statistics. To highlight the role of the computer in contemporary statistics, we have included sample output throughout the book, such as that shown here from pages 185 and 512.

FIGURE 5.21 MINITAB output for the data of Example 5.14

Regression Analysis

The regression equation is
Range of Motion = 108 + 0.871 Age

Predictor	Coef	StDev	T	P
Constant	107.58	11.12	9.67	0.000
Age	0.8710	0.4146	2.10	0.062

S = 10.42   R-Sq = 30.6%   R-Sq(adj) = 23.7%

Analysis of Variance

Source	DF	SS	MS	F	P
Regression	1	479.2	479.2	4.41	0.062
Residual Error	10	1085.7	108.6		
Total	11	1564.9			

SSResid ← 1085.7

SSTo ← 1564.9

9. Because the P-value $> \alpha$, we fail to reject H_0. There is not sufficient evidence to conclude that the mean time spent in personal use of company technology is greater than 75 minutes per day for this company.

MINITAB could also have been used to carry out the test, as shown in the accompanying output.

T-Test of the Mean
Test of mu = 75.00 vs mu > 75.00

Variable	n	Mean	StDev	SE Mean	t	P
Time	10	74.80	9.45	2.99	-0.07	0.53

Although we had to round the computed t value to $-.1$ to use Appendix Table IV, MINITAB was able to compute the P-value corresponding to the actual value of the test statistic.

In addition, numerous exercises contain data that can easily be analyzed by computer, though our exposition firmly avoids a presupposition that students have access to a particular statistical package.

The recent appearance of hand-held calculators with significant statistical and graphing capability has also changed statistics instruction in classrooms where access to computers is still limited. The computer revolution of a previous generation is now being writ small — or, possibly we should say, small*er* — for the youngest generation of investigators. There is not, as we write, anything approaching universal or even wide agreement about the proper role for the calculator in college statistics classes, where access to a computer is virtually assured. At the same time, to tens of thousands of students in Advanced Placement Statistics in our high schools, the calculator is the only dependable access to statistical technology.

This text allows either approach and presents the power of the calculator in a series of Graphing Calculator Explorations. These are placed at the end of each chapter, unobtrusive to those instructors whose technology preference runs to computer packages while accessible to those instructors and students comfortable with this technology. As with computer packages, our exposition firmly avoids pointing to a particular calculator and presents the calculator capabilities in a generic format; specifically, we do not teach particular keystroke sequences, believing that the best source for such specific information is the calculator manual. As much as possible, the calculator explorations are independent of each other, allowing instructors to pick and choose calculator topics that are more relevant to their particular courses.

Advanced Placement Statistics

We have designed this book with a particular eye toward the syllabus of the Advanced Placement Statistics course and the needs of high school teachers and students. Concerns expressed and questions asked in teacher workshops and on the apstat-l listserv have strongly influenced our exposition of certain topics, especially in the area of experimental design and probability. We have taken great care to provide precise definitions and clear examples of concepts that AP Statistics instructors have acknowledged as difficult for their students. We have also expanded the variety of examples and exercises, recognizing the diverse potential futures envisioned by very capable students who have not yet focused on a college major.

Topic Coverage

Our book can be used in courses as short as one quarter or as long as one year in duration. Particularly in shorter courses, an instructor will need to be selective in deciding which topics to include and which to set aside. The book divides naturally into four major sections: collecting data and descriptive methods (Chapters 2–5), probability material (Chapters 6–8), the basic one- and two-sample inferential techniques (Chapters 9–12), and more advanced inferential methodology (Chapters 13–15). We have joined a growing number of books in including an early chapter (Chapter 5) on descriptive methods for bivariate numerical data. This early exposure raises questions and issues that should stimulate student interest in the subject; it is also advantageous for those teaching courses in which time constraints preclude covering advanced inferential material. However, this chapter can easily be postponed until the basics of inference have been covered, and then combined with Chapter 13 for a unified treatment of regression and correlation.

With the possible exception of Chapter 5, Chapters 1–10 should be covered in order. We anticipate that most instructors will then continue with the two-sample material of Chapter 11, although regression could be covered before either of these. Analysis of variance (Chapter 15) and/or categorical data analysis (Chapter 12) can be discussed prior to the regression material of Chapters 13–14. In addition to flexibility in the order in which chapters are covered, material in some sections can be skipped entirely or in part postponed. The authors would be happy to provide more detailed suggestions concerning coverage.

A Note on Probability

The content of the probability chapters is consistent with the Advanced Placement Statistics course description. It includes both a traditional treatment of probability and probability distributions at an introductory level, as well as a section on the use of simulation as a tool for estimating probabilities. For those who prefer a briefer and more informal treatment of probability, the book *Statistics: The Exploration and Analysis of Data,* 4th edition, by Jay Devore and Roxy Peck (also published by Duxbury Press) may be a more appropriate choice. Except for the treatment of probability and the omission of the Graphing Calculator Explorations, it parallels the material in this text.

Acknowledgments

Many people have made valuable contributions to the preparation of this book. Carolyn Crockett, our editor at Duxbury, has been tremendously supportive throughout, as has our wonderful production coordinator, Susan Reiland. Many others at Duxbury have been helpful in bringing this project to completion, and we thank them for their support and help. Steve Rein performed an accuracy check and made many good suggestions for improving our manuscript.

We are also grateful to the constructive suggestions that came from the following manuscript reviewers: Duane C. Hinders, Gunn High School, Palo Alto, California; Lee E. Kucera, Capistrano Valley High School, Mission Viejo, California; Pamela Martin, Northeast Louisiana University; Susan B. Mead, Hockaday School, Dallas, Texas; Paula L. Phillips, North Olmstead High School, North Olmstead, Ohio; and Daren Starnes, The Webb Schools, Claremont, California.

And, as always, we thank our families, friends, and colleagues for their continued support.

Roxy Peck
Chris Olsen
Jay Devore

1

The Role of Statistics

INTRODUCTION

Statistical methods for summary and analysis provide investigators with powerful tools for making sense out of data. Statistical techniques are being employed with increasing frequency in business, medicine, agriculture, social sciences, natural sciences, and applied sciences such as engineering. The pervasiveness of statistical analyses in such diverse fields has led to increased recognition that statistical literacy — a familiarity with the goals and methods of statistics — is a basic component of a well-rounded educational program. In this chapter, we begin by introducing some of the terminology of data analysis and consider the nature and role of variability in statistical analyses.

1.1   Three Reasons to Study Statistics

Because of the widespread use of statistical analysis to organize, summarize, and draw conclusions from data, it is clear that a familiarity with statistical techniques and statistical literacy in general is vital in today's society. Here are three important reasons why it's a good idea for everyone to have a basic understanding of statistics, and why many college majors require a course in statistics.

The First Reason: Being an Informed "Information Consumer"

In today's society, we are bombarded with numerical information in news, in ads, and even in conversation. How do we decide whether claims based on numerical information are reasonable? Here are just a few examples from one day's news (*Los Angeles Times* and *San Francisco Examiner,* May 28, 1995):

■ *A statistical analysis of U.S. public high school reading and math scores.* This analysis showed that, in spite of headlines bemoaning the decline in SAT scores, scores in both reading and math have gone up. The study, conducted by the Rand Corporation, pointed out that SAT scores measure the achievement of college-bound students, who make up less than half of all high school students. U.S. Department of Education Achievement (USDEA) Tests, however, are taken by all students, not just those preparing for college. Analysis of USDEA test scores showed that both reading and math scores were higher in 1995 than in 1970, with the greatest gains reported for blacks and Latinos.

■ *An assessment of the impact of affirmative action on civil service workforce composition in California.* Comparisons based on ethnicity and gender were made using data from 1978 and from 1994. In a separate article, data from a public opinion poll on attitudes toward affirmative action programs was used to support a political candidate's claim that opposition to such programs is growing.

■ *An analysis of change in the average length of nine-inning major-league baseball games.* A graph in the article showed an increase from an average length of 2 hours and 33 minutes in 1981 to an average of 2 hours and 58 minutes in 1994. The analysis prompted some suggestions for shortening game times.

■ *A report on a marketing study that examined the customer base for check-cashing services* (businesses that charge a fee for cashing checks). The study presented data on the reasons people use this type of service rather than a bank, as well as data on client demographics, including race, gender, and age. The reported information is useful in evaluating the potential of a proposed location for such a business.

■ *An annual report on housing prices across the United States.* The average cost of a 2200-square-foot home for 280 cities ranged from a low of $92,125 in Fort Worth, Texas, to a high of $886,000 in Beverly Hills, California. An index for each city that compared the city to the national average also appeared. The article explained how to use these indexes to calculate the cost of replacing a house in one city with a comparable house in another.

To be an informed consumer of such reports, you must be able to do the following: (1) extract information from charts and graphs, (2) follow numerical arguments, and (3) know the basics of how data should be gathered, summarized, and analyzed to draw statistical conclusions.

The Second Reason: Understanding and Making Decisions

No matter what profession you choose, you will almost certainly need to understand statistical information and base decisions on it. Here are some examples:

■ Almost all industries, as well as government and nonprofit organizations, use market research tools, such as consumer surveys, designed to reveal who uses their products or services.

■ Modern science and its applied fields, from astrophysics to zoology, rely on statistical methods for analyzing data and deciding whether various conjectures are supported by observed data. This is true for the social sciences, such as economics and psychology, and increasingly, liberal arts fields, such as literature and history, use statistics as a research tool.

■ In law or government, you may be called on to understand and debate statistical techniques used in another field. Class-action lawsuits can depend on a statistical analysis of whether one kind of injury or illness is more common in a particular group than in the general population. Proof of guilt in a criminal case may rest on statistical interpretation of the likelihood that DNA samples match.

Throughout your professional life, you will have to make informed decisions and assess the risk of various choices. To make these decisions, you must be able to do the following:

1. Decide whether existing information is adequate or whether additional information is required.
2. If necessary, collect more information in a reasonable and thoughtful way.
3. Summarize the available data in a useful and informative manner.
4. Analyze the available data.
5. Draw conclusions, make decisions, and assess the risk of an incorrect decision.

Statistical methods are the tools for accomplishing these steps. You may already informally use these steps to make everyday decisions. Should you go out for a sport that involves the risk of injury? Will your college organization do better by trying to raise funds with a benefit concert or a direct appeal for donations? If you choose a particular major, what are your chances of finding a job when you graduate? How should you select a graduate program based on guidebook ratings that include information on percentage of applicants accepted, time to obtain a degree, and so on?

The Third Reason: Evaluating Decisions That Affect Your Life

An understanding of statistical techniques will allow you to evaluate decisions that affect your well-being. Here are some instances:

■ Insurance companies use statistical techniques to set auto insurance rates, although some states restrict the use of these techniques. Data suggests that young drivers have more accidents than older ones. Should laws or regulations limit how much more the young drivers pay for insurance? What about the common practice of charging higher rates for people who live in inner cities?

■ University financial aid offices survey students on the cost of going to school and collect data on family income, savings, and expenses. The resulting data is used to set criteria used in deciding who receives financial aid. Are the estimates accurate?

■ Medical researchers use statistical methods to make recommendations regarding the choice between surgical and nonsurgical treatment of diseases such as coronary heart disease and cancer. How do they weigh the risks and benefits to reach such a recommendation?

- Many companies now require drug screening as a condition of employment, but with these screening tests there is a risk of a false positive reading (incorrectly indicating drug use) or a false negative reading (failure to detect drug use). What are the consequences of a false result? Given the consequences, is the risk of a false result acceptable?

An understanding of elementary statistical methods can help you to decide whether important decisions like the ones just mentioned are being made in a reasonable way.

We encounter data and conclusions based on data every day. **Statistics** is the scientific discipline that provides methods to help us make sense of data. Some people regard conclusions based on statistical analyses with a great deal of suspicion. Extreme skeptics, usually speaking out of ignorance, characterize the discipline as a subcategory of lying — something used for deception rather than for positive ends. However, we believe that statistical methods, used intelligently, offer a set of powerful tools for gaining insight into the world around us. We hope that this text will help you to understand the logic behind statistical reasoning, prepare you to apply statistical methods appropriately, and enable you to recognize when others are not doing so.

1.2   Statistics and Data Analysis

Statistical methods, used appropriately, allow us to draw conclusions based on data. Data and conclusions based on data appear regularly in a variety of settings: newspapers, advertisements, magazines, and professional journals. In business, industry, and government, informed decisions are often data-driven.

Statistics is the science of collecting, analyzing, and drawing conclusions from data.

Once data has been collected or an appropriate data source identified, the next step in the data analysis process usually involves organizing and summarizing the information. Tables, graphs, and numerical summaries allow increased understanding and provide an effective way to present data. Methods for summarizing data make up the branch of statistics called **descriptive statistics.**

After data has been summarized, we often wish to draw conclusions or make decisions based on the data. This usually involves generalizing from a small group of individuals or objects that we have studied to a much larger group.

Definition

The entire collection of individuals or objects about which information is desired is called the **population** of interest. A **sample** is a subset of the population, selected for study in some prescribed manner.

For example, the admissions director at a large university might be interested in learning why some applicants who were accepted for the fall 2000 term failed to enroll at the university. The population of interest to the director consists of all accepted applicants who did not enroll in the fall 2000 term. Because this population is large and it may be difficult to contact all the individuals, she might be able to collect data from only 300 selected students. These 300 students constitute what is known as a sample.

The second major branch of statistics, **inferential statistics,** involves generalizing from a sample to the population from which it was selected. When we generalize in this way, we run the risk of an incorrect conclusion, since a conclusion about the population will be reached on the basis of incomplete information. An important aspect in the development of inferential techniques involves quantifying the chance of an incorrect conclusion.

Considering some examples will help you develop a preliminary appreciation for the scope and power of statistical methodology. In the following examples, we describe three problems that can be investigated using techniques to be presented in this text.

EXAMPLE 1.1

Suppose that a university has recently implemented a new phone registration system. By entering information from a touch-tone phone, students interact with the computer to select classes for the term. To assess student opinion regarding the effectiveness of the system, a survey of students is to be undertaken. Each student in a sample of 400 will be asked a variety of questions (such as the number of units received, the number of attempts required to get a phone connection, etc.). The survey will yield a rather large and unwieldy data set. To make sense out of the raw data and to describe student responses, it is desirable to summarize the data. This would also make the results more accessible to others. Descriptive techniques can be used to accomplish this task. In addition, inferential methods can be employed to draw various conclusions about the experiences of *all* students who used the registration system.

EXAMPLE 1.2

A study linking depression to low cholesterol levels is described in an Associated Press article (*San Luis Obispo Telegram Tribune,* June 23, 1995). Researchers at a hospital in Italy compared the average cholesterol level for a sample of 331 patients who had been admitted to the hospital after a suicide attempt and diagnosed with clinical depression to the average cholesterol level of 331 patients admitted to the hospital for other reasons. Statistical techniques were used to analyze the data and to show that the average cholesterol level was lower for the depressed group. The article correctly noted that, because of the way in which the data was collected, it was not possible to determine from the statistical analysis alone whether there is a causal relationship between cholesterol level and psychological state — that is, whether low cholesterol levels affect psychological state or vice versa.

EXAMPLE 1.3 A final example comes from the discipline of forestry. When a fire occurs in a forested area, decisions must be made as to the best way to combat the fire. One possibility is to try to contain the fire by building a fire line. If building a fire line requires 4 hours, deciding where the line should be built involves making a prediction of how far the fire will spread during this period. Many factors must be taken into account, including wind speed, temperature, humidity, and time elapsed since the last rainfall. Statistical techniques make it possible to develop a model for the prediction of fire spread, using information available from past fires.

Exercises 1.1 – 1.7

1.1 Give a brief definition of the terms *descriptive statistics* and *inferential statistics.*

1.2 Give a brief definition of the terms *population* and *sample.*

1.3 The student senate at a university with 15,000 students is interested in the proportion of students who favor a change in the grading system to allow for + and − grades (e.g., B−, B, B+, rather than just B). Two hundred students are interviewed to determine their attitude toward this proposed change. What is the population of interest? What group of students constitutes the sample in this problem?

1.4 The supervisors of a rural county are interested in the proportion of property owners who support the construction of a sewer system. Because it is too costly to contact all 7000 property owners, a survey of 500 (selected at random) is undertaken. Describe the population and sample for this problem.

1.5 Representatives of the insurance industry wished to investigate the monetary loss due to earthquake damage to single-family dwellings in Northridge, California, in January of 1994. From the set of all single-family homes in Northridge, 100 homes were selected for inspection. Describe the population and sample for this problem.

1.6 A consumer group conducts crash tests of new-model cars. To determine the severity of damage to 1999 Mazda 626s resulting from a 10-mph crash into a concrete wall, six cars of this type are tested, and the amount of damage is assessed. Describe the population and sample for this problem.

1.7 A building contractor has a chance to buy an odd lot of 5000 used bricks at an auction. She is interested in determining the proportion of bricks in the lot that are cracked and therefore unusable for her current project, but she does not have enough time to inspect all 5000 bricks. Instead, she checks 100 bricks to determine whether each is cracked. Describe the population and sample for this problem.

1.3   The Nature and Role of Variability

Statistics is the science of collecting, analyzing, and drawing conclusions from data. If we lived in a world where all measurements were identical for every individual, all three of these tasks would be simple. Imagine a population consisting of all students at a particular university. Suppose that *every* student is taking exactly the same number of units, spent exactly the same amount of money on textbooks this semester, and favors increasing student fees to support expanding library services (this is fiction!). For this population, there is *no* variability in the values of number of units, amount spent on books, or student opinion on the fee increase.

A researcher studying this population to draw conclusions about these three variables would have a particularly easy task. It would not matter how many students the researcher included in his sample or how the sampled students were selected. In fact, the researcher could collect information on number of units, amount spent on books, and opinion on the fee increase by just stopping the next student who happened to walk by the library. Since there is no variability in the population, this one individual would provide complete and accurate information about the population, and the researcher could draw conclusions based on the sample with no risk of error.

The situation just described is obviously unrealistic. Populations with no variability don't exist — or if they do, they are of little statistical interest because they present no challenge! In fact, variability is almost universal. It is variability that makes life (and the life of a statistician, in particular) interesting. Given the presence of variability, how we collect data and how we analyze and draw conclusions from it require that we understand the nature of variability. One of the primary uses of descriptive statistical methods is to increase our understanding of the nature of variability in a population.

The following two examples illustrate how an understanding of variability is necessary to draw conclusions based on data.

EXAMPLE 1.4   The graphs in Figure 1.1 (page 8) are examples of a type of graph called a histogram. (We will see how to construct histograms in Chapter 3.) Figure 1.1(a) shows the distribution of the heights of female basketball players who played at a particular university between 1990 and 1998. The height of each bar in the graph indicates how many players' heights were in the corresponding interval. For example, 40 basketball players had heights between 72 in. and 74 in., whereas only 2 had heights between 66 in. and 68 in. Figure 1.1(b) shows the distribution of heights for members of the women's gymnastics team over the same period. Both histograms are based on the heights of 100 women.

The first histogram shows that the heights of female basketball players varied, with most heights falling between 68 in. and 76 in. Heights of female gymnasts also varied, with most heights in the range of 60 in. to 72 in. We can also see that there is more variation in the heights of the gymnasts than in the heights of the basketball players, since the heights of gymnasts tended to differ more from one another.

FIGURE 1.1 Heights of female athletes: (a) Basketball players; (b) Gymnasts

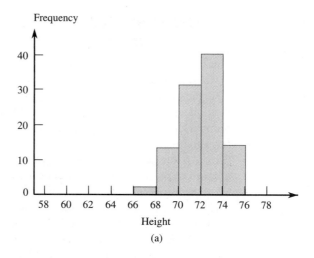

(a)

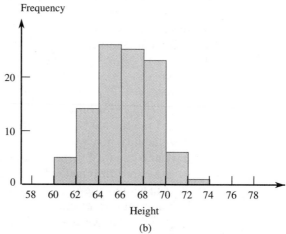

(b)

Now suppose that a tall woman (5 ft 11 in) tells you she is looking for her sister who is practicing with her team at the gym. Would you direct her to where the basketball team is practicing or to where the gymnastics team is practicing? What reasoning would you use to decide? What if you found a pair of size 6 tennis shoes left in the locker room? Would you first try to return them by checking with members of the basketball team or the gymnastics team?

You probably answered that you would send the woman looking for her sister to the basketball practice and try to return the shoes to a gymnastics team member. To reach these conclusions, you informally used statistical reasoning that combined your own knowledge of the relationship between heights of siblings and between shoe size and height with the information about the variability in heights presented in Figure 1.1. You might have reasoned that heights of siblings tend to be similar and that a height as great as 5 ft 11 in, although not impossible, would be unusual for a gymnast. On the other hand, a height as tall as 5 ft 11 in would be a common occurrence for a basketball player. Similarly, you might have reasoned that tall people tend to have bigger feet and short people tend to have smaller feet. The shoes found were a small size, so it is more likely that they belong to a gymnast than to a basketball player, since small heights and small feet are usual for gymnasts and unusual for basketball players.

EXAMPLE 1.5

As part of its regular water quality monitoring efforts, an environmental control board selects five water specimens from a particular well each day. The concentration of contaminants in parts per million (ppm) is measured for each of the five specimens, and then the average of the five measurements is calculated. The histogram in Figure 1.2 summarizes the average contamination values for 200 days.

Suppose a chemical spill has occurred at a manufacturing plant about 1 mile from the well. It is not known whether a spill of this nature would contaminate ground water in the area of the spill and, if so, whether a spill this distance from the well would affect the quality of well water.

One month after the spill, five water specimens are collected from the well and the average contamination is 16 ppm. Would you take this as convincing evidence that the well water was affected by the spill? What if the calculated average was 18 ppm? 22 ppm? How is your reasoning related to the graph in Figure 1.2?

Before the spill, the average contaminant concentration varied from day to day. An average of 16 ppm would not have been an unusual value, and so seeing an average of 16 after the spill isn't necessarily an indication that contamination has increased. On the other hand, an average as large as 18 ppm is less common, and an average as large as 22 is not at all typical of the prespill values. In this case, we would probably conclude that the well contamination level has increased.

FIGURE 1.2 Contaminant concentration in well water

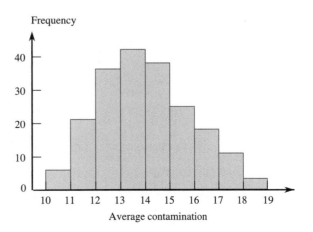

In each of the previous examples, reaching a conclusion required an understanding of variability. Understanding variability allows us to distinguish between usual and unusual values. The ability to recognize unusual values in the presence of variability is the essence of most statistical procedures and is also what enables us to quantify the chance of being incorrect when a conclusion is based on sample data. These concepts will be developed further in subsequent chapters.

Summary of Key Concepts and Formulas

Term or Formula	Comment
Descriptive statistics	Numerical, graphical, and tabular methods for organizing and summarizing data.
Population	The entire collection of individuals or measurements about which information is desired.
Sample	A part of the population selected for study.

References

Moore, David. *Statistics: Concepts and Controversies,* 4th ed. New York: W. H. Freeman, 1997. (A very nice, informal survey of statistical concepts and reasoning.)

Tanur, Judith, ed. *Statistics: A Guide to the Unknown.* Belmont, CA: Duxbury Press, 1989. (Short articles by a number of well-known statisticians and users of statistics, all very nontechnical, on the application of statistics in various disciplines and subject areas.)

2

The Data Analysis Process and Collecting Data Sensibly

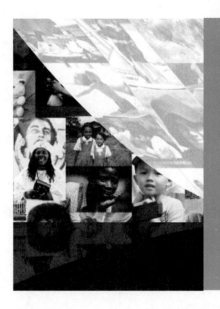

INTRODUCTION

An early step in the data analysis process is data collection. This step is critical because the type of analysis that is appropriate and the conclusions that can be drawn from it depend on how the data is collected. After describing the data analysis process in some detail, this chapter goes on to discuss two methods of data collection — sampling and experimentation.

2.1   Types of Data

Every discipline has its own particular way of using common words, and statistics is no exception. You will recognize some of the terminology from previous math and science courses, but much of the language of statistics will be new to you. We begin by introducing some common terms that will be used throughout this book.

The individuals or objects in any particular population typically possess many characteristics that might be studied. Consider as an example a group of students currently enrolled in a statistics course. One characteristic of the students in the population is the brand of calculator owned (Casio, Hewlett-Packard, Sharp, Texas Instruments, and so on). Another characteristic is the number of textbooks purchased, and yet another is the distance from the university to each student's permanent residence. A **variable** is any characteristic whose value may change from one individual to another. For example, *calculator brand* is a variable, and so are *number of textbooks purchased* and *distance to the university*. **Data** results from making observations either on a single variable or simultaneously on two or more variables. A **univariate data set** consists of observations on a single variable made on individuals in a sample or population.

In the previous example, *calculator brand* is a categorical variable, since each student's response to the query, "What brand of calculator do you own?" is a category. The collection of responses from all these students forms a **categorical data set.** The other two attributes, *number of textbooks purchased* and *distance,* are both numerical in nature. Determining the value of such a numerical variable (by counting or measuring) for each student results in a **numerical data set.**

Definition

A data set consisting of observations on a single attribute is a **univariate data set.** A univariate data set is **categorical** (or **qualitative**) if the individual observations are categorical responses; it is **numerical** (or **quantitative**) if each observation is a number.

EXAMPLE 2.1

The article "Knee Injuries in Women Collegiate Rugby Players" (*Amer. J. of Sports Medicine,* 1997: 360–362) reported the following data on type of injury sustained by 13 female rugby players. MCL, ACL, and the meniscus are different sites in the knee and the patella is the kneecap. (This is a subset of the data given in the article.)

meniscus tear	patella dislocation	MCL tear	MCL tear
meniscus tear	ACL tear	meniscus tear	meniscus tear
MCL tear	meniscus tear	ACL tear	patella dislocation
MCL tear			

Since type of injury is a categorical (nonnumerical) response, this is a categorical data set.

EXAMPLE 2.2

A sample of 20 compact cars, all of the same model, is selected, and the fuel efficiency (miles per gallon) is determined for each one. The resulting numerical data set is

29.8	28.5	27.6	29.5	28.3	27.2	28.7	26.9	27.9	28.4
30.1	28.0	28.0	30.0	28.7	29.6	27.9	29.1	29.9	27.9

In both of the preceding examples, the data sets consisted of observations (categorical responses or numbers) on a single variable, and so they are univariate data sets.

In some studies, attention focuses simultaneously on two different attributes. For example, both the height (in.) and weight (lb) might be recorded for each individual in a group. The resulting data set consists of pairs of numbers, such as

(68, 146). This is called a **bivariate data set. Multivariate data** results from obtaining a category or value for each of two or more attributes (so bivariate data is a special case of multivariate data). For example, multivariate data would result from determining height, weight, pulse rate, and systolic blood pressure for each individual in a group. Much of this book will focus on methods for analyzing univariate data. In the last several chapters, we consider some methods for analyzing bivariate and multivariate data.

Two Types of Numerical Data

With numerical data, it is useful to make a further distinction between *discrete* and *continuous* numerical data. Visualize a number line (Figure 2.1) for locating values of the numerical variable being studied. To every possible number (2, 3.125, −8.12976, etc.) there corresponds exactly one point on the number line. Now suppose that the variable of interest is the number of cylinders of an automobile engine. The possible values of 4, 6, and 8 are identified in Figure 2.2(a) by the dots at the points marked 4, 6, and 8. These possible values are isolated from one another on the line; around any possible value, we can place an interval that is small enough that no other possible value is included in the interval. On the other hand, the line segment in Figure 2.2(b) identifies a plausible set of possible values for the time it takes a car to travel a quarter mile. Here the possible values comprise an entire interval on the number line, and no possible value is isolated from the other possible values.

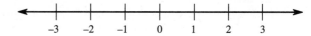

FIGURE 2.1 A number line

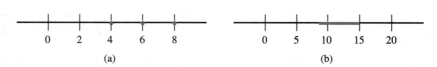

(a)   (b)

FIGURE 2.2 Possible values of a variable: (a) Number of cylinders; (b) Quarter-mile time

---

Definition

Numerical data is **discrete** if the possible values are isolated points on the number line. Numerical data is **continuous** if the set of possible values forms an entire interval on the number line.

Discrete data usually arises when each observation is determined by counting (the number of classes for which a student is registered, the number of petals on a certain type of flower, and so on).

EXAMPLE 2.3

The number of telephone calls per day to a drug hotline is recorded for 12 days. The resulting data set is

3   0   4   3   1   0   6   2   0   0   1   2

Possible values for the *number of calls* are 0, 1, 2, 3, . . .; these are isolated points on the number line, so we have a sample consisting of discrete numerical data.

The sample of fuel efficiencies in Example 2.2 is an example of continuous data. A car's fuel efficiency could be 27.0, 27.13, 27.12796, or any other value in an entire interval. Other examples of continuous data are task completion times, body temperatures, and package weights. In general, data is continuous when observations involve making measurements, as opposed to counting.

In practice, measuring instruments do not have infinite accuracy, so possible measured values, strictly speaking, do not form a continuum on the number line. However, any number in the continuum *could* be a value of the variable. The distinction between discrete and continuous data will be important in our discussion of probability models.

Exercises 2.1 – 2.4

2.1   Classify each of the following attributes as either categorical or numerical. For those that are numerical, determine whether they are discrete or continuous.

a. Number of students in a class of 35 who turn in a term paper before the due date

b. Gender of the next baby born at a particular hospital

c. Amount of fluid (oz) dispensed by a machine used to fill bottles with soda pop

d. Thickness of the gelatin coating of a vitamin E capsule

e. Birth classification (only child, firstborn, middle child, lastborn) of a math major

2.2   Classify each of the following attributes as either categorical or numerical. For those that are numerical, determine whether they are discrete or continuous.

a. Brand of computer purchased by a customer

b. State of birth for someone born in the United States

c. Price of a textbook

d. Concentration of a contaminant (micrograms/cm^3) in a water sample

e. Zip code (Think carefully about this one.)

f. Actual weight of coffee in a 1-lb can

2.3   For the following numerical attributes, state whether each is discrete or continuous.

a. The number of insufficient-fund checks received by a grocery store during a given month

b. The amount by which a 1-lb package of ground beef decreases in weight (because of moisture loss) before purchase

c. The number of New York Yankees during a given year who will not play for the Yankees the next year

d. The number of students in a class of 35 who have purchased a used copy of the textbook

e. The length of a 1-year-old rattlesnake

f. The altitude of a location in California selected randomly by throwing a dart at a map of the state

g. The distance from the left edge at which a 12-in. plastic ruler snaps when bent sufficiently to break

h. The price per gallon paid by the next customer to buy gas at a particular station

2.4   For each of the following situations, give some possible data values that might arise from making the observations described.

a. The manufacturer for each of the next ten automobiles to pass through a given intersection is noted.

b. The grade point average for each of the 15 seniors in a statistics class is determined.

c. The number of gas pumps in use at each of 20 gas stations at a particular time is determined.

d. The actual net weight of each of 12 bags of fertilizer having a labeled weight of 50 lb is determined.

e. Fifteen different radio stations are monitored during a 1-hr period, and the amount of time devoted to commercials is determined for each.

f. The brand of breakfast cereal purchased by each of 16 customers is noted.

g. The number of defective tires is determined for each of the next 20 automobiles stopped for speeding on a certain highway.

2.2   The Data Analysis Process

Statistics involves the collection and analysis of data. Both tasks are critical. Raw data without analysis is of little value, and even a sophisticated analysis cannot extract meaningful information from data that was not collected in a sensible way. In this section, we give an overview of the data analysis process, thus providing a framework for the material covered in this text.

Planning and Conducting a Study

Most scientific studies are undertaken to answer questions about our world. Is a new flu vaccine effective in preventing illness? Is the use of bicycle helmets on the rise? Are injuries that result from bicycle accidents less severe for riders who wear helmets than for those who do not? How many credit cards do college students carry? Do engineering students or psychology students pay more for textbooks? Data collection and analysis allow researchers to answer such questions.

The data analysis process can be organized into six steps: (1) understanding the nature of the problem; (2) deciding what to measure and how to measure it; (3) collecting the data; (4) summarizing the data and making a preliminary analysis; (5) formally analyzing the data; and (6) interpreting the results in the context of the original problem.

Understanding the Nature of the Problem

Effective data analysis begins with an understanding of the research problem. We must know the goal of the researcher and what questions are to be answered. It is important to have clear direction before gathering data in order to avoid being unable to answer the questions of interest using the data collected.

Deciding What to Measure and How to Measure It

The next step in the process is deciding what information is needed to answer the questions of interest. In some cases, the choice is obvious (for example, in a study of the relationship between the weight of a Division I football player and position played), but in other cases it is not as straightforward (for example, in a study of the relationship between preferred learning style and intelligence). It is important to carefully define the variables to be studied and to select an appropriate means of determining these values.

Data Collection

The data collection step in this process is a crucial one. The researcher must first decide whether an existing data source is adequate or whether new data must be collected. Even if a decision is made to use existing data, it is important to understand how the data was collected and for what purpose, so that any resulting limitations are also fully understood and judged to be acceptable. If new data is to be collected, a careful plan must be developed, since the type of analysis that is appropriate and the conclusions that can be drawn from it are dependent on how the data is collected.

Data Summarization and Preliminary Analysis

After data is collected, the next step usually involves a preliminary analysis that includes summarizing the data graphically and numerically. This type of analysis provides insight into important characteristics of the data and can provide guidance in selecting appropriate methods for analysis.

Formal Data Analysis

The data analysis step requires the selection and application of appropriate statistical methods. Much of this text is devoted to methods that may be used to carry out this step.

Interpretation of Results

A critical step in the data analysis process is the interpretation of results. The purpose of this step is to address the following questions:

- What conclusions can be drawn from the analysis?
- How do the results of the analysis inform us about the stated research problem or question?
- How can our results guide future research?

This step in the process often leads to the formulation of new research questions, which, in turn, leads us back to the first step. In this way, good data analysis is often an iterative process.

Evaluating a Research Study

The data analysis steps just described can also be used as a guide for evaluating published research studies. The steps suggest asking the following questions as part of the evaluation of a study:

- What were the researchers trying to learn? What questions motivated their research?
- Was relevant information collected? Were the right things measured?
- Was the data collected in a sensible way?
- Was the data summarized in an appropriate way?
- Was an appropriate method of analysis selected, given the type of data and how the data was collected?
- Are the conclusions drawn by the researchers supported by the data analysis?

The following example illustrates how these questions can guide an evaluation of a research study.

EXAMPLE 2.4   The newspaper article entitled "Spray Away Flu" (*Omaha World-Herald,* June 8, 1998) reported on a study of the effectiveness of a new flu vaccine that is administered by nasal spray rather than by injection. The article states the following:

> Researchers gave the spray to 1070 healthy children 15 months to 6 years old before the flu season two winters ago. One percent developed confirmed influenza, compared with 18 percent of the 532 children who received a placebo. And only one vaccinated child developed an ear infection after coming down with influenza. . . . Typically 30 percent to 40 percent of children with influenza later develop an ear infection.

The researchers concluded that the nasal flu vaccine was effective in reducing the incidence of flu and also in reducing the number of children with flu who subsequently get ear infections.

In this study, the researchers were trying to find out whether the nasal flu vaccine was effective in reducing the number of flu cases and in reducing the number of ear infections in children who do get the flu. The researchers recorded whether a child received the nasal vaccine or a placebo. A **placebo** is a treatment that has no active ingredients. Whether the child developed the flu and subsequent ear infection was also recorded. These are appropriate measurements to make to answer the research question of interest. We typically cannot tell much about the data collection process from a newspaper article. As we will see in Section 2.5, to fully evaluate this study, we would also want to know how the participating children were selected, how it was determined whether a particular child received the vaccine or the placebo, and how the subsequent diagnoses of flu and infection were made.

We will also have to delay discussion of the data analysis and the appropriateness of the conclusions to a later chapter, since we don't yet have the necessary tools to evaluate these aspects of the study.

Other interesting examples of statistical studies can be found in the books *Statistics: A Guide to the Unknown* and *Forty Studies That Changed Psychology: Exploration into the History of Psychological Research* (listed in the chapter references).

2.3   Collecting Data Sensibly: Observation and Experimentation

During the week of August 10, 1998, articles with the following headlines appeared in *USA Today*:

Prayer Can Lower Blood Pressure (August 11, 1998)

Wired Homes Watch 15% Less Television (August 13, 1998)

In each of these articles, a conclusion is drawn from data. In the study on prayer and blood pressure, 2391 people 65 years or older were followed for 6 years. The article states that people who attended a religious service once a week and prayed or studied the Bible at least once a day were less likely to have high blood pressure. The researcher then concluded that "attending religious services lowers blood pressure." The second article reported on a study of 5000 families, some of whom had on-line Internet services. The researcher concluded that TV use was lower in homes that were wired for the Internet, compared to nonwired homes. The researcher who conducted the study cautioned that the study does not answer the question of whether higher family income could be another factor in lower TV usage. Are these conclusions reasonable? As we will see, the answer in each case depends in a critical way on how the data was collected.

Observation and Experimentation

Data collection is an important step in the data analysis process. When we set out to collect information, it is important to keep in mind the questions we hope to answer on the basis of the resulting data. Sometimes we are interested in answering questions about characteristics of an existing population or in comparing two or more well-defined populations. To accomplish this, a sample is selected from each population under consideration, and the sample information is used to gain insight into characteristics of the population(s).

For example, an ecologist might be interested in estimating the average shell thickness of bald eagle eggs. A social scientist studying a rural community may want to determine whether there is a relationship between gender and attitude toward abortion. A city council member in a college town may want to ascertain whether student residents of the city differ from nonstudent residents with respect to their support for various community projects. These are examples of studies that are **observational** in nature. We want to observe characteristics of members of an existing population or several populations and then use the resulting information to draw conclusions.

In an observational study, it is important to obtain a sample that is representative of the corresponding population. To be reasonably certain of this, the way in which the sample is selected must be carefully considered.

Sometimes the questions we are trying to answer deal with the effect of certain explanatory variables on some response and cannot be answered using data from an observational study. Such questions are often of the form, What happens when . . . ? or, What is the effect of . . . ? For example, an educator may wonder what would happen to test scores if the required lab time for a chemistry course were increased from 3 hours to 6 hours per week. To answer questions like this, the researcher conducts an experiment to collect relevant data. The value of some response variable (test score in the chemistry example) is recorded under different experimental conditions (3-hour lab and 6-hour lab in the chemistry example). In an experiment, the researcher manipulates one or more variables, called **factors,** to create the experimental conditions.

A study is an **observational study** if the investigator observes characteristics of a subset of the members of one or more existing populations. The usual goal of an observational study is to draw conclusions about the corresponding population or about differences between two or more populations.

A study is an **experiment** if the investigator observes how a response variable behaves when the researcher manipulates one or more factors. The usual goal of an experiment is to determine the effect of the manipulated factors on some response variable.

The type of conclusion that may be drawn from a research study depends on the study design. Both observational studies and experiments can be used to compare groups, but in an experiment the researcher controls who is in which group, whereas this is not the case in an observational study. This seemingly small difference is critical when it comes to the interpretation of results from the study.

A well-designed experiment can result in data that provides evidence for a cause-and-effect relationship — this is an important difference between an observational study and an experiment. In an observational study, it is impossible to draw cause-and-effect conclusions because we cannot rule out the possibility that the observed effect is due to some other variable rather than to the factor being studied. Such variables are called *confounding variables*.

A **confounding variable** is one that is related both to group membership and to the response variable of interest in the research study.

Both of the studies from the *USA Today* articles discussed previously in this section were observational studies. In the TV-viewing example, the researcher merely observes the TV-viewing behavior of individuals in two groups (wired and nonwired) but does not control which individuals are in which group. A possible confounding variable here is family income. In the prayer example, two groups are compared (those who attend church at least once a week and those who do not), but the researcher does not manipulate this factor to observe the effect on blood pressure by assigning people to church or no-church groups. As a result, cause-and-effect conclusions such as "Prayer Can Lower Blood Pressure," are not reasonable based on the observed data. It is possible that some other variable, such as lifestyle, is related both to church attendance and to blood pressure. In this case, lifestyle is an example of a potential confounding variable.

Both observational studies and experiments must be carefully designed if the resulting data is to be useful. In the next two sections, we will describe the data collection process and discuss what constitutes good practice in this area.

Exercises 2.5 – 2.9

2.5   The Associated Press (Dec. 10, 1999) reported that researchers have found that women who suffer severe morning sickness early in pregnancy are more likely to have a girl. This conclusion was reached by researchers in Sweden based on a "scientific study." How do you think the researchers might have collected data that would have enabled them to reach such a conclusion? Do you think that the scientific study referred to in the article was an experiment or an observational study? Explain.

2.6   "Attending Church Found Factor in Longer Life" is the title of an article that appeared in *USA Today* (Aug. 9, 1999). Based on a "nationally representative survey of 3617 Americans," the article concludes that "Attending services extends the life span about as much as moderate exercise or not smoking." Comment on the validity of this conclusion.

2.7   An article that appeared in the *San Luis Obispo Tribune* (Nov. 11, 1999) was titled "Study Points Out Dangerous Side to SUV Popularity: Half of All 1996 Ejection Deaths Occur in SUVs." This article states that sports utility vehicles (SUVs) have a much higher rate of passengers being thrown from a window during an accident than do automobiles. The article also states that more than half of all deaths caused by ejection involved sports utility vehicles — the basis for the conclusion that SUVs are more dangerous than cars. Later in the article, there is a comment that about 98% of those injured or killed in ejection accidents were not wearing seat belts. Com-

ment on the conclusion that SUVs are more dangerous than cars.

2.8   Does living in the South cause high blood pressure? Data from a group of 6278 whites and blacks questioned in the Third National Health and Nutritional Examination Survey between 1988 and 1994 (see CNN.com Website article of January 6, 2000, entitled "High Blood Pressure Greater Risk in U.S. South, Study Says") indicates that a greater percentage of Southerners have high blood pressure than do people in any other region of the United States. This difference in rate of high blood pressure was found in every race, gender, and age category studied. List at least two possible reasons we cannot conclude that living in the South causes high blood pressure.

2.9   Does eating broccoli reduce the risk of prostate cancer? According to an observational study from the Fred Hutchinson Cancer Research Center (see CNN.com Website article entitled "Broccoli, Not Pizza Sauce, Cuts Cancer Risk, Study Finds," January 5, 2000), men who ate more cruciferous vegetables (broccoli, cauliflower, brussels sprouts, and cabbage) had a lower risk of prostate cancer. This study made separate comparisons for men who ate different levels of vegetables. According to one of the authors, "at any given level of total vegetable consumption, as the percent of cruciferous vegetables increased, the prostate cancer risk decreased." Based on this study, is it reasonable to conclude that eating cruciferous vegetables causes a reduction in prostate cancer risk? Explain.

2.4   Sampling

There are many reasons for selecting a sample rather than obtaining information from an entire population (a **census**). The most common is limited resources: Restrictions on available time or money usually prohibit observation of an entire population. Sometimes the process of measuring the characteristics of interest is destructive, as with the measurement of the breaking strength of soda bottles, the lifetime of flashlight batteries, or the sugar content of oranges, and it would be foolish to study the entire population.

An observational study usually involves an attempt to generalize from a sample to the corresponding population. As a result, it is important that the sample be representative of the population. To be reasonably certain of this, the way in which the sample is selected must be carefully considered. It is sometimes tempting to take the easy way out and gather data in a haphazard way; but if a sample is chosen on the basis of convenience alone, it becomes impossible to interpret the resulting data with confidence. For example, it might be easy to use the students in your statistics class as a sample of students at your university. However, in doing so, you may introduce problems. Not all majors include a statistics course in their curriculum, and most students take statistics in their sophomore or junior year. It is not clear whether or how these factors (and others of which we might not be aware) affect inferences based on information from such a sample. There is no way to tell just by looking at a sample whether it is representative of the population from which it was drawn. Our only assurance comes from the method used to select the sample.

Bias in Sampling

Bias is the tendency for a sample to differ from the corresponding population in some systematic way. The most common types of bias encountered in sampling situations are selection bias, measurement or response bias, and nonresponse bias. **Selection bias** is introduced when some part of the population is systematically excluded from the sample. This problem is sometimes also called *undercoverage*. For example, a researcher may wish to generalize from the results of a study to the population consisting of all residents of a particular city, but the method of selecting individuals may exclude the homeless or those without telephones. If those who are excluded from the sampling process differ in some systematic way from those who are included, the sample will not be representative of the population. If this difference between the included and the excluded occurs on a variable that is important to the study, conclusions based on the sample data may not be valid. Selection bias also occurs if only volunteers or self-selected individuals are used in a study, since self-selected individuals (for example, those who choose to participate in a call-in telephone poll) usually differ from those who choose not to participate.

Measurement or **response bias** occurs when our method of observation tends to produce values that systematically differ from the true population value in some way. This might happen if an improperly calibrated scale is used to weigh items or if questions on a survey are worded in a way that tends to influence the response. For example, a May 1994 Gallup survey sponsored by the American Paper Institute (*Wall Street Journal,* May 17, 1994) included the following question:

> It is estimated that disposable diapers account for less than 2 percent of the trash in today's landfills. In contrast, beverage containers, third-class mail and yard waste are estimated to account for about 21 percent of trash in landfills. Given this, in your opinion, would it be fair to tax or ban disposable diapers?

It is likely that the wording of this question prompted people to respond in a particular way.

Other things that might contribute to response bias are the appearance or behavior of the person asking the question, the group or organization conducting the study, and the tendency for people to lie when asked about illegal behavior or unpopular beliefs.

Although the terms *measurement bias* and *response bias* are often used interchangeably, the term *measurement bias* is usually used to describe systematic deviation from the true value as a result of a faulty measurement instrument (as with the improperly calibrated scale).

Nonresponse bias occurs when responses are not actually obtained from all individuals selected for inclusion in the sample. Like selection bias, nonresponse can distort results if those who respond differ in important ways from those who do not respond. Although some level of nonresponse is unavoidable in most surveys, the biasing effect on the resulting sample is lowest when the response rate is high. It is very important that a serious effort be made to follow up with individuals who do not respond to an initial request for information.

The nonresponse rate for surveys or opinion polls varies dramatically depending on how the data is collected. Surveys are commonly conducted by mail, by phone, or by personal interview. Mail surveys are inexpensive, but often have high nonresponse rates. Telephone surveys can also be inexpensive and can be implemented quickly, but they work well only for short surveys and they can also have high nonresponse rates. Personal interviews are generally expensive but tend to have better response rates.

Types of Bias

Selection Bias
Tendency for samples to differ from the corresponding population as a result of systematic exclusion of some part of the population

Measurement or Response Bias
Tendency for samples to differ from the corresponding population because the method of observation tends to produce values that differ from the true value

Nonresponse Bias
Tendency for samples to differ from the corresponding population because data is not obtained from all individuals selected for inclusion in the sample

---

It is important to note that bias is introduced by the way in which a sample is selected or by the way in which the data is collected from the sample. Increasing the size of the sample, although possibly desirable for other reasons, does nothing to reduce bias. A good discussion of types of bias appears in the book by Lohr listed in the chapter references.

Random Sampling

Most of the inferential methods introduced in this text are based on the idea of random selection. The simplest of these methods is called *simple random sampling*. A simple random sample is one chosen using a method that ensures that each different possible sample of the desired size has an equal chance of being the one chosen. Suppose we want a simple random sample of ten employees chosen from all those who work at a particular department store. For the sample to be a simple random sample, each of the many different sets of ten employees must be equally likely to be the one selected. A sample taken from only full-time employees would not be a simple random sample of *all* employees, since someone who works part-time is still considered an employee but has no chance of being selected. Although a simple random sample may, by chance, include only full-time employees, it must be selected in such a way that *every* employee has the same chance of inclusion in the sample. It is the selection process, not the final sample, that determines whether the sample is a simple random sample.

The letter *n* is used to denote the sample size; it is the number of individuals or objects in the sample.

Definition

A **simple random sample of size *n*** is a sample that is selected from a population in a way that ensures that every different possible sample of the desired size has the same chance of being selected.

The definition implies that every individual member of the population has an equal chance of being selected. **However, the fact that every individual has an equal chance of selection, by itself, is not enough to guarantee that the sample is a simple random sample.** For example, suppose that a class is made up of 100 students, 60 of whom are female. A researcher decides to select 6 of the female students by writing all 60 names on slips of paper, mixing the slips, and then picking 6. She then selects 4 male students from the class using a similar procedure. Even though every student in the class has an equal chance of being included in the sample (6 of 60 females are selected and 4 of 40 males are chosen), the resulting sample is *not* a simple random sample because not all different possible samples of 10 students from the class have the same chance of selection. There are many possible samples of 10 students that consist of 7 females and 3 males or that consist of all females that have no chance of being selected. The sample selection method described here is not necessarily a bad choice (in fact, it is an example of a stratified sample, to be discussed in more detail shortly), but it is not a simple random sample; this must be considered when a method is chosen for analyzing data from such a sample.

There are a number of different methods for selecting a simple random sample. One way is to use a slip of paper to identify each member of the population. Names or numbers are commonly used to label the different population members. The process of thoroughly mixing the slips and then selecting *n* slips one by one (without replacement) yields a random sample of size *n.* This method is easy to understand, but it has obvious drawbacks. The mixing must be adequate, and producing the necessary slips of paper can be extremely tedious, even for populations of moderate size.

A commonly used method for selecting a random sample is to first create a list, called a **sampling frame,** of the objects or individuals in the population. Each item on the list can then be identified by a number, and a table of random digits or a random number generator can be used to select the sample. A random number generator is an algorithm that produces a sequence of numbers that satisfy all reasonable properties associated with the notion of randomness. Most statistics software packages include a random number generator, as do many calculators. A small table of random digits can be found in Appendix Table I.

When selecting a random sample, sampling can be done with or without replacement. **Sampling with replacement** means that after each successive item is selected for the sample, the item is "replaced" back into the population and may therefore be selected again at a later stage. Thus, sampling with replacement allows for the possibility of having the same item or individual appear more than once in the sample. In practice, sampling with replacement is rarely used. Instead, the more common method is to not allow the same individual to be included in the sample more than once. After being included in the sample, an individual or object would not be considered for further selection. Sampling in this manner is called **sampling without replacement.** Although these two forms of sampling are different, when the sample size is small relative to the population size, as is typically the case in most applications, there is little practical difference between them.

EXAMPLE 2.5   Breaking strength is an important characteristic of glass soda bottles. If the strength is too low, a bottle may burst — not a desirable outcome. Suppose that we want to measure the breaking strength of a random sample of three bottles selected from four crates containing a total of 100 bottles (the population). Each crate contains five rows of five bottles each. We can identify each bottle with a number from 1 to 100 by numbering across the rows, starting with the top row of crate 1, as pictured.

Crate 1

1	2	3	4	5
6	...			

Crate 2

26	27	28	...	

⋮

Crate 4

76	77	...		
				100

Using a random number generator from a calculator or statistical software package, we could generate three random numbers between 1 and 100 to determine which bottles would be included in our sample. This might result in bottles 15 (row 3 column 5 of crate 1), 89 (row 3 column 4 of crate 4), and 60 (row 2 column 5 of crate 3) being selected. (Alternately, we could write the numbers from 1 to 100 on slips of paper, place them in a container, mix them well, and then select three.)

The goal of random sampling is to produce a sample that is representative of the population. Although random sampling does not *guarantee* that the sample will be representative, probability-based methods can be used to assess the risk of an unrepresentative sample. It is the ability to quantify this risk that allows us to generalize with confidence from a random sample to the corresponding population.

Other Sampling Methods

Simple random sampling provides researchers with a sampling method that is objective and free of selection bias. In some settings, however, alternative sampling methods may be less costly, may be easier to implement, or may result in more accurate results. When the entire population can be divided into a set of nonoverlapping groups, a method known as stratified sampling often proves to be easier to implement and more cost-effective than random sampling. Stratified random sampling is a method that independently selects a separate simple random sample from each subgroup. For example, for the purpose of estimating average cost of malpractice insurance, it might be convenient to view the population of all doctors practicing in a particular community as being made up of the four subpopulations: (1) surgeons, (2) internists and family practitioners, (3) obstetricians, and (4) a group that includes all other areas of specialization. Rather than taking a simple random sample from the population of all doctors, we could take four separate random samples — one from the group of surgeons, another from the internists and family practitioners, and so on. These four samples would provide information about the four subgroups as well as information about the overall population of doctors.

When the population is divided in this way, the subpopulations are called **strata.** **Stratified sampling** entails selecting a separate simple random sample from each of the strata. Stratified sampling can be used instead of random sampling if it is important to obtain information about characteristics of the individual strata as well as of the entire population.

Stratified sampling usually allows more accurate inferences about a population than does simple random sampling. In general, it is much easier to estimate characteristics of a homogeneous group than of a heterogeneous group. For example, even with a small sample, it is possible to obtain a very accurate estimate of the average grade point average (GPA) of students graduating with high honors from a university. This is a very homogeneous group with respect to GPA. The individual GPAs of group members are all quite similar, and even a sample of three or four individuals from this subpopulation should be representative. On the other hand, estimating the average GPA of *all* seniors at the university, a much more diverse group of GPAs, is a more difficult task.

If a varied population can be divided into strata each of which is more homogeneous than the population with respect to the characteristic of interest, stratified sampling tends to produce more accurate estimates of population characteristics than does simple random sampling. This is because when the strata are relatively homogeneous, a small sample from each stratum can provide reasonably accurate information about strata characteristics. The strata information can then be used to obtain better information about the population as a whole than might have been possible using a simple random sample of the same total size.

Exercises 2.10 – 2.22

2.10 The psychology department at a university produced 140 graduates in 1999. As part of a curriculum review, the department would like to select a simple random sample of 20 1999 graduates to obtain information on how graduates perceived the value of the curriculum. Describe two different methods that might be used to select the sample.

2.11 During the previous calendar year, a county's small-claims court processed 870 cases. A legal researcher would like to select a simple random sample of 50 cases to obtain information regarding the average award in such cases. Describe how a simple random sample of size 50 might be selected from the case files.

2.12 A petition with 500 signatures is submitted to a university's student council. The council president would like to determine the proportion of those who signed the petition who are actually registered students at the university. There is not enough time to check all 500 names with the registrar, so it is decided to select a simple random sample of 30 signatures. Describe how this might be done.

2.13 The financial aid office of a university wishes to estimate the average amount of money that students spend on textbooks each term. They are considering taking a stratified sample. For each of the proposed stratification schemes, discuss whether you think it would be worthwhile to stratify the university students in this manner.

a. Strata corresponding to class standing (freshman, sophomore, junior, senior, graduate student)

b. Strata corresponding to field of study, using the following categories: engineering, architecture, business, other

c. Strata corresponding to first letter of last name: A–E, F–K, etc.

2.14 Devise a procedure for selecting a random sample of words from a dictionary. Explain why your procedure guarantees that, for any *n*, each collection of *n* words has an equal chance of being selected.

2.15 Citrus trees are usually grown in orderly arrangements of rows to facilitate automated farming and harvesting practices. Suppose a group of 1000 trees is laid out in 40 rows of 25 trees each. To determine the sugar content of fruit from a sample of 30 trees, researcher A suggests randomly selecting five rows and then randomly selecting six trees from each sampled row. Researcher B suggests numbering each tree on a map of the trees from

1 to 1000 and using random numbers to select 30 of the trees. Is there a reason for preferring one of these sample selection methods over the other? Explain.

2.16 If liquor stores accept credit cards, does alcohol consumption change? The article "Changes in Alcohol Consumption Patterns Following the Introduction of Credit Cards in Ontario Liquor Stores" (*Journal of Studies on Alcohol* (1999): 378–382) attempted to answer this question with data based on a telephone survey. The sample was obtained by generating random phone numbers.

   a. What are some potential drawbacks of using random phone numbers for sampling? Do any of these drawbacks seem critical for answering this question about alcohol and credit cards?

   b. What time of day would be best for such calls?

2.17 The article "Drinking-Driving and Riding with Drunk Drivers among Young Adults: An Analysis of Reciprocal Effects" (*Journal of Studies on Alcohol* (1999): 615–621) concludes that although 16–24-year olds who drive drunk are more likely to ride with drunk drivers, those who ride with drunk drivers are not more likely to drive drunk themselves. These results are based on a sample of 993 young adults from the state of New York, excluding New York City residents. The article explains the exclusion of New York City by saying that the proportion of young adults is higher but that the driving rate of young adults is far lower.

   a. Describe the population of interest for this study.

   b. Explain how this sample would (or would not) be representative of New York City young adults.

   c. Explain how this sample would (or would not) be representative of adults from the entire United States.

2.18 The study described in Exercise 2.17 actually used a stratified sample with strata determined by age. Why do you think the stratification was desirable?

2.19 According to the article "Effect of Preparation Methods on Total Fat Content, Moisture Content, and Sensory Characteristics of Breaded Chicken Nuggets and Beef Steak Fingers" (*Family and Consumer Sciences Research Journal* (1999): 18–27), sensory tests were conducted using 40 college student volunteers at Texas Women's University. Give three reasons apart from the relatively small sample size why this sample may not be ideal as the basis for generalizing to the population of all college students.

2.20 According to an article on the CNN.com Website dated September 17, 1998, entitled "Majority of U.S. Teens Are Not Sexually Active Study Shows," 52% of surveyed teens had never had sexual intercourse. A very large random sample of 16,262 high school students were the source of these data. If the population of interest is teens in the United States, are there individuals in the population who had no chance of being included in the sample? What is the name for this type of bias?

2.21 A new and somewhat controversial polling procedure that replaces the phone with the Internet is being employed in the conduct of public opinion polls. The article "Pollsters Fear Internet Numbers Offer Warped View" (*Knight Ridder Newspapers,* Oct. 18, 1999) states

> Many pollsters say Internet users as a whole are still too upper-income, highly educated, white and urban to produce results that accurately reflect all Americans. What's more, while 94% of U.S. households had a telephone in 1998, only 26% had Internet access, according to a U.S. Department of Commerce study.

What type of bias is being described in this statement? Do you think that this bias is a serious problem? Explain.

2.22 The article "Study Provides New Data on the Extent of Gambling by College Athletes" (*The Chronicle of Higher Education,* Jan. 22, 1999) reported that "72 percent of college football and basketball players had bet money at least once since entering college." This conclusion was based on a study in which "Copies of the survey were mailed to 3000 athletes at 182 Division I institutions, 25 percent of whom responded." What types of bias might have influenced the results of this study? Explain.

2.5   Simple Comparative Experiments

Sometimes the questions we are trying to answer deal with the effect of certain explanatory variables on some response. Such questions are often of the form, What happens when . . . ? or, What is the effect of . . . ? For example, an educator may wonder what happens to test scores if an activity-based method of instruction is used rather than a traditional lecture method. An industrial engineer may be considering two different workstation designs and might want to know whether the choice of design affects work performance. A medical researcher may want to determine how a proposed treatment for a disease compares to a standard treatment. A nutritionist working for a commercial bakery may want to determine the effect of baking time and baking temperature on the nutritional content of bread.

To address these questions, the researcher conducts an experiment to collect the relevant information. The value of some response variable (test score, assembly time, bread density, etc.) is recorded under different experimental conditions. Such experiments must be carefully planned to obtain information that will give unambiguous answers to questions of interest.

Definition

An **experiment** is a planned intervention undertaken to observe the effects of one or more explanatory variables, often called **factors**, on a response variable. The fundamental purpose of the intervention is to increase understanding of the nature of the relationships between the explanatory and response variables. Any particular combination of values for the explanatory variables is called an **experimental condition** or **treatment**.

The **design** of an experiment is the overall plan for conducting an experiment. A good design minimizes ambiguity in the interpretation of the results.

Suppose that we are interested in determining how student performance on a first-semester calculus exam is affected by room temperature. There are four sections of calculus being offered next term. We might design an experiment in this way: Set the room temperature (in degrees Fahrenheit) to 65° in two of the rooms and to 75° in the other two rooms on test day, and then compare the exam scores for the 65° group and the 75° group. Suppose that the average exam score for the students in the 65° group was noticeably higher than the average for the 75° group. Could we conclude that the increased temperature resulted in a lower average score? The answer is no because many other factors might be related to exam score. Were the sections at different times of the day? Did they have the same instructor? Different textbooks? Was one section required to do more homework than the other sections? Did the sections differ with respect to the abilities of the students? Any of these other factors could be a plausible explanation (having nothing to do with room temperature) of why the average test score was different for the two groups. It is not possible to separate the effect of temperature from the effects of these other factors. As a consequence, simply setting the room temperatures as described makes for a poorly designed experiment. A well-designed experiment requires more than just manipulating the explanatory variables; the design must also eliminate rival explanations or the experimental results will be ambiguous.

The goal of experimentation is to design an experiment that will allow us to determine the effects of the relevant factors on the chosen response variable. To do this, we must take into consideration other factors that, although not of interest in the current study, might also affect the response variable. These are called *extraneous factors*.

Definition

An **extraneous factor** is one that is not of interest in the current study but is thought to affect the response variable.

A researcher can directly **control** some extraneous factors. In the calculus test example, the textbook used is an extraneous factor since part of the differences in test result might be attributed to this factor. We might choose to control this factor directly, by requiring that all sections participating in the study use the same text. If this were the case, any observed differences between temperature groups could not be explained by the use of different texts, since all sections would have used the same text. The extraneous factor *time of day* might also be directly controlled in this way by having all sections meet at the same time.

The effect of some extraneous factors can be filtered out by a process known as **blocking.** Extraneous factors that are addressed through blocking are called *blocking factors*. Blocking creates groups (called *blocks*) that are similar with respect to blocking factors; then all treatments are tried in each block. In our example, we might use *instructor* as a blocking factor. If two instructors are each teaching two sections of calculus, we would make sure that for each instructor, one section was part of the 65° group and the other was part of the 75° group. With this design, if we see a difference in exam scores for the two temperature groups, the factor *instructor* can be ruled out as a possible explanation, because both instructors' students were present in each temperature group. If one instructor taught both 65° sections and the other taught both 75° sections, we would be unable to distinguish the effect of temperature from the effect of instructor. These two factors (temperature and instructor) are said to be *confounded.*

Two factors are **confounded** if their effects on the response variable cannot be distinguished from one another.

We can directly control some extraneous factors by holding them constant, and we can use blocking to create groups that are similar to essentially block out the effect of others. What about factors, such as student ability in our calculus test example, that cannot be controlled by the experimenter and would be difficult to use as blocking factors? These extraneous factors are handled by the use of random assignment to experimental groups — a process called **randomization.** Randomization ensures that our experiment does not favor one experimental condition over any other and attempts to create "equivalent" experimental groups (groups that are as much alike as possible). For example, if the students requesting calculus could be assigned to one of the four available sections using some sort of random mechanism, we would expect the resulting groups to be similar with respect to student ability, as well as similar with respect to other extraneous factors that we are not directly controlling or using as a basis for blocking.

Not all experiments require the use of human subjects to evaluate different experimental conditions. For example, a researcher might be interested in comparing three different gasoline additives with respect to automobile performance as measured by gasoline mileage. The experiment might consist of using a single car with an empty tank. One gallon of gas with one of the additives will be put in the tank and the car driven along a standard route at a constant speed until it runs out of gas. The total distance traveled on the gallon of gas could then be recorded. This could be repeated a number of times, ten for example, with each additive.

The experiment just described can be viewed as consisting of a sequence of trials. Because there are a number of extraneous factors that might have an effect on gas mileage (such as variations in environmental conditions like wind speed or humidity and small variations in the condition of the car), it would not be a good idea to use additive 1 for the first ten trials, additive 2 for the next ten, and so on. An approach that would not unintentionally favor any one of the additives would be to randomly assign additive 1 to 10 of the 30 planned trials, then randomly assign additive 2 to 10 of the remaining 20 trials. The resulting plan for carrying out the experiment might look as follows:

Trial	1	2	3	4	5	6	7	...	30
Additive	2	2	3	3	2	1	2	...	1

When an experiment can be viewed as a sequence of trials, randomization involves the random assignment of treatments to trials. **Just remember that random assignment — either of subjects to treatments or of treatments to trials — is a critical component of a good experiment.**

Randomization can be effective in evening out the effects of extraneous variables only if the number of subjects or observations in each treatment or experimental condition is large enough for each experimental group to reliably reflect variability in the population. For example, if there were only eight students requesting calculus, it is unlikely that we would get equivalent groups for comparison, even with random assignment to the sections. **Replication** is the design strategy of making multiple observations for each experimental condition. Together, replication and randomization allow the researcher to be reasonably confident of comparable experimental groups.

Key Concepts in Experimental Design

Randomization
Random assignment (of subjects to treatments or of treatments to trials) to ensure that the experiment does not systematically favor one experimental condition over another

Blocking
Using extraneous factors to create groups (blocks) that are similar. All experimental conditions are then tried in each block.

Direct Control
Holding extraneous factors constant so that their effects are not confounded with those of the experimental conditions

Replication
Ensuring that there is an adequate number of observations in each experimental condition

These four concepts — randomization, blocking, control, and replication — are the fundamental principles of experimental design. When collecting data through experimentation, you should give careful thought to the role that each of these principles plays in your design.

It is an unremarkable event when we read in our newspapers or see on the news or perhaps read in one of our textbooks that a scientist performed an experiment. The subject matter may be a new therapy for arthritis, a promising drug for asthma control, or a substance that holds promise of longer lasting tires for automobiles, but the stories are very similar. A scientist, doctor, or engineer has performed an experiment and some new or unexpected result has brought new knowledge or refined our understanding of some scientific puzzle or, in some cases, caused rethinking about a heretofore "settled" issue in some area of human endeavor.

Although we may be aware of the results of experiments, the details of carrying out experiments are not usually reported, and, in all likelihood, we have little or no understanding of how experiments are performed or just how they provide the scientific information they do. To illustrate the design of a simple experiment, consider the dilemma of Anna, a waitress in a local restaurant. Anna is saving money for college, and she depends on tips from her customers in the restaurant. She would like to increase the amount of her tips, and her strategy is very simple: In addition to prompt and courteous service, she will project a positive image of herself to her customers by writing "Thank you" on the back of the check before she gives it to the patron. She reasons that friendly waitresses are likely to result in satisfied customers and that this satisfaction would result in higher tips. She will write "Thank you" on some of the checks, and on others she will write nothing. She plans to calculate the percentage of the tip as her measure of success. (For instance, a 15% tip is common.) She will compare the average percentage of the tip calculated from checks with and without thank you.

Writing thank you consumes a bit more time in her already hectic serving activities, and so she has a stake in the outcome of her experiment. If the extra time does not produce higher tips, she may try a different strategy; in any case it is of value to her to have at least a tentative answer to the question: Will writing thank you produce the desired outcome of higher tips? Anna is untrained in the art of planning experiments, but already she has taken some common sense steps in the right direction to answer her question.

1. **Identification of a specific problem**
   Anna has defined a manageable problem, and collecting the appropriate data is feasible. It should be very easy to gather data as a normal part of her work.

2. **Definition of the variables of interest**
   Anna reasons that writing thank you on the customers' bills will have an effect on the amount of tipping. Moreover, it seems logical to her that such a communication could produce this desired effect. In the language of experimentation we would refer to the writing of thank you and the not writing of thank you as **treatments** (the two experimental conditions to be compared in the experiment). The two treatments together are the possible values of the **explanatory** variable. The tipping percentage is the **response** variable. The idea behind this terminology is that the outcome of the experiment (the tipping percentage) is a *response* to the treatments *writing thank you* or *not writing thank you.* Anna's experiment may be thought of as an attempt to explain the variability in the response variable in terms of its presumed cause, the variability in the explanatory variable. That is, as she manipulates the explanatory variable, she expects the response by her customers to vary in a manner consistent with her theory.

3. **Formulation of the measurement**
   Writing or not writing is a categorical variable, which is easily measured. It is possible that Anna may write more or less legibly depending on how hectic her day is, but with practice she should be able to dash off a thank you in a standard manner. The tipping percentage is easily calculated from the tip and the original bill: % tip = (tip/bill total) $\times$ 100.

4. **Interpretation of the results**
   In general, if the tip sizes are higher for the *thank you* treatment, Anna will regard this as evidence in favor of her hypothesis that such activity ought to increase tipping. If the tip sizes are very similar, she will regard the writing as a waste of time with respect to increasing tip percentages.

Anna has a good start, but now she must consider the four fundamental design principles.

Replication   Anna cannot run a successful experiment by gathering tipping information on only one person for each treatment. There is no reason to believe that any single tipping incident is representative of what would happen in other incidents (since customers vary in their tipping practices), and therefore it would be impossible to evaluate the two treatments with only two subjects. To interpret the effects of a particular treatment, she must **replicate** each treatment in the experiment.

Control and Randomization   There are a number of extraneous variables in this example — variables that might have an effect on the size of tip. Some restaurant patrons will be seated near the window with a nice view, some will have to wait for a table while others may be seated immediately, some may be on a fixed income and cannot afford a large tip, and so on. Some of these variables can be controlled. For example, Anna may choose to only use window tables in her experiment, thus eliminating table location as a potential confounding variable. Other variables, such as length of wait and customer income, cannot be easily controlled. As a result, it is important that Anna use randomization to decide which of the window tables will be in the thank you group and which will be in the no thanks group. She might do this by flipping a coin as she prepares the check for each window table. If the coin lands with the head side up, she could write thank you on the bill, omitting the thank you when a tail is observed.

Blocking   Suppose that Anna works on both Thursdays and Fridays. Because day of the week might affect tipping behavior, Anna should block on day of the week and make sure that observations for both treatments are made on each of the two days.

Evaluating an Experimental Design

The key concepts of experimental design provide a framework for evaluating an experimental design, as illustrated in the following two examples.

EXAMPLE 2.6   The article "The Most Powerful Manipulative Messages Are Hiding in Plain Sight" (*Chronicle of Higher Education,* Jan. 29, 1999) reported the results of an experiment on "priming" — the effect of subliminal messages on how we behave. In the experiment, a researcher asked subjects to complete a language test in which the subject was asked to construct a sentence using each word in a list of words. One group of subjects received a list of words related to politeness, and a second group was given a list of words related to rudeness. Subjects were asked to complete the language test and then come into the hall and find the researcher so that he could explain the next part of the test. When each subject came into the hall, he or she found the researcher engaged in conversation. The researcher wanted to see whether the subject would interrupt the conversation. The researcher found that 63% of those primed with words related to rudeness interrupted the conversation, whereas only 17% of those primed with words related to politeness interrupted.

If we assume that the researcher randomly assigned the subjects to the two groups, then this study is an experiment that compares two treatments (primed with words related to rudeness and primed with words related to politeness). The response variable, *politeness,* has the values *interrupted conversation* and *did not interrupt conversation.* The experiment uses replication (many subjects in each treatment group) and randomization to control for extraneous variables that might affect the response.

Many experiments compare a group that receives a particular treatment to a group that receives no treatment. An experimental group that receives no treatment or that receives a placebo treatment (something that looks like the treatment, but which has no active ingredients) is called a **control group.** An experiment that illustrates the use of a control group is described in Example 2.7.

EXAMPLE 2.7
Researchers at the University of North Carolina studied 422 third- and fourth-grade children to determine whether a program of health education and exercise was effective in reducing cholesterol level (*San Luis Obispo Telegram-Tribune,* Aug. 4, 1998). Children with high cholesterol levels were divided into three groups. One group attended classes in healthy nutrition twice a week and exercised three times a week. A second group received individualized instruction from a nurse and exercised three times a week. The third group, a control group, received no specialized instruction and participated in the school's regular physical education classes. After 2 months, cholesterol levels were measured. The researchers concluded that the classroom-based program (group 1) was most effective in lowering cholesterol.

If the researchers randomly assigned the 422 children to the three experimental conditions (treatments), the study was an experiment that employed both replication and randomization.

When planning an experiment or evaluating a design by someone else, be sure to keep the basics in mind:

1. Replication: Design strategy of making multiple observations for each experimental treatment
2. Direct control: Holding the value of an extraneous variable constant so that its effect is not potentially confounded with factors in the experiment
3. Blocking: Using extraneous variables to create groups (blocks) that are similar, and then making sure that all treatments are tried in each block
4. Randomization (**a must!**): The strategy for dealing with all extraneous variables not taken into account through direct control or blocking. We count on randomization to create "equivalent" experimental groups.

Before proceeding with an experiment, you should be able to give a satisfactory answer to each of the following ten questions.

1. What is the research question that data from the experiment will be used to answer?
2. What is the response variable?

3. How are the values of the response variable to be determined?

4. What are the factors (explanatory variables) for the experiment?

5. For each factor, how many different values are there, and what are these values?

6. What are the treatments for the experiment?

7. What extraneous variables might influence the response?

8. How does the design incorporate random assignment of subjects to treatments (or treatments to subjects) or random assignment of treatments to trials?

9. For each extraneous variable listed in question 7, how does the design protect against its potential influence on the response through blocking, direct control, or randomization?

10. Will you be able to answer the research question using the data collected in this experiment?

Exercises 2.23 – 2.32

2.23 The head of the quality control department at a publishing company is interested in factors that affect the strength of the binding on books that it publishes. In particular, she would like to carry out an experiment to determine which of three different glues results in the greatest binding strength. Although they are not of interest in the current investigation, other factors thought to affect binding strength are the number of pages in the book and whether the book is being bound as a paperback or a hardback.
a. What is the response variable in this experiment?
b. What factor will determine the experimental conditions?
c. What two extraneous factors are mentioned in the problem description? Can you think of other extraneous factors that might be considered?

2.24 Based on observing more than 400 drivers in the Atlanta area, two investigators at Georgia State University concluded that people exiting parking spaces did so more slowly when a driver in another car was waiting for the space than when no one was waiting ("Territorial Defense in Parking Lots: Retaliation Against Waiting Drivers," *J. Applied Social Psych.* (1997): 821–834). Describe how you might design a study to collect information that would allow you to determine whether this phenomenon is true for your city. What is the response variable for your study? What extraneous factors might have an effect on the response variable, and how does your design control for them?

2.25 An article in the *San Luis Obispo Tribune* (Sept. 7, 1999) described an experiment designed to investigate the effect of creatine supplements on the development of muscle fibers. The article states that the researchers

... looked at 19 men, all about 25 years of age and similar in weight, lean body mass and capacity to lift weights. Ten were given creatine — 25 grams a day for the first week, followed by 5 grams a day for the rest of the study. The rest were given a fake preparation. No one was told what they were getting. All the men worked out under the guidance of the same trainer.

The response variable measured was gain in fat-free mass (in percent).
a. What extraneous variables are identified in the statement above, and what strategy did the researchers use to deal with them?
b. Do you think it was important that the men participating in the experiment were not told whether they were receiving creatine or the placebo? Explain. This type of study, where subjects do not know what treatment they are receiving, is called a **blinded study.**
c. In a **double-blind study,** not only do the subjects not know which treatment they are receiving, but nei-

ther does the person who measures the response variable. This particular experiment was not conducted in a double-blind manner. Do you think it would have been a good idea to have made this a double-blind experiment? Explain.

2.26 In 1993, researchers proclaimed that listening to Mozart can make you smarter. Dubbed the "Mozart Effect," this conclusion was based on a study that showed college students temporarily gained up to nine IQ points after listening to a Mozart piano sonata. This research has since been criticized by a number of researchers who have been unable to confirm this result in other similar studies. Suppose you wanted to see whether there is a Mozart effect for students at your school.

a. Describe how you might design an experiment for this purpose.
b. Does your experimental design include direct control of any extraneous variables? Explain.
c. Does your experimental design use blocking? Explain why you did or did not include blocking in your design.
d. What role does randomization play in your design?

2.27 Do race and gender influence the type of care that a heart patient receives? The following passage is from the article "Heart Care Reflects Race and Sex, Not Symptoms" (© 1999, *USA Today*, Feb. 25, 1999. Reprinted with permission.):

> Previous research suggested blacks and women were less likely than whites and men to get cardiac catheterization or coronary bypass surgery for chest pain or a heart attack. Scientists blamed differences in illness severity, insurance coverage, patient preference and health care access. The researchers eliminated those differences by videotaping actors — two black men, two black women, two white men and two white women — describing chest pain from identical scripts. They wore identical gowns, used identical gestures and were taped from the same position. Researchers asked 720 primary care doctors at meetings of the American College of Physicians or the American Academy of Family Physicians to watch a tape and recommend care. The doctors thought the study was on clinical decision making.

Evaluate this experimental design. Do you think this is a good design or a poor design, and why? If you were designing such a study, what, if anything, would you propose to do differently?

2.28 Four new word-processing software programs are to be compared by measuring the speed with which various standard tasks can be completed. Before conducting the tests, researchers note that the level of a person's computer experience is likely to have a large influence on the test results. Discuss how you would design an experiment that fairly compares the word-processing programs while simultaneously accounting for possible differences in users' computer proficiency.

2.29 Is status related to a student's understanding of science? The article "From Here to Equity: The Influence of Status on Student Access to and Understanding of Science" (*Culture and Comparative Studies* (1999): 577–602) described a study on the effect of group discussions on learning biology concepts. An analysis of the relationship between status and "rate of talk" (the number of on-task talk speech acts per minute) during group work included gender as a blocking variable. Do you think that gender is a useful blocking variable? Explain.

2.30 In an experiment to determine whether suggesting a number of items to purchase (for example, "limit 12 per person") increases the number of items purchased [see "An Anchoring and Adjustment Model of Purchase Quantity Decision" (*Journal of Marketing Research* (1999): 71–81)], data was collected at three separate stores. At one store the display of sale soup had a sign that said "no limit per person," at another store the sign read "limit four per person," and at the third the sign read "limit 12 per person." On the second and third days of the experiment the signs were switched so that each store had each treatment one day only.

a. Why did the researchers rotate the three treatments across the three different stores for the three days?
b. The results of this study showed that there were no day-to-day or store-to-store differences in the quantity of soup purchased. Also, in the stores with the "limit 12 per person" soup display, the average number of cans purchased per person was twice that of the "no limit" or "limit 4 per person" (7 instead of 3.5). This difference was large enough for the researchers to conclude that the results were not due to chance alone. If a grocery store manager asked you how she should market soup, what would you say?
c. If a grocer asked you about how she should market lettuce, would you say (based on the results of this study) that "limit 12 per person" would increase sales over "no limit per person"? Explain.

2.31 Does alcohol cause increased cravings for cigarettes? Research at Purdue University suggests so (see

CNN.com Website article dated June 13, 1997 entitled "Researchers Find Link Between Cigarette Cravings and Alcohol"). In an experiment, 60 heavy smokers and moderate drinkers were divided into two groups. One group drank vodka tonics and the other group drank virgin tonics (tonic water alone), but all subjects were told they were drinking vodka tonics. The researchers then measured the level of nicotine cravings (by monitoring heart rate, skin conductance, etc.). Those who had consumed the vodka tonics had 35% higher cravings than those who did not. Assuming that the assignment of subjects to treatment (vodka) or control was made at random, do you think that there might be any confounding factors that would make conclusions based on this experiment questionable?

2.32 The level of lead in children's blood should be kept low, and one way to reduce a child's exposure to lead is to control dust in the home. Researchers at the University of Rochester studied the effects of families' dust control on children with low to mild elevations in blood lead levels ("A Randomized Trial of the Effect of Dust Control on Children's Blood Lead Levels," *Pediatrics* (1996): 35). They designed an experiment in which they randomly assigned families whose children had elevated lead levels to two groups: intervention and control. The intervention group was given cleaning supplies and a demonstration of proper cleaning techniques. The control group was given a pamphlet on how to avoid lead exposure. The researchers measured lead levels in the child's blood and in the household dust at the beginning of the study and 7 months later.

a. Identify the population in this experiment.
b. Discuss the role and importance of randomization in this study.
c. Discuss the purpose of a control group.
d. Identify the treatments and response variable(s).
e. Discuss possible extraneous factors and how they could be controlled.
f. Discuss whether and how blocking should be used in this experiment.

Summary of Key Concepts and Formulas

Term or Formula	Comment
Categorical data	Individual observations are categorical responses (nonnumerical).
Numerical data	Individual observations are numerical in nature.
Discrete numerical data	Possible values are isolated points along the number line.
Continuous numerical data	Possible values form an entire interval along the number line.
Bivariate and multivariate data	Each observation consists of two (bivariate) or more (multivariate) responses or values.
Observational study	A study that observes characteristics of an existing population.
Simple random sample of size n	A sample selected in a way that gives every different sample of size n an equal chance of being selected.
Stratified sampling	Dividing a population into subgroups (strata) and then taking a separate random sample from each stratum.
Confounding variable	A variable that is related both to group membership and to the response variable.
Measurement or response bias	The tendency for a sample to differ from the population because the method of observation tends to produce values that differ from the true value.
Selection bias	The tendency for a sample to differ from the population due to systematic exclusion of some part of the population.

Term or Formula	Comment
Nonresponse bias	The tendency for a sample to differ from the population because measurements are not obtained from all individuals selected for inclusion in the sample.
Experiment	A procedure for investigating the effect of an *experimental condition* (which is manipulated by the experimeter) on a *response variable.*
Extraneous factor	A variable that is not of interest in the current study but is thought to affect the response variable.
Direct control	Holding extraneous factors constant so that their effects are not confounded with those of the experimental conditions.
Blocking	Using extraneous factors to create experimental groups that are similar with respect to those factors, thereby filtering out their effect.
Randomization	Random assignment to experimental conditions.
Replication	Ensuring that there is an adequate number of observations on each experimental treatment.
Placebo treatment	A treatment that resembles the other treatments in an experiment, but which has no active ingredients.
Control group	A group that receives no treatment or a placebo treatment.

Supplementary Exercises 2.33 – 2.40

2.33   The article "Heavy Drinking and Problems among Wine Drinkers" (*Journal of Studies on Alcohol* (1999): 467–471) attempts to answer the question of whether wine drinkers tend to drink less excessively than those who drink beer and spirits. A sample of Canadians, stratified by province of residence and other socioeconomic factors, was selected.

a. Why might stratification by province be a good thing?
b. List two socioeconomic factors that would be appropriate to use for stratification. Explain how each would relate to the consumption of alcohol in general and wine in particular.

2.34   According to Social Security Administration information on the names given to all babies born in the United States, the most popular name for girls born in 1982 is Jennifer, with 30.6 per 1000 girls being given that name. By 1998 Jennifer had fallen in popularity to 20th with 8569 Jennifers of the 1,579,399 females born nationwide. The 1982 results are based on a "one-percent sample of Social Security Number card applications" (required for every baby born in the United States within a year of birth). The 1998 results are based on a 100% sample of Social Security records.

a. Calculate the rate per 1000 of Jennifers in 1998.
b. What is another name for a 100% sample, such as the one of 1998 birth names?
c. The one-percent sample from 1992 is "based on the last four digits of the Social Security Number, which are random and not correlated with a particular geographical region." Would you say that this sample is a simple random sample of names of girls? Explain.

2.35   The general manager of a four-star hotel decides to check the quality of housekeeping by inspecting 15 rooms himself. Discuss how the manager should determine which rooms to check. Consider that the hotel offers economy- and business-class rooms, and suites; would a simple random sample or a stratified sample be more appropriate? Also discuss possible extraneous factors.

2.36   The size of a tip that a restaurant server receives is determined, in part, by the behavior of the server. Researchers at the University of Houston ("Effect of Server Posture on Restaurant Tipping," *Journal of Applied Social Psychology* (1993): 678–685) decided to test the hypothesis that servers who squat down to the level of their customers would receive a larger tip. In the experiment, the waiter would flip a coin to determine whether he would stand or squat down next to the table. The waiter would record the amount of the bill and tip, and whether he stood or squatted down.

a. Describe the treatments and response variable.

b. Discuss possible extraneous factors and how they could be controlled.

c. Discuss whether blocking would be necessary.

d. Identify possible confounding variables.

e. Discuss the role of randomization in this experiment.

2.37 You have been asked to determine on what types of grasslands two species of birds, northern harriers and short-eared owls, build nests. The types of grasslands to be used include undisturbed, native grasses; managed, native grasses; undisturbed, nonnative grasses; and managed, nonnative grasses. You are allowed a plot of land 500 meters square to study. Explain how you would determine where to plant the four types of grasses. What role would randomization play in this determination? Identify any confounding variables. Would this study be considered observational or an experiment? (Based on the article "Response of Northern Harriers and Short-Eared Owls to Grassland Management in Illinois," *Journal of Wildlife Management* (1999): 517–523.)

2.38 Clay tiles are fired in a kiln. Unfortunately, some of the tiles crack during the firing process. A manufacturer of clay roofing tiles would like to investigate the effect of clay type on the proportion of cracked tiles. Two different types of clay are to be considered. One hundred tiles can be placed in the kiln at any one time. There will be slight variations in firing temperature at different locations in the kiln, and firing temperature may also affect cracking. Discuss the design of an experiment to collect information that could be used to decide between the two clay types. How does your proposed design deal with the extraneous factor *temperature*?

2.39 A mortgage lender routinely places advertisements in a local newspaper. The advertisements are of three different types: one focusing on low interest rates, one featuring low fees for first-time buyers, and one appealing to people who may want to refinance their homes. The lender would like to determine which advertisement format is most successful in attracting customers to call for more information. Describe an experiment that would provide the information needed to make this determination. Be sure to consider extraneous factors such as the day of the week that the advertisement appears in the paper, the section of the paper in which the advertisement appears, daily fluctuations in the interest rate, and so forth. What role does randomization play in your design?

2.40 The headline "Men Cause More Car Accidents" appeared in the Australian newspaper *The Mercury* (Sept. 12, 1995). The article stated that men were twice as likely as women to be involved in a car accident. This conclusion was based on a sample of 4000 insurance claims. Do you think the headline is justified based on this study? Explain.

References

Cobb, George. *Introduction to the Design and Analysis of Experiments,* Springer-Verlag, 1997. (An interesting and thorough introduction to the design of experiments.)

Freedman, David, Robert Pisani, and Roger Purves. *Statistics,* 3rd ed. New York: W. W. Norton, 1997. (The first two chapters contain some interesting examples of both well-designed and poorly designed experimental studies.)

Hock, Roger R. *Forty Studies That Changed Psychology: Exploration into the History of Psychological Research.* New York: Prentice-Hall, 1995.

Lohr, Sharon. *Sampling Design and Analysis.* Belmont, CA: Duxbury Press, 1996. (A very nice discussion of sampling and sources of bias at an accessible level.)

Moore, David. *Statistics: Concepts and Controversies,* 4th ed. New York: W. H. Freeman, 1997. (Contains an excellent chapter on the advantages and pitfalls of experimentation and another one in a similar vein on sample surveys and polls.)

Scheaffer, Richard L., William Mendenhall, and Lyman Ott. *Elementary Survey Sampling,* 5th ed. Belmont, CA: Duxbury Press, 1996. (An accessible yet thorough treatment of the subject.)

Tanur, Judith, ed. *Statistics: A Guide to the Unknown.* Belmont, CA: Duxbury Press, 1989. (Short articles by a number of well-known statisticians and users of statistics, all very nontechnical, on the application of statistics in various disciplines and subject areas.)

GRAPHING CALCULATOR EXPLORATIONS

2.1 Calculators and the study of statistics

As a student of statistics, you must be able to use your calculator effectively to study statistics effectively. In previous math classes you may have used your calculator for graphing functions, finding solutions to equations, and perhaps some arithmetic calculations (and maybe even playing Tetris!).

You will be using your calculator much differently during your study of statistics — and most likely using calculator keys you didn't notice before and menus you have only seen by accident. To help you get maximum utility from the use of your calculator, we will present these Graphing Calculator Explorations from time to time. With them, we will call your attention to features of your calculators that relate to the statistical topic at hand. The explorations will be generic in nature, not referring to a particular calculator, so that we can speak to the widest possible audience.

There currently exist many calculators at various prices and with varying amounts of statistical capability built in. The characteristics of a graphing calculator that would make it especially useful for the study of statistics are:

1. A built-in capability for statistical calculations
2. A built-in capability to generate statistical graphs, specifically boxplots, histograms, and scatterplots
3. Data entry in a row-and-column format, familiar to users of computer spreadsheets

We suggest the following strategy for calculator evaluation: Consult the manual that came with your calculator. (The calculator manual, by the way, is your BEST source of information!) If your calculator manual has a chapter on statistics, see whether the pictures and descriptions in that chapter seem to match the topics as you scan this book. If there seems to be a reasonable match, and your calculator has the characteristics listed above, it is most likely adequate for this course.

2.2 Generating Random Integers

Procedures for generating random numbers have probably been around for as long as people have been gambling — that is, a very long time. The earliest methods were throwing dice, dealing cards, and selecting numbered balls from a well-mixed barrel. Rapidly turning discs and spinners, and randomly pulsating vacuum tubes and clicking Geiger counters have also been used. With the availability of computers, arithmetic algorithms (a sequence of steps to follow, much like a recipe) for generating random numbers appeared. These numeric algorithms do not, strictly speaking, generate random numbers; they generate what are sometimes referred to as "pseudo-random" numbers. It would be very difficult to justify these methods theoretically, and they are evaluated by studying how well the algorithms actually work. From a practical point of view, the numbers generated by today's computers and calculators are close enough to random for our purposes.

Your calculator can be used to generate random numbers, which can then be used in selecting a random sample. This capability must be already built in your calculator as a function, and to learn the appropriate keystrokes you will need to search the index of the manual that came with your calculator; look for "rand," or possibly "random." On most calculators a single keystroke or a short sequence of keystrokes should produce a random number, which might look something like this: .9435974025. On some calculators the number of digits can be adjusted, and the examples that follow are based on using four digits. When you have ascertained the appropriate keystrokes for your calculator, generate five random numbers. Some calculators will repeat the process each time you press "enter" or "execute." Try pressing this button four times after you get the first random number. If a random number appears each time, smile — you will save many keystrokes! Here are the numbers we obtained (yours will be different): .5147, .4058, .7338, .0440, and .3394. You may have noticed that your numbers are also in the interval $0 - <1$. This is not an accident; random number generators typically deliver a random number, r, such that $0 \le r < 1$. You will have to convert each such number to a more useful form, usually a positive integer corresponding to an individual in the population from which you are sampling. This conversion turns out to be remarkably easy, and some calculators may already have a built-in capability to generate random integers. To see whether your calculator has a built-in capability, look in your index for something like "random integer." In our discussions to follow, we will *not* assume you have a random integer capability, but if you do please use it!

Converting Random Numbers to Integers, 1, 2, 3, . . . , n

To convert a calculator-generated random number r, $0 \le r < 1$, into a random integer in the range $1, 2, 3, \ldots, n$, multiply r by n, add 1, and ignore the digits to the right of the decimal.

This will involve a sequence of keystrokes something like the following, where *rand* denotes the keystrokes needed to generate r:

$rand \times n + 1$

EXAMPLE Generate three random integers between 1 and 100 for purposes of sampling from a population with 100 individuals. The keystrokes needed to generate a random integer between 1 and 100 follow:

int($rand \times 100$) + 1

(Remember, *rand* stands for the sequence of keystrokes needed to get the random number between 0 and 1 and *int* stands for the keystrokes necessary to truncate a decimal to an integer.)

The numbers we obtain (though of course your numbers will probably differ) are

11   1   55

We would then include the individuals numbered 1, 11, and 55 in the sample.

Before we finish, we will mention a couple technical notes. First, there is no magic about adding 1 in the formula above. Because the *rand* keystrokes return random numbers in the range $0 \leq r < 1$, it is possible for *rand* to deliver a 0. If you are not bothered by random integers starting at 0, you need not waste keystrokes by adding 1 each time. Just remember to add 1 or not add 1 consistently as you generate the random integers. Second, notice that we said your numbers would *probably* differ from ours. The arithmetic random number generators are shipped from the factory with a number called a "seed," needed to get the random number generation process going. If two calculators in your class are "right out of the box" you may notice that the random numbers generated by these calculators are suspiciously the same! This will not be a problem for very long; since calculator users will typically press the *rand* sequence different numbers of times, sequences will soon differ.

2.3 Randomization

As we have seen, the process of randomization is critical to properly conducting an experiment. The process of randomly assigning subjects to treatments can be facilitated by using the graphing calculator's ability to generate integers from $1, 2, \ldots, n$. We will walk through a few examples to illustrate how this can be done in some common experimental situations. For these situations we will use very small samples to illustrate the techniques. In real experiments, many more replications would be needed.

In the first experimental situation, we will assign subjects to treatments in a completely random fashion — specifically, we will not necessarily have equal numbers of subjects in our treatments. This experiment involves studying the effect of pizza on performance on a statistics exam. A generous instructor, armed with a generous grant, has decided to use three types of pizza (treatments): sausage pizza, mushroom pizza, and cheese pizza. Twelve randomly selected students will take part in the experiment, getting a free pizza lunch for their troubles. The strategy for assigning treatments is very simple: Generate 12 random integers from 1, 2, and 3. *Before* any students are assigned, we arbitrarily assign mushroom pizza = 1, cheese pizza = 2, and sausage pizza = 3.

We now randomly generate integers between 1 and 3 using the procedure discussed in the previous section:

int(*rand* $\times$ 3) + 1

Our results were (remember that yours will be different):

3, 1, 1, 1, 1, 3, 1, 2, 1, 3, 1, 3

We then use these numbers to assign a treatment (pizza type) to each of the twelve students participating in the experiment, as shown. Entries in the table are (student, treatment number, and treatment).

1 3 (Sausage)	2 1 (Mushroom)	3 1 (Mushroom)	4 1 (Mushroom)
5 1 (Mushroom)	6 3 (Sausage)	7 1 (Mushroom)	8 2 (Cheese)
9 1 (Mushroom)	10 3 (Sausage)	11 1 (Mushroom)	12 3 (Sausage)

We can easily see some problems using this method of assignment: Mushroom pizza accounts for more than 50% of the experimental trials, and there is no replication of the cheese pizza! Even though random assignment was used, an imbalance like the one noted here can be a consequence of the small number of students participating in the experiment. We could avoid this type of imbalance by adding a rule to the randomization process: "Once any treatment has four subjects assigned, do not assign any more to that treatment." One disadvantage of this rule is that it might be necessary to generate many random numbers before finally accomplishing the random assignment, but it is easy to understand and implement. Anticipating the need for more than 12 random numbers, we generated a sequence using int($rand \times 3$) + 1 to obtain: 1, 3, 3, 2, 1, 3, 3, 1, 1, 3, 3, 2, 1, 1, 3, 1, 2, 1, 1, 3, 2. The assignment to treatments now proceeds as shown, with the assignments appearing below the random integers:

1, 3, 3, 2, 1, 3, 3, 1, 1, 3, 3, 2, 1, 1, 3, 1, 2, 1, 1, 3, 2
M   S   S   C   M   S   S   M   M   (Last three assigned C's.)

Students 1, 5, 8, and 9 are assigned to the Mushroom group; students 4, 10, 11, and 12 are assigned to the Cheese group; students 2, 3, 6, and 7 are assigned to the Sausage group. Notice that this particular assignment was essentially finished after nine random numbers were genereated; the only treatments left to be assigned were Cheese.

Randomization in a situation where the experimenter is using blocking to control for an extraneous factor could be implemented in a similar fashion. Suppose the instructor in the previous example has classes in the late morning and early afternoon. It might be possible, he reasons, for the different kinds of pizza to have different effects in the morning and afternoon. Thus he decides to block by time of day. Six students are randomly selected from the morning class, and six students from the afternoon class. We will illustrate the random assignment for the morning class only. There are still three treatments, so a sequence of random integers between 1 and 3 is generated and treatments assigned to the six students in the morning as follows:

Random Integer	Decision
2	Assign morning student 1 to Cheese
3	Assign morning student 2 to Sausage
1	Assign morning student 3 to Mushroom
1	Assign morning student 4 to Mushroom **(no more mushroom)**
1	Ignore
3	Assign morning student 5 to Sausage **(no more sausage)**
	Assign morning student 6 to Cheese

This assignment results in the random assignment of treatments to the individuals in block 1 (the morning class). This process would then be repeated for the afternoon class (second block).

3

Graphical Methods for Describing Data

INTRODUCTION

One interesting characteristic of a book is its length. Some books have very few pages, others are exceedingly long, and most fall between these two extremes. Even within books of a given type, such as science fiction, romance, biography, or cooking, there is variation in length. One of the authors, being addicted to reading mysteries, decided to investigate the lengths of such books. A sample of 40 paperback mysteries was randomly selected from the shelves of a local bookstore — this was only a small fraction of the population of mysteries at the store — and the number of pages in each one was recorded. The resulting data is displayed in the order in which observations were obtained:

229	247	347	246	307	181	198	214	234	340	314
260	202	320	360	320	200	414	262	248	376	211
214	218	276	628	255	352	197	308	203	371	203
406	261	378	223	181	284	196				

Here are some interesting questions about the data. What is a typical or representative book length? Are the observations highly concentrated near the typical value or quite spread out? Do short or long books predominate, or are there roughly equal numbers of these two types? Are there any books whose lengths are somehow unusual when compared to the rest of the data? What proportion or fraction of observed lengths are at most 200 pages? At least 500 pages?

Questions such as these are most easily answered if the data can be organized in a sensible manner. In this chapter, we introduce some useful tabular and graphical techniques for organizing and describing data.

3.1   Displaying Categorical Data: Frequency Distributions, Bar Charts, and Pie Charts

Frequency Distributions for Categorical Data

An appropriate graphical or tabular display of data can be an effective way to summarize and communicate information. When the data set is categorical, a common way to present the data is in the form of a table, called a *frequency distribution.*

Definition

A **frequency distribution for categorical data** is a table that displays the possible categories along with the associated frequencies or relative frequencies.

The **frequency** for a particular category is the number of times the category appears in the data set.

The **relative frequency** for a particular category is the fraction or proportion of the time that the category appears in the data set. It is calculated as

$$\text{relative frequency} = \frac{\text{frequency}}{\text{number of observations in the data set}}$$

When the table includes relative frequencies, it is sometimes referred to as a **relative frequency distribution.**

EXAMPLE 3.1

Many public health efforts are directed toward increasing levels of physical activity. The article "Physical Activity in Urban White, African-American, and Mexican-American Women" (*Medicine and Science in Sports and Exercise* (1997): 1608–1614) reported on physical activity patterns in urban women. The accompanying data set gives the preferred leisure-time physical activity for each of 30 Mexican-American women. The following coding is used: W = walking, T = weight training, C = cycling, G = gardening, A = aerobics.

```
W   T   A   W   G   T   W   W   C   W   T
W   A   T   T   W   G   W   W   C   A   W
A   W   W   W   T   W   W   T
```

The corresponding frequency distribution appears as Table 3.1.

From the frequency distribution, we can see that 15 women indicated a preference for walking and that this was by far the most popular response. Equivalently, using the relative frequencies, we can say that .5 (half or 50%) of the women preferred walking.

A frequency distribution gives a tabular display of a data set. It is also common to display categorical data graphically. Bar charts and pie charts are the two most widely used types of graphic displays to use with categorical data.

Table 3.1 *Frequency Distribution for Type of Injury*

Category	Frequency	Relative Frequency
Walking	15	.500 ← 15/30
Weight training	7	.233 ← 7/30
Cycling	2	.067
Gardening	2	.067
Aerobics	4	.133
	30	1.000
	Total number of observations	*Should total 1, but in some cases may be slightly off due to rounding*

Bar Charts

A bar chart is a graph of the frequency distribution of categorical data. Each category in the frequency distribution is represented by a bar or rectangle, and the picture is constructed in such a way that the *area* of the bar is proportional to the corresponding frequency or relative frequency.

Bar Charts

When to Use: Categorical data

How to Construct:

1. Draw a horizontal line, and write the category names or labels below the line at regularly spaced intervals.

2. Draw a vertical line, and label the scale using either frequency or relative frequency.

3. Place a rectangular bar above each category label. The height is determined by the category's frequency or relative frequency, and all bars should have the same width. With the same width, both the height and the area of the bar are proportional to the relative frequency.

What to Look For: Frequently and infrequently occurring categories

---

EXAMPLE 3.2

Example 3.1 gave data on preferred leisure time physical activity for a sample of 30 Mexican-American women. Figure 3.1 (page 48) shows the bar chart corresponding to the frequency distribution constructed for this data (Table 3.1).

The bar chart provides a visual representation of the information in the frequency distribution. From the bar chart, it is easy to see that walking occurred most often in the data set, followed by weight training. The bar for walking is about twice as tall (and therefore has twice the area) as the bar for weight training, because approximately twice as many women preferred walking to weight training.

FIGURE 3.1 Bar chart of preferred leisure activity from MINITAB

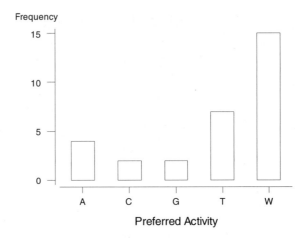

Bar charts can also be used to give a visual comparison of two or more groups. This is done by constructing two or more bar charts that use the same set of horizontal and vertical axes. The article referenced in Example 3.1 also gave data on preferred leisure physical activity for samples of white and African-American women, and included a graph like the one in Figure 3.2. Since the sample sizes for the three groups were not equal, it is important to use relative frequency to construct the scale on the vertical axis so that we can make meaningful comparisons between the groups. It is easy to discern differences among the three groups with respect to preferred leisure activity from the graph.

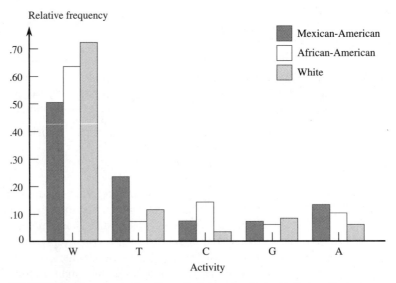

FIGURE 3.2 Bar chart of preferred leisure physical activity by ethnic group

EXAMPLE 3.3

The article "So Close, Yet So Far: Predictors of Attrition in College Seniors" (*J. College Student Development* (1998): 343–348) examined the reasons that college seniors leave their college programs before graduating. Forty-two college seniors at a large public university who dropped out prior to graduation were interviewed and asked the main reason for discontinuing enrollment at the university. Data consistent with that given in the article is summarized in the accompanying frequency distribution.

Reason for Leaving the University	**Frequency**
Academic problems	7
Poor advising or teaching	3
Needed a break	2
Economic reasons	11
Family responsibilities	4
To attend another school	9
Personal problems	3
Other	3

The corresponding bar chart is shown in Figure 3.3. From the bar chart, it is easy to see that more students reported leaving the university for economic reasons or to attend another school than for academic reasons.

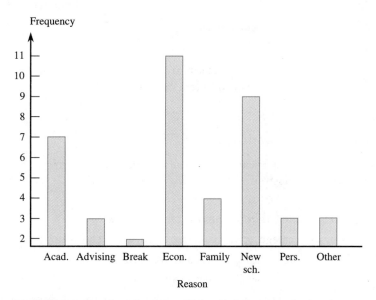

FIGURE 3.3 Bar chart for data of Example 3.3, reason for leaving university

Pie Charts

A categorical data set can also be summarized using a pie chart. In a pie chart, a circle is used to represent the whole data set, with "slices" of the pie representing the possible categories. The size of the slice for a particular category is proportional to the corresponding frequency or relative frequency.

EXAMPLE 3.4

The Chronicle of Higher Education (Jan. 29, 1999) published data collected in a survey of a very large number of students who were college freshmen in the fall of 1998. One question asked whether the student was attending his or her first, second, or third choice of university. Fourth or higher choice were combined in a category called "other." The resulting data is summarized in the pie chart of Figure 3.4.

FIGURE 3.4 Pie chart of data on college choice

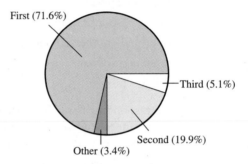

Pie charts are most effective for summarizing data sets when there are not too many different categories.

Pie Chart for Categorical Data

When to Use: Categorical data with a relatively small number of possible categories. Pie charts are most useful for illustrating proportions of the whole data set for various categories.

How to Construct:

1. Draw a circle to represent the entire data set.
2. For each category, calculate the "slice" size. This is done by computing

   slice size = category relative frequency · 360

   (since there are 360 degrees in a circle).
3. Draw a slice of appropriate size for each category. This can be tricky, so most pie charts are generated using a graphing calculator or a statistical software package.

What to Look For: Categories that form large and small proportions of the data set

EXAMPLE 3.5   Night herons and cattle egrets are species of birds that feed on aquatic prey in shallow water. These birds stalk submerged prey while wading in shallow water, and then strike rapidly and downward through the water in an attempt to catch the prey. The article "Cattle Egrets Are Less Able to Cope with Light Refraction Than Are Other Herons" (*Animal Behaviour* (1999): 687–694) gave data on outcome when 240 cattle egrets attempted to capture submerged prey. The data is summarized in the accompanying frequency distribution.

Outcome	Frequency	Relative Frequency
prey caught on first attempt	103	.43
prey caught on second attempt	41	.17
prey caught on third attempt	2	.01
prey not caught	94	.39

To draw a pie chart by hand, we must first compute the slice size for each category. This is done as follows:

Category	Slice Size
first attempt	(.43)(360) = 154.8
second attempt	(.17)(360) = 61.2
third attempt	(.01)(360) = 3.6
not caught	(.39)(360) = 140.4

We would then draw a circle and use a protractor to mark off a slice corresponding to about 155 degrees, as illustrated here:

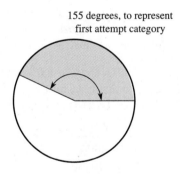

155 degrees, to represent
first attempt category

Continuing to add slices in this way leads to a completed pie chart.

It is much easier to use a statistical software package to construct pie charts. A MINITAB pie chart for the cattle egret data is shown in Figure 3.5(a) on page 52. Figure 3.5(b) shows a pie chart constructed using similar data on outcome for 180 night herons. Although some differences between night herons and cattle egrets can be seen by comparing the pie charts in Figures 3.5(a) and (b), it is difficult to actually compare category proportions using pie charts. A bar chart (Figure 3.6) makes this type of comparison easier.

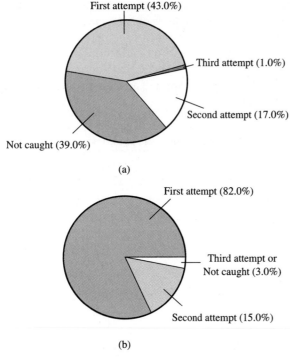

(a)

(b)

FIGURE 3.5 Pie charts for Example 3.5: (a) Cattle egret data; (b) Night heron data

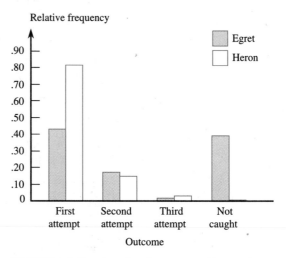

FIGURE 3.6 Bar chart for the egret and heron data

Other Uses of Bar Charts and Pie Charts

As we have seen in previous examples, bar charts and pie charts can be used to summarize categorical data sets. However, they are occasionally also used for other purposes, as illustrated in the following example.

EXAMPLE 3.6   The 1998 Grape Crush Report for San Luis Obispo and Santa Barbara Counties gave the accompanying information on grape production for each of seven different types of grapes used to make wine (*San Luis Obispo Telegram-Tribune,* Feb. 12, 1999).

Type of Grape	Tons Produced
Chardonnay	39,582
Chenin Blanc	1,601
Sauvignon Blanc	5,868
Cabernet Sauvignon	21,656
Merlot	11,210
Pinot Noir	2,856
Zinfandel	7,330
Total	90,103

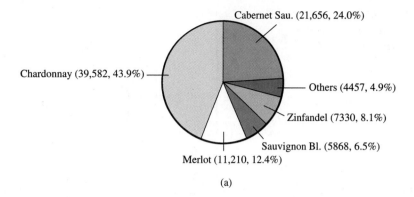

(a)

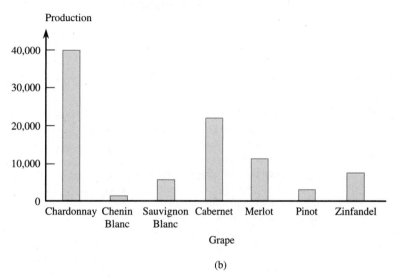

(b)

FIGURE 3.7 Grape production data: (a) Pie chart; (b) Bar chart

Although this table is not a frequency distribution for a categorical data set, it is common to represent information of this type graphically using either a pie chart or a bar chart. A pie chart is shown in Figure 3.7(a). The pie represents the total grape production, and the slices show the proportion of the total production for each of the seven types of grapes. Figure 3.7(b) shows a bar chart representation of the grape production data.

The article also gave grape production information for 1997 and commented on the effect of harsh weather on the counties' grape crop. Figure 3.8 (page 54) is a bar chart of grape production in 1997 and 1998. The bar chart clearly shows decreased production in 1998 for all but two of the seven types of grape.

FIGURE 3.8 Bar chart comparing 1997 and 1998 grape production

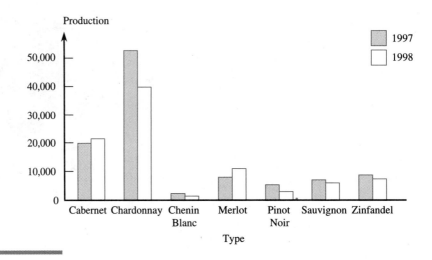

Exercises 3.1 – 3.11

3.1 Many adolescent boys aspire to be a professional athlete. The paper "Why Adolescent Boys Dream of Becoming Professional Athletes" (*Psychological Reports* (1999): 1075–1085) examined some of the reasons. A sample of teenage boys was asked the following question: "Previous studies have shown that more teenage boys say that they are considering becoming professional athletes than any other occupation. In your opinion, why do these boys want to become professional athletes?" The resulting data is shown in the accompanying table.

Response	Frequency
Fame and celebrity	94
Money	56
Attract women	29
Like sports	27
Easy life	24
Don't need an education	19
Other	19

a. Construct a bar chart to display this data.
b. Would a pie chart be a better way to display this data? Why or why not?

3.2 The paper "Profile of Sport/Leisure Injuries Treated at Emergency Rooms of Urban Hospitals" (*Canadian J. of Sports Science* (1991): 99–102) classified non-contact sports injuries by sport, resulting in the accompanying table. Calculate relative frequencies and draw the corresponding bar chart.

Sport	No. of Injuries	Sport	No. of Injuries
Touch football	38	Jogging/running	11
Soccer	24	Bicycling	11
Basketball	19	Volleyball	7
Baseball/softball	11	Others	47

3.3 The paper "Exploring the Factors that Influence Men and Women to Form Medical Career Aspirations" (*J. College Student Development* (1998): 417–421) gave estimates of the percent of college freshmen nationwide who hoped to have a medical career. These percentages were based on "a nationally representative sample of all entering college students" that is conducted each year.

	Percent with Medical Career Aspirations			
	1975	**1980**	**1985**	**1990**
Men	3.9	4.4	4.1	3.9
Women	2.5	2.9	3.4	3.7

Construct a comparative bar graph that shows the proportion with medical career aspirations for men and women over time. (*Hint*: Mark the years on the horizontal axis, and then construct a bar for each gender for each year). Comment on any interesting features of the graph.

3.4 The paper referenced in the previous exercise also gave information on parent's occupation for those who aspire to a medical career. Construct two pie charts, one for men and one for women, to display this information, and comment on the similarities and differences between the two.

Parent Occupation	Male	Female
Only mother is a physician	596	712
Only father is a physician	504	350
Both parents are physicians	734	824
Neither parent is a physician	166	114

3.5 The percentage of the U.S. gross domestic product spent on health care over the period from 1960 to 1995 was given in the paper "Building the Next Generation of Healthy People" (*Public Health Reports* (1999): 213–216). Construct a bar graph for this data and comment on the interesting features of this graph. Be sure to comment on the trend over time.

Year	Percent Spent on Health Care
1960	5.1
1965	5.7
1970	7.1
1975	8.0
1980	8.9
1985	10.3
1990	12.2
1995	13.7

3.6 Nonresponse is a common problem facing researchers who rely on mail questionnaires. In the paper "Reasons for Nonresponse on the Physicians' Practice Survey" (*Proc. Social Statistics Amer. Stat. Assoc.* (1980): 202), 811 doctors who did not respond to the AMA Survey of Physicians were contacted about the reason for their nonparticipation. The results are summarized in the accompanying relative frequency distribution. Draw the corresponding histogram.

Reason	Relative Frequency
1. No time to participate	.264
2. Not interested	.300
3. Don't like surveys in general	.145
4. Don't like this particular survey	.025
5. Hostility toward the government	.054
6. Desire to protect privacy	.056
7. Other reason for refusal	.053
8. No reason given	.103

3.7 The article "Physicians Accessing the Internet: The PAI project" (*J. Amer. Med. Assn.* (1999): 633–634) gave the following information on Internet usage by doctors.

The data is from a survey of 324 randomly selected doctors, who were asked to indicate the category that best described how often they used the Internet.

Internet Usage Pattern	Frequency
Never	31
Rarely (about 3 times per year)	15
Occasionally (about once a month)	52
Often (about once a week)	109
Daily	117

Construct a pie chart for this data and write a brief summary that describes Internet usage for physicians who participated in this survey.

3.8 The paper "Sexual Content of Top-Grossing Motion Pictures" (*J. of Health Education* (1998): 354–357) gave the accompanying information on the ratings of the ten movies that made the most money in the years 1987 and 1992.

	Rating		
	G	PG/PG-13	R
1987	0	3	7
1992	1	5	4

a. Construct a pie chart to show the distribution of ratings for 1987.
b. Construct a pie chart to show the distribution of ratings for 1992, and comment on the differences between this chart and the one based on the 1987 data.

3.9 Each murder committed in Utah during the period 1978–1990 was categorized by day of the week, resulting in the following frequencies: Sunday, 109; Monday, 73; Tuesday, 97; Wednesday, 95; Thursday, 83; Friday, 107; Saturday, 100.
a. Obtain the corresponding frequency distribution.
b. What proportion of these murders were committed on a "weekend day"—i.e., Friday, Saturday, or Sunday?
c. Does this data suggest that a murder is more likely to be committed on some days than on other days? Explain your reasoning.

3.10 The paper "Fire Doors: A Potential Weak Link in the Protection Chain" (*Fire Technology* (1992): 177–179) reported on a survey of companies that use fire doors to subdivide major plant facilities in case of uncontrolled fires. Four types of doors were in use: rolling steel (R), single metal-clad sliding (M), swinging with door closer (C), and single swinging (S). Suppose the data was as shown in the accompanying table. Determine frequencies and relative frequencies for the four categories (percentages agree with those given in the paper).

R	M	M	S	R	R	C	R	M	R	M
M	M	C	S	R	R	C	C	M	R	R
C	S	R	C	C	M	S	M	M	M	R
S	R	R	C	C	S	R	M	C	R	M
C	M	R	C	M	S					

3.11 The use of anabolic steroids (synthetic derivatives of the male sex hormone testosterone) among athletes is believed to have increased dramatically in recent years. A survey of college athletes was carried out to gather relevant data, which was then summarized in the paper "Associations between Academic Performance of Division 1 College Athletes and Their Perceptions of the Effects of Anabolic Steroids" (*Perceptual and Motor Skills* (1995): 284–286). Each athlete was asked for an opinion concerning the severity of the steroid use problem in his or her sport. Responses were also categorized by student grade point average (GPA). (See the accompanying table.)

| | Grade Point Average | | |
Response Category	3.0–4.0	2.0–<3.0	1.0–<2.0
Very great problem	40	73	3
A problem	84	184	11
A small problem	149	170	4
Not a problem	276	317	9
Don't know	146	156	11

a. Construct a separate bar chart of the responses for each GPA category. Compare and contrast the three graphs.

b. Data from any particular GPA category can be summarized in a single rectangular bar, as in the accompanying diagram. The length of each segment is the relative frequency of the corresponding response category. Construct three of these bars above one another for the three different GPA groups. Is it easier to compare and contrast with this picture than with the three separate bar charts?

Figure for Exercise 3.11b

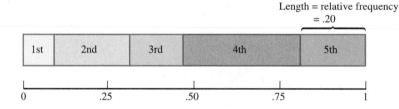

3.2 Displaying Numerical Data: Dotplots and Stem-and-Leaf Displays

Dotplots

A dotplot is a simple way to display numerical data when the data set is reasonably small. Each observation is represented by a dot above the location corresponding to its value on a horizontal measurement scale. When a value occurs more than once, there is a dot for each occurrence and these dots are stacked vertically.

Dotplots

When to Use: **Small numerical data sets**

How to Construct:

1. Draw a horizontal line and mark it with an appropriate measurement scale.

2. Locate each value in the data set along the measurement scale, and represent it by a dot. If there are two or more observations with the same value, stack the dots vertically.

(*continued*)

What to Look For: Dotplots convey information about a representative or typical value in the data set, the extent to which the data values spread out, the nature of the distribution of values along the number line, and the presence of unusual values in the data set.

EXAMPLE 3.7   The accompanying data on gender and birth weight (in kilograms) of foals born to 15 thoroughbred mares appeared in the article "Suckling Behavior Does Not Measure Milk Intake in Horses" (*Animal Behaviour* (1999): 673–678). Figure 3.9 shows a MINITAB dotplot of the weight values.

Foal	1	2	3	4	5	6	7	8	9	10
Gender	F	M	M	F	F	M	F	F	M	F
Weight	129	119	132	123	112	113	95	104	104	93

Foal	11	12	13	14	15
Gender	M	F	M	F	F
Weight	108	95	117	128	127

A typical birth weight is around 113. The 15 observations are quite spread out around this value. A gap separates the three smallest values from the rest of the data. Figure 3.10 shows the dotplots of male and female foal weights. Although the weights for male and female foals seem similar in the main part of the data, it is interesting to note that all three of the foals with low birth weight were female.

FIGURE 3.9 Dotplot of foal birth weights

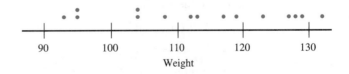

FIGURE 3.10 Dotplot of birth weight for male and female foals

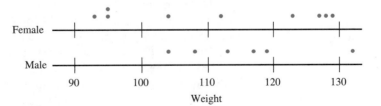

Stem-and-Leaf Displays

A stem-and-leaf display is an effective and compact way to summarize numerical data. Each number in the data set is broken into two pieces called a stem and a leaf.

The **stem** is the first part of the number and consists of the beginning digit(s). The **leaf** is the last part of the number and consists of the final digit(s). For example, the number 213 might be split into a stem of 2 and a leaf of 13 or a stem of 21 and a leaf of 3. The resulting stems and leaves are then used to construct the display.

EXAMPLE 3.8

The use of alcohol by college students is of great concern, not only to those in the academic community, but also, because of potential health and safety consequences, to society at large. The article "Health and Behavioral Consequences of Binge Drinking in College" (*J. of the Amer. Med. Assoc.* (1994): 1672–1677) reported on a comprehensive study of heavy drinking on campuses across the country. A binge episode was defined as five or more drinks in a row for males and four or more for females. Figure 3.11 shows a stem-and-leaf display of 140 values of the percentage of undergraduate students who are binge drinkers at various colleges. (These values were not given in the cited article, but our display agrees with a picture of the data that did appear.)

FIGURE 3.11 Stem-and-leaf display for percentage of binge drinkers at each of 140 colleges

```
0 | 4
1 | 1345678889
2 | 12234566667778899999
3 | 0112233344555666677777888899999          Stem:   Tens digit
4 | 11122222334444556666666777788888999       Leaf:   Ones digit
5 | 001112222334556666667777888899
6 | 01111244455666778
```

The numbers in the vertical column on the left of the display are the **stems.** Each number to the right of the vertical line is a **leaf** corresponding to one of the observations in the data set. The legend

Stem: Tens digit

Leaf: Ones digit

tells us that the observation that had a stem of 2 and a leaf of 1 corresponds to a college where 21% (as opposed to 2.1% or .21%) of the students were binge drinkers.

The display in Figure 3.11 suggests that a typical or representative value is in the stem 4 row, perhaps someplace in the low 40% range. The observations are not highly concentrated about this typical value, as would be the case if all values were between 20% and 49%. The display rises to a single peak as we move downward and then declines, and there are no gaps in the display. The shape of the display is not perfectly symmetric but rather appears to stretch out a bit more in the direction of low stems than in the direction of high stems. The most surprising feature of this data is that at most colleges in the sample, at least one-quarter of the students are binge drinkers.

The leaves on each line of the display in Figure 3.11 have been arranged in order from smallest to largest. Most statistical software packages order the leaves this way, but it is not necessary to do so to get an informative display that still shows many of the important characteristics of the data set, such as shape and spread.

Stem-and-leaf displays can be useful in getting a sense of a typical value for the data set, as well as how spread out the values in the data set are. It is also easy to spot data values that are far away from the rest of the values in the data set. Such values are called **outliers.**

Definition

An **outlier** is an unusually small or large data value. A more formal definition of outliers is given in Chapter 4.

The stem-and-leaf display of the data on binge drinking does not show any outliers.

Stem-and-Leaf Displays

When to Use: Numerical data sets with a small to moderate number of observations (does not work well for very large data sets)

How to Construct:
1. Select one or more leading digits for the stem values. The trailing digits (or sometimes just the first one of the trailing digits) become the leaves.
2. List possible stem values in a vertical column.
3. Record the leaf for every observation beside the corresponding stem value.
4. Indicate the units for stems and leaves someplace in the display.

What to Look For: The display conveys information about a representative or typical value in the data set, the extent of spread about such a value, the presence of any gaps in the data, the extent of symmetry in the distribution of values, the number and location of peaks, and the presence of any outliers.

EXAMPLE 3.9

The chapter introduction gave data on page length for a sample of 40 paperback mysteries. The data is reproduced here. The page lengths ranged from a low value of 181 to a high value of 628.

229	247	347	246	307	181	198	214	234	340	314
260	202	320	360	320	200	414	262	248	376	211
214	218	276	628	255	352	197	308	203	371	203
406	261	378	223	181	284	197				

A natural choice for the stem is the leading (hundreds) digit. This would result in a display with 6 stems (1, 2, 3, 4, 5, 6). Using the first two digits of a number as the stem would result in 45 stems (18, 19, . . ., 62). A stem-and-leaf display with 45 stems would not be a very effective summary of the data. In general, stem-and-leaf displays that use between 5 and 20 stems tend to work well.

If we choose the hundreds digit as the stem, the remaining two digits (the tens and ones) digits would form the leaf. For example, for the first few data values

$$229 \rightarrow \text{stem} = 2, \text{leaf} = 29$$
$$247 \rightarrow \text{stem} = 2, \text{leaf} = 47$$
$$347 \rightarrow \text{stem} = 3, \text{leaf} = 47$$

The leaves have been entered in the display of Figure 3.12 in the order they are encountered in the data set.

Commas are used to separate the leaves only when each leaf has two or more digits. This display shows that most mysteries had page lengths in the two and three hundreds, and that a typical page length is around 300. There is one outlier, 638 pages, that stands out as being unusually large compared to the page lengths of the other books in the sample.

An alternative display (Figure 3.13) results from dropping all but the first digit of the leaf. This is what most statistical computer packages do when generating a display; little information about a typical value, spread, or shape is lost in this truncation.

FIGURE 3.12 Stem-and-leaf display of page length

```
1 | 81,98,97,81,96
2 | 29,47,46,14,34,60,02,00,62,48,11,14,18,76,55,03,03,61,23,84
3 | 47,07,40,14,20,60,20,76,52,08,71,78
4 | 14,06
5 |
6 | 28
```
Stem: Hundreds
Leaf: Tens and ones

FIGURE 3.13 Stem-and-leaf display of the page length data using truncated leaves

```
1 | 89989
2 | 24413600641117500628
3 | 404126275077
4 | 10
5 |
6 | 2
```
Stem: Hundreds
Leaf: Tens

Repeated Stems to Stretch a Display

Sometimes a natural choice of stems gives a display in which too many observations are concentrated on just a few stems. A more informative picture may be obtained by dividing the leaves at any given stem into two groups: those that begin with 0, 1, 2, 3, or 4 (the "low" leaves) and those that begin with 5, 6, 7, 8, or 9 (the "high" leaves). Then each stem is listed twice when constructing the display, once for the low leaves and once again for the high leaves.

EXAMPLE 3.10

The accompanying data on daily protein intake (in grams of protein per kilogram of body weight) for 20 competitive athletes was obtained from a plot in the article "A Comparison of Plasma Glutamine Concentration in Athletes from Different Sports" (*Medicine and Science in Sports and Exercise* (1998): 1693–1697):

| 1.4 | 2.2 | 2.7 | 1.5 | 2.3 | 1.7 | 2.3 | 1.5 | 1.8 | 2.8 |
| 1.8 | 1.9 | 2.0 | 2.3 | 1.5 | 1.9 | 1.7 | 1.8 | 1.6 | 3.0 |

Since each value in the data set has only two digits, we must use the first digit for the stem and the last digit for the leaf. The corresponding stem-and-leaf display is shown in Figure 3.14; the display has only three stems (1, 2, and 3), and all but one of the leaves are located at the 1 and 2 stems.

A more informative display using repeated stems is shown in Figure 3.15.

FIGURE 3.14 Stem-and-leaf display for the protein intake data

```
1 | 457588959786
2 | 2733803        Stem:  Ones
3 | 0              Leaf:  Tenths
```

FIGURE 3.15 Stem-and-leaf display for protein intake data using repeated stems

```
1L | 4
1H | 57588959786
2L | 23303
2H | 78             Stem:  Ones
3L | 0              Leaf:  Tenths
```

It is also possible to repeat a stem more than twice. For example, each stem might be repeated five times, once for each of the leaf groupings {0, 1}, {2, 3}, {4, 5}, {6, 7}, and {8, 9}.

Comparative Stem-and-Leaf Displays

Frequently an analyst wishes to see whether two groups of data differ in some fundamental way. A comparative stem-and-leaf display, in which the leaves for one group extend to the right of the stem values and those from the second group to the left, can provide preliminary visual impressions and insights.

EXAMPLE 3.11

The article "Tobacco and Alcohol Use in G-Rated Children's Animated Films (*J. American Medical Assn.* (1999): 1131–1136) investigated exposure to tobacco and alcohol use in all G-rated animated films released between 1937 and 1997 by five major film studios. The researchers found that tobacco use was shown in 56% of the films reviewed. Data on the total tobacco exposure time (in seconds) for films with tobacco use produced by Walt Disney, Inc., were

| 223 | 176 | 548 | 37 | 158 | 51 | 299 | 37 | 11 | 165 |
| 74 | 9 | 2 | 6 | 23 | 206 | 9 | | | |

Data for 11 G-rated animated films showing tobacco use that were produced by MGM/United Artists, Warner Brothers, Universal, and Twentieth Century Fox was also given. The tobacco exposure times (in seconds) for these films was

205   162   6   1   117   5   91   155   24   55   17

To construct a stem-and-leaf display, think of each observation as a three-digit number. For example, 2 would be 002 and 37 would be 037. We then use the first digit of each number as the stem and the remaining two digits as the leaf. For simplicity, let's truncate the leaves to one digit:

$223 \rightarrow$ stem = 2, leaf = 2   (truncated from 23)

$37 \rightarrow 037 \rightarrow$ stem = 0, leaf = 3   (truncated from 37)

$9 \rightarrow 009 \rightarrow$ stem = 0, leaf = 0   (truncated from 09)

The resulting comparative stem-and-leaf display, using repeated stems, is shown in Figure 3.16.

FIGURE 3.16 Comparative stem-and-leaf display for tobacco exposure times

Other Studios		Disney
12000	0L	33100020
59	0H	57
1	1L	
56	1H	756
All Dogs Go to Heaven 0	2L	20 *Pinocchio, James and the Giant Peach*
	2H	9 *101 Dalmatians*
	3L	
	3H	Stem: Hundreds
	4L	Leaf: Tens
	4H	
	5L	4 *The Three Caballeros*

One first impression is that there are many films where the tobacco exposure time is small, both for Disney films and for those made by the other studios. Disney has more films with longer exposure to tobacco, and there is an obvious outlier in the Disney data (the 548-second exposure in the 1945 film *The Three Caballeros*).

Exercises 3.12 – 3.20

3.12 The article "Can We Really Walk Straight?" (*Amer. J. of Physical Anthropology* (1992): 19–27) reported on an experiment in which each of 20 healthy men was asked to walk as straight as possible to a target 60 meters away at normal speed. Consider the following observations on cadence (number of strides per second):

Figure for Exercise 3.13

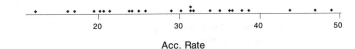

Acc. Rate

| .95 | .85 | .92 | .95 | .93 | .86 | 1.00 | .92 | .85 | .81 |
| .78 | .93 | .93 | 1.05 | .93 | 1.06 | 1.06 | .96 | .81 | .96 |

Construct a dotplot for the cadence data. Do any data values stand out as being unusual?

3.13 Each year *U.S. News and World Report* publishes a ranking of U.S. business schools. The accompanying data gives the acceptance rates (percent of applicants admitted) for the best 25 programs in the most recent survey. Construct a dotplot and compare it to the accompanying MINITAB dotplot (see top of this page).

16.3 12.0 25.1 20.3 31.9 20.7 30.1 19.5 36.2 46.9 25.8
36.7 33.8 24.2 21.5 35.1 37.6 23.9 17.0 38.4 31.2 43.8
28.9 31.4 48.9

3.14 Here is data on both yield and average maturity (in days) for the ten largest tax-free money-market funds, as reported in the Jan. 4, 1996 *Los Angeles Times*:

| Yield | 4.8 | 4.3 | 4.2 | 4.2 | 4.2 | 4.1 | 4.0 | 4.0 | 4.0 | 3.6 |
| Maturity | 26 | 64 | 63 | 48 | 71 | 37 | 55 | 64 | 50 | 48 |

a. Construct a dotplot of the yield data.
b. Construct a dotplot of the maturity data.
c. The first yield and maturity are for the highest-yielding fund, the second yield and maturity are for the second-highest-yielding fund, and so on. Does it appear that decreasing yield is associated either with longer average maturity or with shorter average maturity? That is, does the value of one variable tend to increase or to decrease as the other one decreases? Explain your reasoning. (You might find it helpful to replace the dots in the dotplot of part **b** with the characters 1, 2, . . . , 9, and 0 corresponding to 26, . . . , 48. This type of question will be addressed more formally in Chapter 5.)

3.15 The accompanying stem-and-leaf display shows observations on average shower flow rate (l/min) for a sample of 129 houses in Perth, Australia ("An Application of Bayes Methodology to the Analysis of Diary Records from a Water Use Study," *J. Amer. Stat. Assoc.* (1987): 705–711).

2	23
3	2344567789
4	01356889
5	00001114455666789
6	0000122223344456667789999
7	00012233455555668
8	02233448
9	012233335666788
10	2344455688
11	2335999
12	37
13	8
14	36
15	0035
16	
17	
18	9

Stem: Ones
Leaf: Tenths

a. What is the smallest flow rate in the sample?
b. If one additional house yielded a flow rate of 8.9, where would this observation be placed on the display?
c. What is a typical, or representative, flow rate?
d. Does the display appear to be highly concentrated, or quite spread out?
e. Does the distribution of values in the display appear to be reasonably symmetric? If not, how would you describe the departure from symmetry?
f. Does the data set appear to contain any outliers (observations far removed from the bulk of the data)?

3.16 The accompanying observations are maximum flow rates (at 80 pounds per square inch, or psi) for 34 different shower heads evaluated in a *Consumer Reports* article (July 1990).

2.9	2.8	2.0	3.6	2.7	2.5	2.6	2.9	2.7	2.8	2.5
2.8	2.2	2.5	2.5	2.8	1.8	2.7	2.7	4.7	2.8	2.7
3.1	2.9	3.4	2.6	2.6	2.7	2.4	2.5	5.4	4.9	2.8
2.5										

Construct a stem-and-leaf display using stems 1, 2, 3, 4, and 5. What is the most prominent feature of the display?

3.17 Return to the previous exercise, and construct a stem-and-leaf display using repeated stems (1H, 2L, 2H, 3L, . . .).

3.18 Each observation in the accompanying data set is the number of housing units (homes or condominiums) sold during November 1992 in a region corresponding to a particular Orange County, California, ZIP code. Construct a stem-and-leaf display, and comment on any interesting features.

25	18	16	6	26	11	29	7	5	15	12
37	35	11	16	35	20	27	17	30	10	16
28	13	26	11	12	8	9	29	0	20	30
12	45	26	21	30	18	31	0	46	47	14
13	29	11	18	10	27	5	18	67	21	35
48	42	70	43	0	30	17	35	40	61	18
17	17									

3.19 The accompanying observations are yardages for a sample of golf courses recently listed by *Golf Magazine* as being among the most challenging in the United States. Construct a stem-and-leaf display, and explain why your choice of stems seems preferable to any of the other possible choices.

6526	6770	6936	6770	6583	6464	7005	6927	6790	7209	7040
6850	6700	6614	7022	6506	6527	6470	6900	6605	6873	6798
6745	7280	7131	6435	6694	6433	6870	7169	7011	7168	6713
7051	6904	7105	7165	7050	7113	6890				

3.20 An article on peanut butter in *Consumer Reports* (Sept. 1990) reported the following scores for various brands:

Creamy

56	44	62	36	39	53	50	65	45
40	56	68	41	30	40	50	56	30
22								

Crunchy

62	53	75	42	47	40	34	62	52
50	34	42	36	75	80	47	56	62

Construct a comparative stem-and-leaf display, and discuss similarities and differences for the two types.

3.3   Displaying Numerical Data:
Frequency Distributions and Histograms

A stem-and-leaf display is not always an effective summary technique because it is unwieldy when the data set contains a great many observations. Frequency distributions and histograms are displays that are useful for summarizing even a very large data set in a compact fashion.

Frequency Distributions and Histograms for Discrete Numerical Data

Discrete numerical data almost always results from counting. In such cases, each observation is a whole number. As in the case of categorical data, a frequency distribution for discrete numerical data lists each possible value (either individually or grouped into intervals), the associated frequency, and sometimes the corresponding relative frequency. Recall that relative frequency is calculated by dividing the frequency by the total number of observations in the data set.

EXAMPLE 3.12   The authors of the article "Behavioral Aspects of the Raccoon Mating System: Determinants of Consortship Success" (*Animal Behaviour* (1999): 593–601) monitored raccoons in southern Texas during the 1990–1992 mating seasons in an effort to describe mating behavior. Twenty-nine female raccoons were observed, and the number of male partners during the time the female was accepting partners (generally 1 to 4 days each year) was recorded for each female. The resulting data follows:

1	3	2	1	1	4	2	4	1	1	1	3	1	1	1
1	2	2	1	1	4	1	1	2	1	1	1	1	3	

The corresponding frequency distribution is given in Table 3.2.

Table 3.2 *Frequency Distribution for Number of Partners*

Number of Partners	Frequency	Relative Frequency
1	18	.621
2	5	.172
3	3	.103
4	3	.103
	29	.999

Differs from 1 due to rounding

From the frequency distribution, we can see that 18 of the female raccoons had a single partner. The corresponding relative frequency, .621, tells us that the proportion of female raccoons in the sample with a single partner was .621, or equivalently 62.1% of the females had a single partner. Adding the relative frequencies for the values of 1 and 2 gives

.621 + .172 = .793

indicating that 79.3% of the raccoons had two or fewer partners.

A histogram for discrete numerical data is a graph of the frequency distribution that is very similar to the bar chart for categorical data. Each frequency or relative frequency is represented by a rectangle centered over the corresponding value (or range of values) and with area proportional to the corresponding frequency or relative frequency.

Histogram for Discrete Numerical Data

When to Use: Discrete numerical data. Works well even for large data sets.

How to Construct:

1. Draw a horizontal scale, and mark the possible values.

2. Draw a vertical scale, and mark it with either frequencies or relative frequencies.

3. Above each possible value, draw a rectangle centered at that value (so that the rectangle for 1 is centered at 1, the rectangle for 5 is centered at 5, and so on). The height of each rectangle is determined by the corresponding frequency or relative frequency. Often possible values are consecutive whole numbers, in which case the base width for each rectangle is 1.

What to Look For: Central or typical value, extent of spread or variation, general shape, location and number of peaks, and presence of gaps and outliers

---

EXAMPLE 3.13 The raccoon data of Example 3.12 was summarized in a frequency distribution. The corresponding histogram is shown in Figure 3.17 on page 66.

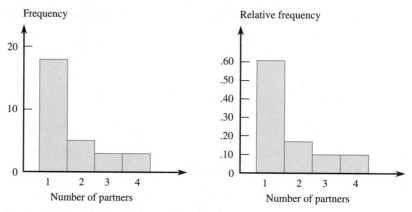

FIGURE 3.17 Histogram and relative frequency histogram of raccoon data

Note that each rectangle in the histogram is centered over the corresponding value. When relative frequency instead of frequency is used for the vertical scale, the scale on the vertical axis is different but all essential characteristics of the graph (shape, location, spread) are unchanged.

Sometimes a discrete numerical data set contains a large number of possible values and may also have a few large or small values that are far away from most of the data. In this case, rather than forming a frequency distribution with a very long list of possible values, it is common to group the observed values into intervals or ranges. This is illustrated in the following example.

EXAMPLE 3.14   Each student in a sample of 176 students at a large public university was asked about alcohol use, and the resulting data appeared in the article "Alcohol, Tobacco, and Marijuana Use: Relationships to Undergraduate Students' Creative Achievement" (*J. College Student Development* (1998): 472–479). The frequency distribution in Table 3.3 summarizes data on number of drinks consumed per week.

Table 3.3 *Frequency Distribution for Number of Drinks per Week*

Drinks per Week	Frequency
0 to 1	52
2 to 5	38
6 to 9	17
10 to 15	35
16 or more	34

The authors of this article chose to group the observed values (rather than list them all: 0, 1, 2, . . .), and they also created an open-ended group of "16 or more." This results in a much more compact table that still communicates one of the important features of the data: the large numbers of individuals at both the low and the high ends. Also note that we could not draw a histogram based on this frequency distribution because of the open-ended group (16 or more). Also, since the number of possible values included differs from group to group (2 in the "0 or 1" group, 4 in the "2 to 5" group, 6 in the "10 to 15" group, etc.), even without the open-ended group drawing a correct histogram is a bit more complicated. You cannot just use frequency or relative frequency for the vertical scale. One method of drawing a histogram that would work, as long as there are no open-ended groups, is described later in this section.

Frequency Distributions and Histograms for Continuous Numerical Data

The difficulty in constructing tabular or graphical displays with continuous data, such as observations on reaction time (seconds) or fuel efficiency (miles per gallon), is that there are no natural categories. The way out of this dilemma is to define our own categories. For fuel efficiency data, suppose that we mark some intervals on a horizontal miles-per-gallon measurement axis, as pictured in Figure 3.18. Each data value should fall in exactly one of these intervals. If the smallest observation were 25.3 and the largest were 29.8, we might use intervals of width .5, with the first one starting at 25.0 and the last one ending at 30.0. The resulting intervals are called **class intervals,** or just **classes.** The class intervals play the same role that the categories or individual values played in frequency distributions for categorical or discrete numerical data.

FIGURE 3.18 Suitable class intervals for miles-per-gallon data

There is one further difficulty. Where should we place an observation such as 27.0, which falls on a boundary between classes? Our convention will be to define intervals so that such an observation is placed in the upper rather than the lower class interval. Thus, in our frequency distribution, a typical class will be 26.5 to <27.0, where the symbol < is a substitute for the phrase *less than.* This class will contain all observations that are greater than or equal to 26.5 and less than 27.0. The observation 27.0 would then fall in the class 27.0 to <27.5.

EXAMPLE 3.15   The trace element zinc is an important dietary constituent, partly because it aids in the maintenance of the immune system. The accompanying data on zinc intake (mg) for a sample of 40 patients with rheumatoid arthritis was read from a graph in the article "Plasma Zinc and Copper Concentrations in Rheumatoid Arthritis: Influence of Dietary Factors and Disease Activity" (*American J. Clinical Nutrition* (1991): 1082–1086):

8.0	12.9	13.0	8.9	10.1	7.3	11.1	10.9	6.2	8.1	8.8
10.4	15.7	13.6	19.3	9.9	8.5	11.1	10.7	8.8	10.7	6.8
7.4	4.8	11.8	13.0	9.5	8.1	6.9	11.5	11.2	13.6	4.9
18.8	15.7	10.8	10.7	11.5	16.1	9.9				

The smallest observation is 4.8, and the largest is 19.3. It seems reasonable to start the first class interval at 3.0 and let each interval have a width of 3.0. This gives the class intervals

3 to <6   6 to <9   9 to <12   12 to <15   15 to <18   18 to <21

Table 3.4 displays the resulting frequency distribution.

Various relative frequencies can be added together to yield other interesting information. For example,

proportion of individuals with intake <12 = (proportion in 3 to <6)
$$+ \text{(proportion in 6 to } <9) + \text{(proportion in 9 to } <12)$$
$$= .050 + .300 + .400$$
$$= .750 \quad (75\%)$$

and

proportion of individuals with intake between 6 and 15 = (proportion in 6 to <9)
$$+ \text{(proportion in 9 to } <12)$$
$$+ \text{(proportion in 12 to } <15)$$
$$= .300 + .400 + .125$$
$$= .825 \quad (82.5\%)$$

Table 3.4 *Frequency distribution for zinc intake (mg)*

Class Interval	Frequency	Relative Frequency
3 to <6	2	.050
6 to <9	12	.300
9 to <12	16	.400
12 to <15	5	.125
15 to <18	3	.075
18 to <21	2	.050
	40	1.000

There are no set rules for selecting either the number of class intervals or the length of the intervals. Using a few relatively wide intervals will bunch the data, whereas using a great many relatively narrow intervals may spread the data over too many intervals, so that no interval contains more than a few observations. Neither type of distribution will give an informative picture of how values are distributed over the range of measurement, and interesting features of the data set may be missed. Generally speaking, with a small amount of data, relatively few intervals, perhaps between five and ten, should be used, whereas with a large amount of data, a distribution based on 15 to 20 (or even more) intervals is often recommended. The quantity

$$\sqrt{\text{number of observations}}$$

often gives a rough estimate for an appropriate number of intervals: 5 intervals for 25 observations, 10 intervals when the number of observations is 100, and so on. Two people making reasonable and similar choices for the number of intervals, their width, and the starting point of the first interval should obtain very similar summaries of the data.

Cumulative Relative Frequencies

Rather than wanting to know what proportion of the data falls in a particular class, we often wish to determine the proportion falling below a specified value. This is easily done when the value is a class boundary. Consider the following classes and relative frequencies:

Class	0 to <25	25 to <50	50 to <75	75 to <100	100 to <125	...
Relative frequency	.05	.10	.18	.25	.20	...

Then

proportion of observations less than 75 = proportion in one of the first three classes

$$= .05 + .10 + .18$$
$$= .33$$

Similarly,

proportion of observations less than 100 = .05 + .10 + .18 + .25 = .33 + .25 = .58

Each such sum of relative frequencies is called a **cumulative relative frequency.** Notice that the cumulative relative frequency .58 is the sum of the previous cumulative relative frequency .33 and the "current'" relative frequency .25. These calculations can also be done for discrete data (e.g., the proportion of observations that are at most 5, at most 6, and so on).

EXAMPLE 3.16   The strength of welds used in aircraft construction has been of great concern to aeronautical engineers in recent years. Table 3.5 (page 70) gives a frequency distribution for shear strengths (force in pounds required to break the weld) of ultrasonic

Table 3.5 *Frequency distribution with cumulative relative frequencies*

Class Interval	Frequency	Relative Frequency	Cumulative Relative Frequency
4000 to <4200	1	.01	.01
4200 to <4400	2	.02	.03
4400 to <4600	9	.09	.12
4600 to <4800	14	.14	.26
4800 to <5000	17	.17	.43
5000 to <5200	22	.22	.65
5200 to <5400	20	.20	.85
5400 to <5600	7	.07	.92
5600 to <5800	7	.07	.99
5800 to <6000	1	.01	1.00
	100	1.00	

spot welds ("Comparison of Properties of Joints Prepared by Ultrasonic Welding and Other Means," *J. Aircraft* (1983): 552–556).

Thus, the proportion of welds with strength values less than 5400 is .85 (that is, 85% of the observations are below 5400). What about the proportion of observations less than 4700? Because 4700 is not a class boundary, we must make an educated guess. Since it is halfway between the boundaries of the 4600–4800 class, let's estimate that half of the relative frequency of .14 for this class, or .07, belongs in the 4600–4700 range. Thus,

estimate of proportion less than 4700 = .01 + .02 + .09 + .07 = .12 + .07 = .19

Similarly, since 5250 is one-fourth of the way from 5200 to 5400,

estimate of proportion less than 5250 = .65 + .25(.20) = .70

In Examples 3.15 and 3.16, the class intervals in the frequency distribution were all of equal width. When this is the case, it is easy to construct a histogram using the information in a frequency distribution.

Histogram for Continuous Numerical Data When the Class Interval Widths Are Equal

When to Use: Continuous numerical data. Works well, even for large data sets.

How to Construct:

1. Mark the boundaries of the class intervals on a horizontal axis.
2. Use either frequency or relative frequency on the vertical axis.

(continued)

3. Draw a rectangle for each class directly above the corresponding interval (so that the edges are at the class boundaries). Use frequency or relative frequency to determine the height of the rectangle.

What to Look For: Central or typical value, extent of spread or variation, general shape, location and number of peaks, and presence of gaps and outliers

EXAMPLE 3.17   Mercury contamination is a serious environmental concern. Mercury levels are particularly high in certain types of fish. Citizens of the Republic of Seychelles, a group of islands in the Indian Ocean, are among those who consume the most fish in the world. The article "Mercury Content of Commercially Important Fish of the Seychelles, and Hair Mercury Levels of a Selected Part of the Population" (*Environ. Research* (1983): 305–312) reported the following observations on mercury content (ppm) in the hair of 40 fishermen:

13.26	32.43	18.10	58.23	64.00	68.20	35.35	33.92	23.94	18.28	22.05	39.14
31.43	18.51	21.03	5.50	6.96	5.19	28.66	26.29	13.89	25.87	9.84	26.88
16.81	37.65	19.63	21.82	31.58	30.13	42.42	16.51	21.16	32.97	9.84	10.64
29.56	40.69	12.86	13.80								

A reasonable choice for class intervals is to start the first interval at zero and set the interval width as 10. The resulting frequency distribution is displayed in Table 3.6, and the corresponding histogram appears in Figure 3.19 (page 72). If it were not for the slight dip in the interval 50 to <60, the histogram would have a single peak; this dip might well disappear with a larger sample size. The upper or right end of the histogram is much more stretched out than the lower or left end. Typical mercury content is somewhere between 20 and 30, but the data exhibits a substantial amount of variability about the center.

Table 3.6 *Frequency distribution for hair mercury content of Seychelles fishermen (ppm)*

Class Interval	Frequency	Relative Frequency
0 to <10	5	.125
10 to <20	11	.275
20 to <30	10	.250
30 to <40	9	.225
40 to <50	2	.050
50 to <60	1	.025
60 to <70	2	.050
	40	1.000

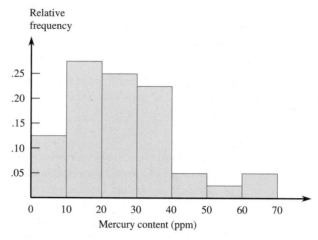

FIGURE 3.19 Histogram for hair mercury content of Seychelles fishermen

Figure 3.20 is a histogram generated by the MINITAB statistical package. Notice that the classes are different from the ones we chose and that the centers of the intervals, rather than the class boundaries, are marked. There is now a gap toward the upper end of the data; this discrepancy in the two graphical displays may well be attributable to the small sample size. With larger samples, the choice of intervals has less impact on the appearance of the histogram.

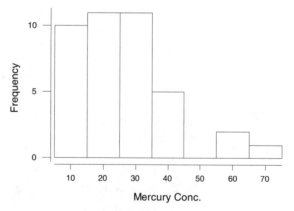

FIGURE 3.20 MINITAB histogram for hair mercury content of Seychelles fishermen

Class Intervals of Unequal Width

Figure 3.21 shows a data set in which a great many observations are concentrated at the center of the set, with only a few outlying, or stray, values both below and above

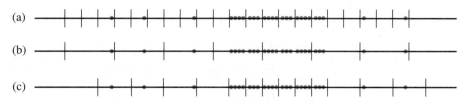

FIGURE 3.21 Three choices of class intervals for a data set with outliers: (a) Many short intervals of equal width; (b) A few wide intervals of equal width; (c) Intervals of unequal width

the main body of data. If a frequency distribution is based on short intervals of equal width, a great many intervals will be required to capture all observations, and many of them will contain no observations (zero frequency). On the other hand, only a few wide intervals will capture all values, but then most of the observations will be grouped into a very few intervals. Neither choice will yield an informative distribution. In such a situation, it is best to use a few, relatively wide class intervals at the ends of the distribution and some shorter intervals in the middle.

Constructing a Histogram for Continuous Data When Class Widths Are Unequal

In this case, frequencies or relative frequencies *should not* be used on the vertical axis. Instead, the height of each rectangle, called the **density** for the class, is given by

$$\text{density} = \text{rectangle height} = \frac{\text{relative frequency of class}}{\text{class width}}$$

The vertical axis is called the **density scale**; it should be marked so that each rectangle can be drawn to the calculated height.

The use of the density scale to construct the histogram ensures that the area of each rectangle in the histogram will be proportional to the corresponding relative frequency. The formula for rectangle height can also be used when class widths are identical; the denominator is then the same for each density calculation. The resulting histogram will look exactly like the one based on relative frequencies, except for the vertical scaling. In the case of equal-width intervals, the extra arithmetic required to obtain the densities is unnecessary.

EXAMPLE 3.18   When people are asked for the values of characteristics such as age or weight, they sometimes shade the truth in their responses. The article "Self-Reports of Academic Performance" (*Soc. Methods and Research* (Nov. 1981): 165–185) focused on such characteristics as SAT scores and grade point average (GPA). For each student in a sample, the difference in GPA (reported − actual) was determined. Positive differences resulted from individuals reporting grade point averages larger than the correct values. Most differences were close to zero, but there were some rather

Table 3.7 *Frequency distribution for errors in reported GPA*

Class Interval	Relative Frequency	Width	Density
−2.0 to <−.4	.023	1.6	.014
−.4 to <−.2	.055	.2	.275
−.2 to <−.1	.097	.1	.970
−.1 to <0	.210	.1	2.100
0 to <.1	.189	.1	1.890
.1 to <.2	.139	.1	1.390
.2 to <.4	.116	.2	.580
.4 to <2.0	.171	1.6	.107

gross errors. Because of this, a frequency distribution based on unequal class widths gives an informative yet concise summary. Table 3.7 displays such a distribution based on classes with boundaries at −2.0, −.4, −.2, −.1, 0, .1, .2, .4, and 2.0.

Figure 3.22 displays two histograms based on this frequency distribution. The histogram of part (a) is correctly drawn, with height = relative frequency/interval width.

The histogram of part (b) has height = relative frequency and is therefore not correct. In particular, this second histogram considerably exaggerates the incidence of grossly overreported and underreported values — the areas of the two most extreme rectangles are much too large. The eye is naturally drawn to large areas, so it is important that the areas correctly represent the relative frequencies.

The formula for rectangle height given in the last box implies that

relative frequency = (rectangle height)(class width) = area of rectangle

That is, a histogram can always be drawn so that the area of each rectangle is the relative frequency of the corresponding class (thus the total area of all rectangles is 1).

Histogram Shapes

The general shape of a histogram is an important characteristic. In describing various shapes, it is convenient to approximate the histogram itself with a smooth curve (called a *smoothed histogram*). This is illustrated in Figure 3.23.

One characterization of general shape relates to the number of peaks, or **modes.** A histogram is said to be **unimodal** if it has a single peak, **bimodal** if it has two peaks, and **multimodal** if it has more than two peaks. These shapes are illustrated in Figure 3.24. Bimodality often occurs when data consists of observations made on two different kinds of individuals or objects. For example, a histogram of heights of college students would show one peak at a typical male height of roughly 70 in. and another at a typical female height of about 67 in. The majority of histograms encountered in practice are unimodal, and multimodality is rather rare.

FIGURE 3.22 Histograms for errors in reporting GPA: (a) A correct picture (height = density); (b) An incorrect picture (height = relative frequency)

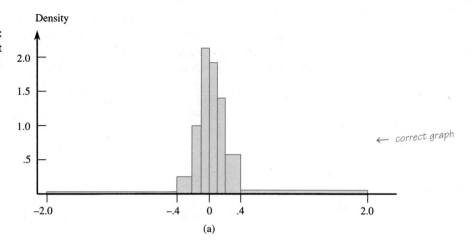

(a)

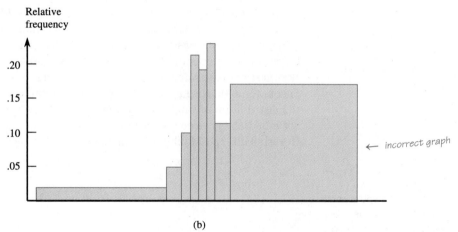

(b)

FIGURE 3.23 Approximating a histogram with a smooth curve

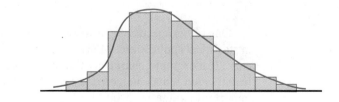

FIGURE 3.24 Smoothed histograms with various numbers of modes: (a) Unimodal; (b) Bimodal; (c) Multimodal

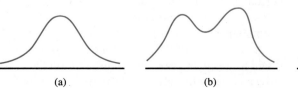
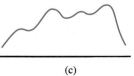

(a)   (b)   (c)

Unimodal histograms come in a variety of shapes. A unimodal histogram is **symmetric** if there is a vertical line of symmetry such that the part of the histogram to the left of the line is a mirror image of the part to the right. (Bimodal and multimodal histograms can also be symmetric in this way.) Several different symmetric smoothed histograms are shown in Figure 3.25. Proceeding to the right from the peak of a unimodal histogram, we move into what is called the **upper tail** of the histogram. Going in the opposite direction moves us into the **lower tail.**

FIGURE 3.25 Several symmetric unimodal smoothed histograms

A unimodal histogram that is not symmetric is said to be **skewed.** If the upper tail of the histogram stretches out much farther than the lower tail, then the distribution of values is **positively skewed.** If, on the other hand, the lower tail is much longer than the upper tail, the histogram is **negatively skewed.** These two types of skewness are illustrated in Figure 3.26. Positive skewness is much more frequently encountered than is negative skewness. An example of positive skewness occurs in the distribution of single-family home prices in Los Angeles County; most homes are moderately priced (at least for California), whereas the relatively few homes in Beverly Hills and Malibu have much higher price tags.

FIGURE 3.26 Two examples of skewed smoothed histograms: (a) Positive skew; (b) Negative skew

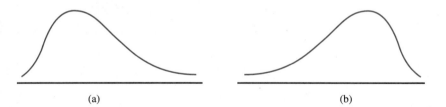

(a)   (b)

One rather specific shape, a **normal curve,** arises more frequently than any other in statistical applications. Many histograms can be well approximated by a normal curve (for example, characteristics such as blood pressure, brain weight, adult male height, adult female height, and IQ score). We briefly mention several of the most important qualitative properties of such a curve, postponing a more detailed discussion until Chapter 7. A normal curve is not only symmetric but also bell shaped; it looks like the curve in Figure 3.27(a). However, not all bell-shaped curves are normal. From the top of the bell, the height of the curve decreases at a well-defined rate when moving into either tail. (This rate of decrease is specified by a certain mathematical function.)

A curve with tails that do not decline as rapidly as the tails of a normal curve is said to specify a **heavy-tailed** distribution (compared to the normal curve). Similarly, a curve with tails that decrease more rapidly than the normal tails is called

FIGURE 3.27 Three examples of bell-shaped histograms: (a) Normal; (b) Heavy-tailed; (c) Light-tailed

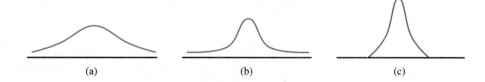

(a)   (b)   (c)

light-tailed. Parts (b) and (c) of Figure 3.27 illustrate these possibilities. Many inferential procedures that work well (that is, result in accurate conclusions) when the population distribution is approximately normal do poorly when the population distribution is heavy tailed. This has spurred development of methods that are not so sensitive to the nature of the tails.

Do Sample Histograms Resemble the Population Histogram?

Sample data is usually collected to make inferences about a population. The resulting conclusions may be in error if the sample is somehow unrepresentative of the population. So how similar might a histogram of sample data be to the histogram of all population values? Will the two histograms be centered at roughly the same place and spread out to about the same extent? Will they have the same number of peaks, and will these occur at approximately the same places?

A related issue concerns the extent to which histograms based on different samples from the same population resemble one another. If two different sample histograms can be expected to differ from one another in obvious ways, then at least one of them will differ substantially from the population histogram. If the sample differs substantially from the population, conclusions about the population based on the sample are likely to be incorrect. **Sampling variability** — the extent to which samples differ from one another and from the population — is a central idea in statistics. We present an example to illustrate such variability in histogram shapes.

EXAMPLE 3.19   A sample of 708 bus drivers employed by public corporations was selected, and the number of traffic accidents in which each bus driver was involved during a 4-year period was determined ("Application of Discrete Distribution Theory to the Study of Noncommunicable Events in Medical Epidemiology," *Random Counts in Biomedical and Social Sciences,* G. P. Patil, ed., University Park, Pa.: Penn. State Univ. Press, 1970). A listing of the 708 sample observations might look like this:

3 0 6 0 0 2 1 4 1 . . . 6 0 2

The frequency distribution (Table 3.8) shows that 117 of the 708 drivers had no accidents, a relative frequency of 117/708 = .165 (or 16.5%). Similarly, the proportion of sampled drivers who had one accident is .222 (or 22.2%). The largest sample observation was 11.

Although the 708 observations actually constituted a sample from the population of all bus drivers, we will regard the 708 observations as constituting the entire population. The first histogram in Figure 3.28 then represents the population histogram. The other four histograms in Figure 3.28 are based on four different samples of 50 observations each from this population. The five histograms certainly

Table 3.8 *Frequency distribution for number of accidents by bus drivers*

Number of accidents	Frequency	Relative frequency
0	117	.165
1	157	.222
2	158	.223
3	115	.162
4	78	.110
5	44	.062
6	21	.030
7	7	.010
8	6	.008
9	1	.001
10	3	.004
11	1	.001
	708	.998

resemble one another in a general way, but there are also some obvious dissimilarities. The population histogram rises to a peak and then declines smoothly, whereas the sample histograms tend to have more peaks, valleys, and gaps. Although the population data set contained an observation of 11, none of the four samples did. In fact, in the first two samples, the largest observations were 7 and 8, respectively. In subsequent chapters, we will see how sampling variability can be described and incorporated into conclusions based on inferential methods.

FIGURE 3.28 A comparison of population and sample histograms for number of accidents

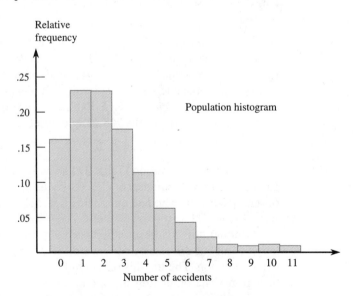

(continued)

FIGURE 3.28 (Continued)

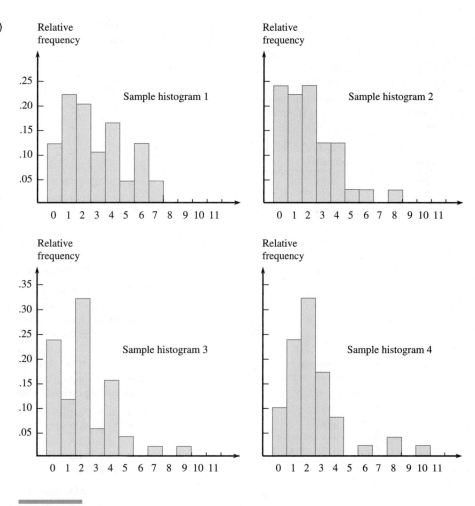

Exercises 3.21 – 3.38

3.21 People suffering from Alzheimer's disease often have difficulty performing basic activities of daily living (ADLs). In one study ("Functional Status and Clinical Findings in Patients with Alzheimer's Disease," *J. of Gerontology* (1992): 177–182), investigators focused on six such activities: dressing, bathing, transferring, toileting, walking, and eating. Here is data on the number of ADL impairments for each of 240 patients.

Number of impairments	0	1	2	3	4	5	6
Frequency	100	43	36	17	24	9	11

a. Determine the corresponding relative frequencies.
b. What proportion of these patients had at most two impairments?
c. Use the result of part **b** to determine what proportion of patients had more than two impairments.
d. What proportion of these patients had at least four impairments?
e. Do you notice anything especially interesting about the frequencies and relative frequencies?

Note: The investigators proposed statistical models from which the number of impairments could be predicted from various patient characteristics and symptoms.

3.22   Is it really the case, as it might seem to an unsuccessful and frustrated angler, that 10% of those fishing reel in 90% of the fish caught? More generally, how is the number of fish caught distributed among those who are trying to catch them? The accompanying table presents data from a survey of 911 anglers done during a particular period on the lower Current River in Canada ("Fisherman's Luck," *Biometrics* (1976): 265–271).

Number of Fish Caught	Number of Anglers (frequency)
0	515
1	65
2	60
3	66
4	53
5	55
6	27
7	25
8	25
9	20
	911

a. Calculate the relative frequencies. (Express each one using four digits of decimal accuracy.)
b. What proportion of those in the sample caught no fish? One fish? At most one fish?
c. What proportion of the 911 anglers caught at least five fish? More than five fish?
d. Calculate the cumulative relative frequencies for the sample of 911 fishermen. Then use them to answer the questions posed in part **c**.
e. Suppose the sample had included an additional four anglers, with numbers of fish caught being 12, 14, 19, and 25. Construct a frequency distribution that has exactly one more row than the one just displayed.

3.23   How unusual is a game with no hits in major-league baseball, and how frequently does a game have more than 10, 15 or even 20 hits? The accompanying table is a frequency distribution for the number of hits per game for all nine-inning games that were played between 1989 and 1993.

Hits/Game	Frequency	Hits/Game	Frequency
0	20	6	1988
1	72	7	2256
2	209	8	2403
3	527	9	2256
4	1048	10	1967
5	1457	11	1509

(continued)

Hits/Game	Frequency	Hits/Game	Frequency
12	1230	20	31
13	834	21	19
14	569	22	13
15	393	23	5
16	253	24	1
17	171	25	0
18	97	26	1
19	53	27	1

a. Calculate the relative frequencies and then construct a relative frequency histogram.
b. Describe the key features of the histogram.
c. What proportion of games had at most two hits?
d. What proportion of games had between 5 and 10 hits?
e. What proportion of games had more than 15 hits?

3.24   The concentration of suspended solids in river water is an important environmental characteristic. The paper "Water Quality in Agricultural Watershed: Impact of Riparian Vegetation during Base Flow" (*Water Resources Bull.* (1981): 233–239) reported on concentration (in parts per million, or ppm) for several different rivers. Suppose that the accompanying 50 observations had been obtained for a particular river.

55.8 60.9 37.0 91.3 65.8 42.3 33.8 60.6 76.0 69.0 45.9
39.1 35.5 56.0 44.6 71.7 61.2 61.5 47.2 74.5 83.2 40.0
31.7 36.7 62.3 47.3 94.6 56.3 30.0 68.2 75.3 71.4 65.2
52.6 58.2 48.0 61.8 78.8 39.8 65.0 60.7 77.1 59.1 49.5
69.3 69.8 64.9 27.1 87.1 66.3

a. Why can't you base a frequency distribution on the class intervals 0–10, 10–20, 20–30, . . . , 90–100?
b. Construct a frequency distribution using class intervals 20–<30, 30–<40, . . . , 90–<100. (The resulting distribution agrees with that for one of the rivers discussed in the paper.)
c. What proportion of the concentration observations were less than 50? At least 60?
d. Just from the frequency distribution, can you determine the proportion of sample observations less than 65? Explain. Can you *estimate* this proportion from the distribution? How does your estimate compare with the actual value of the proportion?

3.25   Refer to the concentration data given in Exercise 3.24.
a. Obtain the cumulative relative frequencies for the class intervals specified in part **b** of that problem.
b. Use the cumulative relative frequencies to calculate the proportions described in part **c** of Exercise 3.24.

c. Use cumulative relative frequencies to calculate the proportion of observations in the interval 40–<70.
d. Just from the frequency distribution, can you determine the proportion of observations that are at most 40? Explain.

3.26 The results of the 1990 census included a state-by-state listing of population density. The accompanying table gives the number of people per square mile for each of the 50 states.

AL	79.6	LA	96.9	OH	264.9
AK	1.0	ME	39.8	OK	45.8
AZ	32.3	MD	489.2	OR	29.6
AR	45.1	MA	767.6	PA	265.1
CA	190.8	MI	163.6	RI	960.3
CO	31.8	MN	55.0	SC	115.8
CT	678.4	MS	54.9	SD	9.2
DE	340.8	MO	74.3	TN	118.3
FL	239.6	MT	5.5	TX	64.9
GA	111.9	NB	20.5	UT	21.0
HI	172.5	NV	10.9	VT	60.8
ID	12.2	NH	123.7	VA	156.3
IL	205.6	NJ	1042.0	WA	73.1
IN	154.6	NM	12.5	WV	74.5
IA	49.7	NY	381.0	WI	90.1
KS	30.3	NC	136.1	WY	4.7
KY	92.8	ND	9.3		

a. Construct a relative frequency distribution for state population density.
b. In your relative frequency distribution, did you use class intervals of equal widths? Why or why not?
c. Use the relative frequency distribution to give an approximate value for the proportion of states that have a population density of more than 100 people per square mile. Is this value close to the actual value?

3.27 The paper "Lessons from Pacemaker Implantations" (*J. Amer. Med. Assoc.* (1965): 231–232) gave the results of a study that followed 89 heart patients who had received electronic pacemakers. The time (in months) to the first electrical malfunction of the pacemaker was recorded.

24	20	16	32	14	22	2	12	24	6
10	20	8	16	12	24	14	20	18	14
16	18	20	22	24	26	28	18	14	10
12	24	6	12	18	16	34	18	20	22
24	26	18	2	18	12	12	8	24	10
14	16	22	24	22	20	24	28	20	22
26	20	6	14	16	18	24	18	16	6
16	10	14	18	24	22	28	24	30	34
26	24	22	28	30	22	24	22	32	

a. Summarize this data in the form of a frequency distribution, using class intervals of 0–<6, 6–<12, and so on.
b. Compute the relative frequencies and cumulative relative frequencies for each class interval of the frequency distribution of part **a**.
c. Show how the relative frequency for the class interval 12–<18 could be obtained from the cumulative relative frequencies.

Use the cumulative relative frequencies to give approximate answers to the following questions.
d. What proportion of those who participated in the study had pacemakers that did not malfunction within the first year?
e. If the pacemaker must be replaced as soon as the first electrical malfunction occurs, approximately what proportion required replacement between 1 and 2 years after implantation?
f. Estimate the time at which about 50% of the pacemakers had failed.
g. Estimate the time at which only about 10% of the pacemakers initially implanted were still functioning.

3.28 A student obtained data on fuel efficiency (mpg) and constructed a frequency distribution using the eight class intervals 27.0–<27.5, 27.5–<28.0, . . . , 30.5–<31.0. He then calculated and reported the following cumulative relative frequencies: .09, .23, .38, .35, .72, .80, .93, 1.00. Comment.

3.29 The clearness index was determined for the skies over Baghdad for each of the 365 days during a particular year ("Contribution to the Study of the Solar Radiation Climate of the Baghdad Environment," *Solar Energy* (1990): 7–12). The accompanying table summarizes the resulting data.

Clearness Index	Number of Days (frequency)
.15–<.25	8
.25–<.35	14
.35–<.45	28
.45–<.50	24
.50–<.55	39
.55–<.60	51
.60–<.65	106
.65–<.70	84
.70–<.75	11

a. Determine the relative frequencies and draw the corresponding histogram. (Be careful here — the intervals do not all have the same width.)
b. Cloudy days are those with a clearness index smaller than .35. What proportion of the days were cloudy?

c. Clear days are those for which the index is at least .65. What proportion of the days were clear?

3.30   The article "Associations between Violent and Nonviolent Criminality" (*Multivariate Behavioral Research* (1981): 237–242) reported the number of previous convictions for 283 adult males arrested for felony offenses. The accompanying frequency distribution is a summary of the data given in the paper. Draw the histogram corresponding to this frequency distribution, and comment on its shape.

Number of Previous Convictions	Frequency
0	0
1	16
2	27
3	37
4	46
5	36
6	40
7	31
8	27
9	13
10	8
11	2

3.31   How does the speed of a runner vary over the course of a marathon (a distance of 42.195 km)? Consider determining both the time to run the first five kilometers and the time to run between the 35 km and 40 km points, and then subtracting the former time from the latter time. A positive value of this difference corresponds to a runner slowing down toward the end of the race. The accompanying histogram is based on times of runners who participated in several different Japanese marathons ("Factors Affecting Runners' Marathon Performance,"

Chance (Fall, 1993): 24–30). What are some interesting features of this histogram? What is a typical difference value? Roughly what proportion of the runners ran the late distance more quickly than the early distance?

3.32   The article "Determination of Most Representative Subdivision" (*J. of Energy Engr.* (1993): 43–55) gave data on various characteristics of subdivisions that could be used in deciding whether to provide electrical power using overhead lines or underground lines. Below is data on the variable x = total length of streets within a subdivision.

1280	5320	4390	2100	1240	3060	4770
1050	360	3330	3380	340	1000	960
1320	530	3350	540	3870	1250	2400
960	1120	2120	450	2250	2320	2400
3150	5700	5220	500	1850	2460	5850
2700	2730	1670	100	5770	3150	1890
510	240	396	1419	2109		

a. Construct a stem-and-leaf display for this data using the thousands digit as the stem. Comment on the various features of the display.
b. Construct a histogram using class boundaries of 0–<1000, 1000–<2000, etc. How would you describe the shape of the histogram?
c. What proportion of subdivisions have total length less than 2000? Between 2000 and 4000?

3.33   The article cited in the previous exercise also gave the following values of the variable y = number of culs-de-sac and z = number of intersections.

y	1	0	1	0	0	2	0	1	1
z	1	8	6	1	1	5	3	0	0

y	1	2	1	0	0	1	1	0	1	1
z	4	4	0	0	1	2	1	4	0	4

(*continued*)

Figure for Exercise 3.31

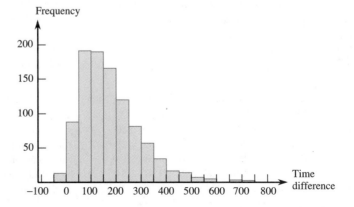

y	1	1	0	0	0	1	1	2	0	
z	0	3	0	1	1	0	1	3	2	
y	1	2	2	1	1	0	2	1	1	0
z	4	6	6	0	1	1	8	3	3	5
y	1	5	0	3	0	1	1	0	0	
z	0	5	2	3	1	0	0	0	3	

a. Construct a histogram for the number of culs-de-sac. What proportion of these subdivisions had no cul-de-sac? At least one?

b. Construct a histogram for the number of intersections. What proportion of these subdivisions had at most five intersections? Fewer than five?

3.34 Disparities among welfare payments by different states have been the source of much political controversy. The accompanying table reports average payment per person (in dollars) in the Aid to Families with Dependent Children Program for the 1990 fiscal year. Construct a relative frequency distribution for this data using equal interval widths. Draw the histogram corresponding to your frequency distribution.

Alaska	244.90	Tennessee	65.93
California	218.31	Wisconsin	155.04
Arizona	93.57	Ohio	115.26
Montana	114.95	Vermont	183.36
Texas	56.79	Connecticut	205.86
Nebraska	115.15	Pennsylvania	127.70
Minnesota	171.75	Maryland	132.86
Arkansas	65.96	South Carolina	71.91
Alabama	39.62	Hawaii	187.71
Illinois	112.28	Oregon	135.99
Indiana	92.43	Nevada	100.25
New Hampshire	164.20	Wyoming	113.84
Rhode Island	179.37	New Mexico	81.87
New Jersey	121.99	Kansas	113.88
Delaware	113.66	North Dakota	130.49
North Carolina	91.95	Missouri	91.93
Florida	95.43	Mississippi	40.22
Washington	160.41	Kentucky	85.21
Idaho	97.93	Michigan	154.75
Utah	118.36	Maine	150.12
Colorado	111.20	Massachusetts	200.99
Oklahoma	96.98	New York	193.48
South Dakota	95.52	West Virginia	82.94
Iowa	129.58	Virginia	97.98
Louisiana	55.81	Georgia	91.31

3.35 The behavior of children watching television has been a much-studied phenomenon. Now the same attention is being given to children engaged in playing with toys. The paper "A Temporal Analysis of Free Toy Play

and Distractibility in Young Children" (*J. of Exp. Child Psychology* (1991): 41–69) reported the accompanying data on play-episode lengths (in seconds) for a particular 5-year-old boy.

Class	Frequency
0–<5	54
5–<10	44
10–<15	28
15–<20	21
20–<40	31
40–<60	15
60–<90	16
90–<120	5
120–<180	8

a. Display this information in a histogram.

b. What proportion of episodes lasted at least 20 sec?

c. Roughly what proportion of episodes lasted between 40 and 75 sec?

3.36 An exam is given to students in an introductory statistics course. What is likely to be true of the shape of the histogram of scores if

a. the exam is quite easy?

b. the exam is quite difficult?

c. half the students in the class have had calculus, the other half have had no prior college math courses, and the exam emphasizes mathematical manipulation?

Explain your reasoning in each case.

3.37 Construct a histogram corresponding to each of the five frequency distributions given in the following table, and state whether each histogram is symmetric, bimodal, positively skewed, or negatively skewed.

	Frequency				
Class Interval	I	II	III	IV	V
0–<10	5	40	30	5	6
10–<20	10	25	10	25	5
20–<30	20	10	8	8	6
30–<40	30	8	7	7	9
40–<50	20	7	7	20	9
50–<60	10	5	8	25	23
60–<70	5	5	30	10	42

3.38 Using the five class intervals 100–<120, 120–<140, ... , 180–<200, devise a frequency distribution based on 70 observations whose histogram could be described as follows.

a. Symmetric

b. Bimodal

c. Positively skewed

d. Negatively skewed

3.4   Interpreting the Results of Statistical Analyses

The use of graphical data displays is quite common in newspapers, magazines, and journals, so it is important to be able to extract information from such displays. For example, data on test scores for a standardized math test given to eighth-graders in 37 states, two territories (Guam and the Virgin Islands), and the District of Columbia was presented in the August 13, 1992, edition of *USA Today*. Figure 3.29 gives both a stem-and-leaf display and a histogram summarizing this data. Careful examination of these displays reveals the following:

1. Most of the participating states had average eighth-grade math scores between 240 and 280. We would describe the shape of this display as negatively skewed, because of the longer tail on the low end of the distribution.

2. Three of the average scores differed substantially from the others. These turn out to be for the Virgin Islands (218), the District of Columbia (229), and Guam (230). These three could be described as outliers. It is interesting to note that the three unusual values are from the areas that are not states.

3. There do not appear to be any outliers on the high side.

4. A "typical" average math score for the 37 states would be somewhere around 260.

5. There is quite a bit of variability in average score from state to state.

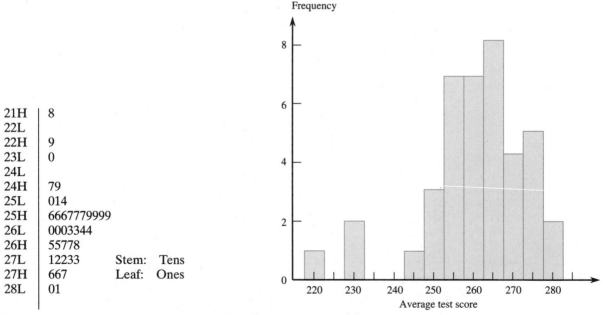

FIGURE 3.29 Stem-and-Leaf display and histogram for math test data

How would the displays have been different if the two territories and the District of Columbia had not participated in the testing? The resulting histogram is shown in Figure 3.30. Note that the display is now more symmetric, with no noticeable outliers. The display still reveals quite a bit of state-to-state variability in average score, and 260 still looks reasonable as a "typical" average score.

Now suppose that the two highest values among the 37 states (Montana and North Dakota) had been even higher. The stem-and-leaf display might then look like the one given in Figure 3.31. In this stem-and-leaf display, two values stand out from the main part of the display. This would catch our attention and might cause us to look carefully at these two states to determine what factors may be related to high math scores. Researchers at Educational Testing Service found math scores to be associated with such factors as amount of time spent reading and amount of time spent watching TV.

FIGURE 3.30 A histogram for the modified math score data

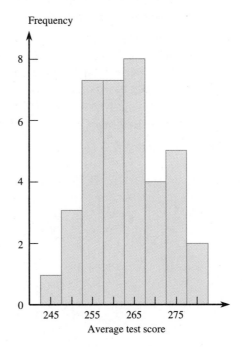

FIGURE 3.31 Stem-and-leaf display for modified math score data

24H	79
25L	014
25H	6667779999
26L	0003344
26H	55778
27L	12233
27H	667
28L	
28H	
29L	
29H	68

Stem: Tens
Leaf: Ones

What to Look for in Published Data

Here are some questions you might ask yourself when attempting to extract information from a graphical data display.

- Is the chosen display appropriate for the type of data collected?
- How would you describe the shape of the distribution, and what does this say about the variable being summarized?
- Are there any outliers (noticeably unusual values) in the data set? Is there any plausible explanation for why these differ from the rest of the data? (This often leads to further avenues of investigation.)
- Where do most of the data values fall? What is a typical value for the data set? What does this say about the variable being summarized?
- Is there much variability in the data values? What does this say about the variable being summarized?

Of course, you should always think carefully about how the data was collected. If the data was not gathered in a reasonable manner (based on sound sampling methods or experimental design principles), you should be cautious in formulating any conclusions based on the data.

Consider the histogram in Figure 3.32, which is based on data published by the National Center for Health Statistics. The data set summarized by this histogram

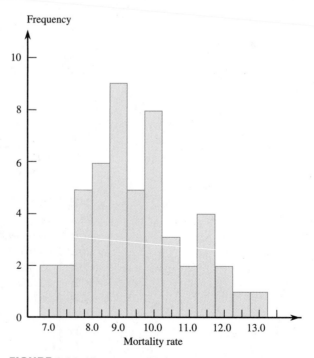

FIGURE 3.32 Histogram of infant mortality rates

consisted of 1990 infant mortality rates (deaths per 1000 live births) for the 50 states in the United States. A histogram is an appropriate way of summarizing this data (although with only 50 observations, a stem-and-leaf display would also have been reasonable). The histogram itself is slightly positively skewed, with most mortality rates between 7.5 and 12. There is quite a bit of variability in infant mortality rate from state to state — perhaps more than we might have expected. This variability might be explained by differences in economic conditions or in access to health care. We may want to look further into these issues. Although there are no obvious outliers, the upper tail is a little longer than the lower tail. The three largest values in the data set are 12.1 (Alabama), 12.3 (Georgia), and 12.8 (South Carolina) — all southern states. Again, this may suggest some interesting questions that deserve further investigation. A typical infant mortality rate would be about 9.5 deaths per 1000 live births. This represents an improvement, since researchers at the National Center for Health Statistics stated that the overall rate for 1988 was 10 deaths per 1000 live births. However, they also point out that the United States still ranked 22nd out of 24 industrialized nations surveyed, with only New Zealand and Israel having higher infant mortality rates.

Summary of Key Concepts and Formulas

Term or Formula	Comment
Frequency distribution	A table that displays frequencies, and sometimes relative and cumulative relative frequencies, for categories (categorical data), possible values (discrete data), or class intervals (continuous data).
Bar chart	A graph of a frequency distribution for a categorical data set. Each category is represented by a bar and the area of the bar is proportional to the corresponding frequency or relative frequency.
Pie chart	A graph of a frequency distribution for a categorical data set. Each category is represented by a slice of the pie and the area of the slice is proportional to the corresponding frequency or relative frequency.
Dotplot	A picture of numerical data in which each observation is represented by a dot on or above a horizontal measurement axis.
Stem-and-leaf display	A method of organizing numerical data in which the stem values (leading digit(s) of the observations) are listed in a column, and the leaf (trailing digit(s)) for each observation is then listed beside the corresponding stem. Sometimes stems are repeated to stretch the display.
Histogram	A picture of the information in a frequency distribution for a numerical data set. A rectangle is drawn above each possible value (discrete data) or class interval. The rectangle's area is proportional to the corresponding frequency or relative frequency.
Histogram shapes	A (smoothed) histogram may be unimodal (a single peak), bimodal (two peaks), or multimodal. A unimodal histogram may be symmetric, positively skewed (a long right or upper tail), or negatively skewed. A frequently occurring shape is the normal curve.

Supplementary Exercises 3.39 – 3.52

3.39   One advantage that margarine has over butter is that it is typically much lower in saturated fat content. (Unfortunately, you can usually taste the difference!) The accompanying MINITAB stem-and-leaf display contains saturated fat contents (in percent) for 37 different margarines (*Consumer Reports,* June 1989). Explain how MINITAB constructed the display. (*Note*: The butters investigated all had fat contents of 64%, 65%, or 66%.)

```
Stem-and-leaf of satfat N = 37
Leaf Unit = 1.0
    2    0   99
    3    1   0
    3    1
    8    1   44444
   10    1   77
    b.   1   8888888888888888899
    7    2   0000
    3    2   2
    2    2   5
    1    2   7
```

3.40   A sample of 50 individuals who recently joined a certain travel club yielded the responses on occupation shown in the accompanying table (C = clerical, M = manager/executive, P = professional, R = retired, S = sales, T = skilled tradesperson, O = other). Summarize this data.

P	R	R	M	S	R	P	R	C	P	M
R	T	O	R	S	R	S	P	M	R	P
C	R	R	S	T	P	P	C	M	S	R
R	P	R	S	R	R	P	M	P	R	R
S	C	P	R	S	M				

3.41   The accompanying frequency distribution of the number of years of continuous service at the time of resignation from a job with an oil company appeared in the paper "The Role of Performance in the Turnover Process" (*Academy of Management J.* (1982): 137–147). Construct the histogram corresponding to this frequency distribution. Which terms introduced in this chapter (*symmetric, skewed,* and so on) would you use to describe the histogram?

Duration of service	Frequency
0–<1	4
1–<2	41
2–<3	67

(continued)

Duration of Service	Frequency
3–<4	82
4–<5	28
5–<6	43
6–<7	14
7–<8	17
8–<9	11
9–<10	7
10–<11	14
11–<12	6
12–<13	14
13–<14	5
14–<15	2

3.42   Many nutritional experts have expressed concern about the high levels of sodium in prepared foods. The accompanying data on sodium content (mg) per frozen meal appeared in the article "Comparison of 'Light' Frozen Meals" (*Boston Globe,* April 24, 1991).

```
720  530  800  690  880  1050  340  810  760  300  400
680  780  390  950  520   500  630  480  940  450  990
910  420  850  390  600
```

MINITAB gave the two accompanying histograms; we selected the classes for the second histogram and let MINITAB do so for the first one.

a. Do the two histograms give different impressions about the distribution of values?

b. Use each histogram to determine approximately the proportion of observations that are less than 800, and compare to the actual proportion.

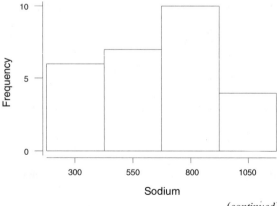

(continued)

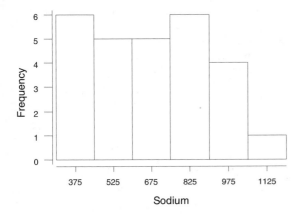

3.43 The paper "Paraquat and Marijuana Risk Assessment" (*Amer. J. of Public Health* (1983): 784–788) reported the results of a 1978 telephone survey on marijuana usage. The accompanying frequency distribution gives the amount of marijuana (in grams) smoked per week for those respondents who indicated that they did use the drug.

Amount Smoked	Frequency
0–<3	94
3–<11	269
11–<18	70
18–<25	48
25–<32	31
32–<39	10
39–<46	5
46–<53	0
53–<60	1
60–<67	0
67–<74	1

a. Display the information given in the frequency distribution in the form of a histogram.
b. What proportion of respondents smoked 25 or more grams per week?
c. Use the histogram to estimate the proportion of respondents who smoked more than 15 g per week.
d. Calculate the cumulative relative frequencies, and use them to determine the proportion of respondents who smoked at least 25 g per week.

3.44 A *Pareto diagram* is a variation of a histogram for categorical data resulting from a quality-control study. Each category represents a different type of product nonconformity or production problem. The categories are or-

dered so that the one with the largest frequency appears on the far left, then the one with the second largest frequency, and so on. Suppose the following information on nonconformities in circuit packs is obtained: failed component, 126; incorrect component, 210; insufficient solder, 67; excess solder, 54; missing component, 131. Construct a Pareto diagram.

3.45 Data on engine emissions for 46 vehicles is given in the accompanying table (*Technometrics* (Nov. 1980): 487).

Vehicle	HC	CO	Vehicle	HC	CO
1	.50	5.01	24	1.10	22.92
2	.65	14.67	25	.65	11.20
3	.46	8.60	26	.43	3.81
4	.41	4.42	27	.48	3.45
5	.41	4.95	28	.41	1.85
6	.39	7.24	29	.51	4.10
7	.44	7.51	30	.41	2.26
8	.55	12.30	31	.47	4.74
9	.72	14.59	32	.52	4.29
10	.64	7.98	33	.56	5.36
11	.83	11.53	34	.70	14.83
12	.38	4.10	35	.51	5.69
13	.38	5.21	36	.52	6.35
14	.50	12.10	37	.57	6.02
15	.60	9.62	38	.51	5.79
16	.73	14.97	39	.36	2.03
17	.83	15.13	40	.49	4.62
18	.57	5.04	41	.52	6.78
19	.34	3.95	42	.61	8.43
20	.41	3.38	43	.58	6.02
21	.37	4.12	44	.46	3.99
22	1.02	23.53	45	.47	5.22
23	.87	19.00	46	.55	7.47

a. Construct a frequency distribution and histogram for the hydrocarbon (HC) emissions data. Do any of the observations appear to be outliers?
b. Construct a frequency distribution and histogram for the carbon monoxide (CO) data.
c. Are the HC and CO histograms symmetric or skewed? If they are both skewed, is the direction of the skew the same?

3.46 The table at the top of page 90 gives data on the cost per ounce (in cents) for 31 shampoos intended for "normal" hair and 29 shampoos intended for "fine" hair (*Consumer Reports,* 1992). Use methods developed in this chapter to describe and compare the two cost distributions.

Normal

79	63	19	9	37	49	20	16	55	69	23
14	9	87	44	13	16	23	20	64	28	18
32	81	85	47	50	8	13	21	9		

Fine

69	9	23	22	8	12	32	12	18	74	19
63	49	37	55	85	44	87	17	11	23	50
65	51	35	14	20	28	8				

3.47 Reconsider the data on lengths of mystery novels given in the chapter introduction.

a. Use various methods developed in this chapter to organize and describe the data.

b. Visit a bookstore with a good selection of science fiction books, select a sample, record page lengths, and compare the data to the mystery novels data.

3.48 The two given frequency distributions of storm duration (in minutes) are based on data appearing in the article "Lightning Phenomenology in the Tampa Bay Area" (*J. of Geophysical Research* (1984): 789–805). Construct a histogram for each of the frequency distributions, and discuss the similarities and differences between the two with respect to shape.

Storm Duration	Single-peak Storms Frequency	Multiple-peak Storms Frequency
0–<25	1	0
25–<50	17	1
50–<75	14	1
75–<100	11	3
100–<125	8	2
125–<150	8	2
150–<175	5	1
175–<200	4	3
200–<225	3	1
225–<250	2	6
250–<275	0	4
275–<300	1	2

3.49 The paper "The Acid Rain Controversy: The Limits of Confidence" (*Amer. Statistician* (1983): 385–394) gave the accompanying data on average SO_2 (sulfur dioxide) emission rates from utility and industrial boilers (lb/million Btu) for 47 states. (Data from Idaho, Alaska, and Hawaii was not given.)

2.3	2.7	1.5	1.5	.3	.6	4.2	1.3	1.2	.4	.5
2.2	4.5	3.8	1.2	.2	1.0	.7	.2	1.4	.7	3.6
1.0	.7	1.7	.5	.1	.6	2.5	2.7	1.5	1.4	2.9
1.0	3.4	2.1	.9	1.9	1.0	1.7	1.8	.6	1.7	2.9
1.8	1.4	3.7								

a. Summarize this set of data by constructing a relative frequency distribution.

b. Draw the histogram corresponding to the frequency distribution in part **a.** Would you describe the histogram as symmetric or skewed?

c. Use the relative frequency distribution of part **a** to compute the cumulative relative frequencies.

d. Use the cumulative relative frequencies to give the approximate proportion of states with SO_2 emission rates that

i. were below 1.0 lb/million Btu.

ii. were between 1.0 and 2.0 lb/million Btu.

iii. were at least 2.0 lb/million Btu.

3.50 Americium 241 (^{241}Am) is a radioactive material used in the manufacture of smoke detectors. The article "Retention and Dosimetry of Injected ^{241}Am in Beagles" (*Radiation Research* (1984): 564–575) described a study in which 55 beagles were injected with a dose of ^{241}Am (proportional to the animals' weights). Skeletal retention of ^{241}Am (μCi/kg) was recorded for each beagle, resulting in the accompanying data.

.196	.451	.498	.411	.324	.190	.489	.300	.346	.448	.188
.399	.305	.304	.287	.243	.334	.299	.292	.419	.236	.315
.447	.585	.291	.186	.393	.419	.335	.332	.292	.375	.349
.324	.301	.333	.408	.399	.303	.318	.468	.441	.306	.367
.345	.428	.345	.412	.337	.353	.357	.320	.354	.361	.329

a. Construct a frequency distribution for this data, and draw the corresponding histogram.

b. Write a short description of the important features of the shape of the histogram.

3.51 In the paper "Reproduction in Laboratory Colonies of Bank Vole" (*Oikos* (1983): 184), the authors presented the results of a study on litter size. (A vole is a small rodent with a stout body, blunt nose, and short ears.) As each new litter was born, the number of babies was recorded, and the accompanying results were obtained.

Size of Litter

3	6	5	6	5	7	5	7	6	6	6
4	6	5	6	4	3	5	6	4	5	9
6	5	6	1	9	7	8	3	7	4	5
5	6	7	3	6	6	9	4	5	7	5
6	8	6	4	7	5	7	4	5	8	6
7	2	7	7	3	3	5	4	6	4	6
3	7	8	5	7	7	7	7	9	8	7
6	7	6	4	7	10	5	2	3	6	6
4	7	6	7	5	5	5	7	5	8	8
4	9	7	5	4	6	5	8	4	5	6
6	3	6	8	6	8	6	5	8	6	11

(*continued*)

4	7	6	8	9	7	3	8	3	4	6
4	5	7	5	6	5	7	6	9	3	5
9	7	5	6	7	5	8	6	8	8	6
5	7	4	8	7	7	7	5	3	8	6
10	4	5	5	5						

a. Construct a relative frequency distribution for this data.

b. What proportion of the litters had more than 6 babies? Between 3 and 8 (inclusive)?

c. Is it easier to answer questions like those posed in part **b** using the relative frequency distribution or the raw data given in the table? Explain.

3.52 Compute the cumulative relative frequencies for the data of Exercise 3.51 and use them to answer the following questions.

a. What proportion of observations are at most 8? At least 8?

b. What proportion of litters contain between 5 and 10 (inclusive) offspring?

References

Chambers, John, William Cleveland, Beat Kleiner, and Paul Tukey. *Graphical Methods for Data Analysis.* Belmont, CA: Wadsworth, 1983. (This is an excellent survey of methods, illustrated with numerous interesting examples.)

Cleveland, William. *The Elements of Graphing Data,* 2d ed. Summit, NJ: Hobart Press, 1994. (An informal and informative introduction to various aspects of graphical analysis.)

Freedman, David, Robert Pisani, and Roger Purves. *Statistics,* 3d ed. New York: W. W. Norton, 1997. (An excellent, very informal introduction to concepts, with some insightful cautionary examples concerning misuses of statistical methods.)

Moore, David. *Statistics: Concepts and Controversies,* 4th ed. New York: W. H. Freeman, 1997. (A nonmathematical yet highly entertaining introduction to our discipline — two thumbs up!)

Velleman, Paul, and David Hoaglin. *Applications, Basics, and Computing of Exploratory Data Analysis.* Boston: Duxbury, 1981. (Subtitled *ABCs of EDA,* this book contains a good treatment of exploratory methods.)

GRAPHING CALCULATOR EXPLORATIONS

3.1 Using Lists on Your Calculator

As is well known, calculators and computers work their arithmetic magic by storing numbers in what are known as "memory cells." When, for instance, an addition is needed, the computer looks in its "memory" for the two numbers, retrieves them, and then adds them. In the early days of calculators, there were very few of these expensive memory cells, and calculations were performed one at a time while the user entered data. Statistical calculations were next to impossible on these calculators; a single keystroke error, usually not detectable by the user, would yield erroneous results. As computer chips have gotten tinier and less expensive, more memory populates computers and calculators. In the modern scientific calculator, the explosion of memory allows a very useful extension of the idea of a single memory cell: a group of memory cells known as a "list." The idea behind a list is that a whole set of data, complete with a "name," can be stored in the calculator and analyzed as a whole. The list capability has transformed the calculator into a very powerful tool for analyzing and displaying data. A data set can be entered into a calculator and stored for future analysis. Furthermore, more than one analysis can be performed after the data set is entered, and since all the numbers are in the calculator at the same time, graphic representations of data such as those presented in this chapter are possible.

The actual capabilities of lists and keystrokes to implement these capabilities will vary from calculator to calculator, so we will not be specific about particular calculator keystrokes. *Your best source for learning about your calculator is the manual that came with it!* The designers of calculators strive to make calculators simple and intuitive, but the increasing power and complexity of the calculator have far outstripped the capability of designers to put everything within reach of a single keystroke. The downside of all this power is that you need a manual to fully realize the potential of the calculator. Your calculator may implement some of its capabilities with special keys or menus, and the menus may present functions that require additional information to be entered in a particular order. You should not expect to be able to memorize all these capabilities! That's why you have the calculator manual to read and use as a reference.

To use your calculator for statistical analysis, you should be able to manipulate lists effectively. We will now present some of the features in calculators which have a "list-based" capability and which will be important to you in performing statistical analysis. In future Graphing Calculator Explorations, we will assume that you have already taught yourself how to perform these list tasks. We strongly suggest that you create a reference table that contains these list operations and either the required keystrokes or the corresponding page reference in the calculator manual. As you use your calculator, many of these skills will become second nature.

Capability	Why it's important
Create a list	Some lists are provided automatically for your use, already labeled "List 1 or L1," or "List 2 or L2," etc. You will want to save data in lists with more informative names than these, such as "height" or "time."
Enter data into a list	This knowledge is, of course, fundamental to all analyses you will be performing.
Delete a list from the calculator	As powerful as your calculator is, there is a limit to its memory — there will come a time when you will need to delete the old data to prepare for the new.
Insert a number in a certain location in the list or Delete a particular number in the list	If you are like everyone else, you will eventually add an extra number you did not intend or leave out a number from where it should be. Correcting these errors is a lot faster than deleting a whole list and starting over.
Copy data from one list to another	This will give you one of those things so precious to everyone who works with a finicky calculator (or finicky fingers?): a backup copy of the data! *(continued)*

Capability	Why it's important
Perform arithmetic operations with lists	Rather than perform the same arithmetic sequence separately on a set of numbers, you can do the calculations one list at a time. For example, to change units from inches to centimeters, you can multiply *all* the numbers in the list by 2.54. Usually, this is done with a statement something like 2.54 × ListName1 → ListName2. (The equal sign, =, is sometimes used in place of the arrow symbol.)

As we proceed, we will be more specific — and more detailed — about how you can utilize your calculator's list capabilities. The procedures given in the table are very general, but very important. You must be familiar with list manipulation to do effective statistical work with your calculator. Your calculator manual probably has a chapter called "Using Lists." We strongly encourage you to read it!

3.2 Setting the Statistics Window

As we move into graphic descriptions of data, we are getting closer to a time when you can use your calculator to produce graphs to help you interpret data. You probably have already used your calculator to graph functions. If so, some of what we will say here will be in the nature of a review. If you are new to the world of graphing calculators, you will need a basic understanding of how to set up your calculator's "viewing window" for displaying graphs.

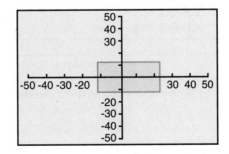

The metaphor of a "window" through which to view the "world" is very apt. If you think of the usual Cartesian x–y axis as the calculator's "world view," and your calculator view window as a portal to view this Cartesian world, you will be in the right frame of mind. As you set up your calculator for graphing, the fundamental problem will be, Where in the world do you put your view window? The quick and easy answer is, you put your view window where your data is! We will illustrate how to do this by constructing a histogram, using the data from Example 3.15. To illustrate some of the problems involved in setting the view window, we will begin with an admittedly terrible graph of a histogram, and then gradually improve it. First, we want you to do something a bit strange, but trust us: Enter the function $y = 10/x$ into your calculator. (This will mean navigating to the "$y =$" screen and entering the function there.) Now enter the data from Example 3.15 into your calculator in List 1. After entering the data, navigate your calculator's menu system to indicate that you want to make a histogram. Actual keystrokes will vary among calculators, but the terms "Stat Graph" and "Stat Plot" are commonly used in calculators. At the end of this navigation, you should have made an explicit choice to plot a histogram.

After making this choice, you can position the view screen where the calculator will draw the graph in its world. In the case of a histogram, your calculator will determine from the data the position of the graph in its Cartesian world. The view window, on the other hand, is based on settings that you will manually supply. (Some calculators have a capability of automatically placing the view window over the appropriate position in its Cartesian system. Pretend, for the moment, that you are unaware of this.) To set up the view window, usually referred to as the "window," navigate to your calculator's menu system, or possibly just press a "window" key. When you find the graph setup screen, it will look something like the accompanying figure.

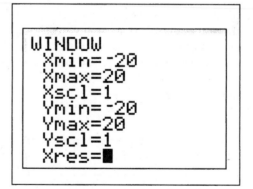

(There may be other information present also, but the numbers here will be our focus for this exploration. Your numbers may be different from these; for ease in following the discussion, change the values on your calculator to match those in the figure.) These numbers determine where the viewing window is placed over the calculator's Cartesian coordinate system. Exit from the setup window and plot a histogram, by using a "graph" key or a menu, depending on your calculator. You should have something like what is shown in the accompanying figure. Keep in mind that we are building a worst-case scenario here, and your graph will get better as we go! Notice that there actually is a histogram there, but it is fairly small, and there is also a pesky function overlaying the histogram. We will, in successive steps, graph a more reasonable histogram of our data!

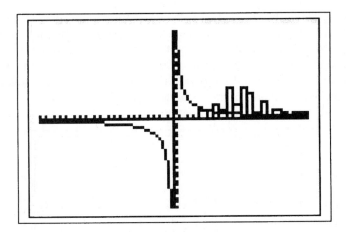

First, remember that the calculator's world is the Cartesian coordinate system. If you have been using the calculator to plot mathematical functions, and they are still "in" the calculator, both the function and the histogram will be drawn. Oddly, this happens by design. Since a user most likely will not graph a mathematical function and a histogram at the same time, the graphing space is "shared" by the two processes to save calculator memory. The solution to this problem is easy: Get rid of that pesky mathematical function. You can do this by deleting the function using the same screen as when you created the function. Do so now, and redraw the histogram. Assuming you have successfully banished the function to cyberspace, we'll take the next step toward a proper histogram.

Your calculator has automatically placed the histogram in the first quadrant of the calculator world, but the view window is centered on the Cartesian origin. We need to manually reposition the viewing window for a better look. Return to the view window setup screen and make the following changes: set Xmin to −1, and Ymin to −1. Now redraw the histogram. The view window will now be positioned so that the left side and bottom of the view screen will contain the *x* and *y* axis, as in the figure above. This is certainly an improvement, but the histogram is still rather small. To correct this problem we will adjust the top of the view window. Return to the graph setup screen and locate the lines, "Xmax = 20" and "Ymax = 20." Change these to "Xmax = 24" and "Ymax = 10" and regraph the histogram. You should see a very well-spaced histogram similar to the one in the figure below.

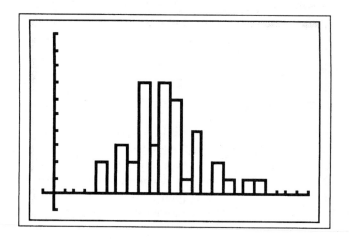

You will find that adjusting the view window is a frequent task in creating effective statistical graphs. While each of the statistical plots has its own individuality, your construction of them will always involve positioning the view window over the Cartesian world view of your calculator. Even if it has an automatic function to position statistical graphs, you will find it necessary sometimes to "improve" on the calculator's automatic choice. When you manually intervene to change your graph, keep in mind the idea of positioning a view window over a Cartesian coordinate system. That idea will help you organize your thoughts about how to change the view window and make the task less frustrating.

3.3 Scaling the Histogram

When we constructed a histogram in the previous exploration, there were some numbers that we ignored in the view screen. We would like to return to those numbers now because they can seriously affect the look of a histogram. When we left the histogram before, the numbers in our view window were set as shown in the accompanying figure. These settings placed the view window over the calculator's Cartesian system for effective viewing of the histogram from the data of Example 3.15. We would now like to do a little experimentation with the "Xscale" and the "Yscale." In all statistical graphs produced by the calculator, the Xscale and Yscale choices will control the placement of the little "tick" marks on the *x* and *y* axis. Return to Graphing Calculator Exploration 3.2, where we constructed the histogram of the

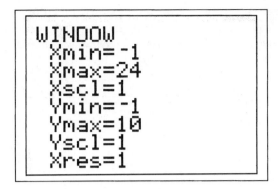

data from Example 3.15. With the X and Yscale defined as they were there, you should notice little tick marks on both axes at the positive integers, 1, 2, 3, 4, ... Now change the Yscale value to 2 and redraw the histogram. You should see a graph similar to the accompanying figure. Notice that the *y*-axis tick marks now appear at 2, 4, 6, 8, ...

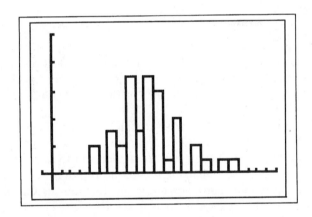

Changing the Xscale to 2 will have the same effect on the *x*-axis tick marks, placing them on the *x*-axis at the values of 2, 4, 6, 8, ... In the case of the histogram, changing the Xscale will have another result, which you can see in the following figure. Changing the Xscale has altered not only the tick marks but also the class intervals for the histogram! This is not usually a problem for statistical graphing, except in the case of the histogram; the choice of class intervals can significantly change the look and feel of the histogram. For this data, the Xscale was previously set equal to 1, and the histogram looked fairly symmetric and mound shaped. When the Xscale is set to equal 2, this picture seems to change. It appears that the data is "tailing out" to the right, and the two leftmost intervals have much smaller values than their immediate neighbors. The choice of Xscale can affect judgments about the shape of the histogram. Because of this possibility, it is wise to look at a histogram with varying choices of the Xscale value. If the shape appears very similar for different choices of Xscale, it is possible to interpret and describe the shape with confidence. However, if different Xscale choices alter the look of the histogram we should probably be more tentative in our descriptions!

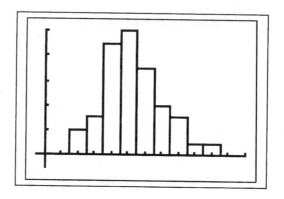

3.4 Drawing a Histogram from a Frequency Distribution

If the data has already been organized as a frequency distribution (such as in Table 3.8), there is no need to enter original data into the calculator to get a histogram. The calculator's list commands can be used to construct a histogram or a relative frequency histogram. We will use the data in Table 3.8 to illustrate how this is done.

First, enter the number of accidents, 0 through 11, in List1. Then enter the frequencies, 117, 157, . . . , 3, 1, in List2. Once these are entered, you can plot the histogram. The methods for doing this can vary from calculator to calculator, so you — yet again! — need to consult your manual. Whatever procedure your calculator uses, you will need to indicate that you want a histogram, that the possible data values are in one list, and that the frequencies are in a second list. Generally speaking, there are two approaches to doing this. If your calculator requires a set of keystrokes to get to a "histogram" menu, you might have to execute a command something like this:

Histogram(List1, List2)

If this is how your calculator works, be sure to get the order of the parameters correct; your manual will tell you whether the list of possible data values or the list of frequencies should come first.

The second approach involves a generic "set-up" screen, where you have a list of graphs, possibly in iconic form similar to the accompanying screen.

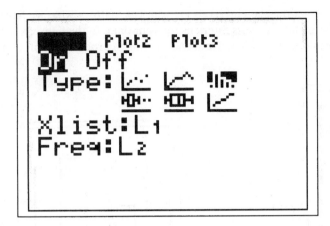

If your calculator has such a set-up screen, you will have to fill in the blanks with the correct information. For this calculator, "Xlist" is the list that contains the set of different possible data values (List1), and "Freq" is the list that contains the frequencies. The resulting plot for the accident data is shown in the accompanying figure.

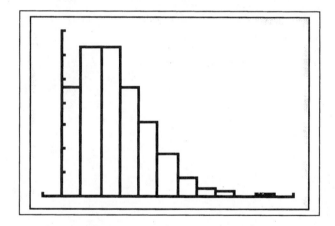

To construct a relative frequency histogram, we need to create a list that contains the relative frequencies. Using the list commands, we divide the frequencies by the total number of elements in the data set (708) and store the results in List3.

List2/708 → List3

Now you can graph the histogram of relative frequencies by substituting List3 for List2 in the procedure above. The shape of the plot will, of course, be very similar to the shape of the original histogram, so we will not show it here.

Sometimes you may have data that is already summarized in class intervals, such as the data in Table 3.6, the hair mercury content of Seychelles fishermen. Your calculator can also handle this situation, *as long as the class intervals have equal widths.*

You will need to set the class intervals for your calculator — what we have been calling Xscale — to match the class intervals for your data. In the figure below, we have used the class midpoints and the frequency counts to produce the plot on the right.

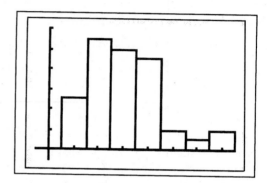

4

Numerical Methods for Describing Data

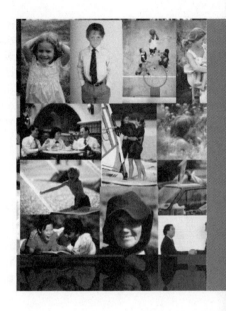

INTRODUCTION

Exposure to various chemical substances can pose substantial health risks. Evidence from recent years suggests that high indoor radon concentration may be linked to the development of childhood cancers. The article "Indoor Radon and Childhood Cancer" (*The Lancet* (1991): 1537–1538) presented the accompanying data on radon concentration (Bq/m^3) in two different samples of houses. The first sample consisted of houses in which a child diagnosed with cancer had been residing. Houses in the second sample had no recorded cases of childhood cancer.

Cancer

10	21	5	23	15	11	9	13	27	13	39	22
7	20	45	12	15	3	8	11	18	16	23	16
34	10	15	11	18	210	22	11	6	17	33	10
9	57	16	21	18	38						

No Cancer

9	38	11	12	29	5	7	6	8	29	24	12
17	11	3	9	33	17	55	11	29	13	24	7
11	21	6	39	29	7	8	55	9	21	9	3
85	11	14									

What is a representative or typical radon concentration value for each sample, and how do these values compare for the two samples? Do the two samples show similar amounts of variability, or are the observations in one sample substantially more spread out than those in the other sample? As we have seen, a stem-and-leaf display, a frequency distribution, or a histogram gives general impressions about where each data set is centered and how much it spreads out about its center. This chapter shows how to calculate numerical summary measures that describe more precisely both the center and the extent of spread. Section 4.1 introduces the mean and median, the two most popular measures of the center of a distribution. The variance and the standard deviation are presented in Section 4.2 as measures of variability. Later sections discuss several techniques for using such summary measures to describe other features of the data.

4.1   Describing the Center of a Data Set

An informative way to describe the location of a numerical data set is to report a value that is representative of the observations in the set. This number describes roughly where the data set is located or "centered" along the number line, and it is called a measure of center. The two most popular measures of center are the *mean* and the *median.*

The Mean

The **mean** of a set of numerical observations is just the familiar arithmetic average, the sum of the observations divided by the number of observations. Values in such a data set are observations on a numerical variable. The variable might be number of traffic accidents, number of pages in a book, reaction time, yield from a chemical reaction, etc. When working with sample data, it is helpful to have concise notation for the variable, sample size, and individual observations. Let

x = the variable for which we have sample data

n = the number of sample observations (sample size)

x_1 = the first sample observation

x_2 = the second sample observation

$\vdots$

x_n = the nth (last) sample observation

For example, we might have a sample consisting of $n = 4$ observations on $x =$ battery lifetime (hr):

$x_1 = 5.9$ $x_2 = 7.3$ $x_3 = 6.6$ $x_4 = 5.7$

Notice that the value of the subscript on x has no relationship to the magnitude of the observation. In this example, x_1 is just the first observation in the data set and not necessarily the smallest observation.

The sum of $x_1, x_2, \ldots, x_n$ can be denoted by $x_1 + x_2 + \cdots + x_n$, but this is cumbersome. The Greek letter Σ is traditionally used in mathematics to denote summation. In particular, $\Sigma\, x$ will denote the sum of all of the x values in the data set under consideration.*

Definition

The **sample mean** of a numerical sample $x_1, x_2, \ldots, x_n$, denoted by $\bar{x}$, is

$$\bar{x} = \frac{\text{sum of all observations in the sample}}{\text{number of observations in the sample}} = \frac{x_1 + x_2 + \cdots + x_n}{n} = \frac{\Sigma\, x}{n}$$

EXAMPLE 4.1

Traumatic knee dislocation often requires surgery to repair ruptured ligaments. One measure of recovery is range of motion (measured by the angle formed when, starting with the leg straight, the knee is bent as far as possible). The article "Reconstruction of the Anterior and Posterior Cruciate Ligaments after Knee Dislocation" (*American Journal of Sports Medicine* (1999): 189–197) reported the following postsurgical range of motion for a sample of 13 patients.

Range of Motion (degrees)

$x_1 = 154 \quad x_2 = 142 \quad x_3 = 137 \quad x_4 = 133 \quad x_5 = 122 \quad x_6 = 126 \quad x_7 = 135$

$x_8 = 135 \quad x_9 = 108 \quad x_{10} = 120 \quad x_{11} = 127 \quad x_{12} = 134 \quad x_{13} = 122$

The sum of these sample values is $154 + 142 + 137 + \cdots + 122 = 1695$, and the sample mean range of motion is

$$\bar{x} = \frac{\Sigma\, x}{n} = \frac{1695}{13} = 130.38$$

We would report 130.38 degrees as a representative value of range of motion for this sample (even though there is no patient in the sample that actually had a range of motion of 130.38 degrees).

The data values in Example 4.1 were all integers, yet the mean was given as 130.38. It is common to use more digits of decimal accuracy for the mean. This allows the value of the mean to fall between possible observable values (for example, the average number of children per family could be 1.8, whereas no single family will have 1.8 children). It could also be argued that the mean carries with it more precision than does any single observation.

The sample mean $\bar{x}$ is computed from sample observations, so it is a characteristic of the particular sample in hand. It is customary to use Roman letters to denote *sample* characteristics, as we have done with $\bar{x}$. Characteristics of the entire

---

*It is also common to see $\Sigma\, x$ written as $\sum_{i=1}^{n} x_i$, but for simplicity we will usually omit the summation indices.

population are usually denoted by Greek letters. One of the most important such characteristics is the population mean.

Definition

The **population mean,** denoted by μ, is the average of all x values in the entire population.

For example, the true average fuel efficiency for all 600,000 cars of a certain type under specified conditions might be $\mu = 27.5$ mpg. A sample of $n = 5$ cars might yield efficiencies of

  27.3   26.2   28.4   27.9   26.5

from which we obtain $\bar{x} = 27.26$ for this particular sample (somewhat smaller than μ). However, a second sample might give $\bar{x} = 28.52$, a third $\bar{x} = 26.85$, and so on. The value of $\bar{x}$ varies from sample to sample, whereas there is just one value for μ. We shall see subsequently how the value of $\bar{x}$ from a particular sample can be used to draw various conclusions about the value of μ.

EXAMPLE 4.2   The 50 states plus the District of Columbia contain a total of 3137 counties. Let x denote the number of residents of a county. Then there are 3137 values of the variable x in the population. The sum of these 3137 values is 248,709,873 (1990 census), so the population average value of x is

$$\mu = \frac{248,709,873}{3137} = 79,282.7 \text{ residents per county}$$

We used *The World Almanac and Book of Facts* to select three different samples at random from this population of counties, with each sample consisting of five counties. The results appear in Table 4.1, along with the sample mean for each sample.

Table 4.1 *Three samples from the population of all U.S. counties (x = number of residents)*

Sample 1		Sample 2		Sample 3	
County	**x value**	**County**	**x value**	**County**	**x value**
Fayette, TX	20,095	Stoddard, MO	28,895	Chattahoochee, GA	16,934
Monroe, IN	108,978	Johnston, OK	10,032	Petroleum, MT	519
Greene, NC	15,384	Sumter, AL	16,174	Armstrong, PA	73,478
Shoshone, ID	13,931	Milwaukee, WI	959,275	Smith, MI	14,798
Jasper, IN	24,960	Albany, WY	30,797	Benton, MO	13,859
$\Sigma x = 183,348$ $\bar{x} = 36,669.6$		$\Sigma x = 1,045,173$ $\bar{x} = 209,034.6$		$\Sigma x = 119,588$ $\bar{x} = 23,917.6$	

Not only are the three $\bar{x}$ values different from one another — because they are based on three different samples and the value of $\bar{x}$ depends on the x values in the sample — but none of the three comes close to the value of the population mean, μ. If we did not know the value of μ but had only sample 1 available, we might use $\bar{x} = 36{,}669.6$ as an *estimate* of μ, but our estimate would be far off the mark. Alternatively, we could combine the three samples into a single sample with $n = 15$ observations:

$$x_1 = 20{,}095, \ldots, x_5 = 24{,}960, \ldots, x_{15} = 13{,}859$$
$$\Sigma x = 1{,}348{,}109$$
$$\bar{x} = \frac{1{,}348{,}109}{15} = 89{,}873.9$$

This is closer to the value of μ but still not very satisfactory as an estimate. The problem here is that there is so much variability in the population of x values (the largest is $x = 8{,}863{,}164$ for Los Angeles County, California, and the smallest is $x = 107$ for Loving County, Texas, which evidently few people love). Therefore, it is difficult for a sample of 15 observations, let alone just 5, to be reasonably representative of the population. But don't lose hope! Once we learn how to measure it, we will also see how to take variability into account in making accurate inferences.

One potential drawback to the mean as a measure of center for a data set is that its value can be greatly affected by the presence of even a single *outlier* (an unusually large or small observation) in the data set.

EXAMPLE 4.3

The following data on $x =$ savings rate for a sample of Southern California financial institutions appeared in the August 15, 1994, *Los Angeles Times*. The values are listed here in increasing order:

2.15	2.20	2.20	2.22	2.22	2.23	2.25	2.25	2.27	2.28
2.28	2.28	2.29	2.30	2.30	2.32	2.32	2.33	2.33	2.33
2.33	2.33	2.33	2.38	2.43	2.55	2.79	3.05	3.68	4.35

The sample mean was reported as $\bar{x} = 2.45$. Figure 4.1 is a MINITAB dotplot of the data. Many would argue that since it is larger than 25 of the 30 observations, 2.45 is not a very representative value for this sample. The two largest rates, 3.68 and 4.35, are unusually large (i.e., outliers) and have a substantial impact on $\bar{x}$.

FIGURE 4.1 A MINITAB dotplot of the data in Example 4.3

We now turn our attention to a measure of center that is not so sensitive to outliers.

The Median

The median strip of a highway divides the highway in half, and the median of a numerical data set does the same thing. Once the data values have been listed in order from smallest to largest, the **median** is the middle value in the list and divides the list into two equal parts. It turns out that the median is found differently, depending on whether the sample size n is even or odd. When n is an odd number (say, 5), the sample median is the single middle value. But when n is even (say, 6), there are two middle values in the ordered list, and we average these two middle values to obtain the sample median.

Definition

The **sample median** is obtained by first ordering the n observations from smallest to largest (with any repeated values included, so that every sample observation appears in the ordered list). Then

$$\text{sample median} = \begin{cases} \text{the single middle value if } n \text{ is odd} \\ \text{the average of the middle two values if } n \text{ is even} \end{cases}$$

EXAMPLE 4.4   (Example 4.3 continued)   The sample size for the savings rate data was $n = 30$, an even number. The median is the average of the 15th and 16th values (the middle two) in the ordered list of the data. Note that the savings rate data is already arranged in numerical order. If this were not the case, we would have had to put the data values in order first before determining the median, as follows:

$$\text{sample median} = \frac{2.30 + 2.32}{2} = 2.31$$

Looking at the dotplot, we see that this value appears to be somewhat more typical of the data than does $\bar{x} = 2.45$.

The **population median** plays the same role for the population as the sample median plays for the sample. It is the middle value in the ordered list consisting of all population observations. We previously noted that the mean — population or sample — is very sensitive to even a single value that lies far above or below the rest of the data. The value of the mean is pulled out toward such an outlying value or values. The median, on the other hand, is quite *in*sensitive to outliers. For example, the largest sample observation (4.35) in Example 4.3 can be increased by an arbitrarily large amount without changing the value of the median. Similarly, an increase in the second or third largest observations does not affect the median, nor would a decrease in several of the smallest observations.

This stability of the median is what sometimes justifies its use as a measure of center in some situations. For example, the July 5, 1999, issue of *The Los Angeles Times* reported that the average earnings for stock brokers in 1997 was $158,201, whereas the median earnings was only $119,010. Income distributions are commonly summarized by reporting the median rather than the mean, since otherwise a few very high salaries could result in a mean that is not representative of a typical salary.

Comparing the Mean and the Median

Figure 4.2 presents several smoothed histograms that might represent either a distribution of sample values or a population distribution. Pictorially, the median is the value on the measurement axis that separates the histogram into two parts, with .5 (50%) of the area under each part of the curve. The mean is a bit harder to visualize. If the histogram were balanced on a triangle with a sharp point (a fulcrum), it would tilt unless the triangle were positioned exactly at the mean. The mean is the balance point for the distribution.

When the histogram is symmetric, the point of symmetry is both the dividing point for equal areas and the balance point, and the mean and median are equal. However, when the histogram is unimodal (single-peaked) with a longer upper tail, the relatively few outlying values in the upper tail pull the mean up, so it generally lies above the median. For example, an unusually high exam score raises the mean but does not affect the median. Similarly, when a unimodal histogram is negatively skewed, the mean is generally smaller than the median (see Figure 4.3).

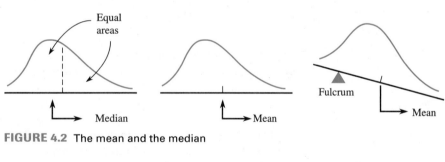

FIGURE 4.2 The mean and the median

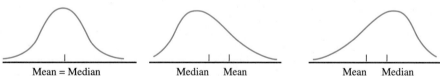

FIGURE 4.3 Relationship between the mean and the median

Categorical Data

The natural numerical summary quantities for a categorical data set are the relative frequencies for the various categories. Each relative frequency is the proportion (fraction) of responses that are in the corresponding category. Often there are only two possible responses (a *dichotomy*) — male or female, does or does not have a driver's license, did or did not vote in the last election, and so on. It is convenient in such situations to label one of the two possible responses S (for success) and the other F (for failure). As long as further analysis is consistent with the labeling, it is immaterial which category is assigned the S label. When the data set is a sample, the fraction of S's in the sample is called the *sample proportion of successes*.

Definition

The **sample proportion of successes,** denoted by p, is

$$p = \text{sample proportion of successes} = \frac{\text{number of } S\text{'s in the sample}}{n}$$

where S is the label used for the response designated as success.

EXAMPLE 4.5   The use of antipollution equipment on automobiles has substantially improved air quality in certain areas. Unfortunately, many car owners have tampered with smog control devices to improve performance. Suppose a sample of $n = 15$ cars was selected, and each was classified as S or F, according to whether or not tampering had taken place. The resulting data was

$$S \quad F \quad S \quad S \quad S \quad F \quad F \quad S \quad S \quad F \quad S \quad S \quad S \quad F \quad F$$

This sample contains nine S's, so

$$p = \frac{9}{15} = .60$$

That is, 60% of the sample responses are S's. In 60% of the cars sampled, there has been tampering with the air pollution control devices.

The Greek letter π is used to denote the **population proportion of S's.*** We will see later how the value of p from a particular sample can be used to make inferences about π.

Trimmed Means

The extreme sensitivity of the mean to even a single outlier and the extreme insensitivity of the median to a substantial proportion of outliers can make both of them suspect as a measure of center. Statisticians have proposed a *trimmed mean* as a compromise between these two extremes.

---

* Note that we are using the symbol π to represent a population proportion and *not* as the mathematical constant $\pi = 3.14 \ldots$. Many other sources use the symbol p for the population proportion and $\hat{p}$ for the sample proportion.

> **Definition**
>
> A **trimmed mean** is computed by first ordering the data values from smallest to largest, deleting a selected number of values from each end of the ordered list, and finally averaging the remaining values. The **trimming percentage** is the percentage of values deleted from each end of the ordered list.

EXAMPLE 4.6   The film industry has been criticized for its depiction of alcohol and tobacco use in children's movies. The article "Tobacco and Alcohol Use in G-Rated Animated Films" (*J. American Medical Assn.* (1999): 1131–1136) gave the following data on the number of seconds showing alcohol use in each of 30 animated films released between 1980 and 1997.

Alcohol Exposure (seconds)

34	414	0	0	76	123	3	0	7	0	46	38	13	73
0	72	0	5	0	0	0	0	74	0	28	0	0	0
0	39

A MINITAB dotplot of this data is shown in Figure 4.4. Since the dotplot shows that the data distribution is not symmetric and that there is an outlier, a trimmed mean is a reasonable choice for describing this data set.

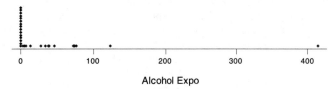

FIGURE 4.4 A MINITAB dotplot of the data in Example 4.6

Since 10% of 30 is 3, a 10% trimmed mean results from deleting the three largest and three smallest data values and then averaging the remaining 24 data values. The ordered data values are

| 0 | 0 | 0 | 0 | 0 | 0 | 0 | 0 | 0 | 0 | 0 | 0 | 0 | 0 | 0 |
| 3 | 5 | 7 | 13 | 28 | 34 | 38 | 39 | 46 | 72 | 73 | 74 | 76 | 123 | 414 |

Deleting three of the 0's (the smallest 10%) and 76, 123, and 414 (the largest 10%) and averaging the remaining values gives

$$\text{10\% trimmed mean} = \frac{0 + 0 + \cdots + 74}{24} = \frac{432}{24} = 18.00$$

The mean, $\bar{x} = 34.53$, is much larger than the trimmed mean because of the unusually large value (414) in the data set. The 10% trimmed mean, which is 1.5 for this data set, falls between the mean and the median.

Sometimes the number of observations to be deleted from each end of the data set is specified and then the corresponding trimming percentage is calculated as

$$\text{trimming percentage} = \left(\frac{\text{number deleted from each end}}{n} \right) \cdot 100$$

In other cases, the trimming percentage is specified and then used to determine how many observations to delete from each end, with

$$\text{number deleted from each end} = (\text{trimming percentage}) \cdot n$$

If the number of observations to be deleted from each end resulting from this calculation is not an integer, it can be rounded to the nearest integer.

A trimmed mean with a small to moderate trimming percentage — between 5% and 25% — is less affected by outliers than the mean, but it is not as insensitive as the median. Trimmed means are therefore being used more and more often.

Exercises 4.1 – 4.18

4.1 The risk of developing iron deficiency is especially high during pregnancy. Detecting such deficiency is complicated by the fact that some methods for determining iron status can be affected by the state of pregnancy itself. Consider the following data on transferrin receptor concentration for a sample of women with laboratory evidence of overt iron-deficiency anemia ("Serum Transferrin Receptor for the Detection of Iron Deficiency in Pregnancy," *Amer. J. of Clinical Nutrition* (1991): 1077–1081): $x_1 = 15.2$, $x_2 = 9.3$, $x_3 = 7.6$, $x_4 = 11.9$, $x_5 = 10.4$, $x_6 = 9.7$, $x_7 = 20.4$, $x_8 = 9.4$, $x_9 = 11.5$, $x_{10} = 16.2$, $x_{11} = 9.4$, $x_{12} = 8.3$. Compute the values of the sample mean and median. Why are these values different here? Which one do you regard as more representative of the sample, and why?

4.2 The paper "The Pedaling Technique of Elite Endurance Cyclists" (*Int. J. of Sport Biomechanics* (1991): 29–53) reported the accompanying data on single-leg power at a high workload.

244 191 160 187 180 176 174 205 211 183 211
180 194 200

a. Calculate and interpret the sample mean and median.
b. Suppose the first observation had been 204, not 244. How would the mean and median change?
c. Calculate a trimmed mean by eliminating the smallest and largest sample observations. What is the corresponding trimming percentage?

d. Suppose that the largest observation had been 204 rather than 244. How would the trimmed mean in part **c** change? What if the largest value had been 284?

4.3 The paper cited in Exercise 4.2 also reported values of single-leg power for a low workload. The sample mean for $n = 13$ observations was $\bar{x} = 119.8$ (actually 119.7692), and the 14th observation, somewhat of an outlier, was 159. What is the value of $\bar{x}$ for the entire sample?

4.4 According to an article about buying Viagra via the Internet (see "Direct Sale of Sildenafil (Viagra) to Consumers over the Internet," *New England Journal of Medicine* (1999): 1389–1392), of 77 Web sites offering to sell Viagra, 55 were in the United States, and 11 were from the United Kingdom. Furthermore, 42 of the 77 provided information about contraindications (reasons a patient should not use Viagra, for example, concomitant use of nitrates).
a. What proportion of the sampled Web sites were in the United Kingdom?
b. What proportion of the sampled Web sites were outside the United States?
c. What proportion of the sampled Web sites did not warn consumers about possible contraindications?

4.5 A sample of 26 offshore oil workers took part in a simulated escape exercise, resulting in the accompanying data on time (sec) to complete the escape ("Oxygen Consumption and Ventilation During Escape from an Offshore Platform," *Ergonomics* (1997): 281–292):

389	356	359	363	375	424	325	394	402
373	373	370	364	366	364	325	339	393
392	369	374	359	356	403	334	397	

a. Construct a stem-and-leaf display of the data. How does it suggest that the sample mean and median will compare?

b. Calculate the values of the sample mean and median.

c. By how much could the largest time be increased without affecting the value of the sample median? By how much could this value be decreased without affecting the sample median?

4.6 Did Civil War veterans have different levels of wealth than those who did not fight after the war was over? The article "The Civil War Generation: Military Service and Mobility in Dubuque, Iowa, 1860–1870" (*Journal of Social History* (1999): 791–820) discusses this issue. One bit of information presented in this article is the average property value (in dollars) of each of 123 non-soldier sons of business-class parents and 14 soldier sons of these same parents in Dubuque, Iowa. The average property value of the 123 non-soldier sons was 14,905.69, whereas the average property value of the 14 soldier sons was 17,678.57.

a. What is the mean income of all 137 sons?

b. If one of the soldier sons had no property at all, what is the mean property value of the remaining 13 sons?

4.7 According to the September 1999 *Survey of Current Business,* the top five metropolitan regions by per capita personal income in 1997 were:

Area Name	Total Personal Income (in millions of dollars)	Per Capita Personal Income	Rank
San Francisco	68,671	41,128	1
New Haven/ Bridgeport/ Stamford/ Danbury/ Waterbury	66,562	40,928	2
West Palm Beach/ Boca Raton	39,269	38,772	3
San Jose	61,345	37,856	4
Bergen/ Passaic	49,111	36,769	5

a. Why would areas be ranked by per capita personal income rather than total personal income?

b. How can an area like West Palm Beach–Boca Raton with such a relatively low personal income still have a very high per capita personal income? Explain.

c. Approximately how many people live in the San Jose area?

4.8 Calculate both the 10% trimmed mean and the 20% trimmed mean for the savings rate data of Example 4.3. Compare these values to $\bar{x}$ and the sample median.

4.9 Consider the following statement: More than 65% of the residents of Los Angeles earn less than the average wage for that city. Could this statement be correct? If so, how? If not, why not?

4.10 Five experimental animals were put on a certain diet for several weeks, and another five animals were put on a second diet. The resulting weight gains (in pounds) at the end of the period were as follows:

Diet 1	12	13	7	5	15
Diet 2	11	14	3	13	4

Without doing any calculation, decide which group of animals achieved the larger average weight gain. (*Hint:* Base your reasoning on the ordered values from each sample.)

4.11 Reconsider the situation described in Exercise 4.10, but suppose now that the weight gains for the second diet are 5, 13, 11, 10, and 3 lb. Without doing any calculation, say how the average gain for the first diet compares to that for the second diet. (*Hint:* Again order the observations from smallest to largest within each sample.)

4.12 A sample consisting of four pieces of luggage was selected from among those checked at an airline counter, yielding the following data on x = weight (lb):

$$x_1 = 33.5 \qquad x_2 = 27.3 \qquad x_3 = 36.7 \qquad x_4 = 30.5$$

Suppose one more piece is selected; denote its weight by x_5. Find a value of x_5 such that $\bar{x}$ = sample median.

4.13 The October 1, 1994, issue of the *San Luis Obispo Telegram-Tribune* reported the following monthly salaries for supervisors from six different counties: $5354 (Kern), $5166 (Monterey), $4443 (Santa Cruz), $4129 (Santa Barbara), $2500 (Placer), and $2220 (Merced). San Luis Obispo County supervisors are supposed to be paid the average of the two counties among these six in the middle of the salary range. Which measure of center determines this salary, and what is its value? Why is the other measure of center featured in this section not as favorable to these supervisors (though it might appeal to taxpayers)?

4.14 Suppose that ten patients with meningitis received treatment with large doses of penicillin. Three days later, temperatures were recorded, and the treatment was considered successful if there had been a reduction in a patient's temperature. Denoting success by S and failure by F, the ten observations are

$$S \quad S \quad F \quad S \quad S \quad S \quad F \quad F \quad S \quad S$$

a. What is the value of the sample proportion of successes?
b. Replace each S with 1 and each F with a 0. Then calculate $\bar{x}$ for this numerically coded sample. How does $\bar{x}$ compare to p?
c. Suppose it is decided to include 15 more patients in the study. How many of these would have to be S's to give $p = .80$ for the entire sample of 25 patients?

4.15 An experiment to study the lifetime (in hours) for a certain type of component involved putting ten components into operation and observing them for 100 hr. Eight of the components failed during that period, and those lifetimes were recorded. Denote the lifetimes of the two components still functioning after 100 hr by 100+. The resulting sample observations were

$$48 \quad 79 \quad 100+ \quad 35 \quad 92 \quad 86 \quad 57 \quad 100+ \quad 17 \quad 29$$

Which of the measures of center discussed in this section can be calculated, and what are the values of those measures? (*Note*: The data from this experiment is said to be "censored on the right.")

4.16 An instructor has graded 19 exam papers submitted by students in a certain class of 20 students, and the average so far is 70. (The maximum possible score is 100.) How high would the score on the last paper have to be to raise the class average by one point? By two points?

4.17 "The Motley Fool," an investment column that appears in newspapers (*San Luis Obispo Tribune*, Oct. 16, 1999), made the following statement about mutual funds:

> Funds with a great three or five or even 10-year average are likely to have that great average because of one amazing year. After all, a five-year average is just an average of five numbers. If one of them is unusually high, the average will be high. Let's consider an example. If in each of five years, a fund earns 8 percent, 11 percent, 4 percent, 12 percent, and 33 percent, its average annual return will be about 13 percent. That might look respectable, but note that in reality it exceeded 13 percent in only one of five years. That 33 percent return (an "outlier" in statistical terms) has skewed the average.

Calculate the median return for the five years mentioned in the above quote. Do you think that the median would be a better indication of a typical return in this case? Explain.

4.18 A certain college has two sections of introductory statistics during a particular semester. One section has 20 students, and the other has 100 students.
a. What is the mean number of students per section (the average over sections)?
b. Let x_1 denote the number of students in the first student's class, x_2 the number of students in the second student's class, and so on (up through x_{120}). What is the average of these x values? (This is the mean class size when averaged over students rather than classes.)
c. From a student perspective, which of the two averages, the one calculated in part **a** or the one in part **b**, is more pertinent?

4.2   Describing Variability in a Data Set

Reporting a measure of center gives only partial information about a data set. It is also important to describe the spread of values about the center. The three different samples displayed in Figure 4.5 all have mean = median = 45. There is much variability in the first sample compared to the third sample. The second sample shows less variability than the first and more than the third; most of the variability in the second sample is due to the two extreme values being so far from the center.

The simplest numerical measure of variability is the **range,** which is defined as the difference between the largest and smallest values. Generally speaking, more variability will be reflected in a larger range. However, variability depends on more than just the distance between the two most extreme values. Variability is a characteristic of the entire data set, and each observation contributes to the variability.

Sample

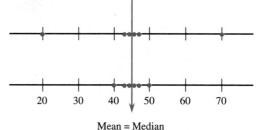

1. 20, 40, 50, 30, 60, 70

2. 47, 43, 44, 46, 20, 70

3. 44, 43, 40, 50, 47, 46

Mean = Median

FIGURE 4.5 Three samples with the same center and different amounts of variability

The first two samples plotted in Figure 4.5 both have a range of 50, but there is substantially less dispersion in the second sample.

Deviations from the Mean

The most common measures of variability describe the extent to which the sample observations deviate from the sample mean $\bar{x}$. Subtracting $\bar{x}$ from each observation gives a set of deviations from the mean.

Definition

The n **deviations from the sample mean** are the differences

$$x_1 - \bar{x}, x_2 - \bar{x}, \ldots, x_n - \bar{x}$$

A particular deviation is positive if the x value exceeds $\bar{x}$ and negative if the x value is less than $\bar{x}$.

EXAMPLE 4.7   The $n = 30$ savings rate observations from Example 4.3 are displayed in Table 4.2 (page 114), along with the corresponding deviations calculated from $\bar{x} = 2.452$.

Only 5 of the deviations are positive, because only 5 of the 30 observations exceed $\bar{x}$. Most of the negative deviations are quite small in magnitude, corresponding to observations slightly smaller than the mean.

Generally speaking, the greater the amount of variability in the sample, the larger the magnitudes (ignoring the sign) of the deviations. Thus, the magnitudes of the deviations for the first sample in Figure 4.5 substantially exceed those for the third sample, implying more variability in sample 1.

Table 4.2 *Deviations of savings rates from the mean*

Observation (x)	Deviation $(x - \bar{x})$	Observation	Deviation $(x - \bar{x})$
2.15	−.302	2.32	−.132
2.20	−.252	2.32	−.132
2.20	−.252	2.33	−.122
2.22	−.232	2.33	−.122
2.22	−.232	2.33	−.122
2.23	−.222	2.33	−.122
2.25	−.202	2.33	−.122
2.25	−.202	2.33	−.122
2.27	−.182	2.38	−.072
2.28	−.172	2.43	−.022
2.28	−.172	2.55	.098
2.28	−.172	2.79	.338
2.29	−.162	3.05	.598
2.30	−.152	3.68	1.228
2.30	−.152	4.35	1.898

We now consider how to combine the deviations into a single numerical measure. Our first thought might be to calculate the average deviation, by adding the deviations together — this sum can be denoted compactly by $\Sigma (x - \bar{x})$ — and dividing by n. This does not work, though, because negative and positive deviations counteract one another in the summation.

Except for the effects of rounding in computing the deviations, it is always true that $\Sigma (x - \bar{x}) = 0$. Since this sum is zero, the average deviation is always zero and so cannot be used as a measure of variability.

As a result of rounding the value of $\bar{x}$, the sum of the 30 deviations in Example 4.7 is $\Sigma (x - \bar{x}) = .01$. If we used even more decimal accuracy in computing $\bar{x}$, the sum would be closer to zero.

The Variance and Standard Deviation

The standard way to prevent negative and positive deviations from counteracting one another is to square them before combining. Then deviations with opposite signs but with the same magnitude, such as +20 and −20, will make identical contributions to variability. The squared deviations are $(x_1 - \bar{x})^2$, $(x_2 - \bar{x})^2$, . . . , $(x_n - \bar{x})^2$, and their sum is

$$(x_1 - \bar{x})^2 + (x_2 - \bar{x})^2 + \cdots + (x_n - \bar{x})^2 = \Sigma (x - \bar{x})^2$$

A common shorthand notation for $\Sigma (x - \bar{x})^2$ is S_{xx}. Dividing this sum by the sample size n gives the average squared deviation. Although this seems to be a reasonable measure of variability, a divisor slightly smaller than n will be used. (The reason for this will be explained subsequently.)

Definition

The **sample variance,** denoted by s^2, is the sum of squared deviations from the mean divided by $n - 1$. That is,

$$s^2 = \frac{\Sigma(x - \bar{x})^2}{n - 1} = \frac{S_{xx}}{n - 1}$$

The **sample standard deviation** is the positive square root of the sample variance and is denoted by **s.**

A large amount of variability in the sample is indicated by a relatively large value of s^2 or s, whereas a small value of s^2 or s indicates a small amount of variability. For most statistical purposes, s is the desired quantity, but s^2 must be computed first. Notice that whatever unit is used for x (such as lb or sec), the squared deviations and therefore s^2 are in squared units. Taking the square root gives a measure expressed in the same units as x. Thus, for a sample of heights, the standard deviation might be $s = 3.2$ in., and for a sample of textbook prices, it might be $s = \$12.43$.

EXAMPLE 4.8   (Example 4.1 continued)   To illustrate the calculation of the sample variance and standard deviation, let's look again at the range of motion data from Example 4.1. Table 4.3 shows squared deviations (obtained by subtracting $\bar{x} = 130.38$ from each observation), which are then combined to compute the values of s^2 and s.

Table 4.3 *Deviations and squared deviations for the range of motion data*

Observation	Deviation $(x - \bar{x})$	Squared Deviation $(x - \bar{x})^2$
154	23.62	557.9044
142	11.62	135.0244
137	6.62	43.8244
133	2.62	6.8644
122	−8.38	70.2244
126	−4.38	19.1844
135	4.62	21.3444
135	4.62	21.3444
108	−22.38	500.8644
120	−10.38	107.7444
127	−3.38	11.4244
134	3.62	13.1044
122	−8.38	70.2244
		Sum = 1579.0772

So,

$$\Sigma (x - \overline{x})^2 = S_{xx} = 1579.0772$$

and

$$s^2 = \frac{\Sigma (x - \overline{x})^2}{n - 1} = \frac{1579.0722}{12} = 131.5898$$

$$s = \sqrt{131.5898} = 11.4713$$

The computation of s^2 can be a bit tedious. Fortunately, many calculators and computer software packages compute the variance and standard deviation. One commonly used statistical computer package is MINITAB. Output resulting from using the MINITAB Describe command with the range of motion data follows. MINITAB gives a variety of numerical descriptive measures, including the mean, median, and standard deviation.

Descriptive Statistics

Variable	N	Mean	Median	TrMean	StDev	SE Mean
Motion	13	130.38	133.00	130.27	11.47	3.18

Variable	Minimum	Maximum	Q1	Q3
Motion	108.00	154.00	122.00	136.00

Interpretation and Properties

A standard deviation may be informally interpreted as the size of a "typical" deviation from the mean. Thus, in Example 4.8, a typical deviation from $\overline{x} = 130.38$ is about 11.47; some observations are closer than 11.47 to $\overline{x}$, and others are farther away. We computed $s = 11.47$ in Example 4.8 without saying whether this value indicated a large or small amount of variability. At this point, it is better to use s for comparative purposes than for an absolute assessment of variability. If we obtained a sample of range of motion values for a second group of patients and computed $s = 9.1$ for those patients, then we would conclude that there is more variability in our original sample than in this second sample. A particular value of s can be judged large or small only in comparison to something else.

There are measures of variability for the entire population that are analogous to s^2 and s for a sample. These measures are called the **population variance** and **population standard deviation,** and are denoted by σ^2 and σ, respectively. (We again use a lowercase Greek letter for a population characteristic.) The population standard deviation σ is expressed in the same units of measurement as are the values in the population. As with s, the value of σ can be used for comparative purposes.

In many statistical procedures, we would like to use the value of σ, but unfortunately it is not usually available. Therefore, in its place we must use a value computed from the sample that we hope is close to σ (that is, a good *estimate* of σ). We use the divisor $n - 1$ in s^2 rather than n because, on average, it tends to be a bit closer to σ^2. We will say more about this in Chapter 9.

An alternative rationale for using $n - 1$ is based on the property $\Sigma (x - \bar{x}) = 0$. Suppose that $n = 5$ and four of the deviations are

$$x_1 - \bar{x} = -4 \qquad x_2 - \bar{x} = 6 \qquad x_3 - \bar{x} = 1 \qquad x_5 - \bar{x} = -8$$

Then, since the sum of these four deviations is -5, the remaining deviation must be $x_4 - \bar{x} = 5$ (so that the sum of all five is 0). Although there are five deviations, only four of them contain independent information about variability. More generally, once any $n - 1$ of the deviations are available, the value of the remaining deviation is determined. The n deviations actually contain only $n - 1$ independent pieces of information about variability. Statisticians express this by saying that s^2 and s are based on $n - 1$ *degrees of freedom* (df). Many inferential procedures encountered in later chapters are based on some appropriate number of df.

The Interquartile Range

As with $\bar{x}$, the value of s can be greatly affected by the presence of even a single unusually small or large observation. The *interquartile range* is a measure of variability that is resistant to the effects of outliers. It is based on quantities called *quartiles*. The *lower quartile* separates the bottom 25% of the data set from the upper 75%, and the *upper quartile* separates the top 25% from the bottom 75%. The *middle quartile* is the median, and it separates the bottom 50% from the top 50%. Figure 4.6 illustrates the locations of these quartiles for a smoothed histogram.

The quartiles for sample data are obtained by dividing the n ordered observations into a lower half and an upper half; if n is odd, the median is excluded from both halves. The two extreme quartiles are then the medians of the two halves.

FIGURE 4.6 The quartiles for a smoothed histogram

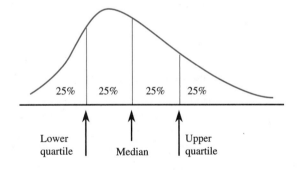

Lower quartile Median Upper quartile

25% 25% 25% 25%

Definition*

    lower quartile = median of the lower half of the sample
    upper quartile = median of the upper half of the sample

(If n is odd, the median of the entire sample is excluded from both halves.)

The **interquartile range** (**iqr**), a resistant measure of variability, is given by

    iqr = upper quartile − lower quartile

* There are several other sensible ways to define quartiles. Some calculators and software packages use an alternate definition.

EXAMPLE 4.9   Cardiac output and maximal oxygen capacity typically decrease with age in seden-
tary individuals, but these decreases are at least partially arrested in middle-aged
individuals who engage in a substantial amount of physical exercise. To understand
better the effects of exercise and aging on various circulatory functions, the article
"Cardiac Output in Male Middle-Aged Runners" (*J. of Sports Medicine* (1982): 17–
22) presented data from a study of 21 middle-aged male runners. Figure 4.7 is a stem-
and-leaf display of oxygen capacity values (mL/kg · min) while pedaling at a specified rate on a bicycle ergometer.

FIGURE 4.7 MINITAB stem-
and-leaf display of oxygen
capacity (mL/kg · min)

```
12 │ 81
13 │
14 │ 95
15 │ 97,83
16 │
17 │ 90
18 │ 34,27
19 │ 94,82
20 │ 99,93,98,62,88
21 │ 15
22 │ 16,24            Stem:   Ones
23 │ 16,56            Leaf:   Hundreths
```

HI: 35.78, 36.73 *These are outliers.*

The sample size $n = 21$ is an odd number, so the median, 20.88, is excluded from
both halves of the sample:

Lower half   12.81   14.95   15.83   15.97   17.90   18.27   18.34   19.82   19.94   20.62
Upper half   20.93   20.98   20.99   21.15   22.16   22.24   23.16   23.56   35.78   36.73

Each half of the sample contains 10 observations, so each quartile is the average of
the fifth and sixth values of the corresponding half. This gives us

$$\text{lower quartile} = \frac{17.90 + 18.27}{2} = 18.085 \qquad \text{upper quartile} = \frac{21.16 + 22.24}{2} = 22.200$$

and $\text{iqr} = 22.200 - 18.085 = 4.115$

The sample mean and standard deviation are 21.10 and 5.75, respectively. If we
were to change the two largest values from 35.78 and 36.73 to 25.78 and 26.73 (so
that they are still the two largest values), the median and the interquartile range
would not be affected, whereas the mean and the standard deviation would change
to 20.14 and 3.44, respectively.

The resistant nature of the iqr follows from the fact that up to 25% of the smallest sample observations and up to 25% of the largest can be made more extreme
without affecting its value.

The **population interquartile range** is the difference between the upper and lower population quartiles. If a histogram of the data set under consideration (whether a population or a sample) can be reasonably well approximated by a normal curve, then the relationship between the standard deviation (sd) and interquartile range is roughly sd = iqr/1.35. A value of the standard deviation much larger than iqr/1.35 suggests a histogram with heavier (or longer) tails than a normal curve. For the oxygen uptake data of Example 4.9, we had $s = 5.75$, whereas iqr/1.35 = 4.115/1.35 = 3.05. This suggests that the distribution of sample values is indeed heavy tailed compared to a normal curve.

A Note Concerning Computation (Optional)

Hand calculation of s^2 using the defining formula can be a bit tedious. An alternative expression for the numerator of s^2 simplifies the arithmetic by eliminating the need to calculate the deviations. In this alternative, it is necessary to distinguish between two quantities that are similar in appearance but very different in value:

1. Sum the x values, then square the sum to obtain $(\Sigma x)^2$.

2. Square each x value, then sum the squares to obtain Σx^2.

If, for example, $x_1 = 3$ and $x_2 = 5$, then

$$(\Sigma x)^2 = (3 + 5)^2 = 8^2 = 64$$
$$\Sigma x^2 = x_1^2 + x_2^2 = 3^2 + 5^2 = 9 + 25 = 34$$

So the order in which the two operations, squaring and summation, are carried out makes a difference!

A computational formula for the sum of squared deviations is

$$S_{xx} = \Sigma (x - \bar{x})^2 = \Sigma x^2 - \frac{(\Sigma x)^2}{n}$$

Thus, a **computational formula for the sample variance** is

$$s^2 = \frac{S_{xx}}{n - 1} = \frac{\Sigma x^2 - \dfrac{(\Sigma x)^2}{n}}{n - 1}$$

According to this formula, after squaring each x value and adding these to obtain Σx^2, the single quantity $(\Sigma x)^2/n$ is subtracted from the result. Instead of the n subtractions required to obtain the deviations, just one subtraction now suffices.

The computed value of s^2, whether obtained from the defining formula or the computational formula, can sometimes be greatly affected by rounding in $\bar{x}$ or $(\Sigma x)^2/n$. Protection against adverse rounding effects can almost always be achieved by using four or five digits of decimal accuracy beyond the decimal accuracy of the data values themselves.

We now illustrate the use of the computational formula with the range of motion data of Example 4.8. For efficient computation, it is convenient to place the x

values in a single column and the x^2 values in a second column. Adding the numbers in these two columns then gives $\Sigma\, x$ and $\Sigma\, x^2$, respectively.

x	x^2
154	23,716
142	20,164
137	18,769
133	17,689
122	14,884
126	15,876
135	18,225
135	18,225
108	11,664
120	14,400
127	16,129
134	17,956
122	14,844
Sum = 1695	Sum = 222,581

Thus,

$$\frac{(\Sigma\, x)^2}{n} = \frac{(1695)^2}{13} = 221{,}001.9231$$

$$s^2 = \frac{222{,}581 - 221{,}001.9231}{13 - 1} = \frac{1579.0769}{12} = 131.5897$$

$$s = \sqrt{131.5897} = 11.4713$$

Note that the values of s^2 and s obtained using the computational formulas agree with those found previously in Example 4.8.

Exercises 4.19 – 4.28

4.19 A sample of $n = 5$ college students yielded the following observations on number of traffic citations for a moving violation during the previous year:

$$x_1 = 1 \qquad x_2 = 0 \qquad x_3 = 0 \qquad x_4 = 3 \qquad x_5 = 2$$

Calculate s^2 and s.

4.20 A study of the relationship between age and various visual functions (such as acuity and depth perception) reported the following observations on area of scleral lamina (mm^2) from human optic nerve heads ("Morphometry of Nerve Fiber Bundle Pores in the Optic Nerve

Head of the Human," *Experimental Eye Research* (1988): 559–568).

2.75	2.62	2.74	3.85	2.34	2.74
3.93	4.21	3.88	4.33	3.46	4.52
2.43	3.65	2.78	3.56	3.01	

It is easily verified that $S_{xx} = 8.025176$. Use this to compute the sample variance s^2 and then the sample standard deviation s. How can the value of s be interpreted?

4.21 a. Give two sets of five numbers that have the same mean but different standard deviations.

b. Give two sets of five numbers that have the same standard deviation but different means.

4.22 Although bats are not known for their eyesight, they are able to locate prey (mainly insects) by emitting high-pitched sounds and listening for echoes. A paper appearing in *Animal Behaviour* ("The Echolocation of Flying Insects by Bats" (1960): 141–154) gave the following distances (in cm) at which a bat first detected a nearby insect:

> 62   23   27   56   52   34   42   40   68   45   83

a. Compute the sample mean distance at which the bat first detects an insect.

b. Compute the sample variance and standard deviation for this data set. Interpret these values.

4.23 For the data in Exercise 4.22, add -10 to each sample observation. (This is the same as subtracting 10.) For the new set of values, compute the mean and the deviations from the mean. How do these deviations compare to the deviations from the mean for the original sample? How does s^2 for the new values compare to s^2 for the old values? In general, what effect does adding the same number to each observation have on s^2 and s? Explain.

4.24 For the data of Exercise 4.22, multiply each data value by 10. How does s for the new values compare to s for the original values?

4.25 The paper "The Pedaling Technique of Elite Endurance Cyclists" (*Int. J. of Sport Biomechanics* (1991): 29–53) reported the accompanying data on single-leg power at a high workload. Calculate and interpret the sample variance and standard deviation.

> 244   191   160   187   180   176   174
> 205   211   183   211   180   194   200

4.26 Refer to the data in Exercise 4.20 on scleral lamina areas.

a. Determine the lower and upper quartiles.

b. Calculate the value of the interquartile range.

c. If the two largest sample values, 4.33 and 4.52, had instead been 5.33 and 5.52, how would this affect the iqr? Explain.

d. By how much could the observation 2.34 be increased without affecting the iqr? Explain.

e. If an eighteenth observation, $x_{18} = 4.60$, is added to the sample, what is the iqr?

4.27 The sample standard deviation of the savings rate data given in Example 4.3 is $s = .469$. Calculate the value of iqr/1.35 and compare this to s. What does this comparison suggest about the shape of the histogram of this data?

4.28 The standard deviation alone does not measure relative variation. For example, a standard deviation of $1 would be considered large if it is describing the variability from store to store in the price of an ice cube tray. On the other hand, a standard deviation of $1 would be considered small if it is describing store-to-store variability in the price of a particular brand of freezer. A quantity designed to give a relative measure of variability is the *coefficient of variation*. Denoted by CV, the coefficient of variation expresses the standard deviation as a percent of the mean. It is defined by the formula

$$CV = 100 \cdot \frac{s}{x}$$

Consider the two samples below. Sample 1 gives the actual weight (in ounces) of the contents of cans of pet food labeled as having a net weight of 8 oz. Sample 2 gives the actual weight (in pounds) of the contents of bags of dry pet food labeled as having a net weight of 50 lb.

Sample 1:
8.3	7.1	7.6	8.1	7.6
8.3	8.2	7.7	7.7	7.5

Sample 2:
52.3	50.6	52.1	48.4	48.8
47.0	50.4	50.3	48.7	48.2

a. For each of the given samples, calculate the mean and the standard deviation.

b. Compute the coefficient of variation for each sample. Do the results surprise you? Why or why not?

4.3   Summarizing a Data Set: Boxplots

In the previous two sections, we looked at ways of describing the center and variability of a data set using numerical measures. It would be nice to have a method of summarizing data that gives more detail than just a measure of center and spread and yet less detail than a stem-and-leaf display or histogram. A *boxplot* is one such technique. It is compact, yet it provides information about the center, spread, and symmetry or skewness of the data. A boxplot is based on the median and interquartile range rather than on the mean and standard deviation. We will consider two types of boxplots: the skeletal and the modified boxplot.

Construction of a Skeletal Boxplot

1. Draw a horizontal (or vertical) measurement scale.

2. Construct a rectangular box whose left (or lower) edge is at the lower quartile and whose right (or upper) edge is at the upper quartile (so box width = iqr).

3. Draw a vertical (or horizontal) line segment inside the box at the location of the median.

4. Extend horizontal (or vertical) line segments from each end of the box to the smallest and largest observations in the data set. (These line segments are called *whiskers*.)

EXAMPLE 4.10   Let's reconsider the savings rate data first given in Example 4.3. Recall that savings rate was reported for a sample of 30 financial institutions in Southern California. The ordered observations are

2.15	2.20	2.20	2.22	2.22	2.23	2.25	2.25	2.27	2.28
2.28	2.28	2.29	2.30	2.30	2.32	2.32	2.33	2.33	2.33
2.33	2.33	2.33	2.38	2.43	2.55	2.79	3.05	3.68	4.35

To construct a boxplot of this data, we need the following information: the smallest observation, the lower quartile, the median, the upper quartile, and the largest observation. This collection of summary measures is often referred to as a five-number summary. For this data set,

smallest observation = 2.15

lower quartile = median of the lower half = 2.25

median = average of the 15th and 16th observations in the ordered list

$$= \frac{2.30 + 2.32}{2} = 2.31$$

upper quartile = median of the upper half = 2.33

largest observation = 4.35

Figure 4.8 shows the corresponding boxplot. The median line is somewhat closer to the upper edge of the box than to the lower edge, suggesting a concentration of values in the upper part of the middle half. The upper whisker is much longer than the lower whisker, giving the impression of positive skewness in the extremes of the data. These conclusions are consistent with the dotplot of Figure 4.1.

FIGURE 4.8 Skeletal boxplot for the savings rate data

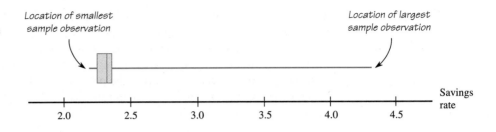

The sequence of steps used to construct a skeletal boxplot is easily modified to give information about outliers.

Definition

An observation is an **outlier** if it is more than 1.5 iqr away from the closest end of the box (the closest quartile). An outlier is **extreme** if it is more than 3 iqr from the closest end of the box, and it is **mild** otherwise. A **modified boxplot** represents mild outliers by shaded circles and extreme outliers by open circles. Whiskers extend on each end to the most extreme observations that are *not* outliers.

EXAMPLE 4.11   The accompanying data came from an anthropological study of rectangular shapes (*Lowie's Selected Papers in Anthropology,* Cora Dubios, ed., Berkeley, Calif.: Univ. of Calif. Press, 1960: 137–142). Observations were made on the variable x = width/length for a sample of n = 20 beaded rectangles used in Shoshoni Indian leather handicrafts.

.553 .570 .576 .601 .606 .606 .609 .611 .615 .628 .654 .662
.668 .670 .672 .690 .693 .749 .844 .933

The quantities needed for constructing the modified boxplot follow:

median = .641   iqr = .681 − .606 = .075
lower quartile = .606   1.5 iqr = .1125
upper quartile = .681   3 iqr = .225

Thus,

upper edge of box (upper quartile) + 1.5 iqr = .681 + .1125 = .7935
lower edge of box (lower quartile) − 1.5 iqr = .606 − .1125 = .4935

so .844 and .933 are both outliers on the upper end, and there are no outliers on the lower end. Since

upper edge of box + 3 iqr = .681 + .225 = .906

.933 is an extreme outlier, and .844 is only a mild outlier. The upper whisker extends to the largest observation that is not an outlier, .749, and the lower whisker extends to .553. The boxplot is presented in Figure 4.9. The median line is not at the center of the box, so there is a slight asymmetry in the middle half of the data. However, the most striking feature is the presence of the two outliers. These two x values considerably exceed the "golden ratio" .618, used since antiquity as an aesthetic standard for rectangles.

FIGURE 4.9 Modified boxplot for rectangle data

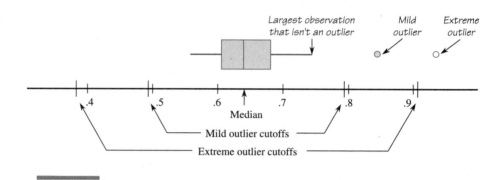

With two or more data sets consisting of observations on the same variable (for example, fuel efficiencies for two types of car, weight gains for a control group and a treatment group, and so on), side-by-side boxplots convey initial impressions concerning similarities and differences between the data sets.

EXAMPLE 4.12

An article in the October 1994 issue of *Consumer Reports* compared various brands of brewed coffee. The study included 31 regular-roast caffeinated brands, 9 decaffeinated brands, and 13 dark-roast brands. Figure 4.10 is a comparative boxplot of cost per cup for the three types. The median marker for the first type (above 6.0) is barely discernible at the lower end of the box, and the same is true at the higher end (above 8.0) for the second type. The box for the third type is much wider than for the other two types (much larger iqr), indicating much more variability in the middle half of the data. However, there are no outliers for the third type, whereas the second type has one extreme outlier and the first type has three extreme and four mild outliers (two of these outlier brands have the same value: 4¢ per cup).

FIGURE 4.10 Comparative boxplot for cost per cup

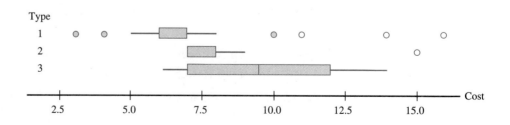

Exercises 4.29 – 4.34

4.29 The percentage of juice lost after thawing for 19 different strawberry varieties appeared in the article "Evaluation of Strawberry Cultivars with Different Degrees of Resistance to Red Scale" (*Fruit Varieties J.* (1991): 12–17):

46	51	44	50	33	46	60	41	55	46
53	53	42	44	50	54	46	41	48	

a. Are there any observations that are mild outliers? Extreme outliers?
b. Construct a boxplot and comment on the important features of the plot.

4.30 A *Consumer Reports* article on peanut butter (Sept. 1990) reported the following scores for various brands:

Creamy 56 44 62 36 39 53 50 65 45 22
        40 56 68 41 30 40 50 56 30

Crunchy 62 53 75 42 47 40 34 62 52
         50 34 42 36 75 80 47 56 62

The statistical package S-Plus was used to obtain the accompanying comparative boxplot. Comment on the similarities and differences in the two data sets.

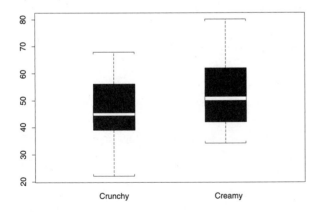

4.31 The paper "Sodium–Calcium Exhange Equilibria in Soils as Affected by Calcium Carbonate and Organic Matter" (*Soil Science* (1984): 109) gave ten observations on soil pH. The data resulted from analysis of ten samples of soil from the Central Soil Salinity Research experimental farm.

8.53	8.52	8.01	7.99	7.93
7.89	7.85	7.82	7.80	7.72

a. Compute the upper quartile, lower quartile, and interquartile range.
b. Construct a skeletal boxplot and comment.

4.32 The amount of aluminum contamination (ppm) in plastic of a certain type was determined for a sample of 26 plastic specimens, resulting in the following data ("The LogNormal Distribution for Modeling Quality Data When the Mean Is Near Zero," *J. of Quality Technology* (1990): 105–110):

30	30	60	63	70	79	87	90	101
102	115	118	119	119	120	125	140	145
172	182	183	191	222	244	291	511	

Construct a boxplot that shows outliers, and comments on the interesting features of this plot.

4.33 The article "Compression of Single-Wall Corrugated Shipping Containers Using Fixed and Floating Test Platens" (*J. of Testing and Evaluation* (1992): 318–320) described an experiment in which several different types of boxes were compared with respect to compressive strength. Consider the following observations on four different types of boxes (this data is consistent with summary quantities given in the paper):

Type of Box	Compression Strength (lb)					
1	655.5	788.3	734.3	721.4	679.1	699.4
2	789.2	772.5	786.9	686.1	732.1	774.8
3	737.1	639.0	696.3	671.7	717.2	727.1
4	535.1	628.7	542.4	559.0	586.9	520.0

Construct a boxplot for each type of box. Use the same scale for all four boxplots. Discuss the similarities and differences in the boxplots.

4.34 Blood cocaine concentration (mg/L) was determined both for a sample of individuals who had died from cocaine-induced excited delirium and for a sample of those who had died from a cocaine overdose without excited delirium. The accompanying data is consistent with summary values given in the paper "Fatal Excited Delirium Following Cocaine Use" (*J. of Forensic Sciences* (1997): 25–31).

Excited Delirium:

0	0	0	0	.1	.1	.1	.1	.2
.2	.3	.3	.3	.4	.5	.7	.7	1.0
1.5	2.7	2.8	3.5	4.0	8.9	9.2	11.7	21.0

No Excited Delirium:

0	0	0	0	0	.1	.1	.1	.1	.2
.2	.2	.3	.3	.3	.4	.5	.5	.6	.8
.9	1.0	1.2	1.4	1.5	1.7	2.0	3.2	3.5	4.1
4.3	4.8	5.0	5.6	5.9	6.0	6.4	7.9	8.3	8.7
9.1	9.6	9.9	11.0	11.5	12.2	12.7	14.0	16.6	17.8

a. Determine the median, quartiles, and iqr for each of the two samples.

b. Are there any outliers in either sample? Any extreme outliers?

c. Construct a boxplot for each of the samples, and use them as a basis for comparing and contrasting the two samples.

4.4   Interpreting Center and Variability: Chebyshev's Rule, the Empirical Rule, and *z* Scores

The mean and standard deviation can be combined to obtain informative statements about how the values in a data set are distributed and about the relative position of a particular value in a data set. To do this, it is useful to be able to describe how far away a particular observation is from the mean in terms of the standard deviation. For example, we might say that an observation is 2 standard deviations above the mean or that an observation is 1.3 standard deviations below the mean. In this way, the standard deviation functions as a unit of measurement for describing distance from the mean. This is illustrated in the following example.

EXAMPLE 4.13

Consider a data set of scores on a standardized test with mean and standard deviation 100 and 15, respectively. We can make the following statements:

1. Because $100 - 15 = 85$, we say that a score of 85 is "1 standard deviation *below* the mean." Similarly, $100 + 15 = 115$ is "1 standard deviation *above* the mean." (See Figure 4.11.)

2. Since 2 standard deviations is $2(15) = 30$ and $100 + 30 = 130$, $100 - 30 = 70$, scores between 70 and 130 are those *within* 2 standard deviations of the mean. (See Figure 4.12.)

3. Because $100 + (3)(15) = 145$, scores above 145 exceed the mean by more than 3 standard deviations.

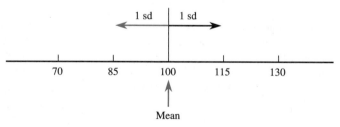

FIGURE 4.11 Values within 1 standard deviation of the mean

FIGURE 4.12 Values within 2 standard deviations of the mean

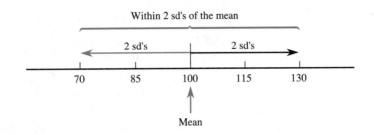

Sometimes in published articles, the mean and standard deviation are reported, but a graphical display of the data is not given. The following result enables us to get a sense of the distribution of data values based on knowledge of the mean and standard deviation only.

Chebyshev's Rule

Consider any number k, where $k \geq 1$. Then *the percentage of observations that are within k standard deviations of the mean is at least* $100(1 - 1/k^2)\%$.

Substituting selected values of k gives the following results.

Number of Standard Deviations, k	$1 - \dfrac{1}{k^2}$	Percentage Within k sd's of the Mean
2	$1 - 1/4 = .75$	at least 75%
3	$1 - 1/9 = .89$	at least 89%
4	$1 - 1/16 = .94$	at least 94%
4.472	$1 - 1/20 = .95$	at least 95%
5	$1 - 1/25 = .96$	at least 96%
10	$1 - 1/100 = .99$	at least 99%

EXAMPLE 4.14   The article "Piecing Together Child Care with Multiple Arrangements: Crazy Quilt or Preferred Pattern for Employed Parents of Preschool Children?" (*J. of Marriage and the Family* (1994): 669–680) examined various modes of care for preschool children. For a sample of families with one such child, it was reported that the mean and standard deviation of child care time per week (hours) were approximately 36 and 12, respectively. Figure 4.13 displays values that are 1, 2, and 3 standard deviations from the mean.

FIGURE 4.13 Measurement scale for child care time

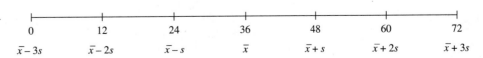

Chebyshev's Rule allows us to assert the following:

1. At least 75% of the sample observations must be between 12 and 60 hours (within 2 sd's of the mean).

2. Because at least 89% of the observations must be between 0 and 72, at most 11% are outside this interval. Time cannot be negative, so we conclude that at most 11% of the observations exceed 72.

3. The values 18 and 52 are 1.5 sd's to either side of $\bar{x}$ ($1.5s = 18$), so using $k = 1.5$ in Chebyshev's Rule implies that at least 55.6% of the observations must be between these two values. Thus, at most 44.4% of the observations are less than 18 — *not* at most 22.2%, since the distribution of values may not be symmetric.

Because Chebyshev's Rule is applicable to any data set (distribution), whether symmetric or skewed, we must be careful when making statements about the proportion above a particular value, below a particular value, or inside or outside an interval that is not centered at the mean. The rule must be used in a conservative fashion. There is another aspect of this conservatism. Whereas the rule states that at least 75% of the observations are within 2 standard deviations of the mean, in many data sets substantially more than 75% of the values satisfy this condition. The same sort of understatement is frequently encountered for other values of k (numbers of standard deviations).

EXAMPLE 4.15

Figure 4.14 gives a stem-and-leaf display of IQ scores of 112 children in one of the early studies that used the Stanford revision of the Binet–Simon intelligence scale (*The Intelligence of School Children*, L. M. Terman, 1919).
Summary quantities include

$$\bar{x} = 104.5 \qquad s = 16.3 \qquad 2s = 32.6 \qquad 3s = 48.9$$

FIGURE 4.14 Stem-and-leaf display of IQ scores

```
 6 | 1
 7 | 25679
 8 | 0000124555668
 9 | 0000112333446666778889
10 | 00011222223333566677778899999
11 | 00001122333344444477899
12 | 01111123445669
13 | 006
14 | 26
15 | 2
```

Stem: Tens
Leaf: Ones

Table 4.4 shows how Chebyshev's Rule considerably understates actual percentages.

Table 4.4 *Summarizing the distribution of IQ scores*

$k =$ no. of sd's	$\bar{x} \pm ks$	Chebyshev	Actual
2	71.9 to 137.1	at least 75%	96% (108)
2.5	63.7 to 145.3	at least 84%	97% (109)
3	55.6 to 153.4	at least 89%	100% (112)

The Empirical Rule

The fact that statements based on Chebyshev's Rule are frequently very conservative suggests that we should look for rules that are less conservative and more precise. The most useful such rule is the **Empirical Rule,** which can be applied whenever the distribution of data values can be reasonably well described by a normal curve. The word *empirical* means deriving from practical experience, and practical experience has shown that a normal curve gives a reasonable fit to many data sets.

Empirical Rule

If the histogram of values in a data set can be reasonably well approximated by a normal curve, then

Approximately 68% of the observations are within 1 standard deviation of the mean.

Approximately 95% of the observations are within 2 standard deviations of the mean.

Approximately 99.7% of the observations are within 3 standard deviations of the mean.

The Empirical Rule makes "approximately" instead of "at least" statements, and the percentages for $k = 1, 2,$ and 3 standard deviations are much higher than those allowed by Chebyshev's Rule. Figure 4.15 illustrates the percentages given in

FIGURE 4.15 Approximate percentages implied by the Empirical Rule

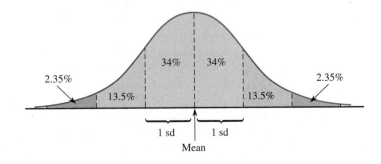

the Empirical Rule. In contrast to Chebyshev's Rule, dividing the percentages in half is permissible, because a normal curve is symmetric.

EXAMPLE 4.16

One of the earliest articles to argue for the wide applicability of the normal distribution was "On the Laws of Inheritance in Man. I. Inheritance of Physical Characters" (*Biometrika* (1903): 375–462). Among the data sets discussed in the article was one consisting of 1052 measurements of the heights of mothers. The mean and standard deviation were

$$\bar{x} = 62.484 \text{ in.} \qquad s = 2.390 \text{ in.}$$

A normal curve did provide a good fit to the data. Table 4.5 contrasts actual percentages with those from Chebyshev's Rule and the Empirical Rule. Clearly, the Empirical Rule is much more successful and informative in this case than Chebyshev's Rule.

Table 4.5 *Summarizing the distribution of mothers' heights*

No. of sd's	Interval	Actual	Empirical Rule	Chebyshev Rule
1	60.094 to 64.874	72.1%	68%	at least 0%
2	57.704 to 67.264	96.2%	95%	at least 75%
3	55.314 to 69.654	99.2%	99.7%	at least 89%

Our detailed study of the normal distribution and areas under normal curves in Chapter 7 will enable us to make statements analogous to those of the Empirical Rule for values other than $k = 1, 2,$ or 3 standard deviations. For now, note that it is rather rare to see an observation from a normally distributed population that is farther than 2 standard deviations from the mean (only 5%), and it is very surprising to see one that is more than 3 standard deviations away. If you encountered a mother whose height was 72 in., you might reasonably conclude that she was not part of the population described by the data set in Example 4.16.

Measures of Relative Standing

When you obtain your score after taking an achievement test, you probably want to know how it compares to the scores of others who have taken the test. Is your score above or below the mean, and by how much? Does your score place you among the top 5% of those who took the test, or only among the top 25%? Questions of this sort are answered by finding ways to measure the position of a particular value in a data set relative to all values in the set. One such measure involves calculating a *z score*.

Definition

The **z score** corresponding to a particular observation in a data set is

$$z \text{ score} = \frac{\text{observation} - \text{mean}}{\text{standard deviation}}$$

The *z* score tells us how many standard deviations the observation is from the mean. It is positive or negative according to whether the observation lies above or below the mean.

The process of subtracting the mean and then dividing by the standard deviation is sometimes referred to as *standardization,* and a *z* score is one example of what is called a *standardized score.*

EXAMPLE 4.17   Suppose that two graduating seniors, one marketing major and one accounting major, are comparing job offers. The accounting major has an offer for $35,000 per year, and the marketing student has one for $33,000 per year. Summary information about the distribution of offers follows:

Accounting: mean = 36,000   standard deviation = 1500

Marketing: mean = 32,500   standard deviation = 1000

Then,

$$\text{accounting } z \text{ score} = \frac{35,000 - 36,000}{1500} = -.67$$

(so $35,000 is .67 standard deviation below the mean), whereas

$$\text{marketing } z \text{ score} = \frac{33,000 - 32,500}{1000} = .5$$

Relative to the appropriate data sets, the marketing offer is actually more attractive than the accounting offer (though this may not offer much solace to the marketing major).

The *z* score is particularly useful when the distribution of observations is approximately normal. In this case, by the Empirical Rule, a *z* score outside the interval from −2 to +2 will occur in about 5% of all cases, whereas a *z* score outside the interval from −3 to +3 will occur only about .3% of the time.

A particular observation can be located even more precisely by giving the percentage of the data that fall at or below that observation. If, for example, 95% of all test scores are at or below 650, whereas only 5% are above 650, then 650 is called the *95th percentile* of the data set (or of the distribution of scores). Similarly, if 10% of all scores are at or below 400 and 90% are above 400, then the value 400 is the 10th percentile.

Definition

For any particular number *r* between 0 and 100, the *r*th percentile is a value such that *r* percent of the observations in the data set fall at or below that value.

Figure 4.16 illustrates the 90th percentile. We have already met several percentiles in disguise. The median is the 50th percentile, and the lower and upper quartiles are the 25th and 75th percentiles, respectively.

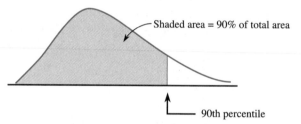

FIGURE 4.16 90th percentile from a smoothed histogram

EXAMPLE 4.18   Variable cost as a percentage of operating revenue is an important indicator of a company's financial health. The article "A Credit Limit Decision Model for Inventory Floor Planning and Other Extended Trade Credit Arrangements" (*Decision Sciences* (1992): 200–220) reported the following summary data for a sample of 350 corporations:

sample mean = 90.1%   sample median = 91.2%
5th percentile = 78.3%   95th percentile = 98.6%

Thus, half of the sampled firms had variable costs that were less than 91.2% of operating revenues because 91.2 is the 50th percentile. If the distribution of values were perfectly symmetric, the mean and median would be identical. Further evidence against symmetry is provided by the two other given percentiles. The 5th percentile is 12.9 below the median, whereas the 95th percentile is only 7.4 above the median. The histogram given in the article had roughly the shape of the negatively skewed smoothed histogram in Figure 4.17. (The total area under the curve is 1.)

FIGURE 4.17 The distribution of variable costs from Example 4.18

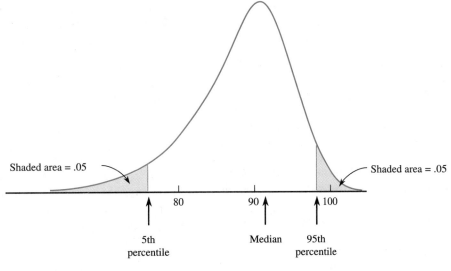

Shaded area = .05

Shaded area = .05

80   90   100

5th
percentile

Median   95th
percentile

Exercises 4.35 – 4.48

4.35 The average playing time of compact discs in a large collection is 35 min, and the standard deviation is 5 min.
  a. What value is 1 standard deviation above the mean? 1 standard deviation below the mean? What values are 2 standard deviations away from the mean?
  b. Without assuming anything about the distribution of times, at least what percentage of the times are between 25 and 45 min?
  c. Without assuming anything about the distribution of times, what can be said about the percentage of times that are either less than 20 min or greater than 50 min?
  d. Assuming that the distribution of times is normal, approximately what percentage of times are between 25 and 45 min? Less than 20 min or greater than 50 min? Less than 20 min?

4.36 In a study investigating the effect of car speed on accident severity, 5000 reports of fatal automobile accidents were examined, and the vehicle speed at impact was recorded for each one. It was determined that the average speed was 42 mph and that the standard deviation was 15 mph. In addition, a histogram revealed that vehicle speed at impact could be described by a normal curve.
  a. Roughly what proportion of vehicle speeds were between 27 mph and 57 mph?

  b. Roughly what proportion of vehicle speeds exceeded 57 mph?

4.37 Mobile homes are very tightly constructed for energy conservation. This may lead to a buildup of pollutants generated indoors. The paper "A Survey of Nitrogen Dioxide Levels Inside Mobile Homes" (*J. Air Pollut. Control Assoc.* (1988): 647–651) discussed various aspects of NO_2 concentration in these structures.
  a. In one sample of homes in the Los Angeles area, the mean NO_2 concentration in kitchens during the summer was 36.92 ppb, and the standard deviation was 11.34. Making no assumptions about the shape of the histogram, what can be said about the percentage of observations between 14.24 and 59.60?
  b. Inside what interval is it guaranteed that at least 89% of the concentration observations will lie?
  c. In a sample of non–Los Angeles homes, the average kitchen concentration during the winter was 24.76 ppb, and the standard deviation was 17.20. Do these values suggest that the histogram of sample observations did not closely resemble a normal curve? (*Hint:* What is $\bar{x} - 2s$?)

4.38 The article "Taxable Wealth and Alcoholic Beverage Consumption in the United States" (*Psychological Reports* (1994): 813–814) reported that the mean annual adult consumption of wine was 3.15 gallons and the standard deviation was 6.09 gallons. Would you use the Empirical Rule to approximate the proportion of adults who consume more than 9.24 gallons (i.e., whose consumption value exceeds the mean by more than 1 standard deviation)? Explain your reasoning.

4.39   A student took two national aptitude tests in the course of applying for admission to colleges. The national average and standard deviation for the first test were 475 and 100, respectively, whereas for the second test the average and standard deviation were 30 and 8, respectively. The student scored 625 on the first test and 45 on the second. Use z scores to determine on which exam the student performed better.

4.40   Suppose that your younger sister is applying for entrance to college and has taken the SAT exams. She scored at the 83rd percentile on the verbal section of the test and at the 94th percentile on the math section of the test. Since you have been studying statistics, she asks you for an interpretation of these values. What would you tell her?

4.41   A sample of concrete specimens of a certain type is selected, and the compressive strength of each one is determined. The mean and standard deviation are calculated as $\bar{x} = 3000$ and $s = 500$, and the sample histogram is found to be very well approximated by a normal curve.

a. Approximately what percentage of the sample observations are between 2500 and 3500, and what result justifies your assertion?

b. Approximately what percentage of sample observations are outside the interval from 2000 to 4000?

c. What can be said about the approximate percentage of observations between 2000 and 2500?

d. Why would you not use Chebyshev's Rule to answer the questions posed in parts **a**–**c**?

4.42   The paper "Modeling and Measurements of Bus Service Reliability" (*Trans. Research* (1978): 253–256) studied various aspects of bus service and presented data on travel times from several different routes. We give here a frequency distribution for bus travel times from origin to destination on one particular route in Chicago during peak morning traffic periods.

Class	Frequency	Relative Frequency
15 –<16	4	.02
16 –<17	0	.00
17 –<18	26	.13
18 –<19	99	.49

(continued)

Class	Frequency	Relative Frequency
19 –<20	36	.18
20 –<21	8	.04
21 –<22	12	.06
22 –<23	0	.00
23 –<24	0	.00
24 –<25	0	.00
25 –<26	16	.08

a. Construct the corresponding histogram.

b. Compute the following (approximate) percentiles.
  i. 86th   **ii.** 15th   **iii.** 90th   **iv.** 95th   **v.** 10th

4.43   An advertisement for the "30-in. wonder" that appeared in the September 1983 issue of the journal *Packaging* claimed that the 30-in. wonder weighs cases and bags up to 110 lb and provides accuracy to $\frac{1}{4}$ oz. Suppose that a 50-oz weight was repeatedly weighed on this scale and the weight readings recorded. The mean value was 49.5 oz, and the standard deviation was .1. What can be said about the proportion of the time that the scale actually showed a weight that was within $\frac{1}{4}$ oz of the true value of 50 oz? (*Hint*: Try to make use of Chebyshev's Rule.)

4.44   Suppose that your statistics professor returned your first midterm exam with only a z score written on it. She also told you that a histogram of the scores was closely described by a normal curve. How would you interpret each of the following z scores?
  a. 2.2   **b.** −.4   **c.** −1.8   **d.** 1.0   **e.** 0

4.45   The paper "Answer Changing on Multiple-Choice Tests" (*J. of Experimental Education* (1980): 18–21) reported that for a group of 162 college students, the average number of responses changed from the correct answer to an incorrect answer on a test containing 80 multiple-choice items was 1.4. The corresponding standard deviation was reported to be 1.5. Based on this mean and standard deviation, what can you tell about the shape of the distribution of the variable *number of answers changed from right to wrong*? What can you say about the number of students who changed at least six answers from correct to incorrect?

4.46   The average reading speed of students completing a speed-reading course is 450 words per minute (wpm). If the standard deviation is 70 wpm, find the z score associated with each of the following reading speeds.
  a. 320 wpm   **b.** 475 wpm   **c.** 420 wpm   **d.** 610 wpm

4.47   The accompanying data values are 1989 per capita expenditures on public libraries for each of the 50 states. (New York is the highest and Arkansas the lowest.)

29.48	24.45	23.64	23.34	22.10	21.16	19.83
18.01	17.95	17.23	16.53	16.29	15.89	15.85
13.64	13.37	13.16	13.09	12.66	12.37	11.93
10.99	10.55	10.24	10.06	9.84	9.65	8.94
7.70	7.56	7.46	7.04	6.58	5.98	19.81
19.25	19.18	18.62	14.74	14.53	14.46	13.83
11.85	11.71	11.53	11.34	8.72	8.22	8.13
8.01						

a. Summarize this data set with a frequency distribution. Construct the corresponding histogram.

b. Use the histogram in part **a** to find approximate values of the following percentiles:

   i. 50th   **ii.** 70th   **iii.** 10th   **iv.** 90th   **v.** 40th

4.48 The accompanying table gives the mean and standard deviation of reaction times (sec) for each of two different stimuli.

	Stimulus 1	Stimulus 2
Mean	6.0	3.6
Standard deviation	1.2	.8

If your reaction time for the first stimulus is 4.2 sec and for the second stimulus 1.8 sec, to which stimulus are you reacting (when compared to all other individuals) relatively more quickly?

4.5 Interpreting the Results of Statistical Analyses

It is relatively rare to find raw data in published sources. Most of the time, only a few numerical summary quantities are reported. We must be able to interpret these values and understand what they tell us about the underlying data set.

For example, a university recently conducted an investigation of the amount of time required to enter the information contained in an application for admission into the university computer system. One of the individuals who performs this task was asked to note starting time and completion time for 50 randomly selected application forms. The resulting entry times (minutes) were summarized using the mean, median, and standard deviation:

$$\bar{x} = 7.854 \qquad \text{median} = 7.423 \qquad s = 2.129$$

What do these summary values tell us about entry times? The average time required to enter admissions data was 7.854 minutes, but the relatively large standard deviation suggests that there was quite a bit of variability among the 50 observed times. The median tells us that half of the applications required less than 7.423 minutes for entry. The fact that the mean exceeds the median suggests that some unusually large values in the data set affected the value of the mean. This last conjecture is confirmed by the stem-and-leaf display of the data given in Figure 4.18.

FIGURE 4.18 Stem-and-leaf display of data entry times

```
 4 | 8
 5 | 0234579
 6 | 00001234566779
 7 | 223556688
 8 | 23334
 9 | 002
10 | 011168
11 | 134
12 | 2              Stem:  Ones
13 |                Leaf:  Tenths
14 | 3
```

Those conducting the study looked at the outlier 14.3 and at the other relatively large values in the data set; they found that the five largest values came from applications that were entered before lunch. After talking with the individual entering the data, administrators speculated that morning entry times might differ from afternoon entry times because there tended to be more distractions and interruptions (phone calls, etc.) during the morning hours, when the admissions office generally was busier. When morning and afternoon entry times were separated, the following summary statistics resulted:

Morning (*based on n = 20 applications*): $\bar{x} = 9.093$ median = 8.743 $s = 2.329$

Afternoon (*based on n = 30 applications*): $\bar{x} = 7.027$ median = 6.737 $s = 1.529$

From these summary values, we see that the average entry time is higher for applications entered in the morning and also that the individual entry times differ more from one another in the mornings than in the afternoons (since the standard deviation for morning entry times is about 1.5 times as large as that for afternoon entry times).

What to Look for in Published Data

Here are a few questions to ask yourself when interpreting numerical summary measures.

- Is the chosen summary measure appropriate for the type of data collected? In particular, watch for inappropriate use of the mean and the standard deviation with categorical data that has simply been coded numerically.

- If both the mean and the median are reported, how do the two values compare? What does this suggest about the distribution of values in the data set? If only the mean or the median was used, was the appropriate measure selected?

- Is the standard deviation large or small? Is the value consistent with your expectations regarding variability? What does the value of the standard deviation tell you about the variable being summarized?

- Can anything of interest be said about the values in the data set by applying Chebyshev's Rule or the Empirical Rule?

The journal article "Smoking During Pregnancy and Lactation and Its Effects on Breast Milk Volume" (*Amer. J. of Clinical Nutrition* (1991): 1011–1016) reported the following summary values for data collected on daily breast milk volume for two groups of nursing mothers. For a group of mothers who smoked, the mean milk volume was 693 g and the standard deviation was 110. Based on data given in the article, the median milk volume was 589.5 g. For a group of nonsmoking mothers, the mean was 961 g, the standard deviation 120, and the median 946 g.

	Mean	Standard Deviation
Smokers	693	110
Nonsmokers	589	120

Since milk volume is a numerical variable, these descriptive measures reasonably summarize the data set. For smoking mothers, the mean and median are similar, indicating that the milk volume distribu̶ely symmetric. The average milk volume for this group is 693, wh̶ng mothers the mean milk volume is 961 — quite a bit higher than ̶others. The standard deviations for the two groups are similar, sugg̶ h the average milk volume may be lower for smokers, the variabi̶ volume is about the same. Since the median for the nonsmokir̶ than the corresponding mean, we suspect that the distributior̶ r nonsmoking mothers is positively skewed.

Summary of Key Concepts and Formulas

Term or Formula	Comment
$x_1, x_2, \ldots, x_n$	Notation for sample data consisting of observations on a variable x, where n is the sample size.
Sample mean, $\bar{x} = \dfrac{\Sigma x}{n}$	The most frequently used measure of center of a sample. It can be very sensitive to the presence of even a single outlier (unusually large or small observation).
Population mean, μ	The average x value in the entire population.
Sample median	The middle value in the ordered list of sample observations. (For n even, the median is the average of the two middle values.) It is very insensitive to outliers.
Trimmed mean	A measure of center in which the observations are first ordered from smallest to largest, one or more observations are deleted from each end, and the remaining ones are averaged. In terms of sensitivity to outliers, it is a compromise between the mean and the median.
Deviations from the mean: $x_1 - \bar{x}, x_2 - \bar{x}, \ldots, x_n - \bar{x}$	Quantities used to assess variability in a sample. Except for rounding effects, $\Sigma (x - \bar{x}) = 0$.
The sample variance $s^2 = \dfrac{\Sigma(x - \bar{x})^2}{n - 1}$ and standard deviation $s = \sqrt{s^2}$	The most frequently used measures of variability for sample data.
$S_{xx} = \Sigma x^2 - \dfrac{(\Sigma x)^2}{n}$	The computing formula for S_{xx}, the numerator of s^2.
The population variance σ^2 and standard deviation σ	Measures of variability for the entire population.
Quartiles and the interquartile range	The lower quartile separates the smallest 25% of the data from the remaining 75%, and the upper quartile separates the largest 25% from the smallest 75%. The interquartile range (iqr), a measure of variability less sensitive to outliers than s, is the difference between the upper and lower quartiles.
Chebyshev's Rule	This rule states that for any number $k \geq 1$, *at least* $100(1 - 1/k^2)\%$ of the observations in *any* data set are within k standard deviations of the mean. It is typically conservative in that the actual percentages often considerably exceed $100(1 - 1/k^2)$.

Empirical Rule	This rule gives the approximate percentage of observations within 1 standard deviation (68%), 2 standard deviations (95%), and 3 standard deviations (99.7%) of the mean when the histogram is well approximated by a normal curve.
z score	This quantity gives the distance between an observation and the mean expressed as a certain number of standard deviations. It is positive (negative) if the observation lies above (below) the mean.
rth percentile	The value such that r percent of the observations in the data set fall at or below that value.
Boxplot	A picture that conveys information about the most important features of a data set: center, spread, extent of skewness, and presence of outliers.

Supplementary Exercises 4.49–4.64

4.49 Reconsider the radon concentration data given in the chapter introduction.

a. Construct a comparative stem-and-leaf display with 210 shown separately as a high value, and comment on similarities and differences.

b. $\Sigma x = 958$ for the cancer sample and 747 for the no-cancer sample. Calculate the two sample means and medians, and comment.

c. $S_{xx} = 41,084.476$ for the cancer sample and 10,969.077 for the no-cancer sample. What do the values of the two sample standard deviations suggest about variability in the two samples?

d. Determine the value of iqr for each sample. What do these values tell you about variability in the two samples? Is the conclusion here different from the conclusion in part **c**? Explain. (*Hint*: Without 210 and 85, cancer $s = 11.4$ and no-cancer $s = 13.3$.)

e. Construct comparative boxplots.

4.50 The article "Can We Really Walk Straight?" (*Amer. J. of Physical Anthropology* (1992): 19–27) reported on an experiment in which each of 20 healthy men was asked to walk as straight as possible to a target 60 meters away at normal speed. Consider the accompanying data on cadence (number of strides per second).

.95 .85 .92 .95 .93 .86 1.00 .92 .85 .81 .78
.93 .93 1.05 .93 1.06 1.06 .96 .81 .96

Use the methods developed in this chapter to summarize the data; include an interpretation or discussion wherever appropriate. (*Note:* The author of the paper used a rather sophisticated statistical analysis to conclude that people cannot walk in a straight line, and suggested several explanations for this.)

4.51 The *San Luis Obispo Telegram-Tribune* (Dec. 18, 1992) reported that for the 1992 season, the average major-league baseball salary was just over $1 million, whereas only 267 of the 772 players made at least this much money. What does this imply about the median salary and the shape of a histogram of salaries?

4.52 The article "Comparing the Costs of Major Hotel Franchises" (*Real Estate Review* (1992): 46–51) gave the accompanying data on franchise cost as a percentage of total room revenue for chains of three different types.

Budget	2.7	2.8	3.8	3.8	4.0	4.1	5.5	5.9	6.7
	7.0	7.2	7.2	7.5	7.5	7.7	7.9	7.9	8.1
	8.2	8.5							
Midrange	1.5	4.0	6.6	6.7	7.0	7.2	7.2	7.4	7.8
	8.0	8.1	8.3	8.6	9.0				
First-class	1.8	5.8	6.0	6.6	6.6	6.6	7.1	7.2	7.5
	7.6	7.6	7.8	7.8	8.2	9.6			

Construct a boxplot for each type of hotel, and comment on interesting features, similarities, and differences.

4.53 In recent years, many teachers have been subject to increased levels of stress, contributing to disenchantment with teaching as a profession. The paper "Professional Burnout Among Public School Teachers" (*Public Personnel Mgmt.* (1988): 167–189) looked at various aspects of this problem. Consider the accompanying information on total psychological effects scores for a sample of 937 teachers.

Burnout Level	Range of Scores	Frequency
Low	0–21	554
Moderate	22–43	342
High	44–65	41

$\bar{x} = 19.93$ $s = 12.89$

a. Draw a histogram.

b. Mark the value of $\bar{x}$ on the measurement scale, and use the histogram to locate the approximate value of the median. Why are the two measures of center not identical?

c. In what interval of values are we guaranteed to find at least 75% of the scores, irrespective of the histogram shape? From the histogram, roughly what percentage of the scores actually fall in this interval?

4.54 The amount of radiation received at a greenhouse plays an important role in determining the rate of photosynthesis. The accompanying observations on incoming solar radiation were read from a graph in the paper "Radiation Components over Bare and Planted Soils in a Greenhouse" (*Solar Energy* (1990): 1–6).

6.3	6.4	7.1	7.7	8.4	8.5	8.8	8.9	9.0
9.1	10.0	10.1	10.2	10.6	10.6	10.7	10.7	10.8
10.9	11.1	11.2	11.2	11.4	11.9	11.9	12.2	13.1

Use some of the methods discussed in this chapter and the previous one to describe this data.

4.55 The accompanying data on milk volume (g/day) was taken from the paper "Smoking during Pregnancy and Lactation and Its Effects on Breast Milk Volume" (*American J. of Clinical Nutrition* (1991): 1011–1016).

Smoking mothers	621	793	593	545	753	655
	895	767	714	598	693	
Nonsmoking mothers	947	945	1086	1202	973	981
	930	745	903	899	961	

Compare and contrast the two samples.

4.56 a. Suppose that n, the number of sample observations, is an odd number. Under what conditions will $\bar{x}$ and the sample median be identical?

b. If $n = 10$, when will $\bar{x}$ and the 10% trimmed mean be identical?

4.57 a. A statistics instructor informed her class that the median exam score was 78 and the mean was 70. Sketch a picture of what the smoothed histogram of scores might look like.

b. Suppose that the majority of students in the class studied for the exam, but a few students did not. How might this be reflected in a smoothed histogram of exam scores?

4.58 Suppose that an auto dealership employs clerical workers, salespeople, and mechanics. The average monthly salaries of the clerical employees, salespeople, and mechanics are $1700, $2200, and $2900, respectively. If the dealership has three clerical employees, ten sales-

people, and eight mechanics, find the average monthly wage for all employees of the dealership.

4.59 The *Los Angeles Times* (July 17, 1995) reported that in a sample of 364 lawsuits in which punitive damages were awarded, the sample median damage award was $50,000, and the sample mean was $775,000. What does this suggest about the distribution of values in the sample?

4.60 The paper "Sodium–Calcium Exchange Equilibria in Soils as Affected by Calcium Carbonate and Organic Matter" (*Soil Science* (1984): 109) gave ten observations on soil pH. The data resulted from analysis of ten samples of soil from the Central Soil Salinity Research Institute experimental farm.

Soil pH 8.53 8.52 8.01 7.99 7.93 7.89 7.85 7.82 7.80 7.72

a. Calculate and interpret the values of the sample mean, variance, and standard deviation.

b. Compute the 10% trimmed mean and the sample median. Do either of these values differ much from the value of the sample mean?

c. Find the upper quartile, the lower quartile, and the interquartile range.

d. Illustrate the location and spread of this sample using a boxplot.

4.61 Age at diagnosis for each of 20 patients under treatment for meningitis was given in the paper "Penicillin in the Treatment of Meningitis" (*J. Amer. Med. Assoc.* (1984): 1870–1874). The ages (in years) were as follows:

18	18	25	19	23	20	69	18	21	18
20	18	18	20	18	19	28	17	18	18

a. Calculate the values of the sample mean and the standard deviation.

b. Calculate the 10% trimmed mean. How does the value of the trimmed mean compare to that of the sample mean? Which would you recommend as a measure of location? Explain.

c. Compute the upper quartile, the lower quartile, and the interquartile range.

d. Are there any mild or extreme outliers present in this data set?

e. Construct the boxplot for this data set.

4.62 Although blood pressure is a continuous variable, its value is often reported to the nearest 5 mm Hg (for example, 100, 105, 110, and so on). Suppose that the actual blood pressure values for nine randomly selected individuals are as follows.

118.6 127.4 138.4 130.0 113.7 122.0 108.3 131.5 133.2

a. If values are reported as suggested (rounded to the nearest 5 mm Hg), what is the sample of the *reported* values, and what is the median of this sample?

b. Suppose that the second individual's blood pressure is 127.6 rather than 127.4 (a small change in a single value). How does this change the median of reported values? What does this say about the median's sensitivity to rounding or grouping of the data?

4.63 The accompanying observations are carbon monoxide levels (ppm) in air samples obtained from a certain region.

9.3   10.7   8.5   9.6   12.2   16.6   9.2   10.5   7.9   13.2   11.0
8.8   13.7   12.1   9.8

a. If a trimmed mean is calculated by first deleting the smallest and largest observations, what is the corresponding trimming percentage? What is it if the two smallest and two largest observations are deleted?

b. Calculate the two trimmed means referred to in part **a.**

c. Using the results of part **b,** how might you calculate a measure of center for this sample that could be regarded as a 10% trimmed mean?

4.64 Suppose that the distribution of scores on an exam is closely described by a normal curve with mean 100. The 16th percentile of this distribution is 80.

a. What is the 84th percentile?

b. What is the approximate value of the standard deviation of exam scores?

c. What z score is associated with an exam score of 90?

d. What percentile corresponds to an exam score of 140?

e. Do you think there were many scores below 40? Explain.

References

See the References at the end of Chapter 3.

GRAPHING CALCULATOR EXPLORATIONS

4.1 Trimmed Mean

The calculation of the trimmed mean is not a built-in feature of most graphing calculators, but it is easy to evaluate on a list-based calculator. To illustrate how this is done, we will use the data from Example 4.6. In Graphing Calculator Exploration 3.1, we discussed common list manipulation capabilities of calculators, and we'll use some of those now. First, enter this data into a list on your calculator in the order it was presented in the example. After the data list is entered, your calculator should look something like part (a) in the accompanying figure. The next thing you should do is to copy the original list into a "working" list. We will be deleting some data in a couple of steps and in case of error you may want to recover the original data. If you work with a copy of the data, you can recover from error quickly by simply making another working copy from the original list.

To calculate the trimmed mean you will need to be able to find the smallest and largest values in the list. To facilitate this, sort the working list; after sorting, the data should look something like part (b) in the accompanying figure.

To find the 10% trimmed mean for Example 4.6, we need to delete the three smallest numbers and the three largest numbers from the working list. After deleting these numbers, the list should appear as indicated in part (c) of the accompanying figure.

(a)	(b)	(c)
34	0	0
414	0	0
0	0	0
0	0	0
76	0	0
123	0	0
...	...	...
0	73	39
0	74	46
0	76	72
0	123	73
39	414	74

The last step in the trimmed mean calculation is to calculate the mean of the remaining numbers in the list, most likely by using a "list function" of your calculator. The actual syntax will vary from calculator to calculator, but will be something like "mean(List1)."

4.2 Quartiles

In this exploration, we bring up a rather delicate statistical point. As it happens, statisticians are not in complete agreement about the best way to calculate quartiles. Thus you may encounter a calculator or computer software package that will give somewhat different quartiles than would be obtained by using the definition we give in this book. The definition of quartiles used in this book will result in boxplots that are consistent with the views of the inventor of the boxplot — Dr. John Tukey — who, after all, ought to know. (Professor Tukey called these numbers "hinges.")

Differing approaches to the calculation of quartiles lead to potentially different values for the quartiles computed by calculators and computer software, and therefore to potentially different results for the interquartile range. That, in turn, may result in different criteria for what constitutes an outlier in a univariate set of data. Don't worry; this is a good news/bad news situation. The bad news is that you may at first be confused by your textbook, software, and calculator giving you different "right" answers when you make boxplots. But, the good news outweighs the bad — unless you have a very small data set, the differences among the different approaches are of little practical importance. The boxplot will be very effective at displaying the data and indicating symmetry, skewness, and outliers, irrespective of the quartile definition used.

Despite the inconsequential nature of the differences in practice, you should be cognizant of what your calculator and software actually do; if your calculator is using a different definition than this book and/or your computer software, you may be more comfortable in the knowledge that an observed discrepancy is likely to be due to the differing definitions, and not an error in data entry or a conceptual error on your part.

To check the agreement among your calculator, your computer software, and this book, use the following sets of data, which are designed to bring out these differences. The answers using the definitions in this book are in parentheses.

Test Data Set #1:   10, 20, 30, 40
(lower quartile = 15, median = 25, upper quartile = 35)

Test Data Set #2:   10, 20, 30, 40, 50
(lower quartile = 15, median = 30, upper quartile = 45)

Test Data Set #3:   10, 20, 30, 40, 50, 60
(lower quartile = 20, median = 35, upper quartile = 50)

Test Data Set #4:   10, 20, 30, 40, 50, 60, 70
(lower quartile = 20, median = 40, upper quartile = 60)

Did you obtain the same values using your calculator? If so, your calculator is probably using the same definition as the text. If not, your calculator is probably using one of the alternate definitions. But don't worry, when analyzing real data, the method used to compute quartiles will not be of concern.

4.3 Boxplots

It may be that your calculator does not provide complete flexibility when graphing a boxplot. Your calculator may graph a skeletal boxplot only, a modified boxplot only, or it may offer you a choice. (If your calculator offers you a choice, it will probably present a menu with a sequence of choices to "set up" a graphics screen.)

But what if you are limited in your choices? Suppose you wish to sketch a boxplot that is different from what your calculator delivers. If your calculator presents a modified boxplot and you need to sketch a skeletal boxplot, simply extend the whiskers of the modified boxplot to include all the outliers.

If your calculator plots only a skeletal boxplot, you have a very different problem — how can you draw the outliers without just starting from scratch on paper? To graph the skeletal boxplot, your calculator will have already calculated the quartiles and the median for your data. To sketch a modified boxplot, you need to locate this information and then decide which, if any, of the data values are outliers.

The first step will be the easiest. Your calculator will have a screen that displays a statistical summary for the data. Such quantities as the minimum and maximum data element, the mean, median, and standard deviation, and the lower and upper quartiles will have been calculated. (You may have to scroll down the screen to bring the ones you want into view.) You must now ascertain whether there are any outliers, and to do that you will need to calculate the following quantities by hand, as discussed in the text:

upper quartile + 1.5 iqr = upper quartile +1.5(upper quartile − lower quartile)

lower quartile − 1.5 iqr = lower quartile −1.5(upper quartile − lower quartile)

If these calculations result in a judgment of no outliers, you are done! The skeletal boxplot and the modified boxplot will be the same. If these calculations indicate there are outliers in your data, you now must find the largest observation that is not an outlier, and the smallest observation that is not an outlier. Any observations larger and smaller than these, respectively, will be outliers. Finding the outliers "by inspection"—that is, by scanning up and down the list of data, would be tedious and error-laden, especially because the observations are not sorted. The solution? Use your calculator to sort the data, making it very easy to spot outliers by looking at either end of the data list.

Your calculator will sort the data from small to large, or large to small; for the purpose of identifying outliers, it does not matter which you choose. The keystroke sequence for sorting data on your particular calculator will be found in your calculator manual; look in the chapter on "lists" and in the index under "sorting." Once the data is sorted, finding the outliers by inspection is very easy.

We need to issue a final caveat before we refer you to your calculator manual. Later in this book we will be entering bivariate data into the calculator. With bivariate data, pairs of numbers are entered into the calculator and stored in two lists. You should be aware that your calculator does not know these numbers are paired! If you sort only one of these lists the pairing will be lost. If sorting is necessary for you to make a boxplot, you should make a copy of the list you want to plot, and work with that. Then any sorting you do on the copy of the list will not affect the original list and the original numbers will still be "paired" together.

4.4 *z* Scores

Despite the usefulness of standardized scores in general, and z scores in particular, there will not be a z score function on your calculator. Suppose you wanted to compute the z scores corresponding to each observation in a data set consisting of 20 values. You might worry that this would require mind-numbing and repetitive calculations — subtracting the mean from every value and then dividing each result by the standard deviation. Fortunately, this is not the case. The list capabilities of your calculator will dramatically save time and effort. Instead of 20 sets of calculations, you can perform one calculation on the whole list of data! Moreover, the keystrokes that will work for lists mirror the keystrokes for working with individual numbers.

You will first need to enter some data into a list we will refer to as List1. We will use the data in column (a) of the figure on page 144. After entering the data, we need to find the mean and sample standard deviation. These values are $\bar{x} = 3.0$, $s = 1.58113883$. (Your calculator may not display all these digits, but it knows they are there!)

At this point you can convert all the raw scores to z scores by performing arithmetic on the whole list. Not only that, but you will also be able to store the results in another list without losing your original data. Again, your required keystrokes may differ, but they will accomplish the following:

1. Subtract the mean from each value in List1 to get the deviations.

2. Divide the deviations by the standard deviation.

3. Store the results of these calculations in List2.

The sequence of keystrokes will probably be very similar to this:

(List1 − 3.0) ÷ 1.58113883 → List2

Some calculators may use the slash "/" to indicate division, and some may use the equal sign "=" to indicate that the results should be stored in a second list. After performing the keystrokes appropriate for your calculator, you should see the results in List2 of column (b) in the figure below. If the lists you have chosen are next to each other on the screen you can easily compare the raw data with the z scores.

You may be able to save a few more keystrokes by "recalling" the mean and standard deviation from the calculator memory, although this is probably a matter of style and individual taste. If you have not done any intervening statistical calculations since you calculated the mean and standard deviation, these values are still stored in the calculator memory. If you wish, you may recall those values from the calculator's memory rather than key in the numbers. These values can be inserted in your calculations in place of the actual numbers by pressing the appropriate keystroke sequence. This capability typically involves using a "VARS" key, short for "variables." (The Texas Instruments TI-80 series calculators use VARS, as does the Casio CFX-9850Ga+. This capability may exist under different names with different calculators; consult your manual!) For example, on the Texas Instruments TI-83, you would press . . .

VARS . . . Statistics . . . S_x . . . ENTER

to recall the sample standard deviation. In the discussion to follow, the keystroke VARS$\overline{x}$ will stand for the particular sequence to recall the sample mean, and VARSs will stand for the particular sequence to recall the sample standard deviation.

With this symbolism in mind, the sequence of keystrokes using the recall capabilities of your calculator would transform (List1 − 3.0) ÷ 1.58113883 → List2 into a keystroke sequence like this: (List1 − VARS$\overline{x}$) ÷ VARSs → List2. The recall capability is a very nice feature, but we have found that individuals differ in their preferences about whether to use the VARS capabilities or just key in the digits when performing arithmetic operations with lists. We encourage you to try both ways and use the method with which you are more comfortable.

List 1
1.0
2.0
3.0
4.0
5.0

(a)

List 2
−1.265
−0.6325
0.0
0.63246
1.2649

(b)

5

Summarizing Bivariate Data

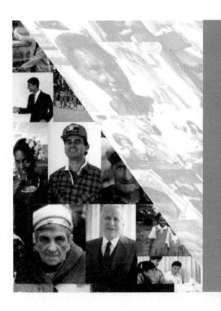

INTRODUCTION

How can the cost-effectiveness of schools be assessed? People inside and outside the education community have been arguing over this issue for years, and the debate has become more heated as spending for education has come under increasing scrutiny. The article "Cost-Effectiveness in Public Education" (*Chance* (1995): 38–41) considered data from various school districts in New Jersey. Part of this data consisted of observations on the two variables x = dollars spent per pupil and y = average SAT scores for 44 districts having only grade levels 7–12 or 9–12. Some representative observations (as read from a graph) are as follows:

x	7750	9900	10,870	12,080	...
y	878	893	966	950	...

What can be said about the general nature of the relationship between expenditure per pupil and SAT score? In particular, are high scores associated with high expenditure levels? Can average SAT score for a district be predicted once expenditure per pupil for the district is known? How accurate are predictions of average SAT score based just on expenditure per pupil?

Chapters 3 and 4 discussed methods for describing and summarizing univariate data. Often, however, an investigator is interested in two or more attributes of individuals or objects in a population, and how they are related to one another. For example, an environmental researcher might wish to know how the lead content of soil varies with distance from a major highway. Researchers in early childhood education might investigate how vocabulary size and age are related. College admissions officers, who must try to predict whether an applicant will succeed in college, might use a model relating college grade point average to high school grades, ACT or SAT scores, and various personal and family characteristics.

In this chapter, we introduce methods for describing relationships between two numerical variables and for assessing the strength of a relationship. Chapter 13 develops methods of statistical inference for drawing conclusions from data consisting of observations on two variables. The techniques introduced in this chapter are also important stepping stones for analyzing data consisting of observations on three or more variables, the topic of Chapter 14.

5.1   Scatter Plots

A **multivariate** data set consists of measurements or observations on each of two or more variables. An important special case, **bivariate** data, involves just two variables x and y. For example, x might be distance from a highway and y lead content of soil at that distance. When both x and y are numerical variables, each observation consists of a pair of numbers, such as (14, 5.2) or (27.63, 18.9). The first number in a pair is the value of x, and the second number is the value of y.

An unorganized list of such pairs yields little information about the distribution of either the x values or the y values separately, and even less information about how the two variables are related to one another. In Chapter 3, we saw how pictures could help make sense of univariate data. Pictures will help us with bivariate data also. The most important picture based on bivariate numerical data is a **scatter plot.** Each observation (pair of numbers) is represented by a point on a rectangular coordinate system, as shown in Figure 5.1(a). The horizontal axis is identified with values of x and is scaled so that any x value can be easily located. Similarly, the vertical or y axis is marked for easy location of y values. The point corresponding to any particular (x, y) pair is placed where a vertical line from the value on the x axis meets a horizontal line from the value on the y axis. Figure 5.1(b) shows the point representing the observation (4.5, 15); it is above 4.5 on the horizontal axis and to the right of 15 on the vertical axis.

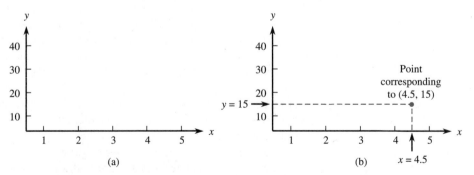

FIGURE 5.1 Constructing a scatter plot: (a) Rectangular coordinate system; (b) Point corresponding to (4.5, 15)

EXAMPLE 5.1

One method used by researchers to estimate the percentage of body fat is based on a measurement of total body electrical conductivity (TOBEC). The authors of the article "Comparing TOBEC-Derived Total Body Fat with Fat Pad Weights" (*Physiology & Behavior* (1995): 319–323) compared TOBEC body fat estimates with the actual weight of two fat deposits (called fat pads) for each rat in a sample of rats. The weight of these fat pads is known to be closely tied to overall body fat. The accompanying data on y = weight of fat pads and x = TOBEC estimate of body fat percentage is a subset of that given in the article. Figure 5.2 gives a scatter plot.

Observation	1	2	3	4	5	6	7	8	9
x	.1	1.9	4.0	4.6	4.2	4.2	7.1	5.4	6.2
y	216	233	251	272	282	295	298	303	311

Observation	10	11	12	13
x	4.6	3.6	7.1	5.4
y	318	318	327	335

FIGURE 5.2 Scatter plot for the data of Example 5.1

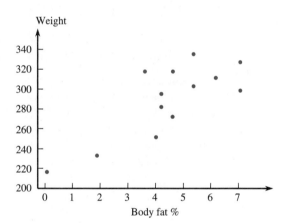

Examination of the data and plot reveals the following:

1. Several observations have identical x values yet different y values (for example, $x_5 = x_6 = 4.2$, but $y_5 = 282$ and $y_6 = 295$). Thus, the value of y is *not* determined *solely* by the value of x but by various other factors as well. That is, y is not a deterministic function of x.

2. There is a tendency for y to increase as x increases. That is, larger values of fat pad weight tend to be associated with larger values of TOBEC estimate of body fat percentage (a positive relationship between the variables).

3. It appears that the value of y could be predicted reasonably well from x by finding a line that is close to the points in the plot.

The horizontal and vertical axes in the scatter plot of Figure 5.2 do not intersect at the point $(0, 0)$. In many data sets, the values of x or of y or of both variables differ considerably from zero relative to the ranges of the values in the data set. For example, a study of how air conditioner efficiency is related to maximum daily outdoor temperature might involve observations at temperatures of $80°$, $82°$, ..., $98°$, and $100°$. In such cases, the plot will be more informative if the axes intersect at some point other than $(0, 0)$ and are marked accordingly. This is illustrated in the following example.

EXAMPLE 5.2

The growth and decline of forests is a matter of great public and scientific interest. The article "Relationships Among Crown Condition, Growth, and Stand Nutrition in Seven Northern Vermont Sugarbushes" (*Canad. J. of Forest Res.* (1995): 386– 397) included a scatter plot of y = mean crown dieback (%), which is one indicator of growth retardation, and x = soil pH (higher pH corresponds to less acidic soil). The following observations were read from the scatter plot:

x	3.3	3.4	3.4	3.5	3.6	3.6	3.7	3.7	3.8	3.8
y	7.3	10.8	13.1	10.4	5.8	9.3	12.4	14.9	11.2	8.0

x	3.9	4.0	4.1	4.2	4.3	4.4	4.5	5.0	5.1
y	6.6	10.0	9.2	12.4	2.3	4.3	3.0	1.6	1.0

Figure 5.3 shows two scatter plots of this data, produced by the statistical computer package MINITAB. In Figure 5.3(a), we let MINITAB select the scale for both axes. We obtained Figure 5.3(b) by specifying minimum and maximum values for x and y such that the axes would intersect at (roughly) the point $(0, 0)$. The second plot is more crowded than the first one, and such crowding can make it more difficult to ascertain the general nature of any relationship. For example, it can be more difficult to spot curvature in a crowded plot.

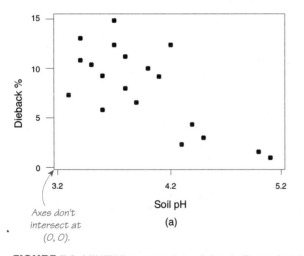

Axes don't
intersect at
$(0, 0)$.

(a)

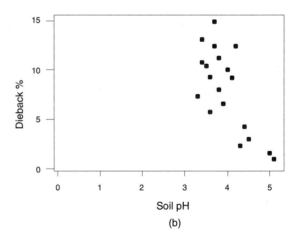

(b)

FIGURE 5.3 MINITAB scatter plots of data in Example 5.2

Large values of crown dieback tend to be associated with low soil pH, indicating a negative or inverse relationship. Furthermore, the two variables appear to be approximately linearly related, although the points would spread out quite a bit about any straight line drawn through the plot.

EXAMPLE 5.3

The focus of many agricultural experiments is to study how the yield of a crop varies with the time at which it is harvested. The accompanying data appeared in the article "Determination of Biological Maturity and Effect of Harvesting and Drying Conditions on Milling Quality of Paddy" (*J. of Agric. Engr. Research* (1975): 353–361). The variables are

x = time between flowering and harvesting (days)

y = yield of paddy, a type of grain farmed in India (kg/hectare)

x	16	18	20	22	24	26	28	30
y	2508	2518	3304	3423	3057	3190	3500	3883

x	32	34	36	38	40	42	44	46
y	3823	3646	3708	3333	3517	3214	3103	2776

Figure 5.4 is a MINITAB scatter plot. Notice that the axes do not intersect at $(0, 0)$. This is because zero lies well outside both the range of x values (16 to 46) and the range of y values (roughly 2500 to 3900). Had the scaling been chosen so that the axes did intersect at this point, all points in the plot would have been squeezed into the upper right-hand corner, obscuring somewhat the general pattern.

Although the pattern of the points in the plot is not linear, the scatter plot shows a fairly strong relationship between x and y: As x increases, y initially tends to increase, but after a certain point, y tends to decrease as x continues to increase. The shape of the plot is much like that of a parabola.

FIGURE 5.4 A scatter plot of the yield data from Example 5.3

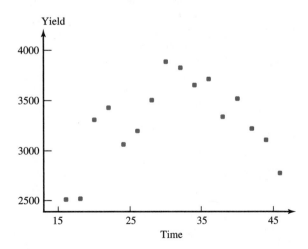

In the next three sections, we will develop methods for summarizing bivariate data when the scatter plot reveals a linear pattern (as in Examples 5.1 and 5.2). A curved pattern, such as the one in Example 5.3, is a bit more complicated to analyze, and methods for summarizing such nonlinear relationships will be discussed in Section 5.5.

Exercises 5.1 – 5.8

5.1 A number of studies concerning muscular efficiency in humans have utilized competitive cyclists as subjects. It had previously been observed that variation in efficiency could not be explained by differences in cycling technique. The paper "Cycling Efficiency Is Related to the Percentage of Type I Muscle Fibers" (*Med. and Sci. in Sports and Exercise* (1992): 782–788) reported the accompanying data on x = percentage of type I (slow twitch) muscle fibers and y = cycling gross efficiency (the ratio of work accomplished per minute to caloric expenditure).

Observation	1	2	3	4	5	6	7
x	32	37	38	40	45	50	50
y	18.3	18.9	19.0	20.9	21.4	20.5	20.1

Observation	8	9	10	11	12	13	14
x	54	54	55	61	63	63	64
y	20.1	20.8	20.5	19.9	20.5	20.6	22.1

Observation	15	16	17	18	19
x	64	70	70	75	76
y	21.9	21.2	20.5	22.7	22.8

a. Construct a scatter plot of the data in which the axes intersect at the point (0, 16). How would you describe the pattern in the plot?
b. Does it appear that the value of cycling gross efficiency is determined solely by the percentage of type I muscle fibers? Explain.

5.2 The article "Exhaust Emissions from Four-Stroke Lawn Mower Engines (*J. of the Air and Water Mgmt. Assoc.* (1997): 945–952) reported data from a study in which both a baseline gasoline and a reformulated gasoline were used, Consider the following observations on age (yr) and NO_x emissions (g/kW-h):

Engine	1	2	3	4	5
Age	0	0	2	11	7
Baseline	1.72	4.38	4.06	1.26	5.31
Reformulated	1.88	5.93	5.54	2.67	6.53

(*continued*)

Engine	6	7	8	9	10
Age	16	9	0	12	4
Baseline	.57	3.37	3.44	.74	1.24
Reformulated	.74	4.94	4.89	.69	1.42

a. Construct a scatter plot of emissions versus age for the baseline gasoline.
b. Construct a scatter plot of emissions versus age for the reformulated gasoline.
c. Do age and emissions appear to be related? If so, in what way?
d. Comment on the similarities and differences in the two scatter plots of parts **a** and **b**.

5.3 A study to assess the capability of subsurface flow wetland systems to remove biochemical oxygen demand (BOD) and various other chemical constituents resulted in the accompanying data on x = BOD mass (kg/ha/d) and y = BOD removal (kg/ha/d) ("Subsurface Flow Wetlands — A Performance Evaluation," *Water Envir. Res.* (1995): 244–247).

x	3	8	10	11	13	16	27	30	35
y	4	7	8	8	10	11	16	26	21

x	37	38	44	103	142
y	9	31	30	75	90

a. Construct boxplots of both mass and mass removal, and comment on any interesting features.
b. Construct a scatter plot of the data and comment on any interesting features.

5.4 The paper "Objective Measurement of the Stretchability of Mozzarella Cheese" (*J. of Texture Studies* (1992): 185–194) reported on an experiment to investigate how the behavior of mozzarella cheese varied with temperature. Consider the data (at the top of page 151) on x = temperature and y = elongation (%) at failure of the cheese. (*Note*: The researchers were Italian and used *real* mozzarella, not the poor cousin widely available in the United States.)

x	59	63	68	72	74	78	83
y	118	182	247	208	197	135	132

a. Construct a scatter plot in which the axes intersect at the point (0, 0). (Mark 0, 20, 40, 60, 80, and 100 on the horizontal scale and 0, 50, 100, 150, 200, and 250 on the vertical scale.)

b. Construct a scatter plot in which the axes intersect at (55, 100) (as was done in the paper). Does this plot seem preferable to the one in part **a**? Explain.

c. What do the plots of parts **a** and **b** suggest about the nature of the relationship between x and y?

5.5 One factor in the development of tennis elbow, a malady that strikes fear into the hearts of all serious players of that sport (including one of the authors), is the impact-induced vibration of the racket-and-arm system at ball contact. It is well known that the likelihood of getting tennis elbow depends on various properties of the racket used. Consider the scatter plot of x = racket resonance frequency (Hz) and y = sum of peak-to-peak accelerations (a characteristic of arm vibration, in m/sec/sec) for n = 23 different rackets ("Transfer of Tennis Racket Vibrations into the Human Forearm," *Med. and Sci. in Sports and Exercise* (1992): 1134–1140). Discuss interesting features of the data and the scatter plot.

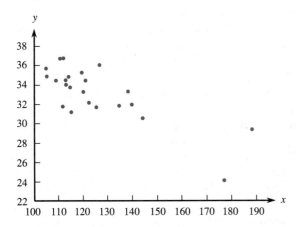

5.6 Stress can affect the physiology and behavior of animals, just as it can with humans (as many of us know all too well). The accompanying data on x = plasma cortisol concentration (ng cortisol/mL plasma) and y = oxygen consumption rate (mg/kg/hr) for juvenile steelhead after three 2-min disturbances was read from a graph in the paper "Metabolic Cost of Acute Physical Stress in Juvenile Steelhead" (*Trans. of Amer. Fish. Soc.* (1987): 257–263). The paper also included data for unstressed fish.

x	25	36	48	59	62
y	155	184	180	220	280

x	72	80	100	100	137
y	163	230	222	241	350

a. Is the value of y determined solely by the value of x? Explain your reasoning.

b. Construct a scatter plot of the data.

c. Does an increase in plasma cortisol concentration appear to be accompanied by a change in oxygen consumption rate? Comment.

5.7 Does the size of a transplanted organ matter? A study that attempted to answer this question ("Minimum Graft Size for Successful Living Donor Liver Transplantation," *Transplantation* (1999): 1112–1116) presented a scatter plot much like the following:

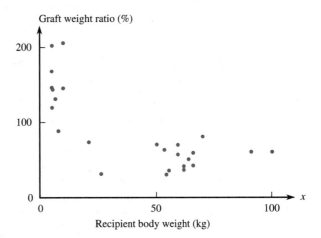

"Graft weight ratio" is the weight of the transplanted liver relative to the ideal size liver for the recipient.

a. Discuss interesting features of this scatter plot.

b. Why do you think the overall relationship is negative?

5.8 An article that appeared in *USA Today* (Dec. 20, 1999) gave the following data on the 25 largest counties in the United States.

County	Population	Number of Death Row Inmates
Los Angeles, CA	8,974,105	156
Cook, IL	5,109,887	81
Harris, TX	2,958,978	140
San Diego, CA	2,576,872	27

(*continued*)

County	Population	Number of Death Row Inmates	County	Population	Number of Death Row Inmates
Orange, CA	2,503,996	41	Broward, FL	1,337,001	26
Maricopa, AZ	2,328,217	53	Alameda, CA	1,323,155	33
Kings, NY	2,283,622	0	Allegheny, PA	1,321,002	8
Dade, FL	1,979,145	50	Nassau, NY	1,297,132	0
Queens, NY	1,958,299	0	Riverside, CA	1,275,338	36
Dallas, TX	1,917,501	37	Bexar, TX	1,245,209	29
King, WA	1,553,110	2	Tarrant, TX	1,227,122	25
Philadelphia, PA	1,541,039	125	Bronx, NY	1,198,220	0
Santa Clara, CA	1,535,916	27			
New York, NY	1,501,014	1			
San Bernardino, CA	1,497,816	28			
Cuyahoga, OH	1,410,021	37			
Suffolk, NY	1,337,486	0			

(continued)

Construct a scatter plot of this data and comment on any unusual features. How would you describe the relationship between county population and number of death row inmates? How are the unusual features of the plot incorporated into your description of the relationship?

5.2   Correlation

A scatter plot of bivariate numerical data gives a visual impression of how strongly x values and y values are related. However, to make precise statements and draw conclusions from data, we must go beyond pictures. A **correlation coefficient** (from *co-relation*) is a quantitative assessment of the strength of relationship between the x and y values in a set of (x, y) pairs. In this section, we introduce the most commonly used correlation coefficient.

Figure 5.5 displays several scatter plots that indicate different types of relationships between the x and y values. The plot in Figure 5.5(a) suggests a strong *positive relationship* between x and y; for every pair of points in the plot, the one with the larger x value also has the larger y value. That is, an increase in x is inevitably paired with an increase in y. The plot in Figure 5.5(b) shows a strong *tendency* for y to increase as x does, but there are a few exceptions. For example, the x and y values of the two points in the extreme upper right-hand corner of the plot go in opposite directions (x increases, but y decreases). Nevertheless, a plot such as this would again indicate a rather strong positive relationship. Figure 5.5(c) suggests that x and y are *negatively related* — as x increases, there is a tendency for y to decrease. Obviously, the negative relationship in this plot is not as strong as the positive relationship in Figure 5.5(b), although both plots show a well-defined linear pattern. The plot of Figure 5.5(d) is indicative of no strong relationship between x and y; there is no tendency for y either to increase or to decrease as x increases. Finally, as illustrated in Figure 5.5(e), a scatter plot can show evidence of a strong positive (or negative) relationship through a pattern that is curved rather than linear in character.

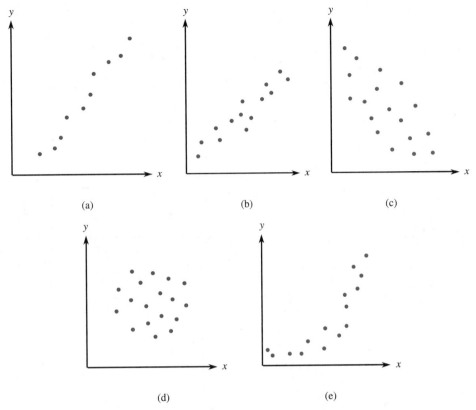

FIGURE 5.5 Scatter plots illustrating various types of relationships: (a) and (b) Positive linear relation; (c) Negative linear relation; (d) No relation; (e) Curved relation

Pearson's Sample Correlation Coefficient

Let $(x_1, y_1), (x_2, y_2), \ldots, (x_n, y_n)$ denote a sample of (x, y) pairs. Consider replacing each x value by the corresponding z score, z_x, (by subtracting $\bar{x}$ and then dividing by s_x) and similarly replacing each y value by its z score. Note that x values that are larger than the mean x value will have a positive z score and those smaller than $\bar{x}$ will have negative z scores. The same will be true for the y values: Those larger than $\bar{y}$ will have positive z scores and those smaller will have negative z scores. A common measure of the strength of linear relationship, called Pearson's sample correlation coefficient, is based on $\Sigma z_x z_y$, the sum of the product of z_x and z_y for each observation in the bivariate data set.

The scatter plot in Figure 5.6(a) on page 154 indicates a substantial positive relationship. A vertical line through $\bar{x}$ and a horizontal line through $\bar{y}$ divide the plot into four regions. In region I, both x and y exceed their mean values, so the z score for x and the z score for y are both positive numbers. It follows that $z_x z_y$ is positive. The product of the z scores is also positive for any point in region III, because both z scores are negative there, and multiplying two negative numbers gives a positive

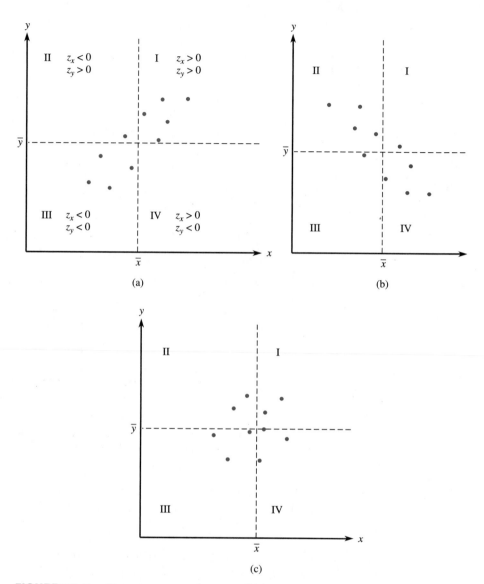

FIGURE 5.6 Breaking up a scatter plot according to the signs of z_x and z_y: (a) A positive relation; (b) A negative relation; (c) No strong relation

number. In each of the other two regions, one z score is positive and the other negative, so $z_x z_y$ is negative. But because almost all points lie in regions I and III, almost all products of z scores are positive. Thus, the *sum* of products, $\Sigma\, z_x z_y$, will be a large positive number.

Similar reasoning for the data displayed in Figure 5.6(b), which exhibits a strong negative relationship, implies that $\Sigma\, z_x z_y$ will be a large negative number. When there is no strong relationship, as in Figure 5.6(c), positive and negative products tend to counteract one another, producing a value of $\Sigma\, z_x z_y$ that is close to zero. In summary, $\Sigma\, z_x z_y$ seems to be a reasonable measure of the degree of association between x and y; it will be a large positive number, a large negative number, or a number close to zero, depending on whether there is a strong positive, a strong negative, or no strong linear relationship.

Pearson's sample correlation coefficient, denoted by r, is obtained by dividing $\Sigma\, z_x z_y$ by $n - 1$.

Definition

Pearson's sample correlation coefficient r is given by

$$r = \frac{\Sigma\, z_x z_y}{n - 1}$$

Although there are several different types of correlation coefficients, Pearson's correlation coefficient is by far the most commonly used, and so the name Pearson's is often omitted and it is referred to as simply "the correlation coefficient."

Hand calculation of the correlation coefficient is illustrated in the following example; such calculation can be quite tedious. Fortunately, all statistical software packages and most scientific calculators will compute r from sample data.

EXAMPLE 5.4

The article "Parents' and Adolescents' Perceptions of a Strong Family" (*Psych. Reports* (1999): 1219–1224) examined the relationship between family strength and other characteristics such as commitment, time spent together, and crisis management skills. Adolescents in what the researchers called "intact families" (those with both of the parents present in the home) rated their families with respect to the degree that the family exhibited effective crisis management and also with respect to family strength. Data consistent with findings that appeared in the article is given here:

Observation	1	2	3	4	5	6	7	8	9	10
Crisis management score	20	13	27	18	19	21	0	21	21	11
Family strength score	50	60	67	57	49	72	52	68	60	58

Figure 5.7 (page 156) is a scatter plot of this data.

Let x denote the crisis management score and y denote the family strength score. It is easy to verify that

$$\bar{x} = 17.10 \qquad s_x = 7.48 \qquad \bar{y} = 59.30 \qquad s_y = 7.82$$

To illustrate the calculation of the correlation coefficient, we begin by computing z scores for each (x, y) pair in the data set. For example, the first observation is $(20, 50)$. The corresponding z scores are

$$z_x = \frac{20 - 17.10}{7.48} = .39 \qquad z_y = \frac{50 - 59.30}{7.82} = -1.19$$

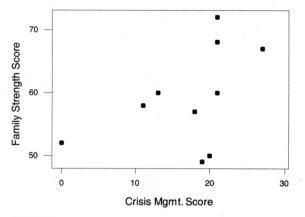

FIGURE 5.7 MINITAB scatter plot for data of Example 5.4

The accompanying table shows the z scores as well as the product $z_x z_y$ for each observation.

x	y	z_x	z_y	$z_x z_y$
20	50	.39	−1.19	−.46
13	60	−.55	.09	−.05
27	67	1.32	.98	1.29
18	57	.12	−.29	−.03
19	49	.25	−1.32	−.33
21	72	.52	1.62	.84
0	52	−2.29	−.93	2.13
21	68	.52	1.11	.58
21	60	.52	.09	.05
11	58	−.82	−.17	.14
				$4.16 = \Sigma\, z_x z_y$

Then, with $n = 10$

$$r = \frac{\Sigma\, z_x z_y}{n - 1} = \frac{4.16}{9} = .46$$

Based on the scatter plot and the properties of the correlation coefficient presented in the discussion that follows this example, there is only a weak positive linear relationship between crisis management score and family strength score.

Properties of *r*

1. *The value of r does not depend on the unit of measurement for either variable.* For example, if x is height, the corresponding z score will be the same whether height is expressed in inches, meters, or miles, and thus the value of the correlation coefficient is not affected. The correlation coefficient measures the inherent strength of the relationship between two numerical variables.

2. *The value of r does not depend on which of the two variables is labeled x.* Thus, if we had let x = family strength score and y = crisis management score in Example 5.4, the same value, $r = .46$, would have resulted.

3. *The value of r is between -1 and $+1$.* A value near the upper limit, $+1$, is indicative of a substantial positive relationship, whereas an r close to the lower limit, -1, suggests a prominent negative relationship. Figure 5.8 shows a useful way to describe the strength of relationship based on r.

FIGURE 5.8 Describing the strength of relationship

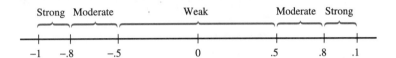

It may seem surprising that a value of r as extreme as $-.5$ or $.5$ should be in the "weak" category; an explanation for this is given later in the chapter. Even a weak correlation may indicate a meaningful relationship.

4. *The correlation coefficient $r = 1$ only when all the points in a scatter plot of the data lie exactly on a straight line that slopes upward. Similarly, $r = -1$ only when all the points lie exactly on a downward-sloping line.* Only when there is a perfect linear relationship between x and y in the sample will r take on one of its two possible extreme values.

5. *The value of r is a measure of the extent to which x and y are linearly related —* that is, the extent to which the points in the scatter plot fall close to a straight line. A value of r close to zero does not rule out any strong relationship between x and y; there could still be a strong relationship but one that is not linear.

EXAMPLE 5.5

College administrators and instructors have long puzzled over what factors affect student performance. The article "The Impact of Athletics, Part-Time Employment, and Other Activities on Academic Achievement" (*J. of College Student Devel.* (1992): 447–453) summarized the results of one large-scale study.

With x = grade point average and y = time spent working at a job (hr/week), it was reported that $r = -.08$ (based on $n = 528$). It appears that the linear relationship between time spent working at a job and grade point average is extremely weak: There is a very slight tendency for those who work more to have lower grades. If time spent working were reexpressed in min/week, r would still be $-.08$.

EXAMPLE 5.6   Economists sometimes use what is called the Misery Index to reflect the gloominess of the U.S. population. The Misery Index is defined as the sum of the inflation rate and the unemployment rate. The article "The Misery Index and Suicide" (*Psych. Reports* (1999): 1086) examined the relationship between suicide rate and both the Misery Index and a proposed revised misery index (the sum of the inflation rate and 2 times the unemployment rate). The revised index was proposed based on the belief that "unemployment inflicts twice the pain of inflation on the general public."

Using inflation, unemployment, and suicide rates for 1958–1992, the authors reported that

Pearson correlations between the Misery Indices and suicide rate were .97 for the original Misery Index and .61 for the revised index.

Based on the correlation coefficient, which measures the strength of any linear relationship, it appears that although there is a positive relationship between suicide rate and both indices, the relationship is much stronger for the original index than for the revised index.

EXAMPLE 5.7   Foal weight at birth is an indicator of health, therefore it is of interest to breeders of thoroughbred horses. Is foal weight related to the weight of the mare (mother)? The accompanying data are from the article "Suckling Behaviour Does Not Measure Milk Intake in Horses" (*Animal Behaviour* (1999): 673–678).

Observation	1	2	3	4	5	6	7	8	9
Mare weight (x, in kg)	556	638	588	550	580	642	568	642	556
Foal weight (y, in kg)	129	119	132	123.5	112	113.5	95	104	104

Observation	10	11	12	13	14	15
Mare weight (x, in kg)	616	549	504	515	551	594
Foal weight (y, in kg)	93.5	108.5	95	117.5	128	127.5

MINITAB was used to compute the value of the correlation coefficient for this data, with the following result.

Correlations (Pearson)
Correlation of Mare Weight and Foal Weight = 0.001, P-Value = 0.996

A correlation coefficient this close to zero indicates no linear relationship between mare weight and foal weight. The conclusion that mare weight and foal weight are unrelated is supported by a scatter plot of this data, which appears as Figure 5.9. From the correlation coefficient alone, we can conclude only that there is no *linear* relationship. We cannot rule out a more complicated curved relationship without examining the scatter plot.

FIGURE 5.9 A MINITAB scatter plot of the mare and foal weight data

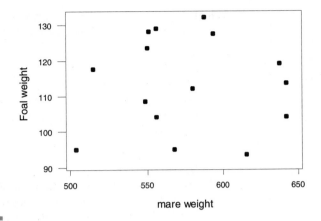

EXAMPLE 5.8   (Example 5.3 continued)   Let's reconsider the data on

x = time between planting and harvesting

and

y = yield

for the grain data of Example 5.3. The scatter plot for this data set appeared in Figure 5.4 and showed a strong curved pattern. Using MINITAB to compute the correlation coefficient between time and yield results in the following:

Correlations (Pearson)
Correlation of Time and Yield = 0.274, P-Value = 0.304

This shows the importance of interpreting r as a measure of the strength of any linear association. Here, r is not large, but there is a strong nonlinear relationship between time and yield. We should not conclude that there is no relationship whatsoever just because r is near 0.

The Population Correlation Coefficient

The sample correlation coefficient r measures how strongly the x and y values in a *sample* of pairs are related to one another. There is an analogous measure of how strongly x and y are related in the entire population of pairs from which the sample $(x_1, y_1), \ldots, (x_n, y_n)$ was obtained. It is called the **population correlation coefficient** and is denoted by **ρ.** (Notice again the use of a Greek letter for a population characteristic and Roman letter for a sample characteristic.) We will never have to calculate ρ from the entire population of pairs, but it is important to know that ρ satisfies properties paralleling those of r:

1. ρ is a number between -1 and $+1$ that does not depend on the unit of measurement for either x or y or on which variable is labeled x and which is labeled y.

2. $\rho = +1$ or -1 if and only if all (x, y) pairs in the population lie exactly on a straight line, so ρ measures the extent to which there is a linear relationship in the population.

In Chapter 13, we show how the sample characteristic r can be used to make an inference concerning the population characteristic ρ. In particular, r can be used to decide whether it is plausible that $\rho = 0$ (no linear relationship in the population).

Correlation and Causation

A value of r close to 1 indicates that relatively large values of one variable tend to be associated with relatively large values of the other variable. This is far from saying that a large value of one variable *causes* the value of the other variable to be large. Correlation (Pearson's or any other) measures the extent of association, but **association does not imply causation.** It frequently happens that two variables are highly correlated not because one is causally related to the other but because they are both strongly related to a third variable. Among all elementary school children, there is a strong positive relationship between the number of cavities in a child's teeth and the size of his or her vocabulary. Yet no one advocates eating foods that result in more cavities to increase vocabulary size (or working to decrease vocabulary size to protect against cavities). Number of cavities and vocabulary size are both strongly related to age, so older children tend to have higher values of both variables than do younger ones.

Scientific experiments can frequently make a strong case for causality by carefully controlling the values of all variables that might be related to the ones under study. Then, if y is observed to change in a "smooth" way as the experimenter changes the value of x, the most plausible explanation would be a causal relationship between x and y. In the absence of such control and ability to manipulate values of one variable, we must admit the possibility that an unidentified underlying third variable is influencing both the variables under investigation. A high correlation in many uncontrolled studies carried out in different settings can marshal support for causality — as in the case of cigarette smoking and cancer — but proving causality is often a very elusive task.

Spearman's Rank Correlation Coefficient (Optional)

Pearson's correlation coefficient r identifies a strong linear relationship between x and y but may miss a strong relationship that is not linear. In addition, the value of r can be greatly affected by the presence of even one or two outlying (x, y) pairs that are far from the main part of the scatter plot. Spearman's correlation coefficient r_S is a measure that is not as sensitive as r to outlying points and identifies both linear and nonlinear relationships. It does so by using the *ranks* of the x and y observations rather than the observations themselves.

To compute r_S, first replace the smallest x value by its rank, 1, the second smallest x value by its rank, 2, and so on. Similarly, the smallest y value is replaced by its rank, 1, the second smallest by its rank, 2, and so on. Suppose, for example, that $n = 4$ and the data pairs are

(110, 24.7)   (125, 24.2)   (116, 22.6)   (95, 23.5)

Then we have

x value	110	125	116	95	y value	24.7	24.2	22.6	23.5
Rank	2	4	3	1	Rank	4	3	1	2

The resulting rank pairs are

(2, 4)   (4, 3)   (3, 1)   (1, 2)

If there is a strong positive relationship between x and y, the x observations with small ranks tend to be paired with y observations having small ranks. An extreme case of this, the rank pairs (1, 1), (2, 2), (3, 3), (4, 4), occurs whenever a larger x value is always associated with a larger y value. Alternatively, large x ranks paired with small y ranks indicate a negative relationship, the most extreme case being (1, 4), (2, 3), (3, 2), and (4, 1) for $n = 4$ observations. Once the rank pairs have been determined, Spearman's r_S is just Pearson's coefficient r applied to these rank pairs. It can be computed by converting the ranks to z scores and then using the formula introduced earlier in this section, or by using the equivalent formula in the accompanying box.

Definition

Spearman's rank correlation coefficient r_S is Pearson's correlation coefficient applied to the rank pairs obtained by replacing each x value by its rank and each y value by its rank. An equivalent computing formula is

$$r_S = \frac{\Sigma (x\ \text{rank})(y\ \text{rank}) - \dfrac{n(n + 1)^2}{4}}{\dfrac{n(n - 1)(n + 1)}{12}}$$

As with r, the value of r_S is between -1 and 1. A value close to 1 or -1 indicates a strong relationship, whereas a value close to 0 indicates a weak relationship.

To illustrate the computation of Spearman's rank correlation coefficient, we will consider data from the paper "The Relation Between Freely Chosen Meals and Body Habits" (*Amer. J. Clinical Nutrition* (1983): 32–40). This article reported results of an investigation into the relationship between body build and energy intake of an individual's diet. A measure of body build is the Quetelet index (x), with a high value of x indicating a thickset individual. The variable reflecting energy intake is y = dietary energy density. There were nine subjects in the investigation, and the resulting (x, y) pairs and corresponding (x rank, y rank) pairs were as shown at the top of page 162.

Subject	x	y	x Rank	y Rank	$(x$ Rank$)\cdot(y$ Rank$)$
1	221	.67	3	3	9
2	228	.86	6	5	30
3	223	.78	4	4	16
4	211	.54	1	2	2
5	231	.91	7	7	49
6	215	.44	2	1	2
7	224	.90	5	6	30
8	233	.94	8	9	72
9	268	.93	9	8	72
					Sum = 282

Then

$$r_S = \frac{282 - \dfrac{9(10)^2}{4}}{\dfrac{(9)(8)(10)}{12}} = \frac{57}{60} = .95$$

This indicates a very strong positive relationship between the x and y values in the sample. The accompanying figure displays a scatter plot of the data. The point (268, .93) is a clear outlier and considerably distorts the linear pattern in the plot. The value of Pearson's r is only .658. Consider the effect on r and r_S of changing x_9. If 268 is replaced by 234, then (234, .93) is no longer an outlier. The value of r_S remains at .95, since 234 has the same x rank as 268, but the value of r for the altered data set rises dramatically to .92.

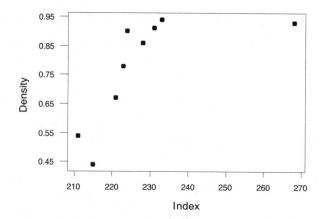

The objective of some experiments is to make comparisons by having individuals rank the objects or entities in a specified group. Thus, some television viewers might be asked to rank a group of programs in order of decreasing preference, or some wine connoisseurs might convene for the purpose of ranking a group of wines in order of descending quality. In such cases, the data values consist of sets of rankings rather than observations on numerical variables. Spearman's r_S can be applied directly to two sets of such rankings to measure the extent to which those doing the rating agree in their judgments. Several examples of this type are given in the exercises.

The procedure for calculating r_S must be modified somewhat if there are ties in either the x values or the y values (for example, three pairs with $x = 27.5$). One of the chapter references can be consulted to find out how this is done.

Exercises 5.9 – 5.25

5.9 For each of the following pairs of variables, indicate whether you would expect a positive correlation, a negative correlation, or no correlation. Explain your choice.
   a. Maximum daily temperature and cooling costs
   b. Interest rate and number of loan applications
   c. Incomes of husbands and wives when both have full-time jobs
   d. Height and IQ
   e. Height and shoe size
   f. Score on the math section of the SAT exam and score on the verbal section of the same test
   g. Time spent on homework and time spent watching television during the same day by elementary school children
   h. Amount of fertilizer used per acre and crop yield (*Hint*: As the amount of fertilizer is increased, yield tends to increase for a while but then tends to start decreasing.)

5.10 Is the following statement correct? Explain why or why not. A correlation coefficient of zero implies that no relationship exists between the two variables under study.

5.11 Draw two scatter plots, one for which $r = 1$ and a second for which $r = -1$.

5.12 Peak heart rate (beats per minute) was determined both during a shuttle run and during a 300-yard run for a sample of $n = 10$ individuals afflicted with Down Syndrome ("Heart Rate Responses to Two Field Exercise Tests by Adolescents and Young Adults with Down Syndrome," *Adapted Phys. Act. Quarterly* (1995): 43–51), resulting in the following data.

Shuttle	168	168	188	172	184
300-yd	184	192	200	192	188

Shuttle	176	192	172	188	180
300-yd	180	182	188	196	196

a. A scatter plot of the data is shown. What does it suggest about the nature of the relationship between the two variables?

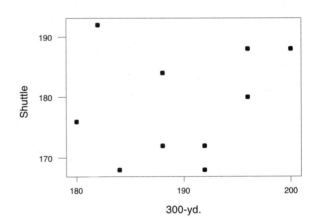

b. With $x = $ shuttle run peak rate and $y = $ 300-yd run peak rate, calculate r. Is the value of r consistent with your answer in part **a**?
c. With $x = $ 300-yd peak rate and $y = $ shuttle run peak rate, how does the value of r compare to what you calculated in part **b**?

5.13 The detection of anemia requires an accurate method for the determination of hemoglobin level. The accompanying scatter plot is for observations on hemoglobin level, determined both by the standard spectrophometric method (x) and by a new, simpler method based on a color scale (y) ("A Simple and Reliable Method for Estimating Hemoglobin," *Bull. World Health Org.* (1995): 369–373).

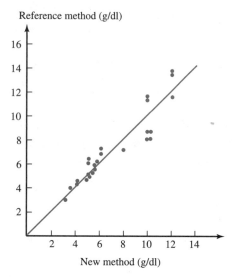

Reference method (g/dl)

New method (g/dl)

a. Does it appear that x and y are highly correlated?
b. The paper reported that r = .9366. How would you describe the relationship between the two variables?
c. The line pictured in the scatter plot has a slope of 1 and passes through (0, 0). If x and y were always identical, all points would lie exactly on this line. The authors of the paper claimed that perfect correlation (r = 1) would result in this line. Do you agree? Explain your reasoning.

5.14 Is there a correlation between test anxiety and exam score performance? Data on x = score on a measure of test anxiety and y = exam score consistent with summary quantities given in the paper "Effects of Humor on Test Anxiety and Performance" (*Psych. Reports* (1999): 1203–1212) appears below.

| x | 23 | 14 | 14 | 0 | 7 | 20 | 20 | 15 | 21 |
| y | 43 | 59 | 48 | 77 | 50 | 52 | 46 | 51 | 51 |

Higher values for x indicate higher levels of anxiety.
a. Construct a scatter plot and comment on the features of the plot.
b. Does there appear to be a linear relationship between the two variables? Based on the scatter plot, would you characterize the relationship as positive or negative? Strong or weak?
c. Compute the value of the correlation coefficient. Is the value of r consistent with your answer to part **b**?
d. Based on the value of the correlation, is it reasonable to conclude that text anxiety caused poor exam performance? Explain.

5.15 According to the article "First-Year Academic Success: A Prediction Combining Cognitive and Psychosocial Variables for Caucasian and African-American Students" (*Journal of College Student Development* (1999): 599–605), there is a mild correlation between high school GPA (x) and first-year college GPA (y). The data, which were collected from a "Southeastern public research university," can be summarized as follows:

$$n = 2600 \qquad \Sigma x = 9620 \qquad \Sigma y = 7436$$
$$\Sigma xy = 27{,}918 \qquad \Sigma x^2 = 36{,}168 \qquad \Sigma y^2 = 23{,}145$$

An alternative formula for computing the correlation coefficient that is algebraically equivalent to the one given in the text is

$$r = \frac{\Sigma xy - \dfrac{(\Sigma x)(\Sigma y)}{n}}{\sqrt{\Sigma x^2 - \dfrac{(\Sigma x)^2}{n}}\sqrt{\Sigma y^2 - \dfrac{(\Sigma y)^2}{n}}}$$

Use this formula to compute the value of the correlation coefficient and interpret this value.

5.16 An employee of an auction house has a list of 25 recently sold paintings. Eight artists were represented in these sales. The sale price of each painting appears on the list. Would the correlation coefficient be an appropriate way to summarize the relationship between artist (x) and sale price (y)? Why or why not?

5.17 Each individual in a sample was asked to indicate on a quantitative scale how willing he or she was to spend money on the environment and also how strongly he or she believed in God ("Religion and Attitudes Toward the Environment," *J. for the Scient. Study of Religion* (1993): 19–28). (This article was authored by Andrew Greeley, who also writes entertaining mysteries with religious themes.) The resulting value of the sample correlation coefficient was r = −.085. Would you agree with the stated conclusion that stronger support for environmental spending is associated with a weaker degree of belief in God? Explain your reasoning.

5.18 A sample of automobiles traversing a certain stretch of highway is selected. Each one travels at roughly a constant rate of speed, though speed does vary from auto to auto. Let x = speed and y = time needed to traverse this segment of highway. Would the sample correlation coefficient be closest to .9, .3, −.3, or −.9? Explain.

5.19 A sample of n = 5 (x, y) pairs gives (1, 1), (2, 2), (3, 3), (4, 4), and (5, y₅). It must be the case that y₅ ≥ 0. Could r be negative?

5.20 Nine students currently taking introductory statistics are randomly selected, and both the first midterm exam score (x) and the second midterm exam score (y) are determined. Three of the students have the class at 8 A.M., another three have it at noon, and the remaining three have a night class. The resulting (x, y) pairs are as follows:

8 A.M.	(70, 60)	(72, 83)	(94, 85)
Noon	(80, 72)	(60, 74)	(55, 58)
Night	(45, 63)	(50, 40)	(35, 54)

a. Calculate the sample correlation coefficient for the nine (x, y) pairs.

b. Let $\bar{x}_1$ = the average score on the first midterm for the 8 A.M. students and $\bar{y}_1$ = the average score on the second midterm for these students. Let $\bar{x}_2$ and $\bar{y}_2$ be these averages for the noon students and $\bar{x}_3$ and $\bar{y}_3$ these averages for the evening students. Calculate r for these three ($\bar{x}$, $\bar{y}$) pairs.

c. Construct a scatter plot of the nine (x, y) pairs and construct another one of the three ($\bar{x}$, $\bar{y}$) pairs. Can you see why r in part **a** is smaller than r in part **b**? Does this suggest that a correlation coefficient based on averages (an "ecological" correlation) might be misleading? Explain.

5.21 An article titled "When it Comes to Sex, We Score" in the newspaper *The Australian* (May 20, 1996) used the results of a large multination survey to rank 15 countries on the variables x = frequency of sex and y = partner's reported satisfaction.

Country	Rank for y	Rank for x
France	5	3
Britain	3	5
United States	11	1
Australia	6	7
Russia	7	2
Italy	4	11
Brazil	10	9
Mexico	2	12
South Africa	9	8
Canada	1	10
Germany	12	4
Thailand	13	15
Spain	8	14
Poland	14	6
Hong Kong	15	13

Is there a correlation between partner satisfaction and frequency? Since only the ranks are given, use Spearman's rank correlation coefficient to describe the strength of the relationship between these two variables.

5.22 In a study of variables thought to be related to urban gang activity, seven Chicago-area communities were ranked according to prevalence of ganging, percent of residents who are African-American, median family income, and percent of families below poverty level ("Youth Gangs," *Pacific Sociol. Rev.* (1981): 366). See the accompanying table. The authors of this paper analyzed this data using Spearman's rank correlation coefficient.

Community	Prevalence of Ganging	Percent African-American	Median Income	% Below Poverty Level
A	2	3	7	1
B	4	6	6	4
C	7	7	1	7
D	5	5	2	5
E	6	4	3	6
F	3	2	5	3
G	1	1	4	2

a. Calculate r_S for the following pairs of variables.
i. Ganging prevalence and percent African-American
ii. Ganging prevalence and median income
iii. Ganging prevalence and percent below poverty level

b. Based on the coefficients in part **a**, which of the variables seems to exhibit the strongest relationship with ganging prevalence?

5.23 The article "Perceived Importance of Self-Determination Factors by Consumers in Substance-Abuse Treatment" (*Perceptual and Motor Skills* (1994): 284–286) reported the accompanying data on ranking of importance of variables related to self-determination. Verify that $r_S = .95$ as stated in the paper, and interpret this value.

Variable	1993 Rank	1994 Rank
Self-confidence	1	1
Positive attitude	2	2
Listening skills	3	3
Responsibility	4	4
Speaking ability	5	7
Know legal rights	6	5
Health and hygiene	7	12
Think about self	8	8
Getting help	9	11
Solve problems	10	6
Planning	11	9
Reading skills	12	14
Persistence	13	16
Getting information	14	10
Strong family	15	15

(continued)

Variable	1993 Rank	1994 Rank
Know needs	16	13
Change plans	17	20
Accepting support	18	17
Negotiate with others	19	19
Writing ability	20	21
Being creative	21	18
Using humor	22	22
Transportation	23	24
Role model	24	23

5.24 Ranking wines is a common practice at wine tastings. Suppose that after tasting nine wines, two judges rank the wines as in the accompanying table. Compute r_S as a measure of agreement between the two judges.

Wine	A	B	C	D	E	F	G	H	I
Judge 1	7	1	3	2	8	5	9	6	4
Judge 2	9	4	1	3	7	5	6	8	2

5.25 Let d_1 denote the difference (x rank $-$ y rank) for the first (x, y) pair, d_2 denote this difference for the second (x, y) pair, and so on. It can be shown that

$$r_S = 1 - \frac{6 \Sigma d^2}{n(n^2 - 1)}$$

a. Use this formula to compute r_S for the Quetelet index data given on page 162.

b. What does this formula imply about the value of r_S when there is a perfect positive relationship? Explain.

5.3   Fitting a Line to Bivariate Data

Given two variables x and y, the general objective of *regression analysis* is to use information about x to draw some type of conclusion concerning y. Often an investigator will want a prediction of the y value that would result from making a single observation at a specified x value — for example, predict product sales y during a given period when shelf space for displaying the product is $x = 6$ ft^2. The different roles played by the two variables are reflected in standard terminology: y is called the **dependent** or **response variable,** and x is referred to as the **independent,** or **predictor,** or **explanatory variable.**

A scatter plot of y versus x (that is, of the (x, y) pairs in a sample) will frequently exhibit a linear pattern. It is natural in such cases to summarize the relationship between the variables by finding a line that is as close as possible to the points in the plot. Before doing so, let's review some elementary facts about lines and linear relationships.

Suppose that a car dealership advertises that a particular model of car can be rented for a flat fee of $25 plus an additional $.30 per mile. If this type of car is rented and driven for 100 miles, the dealer's revenue y is

$$y = 25 + (.30)(100) = 25 + 30 = 55$$

More generally, if x denotes distance driven (miles), then

$$y = 25 + .30x$$

That is, x and y are linearly related.

The general form of a linear relation between x and y is $y = a + bx$. A particular relation is specified by choosing values of a and b. Thus, one such relationship is $y = 10 + 2x$, whereas another is $y = 100 - 5x$. If we choose some x values and compute $y = a + bx$ for each value, the points in the plot of the resulting (x, y) pairs will fall exactly on a straight line.

Definition

The relationship

$$\overset{\text{Intercept}}{\underset{\underset{\text{Slope}}{\uparrow}}{y = a + \overset{\downarrow}{b}x}}$$

is the equation of a straight line. The value of b, called the **slope** of the line, is the amount by which y increases when x increases by 1 unit. The value of a, called the **intercept** (or sometimes the **vertical intercept**) of the line, is the height of the line above the value $x = 0$.

The equation $y = 10 + 2x$ has slope $b = 2$, so each 1-unit increase in x is paired with an increase of 2 in y. When $x = 0$, $y = 10$, and the height at which the line crosses the vertical axis (where $x = 0$) is 10. This is illustrated in Figure 5.10(a). The slope of the line determined by $y = 100 - 5x$ is -5, so y increases by -5 (that is, decreases by 5) when x increases by 1. The height of the line above $x = 0$ is $a = 100$. The resulting line is pictured in Figure 5.10(b).

It is easy to draw the line corresponding to any particular linear equation. First choose any two x values and substitute them into the equation to obtain the corresponding y values. Then plot the resulting two (x, y) pairs as two points. The desired line is the one passing through these points. For the equation $y = 10 + 2x$, substituting $x = 5$ yields $y = 20$, whereas using $x = 10$ gives $y = 30$. The two points are then $(5, 20)$ and $(10, 30)$. The line in Figure 5.10(a) does indeed pass through these points.

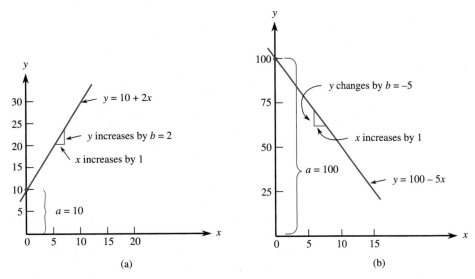

(a)   (b)

FIGURE 5.10 Graphs of two lines: (a) Slope $b = 2$, intercept $a = 10$; (b) Slope $b = -5$, intercept $a = 100$

Fitting a Straight Line: The Principle of Least Squares

Figure 5.11 shows a scatter plot with two lines superimposed on the plot. The line that gives the most effective summary of the approximate linear relationship is the one that in some sense is the best-fitting line, the one closest to the sample data. Line II clearly gives a better fit to the data than does Line I. To measure the extent to which a particular line provides a good fit, we focus on the vertical deviations from the line. For example, Line II in Figure 5.11 has equation $y = 10 + 2x$, and the third and fourth points from the left in the scatter plot are $(15, 44)$ and $(20, 45)$. For these two points, the vertical deviations from this line are

$$\text{3rd deviation} = y_3 - \text{height of the line above } x_3$$
$$= 44 - [10 + 2(15)]$$
$$= 4$$

and

$$\text{4th deviation} = 45 - [10 + 2(20)] = -5$$

FIGURE 5.11 Lines I and II give poor and good fits, respectively, to the data.

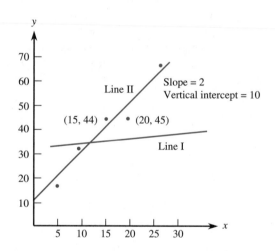

A positive vertical deviation results from a point that lies above the chosen line, and a negative deviation from a point that lies below this line. A particular line is said to be a good fit to the data if the deviations from the line are small in magnitude. Line I in Figure 5.11 fits poorly, because all deviations from that line are larger in magnitude (some are much larger) than the corresponding deviations from Line II.

We now need a way to combine the n deviations into a single measure of fit. The standard approach is to square the deviations (to obtain nonnegative numbers) and sum these squared deviations.

Definition

The most widely used criterion for measuring the goodness of fit of a line $y = a + bx$ to bivariate data $(x_1, y_1), \ldots, (x_n, y_n)$ is the sum of the squared deviations about the line:

$$\Sigma\,[y - (a + bx)]^2 = [y_1 - (a + bx_1)]^2 + [y_2 - (a + bx_2)]^2 + \cdots + [y_n - (a + bx_n)]^2$$

The line that gives the best fit to the data is the one that minimizes this sum; it is called the **least squares line** or **sample regression line.**

Fortunately, the equation of the least squares line can be obtained without having to calculate deviations from any particular line. This is because mathematical techniques can be applied to obtain relatively simple formulas for the slope and vertical intercept of the least squares line. The formula for b involves $\Sigma\,(x - \bar{x})(y - \bar{y})$, which is the sum of products of the x and y deviations from their respective means.

The slope of the least squares line is

$$b = \frac{\Sigma\,(x - \bar{x})(y - \bar{y})}{\Sigma\,(x - \bar{x})^2}$$

and the y intercept is

$$a = \bar{y} - b\bar{x}$$

We write the equation of the least squares line as

$$\hat{y} = a + bx$$

where the $\hat{}$ above y emphasizes that $\hat{y}$ (read as y-hat) is a prediction of y resulting from the substitution of a particular x value into the equation.

Statistical software packages and many calculators will compute the slope and intercept of the least squares line. If the slope and intercept are to be computed by hand, the following computational formula can be used to reduce the amount of time required to perform the calculations.

Calculating Formula for the Slope of the Least Squares Line

$$b = \frac{\Sigma\,xy - \dfrac{(\Sigma\,x)(\Sigma\,y)}{n}}{\Sigma\,x^2 - \dfrac{(\Sigma\,x)^2}{n}}$$

EXAMPLE 5.9   Corrosion of the steel frame is the most important factor affecting the durability of reinforced concrete buildings. It is believed that carbonation of the concrete (caused by a chemical reaction that lowers the pH of the concrete) leads to corrosion of the steel frame and thus reduces the strength of the concrete.

Representative data on

$$x = \text{carbonation depth (mm)}$$

and $y = \text{strength of concrete (Mpa)}$

for a sample of core specimens taken from a particular building were read from a plot in the article "The Carbonation of Concrete Structures in the Tropical Environment of Singapore" (*Magazine of Concrete Research* (1996): 293–300):

Depth, x	8.0	20.0	20.0	30.0	35.0	40.0	50.0	55.0	65.0
Strength, y	22.8	17.1	21.5	16.1	13.4	12.4	11.4	9.7	6.8

A scatter plot of this data (Figure 5.12) shows that the relationship between corrosion depth and strength could be summarized by a line.

FIGURE 5.12 MINITAB scatter plot for data of Example 5.9

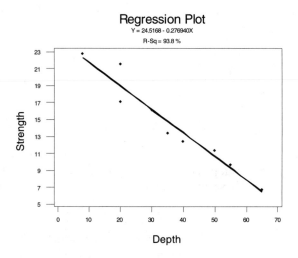

The summary quantities necessary to compute the equation of the least squares line are

$$\Sigma x = 323 \quad \Sigma x^2 = 14{,}339 \quad \Sigma y = 131.2 \quad \Sigma y^2 = 2137.3 \quad \Sigma xy = 3947.9$$

From these quantities, we compute

$$\bar{x} = 35.89 \qquad \bar{y} = 14.58$$

$$b = \frac{\Sigma xy - \dfrac{(\Sigma x)(\Sigma y)}{n}}{\Sigma x^2 - \dfrac{(\Sigma x)^2}{n}} = \frac{3947.9 - \dfrac{(323)(131.2)}{9}}{14{,}339 - \dfrac{(323)^2}{9}} = -.277$$

and

$$a = \bar{y} - b\bar{x} = 14.58 - (-.277)(35.89) = 24.5$$

The least squares line is then

$$\hat{y} = 24.5 - .277x$$

This line is also shown on the scatter plot of Figure 5.12.

If we wanted to predict the strength of concrete that had a depth of corrosion of 25, we could use the point on the least squares line above $x = 25$:

$$\hat{y} = 24.5 - (.277)(25) = 17.575$$

Predicted strengths for other corrosion depths could be obtained in a similar way. **However, the least squares line should not be used to predict strength for corrosion depths much outside the range 8 to 59 (the range of x values in the data set) since we do not know whether the linear pattern observed in the scatter plot continues outside this range.** This is sometimes referred to as the **danger of extrapolation.**

Calculations involving the least squares line can obviously be quite tedious, and fitting a function to multivariate data is even worse. This is where the computer comes to our rescue. All the standard statistical packages fit a straight line to bivariate data when asked to do so. To accomplish this, a regression command or option is used. (This terminology is explained shortly.)

EXAMPLE 5.10

Millions of Muslims all over the world participate in controlled or partial fasting during the month of Ramadan. The article "Body Weight Loss and Changes in Blood Lipid Levels in Normal Men on Hypo-caloric Diets During Ramadan Fasting" (*Amer. J. of Clinical Nutrition* (1988): 1197–1210) reported on a study carried out to assess the physiological effects of such fasting. The accompanying table gives selected data (read from a graph in the article) on x = percentage change in body weight during phase two of the fasting period (in this phase, experimental subjects received a high-fat, low-carbohydrate diet) and y = high-density lipoprotein cholesterol level (mg/dl):

x	-5	-3.2	-2.2	-1.7	-1.6	-1.5	$-.9$	0	0	1.2	1.6	1.7	2.8
y	43	50	61	63	47	57	51	60	67	76	70	51	74

Figure 5.13 (page 172) is a scatter plot. The predominant pattern is linear, though the points in the plot spread out about *any* line (even the least squares line).

MINITAB was used to fit the least squares line, and Figure 5.14 shows part of the resulting output. Instead of x and y, the variable labels **perchgwt** (*percent change in weight*) and **chollevl** (*cholesterol level*) are used. The equation at the top is that of the least squares line. In the rectangular table just below the equation, the first row gives information about the intercept, a, and the second row gives information concerning the slope, b. In particular, the coefficient column labeled **Coef** contains the values of a and b using more significant figures than appear in the rounded equation.

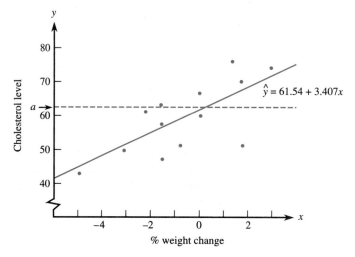

FIGURE 5.13 Scatter plot and least squares line for the data of Example 5.10

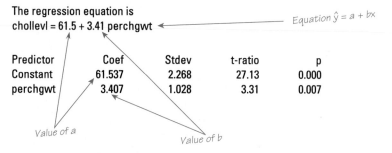

FIGURE 5.14 Partial MINITAB output for Example 5.10

The least squares line should not be used here to predict cholesterol level for values of percentage change in body weight such as $x = -10$ or $x = 8$. These x values are well outside the range of the data, and there is no evidence to support extrapolation of the linear relationship.

Regression

The least squares line is often called the **sample regression line.** This terminology comes from the relationship between the least squares line and Pearson's correlation coefficient. To understand this relationship, we first need alternative expressions for the slope b and the equation of the line itself. With s_x and s_y denoting the sample standard deviations of the x's and y's, respectively, a bit of algebraic manipulation gives

$$b = r \cdot \frac{s_y}{s_x}$$

$$\hat{y} = \bar{y} + r \cdot \frac{s_y}{s_x}(x - \bar{x})$$

You do not need to use these formulas in any computations, but several of their implications are important for appreciating what the least squares line does.

1. When $x = \bar{x}$ is substituted in the equation of the line, $\hat{y} = \bar{y}$ results. That is, the least squares line passes through the *point of averages* $(\bar{x}, \bar{y})$.

2. Suppose for the moment that $r = 1$, so that all points lie exactly on the line whose equation is

$$\hat{y} = \bar{y} + \frac{s_y}{s_x}(x - \bar{x})$$

Now substitute $x = \bar{x} + s_x$, which is 1 standard deviation above $\bar{x}$:

$$\hat{y} = \bar{y} + \frac{s_y}{s_x}(\bar{x} + s_x - \bar{x}) = \bar{y} + s_y$$

That is, with $r = 1$, when x is 1 standard deviation above its mean, we predict that the associated y value will be 1 standard deviation above its mean. Similarly, if $x = \bar{x} - 2s_x$ (2 standard deviations below its mean), then

$$\hat{y} = \bar{y} + \frac{s_y}{s_x}(\bar{x} - 2s_x - \bar{x}) = \bar{y} - 2s_y$$

also 2 standard deviations below the mean. If $r = -1$, then $x = \bar{x} + s_x$ results in $\hat{y} = \bar{y} - s_y$, so the predicted y is also 1 standard deviation from its mean but on the opposite side of $\bar{y}$ from where x is relative to $\bar{x}$. In general, if x and y are perfectly correlated, the predicted y value associated with a given x value will be the same number of standard deviations (of y) from its mean, $\bar{y}$, as x is from its mean, $\bar{x}$.

3. Suppose that x and y are not perfectly correlated. To be specific, take $r = .5$, so the least squares line has the equation

$$\hat{y} = \bar{y} + .5\left(\frac{s_y}{s_x}\right)(x - \bar{x})$$

Then substituting $x = \bar{x} + s_x$ gives

$$\hat{y} = \bar{y} + .5\left(\frac{s_y}{s_x}\right)(\bar{x} + s_x - \bar{x}) = \bar{y} + .5s_y$$

That is, for $r = .5$, when x lies 1 standard deviation above its mean, we predict that y will be only .5 standard deviation above its mean.

Similarly, we can predict y when r is negative. If $r = -.5$, then the predicted y value will be only half the number of standard deviations from $\bar{y}$ that x is from $\bar{x}$, but x and the predicted y will now be on opposite sides of their respective means.

Consider using the least squares line to predict the value of y associated with an x value some specified number of standard deviations away from $\bar{x}$. Then the predicted y value will be only r times this number of standard deviations from $\bar{y}$. In terms of standard deviations, except when $r = 1$ or -1, the predicted y will always be closer to $\bar{y}$ than x is to $\bar{x}$.

Using the least squares line for prediction results in a predicted y that is pulled back in, or regressed, toward its mean compared to where x is relative to its mean. This regression effect was first noticed by Sir Francis Galton (1822–1911), a famous biologist, while studying the relationship between the heights of fathers and their sons. He found that the predicted height of a son whose father was above average in height would also be above average (because r is positive here) but not by as much as the father's; he found a similar relationship for a father whose height was below average. This regression effect has led to the term **regression analysis** for the collection of methods involving the fitting of lines, curves, and more complicated functions to bivariate and multivariate data.

The alternate form of the regression (least squares) line emphasizes that predicting y from knowledge of x is not the same problem as predicting x from knowledge of y. The slope of the least squares line for predicting x is $r \cdot (s_x/s_y)$ rather than $r \cdot (s_y/s_x)$, and the intercepts of the lines are also usually different. It makes a difference whether y is regressed on x, as we have done, or whether x is regressed on y. The regression line of y on x should not be used to predict x, since it is not the line that minimizes the sum of squared x deviations.

Exercises 5.26 – 5.36

5.26 In the article "Reproductive Biology of the Aquatic Salamander *Amphiuma tridactylum* in Louisiana (*Journal of Herpetology* (1999): 100–105), 14 female salamanders were studied. Using regression, the author predicted y = clutch size (number of salamander eggs) from x = snout-vent length (measured in cm) as follows:

$$\hat{y} = -147 + 6.175x$$

It is interesting to note that for the salamanders studied, the range of snout-vent lengths was from approximately 30 to 70. Because the scatter plot looked linear, the authors thought a simple linear regression was a reasonable way of describing the relationship between x and y.

a. What is the value of the y intercept of the least squares line? What is the value of the slope of the least squares line? Give an interpretation of the slope in the context of this problem.

b. Would you be reluctant to predict the clutch size when snout-vent length is 22 cm? Explain.

5.27 A number of studies have shown lichens (certain plants composed of an alga and a fungus) to be excellent bioindicators of air pollution. The article "The Epiphytic Lichen Hypogymnia Physodes as a Bioindicator of Atmospheric Nitrogen and Sulphur Deposition in Norway" (*Envir. Monitoring and Assessment* (1993): 27–47) gives the following data (read from a graph in the paper) on x = NO_3 wet deposition (g/m^3) and y = lichen (% dry weight):

x	.05	.10	.11	.12	.31	.37	.42
y	.48	.55	.48	.50	.58	.52	1.02

x	.58	.68	.68	.73	.85	.92
y	.86	.86	1.00	.88	1.04	1.70

a. What is the equation of the least squares regression line?

b. Predict lichen for an NO_3 deposition of .5.

5.28 According to the article "First-Year Academic Success: A Prediction Combining Cognitive and Psychosocial Variables for Caucasian and African-American Students" (*Journal of College Student Development* (1999): 599–

605) there is a mild correlation between high school GPA (x) and first-year college GPA (y). The data, which was collected from a "Southeastern public research university," can be summarized:

$$n = 2600 \qquad \Sigma x = 9620 \qquad \Sigma y = 7436$$
$$\Sigma xy = 27{,}918 \qquad \Sigma x^2 = 36{,}168 \qquad \Sigma y^2 = 23{,}145$$

a. Find the equation of the least squares regression line.
b. Interpret the value of b, the slope of the least squares line, in the context of this problem.
c. What first-year GPA would you predict for a student with a 4.0 high school GPA?

5.29 Shells of molluscs function as both part of the skeletal system and as protective armor. It has been argued that many features of these shells were the result of natural selection in the constant battle against predators. The paper "Postmortem Changes in Strength of Gastropod Shells" (*Paleobiology* (1992): 367–377) included scatter plots of data on x = shell height (cm) and y = breaking strength (newtons). The least squares line for a sample of $n = 38$ hermit crab shells was $\hat{y} = -275.1 + 244.9x$.

a. What are the slope and intercept of this line?
b. When shell height increases by 1 cm, by how much does breaking strength tend to change?
c. What breaking strength would you predict when shell height is 2 cm?
d. Does this approximate linear relationship appear to hold for shell heights as small as 1 cm? Explain your reasoning.

5.30 It has been observed that Andean high-altitude natives have larger chest dimensions and lung volumes than do sea-level residents. Is this also true of lifelong Himalayan residents? The paper "Increased Vital and Total Lung Capacities in Tibetan Compared to Han Residents of Lhasa" (*Amer. J. Phys. Anthro.* (1991): 341–351) reported on the results of an investigation into this question. Included in the paper was a plot of vital capacity (y) versus chest circumference (x) for a sample of 16 Tibetan natives, from which the accompanying data was read.

x	79.4	81.8	81.8	82.3	83.7	84.3	84.3	85.2
y	4.3	4.6	4.8	4.7	5.0	4.9	4.4	5.0

x	87.0	87.3	87.7	88.1	88.1	88.6	89.0	89.5
y	6.1	4.7	5.7	5.7	5.2	5.5	5.0	5.3

a. Construct a scatter plot. What does it suggest about the nature of the relationship between x and y?
b. The summary quantities are

$$\Sigma x = 1368.1 \qquad \Sigma y = 80.9 \qquad \Sigma x^2 = 117{,}123.85$$
$$\Sigma y^2 = 412.81 \qquad \Sigma xy = 6933.48$$

Verify that the equation of the least squares line is $\hat{y} = -4.54 + .1123x$; draw this line on your scatter plot.
c. Roughly what change in vital capacity is associated with a 1-cm increase in chest circumference? with a 10-cm increase?
d. What vital capacity would you predict for a Tibetan native whose chest circumference is 85 cm?
e. Is vital capacity completely determined by chest circumference? Explain.

5.31 Example 5.2 gave data on x = soil pH and y = crown dieback (an indicator of forest growth retardation). Consider the accompanying MINITAB output.

The regression equation is
dieback = 31.0 - 5.79 soil ph

Predictor	Coef	Stdev	t-ratio	p
Constant	31.040	5.445	5.70	0.000
soil ph	−5.792	1.363	-4.25	0.001
s = 2.981	R-sq = 51.5%			

a. What is the equation of the least squares line?
b. Roughly what change in crown dieback would be associated with an increase of 1 in soil pH (for pH within the range of the data)?
c. What value of crown dieback would you predict when soil pH = 4.0?
d. Would it be sensible to use the least squares line to predict crown dieback when soil pH = 5.6? Explain.

5.32 Athletes competing in a triathlon participated in a study described in the paper "Myoglobinemia and Endurance Exercise" (*Amer. J. of Sports Med.* (1984): 113–118). The accompanying data on finishing time x (hr) and myoglobin level y (ng/mL) was read from a scatter plot in the paper.

x	4.90	4.70	5.35	5.22	5.20
y	1590	1550	1360	895	865

x	5.40	5.70	6.00	6.20	6.10
y	905	895	910	700	675

x	5.60	5.35	5.75	5.35	6.00
y	540	540	440	380	300

$$\Sigma (x - \bar{x})(y - \bar{y}) = -1561.227 \qquad \Sigma (x - \bar{x})^2 = 2.702$$

a. Obtain the equation of the least squares line.
b. Interpret the value of b.
c. What happens if the line in part **a** is used to predict the myoglobin level for a finishing time of 8 hr? Is this reasonable? Explain.

5.33 Explain why it can be dangerous to use the least squares line to obtain predictions for x values that are substantially larger or smaller than those contained in the sample.

5.34 The sales manager of a large company selected a random sample of $n = 10$ salespeople and determined for each one the values of $x =$ years of sales experience and $y =$ annual sales (in thousands of dollars). A scatter plot of the resulting (x, y) pairs showed a marked linear pattern.

a. Suppose that the sample correlation coefficient is $r = .75$ and average annual sales is $\bar{y} = 100$. If a particular salesperson is 2 standard deviations above the mean in terms of experience, what would you predict for that person's annual sales?

b. If a particular person whose sales experience is 1.5 standard deviations below average experience is predicted to have an annual sales value that is 1 standard deviation below average annual sales, what is the value of r?

5.35 Explain why the slope b of the least squares line always has the same sign (positive or negative) as does the sample correlation coefficient r.

5.36 The accompanying data resulted from an experiment in which weld diameter, x, and shear strength, y (lb), were determined for five different spot welds on steel. A scatter plot shows a pronounced linear pattern. With $\Sigma (x - \bar{x})^2 = 1000$ and $\Sigma (x - \bar{x})(y - \bar{y}) = 8577$, the least squares line is $\hat{y} = -936.22 + 8.577x$.

x	200	210	220	230	240
y	813.7	785.3	960.4	1118.0	1076.2

a. Since 1 lb = .4536 kg, strength observations can be reexpressed in kilograms through multiplication by this conversion factor: new $y = .4536 \cdot (\text{old } y)$. What is the equation of the least squares line when y is expressed in kilograms?

b. More generally, suppose that each y value in a data set consisting of n (x, y) pairs is multiplied by a conversion factor c (which changes the units of measurement for y). What effect does this have on the slope b (that is, how does the new value of b compare to the value before conversion), on the intercept a, and on the equation of the least squares line? Verify your conjectures by using the given formulas for b and a. (*Hint*: Replace y with cy and see what happens — and remember, this conversion will affect $\bar{y}$.)

5.4 Assessing the Fit of a Line

Once the least squares regression line has been obtained, it is natural to examine how effectively the line summarized the relationship between x and y. Important questions to consider are

1. Is a line an appropriate way to summarize the relationship between the two variables?

2. Are there any unusual aspects of the data set that we need to consider before proceeding to use the regression line to make predictions?

3. If we decide that it is reasonable to use the regression line as a basis for prediction, how accurate can we expect predictions based on the regression line to be?

In this section, we look at graphical and numerical methods that will allow us to answer these questions. Most of these methods are based on the vertical deviations of the data points from the regression line. These vertical deviations are called *residuals,* and each represents the difference between an actual y value and the corresponding predicted value, $\hat{y}$, that would have resulted from using the regression line to make a prediction.

Predicted Values and Residuals

If the x value for the first observation is substituted in the equation for the least squares line, the result is $a + bx_1$, the height of the line above x_1. The point (x_1, y_1) in the scatter plot also lies above x_1, so the difference

$$y_1 - (a + bx_1)$$

is the vertical deviation from this point to the line (see Figure 5.15). A point lying above the line gives a positive deviation, and a point below the line results in a negative deviation. The remaining vertical deviations come from repeating this process for $x = x_2$, then $x = x_3$, and so on.

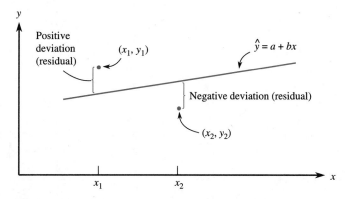

FIGURE 5.15 Positive and negative deviations (residuals) from the least squares line

Definition

The **predicted** or **fitted values** result from substituting each sample x value in turn into the equation for the least squares line. This gives

$\hat{y}_1$ = 1st predicted value $= a + bx_1$

$\hat{y}_2$ = 2nd predicted value $= a + bx_2$

$\vdots$

$\hat{y}_n$ = nth predicted value $= a + bx_n$

The residuals from the least squares line are the n quantities

$$y_1 - \hat{y}_1, y_2 - \hat{y}_2, \ldots, y_n - \hat{y}_n$$

Each residual is the difference between an observed y value and the corresponding predicted y value.

EXAMPLE 5.11　One measure of the success of knee surgery is postsurgical range of motion for the knee joint. Postsurgical range of motion was recorded for 12 patients who had surgery following a knee dislocation. The age of each patient was also recorded ("Reconstruction of the Anterior and Posterior Cruciate Ligaments after Knee

Dislocation," *Amer. J. of Sports Medicine* (1999): 189–194). The data is given in Table 5.1.

Table 5.1 *Predicted values and residuals for the data of Example 5.11*

Patient	Age x	Range of Motion y	Predicted Range of Motion $\hat{y}$	Residual $y - \hat{y}$
1	32	154	138.07	15.93
2	24	142	128.49	13.51
3	40	137	142.42	−5.42
4	31	133	134.58	−1.58
5	28	122	131.97	−9.97
6	25	126	129.36	−3.36
7	26	135	130.23	4.77
8	16	135	121.52	13.48
9	14	108	119.78	−11.78
10	20	120	125.00	−5.00
11	21	127	125.87	1.13
12	30	122	133.71	−11.71

MINITAB was used to fit the least squares regression line; partial computer output is shown. The resulting least squares regression line is $\hat{y} = 107.58 + .871$age.

Regression Analysis

The regression equation is

Range of Motion = 108 + 0.871 Age

Predictor	Coef	StDev	T	P
Constant	107.58	11.12	9.67	0.000
Age	0.8710	0.4146	2.10	0.062

S = 10.42   R-Sq = 30.6%   R-Sq(adj) = 23.7%

A scatter plot that also includes the regression line is shown in Figure 5.16. The residuals for this data set are the vertical distances from the points to the regression line.

For the youngest patient (patient 9 with $x_9 = 14$ and $y_9 = 108$), the corresponding predicted value and residual are

$$\text{predicted value} = \hat{y}_9 = 107.58 + .871\text{age} = 119.78$$

$$\text{residual} = y_9 - \hat{y}_9 = 108 - 119.78 = -11.78$$

The other predicted values and residuals are computed in a similar manner and are included in Table 5.1.

Computing the predicted values and residuals by hand can be tedious, but MINITAB and other statistical software packages, as well as many graphing calculators, include them as part of the output, as shown in Figure 5.17. The predicted

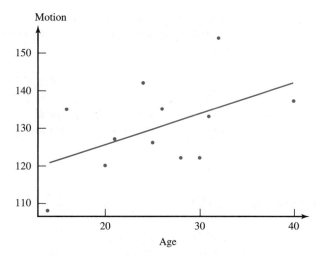

FIGURE 5.16 Scatter plot of the data from Example 5.11

The regression equation is

Range of Motion = 108 + 0.871 Age

Predictor	Coef	StDev	T	P
Constant	107.58	11.12	9.67	0.000
Age	0.8710	0.4146	2.10	0.062

S = 10.42   R-Sq = 30.6%   R-Sq(adj) = 23.7%

Analysis of Variance

Source	DF	SS	MS	F	P
Regression	1	479.2	479.2	4.41	0.062
Residual Error	10	1085.7	108.6		
Total	11	1564.9			

Obs	Age	Range of	Fit	StDev Fit	Residual	St Resid
1	35.0	154.00	138.07	4.85	15.93	1.73
2	24.0	142.00	128.49	3.10	13.51	1.36
3	40.0	137.00	142.42	6.60	-5.42	-0.67
4	31.0	133.00	134.58	3.69	-1.58	-0.16
5	28.0	122.00	131.97	3.14	-9.97	-1.00
6	25.0	126.00	129.36	3.03	-3.36	-0.34
7	26.0	135.00	130.23	3.01	4.77	0.48
8	16.0	135.00	121.52	5.07	13.48	1.48
9	14.0	108.00	119.78	5.75	-11.78	-1.36
10	20.0	120.00	125.00	3.86	-5.00	-0.52
11	21.0	127.00	125.87	3.61	1.13	0.12
12	30.0	122.00	133.71	3.47	-11.71	-1.19

FIGURE 5.17 MINITAB output for the data of Example 5.11

values and residuals can be found in the table at the bottom of the MINITAB output in the columns labeled **Fit** and **Residual,** respectively.

─────────

Plotting the Residuals

A careful look at residuals can reveal many potential problems. A *residual plot* is a good place to start when assessing the appropriateness of the regression line.

Definition

A **residual plot** is a scatter plot of the (*x*, residual) pairs.

Isolated points or a pattern of points in the residual plot indicate potential problems. A desirable plot is one that exhibits no particular pattern, such as curvature. Curvature in the residual plot is an indication that the relationship between *x* and *y* is not linear and that a curve would be a better choice than a line for describing the relationship between *x* and *y*. This is sometimes easier to see in a residual plot than in a scatter plot of *y* versus *x*, as illustrated in the next example.

EXAMPLE 5.12

Consider the accompanying data on *x* = height (in) and *y* = average weight (lb) for American females aged 30–39 (from *The World Almanac and Book of Facts*). The scatter plot displayed in Figure 5.18(a) appears rather straight. However, when the

FIGURE 5.18 Plots for data from Example 5.12: (a) Scatter plot; (b) Residual plot

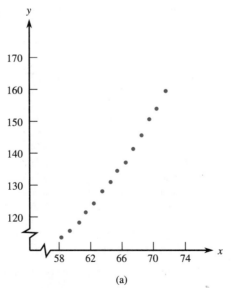

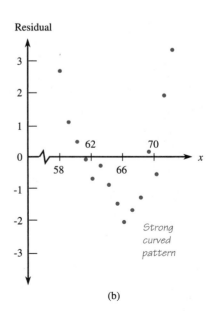

residuals from the least squares line ($\hat{y} = 98.23 + 3.596x$) are plotted, substantial curvature is apparent (even though $r \approx .99$). It is not accurate to say that weight increases in direct proportion to height (linearly with height). Instead, average weight increases somewhat more rapidly for relatively large heights than it does for relatively small heights.

x	58	59	60	61	62	63	64	65
y	113	115	118	121	124	128	131	134

x	66	67	68	69	70	71	72
y	137	141	145	150	153	159	164

There is another common type of residual plot — one that plots the residuals versus the corresponding $\hat{y}$ values rather than versus the x values. Since $\hat{y} = a + bx$ is simply a linear function of x, the only real difference between the two types of residual plots is the scale on the horizontal axis. The pattern of points in the residual plots will be the same, and it is this pattern of points that is important, not the scale. Thus the two plots give equivalent information, as can be seen in Figure 5.19, which gives both plots for the data of Example 5.11.

FIGURE 5.19 (a) Plot of residuals versus x; (b) Plot of residuals versus $\hat{y}$

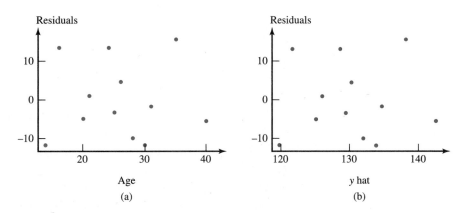

It is also important to look for unusual values in the scatter plot or the residual plot. A point falling far above or below the horizontal line at height zero corresponds to a large residual, which may indicate some type of unusual behavior, such as a recording error, nonstandard experimental condition, or atypical experimental subject. A point whose x value differs greatly from others in the data set may have exerted excessive influence in determining the fitted line. One method for assessing the impact of such an isolated point on the fit is to delete it from the data set and then recompute the best-fit line.

EXAMPLE 5.13   One factor in the development of tennis elbow is the impact-induced vibration of the racket and arm at ball contact. The likelihood of developing tennis elbow is known to depend on various properties of the tennis racket used. The accompanying data is a subset of that analyzed in the article "Transfer of Tennis Racket Vibrations into the Human Forearm" (*Med. and Sci. in Sports and Exercise* (1992): 1134–1140). Measurements on

x = racket resonance frequency (Hz)

and

y = sum of peak-to-peak accelerations (a characteristic of arm vibration in m/sec/sec) are given for $n = 14$ different rackets.

Racket	Resonance (x)	Acceleration (y)
1	105	36.0
2	106	35.0
3	110	34.5
4	111	36.8
5	112	37.0
6	113	34.0
7	113	34.2
8	114	33.8
9	114	35.0
10	119	35.0
11	120	33.6
12	121	34.2
13	126	36.2
14	189	30.0

A scatter plot and residual plot are shown in Figures 5.20(a) and (b), respectively.

One observation in the data set is far to the right of the other points in the scatter plot. Because the least squares line minimizes the sum of squared residuals, the least squares line is pulled down toward this discrepant point. This single observation plays a big role in determining the slope of the least squares line, and it is therefore called an *influential* observation. Notice that an influential observation is not necessarily the one with the largest residual, since the least squares line actually passes very near this point.

Figure 5.20(c) and (d) show what happens when the influential observation is removed from the sample. Both the slope and intercept of the least squares line are quite different from the slope and intercept of the line when this observation is included.

Deletion of the observation corresponding to the largest residual (1.955 for observation 13) also changes the value of the slope and intercept of the least squares line, but these changes are not profound. For example, the least squares line for the

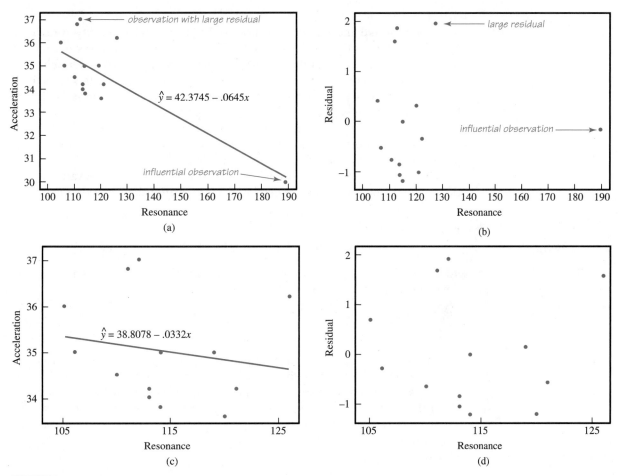

FIGURE 5.20 Plots of data from Example 5.13: (a) Scatter plot for full sample; (b) Residual plot for full sample; (c) Scatter plot when influential observation is deleted; (d) Residual plot when influential observation is deleted

data set consisting of the 13 points that remain when observation 13 is deleted is $\hat{y} = 42.5 - .0670x$. This observation does not appear to be all that influential.

Careful examination of a scatter plot and a residual plot can help us determine the appropriateness of a line for summarizing a relationship. If we decide that a line is appropriate, the next step is to think about assessing the accuracy of predictions to be based on the least squares line and whether these predictions (based on the value of x) are better in general than those made without knowledge of the value of x. Two numerical measures that are helpful in this assessment are the coefficient of determination and the standard deviation about the regression line.

The Coefficient of Determination

Suppose we would like to predict the price of homes in a particular city. A random sample of 20 homes that are for sale is selected and y = price and x = size (sq ft) is recorded for each house in the sample. Undoubtedly, there will be variability in house price (the houses will differ with respect to price), and it is this variability that makes accurate prediction of price a challenge. How much of the variability in house price can be explained by the fact that price is related to house size and houses differ in size? If differences in size account for a large proportion of the variability in price, a price prediction that takes house size into account is a big improvement over a prediction that is not based on size.

The **coefficient of determination** is a measure of the proportion of variability in the y variable that can be "explained by" a linear relationship between x and y.

Definition

The **coefficient of determination,** denoted by r^2, gives the proportion of variation in y that can be attributed to an approximate linear relationship between x and y.

Variation in y can effectively be explained by an approximate straight-line relationship when the points in the scatter plot fall close to the least squares line — that is, when the residuals are small in magnitude. A natural measure of variation about the least squares line is the sum of the squared residuals. (Squaring before combining prevents negative and positive residuals from counteracting one another.) A second sum of squares assesses the total amount of variation in observed y values.

Definition

The **total sum of squares,** denoted by **SSTo,** is defined as

$$SSTo = (y_1 - \bar{y})^2 + (y_2 - \bar{y})^2 + \cdots + (y_n - \bar{y})^2 = \Sigma (y - \bar{y})^2$$

The **residual sum of squares** (sometimes referred to as the error sum of squares), denoted by **SSResid,** is defined as

$$SSResid = (y_1 - \hat{y}_1)^2 + (y_2 - \hat{y}_2)^2 + \cdots + (y_n - \hat{y}_n)^2 = \Sigma (y - \hat{y})^2$$

These sums of squares can be found as part of the regression output from most standard statistical packages or can be obtained using the following computational formulas:

$$SSTo = \Sigma y^2 - \frac{(\Sigma y)^2}{n}$$

$$SSResid = \Sigma y^2 - a \Sigma y - b \Sigma xy$$

EXAMPLE 5.14   (Example 5.11 continued)   Figure 5.21 displays part of the MINITAB output that results from fitting the least squares line to the data on range of motion and age of Example 5.11. From the output,

$$SSTo = 1564.9 \quad and \quad SSResid = 1085.7$$

Notice that SSResid is relatively large compared to SSTo.

FIGURE 5.21 MINITAB output for the data of Example 5.14

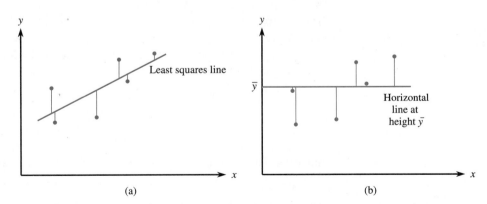

Regression Analysis

The regression equation is
Range of Motion = 108 + 0.871 Age

Predictor	Coef	StDev	T	P
Constant	107.58	11.12	9.67	0.000
Age	0.8710	0.4146	2.10	0.062

S = 10.42   R-Sq = 30.6%   R-Sq(adj) = 23.7%

Analysis of Variance

Source	DF	SS	MS	F	P
Regression	1	479.2	479.2	4.41	0.062
Residual Error	10	1085.7 ⟵ *SSResid*	108.6		
Total	11	1564.9			

↗ *SSTo*

The residual sum of squares is the sum of squared vertical deviations from the least squares line. As Figure 5.22 illustrates, SSTo is also a sum of squared vertical

FIGURE 5.22 Interpreting sums of squares: (a) SSResid = sum of squared vertical deviations from the least squares line; (b) SSTo = sum of squared vertical deviations from the horizontal line at height $\bar{y}$

deviations from a line — the horizontal line at height $\bar{y}$. The least squares line is by definition the one having the smallest sum of squared deviations. It follows that SSResid ≤ SSTo. The two sums of squares will be equal only when the least squares line *is* the horizontal line.

SSResid is often referred to as a measure of "unexplained" variation — the amount of variation in y that cannot be attributed to the linear relationship between x and y. The more points in the scatter plot deviate from the least squares line, the larger will be the value of SSResid, and the greater the amount of y variation that cannot be explained by the approximate linear relationship. Similarly, SSTo is interpreted as a measure of total variation. The larger the value of SSTo, the greater the amount of variability in $y_1, y_2, \ldots, y_n$. The ratio SSResid/SSTo is the fraction or proportion of total variation that is unexplained by a straight-line relation. Subtracting this ratio from 1 gives the proportion of total variation that *is* explained.

The coefficient of determination, r^2, can be computed as

$$r^2 = 1 - \frac{\text{SSResid}}{\text{SSTo}}$$

Multiplying r^2 by 100 gives the percentage of y variation attributable to the approximate linear relationship. The closer this percentage is to 100%, the more successful is the relationship in explaining variation in y.

EXAMPLE 5.15

(Example 5.14 continued)   For the data on range of motion and age of Example 5.14, we found SSTo = 1564.9 and SSResid = 1085.7, so

$$r^2 = 1 - \frac{\text{SSResid}}{\text{SSTo}} = 1 - \frac{1085.7}{1564.9} = .306$$

This means that only 30.6% of the observed variability in postsurgical range of motion can be explained by an approximate linear relationship between range of motion and age. Note that the r^2 value can be found in the MINITAB output of Figure 5.21 labeled as **R-Sq.**

The symbol r was used in Section 5.2 to denote Pearson's sample correlation coefficient. It is not coincidental that r^2 is used to represent the coefficient of determination. The notation suggests how these two quantities are related:

(correlation coefficient)^2 = coefficient of determination

Thus, if $r = .8$ or $r = -.8$, then $r^2 = .64$, so 64% of the observed variation in the dependent variable can be attributed to the linear relationship. Because the value of r does not depend on which variable is labeled x, the same is true of r^2. The coefficient of determination is one of the few quantities calculated in the course of a regression analysis whose value remains the same when the role of dependent and independent variables are interchanged. When $r = .5$, we get $r^2 = .25$, so only 25% of the observed variation is explained by a linear relation. This is why a value of r between $-.5$ and $.5$ is not considered as evidence of a strong linear relationship.

Standard Deviation About the Least Squares Line

The coefficient of determination measures the extent of variation about the best-fit line *relative* to overall variation in y. A high value of r^2 does not by itself promise that the deviations from the line are small in an absolute sense. A typical observation could deviate from the line by quite a bit, yet these deviations might still be small relative to overall y variation. Recall that in Chapter 4 the sample standard deviation

$$s = \sqrt{\frac{\Sigma (x - \bar{x})^2}{n - 1}}$$

was used as a measure of variability in a single sample; roughly speaking, s is the typical amount by which a sample observation deviates from the mean. There is an analogous measure of variability when a line is fit by least squares.

Definition

The **standard deviation about the least squares line** is given by

$$s_e = \sqrt{\frac{\text{SSResid}}{n - 2}}$$

Roughly speaking, s_e is the typical amount by which an observation deviates from the least squares line. Justification for division by $n - 2$ and the use of the subscript e are given in Chapter 13.

EXAMPLE 5.16   The values of $x =$ commuting distance and $y =$ commuting time were determined for workers in samples from three different regions. The data is given in Table 5.2; the three scatter plots are displayed in Figure 5.23 on page 188.

Table 5.2 *Data for Example 5.16*

Region 1		Region 2		Region 3	
x	y	x	y	x	y
15	42	5	16	5	8
16	35	10	32	10	16
17	45	15	44	15	22
18	42	20	45	20	23
19	49	25	63	25	31
20	46	50	115	50	60

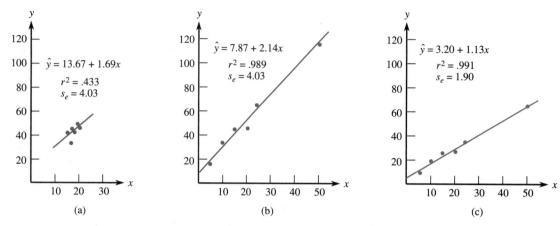

FIGURE 5.23   Scatter plots for Example 5.16: (a) Region 1; (b) Region 2; (c) Region 3

For sample 1, a rather small proportion of variation in y can be attributed to an approximate linear relationship, and a typical deviation from the least squares line is roughly 4. The amount of variability about the line for sample 2 is the same as for sample 1, but the value of r^2 is much higher, because y variation is much greater overall in sample 2 than in sample 1. Sample 3 yields roughly the same high value of r^2 as does sample 2, but the typical deviation from the line for sample 3 is only half that for sample 2. For a complete picture of variation, both r^2 and s_e must be computed.

Exercises 5.37 – 5.51

5.37 Exercise 5.26 gave the least squares regression line for predicting y = clutch size from x = snout-vent length ("Reproductive Biology of the Aquatic Salamander *Amphiuma tridactylum* in Louisiana," *Journal of Herpetology* (1999): 100–105). The paper also reported $r^2 = .7664$ and SSTo = 43,951.
   a. Interpret the value of r^2.
   b. Find and interpret the value of s_e. (The sample size was $n = 14$.)

5.38 The article "Characterization of Highway Runoff in Austin, Texas, Area" (*J. of Envir. Engr.* (1998)) gave a scatter plot, along with the least squares line for x = rainfall volume (m³) and y = runoff volume (m³) for a particular location. The accompanying data was read from the plot in the paper.

x	5	12	14	17	23	30	40	47
y	4	10	13	15	15	25	27	46

x	55	67	72	81	96	112	127
y	38	46	53	70	82	99	100

a. Does a scatter plot of the data suggest a linear relationship between x and y?
b. Estimate the slope and intercept of the least squares line.
c. Compute an estimate of the average runoff volume when rainfall volume is 80.
d. Compute the estimated standard deviation about the regression line.
e. What proportion of the observed variation in runoff volume can be explained by the linear relationship to rainfall volume?

5.39 The article cited in the chapter introduction reported that for a regression of y = average SAT score on x = expenditure per pupil, based on data from $n = 44$ New Jersey school districts, $a = 766$, $b = .015$, $r^2 = .160$, and $s_e = 53.7$.
   a. One observation in the sample was (9900, 893). What average SAT score would you predict for this district, and what is the corresponding residual?
   b. Interpret the value of s_e.

c. How effectively do you think the least squares line summarizes the relationship between x and y? Explain your reasoning.

5.40 The paper "Root Dentine Transparency: Age Determination of Human Teeth Using Computerized Densitometric Analysis" (*Amer. J. Phys. Anthro.* (1991): 25–30) reported on an investigation of methods for age determination based on tooth characteristics. With $y =$ age (years) and $x =$ percentage of root with transparent dentine, a regression analysis for premolars using an image-analyser method to determine x gave $n = 36$, SSResid $= 5987.16$, and SSTo $= 17,409.60$. Calculate and interpret the values of r^2 and s_e.

5.41 The paper cited in Exercise 5.40 gave a scatter plot in which x values were for anterior teeth. Consider the accompanying representative subset of the data.

x	15	19	31	39	41
y	23	52	65	55	32

x	44	47	48	55	65
y	60	78	59	61	60

$$\Sigma x = 404 \qquad \Sigma x^2 = 18,448 \qquad \Sigma y = 545$$
$$\Sigma y^2 = 31,993 \qquad \Sigma xy = 23,198$$
$$a = 32.080888 \qquad b = .554929$$

a. Calculate the predicted values and residuals.
b. Use the results of part **a** to obtain SSResid and r^2.
c. Does the least squares line appear to give very accurate predictions? Explain your reasoning.

5.42 The decline of salmon fisheries along the Columbia River in Oregon has caused great concern among commercial and recreational fishermen. The paper "Feeding of Predaceous Fishes on Out-Migrating Juvenile Salmonids in John Day Reservoir, Columbia River" (*Trans. Amer. Fisheries Soc.* (1991): 405–420) gave the accompanying data on $y =$ maximum size of salmonids consumed by a northern squawfish (the most abundant salmonid predator) and $x =$ squawfish length, both in mm.

x	218	246	270	287	318	344
y	82	85	94	127	141	157

x	375	386	414	450	468
y	165	216	219	238	249

Use the accompanying output from MINITAB to answer the following questions.

The regression equation is
size = -89.1 + 0.729 length

Predictor	Coef	Stdev	t ratio	p
Constant	-89.09	16.83	-5.29	0.000
length	0.72907	0.04778	15.26	0.000

$s = 12.56$ R-sq $= 96.3\%$ R-sq(adj) $= 95.9\%$

Analysis of Variance

SOURCE	DF	SS	MS	F	p
Regression	1	36736	36736	232.87	0.000
Error	9	1420	158		
Total	10	38156			

a. What maximum size would you predict for a squawfish whose length is 375 mm, and what is the residual corresponding to the observation (375, 165)?
b. What proportion of observed variation in y can be attributed to the approximate linear relationship between the two variables?

5.43 There have been numerous studies on the effects of radiation. The accompanying data on the relationship between degree of exposure to ^{242}Cm alpha particles (x) and the percentage of exposed cells without aberrations (y) appeared in the paper "Chromosome Aberrations Induced in Human Lymphocytes by D-T Neutrons" (*Radiation Research* (1984): 561–573).

x	.106	.193	.511	.527	1.08	1.62
y	98	95	87	85	75	72

x	1.73	2.36	2.72	3.12	3.88	4.18
y	64	55	44	41	37	40

Summary quantities are

$$n = 12 \qquad \Sigma x = 22.027 \qquad \Sigma y = 793$$
$$\Sigma x^2 = 62.600235 \qquad \Sigma xy = 1114.5 \qquad \Sigma y^2 = 57,939$$

a. Obtain the equation of the least squares line.
b. Calculate SSResid and SSTo.
c. What percentage of observed variation in y can be explained by the approximate linear relationship between the two variables?
d. Calculate and interpret the value of s_e.
e. Using just the results of parts **a** and **c,** what is the value of Pearson's sample correlation coefficient?

5.44 The paper "Crop Improvement for Tropical and Subtropical Australia: Designing Plants for Difficult Climates" (*Field Crops Research* (1991): 113–139) gave the accompanying data on $x =$ crop duration (days) for soybeans and $y =$ crop yield (tons/ha).

x	92	92	96	100	102
y	1.7	2.3	1.9	2.0	1.5

x	102	106	106	121	143
y	1.7	1.6	1.8	1.0	.3

$$\Sigma x = 1060 \qquad \Sigma y = 15.8 \qquad \Sigma xy = 1601.1$$
$$\Sigma x^2 = 114,514 \qquad \Sigma y^2 = 27.82$$
$$a = 5.20683380 \qquad b = -0.3421541$$

a. Construct a scatter plot of the data. Do you think the least squares line will give accurate predictions?

b. The largest x value in the sample greatly exceeds the remaining ones. Delete the corresponding observation from the sample and recalculate the equation of the least squares line. Does this observation greatly affect the equation of the line?

c. What effect does the deletion suggested in part **b** have on the value of r^2? Can you explain why this is so?

5.45 A study was carried out to investigate the relationship between the hardness of molded plastic (y, in Brinell units) and the amount of time elapsed since termination of the molding process (x, in hours). Summary quantities include

$$n = 15 \qquad \text{SSResid} = 1235.470 \qquad \text{SSTo} = 25{,}321.368$$

Calculate and interpret the coefficient of determination.

5.46 There has been an ongoing debate in recent years as to whether various species of wolves are endangered and what steps should be taken to prevent their demise. Disease can certainly pose a serious threat to such wildlife populations. The paper "Effects of Canine Parvovirus (CPV) on Gray Wolves in Minnesota" (*J. of Wildlife Manage.* (1995): 565–570) summarized a regression of y = percentage of pups in a capture on x = percentage of CPV prevalence among adults and pups. The equation of the least squares line, based on $n = 10$ observations, was $\hat{y} = 62.9476 - .54975x$, with $r^2 = .57$.

a. One observation was (25, 70). What is the corresponding residual?

b. What is the value of the sample correlation coefficient?

c. Suppose that SSTo = 2520.0 (this value was not given in the paper). What is the value of s_e?

5.47 Construct a residual plot for the data of Exercise 5.43, and comment on any interesting features.

5.48 Anthropologists often study soil composition for clues as to how the land was used during different periods. The accompanying data on x = soil depth (cm) and y = percent montmorillonite in the soil was taken from a scatter plot in the paper "Ancient Maya Drained Field Agriculture: Its Possible Application Today in the New River Floodplain, Belize, C. A." (*Ag. Ecosystems and Environ.* (1984): 67–84).

x	40	50	60	70	80	90	100
y	58	34	32	30	28	27	22

a. Draw a scatter plot of y versus x.

b. The equation of the least squares line is $\hat{y} =$ 64.50 − .45x. Draw this line on your scatter plot. Do there appear to be any large residuals?

c. Compute the residuals and construct a residual plot. Are there any unusual features in the plot?

5.49 a. Is it possible that both r^2 and s_e could be large for a bivariate data set? Explain. (A picture might be helpful.)

b. Is it possible that a bivariate data set could yield values of r^2 and s_e that are both small? Explain. (Again, a picture might be helpful.)

c. Explain why it is desirable to have r^2 large and s_e small if the relationship between two variables x and y is to be described using a straight line.

5.50 a. Show that the sum of the residuals $\Sigma [y - (a + bx)]$ is zero. [*Hint:* Substitute $a = \bar{y} - b\bar{x}$ and then use facts about $\Sigma (x - \bar{x})$ and $\Sigma (y - \bar{y})$.]

b. Let the residuals be denoted by $e_1, e_2, \ldots, e_n$ and consider fitting a line to the (x, e) pairs — that is, to the points in the residual plot — using the method of least squares. Then the slope of the least squares line is

$$\frac{\Sigma (x - \bar{x})(e - \bar{e})}{\Sigma (x - \bar{x})^2}$$

From part **a**, $\bar{e} = 0$, so $e - \bar{e} = e = y - (a + bx) = y - \bar{y} - b(x - \bar{x})$. Use this to show that the slope in the given expression is zero, so the residual plot has no tilt (though it may still have a nonlinear pattern if a straight-line fit is not appropriate).

5.51 Some straightforward but slightly tedious algebra shows that

$$\text{SSResid} = (1 - r^2) \Sigma (y - \bar{y})^2$$

from which it follows that

$$s_e = \sqrt{\frac{n - 1}{n - 2}} \sqrt{1 - r^2} s_y$$

Unless n is quite small, $(n - 1)/(n - 2) \approx 1$, so $s_e \approx \sqrt{1 - r^2} s_y$.

a. For what value of r is s_e as large as s_y? What is the least squares line in this case?

b. For what values of r will s_e be much smaller than s_y?

c. A study by the Berkeley Institute of Human Development (see the book *Statistics* by Freedman et al., listed in the Chapter 2 references) reported the following summary data for a sample of $n = 66$ California boys:

$$r \approx .80$$

At age 6, average height $\approx$ 46 in., standard deviation ≈ 1.7 in.

At age 18, average height $\approx$ 70 in., standard deviation $\approx$ 2.5 in.

What would s_e be for the least squares line used to predict 18-year-old height from 6-year-old height?

d. Referring to part **c,** suppose you wanted to predict the past value of 6-year-old height from knowledge of 18-year-old height. Find the equation for the appropriate least squares line. What is the corresponding value of s_e?

5.5 Nonlinear Relationships and Transformations

As we have seen in the previous sections, when the points in a scatter plot exhibit a linear pattern and the residual plot does not reveal any problems with the linear fit, the least squares line is a sensible way to summarize the relationship between x and y. A linear relationship is easy to interpret, departures from the line are easily detected, and using the line to predict y from our knowledge of x is straightforward. Often, though, a scatter plot or residual plot exhibits a curved pattern, indicating a more complicated relationship between x and y. In this case, finding a curve that fits the observed data well is a more complex task. In this section, we will consider two common approaches to fitting nonlinear relationships: polynomial regression and transformations.

Polynomial Regression

Let's reconsider the data on

x = time between flowering and harvesting

and

y = yield of paddy (a type of grain)

from Example 5.3. The scatter plot of this data is reproduced here as Figure 5.24. Since this plot shows a marked curved pattern, it is clear that no straight line would do a reasonable job of describing the relationship between x and y. However, the

FIGURE 5.24 Scatter plot for the paddy data

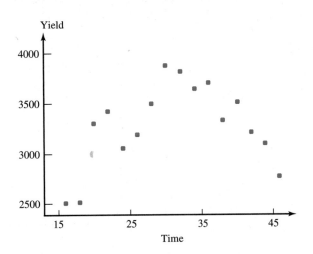

relationship could be described by a curve, and in this case the curved pattern in the scatter plot looks like a parabola (the graph of a quadratic function). This suggests trying to find a quadratic function of the form

$$\hat{y} = a + b_1 x + b_2 x^2$$

that would reasonably describe the relationship. That is, the values of the coefficients a, b_1, and b_2, in this equation must be selected to obtain a good fit to the data.

What are the best choices for the values of a, b_1, and b_2? In fitting a line to data, we used the principle of least squares to guide our choice of slope and intercept. Least squares can be used to fit a quadratic function as well. The deviations, $y - \hat{y}$, are still represented by vertical distances in the scatter plot, but now they are vertical distances from the points to a parabolic curve (the graph of a quadratic equation) rather than to a line, as shown in Figure 5.25. We then choose values for the coefficients in the quadratic equation so that the sum of squared deviations is as small as possible.

For a quadratic regression, the least squares estimates of a, b_1, and b_2 are those values that minimize the sum of squared deviations $\Sigma (y - \hat{y})^2$.

For quadratic regression, a measure that is useful for assessing fit is $R^2 = 1 - \text{SSResid}/\text{SSTo}$, where $\text{SSResid} = \Sigma (y - \hat{y})^2$. The measure R^2 is defined in a way similar to r^2 for simple linear regression and is interpreted in a similar fashion. The notation r^2 is used only with linear regression to emphasize the relationship between r^2 and the correlation coefficient, r, in the linear case.

The general expressions for computing the least squares estimates are somewhat complicated, so we will rely on a statistical software package or graphing calculator to do the computations for us.

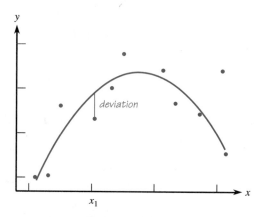

FIGURE 5.25 Deviations for a quadratic function

EXAMPLE 5.17

(Example 5.3 continued)   For the paddy yield data, the scatter plot (see Figure 5.24) showed a marked curved pattern. If the least squares line is fit to this data, it is no surprise that the line does not do a good job of describing the relationship ($r^2 = .075$ or 7.5% and $s_e = 416.7$), and the residual plot shows a distinct curved pattern as well (Figure 5.26).

FIGURE 5.26 Paddy data of Example 5.17: (a) Fitted line plot; (b) Residual plot

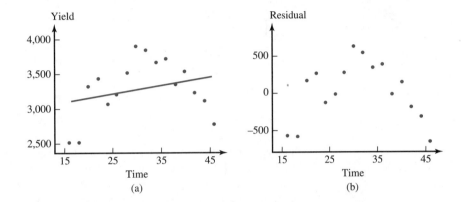

Partial MINITAB output from fitting a quadratic function to this data follows.

Polynomial Regression

Y = -1074.63 + 293.924X - 4.54644X**2

R-Sq = 79.4 %

Analysis of Variance

SOURCE	DF	SS	MS	F	P
Regression	2	2086388	1043194	25.0334	3.48E-05
Error	13	541738	41672		
Total	15	2628126			

The least squares coefficients are

$$a = -1074.63 \qquad b_1 = 293.924 \qquad b_2 = -4.54644$$

and the least squares quadratic is $\hat{y} = -1074.63 + 293.924x - 4.54644x^2$.

A plot showing the curve and the corresponding residual plot for the quadratic regression are given in Figure 5.27 (page 194). Notice that there is no strong pattern in the residual plot for the quadratic case as there was in the linear case. For the quadratic regression, $R^2 = .794$ (as opposed to .075 for the least squares line), which means that 79.4% of the variability in yield can be explained by an approximate quadratic relationship between yield and time between flowering and harvesting.

Linear and quadratic regression are special cases of polynomial regression. A polynomial regression curve is described by a function of the form

$$\hat{y} = a + b_1x + b_2x^2 + b_3x^3 + \cdots + b_kx^k$$

FIGURE 5.27 Quadratic regression of Example 5.17: (a) Scatter plot; (b) Residual plot

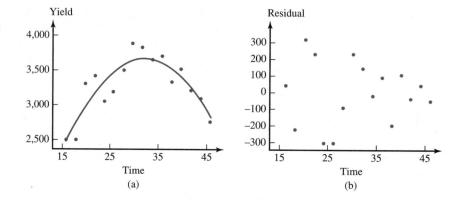

which is called a kth degree polynomial. The case of $k = 1$ results in linear regression ($\hat{y} = a + b_1 x$), and $k = 2$ yields a quadratic regression ($\hat{y} = a + b_1 x + b_2 x^2$). A quadratic curve has only one bend (see Figure 5.28(a) and (b)). A less frequently encountered special case is for $k = 3$, $\hat{y} = a + b_1 x + b_2 x^2 + b_3 x^3$, which is called a cubic regression curve. Cubic curves have two bends, as shown in Figure 5.28(c).

FIGURE 5.28 Polynomial regression curves: (a) Quadratic curve with $b_2 < 0$; (b) Quadratic curve with $b_2 > 0$; (c) Cubic curve

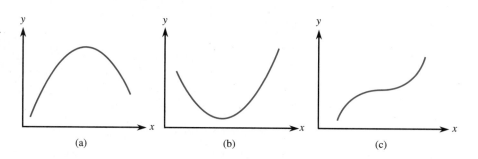

EXAMPLE 5.18

Researchers have examined a number of climatic variables in an attempt to understand the mechanisms that govern rainfall runoff. The article "The Applicability of Morton's and Penman's Evapotranspiration Estimates in Rainfall-Runoff Modeling" (*Water Resources Bull.* (1991): 611–620) reported on a study that examined the relationship between

x = cloud cover index   and y = sunshine index

Suppose that cloud cover index can have values between 0 and 1. Consider the accompanying data that is consistent with some of the summary quantities in the article:

Cloud Cover Index (x)	Sunshine Index (y)
.2	10.98
.5	10.94
.3	10.91
.1	10.94
.2	10.97
.4	10.89
0	10.88
.4	10.92
.3	10.86

The authors of the article used a cubic regression to describe the relationship between cloud cover and sunshine. Using the given data, MINITAB was instructed to fit a cubic regression to this data, resulting in the following output.

Polynomial Regression

Y = 10.8801 + 1.43140X - 7.20079X**2 + 9.19225X**3

R-Sq = 57.9 %

Analysis of Variance

SOURCE	DF	SS	MS	F	P
Regression	3	8.02E-03	2.67E-03	2.28796	0.195998
Error	5	5.84E-03	1.17E-03		
Total	8	1.39E-02			

The least squares cubic regression is then $\hat{y} = 10.8801 + 1.4314x - 7.20079x^2 + 9.19225x^3$. To predict the sunshine index for a day when the cloud cover index is .45, we use the regression equation to obtain

$$\hat{y} = 10.8801 + 1.4314x - 7.20079x^2 + 9.19225x^3$$
$$= 10.8801 + 1.4314(.45) - 7.20079(.45)^2 + 9.19225(.45)^3 = 10.91$$

The least squares cubic regression and the corresponding residual plot are shown in Figure 5.29 (page 196). The residual plot does not reveal any troublesome patterns that would suggest modification of the choice of cubic regression.

Transformations

An alternative to finding a curve to fit the data is to find a way to "transform" the x values and/or y values so that a scatter plot of the transformed data has a linear appearance. A **transformation** (sometimes called a reexpression) involves using a simple function of a variable in place of the variable itself. For example, instead of trying to describe the relationship between x and y, it might be easier to describe the relationship between $\sqrt{x}$ and y or between x and log(y). And, if we can describe the relationship between, say, $\sqrt{x}$ and y, it will still be possible for us to give a prediction

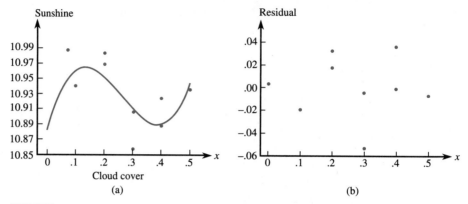

FIGURE 5.29 (a) Least squares cubic regression; (b) Residual plot

of the value of y for a given x value. Common transformations involve taking square roots, logs, or reciprocals.

EXAMPLE 5.19

As fans of white-water rafting know, a river flows more slowly close to its banks (due to friction between the river bank and the water). To study the nature of the relationship between water velocity and the distance from the shore, data is gathered on velocity (cm/sec) of a river at different distances (meters) from the bank. Suppose that the resulting data was as follows:

Distance	.5	1.5	2.5	3.5	4.5	5.5	6.5	7.5	8.5	9.5
Velocity	22.00	23.18	25.48	25.25	27.15	27.83	28.49	28.18	28.50	28.63

A graph of the data exhibits a curved pattern, as seen in both the scatter plot of Figure 5.30(a) and the residual plot from a linear fit shown in Figure 5.30(b).

Let's try transforming the x values by replacing each x value by its square root. We define

$$x' = \sqrt{x}$$

The resulting transformed data is given in Table 5.3.

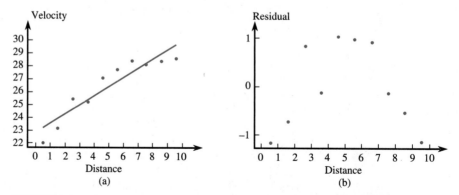

FIGURE 5.30 (a) Scatter plot of the river data; (b) Residual plot from linear fit

Table 5.3 *Original and transformed data of Example 5.19*

Original Data		Transformed Data	
x	*y*	*x'*	*y*
.5	22.00	0.7071	22.00
1.5	23.18	1.2247	23.18
2.5	25.48	1.5811	25.48
3.5	25.25	1.8708	25.25
4.5	27.15	2.1213	27.15
5.5	27.83	2.3452	27.83
6.5	28.49	2.5495	28.49
7.5	28.18	2.7386	28.18
8.5	28.50	2.9155	28.50
9.5	28.63	3.0822	28.63

Figure 5.31(a) shows a scatter plot of y versus x' (or equivalently y versus $\sqrt{x}$). The pattern of points in this plot looks linear, and so we can fit a least squares line using the transformed data. The MINITAB output from this regression follows.

Regression Analysis

The regression equation is
velocity = 20.1 + 3.01 sqrt distance

Predictor	Coef	StDev	T	P
Constant	20.1102	0.6097	32.99	0.000
sqrt dis	3.0085	0.2726	11.03	0.000

S = 0.6292   R-Sq = 93.8%   R-Sq(adj) = 93.1%

Analysis of Variance

Source	DF	SS	MS	F	P
Regression	1	48.209	48.209	121.76	0.000
Residual Error	8	3.168	0.396		
Total	9	51.376			

FIGURE 5.31 (a) Scatter plot of y versus x' for data of Example 5.19; (b) Residual plot resulting from a linear fit to transformed data

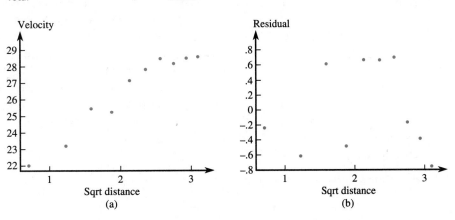

The residual plot in Figure 5.31(b) shows no indication of a pattern. The resulting regression equation is

$$\hat{y} = 20.1 + 3.01x'$$

or, equivalently,

$$\hat{y} = 20.1 + 3.01\sqrt{x}$$

The values of r^2 and s_e (see the MINITAB output) indicate that a line is a reasonable way to describe the relationship between y and x'. To predict velocity of the river at a distance of 9 meters from shore, we first compute $x' = \sqrt{x} = \sqrt{9} = 3$ and then use the sample regression line to obtain a prediction of y:

$$\hat{y} = 20.1 + 3.01x' = 20.1 + (3.01)(3) = 29.14$$

In Example, 5.19, transforming the x values using the square root function worked well. In general, how can we choose a transformation that will result in a linear pattern? Table 5.4 gives some guidance and summarizes some of the properties of the most commonly used transformations.

Table 5.4 *Commonly used transformations*

Transformation	Mathematical Description	Try This Transformation When
No transformation	$\hat{y} = a + bx$	The change in y is constant as x changes. A 1-unit increase in x is associated with, on average, an increase of b in the value of y.
Square root of x	$\hat{y} = a + b\sqrt{x}$	The change in y is not constant. A 1-unit increase in x is associated with smaller increases or decreases in y for larger x values.
Log of x*	$\hat{y} = a + b\log_{10}(x)$ or $\hat{y} = a + b\ln(x)$	The change in y is not constant. A 1-unit increase in x is associated with smaller increases or decreases in the value of y for larger x values.
Reciprocal of x	$\hat{y} = a + b(\frac{1}{x})$	The change in y is not constant. A 1-unit increase in x is associated with smaller increases or decreases in the value of y for larger x values. In addition, y has a limiting value of a as x increases.
Log of y* (Exponential growth or decay)	$\log(\hat{y}) = a + bx$ or $\ln(\hat{y}) = a + bx$	The change in y associated with a 1-unit change in x is proportional to x.

*The values of a and b in the regression equation will depend on whether $\log_{10}$ or ln is used, but the $\hat{y}$'s and r^2 values will be identical.

EXAMPLE 5.20   No tortilla chip lover likes soggy chips, so it is important to find characteristics of the production process that produce chips with an appealing texture. The following data on

x = frying time (sec)

and

y = moisture content (%)

appeared in the article "Thermal and Physical Properties of Tortilla Chips as a Function of Frying Time" (*J. of Food Processing and Preservation* (1995): 175–189).

Frying time, x	5	10	15	20	25	30	45	60
Moisture content, y	16.3	9.7	8.1	4.2	3.4	2.9	1.9	1.3

Figure 5.32(a) on page 200 shows a scatter plot of the data. The pattern in this plot is typical of exponential decay, with the change in y as x increases much smaller for large x values than for small x values. You can see that a 5-second change in frying time is associated with a much larger change in moisture content in the part of the plot where the x values are small than in the part of the plot where the x values are large. Table 5.4 suggests transforming the y values (moisture content in this example) by taking logarithms.

Two standard logarithmic functions are commonly used for such transformations — the common log (log base 10, denoted by log) and the natural log (log base e, denoted by ln). Either the common or natural log can be used; the only difference in the resulting scatter plots is the scale of the transformed y variable. This can be seen in Figures 5.32(b) and (c). These two scatter plots show the same pattern, and it looks like a line would be appropriate to describe this relationship.

Table 5.5 displays the original data along with the transformed y values using $y' = \log(y)$.

The MINITAB output at the top of page 200 shows the result of fitting the least squares line to the transformed data.

Table 5.5 *Transformed data from Example 5.20*

Frying Time x	Moisture y	Log(moisture) y'
5	16.3	1.21219
10	9.7	.98677
15	8.1	.90849
20	4.2	.62325
25	3.4	.53148
30	2.9	.46240
45	1.9	.27875
60	1.3	.11394

Regression Analysis

The regression equation is
log(moisture) = 1.14 - 0.0192 frying time

Predictor	Coef	StDev	T	P
Constant	1.14287	0.08016	14.26	0.000
frying t	-0.019170	0.002551	-7.52	0.000

S = 0.1246   R-Sq = 90.4%   R-Sq(adj) = 88.8%

Analysis of Variance

Source	DF	SS	MS	F	P
Regression	1	0.87736	0.87736	56.48	0.000
Residual Error	6	0.09320	0.01553		
Total	7	0.97057			

The resulting regression equation is

$$y' = 1.14 - .0192x$$

or, equivalently,

$$\log(y) = 1.14 - .0192x$$

FIGURE 5.32 (a) Scatter plot of the tortilla chip data; (b) Scatter plot of transformed data with $y = \log(y)$

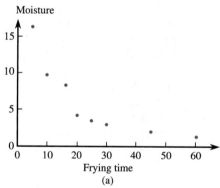

(a)

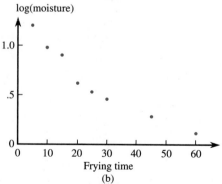

(b)

(*continued*)

FIGURE 5.32 (continued)
(c) Scatter plot of transformed data with $y' = \ln(y)$

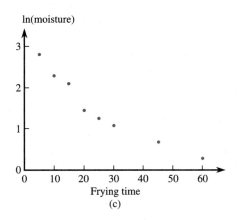

(c)

Fitting a Curve Using Transformations

The objective of a regression analysis is usually to describe the approximate relationship between x and y with an equation of the form

$$y = \text{some function of } x$$

If we have transformed only x, fitting a least squares line to the transformed data will result in an equation of the desired form, for example, something like

$$\hat{y} = 5 + 3x' = 5 + 3\sqrt{x} \qquad \text{where } x' = \sqrt{x}$$

or

$$\hat{y} = 4 + .2x' = 4 + .2\left(\frac{1}{x}\right) \qquad \text{where } x' = \frac{1}{x}$$

These functions specify lines when graphed using y and x', and they specify curves when graphed using y and x, as illustrated in Figure 5.33 for the case of the square-root transformation.

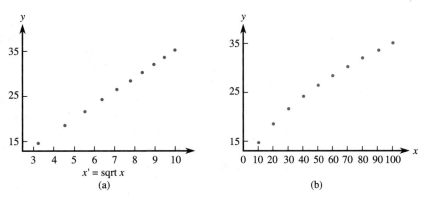

FIGURE 5.33 (a) A plot of $\hat{y} = 5 + 3x'$, where $x' = \sqrt{x}$ (b) A plot of $\hat{y} = 5 + 3\sqrt{x}$

If the y values have been transformed, after obtaining the least squares line, the transformation can be "undone" to yield an expression of the form y = some function of x (as opposed to y' = some function of x). For example, to reverse a log transformation ($y' = \log(y)$), we can take the antilog of each side of the equation. To reverse a square-root transformation ($y' = \sqrt{y}$), we can square both sides of the equation, and to reverse a reciprocal transformation ($y' = 1/y$), we can take the reciprocal of each side of the equation. This is illustrated in the following example.

EXAMPLE 5.21

(Example 5.20 continued)   For the tortilla chip data of Example 5.20, $y' = \log(y)$ and the least squares line relating y' and x was

$$y' = 1.14 - .0192x$$

or, equivalently,

$$\log(y) = 1.14 - .0192x$$

To reverse this transformation, we take the antilog of both sides of the equation:

$$10^{\log(y)} = 10^{1.14-.0192x}$$

Using the properties of logs and exponents

$$10^{\log(y)} = y$$
$$10^{1.14-.0192x} = (10^{1.14})(10^{-.0192x})$$

We get

$$\hat{y} = (10^{1.14})(10^{-.0192x}) = (13.8038)(10^{-.0192x})$$

This equation can now be used to predict the y value (moisture content) for a given x (frying time). For example, the predicted moisture content when frying time is 35 seconds is

$$\hat{y} = (13.8038)(10^{-.0192x}) = (13.8038)(.2128) = 2.9376$$

It should be noted that the process of transforming data, fitting a line to the transformed data, and then "undoing" the transformation to get an equation for a curved relationship between x and y usually results in a curve that provides a reasonable fit to the sample data, but it is not the "least squares" curve for the data. For example, in Example 5.21, a transformation was used to fit the curve $\hat{y} = (13.8038)(10^{-.0192x})$. However, there may be another equation of the form $\hat{y} = a(10^{bx})$ that has a smaller sum of squared residuals for the *original* data than the one we obtained using transformations. Finding the least squares estimates for a and b in an equation of this form is complicated. Fortunately, the curves found using transformations usually provide reasonable predictions of y.

Power Transformations

Frequently, an appropriate transformation will be suggested by the data. One type of transformation that statisticians have found useful for straightening a plot is a

Table 5.6 *Power transformation ladder*

Power	Transformed Value*	Name
3	(Original value)^3	Cube
2	(Original value)^2	Square
1	Original value	No transformation
$\frac{1}{2}$	$\sqrt{\text{Original value}}$	Square root
$\frac{1}{3}$	$\sqrt[3]{\text{Original value}}$	Cube root
0	Log(original value)	Logarithm
−1	1/(original value)	Reciprocal

*Transformed value = (original value)^{power}

power transformation. A power (exponent) is first selected, and each original value is raised to that power to obtain the corresponding transformed value. Table 5.6 displays a "ladder" of the most frequently used power transformations. The power 1 corresponds to no transformation at all. Using the power 0 would transform every value to 1, which is certainly not informative, so statisticians use the logarithmic transformation in its place in the ladder of transformations. Other powers intermediate to or more extreme than those listed can be used, of course, but they are less frequently needed than those on the ladder. Notice that all of the transformations previously presented are included in this ladder.

Figure 5.34 is designed to suggest where on the ladder we should go to find an appropriate transformation. The four curved segments labeled 1, 2, 3, and 4 represent shapes of curved scatter plots that are commonly encountered. Suppose that a scatter plot looks like the curve labeled 1. Then to straighten the plot, we should use a power on x that is up the ladder from the no-transformation row (x^2 or x^3) and/or

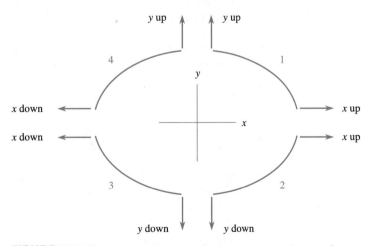

FIGURE 5.34 Scatter plot shapes and where to go on the transformation ladder to straighten the plot

a power on y that is also up the ladder from the power 1. Thus, we might be led to squaring each x value, cubing each y, and plotting the transformed pairs. If the curvature looks like curved segment 2, a power up the ladder from no transformation for x and/or a power down the ladder for y (for example, $\sqrt{y}$ or $\log(y)$) should be used.

Note that the scatter plot for the tortilla chip data (Figure 5.32(a)) has the pattern of segment 3 in Figure 5.34. This suggests going down the ladder of transformations for x and/or for y. We found that transforming the y values in this data set by taking logs worked well, and this is consistent with the suggestion of going down the ladder of transformations for y.

EXAMPLE 5.22

In many parts of the world, a typical diet consists mainly of cereals and grains, and many individuals suffer from a substantial iron deficiency. The article "The Effects of Organic Acids, Phytates, and Polyphenols on the Absorption of Iron from Vegetables" (*British J. Nutrition* (1983): 331–342) reported the data in Table 5.7 on

x = proportion of iron absorbed

and

y = polyphenol content (mg/g)

when a particular vegetable is consumed. The scatter plot of the data in Figure 5.35(a) shows a clear curved pattern, which resembles the curved segment 3 in

Table 5.7 *Original and transformed data for Example 5.22*

Vegetable	x	y	$\sqrt{x}$	$\sqrt{y}$
Wheat germ	.007	6.4	.083666	2.52982
Aubergine	.007	3.0	.083666	1.73205
Butter beans	.012	2.9	.109545	1.70294
Spinach	.014	5.8	.118322	2.40832
Brown lentils	.024	5.0	.154919	2.23607
Beetroot greens	.024	4.3	.154919	2.07364
Green lentils	.032	3.4	.178885	1.84391
Carrot	.096	.7	.309839	.83666
Potato	.115	.2	.339116	.44721
Beetroot	.185	1.5	.430116	1.22474
Pumpkin	.206	.1	.453872	.31623
Tomato	.224	.3	.473286	.54772
Broccoli	.260	.4	.509902	.63246
Cauliflower	.263	.7	.512835	.83666
Cabbage	.320	.1	.565685	.31623
Turnip	.327	.3	.571839	.54772
Sauerkraut	.327	.2	.571839	.44721

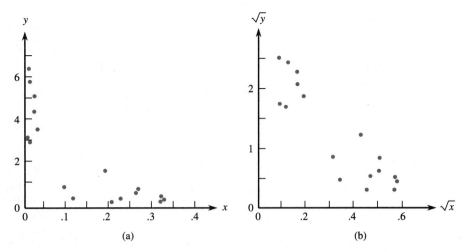

FIGURE 5.35 Scatter plot of data from Example 5.22: (a) Original data; (b) Data transformed by taking square roots

Figure 5.34. This suggest that x and/or y should be transformed by a power transformation down the ladder from 1. The authors of the article applied a square-root transformation to both x and y. The resulting scatter plot in Figure 5.35(b) is reasonably straight.

Using MINITAB to fit a least squares line to the transformed data ($y' = \sqrt{y}$ and $x' = \sqrt{x}$) results in the following output:

Regression Analysis

The regression equation is
sqrt y = 2.45 - 3.72 sqrt x

Predictor	Coef	StDev	T	P
Constant	2.4483	0.1833	13.35	0.000
sqrt x	-3.7247	0.4836	-7.70	0.000

S = 0.3695   R-Sq = 79.8%   R-Sq(adj) = 78.5%

The corresponding least squares line is

$$\hat{y}' = 2.45 - 3.72x'$$

This transformation can be reversed by squaring both sides (to obtain an equation of the form $y =$ some function of x) to obtain

$$\hat{y}'^2 = (2.45 - 3.72x')^2$$

Since $y'^2 = y$ and $x' = \sqrt{x}$, we get

$$\hat{y} = (2.45 - 3.72\sqrt{x})^2$$

Exercises 5.52 – 5.59

5.52 The following data on x = frying time (sec) and y = moisture content (%) appeared in the paper "Thermal and Physical Properties of Tortilla Chips as a Function of Frying Time" (*J. of Food Processing and Preservation* (1995): 175–189).

x	5	10	15	20	25	30	45	60
y	16.3	9.7	8.1	4.2	3.4	2.9	1.9	1.3

a. Construct a scatter plot of the data. Would a straight line provide an effective summary of the relationship?
b. Here are the values of $x' = \log(x)$ and $y' = \log(y)$. Construct a scatter plot of this transformed data and comment on the pattern.

x'	.70	1.00	1.18	1.30	1.40	1.48	1.65	1.78
y'	1.21	.99	.91	.62	.53	.46	.28	.11

c. Based on the accompanying MINITAB output, does the least squares line effectively summarize the relationship between y' and x'?

```
The regression equation is
logmoist = 2.01 - 1.05 logtime

Predictor      Coef     Stdev    t-ratio      p
Constant    2.01444   0.09166     21.98   0.000
logtime    -1.04920   0.06786    -15.46   0.000

s = 0.06293   R-sq = 97.6%    R-sq(adj) = 97.1%

Analysis of Variance
SOURCE        DF       SS        MS        F       p
Regression     1   0.94680   0.94680   239.06   0.000
Error          6   0.02376   0.00396
Total          7   0.97057
```

d. Use the MINITAB output in part **c** to predict moisture content when frying time is 35 seconds.
e. Do you think that predictions of moisture content using the model in part **c** will be better than those using the model fit in Example 5.20, which used transformed y values but did not transform x? Explain.

5.53 The article "Reduction in Soluble Protein and Chlorophyll Contents in a Few Plants as Indicators of Automobile Exhaust Pollution" (*Int. J. of Environ. Studies* (1983): 239–244) reported the accompanying data on x = distance from a highway (m) and y = lead content of soil at that distance (ppm).

x	.3	1	5	10	15	20
y	62.75	37.51	29.70	20.71	17.65	15.41

x	25	30	40	50	75	100
y	14.15	13.50	12.11	11.40	10.85	10.85

a. Use a statistical computer package to construct scatter plots of y versus x, y versus $\log(x)$, $\log(y)$ versus $\log(x)$, and $1/y$ versus $1/x$.
b. Which transformation considered in part **a** does the best job of producing an approximately linear relationship? Use the selected transformation to predict lead content when distance is 25 m.

5.54 The paper "Aspects of Food Finding by Wintering Bald Eagles" (*The Auk* (1983): 477–484) examined the relationship between the time that eagles spend aerially searching for food (indicated by the percentage of eagles soaring) and relative food availability. The accompanying data is taken from a scatter plot that appeared in this paper. Let x denote salmon availability and y denote percent of eagles in the air.

x	0	0	.2	.5	.5	1.0
y	28.2	69.0	27.0	38.5	48.4	31.1

x	1.2	1.9	2.6	3.3	4.7	6.5
y	26.9	8.2	4.6	7.4	7.0	6.8

a. Draw a scatter plot for this data set. Would you describe the plot as linear or curvilinear?
b. One possible transformation that might lead to a straighter plot involves taking the square root of both the x and y values. Use Figure 5.34 to explain why this might be a reasonable transformation.
c. Construct a scatter plot using the variables $\sqrt{x}$ and $\sqrt{y}$. Is this scatter plot straighter than that in part **a**?
d. Using Table 5.6, suggest another transformation that might be used to straighten the original plot.

5.55 Data on salmon availability (x) and the percent of eagles in the air (y) is given in the previous exercise.
a. Calculate the correlation coefficient for this data.
b. Since the scatter plot of the original data appeared curved, transforming both the x and y values by taking square roots was suggested. Calculate the correlation coefficient for the variables $\sqrt{x}$ and $\sqrt{y}$. How does this value compare with that of part **a**? Does this indicate that the transformation was successful in straightening the plot?

5.56 Penicillin was administered orally to five horses, and the concentration of penicillin in the blood was determined after five different lengths of time (a different horse was used each time). The accompanying data appeared in the paper "Absorption and Distribution Patterns of Oral Phenoxymethyl Penicillin in the Horse" (*Cornell Veterinarian* (1983): 314–323).

x (Time elapsed, hr)	1	2	3	6	8
y (Penicillin concentration, mg/mL)		1.8	1.0	.5	.1	.1

Construct scatter plots using the following variables. Which transformation, if any, would you recommend?

a. x and y

b. $\sqrt{x}$ and y

c. x and $\sqrt{y}$

d. $\sqrt{x}$ and $\sqrt{y}$

e. x and $\log(y)$ (Values of $\log(y)$ are .26, 0, −.30, −1, −1.)

5.57 The paper "Population Pressure and Agricultural Intensity" (*Annals of the Assoc. of Amer. Geog.* (1977): 384–396) reported a positive association between population density and agricultural intensity. The accompanying data consists of measures of population density (x) and agricultural intensity (y) for 18 different subtropical locations.

x	1.0	26.0	1.1	101.0	14.9	134.7
y	9	7	6	50	5	100

x	3.0	5.7	7.6	25.0	143.0	27.5
y	7	14	14	10	50	14

x	103.0	180.0	49.6	140.6	140.0	233.0
y	50	150	10	67	100	100

a. Construct a scatter plot of y versus x. Is the scatter plot compatible with the statement of positive association made in the paper?

b. The scatter plot in part **a** is curved upward like segment 2 in Figure 5.34, suggesting a transformation that is up the ladder for x or down the ladder for y. Try a scatter plot that uses y and x^2. Does this transformation straighten the plot?

c. Draw a scatter plot that uses $\log(y)$ and x. The $\log(y)$ values, given in order corresponding to the y values, are .95, .85, .78, 1.70, .70, 2.00, .85, 1.15, 1.15, 1.00, 1.70, 1.15, 1.70, 2.18, 1.00, 1.83, 2.00, and 2.00. How does this scatter plot compare with that of part **b**?

d. Next consider a scatter plot that uses transformations on both x and y: $\log(y)$ and x^2. Is this effective in straightening the plot? Explain.

5.58 Determining the age of an animal can sometimes be a difficult task. One method of estimating the age of harp seals is based on the width of the pulp canal in the seal's canine teeth. To investigate the relationship between age and the width of the pulp canal, age and canal width were measured for seals of known age. The accompanying data is a portion of a larger data set that appeared in the paper "Validation of Age Estimation in the Harp Seal Using Dentinal Annuli" (*Canadian J. of Fisheries and Aquatic Sci.* (1983): 1430–1441). Let x denote age (years) and y denote canal length (mm).

x	.25	.25	.50	.50	.50	.75	.75	1.00
y	700	675	525	500	400	350	300	300

x	1.00	1.00	1.00	1.00	1.25	1.25	1.50	1.50
y	250	230	150	100	200	100	100	125

x	2.00	2.00	2.50	2.75	3.00	4.00	4.00	5.00
y	60	140	60	50	10	10	10	10

x	5.00	5.00	5.00	6.00	6.00
y	15	10	10	15	10

Construct a scatter plot for this data set. Would you describe the relationship between age and canal length as linear? If not, suggest a transformation that might straighten the plot.

5.59 A frequently encountered problem in crop planting situations involves deciding when to harvest in order to maximize yield. The accompanying data on x = date of harvesting (number of days after flowering) and y = yield (kg/ha) of paddy, a grain farmed in India, appeared in the paper "Determination of Biological Maturity and Effect of Harvesting and Drying Conditions on Milling Quality of Paddy" (*J. of Ag. Eng.* (1975): 353–361). Construct a scatter plot of this data. Can the methodology discussed in this section be used to select a straightening transformation? Why or why not? What kind of a curve might provide a reasonable fit to the plot?

x	16	18	20	22	24	26	28	30
y	2508	2518	3304	3423	3057	3190	3500	3883

x	32	34	36	38	40	42	44	46
y	3823	3646	3708	3333	3517	3241	3103	2776

5.6   Interpreting the Results of Statistical Analyses

Using either a least squares line to summarize a relationship or a correlation coefficient to describe the strength of a linear relationship is common in investigations that focus on more than a single variable. In fact, the methods described in this chapter are among the most widely used of all statistical tools. When numerical bivariate data is analyzed in journal articles and other published sources, it is common to find a scatter plot of the data and a least squares line or a correlation coefficient.

What to Look for in Published Data

Here are a few things to consider when you read an article that includes an analysis of bivariate data:

- What two variables are being studied? Are they both numerical? Is a distinction made between a dependent variable and an independent variable?

- Does the article include a scatter plot of the data? If so, does the relationship between the two variables appear to be linear or curved?

- Does the relationship between the two variables appear to be weak or strong? Is the value of a correlation coefficient reported?

- If the least squares line is used to summarize the relationship between the dependent and independent variables, is any measure of goodness of fit reported, such as r^2 or s_e? How are such values interpreted, and what do they imply about the usefulness of the least squares line?

- If a correlation coefficient is reported, is it interpreted properly? Be cautious of interpretations that claim a causal relationship.

The article "Rubbish Regression and the Census Undercount" (*Chance* (1992): 33) describes work done by the Garbage Project at the University of Arizona. Project researchers had analyzed different categories of garbage for a number of households. They were asked by the Census Bureau to see whether any of the garbage data variables were related to household size. They reported the following:

> The weight data for different categories of garbage were plotted on graphs against data on household size, dwelling by dwelling, and the resulting scatter plots were analyzed to see in which categories the weight showed a steady, monotonic rise relative to household size.

They determined that the strongest linear relationship appeared to be that between the amount of plastic discarded and household size. The line used to summarize this relationship was stated to be $y = .2815x$, where y = household size and x = weight (in lb) of plastic during a 5-week collection. Note that the line reported has an intercept of zero. Scientists at the Census Bureau believed that this relationship would extend to entire neighborhoods, and so the amount of plastic discarded by a neighborhood could be measured (rather than having to measure house to house) and then used to approximate the neighborhood size.

An example of the use of r^2 and the correlation coefficient is found in the article "Affective Variables Related to Mathematics Achievement Among High-Risk College Freshmen" (*Psychological Reports* (1991): 399–403). The authors write the following:

> An examination of the Pearson correlations indicated that the two scales, Confidence in Learning Mathematics and Mathematics Anxiety, were highly related, with values ranging from .87 to .91 for the men, women, and for the total group. Such a high relationship indicates that the two scales are measuring essentially the same concept, so for further analysis, one of the two scales could be eliminated.

The correlation coefficients reported (from .87 to .91) indicate a strong positive linear relationship between scores on the two scales. Students who scored high on the Confidence in Learning Mathematics scale tended also to score high on the Mathematics Anxiety scale. (The authors explain that a high score on the Mathematics Anxiety scale corresponds to low anxiety toward mathematics.) We should be cautious of the statement that the two scales are measuring essentially the same concept. Although that interpretation may make sense in this context given what the researchers know about these two scales, it does not follow solely from the fact that the two are highly correlated. (For example, there might be a correlation between number of years of education and income, but it would be hard to argue that these two variables are measuring the same thing!)

The same article also states that Mathematics Anxiety score accounted for 27% of the variability in first-quarter math grades. The value of 27% reported is the value of r^2 converted to a percentage. Although a value of 27% ($r^2 = .27$) may not seem particularly large, it is in fact surprising that the variable *Math Anxiety score* would be able to explain more than one-fourth of the student-to-student variability in first-quarter math grades.

Summary of Key Concepts and Formulas

Term or Formula	Comment
Scatter plot	A picture of bivariate numerical data in which each observation (x, y) is represented as a point located with respect to a horizontal x axis and a vertical y axis.
Pearson's sample correlation coefficient $$r = \frac{\Sigma\, z_x z_y}{n - 1}$$	A measure of the extent to which sample x and y values are linearly related; $-1 \leq r \leq 1$, so values close to 1 or -1 indicate a strong linear relationship.
Spearman's correlation coefficient r_S	Pearson's r applied to the *ranks* of the x and y values. It will detect both linear and nonlinear relationships and is not as sensitive to outliers as is r.
Principle of least squares	The method used to select a line that summarizes an approximate linear relationship between x and y. The least squares line is the line that minimizes the sum of the squared vertical deviations from the points in the scatter plot.

Term or Formula	Comment
$b = \dfrac{\Sigma (x - \bar{x})(y - \bar{y})}{\Sigma (x - \bar{x})^2} = \dfrac{\Sigma xy - \dfrac{(\Sigma x)(\Sigma y)}{n}}{\Sigma x^2 - \dfrac{(\Sigma x)^2}{n}}$ $a = \bar{y} - b\bar{x}$	The slope and intercept of the least squares line.
Predicted (fitted) values $\hat{y}_1, \ldots, \hat{y}_n$	Obtained by substituting the x value for each observation into the least squares line; $\hat{y}_1 = a + bx_1, \ldots, \hat{y}_n = a + bx_n$
Residuals	Obtained by subtracting each predicted value from the corresponding observed y value: $y_1 - \hat{y}_1, \ldots, y_n - \hat{y}_n$. These are the vertical deviations from the least squares line.
Residual plot	Scatter plot of the $(x, \text{residual})$ pairs. Isolated points or a pattern of points in a residual plot are indicative of potential problems.
Residual (error) sum of squares SSResid $= \Sigma (y - \hat{y})^2$	The sum of the squared residuals is a measure of y variation that cannot be attributed to an approximate linear relationship (unexplained variation).
Total sum of squares SSTo $= \Sigma (y - \bar{y})^2$	The sum of squared deviations from the sample mean $\bar{y}$ is a measure of total variation in the observed y values.
Coefficient of determination $r^2 = 1 - \dfrac{\text{SSResid}}{\text{SSTo}}$	The proportion of variation in observed y's that can be attributed to an approximate linear relationship.
Standard deviation about the least squares line, $s_e = \sqrt{\dfrac{\text{SSResid}}{n - 2}}$	The size of a "typical" deviation from the least squares line.
Transformation	A simple function of the x and/or y variable which is then used in a regression.
Power transformation	An exponent, or power, p, is first specified, and then new (transformed) data values are calculated as *transformed value* = (original value)^p. A logarithmic transformation is identified with $p = 0$. When the scatter plot of original data exhibits curvature, a power transformation of x and/or y will often result in a scatter plot that has a linear appearance.

Supplementary Exercises 5.60 – 5.71

5.60 The sample correlation coefficient between annual raises and teaching evaluations for a sample of $n = 353$ college faculty was found to be $r = .11$ ("Determination of Faculty Pay: An Agency Theory Perspective," *Academy of Mgmt. J.* (1992): 921–955).
   a. Interpret this value.
   b. If a straight line were fit to the data using least squares, what proportion of variation in raises could be attributed to the approximate linear relationship between raises and evaluations?

5.61 An accurate assessment of oxygen consumption provides important information for determining energy-expenditure requirements for physically demanding tasks. The paper "Oxygen Consumption During Fire Suppression: Error of Heart Rate Estimation" (*Ergonomics* (1991): 1469–1474) reported on a study in which oxygen consumption (mL/kg/min) during a treadmill test was determined for a sample of ten firefighters. Then oxygen consumption at a comparable heart rate was measured for each of the ten individuals while performing a fire-suppression simulation. This resulted in the following data and scatter plot.

Firefighter	1	2	3	4	5
x = treadmill consumption	51.3	34.1	41.1	36.3	36.5
y = fire-simulation consumption	49.3	29.5	30.6	28.2	28.0

Firefighter	6	7	8	9	10
x = treadmill consumption	35.4	35.4	38.6	40.6	39.5
y = fire-simulation consumption	26.3	33.9	29.4	23.5	31.6

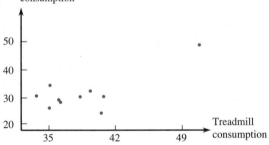

a. Does the scatter plot suggest an approximate linear relationship?

b. The investigators fit a least squares line. The resulting MINITAB output is shown. Predict fire-simulation consumption when treadmill consumption is 40.

The regression equation is
firecon = -11.4 + 1.09 treadcon

Predictor	Coef	Stdev	t-ratio	p
Constant	-11.37	12.46	-0.91	0.388
treadcon	1.0906	0.3181	3.43	0.009

s = 4.70   R-sq = 59.5%   R-sq(adj) = 54.4%

c. How effectively does a straight line summarize the relationship?

d. Delete the observation (51.3, 49.3) and calculate the new equation of the least squares line and the value of r^2. What do you conclude? (*Hint*: For the original data, $\Sigma x = 388.8$, $\Sigma y = 310.3$, $\Sigma x^2 = 15{,}338.54$, $\Sigma xy = 12{,}306.58$, and $\Sigma y^2 = 10{,}072.41$.)

5.62 Silane coupling agents have been used in the rubber industry to improve the performance of fillers in rubber compounds. The accompanying data on y = tensile modulus (in MPa, a measure of silane coupling effectiveness) and x = bound rubber content (%) appeared in the paper "The Effect of the Structure of Sulfur-Containing Silane Coupling Agents on Their Activity in Silica-Filled SBR" (*Rubber Chem. and Tech.* (1984): 675–685).

x	16.1	31.5	21.5	22.4	20.5	28.4
y	4.41	6.81	5.26	5.99	5.92	6.14

x	30.3	25.6	32.7	29.2	34.7
y	6.84	5.87	7.03	6.89	7.87

$n = 11$ $\Sigma y = 69.03$ $\Sigma x = 292.9$
$\Sigma y^2 = 442.1903$ $\Sigma x^2 = 8141.75$ $\Sigma xy = 1890.200$

a. Construct a scatter plot. Does it look linear?
b. Find the equation of the least squares line.
c. Compute and interpret the value of r^2.

5.63 The relationship between the depth of flooding and the amount of flood damage was examined in the paper "Significance of Location in Computing Flood Damage" (*J. of Water Resources Planning and Mgmt.* (1985): 65–81). The accompanying data on x = depth of flooding (feet above first-floor level) and y = flood damage (as a percent of structure value) was obtained using a sample of flood insurance claims.

x	1	2	3	4	5	6
y	10	14	26	28	29	41

x	7	8	9	10	11	12	13
y	43	44	45	46	47	48	49

a. Obtain the equation of the least squares line.
b. Construct a scatter plot and draw the least squares line on the plot. Does it look as though a straight line provides an adequate description of the relationship between y and x? Explain.
c. Predict flood damage for a structure subjected to 6.5 ft of flooding.
d. Would you use the least squares line to predict flood damage when depth of flooding is 18 ft? Explain.

5.64 An investigation was carried out to study the relationship between speed (ft/sec) and stride rate (number of steps taken/sec) among female marathon runners. Resulting summary quantities included

$n = 11$ $\Sigma \,(\text{speed}) = 205.4$ $\Sigma \,(\text{speed})^2 = 3880.08$
$\Sigma \,(\text{rate}) = 35.16$ $\Sigma \,(\text{rate})^2 = 112.681$
$\Sigma \,(\text{speed})(\text{rate}) = 660.130$

a. Calculate the equation of the least squares line that you would use to predict stride rate from speed.
b. Calculate the equation of the least squares line that you would use to predict speed from stride rate.
c. Calculate the coefficient of determination for the "stride rate on speed" regression of part **a** and for the "speed on stride rate" regression of part **b**. How are these related?

5.65 Consider the four (x, y) pairs $(0, 0)$, $(1, 1)$, $(1, -1)$, and $(2, 0)$.

a. What is the value of the sample correlation coefficient r?

b. If a fifth observation is made at the value $x = 6$, find a value of y for which $r > .5$.

c. If a fifth observation is made at the value $x = 6$, find a value of y for which $r < -.5$.

5.66 It is certainly plausible that workers are less likely to quit their jobs when wages are high than when they are low. The paper "Investigating the Causal Relationship Between Quits and Wages: An Exercise in Comparative Dynamics" (*Economic Inquiry* (1986): 61–83) presented the accompanying data on x = average hourly wage and y = quit rate (number of employees per 100 who left jobs during 1986). Each observation is for a different industry.

x	8.20	10.35	6.18	5.37	9.94	9.11	10.59	13.29
y	1.4	.7	2.6	3.4	1.7	1.7	1.0	.5

x	7.99	5.54	7.50	6.43	8.83	10.93	8.80
y	2.0	3.8	2.3	1.9	1.4	1.8	2.0

a. Does the data demonstrate unequivocally that quit rate is determined at least in part by factors other than wages? What does your intuition suggest?

b. Construct a scatter plot of the data. What does the plot suggest about the nature of the relationship?

c. Use the summary quantities $\Sigma x = 129.05$, $\Sigma x^2 = 1178.9601$, $\Sigma y = 28.2$, $\Sigma y^2 = 64.34$, and $\Sigma xy = 218.806$ to obtain the equation of the least squares line.

d. Predict the quit rate for an industry with an average hourly wage of $7.50, and calculate the corresponding residual.

e. What proportion of observed variation in quit rate can be attributed to the linear relationship?

5.67 The paper "Biomechanical Characteristics of the Final Approach Step, Hurdle, and Take-Off of Elite American Springboard Divers" (*J. Human Movement Studies* (1984): 189–212) gave the accompanying data on y = judge's score and x = length of final step (m) for a sample of seven divers performing a forward pike with a single somersault.

y	7.40	9.10	7.20	7.00	7.30	7.30	7.90
x	1.17	1.17	.93	.89	.68	.74	.95

a. Construct a scatter plot.

b. Calculate the slope and intercept of the least squares line. Draw this line on your scatter plot.

c. Calculate and interpret the value of Pearson's sample correlation coefficient.

d. Compute the value of Spearman's rank correlation coefficient, r_S. [For tied scores, assign the av-

erage of the ranks that would have been assigned had they differed slightly; for example, if the values of 7.3 in the score row were slightly different, they would be ranked 4 and 5, so each is assigned rank $(4 + 5)/2 = 4.5$.] How does the value of r_S compare to that of r from part **c**?

5.68 The accompanying data on y = concentration of penicillin-G in pig's blood plasma (units/mL) and x = time (min) from administration of a dose of penicillin (22 mg/kg body weight) appeared in the paper "Calculation of Dosage Regimens of Antimicrobial Drugs for Surgical Prophylaxis" (*J. Amer. Vet. Med. Assoc.* (1984): 1083–1087).

x	5	15	45	90	180	240	360	480	1440
y	32.6	43.3	23.1	16.7	5.7	6.4	9.2	.4	.2

a. Construct a scatter plot for this data.

b. Using the ladder of transformations of Section 5.5, suggest a transformation that might straighten the plot. Give reasons for your choice of transformation.

5.69 The accompanying data resulted from an experiment in which x was the amount of catalyst added to accelerate a chemical reaction and y was the resulting reaction time.

x	1	2	3	4	5
y	49	46	41	34	25

a. Calculate r. Does the value of r suggest a strong linear relationship?

b. Construct a scatter plot. From the plot, does the word *linear* really provide the most effective description of the relationship between x and y? Explain.

5.70 The least squares intercept a and slope b come from solving a system of two linear equations called the **normal equations**:

$$na + (\Sigma x)b = \Sigma y$$
$$(\Sigma x)a + (\Sigma x^2)b = \Sigma xy$$

a. Verify by direct substitution that

$$a = \bar{y} - b\bar{x} = \frac{\Sigma y}{n} - b\frac{\Sigma x}{n}$$

$$b = \frac{\Sigma xy - \frac{(\Sigma x)(\Sigma y)}{n}}{\Sigma x^2 - \frac{(\Sigma x)^2}{n}}$$

is the solution to this system of equations.

b. Show that the first normal equation implies that the sum of the residuals from the least squares line is zero — that is, that $\Sigma (y - \hat{y}) = 0$. (*Hint*: Recall that $\hat{y} = a + bx$.)

5.71 **a.** Show that

SSResid = SSTo − b(numerator of b)

(*Hint*: Use the computational formula for SSResid and the fact that $a = \bar{y} - b\bar{x}$.)

b. Argue that the expression for SSResid in part **a** implies that SSResid ≤ SSTo. (*Hint*: Both b and the numerator of b have the same sign.)

References

Chatterjee, Samprit and Bertram Price. *Regression Analysis by Example,* 2nd edition. New York, Wiley Inter-Science, 1991.

Hamilton, Lawrence C. *Regression with Graphics: A Second Course in Applied Statistics.* Belmont, CA, Duxbury Press, 1992.

Neter, John, William Wasserman, and Michael Kutner. *Applied Linear Statistical Models,* 4th edition. New York: McGraw Hill, 1996. (The first half of this book gives a comprehensive treatment of regression analysis without overindulging in mathematical development; a highly recommended reference.)

GRAPHING CALCULATOR EXPLORATIONS

5.1 The Scatter Plot

In Chapter 3 we addressed the basics of setting up a view window and scaling graphs appropriately. You may wish to review those Graphing Calculator Explorations. We will expand on them now in the context of plotting a scatter plot of bivariate data. Here are the steps for creating a scatter plot.

1. You must navigate your calculator's menu system to select the type of graph you want.
2. You must select an appropriate viewing window.
3. You must select data from two lists rather than one.
4. And finally, you must tell the calculator which list will correspond to the horizontal axis and which will correspond to the vertical axis.

Using the data from Example 5.1, we have selected our graphing parameters and scale information as indicated in the accompanying figures. We have set the x and y scales so that the points will fill the view screen, and set the tick marks to be consistent with Figure 5.2. Notice also that we set our plot to be made up of small squares — they are more easily seen than small dots. The resulting plot appears on page 214.

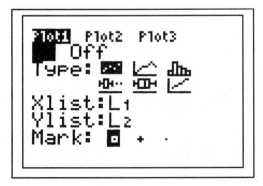

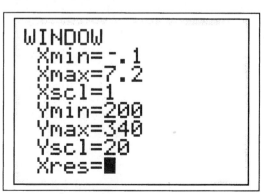

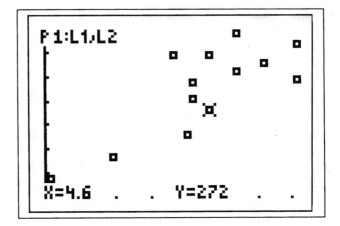

One problem that frequently arises with scatter plots on calculators is the lack of axes (in this case, the horizontal axis) and the lack of a discernible scale on either axis. This occurs because the calculator has been hardwired to plot the *x* axis and *y* axis, and these will not show up if — as with our data — the view screen is not set up for both positive and negative values on each axis. Unfortunately, the demand for axes as reference points competes with the desire for detail in a scatter plot. Although selecting a *y* scale from −1 to 200 would show each axis, the points would cluster close to the top of the screen, possibly hiding some detail or pattern from the data analyst.

There are no easy solutions to this problem, but use of the "trace" capability on your calculator can help a little. You may have encountered the trace button in your math classes when you graphed functions. Pressing the trace button causes the calculator to display the coordinates of the location of a little crosshair icon. You can move this icon about the screen by pressing special "arrow" keys on your calculator. If you are displaying data rather than a function, pressing the trace button and then arrow keys may move the crosshair icon from point to point, displaying the coordinates of the points of your scatter plot. The scatter plot above shows the crosshair icon on the point (4.6, 272).

Your instructor may want you to sketch a scatter plot on a particular assignment, and most likely will want axes and scales as in Figure 5.2. The scaling information is easily found in the viewing window and translated to the sketch of the plot on your paper. And don't forget to put a title on your graph, as well as indicate the variable on each axis of your sketch!

5.2 Linear Regression

It is easy to fall into thinking that a scatter plot and linear regression are always used together, but such is not the case. The scatter plot is a very versatile graph, not limited to its use in fitting a model to data. Your calculator will most likely separate the scatter plot and the regression procedures, requiring that you do two things as you perform a regression and plot the results. The construction of a scatter plot was considered in the previous Graphing Calculator Exploration. Here we point out some of the things you will encounter as you construct a best-fit line.

In general, you will need to perform three tasks to get the best-fit line:

1. Prepare for the plotting of the scatter plot, as discussed previously.
2. Navigate the menus to select linear regression.
3. Transfer the best-fit equation information back to the graphing screen.

Navigating your calculator's menu system means yet another session with your calculator manual. The procedure should end with a choice that looks like "Linear Regression," or possibly "LinReg" for short. When you arrive at that menu option, you should be able to just press the appropriate key, and — voilà! — the best-fit line will be calculated for you. The reason it is so easy is that your calculator has built-in "defaults" to save you keystrokes when possible. So, when you type the final keystroke or select the final menu item, you see

LinReg

What your calculator sees, however, is,

LinReg, List1, List2

which it interprets as "*find the best-fit line assuming that the x values are in List1, and the y values are in List2.*" If you have set up your data in different lists, possibly because you wish to save the data from a previous analysis, you will need to key in the "complete" LinReg command to override the default lists, possibly something like this:

LinReg, List3, List4

Again, you will need to check your manual for the exact syntax, especially to see which variable is assigned the first list in the command. The figures below show the two equations found from the data in Example 5.1. The first equation is found by using the default LinReg command without any parameters — that is, assuming the x data (the TOBEC body fat percentage estimate) is in List1, and the y data (weight of fat pads) is in List2. The second equation comes from reversing the roles of the variables and predicting the TOBEC body fat percentage from the weight of the fat pads. Use these two equations to check and make sure you are entering the List parameters in the correct order.

```
LinReg
  y=a+bx
  a=222.3947689
  b=14.86075349

■
```

```
LinReg
  y=a+bx
  a=-7.489096611
  b=.0414360883

■
```

Now that we have calculated the best-fit line, we want to show it on the scatter plot. There are two ways to do this — one easy and one even easier. The easy way is to simply enter the equation as you would any other function. Then, when the graph is made, the line will appear along with the scatter plot. This is because the "data plotting screen" and the "function plotting screen" share the same memory and the same coordinate system. The easier way will differ from calculator to calculator, but will in general be one of these three possibilities:

1. There may be a "draw" menu option on the screen with the scatter plot. Pressing this will cause the best-fit line to be drawn.

2. You may be able to "copy and paste" the equation stored by the best-fit procedure. This would be similar to how a copy-and-paste operation is done on modern desktop computers.

3. On many calculators this copy-and-paste operation can be directed on the LinReg line when you describe the regression. For example, the LinReg line might look like

   LinReg List1, List2, Y1

which your calculator would interpret as "*find the best-fit line assuming that the x values are in List1, and the y values are in List2. Then, paste that best-fit line into function Y1 for graphing purposes.*"

With these hints you should be able to get the maximum benefit from the best-fit capabilities of your calculator. As your study of bivariate data continues, you will find an astounding number of analyses that your calculator can perform for you. Keep that calculator manual handy, and enjoy the trip!

5.3 The Calculator as Communicator

As many aficionados of both movies and television are aware, when movies are later shown on television there are some compromises made because of the space limitations of TV. *Star Wars* simply doesn't look the same on television! There are also compromises made when the designers of calculators create displays on calculator screens.

When we do "mathematical" work on paper, our instructors usually have certain requirements, which will vary from instructor to instructor, but those requirements usually assume access to 8.5″ by 11″ paper. It can safely be said that not very many calculators will have a screen that size, and this creates problems for calculator manufacturers: What is the best way to present information on a calculator screen? Calculator screens have only so much space for effective display of text characters, and only so much space for graphics. Although the designers of the calculators are very knowledgeable about how mathematics is written, they cannot work miracles and convert from 8.5″ by 11″ paper to a much smaller area. And in any case they cannot anticipate what each instructor will want shown for responses to different questions.

In general, it is our belief that how you solve a problem and report your method of solution should closely mirror what you would write if you were not using a calculator. We believe that when answering a question, your fundamental task is not only to provide an answer, but to communicate the method by which you got that answer. The calculator, because of the compromises discussed above, is singularly unable to communicate effectively. We would like to use the context of regression to discuss some of the differences between what the calculator gives you, and what you should give to your instructor. (You should check with your instructor to make sure what his or her requirements are — we will merely point out that there may well be a mismatch between instructors' requirements and calculators' screen.) When doing linear regression on your calculator, you will probably see something like the accompanying figures for text and graphic output.

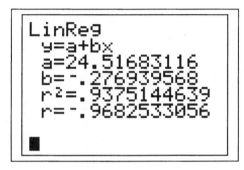 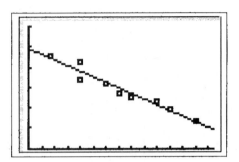

Consider first the regression equation. Notice all those digits? You probably don't want to report all these digits! Also, notice that the equation is reported very generically. To the designer of the calculator screen this makes perfect sense, because with a generic presentation the amount of screen area needed remains constant, irrespective of what variables you are using. However, you should indicate what the variables are. We recommend presenting the regression information as follows:

$$\text{Strength} = 24.52 - .277 \cdot \text{Depth}$$

or

$$\hat{S} = 24.52 - .277D$$

where S and D have been previously defined. Also, report $r = -.97$ or $r = -.968$, a more reasonable choice for number of significant digits.

Suppose now you were asked to predict the strength of a core taken from a depth of $D = 25.0$ mm. We suggest that you communicate a method, not just answer the question! Here's a possible template with notes about what each line is saying:

$\hat{S} = 24.52 - 0.277D$ *("Here's the formula I'm using . . .")*

$= 24.52 - 0.277(25.0)$ *("Here are numbers I'm substituting . . .")*

$\hat{S} = 17.60$ MPa   *("Here's my result.")*

The advantages of this presentation are:

1. If you err in your keystrokes, your instructor can still evaluate your method and assess your understanding.

2. If your answer differs from that of your fellow statistics students, you can easily tell whether you made a conceptual error or a keystroke error without having to take time to duplicate your work.

3. Most important from our perspective, you are communicating in the language of mathematics, where reliable communication demands a standard symbolism.

Now consider the scatter plot as it appears on the calculator screen. The most serious omission is the scale on the two axes. From looking at the graph, one cannot tell how large or small the numbers are. Also, there is no indication of what the variables are. This is certainly understandable on a calculator screen, where all this information would take up precious space, but it is not sufficient if you are sketching a plot on a homework assignment. We recommend that every graph have a title, indicate units, and show any other relevant information, as in the accompanying figure.

$$\text{Strength} = 24.52 - .277 \, \text{Depth}$$
$$r = -.97$$

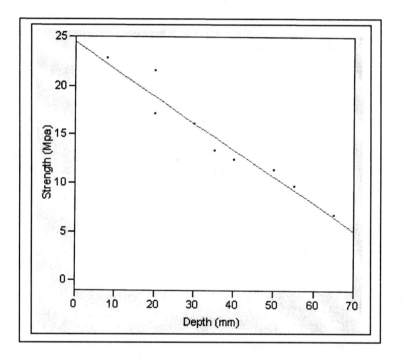

Our message to the student using a graphing calculator is really very simple: You must realize not only the power of the calculator to save you time and effort, but also its limitations of communication.

5.4 In the Matter of Residuals

Different calculators handle residuals in a variety of ways. One calculator may store the residuals but not let you work with them; another calculator may not even store the residuals. In this Graphing Calculator Exploration we would like to offer several "workarounds" related to residuals. Analyzing the residuals is, of course, an essential part of regression analysis. We assume that your calculator is list-based, and that you can perform calculations using those lists.

First, suppose that your calculator does not store the residuals. Let's see how we might get around this problem. We will suppose that the x variable data has been stored in List1, and the y variable data in List2, and that you have successfully performed the appropriate regression calculations, giving the best-fit line $\hat{y} = a + bx$. (We will use simple linear regression, but the workarounds we discuss here will work just as well for polynomial regression and nonlinear regression.)

Once you have the best-fit line, you can calculate the residuals and put them in, say, List3 as follows:

List2 $-$ (a + b * (List1)) $\rightarrow$ List3

As you have probably already seen, we are simply using the definition of the residual in our calculations. This "List algebra" corresponds to our usual algebra, which would be

$$y_i - \hat{y}_i \rightarrow \text{residual list}$$

or

$$y_i - (a + bx_i) \rightarrow \text{residual list}$$

Once we have the residuals stored in this list, we can plot them and use them in other ways as well.

A second problem we would like to address is the calculation of the standard deviation about the least squares line, $s_e = \sqrt{SSResid/(n - 2)}$. If your calculator does not compute this quantity for you, it turns out that you can obtain it directly from information at hand: the number of points, n, the sample standard deviation of y, s_y, and the sample correlation coefficient, r. We will skip the straightforward but slightly tedious algebra and cut directly to the result:

$$s_e = \sqrt{\frac{n - 1}{n - 2}}\sqrt{1 - r^2}s_y$$

With these two workarounds, you should be able to circumvent a slightly statistically challenged calculator and work with residuals with the best of them!

5.5 How Does Your Calculator Perform Nonlinear Regression?

Scientific graphing calculators come with built-in significant statistical capability, including linear regression. Some will also have options for nonlinear regression, such as exponential regression ($\hat{y} = ab^x$), power regression ($\hat{y} = ax^b$), logarithmic regression ($\hat{y} = a + \ln b$), as well as polynomial regression. As we have mentioned, the process of transforming data, fitting a line to the transformed data, and then "undoing" the transformation is used to get a good — though not "best" in the least squares sense — fit of a function to the data.

Depending on which calculator and which nonlinear fit you have chosen to utilize, your calculator may transform variable(s) and fit a line to the transformed data, or it may use more complicated methods to estimate the least squares solution. It is also possible that your calculator may produce unusual residual plots. This is not necessarily a bad thing, but does place some responsibility on your shoulders. Although your calculator does the arithmetic, it is *you* who are responsible for understanding and interpreting what your calculator is telling you.

We will use exponential regression to show how you might explore the calculator's output. We will compare two approaches to the regression: using linear regression on the transformed data, and using the "exp reg" button on the calculator. Since your calculator may differ from ours, we will continue to avoid specific keystrokes. You should be able to reproduce this analysis for different nonlinear regressions. If you choose to do reexpression "by hand" (transforming the data, fitting a line to the transformed data, and then back-transforming), you do not need to worry about what the calculator "button" does. This analysis needs to be done only if you wish to get quicker answers by using your calculator's built-in nonlinear regression capabilities.

The strategy we will use is to pick pairs of points with the same x values, and choose y values on each side of our model, $f(x) - k$ and $f(x) + k$. Choosing points in this manner will produce data that will force a least squares solution through $f(x)$. Generating data using the function $f(x) = e^x$, x values from 2 to 5, and $k = 5$ results in the data shown in the x and y columns of the accompanying table. Because of the way the data was generated, the least squares exponential function is $\hat{y} = e^x$.

Let's transform the data, fit a line, and back-transform, and then compare the result to what the calculator gives when an exponential regression is requested. If the relationship between x and y is exponential (as it is here because of the way in which the data was generated), a log transformation would be an appropriate transformation to create a linear relationship. Taking the natural logarithm of y results in the transformed y values in the $\ln(y)$ column of the table. Using the calculator to fit a line to the $(x, \ln(y))$ data gives

$$\ln y = -.416303559 + 1.094450382x \quad \text{with } r^2 = .885784757$$

Back-transforming gives the nonlinear relationship

$$y = .659480049 * 2.987540226^x$$

Note that this is different from the least squares solution of $y = e^x$, but it still provides a reasonable fit to the data.

x	y (Formula)	y (Decimal)	$\ln(y)$	Residual (Calculator)	Residual (Formula)
2	$e^2 + 5$	12.38905610	2.516813510	.7442163052	6.502935110
2	$e^2 - 5$	2.38905610	.870898350	−.901698854	−3.4970648
3	$e^3 + 5$	25.08553692	3.22229146	.3552438748	7.50051369
3	$e^3 - 5$	15.08553692	2.71373646	−.153311121	−2.4994863
4	$e^4 + 5$	59.59815003	4.08762453	.1261265648	7.06218576
4	$e^4 - 5$	49.59815003	3.90395353	−.057544433	−2.9378142
5	$e^5 + 5$	153.4131591	5.03313466	−.022813682	−3.5401474
5	$e^5 - 5$	143.4131591	4.96572968	−.090218662	−13.540147

The accompanying figure shows a scatter plot of ln(y) versus x and a plot of the residuals for the linear fit to the transformed data. The residual plot certainly does not look "random," but for our present purposes this is not a problem. The calculator reports the residuals from this regression, and we have reproduced these in the Residual (Calculator) column in the previous table.

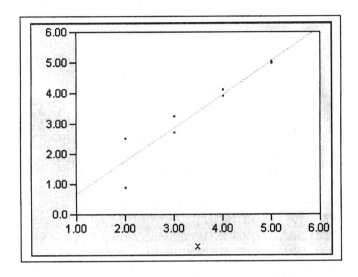

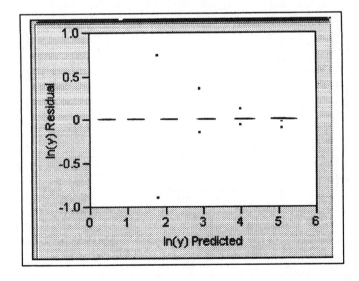

Now try the "exp reg" function on your calculator. You should find that the reported regression is $y = .659480049 \cdot 2.987540226^x$ with $r^2 = .886$, the same values that were obtained by transforming and then fitting a linear function to the transformed data. Note that the calculator did not produce the least squares fit of $y = e^x$.

If you now request a scatter plot and a residual plot, you will see something like the plots below.

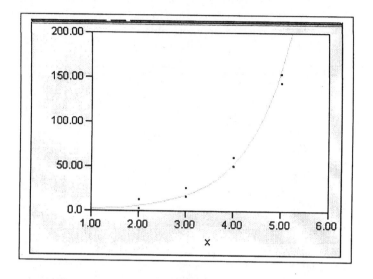

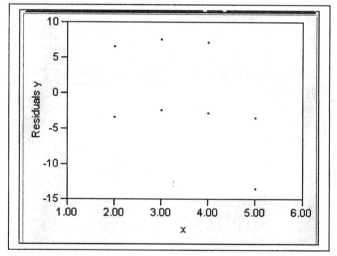

The scatter plots of the data look different, which would be expected because the second plot is of the untransformed data. However, the pattern of residuals is very different. This is because the calculator is computing the residuals by using the fitted equation to get a predicted y and then computing residual = y − predicted y, rather than using the residuals from the linear fit to the transformed data, $\ln(y)$ − predicted $\ln(y)$. Another fact to note is that the r^2 reported by the exp reg button analysis is the r^2 for the *transformed* fit, even though the residuals reported are in the *untransformed* scale. The calculator is doing the transformations for us, which seems reasonable. However, the residuals from the linear regression are not the ones presented; moreover, the residuals that are reported have a very different pattern than those from the linear regression of the transformed data.

The lesson from all these calculations is a very simple one: If you want to use the built-in capabilities of the calculator, you need to understand what it is actually doing. The book that comes with the calculator may not be particularly clear, and you should perform an example such as the one above to be sure what is happening under the buttons! Remember, *you* are ultimately responsible for your analyses. Your calculator can only perform really quick arithmetic — you, the analyst, must provide the careful thinking and clear understanding of that arithmetic.

6

Probability

INTRODUCTION

Every day, each of us makes decisions based on uncertainty. Should you buy an extended warranty for your new CD player? It depends on the likelihood that it will fail during the warranty period. Should you put off taking a required course until the last term before graduation? It depends on the likelihood that all sections will be full before it is your turn to register and whether the course will be offered during that term. Should you allow 45 minutes to get to your 8 A.M. class, or is 35 minutes enough? From experience, you may know that most mornings you can drive to school and park in 25 minutes or less. Most of the time, the walk from your parking space to class is 5 minutes or less. But how often will the drive to school or the walk to class take longer than you expect? How often will both take longer? When it takes longer than usual to drive to campus, is it more likely that it will also take longer to walk to class? Less likely? Or are the driving and walking times unrelated? Some questions involving uncertainty are more serious: If an artificial heart has four key systems, how likely is each one to fail? How likely is it that at least one will fail? If a satellite has a backup solar power system, how likely is it that both the main and the backup components will fail? We can answer questions like these using the ideas and methods of probability, the systematic study of uncertainty.

6.1   Chance Experiments and Events

The basic ideas and terminology of probability are most easily introduced in situations that are both familiar and reasonably simple. Thus some of our initial examples

225

will involve such elementary activities as tossing a coin once or several times, selecting one or more cards from a deck, and rolling a single die or several dice. It may seem to you at times that cards and dice are almost the only subjects of concern to those who study probability. We will attempt to convince you otherwise by interweaving practical probability examples with the intuitively clear examples involving games of chance. Then we will move to more interesting and realistic situations.

Chance Experiments

When a single coin is tossed, it can land with its head side up or its tail side up. The selection of a single card from a well-mixed standard deck may result in the ace of spades, the five of diamonds, the jack of hearts, or any one of the other 49 possibilities. Consider rolling both a red die and a green die. One possible outcome is the red die lands with four dots facing up and the green die shows one dot on its upturned face. Another outcome is the red die lands with three dots facing up and the green die also shows three dots on its upturned face. There are 36 possible outcomes in all, and we do not know in advance what the result of a particular roll will be. Situations such as this are referred to as *chance experiments*.

Definition

A **chance experiment** is any activity or situation in which there is uncertainty about which of two or more possible outcomes will result.

To someone with visions of scientists in lab coats skillfully injecting test mice with new miracle drugs and carefully recording the reactions, rolling a die might not seem like much of an experiment. But the previous definition encompasses experiments both trivial and significant. In an opinion poll or survey, there is uncertainty about whether an individual in a sample supports a school bond. In a laboratory investigation, there is uncertainty about what effect varying the amount of an active ingredient in a drug has on the mice. And in an experiment in which a die is rolled, there is uncertainty about which face will land upturned. Each of these fits the definition of a chance experiment.

Consider an experiment designed to see whether men and women have different shopping preferences when buying a CD at a music store. Four types of music are sold at the store: classical, rock, country, and "other" (all other types of music). For this experiment, a randomly selected shopper will be asked what type of CD has been purchased, and the shopper's gender will be noted by the investigator. Before we interact with the shopper, the outcome of this experiment is unknown to us. We do know, however, what the possible outcomes are. This set of possible outcomes is called the *sample space*.

Definition

The collection of all possible outcomes of a chance experiment is the **sample space** for the experiment.

There are many ways to represent the sample space of an experiment. An obvious representation is a simple list of all the possible outcomes. For the CD purchase experiment, these are the possible outcomes:

1. A male buying classical
2. A female buying classical
3. A male buying rock
4. A female buying rock
5. A male buying country
6. A female buying country
7. A male buying "other"
8. A female buying "other"

For brevity, we can use set notation and ordered pairs. A male purchasing a classical CD would be represented as (male, classical). The sample space would then be

$$\text{sample space} = \left\{ \begin{array}{l} \text{(male, classical), (female, classical), (male, rock), (female, rock),} \\ \text{(male, country), (female, country), (male, other), (female, other)} \end{array} \right\}$$

Another representation of the sample space, this time as a picture, is a "tree" diagram. A tree diagram (shown in Figure 6.1) for the outcomes of the CD shopper experiment has two sets of "branches" corresponding to the two pieces of

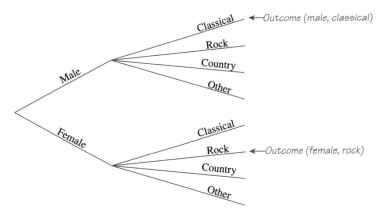

FIGURE 6.1 Tree diagram for the CD purchase example

information that were gathered. To identify any particular element of the sample space, you traverse the tree by first selecting a branch corresponding to gender and then a branch identified with a type of CD.

In the tree diagram of Figure 6.1, there is no particular reason for having the gender in the first branch and the type of CD in the second branch. Some experiments involve making observations in a particular order, in which case the order of the branches in the tree will matter. In this example, however, the order of appearance does not have any particular interpretation or meaning. It would be perfectly acceptable to represent the sample space with the tree diagram shown in Figure 6.2.

As we have seen, the sample space can be represented in several ways, but they all have one thing in common: *Every* element of the sample space is listed, represented, or drawn in some form.

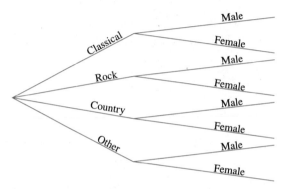

FIGURE 6.2 Another tree diagram for the CD purchase example

Events

In the survey of CD purchasers, we might be interested in which particular outcome will result. Or we may focus on whether the outcomes will involve the purchase of classical music, i.e., in the group of outcomes consisting of (male, classical) and (female, classical). Or we may focus attention on those outcomes where the shopper is male. When we combine one or more individual outcomes in a collection, we are creating what is known as an *event*.

Definition

An **event** is any collection of outcomes from the sample space of a chance experiment.

A **simple event** is an event consisting of exactly one outcome.

We usually represent an event by an uppercase letter, such as *A*. Sometimes different letters are used to denote different events — *A, B, C,* and so on — but on other occasions, the same letter with different numerical subscripts, such as E_1, E_2, E_3, . . . , is used for this purpose.

EXAMPLE 6.1   Reconsider the situation where shoppers were categorized by gender (M or F) and type of music purchased (C = classical, R = rock, K = country, O = other). Using this notation, one possible representation of the sample space is

sample space = {MC, FC, MR, FR, MK, FK, MO, FO}

Because there are eight outcomes, there are eight simple events:

$$O_1 = MC \qquad O_2 = FC \qquad O_3 = MR \qquad O_4 = FR$$
$$O_5 = MK \qquad O_6 = FK \qquad O_7 = MO \qquad O_8 = FO$$

One event of interest might be the event consisting of all outcomes for which a classical CD was purchased. A symbolic description of the event *classical* is

classical = {MC, FC}

Another event is the event that the purchaser is male,

male = {MC, MR, MK, MO}

EXAMPLE 6.2   Suppose you are interested in studying "demoralization" in golf. You believe that after losing a game to an opponent, a golfer is more likely to lose the next game because he or she is demoralized after losing the first game. Your experiment might consist of watching two consecutive games of play, and observing whether a player won, tied, or lost each of the two games. In this case (using W, T, and L to represent win, tie, lose), the sample space can be represented as

sample space = {WW, WT, WL, TW, TT, TL, LW, LT, LL}

The event *lose exactly one game of the pair,* denoted by L_1, could then be defined as

$$L_1 = \{WL, TL, LW, LT\}$$

In general, only one outcome, and thus one simple event, occurs when an experiment is performed. We say that a given event occurs whenever one of the outcomes making up the event occurs. If the outcome in Example 6.1 is MC, then the simple event *a male purchasing classical music* has occurred, and so has the non-simple event *classical music purchased.*

Forming New Events

Once some events have been specified, there are several useful ways of manipulating them to create new events. Some of these are presented in the box on page 230.

Definition

Let A and B denote two events.

1. The event **not A** consists of all experimental outcomes that are not in event A. *Not A* is sometimes called the *complement* of A and is usually denoted by A^c, A', or possibly $\overline{A}$.

2. The event **A or B** consists of all experimental outcomes that are in at least one of the two events, that is, in A or in B or in both of these. *A or B* is called the *union* of the two events and is denoted by $A \cup B$.

3. The event **A and B** consists of all experimental outcomes that are in both of the events A and B. *A and B* is called the *intersection* of the two events and is denoted by $A \cap B$.

EXAMPLE 6.3 An observer stands at the bottom of a freeway off-ramp and records the turning direction (L or R) of each of three successive vehicles. The sample space contains eight outcomes:

$$\{LLL, RLL, LRL, LLR, RRL, RLR, LRR, RRR\}$$

Each of these outcomes determines a simple event. Other events include

A = event that exactly one of the cars turn right
 = $\{RLL, LRL, LLR\}$

B = event that at most one of the cars turns right
 = $\{LLL, RLL, LRL, LLR\}$

C = event that all cars turn in the same direction
 = $\{LLL, RRR\}$

Some other events that can be formed from those just defined are

not C = event that not all cars turn in the same direction
 = C^c
 = $\{RLL, LRL, LLR, RRL, RLR, LRR\}$

A or C = event that either exactly one of the cars turns right or all cars turn in the same direction
 = $A \cup C$
 = $\{RLL, LRL, LLR, LLL, RRR\}$

B and C = event that at most one car turns right and all cars turn in the same direction
 = $B \cap C$
 = $\{LLL\}$

EXAMPLE 6.4 (Example 6.2 continued)   In the golf example, in addition to the event that a golfer loses exactly one of the two games $L_1 = \{WL, TL, LW, LT\}$, we could also define events corresponding to L_0 = neither game is lost and L_2 = both games are lost. Then

$$L_0 = \{WW, WT, TW, TT\} \quad \text{and} \quad L_2 = \{LL\}$$

Some other events that can be formed from those just defined include

$$L_2^c = \text{event that at most one game was lost}$$
$$= \{\text{WW, WT, WL, TW, TT, TL, LW, LT}\}$$

$$L_1 \cup L_2 = \text{event that at least one game was lost}$$
$$= \{\text{WL, TL, LW, LT, LL}\}$$

$$L_1 \cap L_2 = \text{event that exactly one game was lost } and \text{ two games were lost} = \{\ \},$$

where $\{\ \}$ denotes the empty set.

It frequently happens, as was the case for events A and C in Example 6.3, and L_1 and L_2 in the golf example, that two events have no common outcomes. Such situations are described by special terminology.

> **Definition**
>
> Two events that have no common outcomes are said to be **disjoint** or **mutually exclusive.**

Two events are disjoint if their intersection is the empty set, so any two *simple* events are disjoint.

It is sometimes useful to draw an informal picture of events to visualize relationships. In a **Venn diagram,** the collection of all possible outcomes is typically shown as the interior of a rectangle. Other events are then identified with specified regions inside this rectangle. Figure 6.3 illustrates several Venn diagrams.

FIGURE 6.3 Venn diagrams

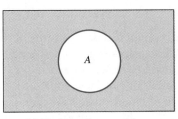

(a) Shaded region = *not A*

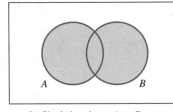

(b) Shaded region = *A or B*

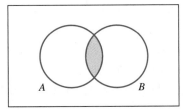

(c) Shaded region = *A and B*

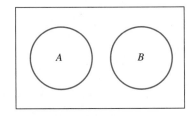

(d) Two disjoint events

The use of the *or* and *and* operations can be extended to form new events from more than two initially specified events.

Definition

Let $A_1, A_2, \ldots, A_k$ denote k events.

1. The event A_1 *or* A_2 *or* ... *or* A_k consists of all outcomes in at least one of the individual events $A_1, A_2, \ldots, A_k$.

2. The event A_1 *and* A_2 *and* ... *and* A_k consists of all outcomes that are simultaneously in every one of the individual events $A_1, A_2, \ldots, A_k$.

These k events are disjoint if no two of them have any common outcomes.

Venn diagrams illustrating these concepts are shown in Figure 6.4.

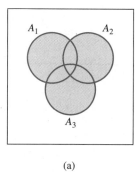   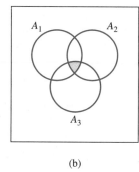   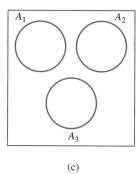

(a)   (b)   (c)

FIGURE 6.4 Venn diagrams: (a) A_1 *or* A_2 *or* A_3; (b) A_1 *and* A_2 *and* A_3; (c) Three disjoint events

EXAMPLE 6.5   The instructor in a seminar class consisting of four students has an unusual way of asking questions. Four slips of paper numbered 1, 2, 3, and 4, respectively, are placed in a box. The student to whom any particular question is to be addressed is then determined by selecting one of these four slips. Suppose that one question is to be posed during each of the next two class meetings. One possible outcome could be represented as (3, 1) — the first question is addressed to student 3 and the second to student 1. There are 15 other possibilities. Consider the following events:

$A = \{(1, 1), (2, 2), (3, 3), (4, 4)\}$

$B = \{(1, 1), (1, 2), (1, 3), (1, 4), (2, 1), (3, 1), (4, 1)\}$

$C = \{(3, 1), (2, 2), (1, 3)\}$

$D = \{(3, 3), (3, 4), (4, 3)\}$

$E = \{(1, 1), (1, 3), (2, 2), (3, 1), (4, 2), (3, 3), (2, 4), (4, 4)\}$

$F = \{(1, 1), (1, 2), (2, 1)\}$

Then

$$A \text{ or } C \text{ or } D = \{(1, 1), (2, 2), (3, 3), (4, 4), (3, 1), (1, 3), (3, 4), (4, 3)\}$$

The outcome $(3, 1)$ is contained in each of the events B, C, and E, as is the outcome $(1, 3)$. These are the only two common outcomes, so

$$B \text{ and } C \text{ and } E = \{(3, 1), (1, 3)\}$$

The events C, D, and F are disjoint because no outcome in any one of these events is contained in either of the other two events.

▬▬▬▬▬▬▬

Exercises 6.1 – 6.13

▬▬▬▬

6.1 Define the term *chance experiment* and give an example of an experiment with four possible outcomes.

6.2 Define the term *sample space* and then give the sample space for the experiment you described in Exercise 6.1.

6.3 Consider the experiment in which the type of transmission — automatic (A) or manual (M) — is recorded for each of the next two cars purchased from a certain dealer.
  a. What is the set of all possible outcomes (the sample space)?
  b. Display the possible outcomes in a tree diagram.
  c. List the outcomes in each of the following events: B = the event that at least one car has an automatic transmission; C = the event that exactly one car has an automatic transmission; D = the event that neither car has an automatic transmission. Which of these events are simple events?
  d. What outcomes are in the event B *and* C? In the event B *or* C?

6.4 A tennis shop sells five different brands of rackets, each of which comes in either a midsize version or an oversize version. Consider the experiment in which brand and size are noted for the next racket purchaser. Possible outcomes correspond to cells in the accompanying rectangular table.

For example, one possible outcome is Head midsize, and another is Prince oversize.
  a. Let A denote the event that an oversize racket is purchased. List the outcomes in A.
  b. Let B denote the event that the name of the brand purchased begins with a W. List the outcomes in B.
  c. Refer to part **b,** and list the outcomes in the event *not B.*
  d. Head, Prince, and Wilson are U.S. companies. Let C denote the event that the racket purchased is made by a U.S. company. List outcomes in the event B *or* C.
  e. Refer to part **d,** and list outcomes in B *and* C.
  f. Display the possible outcomes on a tree diagram, with a first-generation branch for each brand.
  g. Suppose that the store stocks each racket in three grip sizes: 3, 4, and 5 (for $4\frac{3}{8}$ in., $4\frac{1}{2}$ in., and $4\frac{5}{8}$ in.). If brand, head size, and grip size are all noted, display possible outcomes on a tree diagram.

6.5 Consider the experiment in which an automobile is selected and both the number of defective headlights (0, 1, or 2) and the number of defective tires (0, 1, 2, 3, or 4) are determined.
  a. Display possible outcomes using a tree diagram.
  b. Let A be the event that at most one headlight is defective and B be the event that at most one tire is defective. What outcomes are in A^c? in $A \cup B$? in $A \cap B$?

Table for Exercise 6.4

	Head	Prince	Slazenger	Wimbledon	Wilson
Midsize					
Oversize					

c. Let C denote the event that all four tires are defective. Are A and C disjoint events? Are B and C disjoint?

6.6 A college library has four copies of a certain book; the copies are numbered 1, 2, 3, and 4. Two of these are randomly selected. The first book selected will be placed on 2-hour reserve, and the second one may be checked out on an overnight basis.
a. Construct a tree diagram to display the 12 outcomes in the sample space.
b. Let A denote the event that at least one of the books selected is an even-numbered copy. What outcomes are in A?
c. Suppose that copies 1 and 2 are first printings, whereas copies 3 and 4 are second printings. Let B denote the event that exactly one of the copies selected is a first printing. What outcomes are contained in B?

6.7 A library has five copies of a certain text on reserve of which two copies (1 and 2) are first printings and the other three (3, 4, and 5) are second printings. A student examines these books in random order, stopping only when a second printing has been selected.
a. Display the possible outcomes in a tree diagram.
b. What outcomes are contained in the event A, that exactly one book is examined before the experiment terminates?
c. What outcomes are contained in the event C, that the experiment terminates with the examination of book 5?

6.8 Suppose that starting at a certain time, batteries coming off an assembly line are examined one by one to see whether they are defective (let D = defective and N = not defective). The experiment terminates as soon as a nondefective battery is obtained.
a. Give five possible experimental outcomes.
b. What can be said about the number of outcomes in the sample space?
c. What outcomes are in the event E, that the number of batteries examined is an even number?

6.9 Refer to Exercise 6.8, and now suppose that the experiment terminates only when two nondefective batteries have been obtained.
a. Let A denote the event that at most three batteries must be examined in order to terminate the experiment. What outcomes are contained in A?
b. Let B be the event that exactly four batteries must be examined before the experiment terminates. What outcomes are in B?
c. What can be said about the number of possible outcomes?

6.10 A family consisting of three people — P_1, P_2, and P_3 — belongs to a medical clinic that always has a physician at each of stations 1, 2, and 3. During a certain week, each member of the family visits the clinic exactly once and is randomly assigned to a station. One experimental outcome is (1, 2, 1), which means that P_1 is assigned to station 1, P_2 to station 2, and P_3 to station 1.
a. There are 27 possible outcomes; list them. (*Hint*: First list the nine outcomes in which P_1 goes to station 1, then the nine in which P_1 goes to station 2, and finally the nine in which P_1 goes to station 3; a tree diagram might help.)
b. List all outcomes in the event A, that all three people go to the same station.
c. List all outcomes in the event B, that all three people go to different stations.
d. List all outcomes in the event C, that no one goes to station 2.
e. Identify outcomes in the following events: B^c, C^c, $A \cup B, A \cap B, A \cap C$.

6.11 An engineering construction firm is currently working on power plants at three different sites. Define events E_1, E_2, and E_3 as follows:

E_1 = the plant at site 1 is completed by the contract date

E_2 = the plant at site 2 is completed by the contract date

E_3 = the plant at site 3 is completed by the contract date

The accompanying Venn diagram pictures the relationships among these events.

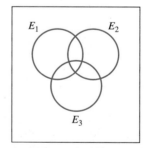

Shade the region in the Venn diagram corresponding to each of the following events (redraw the Venn diagram for each part of the problem).
a. At least one plant is completed by the contract date.
b. All plants are completed by the contract date.
c. None of the plants is completed by the contract date.
d. Only the plant at site 1 is completed by the contract date.
e. Exactly one of the three plants is completed by the contract date.

f. Either the plant at site 1 or both of the other two plants are completed by the contract date.

6.12 Consider a Venn diagram picturing two events *A* and *B* that are not disjoint.
  a. Shade the event $(A \cup B)^c$. On a separate Venn diagram shade the event $A^c \cap B^c$. How are these two events related?
  b. Shade the event $(A \cap B)^c$. On a separate Venn diagram shade the event $A^c \cup B^c$. How are these two events related?
(*Note*: These two relationships together are called De-Morgan's laws.)

6.13 With the marked increase in the use of word processors in recent years, there is some question about a decline in typing skills — since errors are easy to correct, people may spend less energy avoiding them. The Ace Secretary School would like to study this question, and has developed a typing test and a rating system as follows:

Excellent: at most 5 errors per thousand keystrokes

Good: between 5 and 10 errors per thousand keystrokes

Fair: between 10 and 20 errors per thousand keystrokes

a. If we consider the administration of a test to a single individual to be a chance experiment and the outcomes to be the number of errors made per thousand keystrokes (rounded to the nearest integer), what is the sample space for this experiment?
b. Define events *E* (Excellent), *G* (Good), and *F* (Fair) consistent with the above rating system, and list the outcomes that make up these events.
c. There is a certain ambiguity in the meaning of the word *between* in the classification above. In your solution to part **b**, did you assign the outcomes of the sample space to events *E*, *G*, and *F* such that these events are disjoint? How would you reword the categories above to eliminate the ambiguity and make the events disjoint?
d. Make a Venn diagram and carefully label the sample space and the events *E*, *G*, *F*. If we consider a newly defined event, *P* (Poor) where poor is defined to be more than 20 errors per thousand keystrokes, what region in the Venn diagram corresponds to the event *P*?

6.2   The Definition of Probability

Having introduced the basic ideas and definitions of sample space and events in the previous section, we are now in a position to define the notion of probability. If you have previously encountered probability, you may have approached probability as a mathematical subject, complete with definitions, axioms, and theorems. We can also approach probability from an empirical point of view, in the context of collecting data.

Probabilistic reasoning of some sort may well be as old as man. Archeologists have found evidence that Egyptians used a small bone in mammals, the astragalus, as a sort of four-sided die as early as 3500 B.C. Games of chance were common in Greek and Roman times, as well as during the Renaissance of western Europe. Girolamo Cardano, a character of some ill-repute in early mathematical circles, wrote about such games and did some calculations of odds in the 16th century; Galileo mentioned dice in his writings in the early 17th century.

Probability is traditionally thought of as beginning with the correspondence between Blaise Pascal (1623–62) and Pierre de Fermat (1601–65) in 1654. Their exchange of letters discussed some problems related to gambling that were posed by a French nobleman Chevalier de Mère. The methods and solutions that resulted from this exchange, especially techniques of combinatorics (advanced counting), greatly enhanced the study of probability. From these beginnings to the later writings of the mathematician Pierre Simon de Laplace (1749–1827), the techniques, methods, and theory now known as classical probability grew to be a major contribution to the quantification of uncertainty.

From this early work, new interpretations of probability have evolved, including an empirical approach preferred by many statisticians today. We will begin our discussion of probability with a look at some of the different approaches to probability.

The Classical Approach to Probability

The mathematical study of probability developed in response to questions about gambling. Early mathematicians' development of the theory of probability was primarily in that context, and their evolving understanding reflected the peculiarities of games of chance. Perhaps the characteristic most common to games of chance is a physical device used to generate the different outcomes. In children's games, a "spinner" might be used to create chance outcomes. In Monopoly, dice are used to determine by chance the number of squares to move on the board. In other games, the sequence of events is determined by playing cards. In the 17th century, the most popular gaming devices were dice. Dice were constructed so that the physical characteristics of each face were alike (except of course for the number of dots), ensuring that the different outcomes for an individual die were very close to equally likely. Therefore, it seemed quite natural for mathematicians of the time to assume, for example, that the probability of getting a "five" when a single six-sided die was rolled was one-sixth (one chance in six). Generally, if there are N *equally likely* outcomes in a game of chance, the probability of each of the outcomes is defined as $1/N$. The early probabilists often referred to "honest" dice or "fair" coins as physical devices that seemed to obey this assumption of equal probabilities. These early efforts led to what is known as the "classical" definition of probability.

Classical Approach to Probability

The **probability of an event E,** denoted by $P(E)$, is ratio of the number of outcomes favorable to E to the total number of outcomes in the sample space:

$$P(E) = \frac{\text{number of outcomes favorable to } E}{\text{number of outcomes in the sample space}}$$

According to this definition, the calculation of a probability consists of counting the number of outcomes that make up an event, counting the number of outcomes in the sample space, and then dividing.

Important: This method for calculating probabilities is only appropriate when the outcomes of an experiment are equally likely.

Experiments involving tossing fair coins, rolling fair dice, or selecting cards from a well-mixed deck have equally likely outcomes. For example, if a fair die is rolled once, each outcome (simple event) has probability 1/6. With E denoting the event that the number rolled is even, $P(E) = 3/6$. This is just the number of outcomes in E divided by the total number of possible outcomes.

EXAMPLE 6.6

On some football teams, the honor of calling the toss at the beginning of a football game is determined by random selection. Suppose that this week a member of the offensive team will call the toss. There are 5 interior linemen on the 11-player team. If we define the event L as the event that a lineman is selected to call the toss, 5 of the 11 possible outcomes would be included in L. The probability that a lineman will be selected is then

$$P(L) = \frac{5}{11}$$

EXAMPLE 6.7

Four students (Adam, Betina, Carlos, and Debra) have submitted correct solutions to a math contest with two prizes. The contest rules specify that if more than two correct responses are submitted, the winners will be selected at random from those submitting correct responses. In this case, possible outcomes for the experiment that consists of selecting the two winners from the four correct responses are:

$$\{(A, B), (A, C), (A, D), (B, C), (B, D), (C, D)\}$$

Since the winners are selected at random, the six outcomes are equally likely, and the probability of each individual outcome is 1/6.

Let E be the event that both selected winners are the same gender. Then

$$E = \{(A, C), (B, D)\}$$

Since E contains two outcomes, $P(E) = 2/6 = .333$. If F denotes the event that at least one of the selected winners is female, then F consists of all outcomes except (A, C) and $P(F) = 5/6 = .833$.

Limitations of the Classical Approach to Probability

The classical approach to probability works very well with games of chance or other situations with a finite set of outcomes that may be regarded as equally likely. However, some situations that are clearly probabilistic in nature do not fit the classical model. Consider the accident rates of young adults driving cars. Information of this kind is used by insurance companies to set car insurance rates. Suppose we have two individuals, one 18 and one 28. A person purchasing a standard accident insurance policy at age 28 should have a lower annual premium than a person purchasing the same policy at 18, because the 28-year-old would seem to have a much lower chance of an accident, based on prior experience. However, from the classical probability standpoint, there is no reasonable basis for calculating a probability of an accident during the policy year for an individual at age 28. The classical probabilist, playing by the classical rules, would have to consider that there are two outcomes for the year — having an accident or not having an accident — *and those outcomes are equally likely for both the 28-year-old and the 18-year-old.* Moreover, by the assumption of equal probabilities, those probabilities would be equal to $\frac{1}{2}$! This certainly seems to defy reason and experience.

Another example of a problem that cannot be handled using the classical approach is one with a geometric flavor. Suppose that you have a square that is 8 inches on each side and that a circle with a diameter of 8 inches is drawn inside the square, touching the square on all four sides. This will be used as a target in a dart game. A dart will be thrown at the target from a specified distance, and any that completely miss the square are "do-overs" and will be thrown again. It seems reasonable to think about the probability of a dart landing in the circle as a ratio of the area of the circle to the area of the square. That ratio is

$$\frac{\text{area of circle}}{\text{area of square}} = \frac{\pi r^2}{s^2} = \frac{\pi(4)^2}{8^2} = \frac{\pi}{4} \approx .785$$

A practiced dart player who was aiming for the center of the target might have an even greater probability of hitting the inside of the circle. In any case, it is not reasonable to think that the two outcomes *inside the circle* and *not inside the circle* are equally likely, so the classical approach to probability will not help here.

From the standpoint of statistics, a major limitation of the classical approach to probability is that there seems to be no way to convert past experience into an expectation about the future. To return to the car insurance example, suppose that in the experience of the insurance companies, .5% of 18-year-olds have accidents during their 18th year. It seems reasonable that, all things being equal, we could expect this percentage to be stable enough to set the prices for an insurance policy. Although the assumption of equal probabilities is based on and consistent with the experience with dice, classical probability had no method for observing a proportion in real life and constructing a probability based on that observation. For a statistician, this is a crucial defect. Statisticians routinely convert information from a sample survey into statements about the whole population. Without an understanding of probability that allows such a generalization, statistical inference would be impossible.

The Relative Frequency Approach to Probability

After the revolutionary writings of Sir Isaac Newton (1642–1727), modern science became solidly based on observation and the recording of data. Early scientific investigators were certainly aware that experiments and observations do not always give the same results when repeated. Even in the most carefully replicated experiment, there is variation! The idea that rational scientific activity would exhibit the behaviors of chance events was not surprising to scientists. After all, series of dice rolls were not *perfectly* consonant with the calculated classical probabilities, they were only very close.

Early on, it was noticed that the unpredictability of an individual chance experiment gave rise to a dependably stable regularity when an experiment was repeated many times. This was the basis for "fair" wagering in games of chance. For example, suppose two friends, Chris and Jay, meet to play a game. A fair coin is flipped. If it lands heads up, Chris pays Jay $1, otherwise Jay pays the same amount to Chris. After many repetitions, the proportion of the time that Chris wins will be very close to one-half, and the two friends will have simply enjoyed the pleasure of each other's company for a few hours. This happy circumstance occurs because in the long run (i.e., over many repetitions) the proportion of heads would be .5; half the time Chris would win, half the time Jay.

When any given chance experiment is performed, some events are relatively likely to occur, whereas others are not so likely to occur. For a specified event E, we want to assign a number to this event that gives a precise indication of how likely it is that E will occur. In the empirical approach to probability, the probability $P(E)$ will depend on how frequently E occurs when the experiment is performed repeatedly.

EXAMPLE 6.8 Consider the simple chance experiment of tossing a coin just once. Define events H and T for this chance experiment by

H = event that the coin lands with its head side facing up

T = event that the coin lands with its tail side facing up

Frequently, we hear a coin described as fair, or we are told that there is a 50% chance of a coin's landing head up. What might be the meaning of the expression *fair*? Such a description cannot refer to the result of a single toss, since a single toss cannot result in both a head and a tail. Might "fairness" and "50%" refer to ten successive tosses yielding exactly five heads and five tails? Not really, since it is easy to imagine a "fair" coin landing heads up on only three or four of the ten tosses.

Suppose that we take a fair coin and begin to toss it over and over. After each toss, we compute the relative frequency of heads observed so far. This calculation gives the value of the ratio

$$\frac{\text{number of times event } H \text{ occurs}}{\text{number of tosses}}$$

The results of the first ten tosses might be as follows:

Toss	1	2	3	4	5	6	7	8	9	10
Cumulative number of H's	0	1	2	3	3	3	4	5	5	5
Relative frequency of H's	0	.5	.667	.75	.6	.5	.571	.625	.556	.5

Figure 6.5 (page 240) illustrates how the relative frequency of heads fluctuates during a sample sequence of 50 tosses. *As the number of tosses increases, the relative frequency of heads does not continue to fluctuate wildly but instead stabilizes and approaches some fixed number (limiting value).* This stabilization is illustrated for a sequence of 1000 tosses in Figure 6.6.

FIGURE 6.5 Relative frequency of heads in the first 50 of a long series of tosses

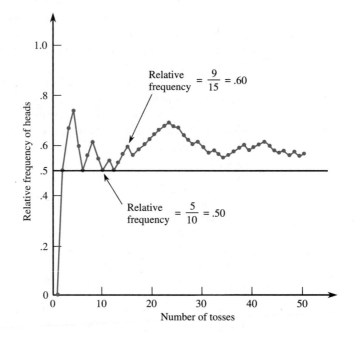

FIGURE 6.6 Stabilization of the relative frequency of heads in coin tossing

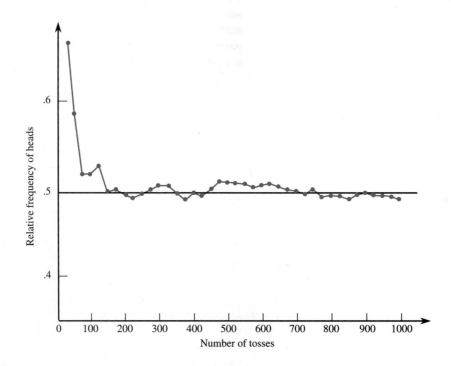

It may seem natural to you that the proportion of H's would get closer and closer to the "real" probability of .5. That an empirically observed proportion could behave like this in real life also seemed reasonable to James Bernoulli in 1713. He focused his considerable mathematical power on this topic and was able to prove mathematically what we now know as the law of large numbers.*

A Law of Large Numbers

As the number of repetitions of a chance experiment increases, the chance that the relative frequency of occurrence for an event will differ from the true probability of the event by more than any very small number approaches zero.

You may still be somewhat skeptical about why such an intuitively pleasing result, that is, the proportion of H's getting closer and closer to the "real" probability of .5, would occupy the interest of a busy mathematician. Note that, as the number of repetitions of the chance experiment increases, the proportion of H's gets closer and closer to the real probability of H occurring in a single experiment *even if the value of this probability is not known.* The law of large numbers is the logical basis for this fundamental scientific activity: Observe the outcomes of repetitions of chance experiments, and then use the observed outcomes to estimate probabilities associated with the natural processes underlying the observations. The most important aspect of this activity is that a finite number of observations will enable the scientist to estimate with confidence. This confidence comes from the long-run stability of repeated observations.

Relative Frequency Approach to Probability

The **probability of an event E,** denoted by $P(E)$, is defined to be the value approached by the relative frequency of occurrence of E in a very long series of trials of a chance experiment. Thus, if the number of trials is quite large,

$$P(E) \approx \frac{\text{number of times } E \text{ occurs}}{\text{number of trials}}$$

Because the probability of an event is the limiting value of its relative frequency of occurrence, a probability is a number between 0 and 1. Thus an event E can be judged relatively unlikely or likely according to where the value of $P(E)$ lies compared to the extremes of 0 and 1. When we informally speak of a 10% chance of occurrence, we mean a relative frequency of .10 in the long run, or $P(E) = .1$.

The relative frequency definition of probability depends on being able to repeat a chance experiment under identical conditions. Suppose we perform a chance

---

*Technically, this should be referred to more specifically as the weak law of large numbers for Bernoulli trials. After Bernoulli's proof, mathematical statisticians proved more general laws of large numbers.

experiment that consists of flipping a cap from a 20-oz bottle of soda and noting whether the cap lands with the open side up or down. The crucial difference from the previous coin-tossing experiment is that there is no particular reason to believe the cap is equally likely to land top up or top down. If we assume that the experiment can be repeated under similar conditions (which seems reasonable), we can flip the cap a number of times and compute the relative frequency of the event *top up* observed so far,

$$\frac{\text{number of times the event } \textit{top up} \text{ occurs}}{\text{number of tosses}}$$

The results of the first ten flips follow, with U indicating top up and D indicating top down:

Flip	1	2	3	4	5	6	7	8	9	10
Outcome	U	U	D	U	D	D	U	D	U	U
Cumulative number of "ups"	1	2	2	3	3	3	4	4	5	6
Relative frequency of "ups"	1.0	1.0	.67	.75	.6	.5	.57	.5	.55	.6

Figure 6.7 illustrates how the relative frequency of heads fluctuates during a sample sequence of 100 flips. Based on these results and faith in the law of large numbers, it is reasonable to think that the probability of the cap landing top up is about .7.

The relative frequency approach to probability is based on observation. From repeated observation, we can get stable relative frequencies that will, in the long run, provide good estimates of the probabilities of different events. The relative frequency interpretation of probability is intuitive, very widely used, and most relevant for the inferential procedures introduced in later chapters.

FIGURE 6.7 Stabilization of the relative frequency of *top up*

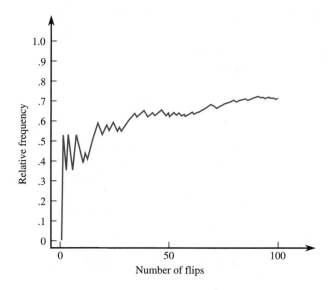

The Subjective Approach to Probability

A third distinct approach to probability is based on subjective judgments. In this view, probability can be interpreted as a personal measure of the strength of belief that a particular outcome will occur. A probability of 1 represents a belief that the outcome will certainly occur. A probability of 0 then represents a belief that the outcome will certainly not occur — that it is impossible. Other probabilities are placed somewhere between 0 and 1 based on the strength of one's beliefs. For example, an airline passenger might report her subjective assessment of the probability of being denied a seat on a particular flight as .01. Since this probability is close to 0, she believes it is very unlikely that she will be denied a seat. The subjective interpretation of probability presents some difficulties. For one thing, different people may assign different probabilities to the same outcome, since each could have a different subjective belief. It seems clear from existing scientific research that people subjectively assess risk in uncertain situations, and some people seem to be very good at it. Although useful in studying and analyzing decision making, subjective probabilities are of limited use because they are personal and not generally replicable by others.

How Are Probabilities Determined?

Probability statements are frequently encountered in newspapers and magazines (and, of course, in this text). What is the basis for these probability statements? The reported probabilities of being struck by lightning, getting some dreaded disease, winning the lottery, and passing a statistics test by guessing alone are *not* delivered by a probability stork, and it is important to understand where and how such probabilities are acquired. We confess that in some cases probabilities are "cooked up" for the purpose of teaching the concept of the moment, and the real-life numbers are unimportant. In other cases, research allows reasonable estimates based on observation and analysis. In general, if a probability is stated, it is based on one of the following approaches:

1. *The classical approach:* This approach is appropriate only for modeling experiments with equally likely outcomes.

2. *The subjective approach:* In this case, the probability represents an individual's judgment based on facts combined with personal evaluation of other information.

3. *The relative frequency approach:* An estimate is based on an accumulation of experimental results. This estimate, usually derived empirically, presumes a replicable chance experiment.

When you see probabilities in this book and in other published sources, you should consider how the probabilities were produced. Be particularly cautious in the case of subjective probabilities.

6.3 Basic Properties of Probability

The view of probabilities as long-run relative frequencies is intuitively pleasing but has some serious limitations when it comes to calculating probabilities in real life.

The most obvious problem is time. If computing even simple probabilities were to require hundreds or thousands of repeated trials of an experiment, even the most ardent statistics students would avoid this task! An alternate approach that alleviates this difficulty in some situations is to study the axioms (fundamental properties) of probability. After presenting some basic properties, we will develop some propositions that follow from these basic properties and that can be used to find probabilities of complex events. These fundamental properties are stated in the box and discussed subsequently.

Basic Properties of Probability

1. For any event E, $0 \leq P(E) \leq 1$.

2. If S is the sample space for an experiment, $P(S) = 1$.

3. If two events E and F are disjoint, then

$$P(E \text{ or } F) = P(E) + P(F)$$

4. For any event E,

$$P(E) + P(not\ E) = 1$$

so

$$P(not\ E) = 1 - P(E) \quad \text{and} \quad P(E) = 1 - P(not\ E)$$

Property 1:   For any event E, $0 \leq P(E) \leq 1$.

To understand the first property of probability, recall the previous bottle cap experiment. You may remember that we were keeping track of the number of successes, which were defined as those landing with the inside of the bottle cap facing up. Suppose that, after flipping the bottle cap N times, we have observed x successes. What are the possible values of x? The fewest that could have been counted would be 0, and the most that could have been counted is N. Therefore, the relative frequency falls between two numbers:

$$\frac{0}{N} \leq \text{relative frequency} \leq \frac{N}{N}$$

Thus, $0 \leq$ relative frequency ≤ 1. As N increases, the long-run value of the relative frequency, which defines the probability, must also lie between 0 and 1.

Property 2:   If S is the sample space for an experiment, $P(S) = 1$.

Since the probability of any event is the proportion of time an outcome in the event will occur in the long run and since the sample space consists of all possible outcomes for an experiment, in the long run an outcome in S must occur 100% of the time. Thus $P(S) = 1$.

Property 3:   If two events E and F are disjoint, then $P(E \text{ or } F) = P(E) + P(F)$.

This is one of the most important properties of probability because if there are a finite number of possible outcomes (simple events) in the sample space, it provides a method for computing probabilities. Any nonsimple event is just a collection of outcomes from the sample space — some outcomes are in the event and others are not. Since simple events are disjoint, any event can be viewed as a union of disjoint events. Suppose, for example, that an experiment has a sample space that consists of six outcomes, $O_1, O_2, \ldots, O_6$. Then $P(O_1)$ can be interpreted as the proportion of time that the outcome O_1 will occur, and the probabilities of the other outcomes can be interpreted in a similar way. Suppose that an event E is made up of outcomes O_1, O_2, and O_4. Since E will occur whenever O_1, O_2, or O_4 occur and O_1, O_2, and O_4 are disjoint, the long-run proportion of time that E will occur is just the sum of the proportion of time each of the outcomes O_1, O_2, and O_4 will occur. Because it is always possible to express any event as a collection of disjoint simple events in this way, finding the probability of an event can be reduced to finding the probability of the simple events (outcomes) that make up the event, and then adding. If the probabilities of all the simple events are known, it is then easy to compute the probability of more complex events constructed from the simple events.

Property 4:   For any event E, $P(E) + P(not\ E) = 1$, so $P(not\ E) = 1 - P(E)$ and $P(E) = 1 - P(not\ E)$.

Property 4 follows from properties 2 and 3. Property 2 tells us that $P(E)$ is the sum of the probabilities in E and that $P(not\ E)$ is the sum of the probabilities for simple events corresponding to outcomes in *not E.* Every outcome in the sample space is in either E or *not E,* and we know from property 3 that the sum of the probabilities of all the simple events is 1. It follows that $P(E) + P(not\ E)$ must equal 1. The implication of property 4, namely, that $P(E) = 1 - P(E^c)$, is surprisingly useful. There are many situations in which calculation of $P(E^c)$ is much easier than direct determination of $P(E)$.

EXAMPLE 6.9   Customers at a certain department store pay for purchases either with cash or with one of four types of credit card. Store records, kept for a long period of time, indicate that 30% of all purchases involve cash, 25% are made with the store's own credit card, 18% with MasterCard (MC), 15% with Visa (V), and the remaining 12% of purchases with an American Express card (AE). The accompanying table displays the probabilities of the simple events for the experiment in which the mode of payment for a randomly selected transaction is observed.

Simple event	O_1 (Cash)	O_2 (Store)	O_3 (MC)	O_4 (V)	O_5 (AE)
Probability	.30	.25	.18	.15	.12

Let's create event E, the event *a randomly selected purchase is made with a nationally distributed credit card.* This event consists of the outcomes MC, V, and AE, so

$$P(E) = P(O_3 \cup O_4 \cup O_5)$$
$$= P(O_3) + P(O_4) + P(O_5)$$
$$= .18 + .15 + .12$$
$$= .45$$

That is, in the long run, 45% of all purchases are made using one of the three national cards. Additionally,

$$P(not\ E) = 1 - P(E)$$
$$= 1 - .45$$
$$= .55$$

which could also have been obtained by noting that *not E* consists of outcomes corresponding to the simple events O_1 and O_2.

The discussion of these properties has been nonmathematical; instead it appeals to your intuition. Students interested in a more formal mathematical treatment of these properties in particular and probability in general are encouraged to consult a more advanced text (see the chapter references).

Having established the fundamental properties of probability, we now present some practical probability rules that can be used to evaluate probabilities in some situations. One important aspect of each rule is its associated assumptions. A common error for beginning statistics students is to believe that the rules can be used in *any* probability calculation. This is NOT the case. You must be careful to use the rules only after verifying that the assumptions are met.

Equally Likely Outcomes

The first probability rule that we consider applies only when events are equally likely and is *not* always true for all events in all situations. As you will recall, experiments involving tossing fair coins, rolling fair dice, or selecting cards from a well-mixed deck have equally likely outcomes. For example, if a fair die is rolled once, each outcome has probability 1/6. With E denoting the event that the outcome is an even number, $P(E) = \frac{1}{6} + \frac{1}{6} + \frac{1}{6} = \frac{3}{6}$. This is just the ratio of the number of outcomes in E to the total number of possible outcomes. The following box presents the generalization of this result.

Calculating Probabilities When Outcomes Are Equally Likely

Consider an experiment that can result in any one of N possible outcomes. Denote the corresponding simple events by $O_1, O_2, \ldots, O_N$. If these simple events are equally likely to occur, then

1. $P(O_1) = \dfrac{1}{N}, P(O_2) = \dfrac{1}{N}, \ldots, P(O_N) = \dfrac{1}{N}$

2. For *any* event E,

   $$P(E) = \frac{\text{number of outcomes in } E}{N}$$

Addition Rule for Disjoint Events

We have seen previously that the probability of an event can be calculated by adding together probabilities of the simple events that correspond to the outcomes making up the event. Simple events are, by definition, disjoint. The reason we can find the probability of an event by adding the corresponding simple event probabilities is that the probability associated with each outcome in the event is included exactly once. This addition process is also legitimate when calculating the probability of the union of two events that are disjoint but not necessarily simple.

The Addition Rule for Disjoint Events

Let E and F be two disjoint events. One of the basic properties (axioms) of probability is

$$P(E \text{ or } F) = P(E \cup F) = P(E) + P(F)$$

This property of probability is known as the addition rule for disjoint events.

More generally, if events $E_1, E_2, \ldots, E_k$ are disjoint, then

$$P(E_1 \text{ or } E_2 \text{ or } \ldots \text{ or } E_k) = P(E_1 \cup E_2 \cup \cdots \cup E_k) = P(E_1) + P(E_2) + \cdots + P(E_k)$$

In words, the probability that any of these k events occurs is the sum of the probabilities of the individual events.

Consider the experiment that consists of rolling a pair of fair dice. There are 36 possible outcomes for this chance experiment ((1, 1), (1, 2), etc.). Since these outcomes are equally likely, we can compute the probability of observing a total of 5 on the two dice by counting the number of outcomes that result in a total of 5. There are a total of 4 such outcomes — (1, 4), (2, 3), (3, 2), and (4, 1) — giving

$$P(\text{total of 5}) = \frac{4}{36}$$

The probabilities for the other totals can be computed in a similar fashion; they are shown in the accompanying table.

Total	Probability	Total	Probability
2	1/36	8	5/36
3	2/36	9	4/36
4	3/36	10	3/36
5	4/36	11	2/36
6	5/36	12	1/36
7	6/36		

What is the probability of getting a total of 3 or 5? Consider the two events

$$E = \text{total is 3} \quad \text{and} \quad F = \text{total is 5}$$

Clearly E and F are disjoint since the sum cannot silmultaneously be both 3 and 5. Notice also that neither of these events is simple, since neither consists of only a single outcome from the sample space. We can apply the addition rule for disjoint events as follows:

$$P(E \text{ or } F) = P(E \cup F)$$
$$= P(E) + P(F)$$
$$= \frac{2}{36} + \frac{4}{36}$$
$$= \frac{6}{36}$$

EXAMPLE 6.10

A large auto center has dealers that sell cars made by a number of different manufacturers. Three of these are Japanese: Honda, Nissan, and Toyota. Consider the manufacturer and models of the next car purchased, and define events E_1, E_2, and E_3 by

$$E_1 = \text{Honda} \qquad E_2 = \text{Nissan} \qquad E_3 = \text{Toyota}$$

Notice that E_1 is not a simple event, since there are Honda Civics, Honda Accords, and Honda Preludes.

Suppose that $P(E_1) = .25$, $P(E_2) = .18$, and $P(E_3) = .14$. Since E_1, E_2, and E_3 are disjoint, the addition rule gives

$$P(\text{Honda } or \text{ Nissan } or \text{ Toyota}) = P(E_1 \cup E_2 \cup E_3)$$
$$= P(E_1) + P(E_2) + P(E_3)$$
$$= .25 + .18 + .14$$
$$= .57$$

The probability that the next car purchased is *not* one of these three types is

$$P(not(E_1 \text{ or } E_2 \text{ or } E_3)) = P((E_1 \cup E_2 \cup E_3)^c)$$
$$= 1 - .57$$
$$= .43$$

In a subsequent section, we will show how $P(E \cup F)$ can be calculated when the two events are not known to be disjoint.

Exercises 6.14 – 6.28

6.14 An article in the *New York Times* (Mar. 2, 1994) reported that people who suffer cardiac arrest in New York City have only a 1 in 100 chance of survival. Using probability notation, an equivalent statement would be

$$P(\text{survival}) = .01$$

for people who suffer a cardiac arrest in New York City. (The article attributed this poor survival rate to factors common in large cities: traffic congestion and the difficulty of finding victims in large buildings. Similar studies in smaller cities showed higher survival rates.)

a. Give a relative frequency interpretation of the given probability.
b. The research that was the basis for the *New York Times* article was a study of 2329 consecutive cardiac arrests in New York City. To justify the "1 in 100 chance of survival" statement, how many of the 2329 cardiac arrest sufferers do you think survived? Explain.

6.15 Consider the following five outcomes for the experiment in which the type of ice cream purchased by the next customer at a certain store is noted.

Data for Exercise 6.15

Outcome (simple event)	O_1	O_2	O_3	O_4	O_5
Brand	Steve's	Ben & Jerry's	Dreyer's	Dreyer's	Von's
Size of container	pint	pint	quart	half-gallon	half-gallon
Probability	.10	.15	.20	.25	.30

a. What is the probability that Dreyer's ice cream is purchased?
b. What is the probability that Von's brand is not purchased?
c. What is the probability that the size purchased is larger than a pint?

6.16 Insurance status — covered (C) or not covered (N) — is determined for each individual arriving for treatment at a hospital's emergency room. Consider the experiment in which this determination is made for two randomly selected patients. The simple events are $O_1 = $ (C, C), $O_2 = $ (C, N), $O_3 = $ (N, C), and $O_4 = $ (N, N). Suppose that probabilities are $P(O_1) = .81$, $P(O_2) = .09$, $P(O_3) = .09$, and $P(O_4) = .01$.
a. What outcomes are contained in the event A, that at most one patient is covered, and what is $P(A)$?
b. What outcomes are contained in the event B, that the two patients have the same status with respect to coverage, and what is $P(B)$?

6.17 Refer to the accompanying information on births in the United States over a given period of time.

Type of Birth	Number of Births
Single birth	41,500,000
Twins	500,000
Triplets	5,000
Quadruplets	100

Use this information to approximate the probability that a randomly selected pregnant woman who reaches full term
a. Delivers twins
b. Delivers quadruplets
c. Gives birth to more than a single child

6.18 Consider the experiment in which both tennis racket head size and grip size are noted for a randomly selected customer at a particular store. The six possible outcomes (simple events) and their probabilities are displayed in the accompanying table.

	Grip Size		
Head Size	$4\frac{3}{8}$ in.	$4\frac{1}{2}$ in.	$4\frac{5}{8}$ in.
Midsize	O_1 (.10)	O_2 (.20)	O_3 (.15)
Oversize	O_4 (.20)	O_5 (.15)	O_6 (.20)

a. The probability that grip size is $4\frac{1}{2}$ in. (the event D) is

$$P(D) = P(O_2 \text{ or } O_5)$$
$$= .20 + .15 = .35$$

How would you interpret this probability?
b. Use the result of part **a** to calculate the probability that grip size is not $4\frac{1}{2}$ in.
c. What is the probability that the racket purchased has an oversize head (the event B), and how would you interpret this probability?
d. What is the probability that grip size is *at least* $4\frac{1}{2}$ in.?

6.19 A mutual fund company offers its customers several different funds: a money-market fund, three different bond funds (short, intermediate, and long-term), two stock funds (moderate and high-risk), and a balanced fund. Among customers who own shares in just one fund, the percentages of customers in the different funds are as follows.

Money-market	20%	High-risk stock	18%
Short bond	15%	Moderate-risk stock	25%
Intermediate bond	10%	Balanced	7%
Long bond	5%		

A customer who owns shares in just one fund is randomly selected.
a. What is the probability that the selected individual owns shares in the balanced fund?
b. What is the probability that the individual owns shares in a bond fund?
c. What is the probability that the selected individual does not own shares in a stock fund?

6.20 A radio station that plays classical music has a "by request" program each Saturday evening. The percentages of requests for composers on a particular night are as follows:

Bach	5%	Mozart	21%
Beethoven	26%	Schubert	12%
Brahms	9%	Schumann	7%
Dvorak	2%	Tchaikovsky	14%
Mendelssohn	3%	Wagner	1%

Suppose that one of these requests is randomly selected.
a. What is the probability that the request is for one of the three B's?

b. What is the probability that the request is not for one of the two S's?

c. Neither Bach nor Wagner wrote any symphonies. What is the probability that the request is for a composer who wrote at least one symphony?

6.21 Refer to Exercise 6.18. Adding probabilities in the first row of the given table yields $P(\text{midsize}) = .45$, whereas from the first column, $P(4\frac{3}{8}\text{-in. grip}) = .30$. Is the following true?

$$P(\text{midsize or } 4\tfrac{3}{8}\text{-in. grip}) = .45 + .30 = .75$$

Explain.

6.22 A deck of 52 cards is mixed very well, and 5 cards are dealt out.

a. It can be shown that (disregarding the order in which the cards are dealt) there are 2,598,960 possible hands, of which only 1287 are hands consisting entirely of spades. What is the probability that a hand will consist entirely of spades? What is the probability that a hand will consist entirely of a single suit?

b. It can be shown that exactly 63,206 hands contain only spades and clubs, with both suits represented. What is the probability that a hand consists entirely of spades and clubs with both suits represented?

c. Using the result of part **b,** what is the probability that a hand contains cards from exactly two suits?

6.23 After all students have left the classroom, a statistics professor notices that four copies of the text were left under desks. At the beginning of the next lecture, the professor distributes the four books at random to the four students (1, 2, 3, and 4) who claim to have left books. One possible outcome is that 1 receives 2's book, 2 receives 4's book, 3 receives his or her own book, and 4 receives 1's book. This outcome can be abbreviated (2, 4, 3, 1).

a. List the 23 other possible outcomes.

b. Which outcomes are contained in the event that exactly two of the books are returned to their correct owners? Assuming equally likely outcomes, what is the probability of this event?

c. What is the probability that exactly one of the four students receives his or her own book?

d. What is the probability that exactly three receive their own books?

e. What is the probability that at least two of the four students receive their own books?

6.24 An individual is presented with three different glasses of cola, labeled C, D, and P. He is asked to taste all three and then list them in order of preference. Suppose that the same cola has actually been put into all three glasses.

a. What are the simple events in this ranking experiment, and what probability would you assign to each one?

b. What is the probability that C is ranked first?

c. What is the probability that C is ranked first *and* D is ranked last?

6.25 The student council for a School of Science and Math has one representative from each of the five academic departments: biology (B), chemistry (C), mathematics (M), physics (P), and statistics (S). Two of these students are to be randomly selected for inclusion on a university-wide student committee (by placing five slips of paper in a bowl, mixing, and drawing out two of them).

a. What are the ten possible outcomes (simple events)?

b. From the description of the selection process, all outcomes are equally likely; what is the probability of each simple event?

c. What is the probability that one of the committee members is the statistics department representative?

d. What is the probability that both committee members come from "laboratory science" departments?

6.26 A video store sells two different brands of VCRs, each of which comes with either two heads or four heads. The accompanying table gives the percentages of recent purchasers buying each type of VCR.

	Number of Heads	
Brand	2	4
M	25%	16%
Q	32%	27%

Suppose that a recent purchaser is randomly selected and both the brand and the number of heads are determined.

a. What are the four simple events?

b. What is the probability that the selected purchaser bought brand Q with 2 heads?

c. What is the probability that the selected purchaser bought brand M?

6.27 A student placement center has requests from five students for interviews regarding employment with a particular consulting firm. Three of these students (1, 2, and 3) are math majors, and the other two are statistics majors. Unfortunately, the interviewer has time to talk to only two of the students; these will be randomly selected from among the five.

a. What is the probability that both selected students are statistics majors?

b. What is the probability that both students are math majors?

c. What is the probability that at least one of the students selected is a statistics major?

d. What is the probability that the selected students have different majors?

6.28 Suppose that a six-sided die is "loaded" so that any particular even-numbered face is twice as likely to be observed as any particular odd-numbered face.

    a. What are the probabilities of the six simple events? (*Hint*: Denote these events by $O_1, \ldots, O_6$. Then $P(O_1) = p$, $P(O_2) = 2p$, $P(O_3) = p$, $\ldots$, $P(O_6) = 2p$. Now use a condition on the sum of these probabilities to determine p.)

b. What is the probability that the number showing is an odd number? At most 3?

c. Now suppose the die is loaded so that the probability of any particular simple event is proportional to the number showing on the corresponding upturned face; that is, $P(O_1) = c$, $P(O_2) = 2c$, $\ldots$, $P(O_6) = 6c$. What are the probabilities of the six simple events? Calculate the probabilities of part **b** for this die.

6.4   Conditional Probability

Sometimes the knowledge that one event has occurred changes the likelihood that another event will occur. One example involves a population in which .1% of all individuals have a certain disease. The presence of the disease cannot be discerned from outward appearances, but there is a diagnostic test available. Unfortunately, the test is not infallible: 80% of those with positive test results actually have the disease; the other 20% who show positive test results are false positives. To cast this in probability terms, consider the experiment in which an individual is randomly selected from the population. Define the following events:

    E = event that the individual has the disease

    F = event that the individual's diagnostic test is positive

Let $P(E \mid F)$ denote the probability of the event E *given that* the event F is known to have occurred. A new symbol has been used to indicate that a probability calculation has been made conditional on the occurrence of another event. The standard symbol for this is $\mid$ and it is read "given." We would say, The probability that an individual has the disease *given* the diagnostic test is positive, and represent this symbolically as $P(\text{has disease} \mid \text{positive test})$ or $P(E \mid F)$. This probability is called a **conditional probability.**

    The information provided then implies that

$$P(E) = .001 \qquad P(E \mid F) = .8$$

That is, before we have diagnostic test information, the occurrence of E is unlikely, whereas once it is known that the test result is positive, the likelihood of the disease increases dramatically. (If this were not so, the diagnostic test would not be very useful!)

EXAMPLE 6.11   At the fictional high school George Washington High (GWH), after-school activities can be classified into three types: athletics, fine arts, and other. The accompanying table gives the number of students participating in each of these types of activities by grade. Row and column totals have also been added to the table.

	9th	10th	11th	12th	Total
Athletics	150	160	140	150	600
Fine arts	100	90	120	125	435
Other	125	140	150	150	565
Total	375	390	410	425	1600

For the purposes of this example, we will assume that any given student is in exactly one of these after-school activities. Consider each of the following statements and make sure that you see how each follows from the information in the table:

1. There are 160 10th grade students participating in athletics.
2. The number of seniors participating in fine arts activities is 125.
3. There are 435 students in fine arts activities.
4. GWH has 410 juniors.
5. The total number of students is 1600.

The principal at GWH selects students at random and invites them to have lunch with her to discuss various issues that might be of concern to them. She feels that random selection will give her the greatest chance of listening to the concerns of a cross-section of the student body. What is the probability that a randomly selected student is a senior athlete? Assuming each student is equally likely to be selected, we can calculate this probability as follows:

$$P(\text{senior athlete}) = \frac{\text{number of senior athletes}}{\text{total number of students}}$$
$$= \frac{150}{1600}$$
$$= .09375$$

Now suppose that the principal's secretary records not only the student's name but also the student's grade level. The secretary has indicated that the selected student is a senior. Does this information change our assessment of the likelihood that the selected student is an athlete? Since 150 of the 425 seniors participate in athletics, this suggests that

$$P(\text{athlete} \mid \text{senior}) = \frac{\text{number of senior athletes}}{\text{total number of seniors}}$$
$$= \frac{150}{425}$$
$$= .3529$$

The probability is calculated in this way because we know that the selected student is one of 425 seniors, each of whom is equally likely to have been the one selected. The interpretation of this conditional probability is that if we were to repeat the experiment of selecting a student at random, about 35.29% of the trials that resulted in a senior being selected would also result in the selection of someone who participates in athletics.

EXAMPLE 6.12   A GFI (ground fault interrupt) switch will turn off power to a system in the event of an electrical malfunction. A spa manufacturer currently has 25 spas in stock, each equipped with a singe GFI switch. The switches are supplied by two different companies, and some of them are defective, as summarized in the accompanying table.

	Nondefective	Defective	Total
Company 1	10	5	15
Company 2	8	2	10
Total	18	7	

A spa is randomly selected for testing. Let

   E = event that GFI switch in selected spa is from company 1

   F = event that GFI switch in selected spa is defective

The tabulated information implies that

$$P(E) = \frac{15}{25} = .60 \qquad P(F) = \frac{7}{25} = .28 \qquad P(E \text{ and } F) = P(E \cap F) = \frac{5}{25} = .20$$

Now suppose that testing reveals a defective switch. (Thus the chosen spa is one of the seven in the "defective" column.) How likely is it that the switch came from the first company? Since five of the seven defective switches are from company 1,

$$P(E \mid F) = P(\text{company 1} \mid \text{defective}) = \frac{5}{7} = .714$$

Notice that this is larger than the unconditional probability $P(E)$. This is because company 1 has a much higher defective rate than does company 2.

An alternative expression for the conditional probability is

$$P(E \mid F) = \frac{5}{7} = \frac{5/25}{7/25} = \frac{P(E \text{ and } F)}{P(F)} = \frac{P(E \cap F)}{P(F)}$$

That is, $P(E \mid F)$ is a ratio of two previously specified probabilities: the probability that both events occur divided by the probability of the "conditioning event" F. Additional insight comes from the Venn diagram of Figure 6.8. Once it is known that the outcome lies in F, the likelihood of E (also) occurring is the "size" of $(E \text{ and } F)$ relative to that of F.

FIGURE 6.8 Venn diagram for Example 6.12 (each dot represents one GFI switch)

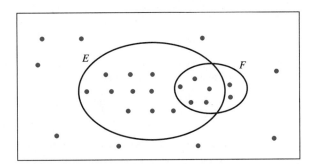

The results of the previous example lead us to a general definition of conditional probability.

Definition

Let E and F be two events with $P(F) > 0$. The **conditional probability of the event E given that the event F has occurred,** denoted by **$P(E \mid F)$,** is

$$P(E \mid F) = \frac{P(E \cap F)}{P(F)}$$

Notice the requirement that $P(F) > 0$. In addition to the standard warning about division by 0, there is another reason for requiring $P(F)$ to be positive. If the probability of F were zero, the event F would never occur; therefore, it would be unreasonable to calculate the probability of another event conditional on F having occurred.

EXAMPLE 6.13   One exciting aspect of increased communication using the Internet is that diverse individuals from widely scattered places all over the world can form an electronic chat room and exchange opinions on various topics of interest. A side effect of such conversation is "flaming," that is, negative criticism of others' contributions to the conversation. M. Dsilva and her colleagues studied this phenomenon ("Criticism on the Internet: An Analysis of Participant Reactions," *Communication Research Reports,* (1999)). The investigators were interested in the effect that personal criticism has on an individual. Would being criticized make one more likely to criticize others? Data from this study is reproduced here.

	Have Been Personally Criticized	Have Not Been Personally Criticized	Total
Have Criticized Others	19	8	27
Have Not Criticized Others	23	143	166
Total	42	151	193

We will assume that the table from the article is indicative of the larger group of chat room users. Suppose a chat room user is randomly selected. Let

  C = event that the individual has criticized others

  O = event that the individual has been personally criticized by others

Assuming that the frequencies in the previous table are indicative of the long-run behavior of events in this situation, we can use the tabulated information to calculate

$$P(C) = \frac{27}{193} = .1400$$

$$P(O) = \frac{42}{193} = .2176$$

$$P(C \cap O) = \frac{19}{193} = .0984$$

Now suppose that it is known that the selection resulted in an individual that has been criticized. (Thus the individual is one of the 42 in the column labeled "Has been personally criticized.") How likely is it that he or she has criticized others? Using the probabilities previously computed, we get

$$P(C \mid O) = \frac{P(C \cap O)}{P(O)} = \frac{19/193}{42/193} = .4524$$

Notice again that this is larger than the original probability $P(C)$. We conclude that there is a higher probability of criticizing others among those who have been personally criticized.

It is also possible to compute $P(O \mid C)$, the probability that an individual has been criticized by others given that he or she has been critical of someone else:

$$P(O \mid C) = \frac{P(O \cap C)}{P(C)} = \frac{P(C \cap O)}{P(C)} = \frac{19/193}{27/193} = \frac{19}{27} = .7037$$

Let's look carefully at the interpretation of each of the following probabilities:

1. $P(C) = .1400$
2. $P(O) = .2176$
3. $P(C \cap O) = .0984$
4. $P(C \mid O) = .4419$
5. $P(O \mid C) = .7037$

1. $P(C) = .1400$ is interpreted as the proportion of chat room users who have been critical of others. Approximately 14% of chat room users have criticized others.
2. $P(O) = .2176$ gives the proportion of chat room users who have been criticized.
3. $P(C \cap O) = .0984$ is the proportion of chat room users who have been critical of others *and* who have also been criticized.
4. $P(C \mid O) = .4419$ is the proportion *of those who have been criticized* who have also been critical of others.
5. $P(O \mid C) = .7037$ is the proportion *of those who have been critical of others* who have themselves been criticized.

In particular, notice the difference between the probabilities in (1)–(3) and the conditional probabilities in (4) and (5). The reference point for the unconditional probabilities is the entire group of interest (all chat room users), whereas the conditional probabilities are interpreted in a more restricted context defined by the "given" event.

▬▬▬▬▬▬

The following example demonstrates the calculation of conditional probabilities and also makes the point that we must be very careful when translating real-world problems — especially probability problems — into mathematical form. Not only are probability problems sometimes difficult to formulate precisely, but the answers are frequently nonintuitive and occasionally counterintuitive.

EXAMPLE 6.14   Consider the population of all families with two children. Representing the gender of each child using G for girl and B for boy results in four possibilities: BB, BG, GB, GG. The gender information is sequential, with the first letter indicating the gender of the older sibling. Thus, a family having a girl first and then a boy would be denoted as GB.

If we assume that a child is equally likely to be male or female, each of the four possibilities in the sample space for the experiment that selects at random from families with two children is equally likely. Consider the following two questions:

1. What is the probability of obtaining a family with two girls, given that the family has at least one girl?

2. What is the probability of drawing a family with two girls, given that the older sibling is a girl?

To many people, these questions *appear* to be identical. However, by computing the appropriate probabilities, we can see that they are indeed different.

In problem 1,

P(family with two girls | family has at least one girl)

$$= \frac{P(\text{family with two girls } and \text{ family with at least one girl})}{P(\text{family with at least one girl})}$$

$$= \frac{P(\text{GG})}{P(\text{GG } or \text{ BG } or \text{ GB})}$$

$$= \frac{1/4}{3/4} = \frac{.25}{.75} = .3333$$

For problem 2:

P(family with two girls | family with older sibling a girl)

$$= \frac{P(\text{family with two girls } and \text{ family with older sibling a girl})}{P(\text{family with older sibling a girl})}$$

$$= \frac{P(\text{GG})}{P(\text{GB } or \text{ GG})}$$

$$= \frac{1/4}{1/2} = \frac{.25}{.50} = .50$$

The moral of this story is: Correct solutions to probability problems, especially conditional probability problems, are much more the result of careful consideration of the sample spaces and the probabilities than they are a product of what your intuition tells you!

As we mentioned in the opening paragraphs of this section, one of the most important practical uses of conditional probability is in making diagnoses. Your teachers diagnose your knowledge by giving you tests. Your mechanic diagnoses your car by hooking it up to a machine and reading the pressures and speeds of the various components of your car. A meteorologist diagnoses the weather by looking at temperatures, isobars, and wind speeds. And of prime importance to all of us, medical doctors diagnose the state of a person's health by performing various tests and gathering information, such as weight and blood pressure.

Doctors observe characteristics of their patients in an attempt to ascertain the "reality" of whether their patients have a certain disease. Many diseases are not actually observable — or at least not easily so — and the doctor must make a probabilistic judgment based on partial information; that is, she must assess conditional probabilities. Because medical diagnosis is very important, and not well understood by most patients, it will be instructive to focus our attention on the probabilistic analysis that underpins diagnosis. As we will see, conditional probability plays a large role in evaluating diagnostic techniques.

We begin with the commonsense notion that a randomly selected individual either has or does not have a particular disease, for example, toxoplasmosis. Toxoplasmosis is a disease caused by parasites and can be brought about by the ingestion of improperly cooked meat. Toxoplasmosis usually does not result in any symptoms, but it may pose a significant risk to the developing fetus of a pregnant woman, who might pass on the disease with serious consequences for the baby: mental retardation, hearing disability, or seizures. The problem here is that the patient will feel no symptoms and the doctor will observe no symptoms in a routine checkup; the disease must be detected by prenatal screening. Normally, a criterion or standard exists by which it would be unequivocally decided whether a person has a disease; this criterion is colloquially known as the "gold standard." The gold standard might be an invasive surgical procedure, expensive and dangerous. The gold standard might be a test that takes a long time to perform in the lab or that is very costly. In some cases, definitive detection may be possible only postmortem. For a diagnostic test to improve on the gold standard, it would need to be faster, less expensive, or less invasive and yet still produce results that agree with the gold standard. In this context, "agreement" means (1) the test generally comes out positive when the patient has the disease, and (2) the test generally comes out negative when the patient does not have the disease. As you may have noticed, these statements will lead us to consideration of conditional probabilities.

EXAMPLE 6.15   To illustrate the calculations involved in evaluating a diagnostic test, we consider the case of tuberculosis (TB), an infectious disease that typically attacks lung tissue.

Before 1998, culturing (that is, growing the tuberculosis bacteria from cells taken from individual patients) was the existing gold standard for diagnosing TB. This method took 10 to 15 days to yield a positive or negative result. In 1998, investigators evaluated a technique that turned out to be much faster ("LCx: A Diagnostic Alternative for the Early Detection of *Mycobacterium tuberculosis* Complex," *Diagnostic Microbiology and Infections Diseases* (1998): 259–264). We will use numbers from that study, but we will not attempt to describe the technique other than to call it a DNA technique. The DNA technique for detecting tuberculosis was evaluated by comparing results from the test to the existing gold standard, with the following results for 207 patients exhibiting respiratory symptoms:

	Has Tuberculosis (gold standard)	**Does Not Have Tuberculosis (gold standard)**
DNA Positive Indication	14	0
DNA Negative Indication	12	181

Converting these data to proportions and inserting the column and row totals into the table, we get the following information:

	Has TB	**Does Not Have TB**	**Total**
DNA+	.0676	.0000	.0676
DNA−	.0580	.8744	.9324
Total	.1256	.8744	1.0000

A cursory look at the table indicates that the DNA technique seems to be working in a manner consistent with what we expect from a diagnostic test. Samples that tested positive with the technique agreed with the gold standard in every case. Samples that tested negative were generally in agreement with the gold standard, but the table also indicates some false negatives. Consider a randomly selected individual who is tested for TB. Define the following events:

T = event that the individual has tuberculosis

N = event that the DNA test is negative

Let $P(T \mid N)$ denote the probability of the event T *given* the event N has occurred. We calculate this probability as follows:

$$P(T \mid N) = P(\text{tuberculosis} \mid \text{a negative DNA test})$$
$$= \frac{P(\text{tuberculosis} \cap \text{a negative DNA test})}{P(\text{a negative DNA test})}$$
$$= \frac{.0580}{.9324}$$
$$= .0622$$

Notice that .1256 of those tested had tuberculosis. The added information provided by the diagnostic test has altered the probability — and provided some measure of relief for the patients who test negative. Once it is known that the test result is negative, the estimated likelihood of the disease is cut in half. If the diagnostic test did not significantly alter the probability, the test would not be very useful to the doctor or patient.

Exercises 6.29 – 6.35

6.29 Two different airlines have a flight from Los Angeles to New York that departs each weekday morning at a certain time. Let E denote the event that the first airline's flight is fully booked on a particular day, and let F denote the event that the second airline's flight is fully booked on that same day. Suppose that $P(E) = .7$, $P(F) = .6$, and also that $P(E \cap F) = .54$.

a. Calculate $P(E \mid F)$, the probability that the first airline's flight is fully booked given that the second airline's flight is fully booked.

b. Calculate $P(F \mid E)$.

6.30 Of the 60 movies reviewed last year by two critics on their joint television show, critic 1 gave a "thumbs up" rating to 15, critic 2 gave this rating to 20, and 10 of the movies were rated "thumbs up" by both critics. Suppose one of these 60 movies is randomly selected.

a. Given that the movie rated "thumbs up" from critic 1, what is the probability that it also had this rating from critic 2?

b. If the movie did not receive a "thumbs up" rating from critic 2, what is the probability that it also received this rating from critic 1? (*Hint:* Construct a table with two rows for the first critic (for "up" and "not up") and two columns for the second critic; then enter the relevant probabilities.)

6.31 The newspaper article "Folic Acid Might Reduce Risk of Down Syndrome" (*USA Today,* Sept. 29, 1999) makes the following statement:

> Older women are at a greater risk of giving birth to a baby with Down Syndrome than are younger women. But younger women are more fertile, so most children with Down Syndrome are born to mothers under 30.

Let D and Y denote the following events:

> D = event that a randomly selected baby is born with Down Syndrome
>
> Y = event that a randomly selected baby is born to a young mother (under age 30)

For each of the following probability statements, indicate whether it is consistent with the quote from the article, and if not, explain why not.

- I. $P(D \mid Y) = .001$; $P(D \mid Y^c) = .004$; $P(Y) = .7$
- II. $P(D \mid Y) = .001$; $P(D \mid Y^c) = .001$; $P(Y) = .7$
- III. $P(D \mid Y) = .004$; $P(D \mid Y^c) = .004$; $P(Y) = .7$
- IV. $P(D \mid Y) = .001$; $P(D \mid Y^c) = .004$; $P(Y) = .4$
- V. $P(D \mid Y) = .001$; $P(D \mid Y^c) = .001$; $P(Y) = .4$
- VI. $P(D \mid Y) = .004$; $P(D \mid Y^c) = .004$; $P(Y) = .4$

6.32 Suppose that an individual is randomly selected from the population of all adult males living in the United States. Let A be the event that the selected individual is over 6 feet in height, and let B be the event that the selected individual is a professional basketball player. Which do you think is larger, $P(A \mid B)$ or $P(B \mid A)$, and why?

6.33 A certain model of car comes in a two-door version, a four-door version, and a hatchback version. Each version can be equipped with either an automatic transmission or a manual transmission. The accompanying table gives the relevant proportions.

	Version		
Transmission Type	**TD**	**FD**	**HB**
A	.32	.27	.18
M	.08	.04	.11

A customer who has purchased one of these cars was randomly selected.

a. What is the probability that this customer purchased a car with an automatic transmission? A four-door car?

b. Given that the customer purchased a four-door car, what is the probability that it has an automatic transmission?

c. Given that the customer did not purchase a hatchback, what is the probability that the car has a manual transmission? How does this conditional probability compare to the (unconditional) probability that the car has a manual transmission?

6.34 An insurance company offers four different deductible levels — none, low, medium, and high — for its homeowner's policy holders, and three different levels — low, medium, and high — for its automobile policy holders. The accompanying table gives proportions for the various categories of policy holders who have both types of insurance. For example, the proportion of individuals with both low homeowner's deductible and low auto deductible is .06 (6% of all such individuals having both types of policy fall into this category).

		Homeowner's			
		N	**L**	**M**	**H**
	L	.04	.06	.05	.03
Auto	**M**	.07	.10	.20	.10
	H	.02	.03	.15	.15

Suppose an insured individual is randomly selected.

a. What is P(low auto deductible | low homeowner's deductible)?

b. Given that the selected individual does not have a high auto deductible, what is the probability that he or she does not have a high homeowner's deductible?

6.35 The *Cedar Rapids Gazette* (Nov. 20, 1999) reported seat belt and car seat usage for cities in Iowa. Iowa law requires that children under 3 be restrained in a car seat and that children aged 3 to 6 be restrained in a car seat or seat belt when riding in an automobile.

City	Number of Children Observed	Number Properly Restrained
Cedar Falls	210	173
Cedar Rapids	231	206
Dubuque	182	135
Iowa City (city)	175	140
Iowa City (interstate)	63	47

a. Use the information provided to estimate the following probabilities:

i. The probability that a randomly selected child is properly restrained given that the child is observed in Dubuque.

ii. The probability that a randomly selected child is properly restrained given that the child is observed in a city that has "Cedar" in its name.

b. Suppose you are observing children in the Iowa City area. Use a tree diagram to illustrate the possible outcomes of an observation that considers both the location of the observation (city or interstate) and whether the child observed was properly restrained.

6.5   Independence

In the previous section, we saw that knowledge of the occurrence of one event can alter our assessment of the likelihood that some other event has occurred. For example, the probability that a randomly selected student participated in athletics in Example 6.11 was altered by learning that the selected student was a senior. We also saw how information about conditional probabilities could be used in medical diagnosis to revise assessments of patients in light of the outcome of a diagnostic procedure. It is also entirely possible, however, that knowledge that one event has occurred will not change the probability of occurrence of a second event.

EXAMPLE 6.16   A large lending institution issues both adjustable-rate and fixed-rate mortgage loans on residential property, which it classifies into three categories: single-family houses, condominiums, and multifamily dwellings. The accompanying table, sometimes called a *joint probability table,* displays probabilities based on the bank's long-run lending behavior.

	Single-family	**Condo**	**Multifamily**	**Total**
Adjustable	.40	.21	.09	.70
Fixed	.10	.09	.11	.30
Total	.50	.30	.20	

From the table, we see that 70% of all mortgages are adjustable rate, 50% of all mortgages are for single-family properties, 40% of all mortgages are adjustable rate for single-family properties (both adjustable-rate *and* single-family), and so on. Define the events E and F by

E = event that a mortgage is adjustable-rate

F = event that a mortgage is for a single-family property

Then

$$P(E \mid F) = \frac{P(E \text{ and } F)}{P(F)} = \frac{.40}{.50} = .80$$

That is, 80% of loans made for single-family properties are for adjustable rate loans. Notice that $P(E \mid F)$ is larger than the original (unconditional) probability $P(E) = .70$. Also,

$$P(F \mid E) = \frac{P(E \text{ and } F)}{P(E)} = \frac{.40}{.70} = .571 > .5 = P(F)$$

If we define another event C by

C = event that a mortgage is for a condominium

we have

$$P(E \mid C) = \frac{P(E \text{ and } C)}{P(C)} = \frac{.21}{.30} = .70$$

Notice that $P(E \mid C) = P(E)$, so if we are told that a mortgage is for a condominium, the probability that it is adjustable remains unchanged.

When two events E and F are such that $P(E \mid F) = P(E)$, the likelihood that event E has occurred is the same after we learn that F has occurred as it was before we had information about F's occurrence. That is, the chance of E having occurred is unaffected by our knowledge that F has occurred. We then say that E and F are independent of one another.

Definition

Two events E and F are said to be **independent** if
$$P(E \mid F) = P(E)$$
If E and F are not independent, they are said to be **dependent** events.
If $P(E \mid F) = P(E)$, it is also true that $P(F \mid E) = P(F)$ and vice versa.

In words, two events are independent if the chance that one of them has occurred is unchanged once it is known that the other has occurred. Moreover, independence of events E and F implies the following additional three relationships:

$$P(not\ E \mid F) = P(not\ E)$$
$$P(E \mid not\ F) = P(E)$$
$$P(not\ E \mid not\ F) = P(not\ E)$$

That is, to say that E and F are independent implies that nothing we learn about F will change the likelihood of E or of $not\ E$.

The formula for conditional probability

$$P(E \mid F) = \frac{P(E \cap F)}{P(F)}$$

can be rearranged to give

$$P(E \cap F) = P(E \mid F)P(F)$$

When E and F are independent, $P(E \mid F) = P(E)$, giving the result in the following box.

Multiplication Rule for Two Independent Events

The events E and F are independent if and only if
$$P(E \cap F) = P(E)P(F)$$
That is, independence implies the relation $P(E \cap F) = P(E)P(F)$, and this relation implies independence.

So, if E occurs 50% of the time when an experiment is repeatedly performed, F occurs 20% of the time, and E and F are independent, then E and F will occur together 10% of the time in the long run, since $(.5)(.2) = .1$.

EXAMPLE 6.17   As an example of independence, consider the study of genetics. In humans, a blueprint for the development of offspring is passed along by parents in DNA material called chromosomes. One such chromosome controls a characteristic known as "hitchhiker's thumb," the ability to bend the last joint of the thumb back at an angle of 60 degrees or more. Whether an offspring has hitchhiker's thumb is determined by two random events: which of two alleles is contributed by the father and which of two alleles is contributed by the mother. (You can think of these alleles as a parental vote of Yes or No for hitchhiker's thumb. In case of a disagreement in votes by the two parents, the dominant allele wins.) These two random events, the results of cell division in two different biological parents, are independent of each other. Suppose there is a .10 probability that a parent will contribute a positive hitchhiker's thumb allele. Since the events are independent, the probability that each parent will contribute a positive hitchhiker's thumb allele, H+, to the offspring is

$P(\text{mother contributes H+} \cap \text{father contributes H+})$

$\quad = P(\text{mother contributes H+}) \cdot P(\text{father contributes H+})$

$\quad = (.10) \cdot (.10)$

$\quad = .01$

EXAMPLE 6.18

Let's look at another example, this time from the field of animal behavior, to illustrate how an investigator could judge whether two events are independent. In the typical kill or be killed aquatic environment, there are predators and there are prey. In a number of fish species, a phenomenon known as "predator inspection" has been reported in the scientific literature. It is thought that predator inspection allows the prey to assess the risk posed by a potential predator. In a typical inspection, a guppy moves toward a predator, presumably to acquire the necessary information and then (hopefully) depart to inspect another day.

Investigators have observed that guppies will sometimes approach and inspect a predator in pairs. Suppose it is not known whether these predator inspections are independent or whether the guppies may be operating as a team. Denote the probability that an individual fish will inspect a predator by p. Let event E_i be the event that fish i would approach and inspect a predator. Then the probability of two guppies, i and j, approaching the predator by chance *independently* would be

$P(E_i \cap E_j) = P(E_i) \cdot P(E_j)$

$\quad\quad\quad\quad = p^2$

Based on our analysis, if the inspections are in fact independent, we would expect that the proportion of times that two guppies happen to simultaneously inspect a predator is equal to the square of the proportion of times that a single fish does so. Based on observing the inspection behavior of a large number of guppies, scientists have concluded that inspection behavior of guppies does not appear to be independent.

The concept of independence extends to more than two events. Consider three events, E_1, E_2, and E_3. Then independence means not only that

$P(E_1 \mid E_2) = P(E_1)$
$P(E_3 \mid E_2) = P(E_3)$

and so on, but also that

$P(E_1 \mid E_2 \text{ and } E_3) = P(E_1)$
$P(E_1 \text{ and } E_3 \mid E_2) = P(E_1 \text{ and } E_3)$

and so on. Furthermore, independence again implies the validity of a multiplication rule.

Events $E_1, E_2, \ldots, E_k$ are **independent** if knowledge that some of the events have occurred does not change the probabilities that any particular one or more of the other events has occurred. Independence implies that

$$P(E_1 \cap E_2 \cap \cdots \cap E_k) = P(E_1) \cdot P(E_2) \cdots P(E_k)$$

Thus when events are independent, the probability that all occur together is the product of the individual probabilities. Furthermore, this relationship remains valid if one or more E_i's is replaced by the event E_i^c.

The independence of more than two events is an important concept in studying complex systems with many components. If these components are critical to the operation of a machine, an examination of the probability of the machine's failure is undertaken by analyzing the failure probabilities of the components. We will take a rather simplified view of a desktop computer to illustrate the role of independence of many events.

EXAMPLE 6.19   Suppose that a desktop computer system consists of a monitor, a mouse, a keyboard, the computer processor itself, and storage devices such as a disk drive. It is conventional wisdom that if a computer is going to fail, it will fail very soon, a phenomenon known as "infant mortality." Purchasers of new computer systems are advised to turn their computers on as soon as they are purchased, and then let them run for a few hours. The theory is that any problems will show up very quickly, and if no problems crop up in the first few hours, the computer system should be OK. Define the following events:

E_1 = event that a newly purchased monitor operates properly

E_2 = event that a newly purchased mouse operates properly

E_3 = event that a newly purchased disk drive operates properly

E_4 = event that a newly purchased computer processor operates properly

Suppose the four events are independent, with

$$P(E_1) = P(E_2) = .98 \qquad P(E_3) = .95 \qquad P(E_4) = .99$$

The probability that all these components operate properly is then

$$\begin{aligned}
P(E_1 \cap E_2 \cap E_3 \cap E_4) &= P(E_1) \cdot P(E_2) \cdot P(E_3) \cdot P(E_4) \\
&= (.98) \cdot (.98) \cdot (.94) \cdot (.99) \\
&= .89
\end{aligned}$$

We interpret this probability as follows: In the long run, 89% of such systems will run properly when tested shortly after purchase. (In reality, the reliability of these parts is much higher than the numbers used in this example!) The probability that all except the monitor will run properly is

$$P(E_1^c \cap E_2 \cap E_3 \cap E_4) = P(E_1^c) \cdot P(E_2) \cdot P(E_3) \cdot P(E_4)$$
$$= [1 - P(E_1)] \cdot P(E_2) \cdot P(E_3) \cdot P(E_4)$$
$$= (.02) \cdot (.98) \cdot (.94) \cdot (.99)$$
$$= .018$$

Sampling With and Without Replacement

One area of statistics where the rules of probability are very important is that of sampling. In previous chapters, we have mentioned that investigators will make inferences about a population based on information from a sample. An important aspect of such studies is the method of sample selection. Sampling methods can be classified into two categories: sampling with replacement and sampling without replacement. Most methods presented in an introductory statistics course are based on the assumption of sampling *with* replacement, but when sampling from real populations, we almost always sample *without* replacement. This seemingly contradictory practice can be a source of confusion. Fortunately, in many circumstances, the distinction between sampling with and without replacement is unimportant for *practical* purposes.

Definition

Sampling is **with replacement** if, once selected, an individual or object is put back into the population before the next selection. Sampling is **without replacement** if, once selected, an individual or object is *not* returned to the population prior to subsequent selections.

EXAMPLE 6.20

Consider the process of selecting three cards from a standard deck. This selection can be made in two ways. One method is to deal three cards off the top of the deck. This method would constitute sampling without replacement. A second method, rarely seen in real games, is selecting a card at random, observing what it is, replacing it in the deck, and shuffling before selecting the next card. This method, as you might have guessed, is sampling with replacement. From the standpoint of probability, these procedures are analyzed very differently. To see this, define these events:

E_1 = event that the first card is a heart

E_2 = event that the second card is a heart

E_3 = event that the third card is a heart

For sampling with replacement, the probability of E_3 is .25, irrespective of whether either E_1 or E_2 occurs, since replacing selected cards gives the same deck for the third selection as for the first two. Whether either of the first two cards is a heart has no bearing on the third card selected, so the three events, E_1, E_2, and E_3 are independent. The chance of getting a heart on the third try is exactly the same as the chance of getting a heart on the first draw and the chance of getting a heart on the second draw.

When sampling is without replacement, the chance of getting a heart on the third draw very definitely depends on the results of the first two draws. If both E_1 and E_2 occur, only 11 of the 50 remaining cards are hearts. Since any one of these 50 has the same chance of being selected, the probability of E_3 in this case is

$$P(E_3 \mid E_1 \text{ and } E_2) = \frac{11}{50} = .22$$

Alternatively, if neither of the first two cards is a heart, then all 13 hearts remain in the deck for the third draw, so

$$P(E_3 \mid \text{not } E_1 \text{ and not } E_2) = \frac{13}{50} = .26$$

Information about the occurrence of E_1 and E_2 affects the chance that E_3 has occurred; for sampling without replacement, the events are not independent.

In opinion polls and other types of surveys, sampling is virtually always done without replacement. For this method of sampling, the results of successive selections are not independent of one another. This is unfortunate, because many of the results from probability and statistics are much easier to state and use when independence can be assumed. The next example suggests that, *under certain circumstances*, the fact that selections in sampling without replacement are not independent is not a cause for concern.

EXAMPLE 6.21 A lot of 10,000 industrial components consists of 2500 manufactured by one firm and 7500 manufactured by a second firm, all mixed together. Three of the components are to be selected at random without replacement. Let

E_1 = event that first component selected was manufactured by firm 1

E_2 = event that second component selected was manufactured by firm 1

E_3 = event that third component selected was manufactured by firm 1

Following the reasoning we used in Example 6.20, we get

$$P(E_3 \mid E_1 \text{ and } E_2) = \frac{2498}{9998} = .24985$$

$$P(E_3 \mid \text{not } E_1 \text{ and not } E_2) = \frac{2500}{9998} = .25005$$

Although these two probabilities differ slightly, to three decimal places they are both .250. We conclude that the occurrence or nonoccurrence of E_1 or E_2 has virtually no effect on the chance that E_3 will occur. *For practical purposes,* the three events can be considered independent.

The essential difference between the situations of Example 6.20 and Example 6.21 is the size of the sample relative to the size of the population. In Example 6.20, a relatively large proportion of the population was sampled (3 out of 52), whereas in Example 6.21, the proportion of the population sampled was quite small (only 3 out of 10,000).

You might think that the differences between sampling without replacement and sampling with replacement should be a source of concern to the practicing statistician. If the inferential methods we will study assume sampling with replacement, and in practice sampling is done without replacement, won't this undermine the credibility of the results? It turns out that the answer is no. The size of the sample is typically very small compared to the size of the population. Whether we are sampling cards, people, light bulbs, or concentrations of metals in rivers and streams, the theory of sampling with replacement can coexist with the practice of sampling without replacement because of the following principle.

If a random sample of size n is taken from a population of size N, the theoretical probabilities of successive selections calculated on the basis of sampling with replacement and on the basis of sampling without replacement differ by insignificant amounts. In practice, therefore, independence can be assumed for the purpose of calculating probabilities. As a reasonable rule of thumb for this practice, n should be no larger than 5% of N.

This principle justifies the assumption of independence in many statistical problems. The phrase *assumption of independence* does not signify that the investigators are in some sense fooling themselves; they are recognizing that the results will not differ from the "right" answers for any practical purpose. In some sampling situations, the sample size might be a significant fraction of the population. For example, a school newspaper editor might easily sample more than 5% of the student body. In that instance, the editors would be wise to consult a statistician before proceeding. Does this mean that an investigator should not sample more than 5% of a population? Certainly not! Except for the time and resources used in sampling a population, it is almost always the case that a larger sample will result in better inferences about the population. The only disadvantage of sampling more than 5% of the population is that the analysis of the resulting data will be slightly more complicated.

Exercises 6.36 – 6.52

6.36 Many fire stations handle emergency calls for medical assistance as well as those requesting firefighting equipment. A particular station says that the probability that an incoming call is for medical assistance is .85. This can be expressed as

$P(\text{call is for medical assistance}) = .85$

a. Give a relative frequency interpretation of the given probability.

b. What is the probability that a call is not for medical assistance?

c. Assuming that successive calls are independent of one another (that is, knowing that one call is for medical assistance doesn't influence our assessment of the probability that the next call will be for medical assistance), calculate the probability that two successive calls will both be for medical assistance.

d. Still assuming independence, calculate the probability that for two successive calls, the first is for medical assistance and the second is not for medical assistance.

e. Still assuming independence, calculate the probability that exactly one of the next two calls will be for medical assistance. (*Hint*: There are two different possibilities that you should consider. The one call for medical assistance might be the first call, or it might be the second call.)

f. Do you think it is reasonable to assume that the requests made in successive calls are independent?

6.37   The Associated Press (*San Luis Obispo Telegram-Tribune,* Aug. 23, 1995) reported on the results of mass screening of school children for tuberculosis (TB). It was reported that for Santa Clara County, California, the proportion of all kindergartners tested who were found to have TB was .0006. The corresponding proportion for recent immigrants (thought to be a high-risk group) was .0075. Suppose that a Santa Clara County kindergartner is to be selected at random. Are the outcomes *selected student is a recent immigrant* and *selected student has TB* independent or dependent outcomes? Justify your answer using the given information.

6.38   The article "Men, Women at Odds on Gun Control" (*Cedar Rapids Gazette,* Sept. 8, 1999) includes the following statement:

> The survey found that 56 percent of American adults favored stricter gun control laws. Sixty-six percent of women favored the tougher laws, compared with 45 percent of men.

These figures are based on a large telephone survey conducted by Associated Press Polls. If an adult is selected at random, are the outcomes *selected adult is female* and *selected adult favors stricter gun control* independent or dependent outcomes? Explain.

6.39   The Australian newspaper *The Mercury* (May 30, 1995) reported that, based on a survey of 600 reformed and current smokers, 11.3% of those who had attempted to quit smoking in the previous two years had used a nicotine aid (such as a nicotine patch). It also reported that 62% of those who quit smoking without a nicotine aid began smoking again within two weeks and 60% of those who used a nicotine aid began smoking again within two weeks. If a smoker who is trying to quit smoking is selected at random, are the outcomes *selected smoker who is trying to quit uses a nicotine aid* and *selected smoker who has attempted to quit begins smoking again within two weeks* independent or dependent outcomes? Justify your answer using the given information.

6.40   In a small city, approximately 15% of those eligible are called for jury duty in any one calendar year. People are selected for jury duty at random from those eligible, and the same individual cannot be called more than once in the same year. What is the probability that a particular eligible person in this city is selected two years in a row? Three years in a row?

6.41   Jeanie is a bit forgetful, and if she doesn't make a "to do" list, the probability that she forgets something she is supposed to do is .1. Tomorrow she intends to run three errands, and she fails to write them on her list.

a. What is the probability that Jeanie forgets all three errands? What assumptions did you make in order to calculate this probability?

b. What is the probability that Jeanie remembers at least one of the three errands?

c. What is the probability that Jeanie remembers the first errand but not the second or third?

6.42   Approximately 30% of the calls to an airline reservation phone line result in a reservation being made.

a. Suppose that an operator handles 10 calls. What is the probability that none of the 10 results in a reservation?

b. What assumption did you make in order to calculate the probability in part **a**?

c. What is the probability that at least one call results in a reservation being made?

6.43   Of the 10,000 students at a certain university, 7000 have Visa cards, 6000 have MasterCards, and 5000 have both. Suppose that a student is randomly selected.

a. What is the probability that the selected student has a Visa card?

b. What is the probability that the selected student has both cards?

c. Suppose you learn that the selected individual has a Visa card (was one of the 7000 with such a card). Now what is the probability that this student has both cards?

d. Are the outcomes *has a Visa card* and *has a Master-Card* independent? Explain.

e. Answer the question posed in part **d** if only 4200 of the students have both cards.

6.44   Consider a system consisting of four components, as pictured in the accompanying diagram. Components 1 and 2 form a series subsystem, as do components 3 and 4. The two subsystems are connected in parallel. Suppose $P(1 \text{ works}) = .9$, $P(2 \text{ works}) = .9$, $P(3 \text{ works}) = .9$, and $P(4 \text{ works}) = .9$, and that these four outcomes are independent (the four components work independently of one another).

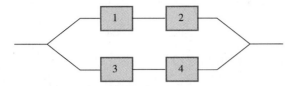

a. The 1–2 subsystem works only if both components work. What is the probability of this happening?
b. What is the probability that the 1–2 subsystem doesn't work? That the 3–4 subsystem doesn't work?
c. The system won't work if the 1–2 subsystem doesn't work and the 3–4 subsystem also doesn't work. What is the probability that the system won't work? That it will work?
d. How would the probability of the system working change if a 5–6 subsystem were added in parallel with the other two subsystems?
e. How would the probability that the system works change if there were three components in series in each of the two subsystems?

6.45 Information from a poll of registered voters in Cedar Rapids, Iowa, to assess voter support for a new school tax (*Cedar Rapids Gazette,* Aug. 28, 1999) was the basis for the following statements:

The poll showed 51 percent of the respondents in the Cedar Rapids school district are in favor of the tax. The approval rating rises to 56 percent for those with children in public schools. It falls to 45 percent for those with no children in public schools. The older the respondent, the less favorable the view of the proposed tax: 36 percent of those over age 56 said they would vote for the tax compared with 72 percent of 18- to 25-year-olds.

Suppose that a registered voter from Cedar Rapids is selected at random, and define the following events

F = event that the selected individual favors the school tax

C = event that the selected individual has children in the public schools

O = event that the selected individual is over 56 years old

Y = event that the selected individual is 18 to 24 years old

a. Use the given information to estimate the values of the following probabilities:

i. $P(F)$ **ii.** $P(F\mid C)$ **iii.** $P(F\mid C^c)$ **iv.** $P(F\mid O)$ **v.** $P(F\mid Y)$
b. Are F and C independent? Justify your answer.
c. Are F and O independent? Justify your answer.

6.46 This case study is reported in the article "Parking Tickets and Missing Women," which appears in the book *Statistics: A Guide to the Unknown,* Judith Tanur, ed. (see Chapter 1 references). In a Swedish trial on a charge of overtime parking, a police officer testified that he had noted the position of the two air valves on the tires of a parked car: To the closest hour, one was at the one o'clock position and the other was at the six o'clock position. After the allowable time for parking in that zone had passed, the policeman returned, noted that the valves were in the same position, and ticketed the car. The owner of the car claimed that he had left the parking place in time and had returned later. The valves just happened by chance to be in the same positions. An "expert" witness computed the probability of this occurring as (1/12)(1/12) = 1/144.
a. What reasoning did the expert use to arrive at the probability of 1/144?
b. Can you spot the error in the reasoning that leads to the stated probability of 1/144? What effect does this error have on the probability of occurrence? Do you think that 1/144 is larger or smaller than the correct probability of occurrence?

6.47 Three friends (A, B, and C) will participate in a round-robin tournament in which each one plays both of the others. Suppose that $P(A$ beats $B) = .7$, $P(A$ beats $C) = .8$, $P(B$ beats $C) = .6$, and that the outcomes of the three matches are independent of one another.
a. What is the probability that A wins both her matches and B beats C?
b. What is the probability that A wins both her matches?
c. What is the probability that A loses both her matches?
d. What is the probability that each person wins one match? (*Hint:* There are two different ways for this to happen. Calculate the probability of each separately, and then add.)

6.48 A shipment of 5000 printed circuit boards contains 40 that are defective. Two boards are chosen at random, without replacement. Consider the two events

E_1 = event that first board selected is defective

E_2 = event that second board selected is defective

a. Are E_1 and E_2 dependent events? Explain in words.
b. Let *not* E_1 be the event that the first board selected is not defective (the event E_1^c). What is $P(not\ E_1)$?

c. How do the two probabilities $P(E_2 \mid E_1)$ and $P(E_2 \mid not\ E_1)$ compare?

d. Based on your answer to part **c,** would it be reasonable to view E_1 and E_2 as approximately independent?

6.49 A store sells two different brands of dishwasher soap, B_1 and B_2. Each brand comes in three different sizes: small (S), medium (M), and large (L). The proportions of the two brands and of the three sizes purchased are displayed as marginal totals in the accompanying table.

Size

	S	M	L	
Brand B_1				.40
B_2				.60
	.30	.50	.20	

Suppose that any event involving brand is independent of any event involving size. What is the probability of the event that a randomly selected purchaser buys the small size of brand B_1 (the event $B_1 \cap S$)? What are the probabilities of the other brand-size combinations?

6.50 Let E_1 denote the event that the next fire to occur in a certain city is a single-alarm blaze. Similarly, let E_2 be the event that the next fire is in the two-alarm category. Suppose that $P(E_1) = .4$ and $P(E_2) = .3$. What is $P(E_1 \mid E_2)$? What is $P(E_2 \mid E_1)$? Are disjoint events independent? Explain.

6.51 On the National Public Radio show "Car Talk," the hosts have a feature called "The Puzzler." Listeners are asked to send in answers to some puzzling questions — usually about cars but sometimes about probability (which, of course, must account for the incredible popularity of the program!). Suppose that for a car question, 800 answers are submitted, of which 50 are correct. Suppose also that the hosts randomly select two answers from those submitted *with replacement.* That is, they randomly select an answer, note whether it is correct, replace it, and then select the second answer at random.

a. What are the possible outcomes of this experiment? Use "C" to represent an answer that is correct and "N" to represent an answer that is not correct.

b. Calculate the probability that both answers selected are correct. (For purposes of this problem, keep at least five digits to the right of the decimal.)

c. Suppose now that the hosts select the answers at random but *without replacement.* Use conditional probability to evaluate the probability that both answers selected are correct. How does this probability compare to the one computed in part **b**?

6.52 Reconsider the problem posed in Exercise 6.51. Suppose now that for a probability question, 100 answers are submitted of which 50 are correct. Calculate the probabilities in parts **b** and **c** of Exercise 6.51 for a probability question rather than a car question.

6.6 Some General Probability Rules

In previous sections, we saw how the probability of $P(E \cup F)$ could be easily computed when E and F are disjoint (mutually exclusive) and how $P(E \cap F)$ could be computed when E and F are independent. In this section, we will develop more general rules: an addition rule that can be used when events are not necessarily disjoint and a multiplication rule that can be used when events are not necessarily independent.

The computation of $P(E \cup F)$ when the two events are not disjoint is a bit more complicated than in the case of disjoint events. Consider Figure 6.9, in which E and

FIGURE 6.9 The shaded region is $P(E \cup F)$, and $P(E \cup F) \neq P(E) + P(F)$.

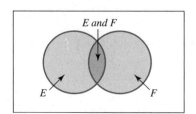

F overlap. The area of the shaded region $(E \cup F)$ is not the sum of the area of *E* and the area of *F*. This is because when the two individual areas are added, the area of the intersection $(E \cap F)$ is counted twice. Similarly, $P(E) + P(F)$ includes $P(E \cap F)$ twice, so this intersection probability must then be subtracted to obtain $P(E \cup F)$. This reasoning leads to the general addition rule that follows.

General Addition Rule

General Addition Rule for Two Events

For any two events *E* and *F*,

$$P(E \cup F) = P(E) + P(F) - P(E \cap F)$$

In words, to calculate the probability that at least one of the two events occurs (the event $E \cup F$), we add the two individual probabilities together and then subtract the probability that both events occur. When *E* and *F* are disjoint, the addition rule simplifies to the previous rule for disjoint events. This is because when $(E \cap F)$ contains no outcomes, $P(E \cap F) = 0$. The general addition rule can be used to determine any one of the four probabilities $P(E), P(F), P(E \cap F)$, or $P(E \cup F)$, provided that the other three probabilities are known.

EXAMPLE 6.22 Suppose that 60% of all customers of a large insurance agency have automobile policies with the agency, 40% have homeowner's policies, and 25% have both types of policies. If a customer is randomly selected, what is the probability that he or she has at least one of these two types of policies with the agency? Let

E = event that a selected customer has auto insurance with the agency

F = event that a selected customer has homeowner's insurance with the agency

The given information implies that

$$P(E) = .60 \qquad P(F) = .40 \qquad P(E \cap F) = .25$$

from which we obtain

$$
\begin{aligned}
P(\text{customer has at least one of the two types of policy}) &= P(E \cup F) \\
&= P(E) + P(F) - P(E \cap F) \\
&= .60 + .40 - .25 \\
&= .75
\end{aligned}
$$

The event that the customer has neither type of policy is $(E \cup F)^c$, so

$$P(\text{customer has neither type of policy}) = 1 - P(E \cup F) = .25$$

Now let's determine the probability that the selected customer has exactly one type of policy. Referring to the Venn diagram in Figure 6.10 (page 272), we see that the

FIGURE 6.10 Representing $P(E \cup F)$ as the union of two disjoint events

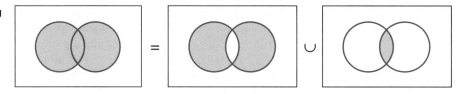

event *at least one*, $(E \cup F)$, can be thought of as consisting of two disjoint parts: *exactly one* and *both*. Thus

$$P(\text{at least one}) = P(E \cup F)$$
$$= P(\text{exactly one}) + P(\text{both})$$
$$= P(\text{exactly one}) + P(E \cap F)$$

so

$$P(\text{exactly one}) = P(E \cup F) - P(E \cap F)$$
$$= .75 - .25$$
$$= .50$$

The addition rule for more than two nondisjoint events is rather complicated. For example, in the case of three events,

$$P(E \cup F \cup G) = P(E) + P(F) + P(G) - P(E \cap F) - P(E \cap G) - P(F \cap G) + P(E \cap F \cap G)$$

A more advanced treatment of probability can be consulted for examples and extensions of these methods.

A General Multiplication Rule

Recall that the definition of conditional probability states that when $P(F) > 0$,

$$P(E \mid F) = \frac{P(E \cap F)}{P(F)}$$

Up to this point, we have used this formula to compute conditional probabilities when $P(E \cap F)$ is known. Sometimes, however, conditional probabilities are known or can be estimated. When this is the case, they can be used to calculate the probability of the intersection of two events. Multiplying both sides of the conditional probability formula by $P(F)$ gives a useful expression for the probability that both events will occur.

General Multiplication Rule

For any two events E and F,

$$P(E \cap F) = P(E \mid F)P(F)$$

The right-hand side of this equation is the product of two probabilities, the first conditional and the second unconditional.

EXAMPLE 6.23

Suppose that 20% of all teenage drivers in a certain county received a citation for a moving violation within the past year. Assume in addition that 80% of those receiving such a citation attended traffic school so that the citation would not appear on their permanent driving records. If a teenage driver from this county is randomly selected, what is the probability that he or she received a citation and attended traffic school?

Let's define two events E and F as follows:

F = selected driver received such a citation

E = selected driver attended traffic school

The question posed can then be answered by calculating $P(E \cap F)$. The percentages given in the problem imply that

$$P(F) = .20 \qquad P(E \mid F) = .80$$

Using the multiplication rule, we calculate

$$
\begin{aligned}
P(E \text{ and } F) &= P(E \mid F)P(F) \\
&= (.80)(.20) \\
&= .16
\end{aligned}
$$

Thus 16% of all teenage drivers in this county received a citation and attended traffic school.

EXAMPLE 6.24

The accompanying table gives information on VCRs sold by a certain appliance store.

	Percentage of Customers Purchasing	Of Those Who Purchase, Percentage Who Purchase Extended Warranty
Brand 1	70	20
Brand 2	30	40

A purchaser is randomly selected from among all those having bought a VCR from the store. What is the probability that the selected customer purchased a brand 1 model and an extended warranty?

To answer this question, we first define the following events:

B_1 = event that brand 1 is purchased

B_2 = event that brand 2 is purchased

E = event that an extended warranty is purchased

The tabulated information implies that

$$P(\text{brand 1 purchased} = P(B_1) = .70$$

$$P(\text{extended warranty} \mid \text{brand 1 purchased}) = P(E \mid B_1) = .20$$

Notice that the 20% is identified with a *conditional* probability; *among purchasers of brand 1,* this is the percentage opting for an extended warranty. Substituting these numbers into the multiplication rule yields

$$
\begin{aligned}
P(B_1 \text{ and } E) &= P(E \mid B_1) \cdot P(B_1) \\
&= (.20)(.70) \\
&= .14
\end{aligned}
$$

The tree diagram of Figure 6.11 gives a nice visual display of how the multiplication rule is used here. The two "first-generation" branches are labeled with events B_1 and B_2 along with their probabilities. Two "second-generation" branches extend from each first-generation branch; these correspond to the two events E and *not E*. The *conditional* probabilities $P(E \mid B_1)$, $P(not\ E \mid B_1)$, $P(E \mid B_2)$, and $P(not\ E \mid B_1)$ appear on these branches. Application of the multiplication rule then consists of multiplying probabilities across the branches of the tree diagram. For example,

$$
\begin{aligned}
P(\text{brand 2 } and \text{ warranty purchased}) &= P(B_2 \text{ and } E) \\
&= P(B_2 \cap E) \\
&= P(E \mid B_2) \cdot P(B_2) \\
&= (.4)(.3) \\
&= .12
\end{aligned}
$$

and this probability is displayed to the right of the E branch that comes from the B_2 branch.

We can now easily calculate $P(E)$, the probability that an extended warranty is purchased. The event E can occur in two different ways: buy brand 1 *and* warranty, or buy brand 2 *and* warranty. Symbolically, these events are $B_1 \cap E$ and $B_2 \cap E$. Furthermore, since each customer purchased a single VCR, he or she could not have

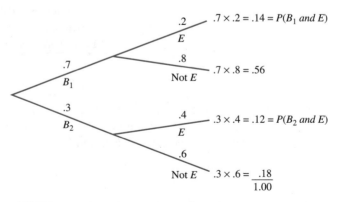

FIGURE 6.11 A tree diagram for probability calculations

simultaneously purchased both brand 1 and brand 2, so the two events $B_1 \cap E$ and $B_2 \cap E$ are disjoint. Thus

$$
\begin{aligned}
P(E) &= P(B_1 \cap E) + P(B_2 \cap E) \\
&= P(E \mid B_1) \cdot P(B_1) + P(E \mid B_2) \cdot P(B_2) \\
&= (.2)(.7) + (.4)(.3) \\
&= .14 + .12 \\
&= .26
\end{aligned}
$$

This probability is the sum of two of the probabilities shown on the right side of the tree diagram. Thus 26% of all VCR purchasers selected an extended warranty.

The multiplication rule can be extended to give an expression for the probability that several events occur together. In the case of three events E, F, and G, we have

$$
P(E \cap F \cap G) = P(E \mid F \cap G) \cdot P(F \mid G) \cdot P(G)
$$

When the events are all independent, $P(E \mid F \cap G) = P(E)$ and $P(F \mid G) = P(F)$, so the right-hand side is simply the product of the three unconditional probabilities.

EXAMPLE 6.25

Twenty percent of all passengers who fly from Los Angeles (LA) to New York (NY) do so on airline G. This airline misplaces luggage for 10% of its passengers, and 90% of this lost luggage is subsequently recovered. If a passenger who has flown from LA to NY is randomly selected, what is the probability that the selected individual flew on airline G (event G), had luggage misplaced (event F), and subsequently recovered the misplaced luggage (event E)? The given information implies that

$$
P(G) = .20 \qquad P(F \mid G) = .10 \qquad P(E \mid F \cap G) = .90
$$

Then

$$
P(E \cap F \cap G) = P(E \mid F \cap G) \cdot P(F \mid G) \cdot P(G) = (.90)(.10)(.20) = .018
$$

That is, about 1.8% of passengers flying from LA to NY fly on airline G have their luggage misplaced and subsequently recover the lost luggage.

The Law of Total Probability and Bayes' Rule

Let's reconsider the information on VCR sales from Example 6.24. In this example, the following events were defined:

B_1 = event that brand 1 is purchased

B_2 = event that brand 2 is purchased

E = event that extended warranty is purchased

Based on the information given in Example 6.24, the following probabilities are known:

$$P(B_1) = .70 \qquad P(B_2) = .30 \qquad P(E \mid B_1) = .20 \qquad P(E \mid B_2) = .4$$

Notice that the conditional probabilities $P(E \mid B_1)$ and $P(E \mid B_2)$ are known, but that the unconditional probability $P(E)$ is not known.

To find $P(E)$, we note that the event E can occur in two ways: (1) a customer purchases an extended warranty *and* buys brand 1 ($E \cap B_1$) or (2) a customer purchases an extended warranty *and* buys brand 2 ($E \cap B_2$). Since these are the only ways in which E can occur, we can write the event E as

$$E = (E \cap B_1) \cup (E \cap B_2)$$

Again, the two events $(E \cap B_1)$ and $(E \cap B_2)$ are disjoint (since B_1 and B_2 are disjoint), so using the addition rule for disjoint events gives

$$\begin{aligned} P(E) &= P((E \cap B_1) \cup (E \cap B_2)) \\ &= P(E \cap B_1) + P(E \cap B_2) \end{aligned}$$

Finally, using the general multiplication rule to evaluate $P(E \cap B_1)$ and $P(E \cap B_2)$ results in

$$\begin{aligned} P(E) &= P(E \cap B_1) + P(E \cap B_2) \\ &= P(E \mid B_1)P(B_1) + P(E \mid B_2)P(B_2) \end{aligned}$$

Substituting in the known probabilities gives

$$\begin{aligned} P(E) &= P(E \mid B_1)P(B_1) + P(E \mid B_2)P(B_2) \\ &= (.2)(.7) + (.4)(.3) \\ &= .26 \end{aligned}$$

We would then conclude that 26% of the VCR customers purchased an extended warranty.

As we've just illustrated, when conditional probabilities are known, they can sometimes be used to compute unconditional probabilities. The **law of total probability** formalizes this use of conditional probabilities.

The Law of Total Probability

If B_1 and B_2 are disjoint events with $P(B_1) + P(B_2) = 1$, then for any event E

$$\begin{aligned} P(E) &= P(E \cap B_1) + P(E \cap B_2) \\ &= P(E \mid B_1)P(B_1) + P(E \mid B_2)P(B_2) \end{aligned}$$

More generally, if $B_1, B_2, \ldots, B_k$ are disjoint events with $P(B_1) + P(B_2) + \cdots + P(B_k) = 1$, then for any event E

$$\begin{aligned} P(E) &= P(E \cap B_1) + P(E \cap B_2) + \cdots + P(E \cap B_k) \\ &= P(E \mid B_1)P(B_1) + P(E \mid B_2)P(B_2) + \cdots + P(E \mid B_k)P(B_k) \end{aligned}$$

EXAMPLE 6.26

The article entitled "Association Between Bicycle Helmet Legislation, Bicycle Safety Education, and Use of Bicycle Helmets in Children," (*Archives of Pediatric & Adolescent Medicine* (1994)) gave information on bicycle helmet usage in some Cleveland suburbs. In Beachwood, Ohio, a safety education program as well as a helmet law was in place, whereas Morland Hills had neither a helmet law nor a safety education program. The article reported that 68% of elementary school students from Beachwood said that they always wear a helmet when bicycling, but only 21% of the elementary school students from Morland Hills reported that they always wear a helmet. For purposes of this example, suppose that these two communities have the same number of elementary school students. A student from one of these two communities is to be selected at random. Let's define

B = selected student is from Beachwood

M = selected student is from Morland Hills

H = selected student reports that he or she always wears a helmet when bicycling

From the information given, we can reason that $P(B) = P(M) = .5$ (since there are the same number of elementary school students in each community) and

$$P(H \mid B) = .68$$
$$P(H \mid M) = .21$$

What proportion of elementary school students always wears helmets? We can use the law of total probability to answer this question. Substituting B and M for B_1 and B_2, respectively, in the formula, we get

$$\begin{aligned} P(H) &= P(H \mid B)P(B) + P(H \mid M)P(M) \\ &= (.68)(.5) + (.21)(.5) \\ &= .34 + .105 \\ &= .445 \end{aligned}$$

That is, about 44.5% of the elementary school children in these two communities always wear a helmet when cycling.

Bayes' Rule

We conclude our discussion of probability rules by considering a formula discovered by the Reverend Thomas Bayes (1702–1761), an English Presbyterian minister. He discovered what is now known as Bayes' rule (or Bayes' theorem). Bayes' rule is a solution to what Rev. Bayes called the "converse problem." To see what he means by this, we return to the field of medical diagnosis. Lyme disease is the leading tick-borne disease in the United States and Europe. Diagnosis of the disease is very difficult and is aided by a test that detects particular antibodies in the blood. The article "Laboratory Considerations in the Diagnosis and Management of Lyme Borreliosis" (*Amer. J. of Clinical Pathology* (1993)) used the following notation to calculate probabilities:

+	represents a positive result on the blood test
−	represents a negative result on the blood test
L	represents the event that the patient actually has Lyme disease
L^c	represents the event that the patient actually does not have Lyme disease

The following probabilities were reported in the article:

$P(L) = .00207$ — The prevalence of Lyme disease in the population; .207% of the population actually has Lyme disease.

$P(L^c) = .99793$ — 99.793% of the population does not have Lyme disease.

$P(+ \mid L) = .937$ — 93.7% of those with Lyme disease test positive.

$P(- \mid L) = .063$ — 6.3% of those with Lyme disease test negative.

$P(+ \mid L^c) = .03$ — 3% of those who do not have Lyme disease test positive.

$P(- \mid L^c) = .97$ — 97% of those who do not have Lyme disease test negative.

Notice the form of the known conditional probabilities; for example, $P(+ \mid L)$ is the probability of a positive test given that a person selected at random from the population actually has Lyme disease. Bayes' converse problem poses a question of a different form: Given that a person tested positive for the disease, what is the probability that he or she actually has Lyme disease? This converse problem is the one that is of primary interest in medical diagnosis problems.

Bayes reasoned as follows to obtain the answer to the converse problem of finding $P(L \mid +)$. We want to calculate $P(L \mid +)$, a conditional probability. We know from the definition of conditional probability that

$$P(L \mid +) = \frac{P(L \cap +)}{P(+)}$$

Since $P(L \cap +) = P(+ \cap L)$, we can use the general multiplication rule to get

$$P(L \cap +) = P(+ \mid L)P(L)$$

This helps, since each of these probabilities is known. We now have

$$P(L \mid +) = \frac{P(+ \mid L)P(L)}{P(+)}$$

What about the denominator, $P(+)$? Fortunately, it can be evaluated using the law of total probability, since L and L^c are disjoint with $P(L) + P(L^c) = 1$. Applying the law of total probability to the denominator, we obtain

$$P(+) = P(+ \cap L) + P(+ \cap L^c)$$
$$= P(+ \mid L)P(L) + P(+ \mid L^c)P(L^c)$$

We now have all we need to answer the converse problem:

$$P(L \mid +) = \frac{P(+ \mid L)P(L)}{P(+ \mid L)P(L) + P(+ \mid L^c)P(L^c)} = \frac{(.937)(.00207)}{(.937)(.00207) + (.03)(.97)} = \frac{.0019}{.1219} = .0156$$

The probability $P(L \mid +)$ is a conditional probability and can be interpreted as such. So, $P(L \mid +) = .0156$ means that, in the long run, 1.56% of those who test positive actually have the disease. Note the difference between $P(L \mid +)$ and the previously reported conditional probability $P(+ \mid L) = .937$, which means that 93.7% of those with Lyme disease test positive.

The accompanying box formalizes this reasoning in the statement of Bayes' rule.

Bayes' Rule

If B_1 and B_2 are disjoint events with $P(B_1) + P(B_2) = 1$, then for any event E

$$P(B_1 \mid E) = \frac{P(E \mid B_1)P(B_1)}{P(E \mid B_1)P(B_1) + P(E \mid B_2)P(B_2)}$$

More generally, if $B_1, B_2, \ldots, B_k$ are disjoint events with $P(B_1) + P(B_2) + \cdots + P(B_k) = 1$, then for any event E,

$$P(B_i \mid E) = \frac{P(E \mid B_i)P(B_i)}{P(E \mid B_1)P(B_1) + P(E \mid B_2)P(B_2) + \cdots + P(E \mid B_k)P(B_k)}$$

EXAMPLE 6.27

Two shipping services offer overnight delivery of parcels, and both promise delivery before 10 A.M. A mail-order catalog company ships 30% of its overnight packages using shipping service 1 and 70% using service 2. Service 1 fails to meet the 10 A.M. delivery promise 10% of the time, whereas service 2 fails to deliver by 10 A.M. 8% of the time. Suppose that you made a purchase from this company and were expecting your package by 10 A.M., but it is late. Which shipping service is more likely to have been used?

Let's define the following events:

S_1 = event that package was shipped using service 1

S_2 = event that package was shipped using service 2

L = event that the package is late

Then the following probabilities are known:

$$P(S_1) = .3 \qquad P(S_2) = .7 \qquad P(L \mid S_1) = .1 \qquad P(L \mid S_2) = .08$$

Since you know that your package is late, you can use Bayes' rule to evaluate $P(S_1 \mid L)$ and $P(S_2 \mid L)$ as follows:

$$P(S_1 \mid L) = \frac{P(L \mid S_1)P(S_1)}{P(L \mid S_1)P(S_1) + P(L \mid S_2)P(S_2)} = \frac{(.1)(.3)}{(.1)(.3) + (.08)(.7)} = \frac{.03}{.086} = .3488$$

$$P(S_2 \mid L) = \frac{P(L \mid S_2)P(S_2)}{P(L \mid S_1)P(S_1) + P(L \mid S_2)P(S_2)} = \frac{(.08)(.7)}{(.1)(.3) + (.08)(.7)} = \frac{.056}{.086} = .6512$$

So you should call service 2. Even though service 2 has a lower percentage of late packages, it is more likely that a late package was shipped using service 2. This is because many more packages (70%) are sent using service 2.

Exercises 6.53 – 6.69

6.53   A certain university has ten cars available for use by faculty and staff upon request. Six of these are Chevrolets and four are Plymouths. On a particular day, only two requests for cars have been made. Suppose that the two cars to be assigned are chosen in a completely random fashion from among the ten.

    a. Let E denote the event that the first car assigned is a Chevrolet. What is $P(E)$?

    b. Let F denote the probability that the second car assigned is a Chevrolet. What is $P(F \mid E)$?

    c. Use the results of parts **a** and **b** to calculate $P(E \text{ and } F)$. (*Hint*: Use the definition of $P(F \mid E)$.)

6.54   A construction firm has bid on two different contracts. Let E_1 be the event that the bid on the first contract is successful, and define E_2 analogously for the second contract. Suppose that $P(E_1) = .4$, $P(E_2) = .3$, and that E_1 and E_2 are independent events.

    a. Calculate the probability that both bids are successful (the probability of the event $E_1 \text{ and } E_2$).

    b. Calculate the probability that neither bid is successful (the probability of the event $(not\ E_1) \text{ and } (not\ E_2)$).

    c. What is the probability that the firm is successful in at least one of the two bids?

6.55   Let E denote the event that a randomly selected student at a certain university has a MasterCard credit card, and let F be the event that such a student has a Visa credit card.

    a. If $P(E) = .40$, $P(F) = .50$, and $P(E \cap F) = .30$, calculate $P(E \cup F)$.

    b. If $P(E) = .25$, $P(F) = .40$, and $P(E \cup F) = .50$, use the general addition rule to calculate $P(E \cap F)$.

6.56   There are two traffic lights on the route used by a certain individual to go from home to work. Let E denote the event that the individual must stop at the first light, and define the event F in a similar manner for the second light. Suppose that $P(E) = .4$, $P(F) = .3$, and $P(E \cap F) = .15$.

    a. What is the probability that the individual must stop at at least one light; that is, what is the probability of the event $E \cup F$?

    b. What is the probability that the individual needn't stop at either light?

    c. What is the probability that the individual must stop at exactly one of the two lights?

    d. What is the probability that the individual must stop just at the first light? (*Hint*: How is the probability of

this event related to $P(E)$ and $P(E \cap F)$? A Venn diagram might help.)

6.57   A family with two automobiles is to be randomly selected. Let A be the event that the older car has an automatic transmission, and let B be the event that the newer car has an automatic transmission. Suppose that $P(A) = .7$ and $P(B) = .8$.

    a. Can $P(A \cup B)$ be calculated from the given information? Explain.

    b. Could it be the case that $P(A \cap B) = .75$? Explain.

    c. If the probability is .9 that the selected family owns at least one car with an automatic transmission, what is the probability that both cars owned by the selected family have automatic transmissions? That at least one car has a manual transmission?

6.58   Exercise 6.34 presented the accompanying table giving proportions of insured individuals with various automobile and homeowner deductible levels.

		Homeowner's			
		N	**L**	**M**	**H**
	L	.04	.06	.05	.03
Auto	**M**	.07	.10	.20	.10
	H	.02	.03	.15	.15

Suppose an individual who has both types of policies is randomly selected.

    a. What is the probability that the individual has a medium auto deductible and a high homeowner's deductible?

    b. What is the probability that the individual has a low auto deductible? A low homeowner's deductible?

    c. What is the probability that the individual is in the same category for both auto and homeowner's deductibles?

    d. Based on your answer to part **c**, what is the probability that the two categories are different?

    e. What is the probability that the individual has at least one low deductible level?

    f. Using the answer in part **e**, what is the probability that neither deductible level is low?

6.59   Let F denote the event that a randomly selected registered voter in a certain city has signed a petition to recall the mayor. Also let E denote the event that a randomly selected registered voter actually votes in the recall election. Describe the event $E \cap F$ in words. If $P(F) = .10$ and $P(E \mid F) = .80$, determine $P(E \cap F)$.

6.60 Suppose a new Internet company Mumble.com was to require all employees to take a drug test. Mumble.com can afford only the inexpensive drug test — the one with a 5% false positive and a 10% false negative rate. (That means that five percent of those who are not using drugs will incorrectly test positive, and ten percent of those who are actually using drugs will test negative.) Suppose that 10% of those who work for Mumble.com are using the drugs for which Mumble is checking. (*Hint*: It may be helpful to draw a tree diagram.)

a. If one employee is chosen at random, what is the probability the employee both uses drugs and tests positive?

b. If one employee is chosen at random, what is the probability the employee does not use drugs but tests positive anyway?

c. If one employee is chosen at random, what is the probability the employee tests positive?

d. If we know that a randomly chosen employee has tested positive, what is the probability he or she uses drugs?

6.61 Refer to the previous exercise. Suppose that because of the high rate of false positives for the drug test, Mumble.com has instituted a mandatory second test for those who test positive on the first test.

a. If one employee is selected at random, what is the probability the selected employee uses drugs and tests positive twice?

b. If one employee is selected at random, what is the probability the employee tests positive twice?

c. If we know the randomly chosen employee has tested positive twice, what is the probability he uses drugs?

d. What is the chance that an individual who does use drugs doesn't test positive twice (either this employee tests negative on the first round and doesn't need a retest or tests positive the first time but that is followed by a negative result) and is thus assumed to be clean?

e. Discuss the benefits and drawbacks of using a retest scheme such as this one.

6.62 According to a study released by the federal Substance Abuse and Mental Health Services Administration (*Knight Ridder Tribune,* Sept. 9, 1999), approximately 8% of all adult full-time workers are drug users and approximately 70% of adult drug users are employed full-time. (These figures are estimates based on a large national survey of nearly 8000 households.)

a. Is it possible for both of the reported percentages to be correct? Explain.

b. Define the events D and E as

D = event that a randomly selected adult is a drug user

E = event that a randomly selected adult is employed full-time

What are the estimated values of $P(D \mid E)$ and $P(E \mid D)$?

c. Is it possible to determine $P(D)$, the probability that a randomly selected adult is a drug user, from the information given? If not, what additional information would be needed?

6.63 A company that manufactures video cameras produces a basic model and a deluxe model. Over the past year, 40% of the cameras sold have been the basic model. Of those buying the basic model, 30% purchase an extended warranty, whereas 50% of all purchasers of the deluxe model buy an extended warranty. If you learn that a randomly selected purchaser bought an extended warranty, what is the probability that he or she has a basic model?

6.64 At a large university, in the never-ending quest for a satisfactory textbook, the Statistics Department has tried a different text during each of the last three quarters. During the fall quarter, 500 students used a book by Professor Mean; during the winter quarter, 300 students used a book by Professor Median, and during the spring quarter, 200 students used a book by Professor Mode. A survey at the end of each quarter showed that 200 students were satisfied with the text in fall quarter, 150 in winter quarter, and 160 in spring quarter.

a. If a student who took statistics during one of these three quarters is selected at random, what is the probability that the student was satisfied with the textbook?

b. If a randomly selected student reports being satisfied with the book, is the student most likely to have used the book by Mean, Median, or Mode? Who is the least likely author? (*Hint*: Use Bayes' rule to compute three probabilities.)

6.65 A friend who works in a big city owns two cars, one small and one large. Three-quarters of the time he drives the small car to work, and one-quarter of the time he takes the large car. If he takes the small car, he usually has little trouble parking, and so is at work on time with probability .9. If he takes the large car, he is on time to work with probability .6. Given that he was on time to work on a particular morning, what is the probability that he drove the small car?

6.66   Only one-tenth of 1% of the individuals in a certain population have a particular disease (an incidence rate of .001). Of those having the disease, 95% test positive when a certain diagnostic test is applied. Of those not having the disease, 90% test negative when the test is applied. Suppose that an individual from this population is randomly selected and given the test.

a. Construct a tree diagram having two first-generation branches, for "has disease" and "doesn't have disease," and two second-generation branches leading out from each of these, for "positive test" and "negative test." Then enter appropriate probabilities on the four branches.

b. Use the multiplication rule to calculate P(has disease *and* positive test).

c. Calculate P(positive test).

d. Use the definition of conditional probability along with the results of parts **b** and **c** to calculate P(has disease | positive test). Does the result surprise you? Give an intuitive explanation for the size of this probability.

6.67   A company uses three different assembly lines — A_1, A_2, and A_3— to manufacture a particular component. Of those manufactured by line A_1, 5% need rework to remedy a defect, whereas 8% of A_2's components need rework and 10% of A_3's need rework. Suppose that 50% of all components are produced by line A_1, whereas 30% are produced by line A_2 and 20% come from line A_3.

a. Construct a tree diagram with first-generation branches corresponding to the three lines. Leading from each one, draw one branch for rework (R) and another for no rework (N). Then enter appropriate probabilities on the branches.

b. What is the probability that a randomly selected component came from line 1 and needed rework?

c. What is the probability that a randomly selected component needed rework?

6.68   A certain company sends 40% of its overnight mail parcels via express mail service A_1. Of these parcels, 2% arrive after the guaranteed delivery time (denote by L the event *late delivery*). If a record of an overnight mailing is randomly selected from the company's files, what is the probability that the parcel went via A_1 and was late?

6.69   Return to Exercise 6.68, and suppose that 50% of the overnight parcels are sent via express mail service A_2 and the remaining 10% are sent via A_3. Of those sent via A_2, only 1% arrived late, whereas 5% of the parcels handled by A_3 arrive late.

a. What is the probability that a randomly selected parcel arrived late? (*Hint*: A tree diagram should help.)

b. Suppose that the selected record shows that the parcel arrived late, but the name of the service does not appear on the record. What is the probability that the parcel was handled by A_1? That is, what is the probability of A_1 given L, denoted $P(A_1 \mid L)$? What is $P(A_2 \mid L)$? $P(A_3 \mid L)$?

6.7   Estimating Probabilities Empirically and Using Simulation

In the examples presented so far, reaching conclusions required knowledge of the probabilities of various outcomes. In some cases, this is reasonable, and we know the true long-run proportion of the time that each outcome will occur. In other situations, these probabilities are not known and must be determined. Sometimes probabilities can be determined analytically, by employing mathematical rules and probability properties, including the basic ones introduced in this chapter. When an analytical approach is impossible or impractical, we can *estimate* probabilities empirically through observation or by using simulation.

Estimating Probabilities Empirically

It is fairly common practice to use observed long-run proportions to estimate probabilities, as illustrated in the following examples.

EXAMPLE 6.28   The Biology Department at a university plans to recruit a new faculty member and intends to advertise for someone with a Ph.D. in biology and at least 10 years of college-level teaching experience. A member of the department expresses the belief that requiring at least 10 years of teaching experience will exclude most potential applicants and will exclude far more of the female applicants than of the male applicants. The Biology Department would like to determine the probability that a Ph.D. in biology who is looking for an academic position would be eliminated from consideration because of the experience requirement.

A similar university just completed a search in which there was no requirement for prior teaching experience, but the information about prior teaching experience was recorded. The 410 applications yielded the following:

	Number of Applicants		
	Less Than 10 Years Experience	**10 Years Experience or More**	**Total**
Male	178	112	290
Female	99	21	120
Total	277	133	410

Let's assume that the populations of applicants for the two positions can be regarded as the same. We will use the available information to approximate the probability that an applicant will fall into each of the four gender–experience combinations. The estimated probabilities (obtained by dividing the number of applicants for each gender–experience combination by 410) are given in Table 6.1. From Table 6.1,

estimate of P(candidate excluded) = .4341 + .2415 = .6756

We can also assess the impact of the experience requirement separately for male and for female applicants. From the given information, we calculate that the proportion of male applicants who have less than 10 years of experience is 178/290 = .6138, whereas the corresponding proportion for females is 99/120 = .8250. Therefore, approximately 61% of the male applicants would be eliminated by the experience requirement, and about 83% of the female applicants would be eliminated.

These subgroup proportions — .6138 for males and .8250 for females — are examples of *conditional probabilities*. As discussed in Section 6.5, outcomes are dependent if the occurrence of one outcome changes our assessment of the probability

Table 6.1 *Estimated probabilities for Example 6.28*

	Less Than 10 Years Experience	**10 Years Experience or More**
Male	.4341	.2732
Female	.2415	.0512

that the other outcome will occur. A conditional probability shows how the original probability changes in light of new information. In this example, the probability that a potential candidate has less than 10 years experience is .6756, but this probability changes to .8250 if we know that a candidate is female. These probabilities can be expressed as an unconditional probability

P(less than 10 years experience) = .6756

and a conditional probability

P(less than 10 years experience | female) = .8250

EXAMPLE 6.29

Consider a state lottery in which six numbers are drawn from the numbers $1, 2, \ldots,$ 54. Suppose that we want to determine the probability that our two favorite numbers (18 and 36) are both drawn, that is, to determine the proportion of the time, in the long run, that 18 and 36 will both be among the six numbers selected. We will estimate this probability empirically. Suppose that we have access to lottery records for the last 5 years and that drawings occur twice each week (for a total of 520 drawings). For each drawing, we determine whether the numbers 18 and 36 were both among the six selected. For example, consider the first few drawings:

Drawing	Numbers Selected	18	36	Both
1	6 17 23 36 38 51	no	yes	no
2	4 12 22 29 48 53	no	no	no
3	14 21 27 32 33 41	no	no	no

Summarizing the data year by year results in the following information:

Year	Number of Drawings	Number with 18 and 36	Cumulative Proportion with 18 and 36
1	104	0	0/104 = 0
2	104	3	3/208 = .0144
3	104	0	3/312 = .0096
4	104	2	5/416 = .0120
5	104	1	6/520 = .0115

Based on the available data (the 520 observed drawings), we estimate the probability that both 18 and 36 appear to be .0115. This is reasonably close to the actual value of .0105.*

If we had used only the first 104 observations, our estimate would have been 0. Increasing the number of observations tends to increase the reliability of an estimated probability.

---

*If the lottery selection equipment were fair, this probability could be determined analytically to be .0105.

Estimating Probabilities Using Simulation

Simulation provides a means of estimating probabilities when we are unable (or do not have the time or resources) to determine probabilities analytically and it is impractical to estimate them empirically by observation. Simulation is a method that generates "observations" by performing an experiment that is as similar as possible in structure to the real situation of interest.

To illustrate the idea of simulation, consider the situation in which a professor wishes to estimate the probabilities of different possible scores on a 20-question true–false quiz when students are merely guessing at the answers. Observations could be collected by having 500 students actually guess at the answers to 20 questions and then scoring the resulting papers. However, obtaining the probability estimates in this way requires considerable time and effort; simulation provides an alternative approach.

Since each question on the quiz is a true–false question, a person who is guessing should be equally likely to answer correctly or incorrectly on any given question. Rather than asking a student to select T or F and then comparing the choice to the correct answer, an equivalent process would be to pick a ball at random from a box that contains half red balls and half blue balls, with a blue ball representing a correct answer. Making 20 selections from the box (replacing each ball selected before picking the next one) and then counting the number of correct choices (the number of times a blue ball is selected) is a physical substitute for an observation from a student who has guessed at the answers to 20 true–false questions. The number of blue balls in 20 selections, with replacement, from a box that contains half red and half blue balls should have the same probability as the number of correct responses to the quiz when a student is guessing.

For example, 20 selections of balls might yield the following results:

Selection	1	2	3	4	5	6	7	8	9	10
	R	R	B	R	B	B	R	R	R	B
Selection	11	12	13	14	15	16	17	18	19	20
	R	R	B	R	R	B	B	R	R	B

This would correspond to a quiz with eight correct responses, and it would provide us with one observation for estimating the probabilities of interest. This process could then be repeated a large number of times to generate additional observations. For example, we might find the following:

Repetition	Number of "Correct" Responses
1	8
2	11
3	10
4	12
⋮	⋮
1000	11

The 1000 simulated quiz scores could then be used to construct a table of estimated probabilities.

Taking this many balls out of a box and writing down the results would be cumbersome and tedious. The process can be simplified by using random digits to substitute for drawing balls from the box. For example, a single digit could be selected at random from the ten digits 0, 1, 2, 3, 4, 5, 6, 7, 8, 9. When using random digits, each of the ten possibilities is equally likely to occur, so we can use the even digits (including 0) to indicate a correct response and the odd digits to indicate an incorrect response. This would maintain the important property that a correct response and an incorrect response are equally likely, since correct and incorrect are each represented by five of the ten digits.

To aid in carrying out such a simulation, tables of random digits (such as Appendix Table I) or computer-generated random digits can be used. The numbers in Table I were generated using a computer's random number generator. You can think of the table as being produced by repeatedly drawing a chip from a box containing ten chips numbered 0, 1, . . . , 9. After each selection, the result is recorded, the chip returned to the box, and the chips mixed. Thus, any of the digits is equally likely to occur on any of the selections.

To see how a table of random numbers can be used to carry out a simulation, let's reconsider the quiz example. We will use a random digit to represent the guess on a single question, with an even digit representing a correct response. A series of 20 digits will represent the answers to the 20 quiz questions. We pick an arbitrary starting point in Table I. Suppose that we start at row 10 and take the 20 digits in a row to represent one quiz. The first five "quizzes" and the corresponding number correct are

Quiz	Random Digits	Number Correct
	* * * * * * * * * * * * *	
1	9 4 6 0 6 9 7 8 8 2 5 2 9 6 0 1 4 6 0 5	13
2	6 6 9 5 7 4 4 6 3 2 0 6 0 8 9 1 3 6 1 8	12
3	0 7 1 7 7 7 2 9 7 8 7 5 8 8 6 9 8 4 1 0	9
4	6 1 3 0 9 7 3 3 6 6 0 4 1 8 3 2 6 7 6 8	11
5	2 2 3 6 2 1 3 0 2 2 6 6 9 7 0 2 1 2 5 8	13

Correct response

This process would be repeated to generate a large number of observations, which would then be used to construct a table of estimated probabilities.

The method for generating observations must preserve the important characteristics of the actual process being considered if simulation is to be successful. For example, it would be easy to adapt the simulation procedure for the true–false quiz to one for a multiple-choice quiz. Suppose that each of the 20 questions on the quiz has five possible responses, only one of which is correct. For any particular question, we would expect a student to be able to guess the correct answer only 1/5 of the time in the long run. To simulate this situation, we could select at random from a box that contained four red balls and only one blue ball (or, more generally, four times as many reds as blues). If we are using random digits for the simulation, we could use 0 and 1 to represent a correct response and 2, 3, . . . , 9 to represent an incorrect response.

Using Simulation to Approximate a Probability

1. Design a method that uses a random mechanism (such as a random number generator or table, the selection of a ball from a box, the toss of a coin, etc.) to represent an observation. Be sure that the important characteristics of the actual process are preserved.

2. Generate an observation using the method from step 1, and determine whether the outcome of interest has occurred.

3. Repeat step 2 a large number of times.

4. Calculate the estimated probability by dividing the number of observations for which the outcome of interest occurred by the total number of observations generated.

The simulation process is illustrated in the three examples that follow.

EXAMPLE 6.30   Many California cities limit the number of building permits that are issued each year. Because of limited water resources, one such city plans to issue permits for only ten dwelling units in the upcoming year. The city will decide who is to receive permits by holding a lottery. Suppose that you are one of a total of 39 individuals who apply for permits. Thirty of these individuals are requesting permits for a single-family home, eight are requesting permits for a duplex (which counts as two dwelling units), and one person is requesting a permit for a small apartment building with eight units (which counts as eight dwelling units). Each request will be entered into the lottery. Requests will be selected at random one at a time, and if there are enough permits remaining, the request will be granted. This process will continue until all ten permits have been issued. If your request is for a single-family home, what are you chances of receiving a permit? Let's use simulation to estimate this probability. (It is not easy to determine analytically.)

To carry out the simulation, we can view the requests as being numbered from 1 to 39 as follows:

01–30	Requests for single-family homes
31–38	Requests for duplexes
39	Request for 8-unit apartment

For ease of discussion, let's assume that your request is number 1.

One method for simulating the permit lottery consists of these three steps:

1. Choose a random number between 1 and 39 to indicate which permit request is selected first, and grant this request.

2. Select another random number between 1 and 39 to indicate which permit request will be considered next. Determine the number of dwelling units for the

selected request. Grant the request only if there are enough permits remaining to satisfy the request.

3. Repeat step 2 until permits for 10 dwelling units have been granted.

We used MINITAB to generate random numbers between 1 and 39 to imitate the lottery drawing. (The random number table in Appendix Table I could also be used, by selecting two digits and ignoring 00 and any value over 39). For example, the first sequence generated by MINITAB is

Random Number	Type of Request	Total Number of Units so Far
25	Single-family home	1
7	Single-family home	2
38	Duplex	4
31	Duplex	6
26	Single-family home	7
12	Single-family home	8
33	Duplex	10

We would stop at this point, since permits for 10 units would have been issued. In this simulated lottery, request number 1 was not selected, so you would not have received a permit.

The next simulated lottery (using MINITAB to generate the selections) is as follows.

Random Number	Type of Request	Total Number of Units so Far
38	Duplex	2
16	Single-family home	3
30	Single-family home	4
39	Apartment — not granted, since there are not 8 permits remaining	4
14	Single-family home	5
26	Single-family home	6
36	Duplex	8
13	Single-family home	9
15	Single-family home	10

Again, request number 1 was not selected, so you would not have received a permit in this simulated lottery.

Now that a strategy for simulating a lottery has been devised, the tedious part of the simulation begins. We would now have to simulate a large number of lottery drawings, determining for each whether request number 1 was granted. We simulated 500 such drawings and found that request number 1 was selected in 85 of the lotteries. Thus,

$$\text{estimated probability of receiving a permit} = \frac{85}{500} = .17$$

EXAMPLE 6.31   Suppose that couples who wanted children were to continue having children until a boy is born. Assuming that each newborn child is equally likely to be a boy or a girl, would this behavior change the proportion of boys in the population? This question was posed in an article that appeared in *The American Statistician* (1994: 290–293), and many people answered the question incorrectly. We will use simulation to estimate the long-run proportion of boys in the population if families were to continue to have children until they have a boy. This proportion is an estimate of the probability that a randomly selected child from this population is a boy. Note that every sibling group would have exactly one boy.

We will use a single-digit random number to represent a child. The odd digits (1, 3, 5, 7, 9) will represent a male birth, and the even digits will represent a female birth. An observation will be constructed by selecting a sequence of random digits. If the first random number obtained is odd (a boy), the observation is complete. If the first selected number is even (a girl), another digit will be chosen. We continue in this way until an odd digit is obtained. For example, reading across row 15 of the random number table (Appendix Table I), the first 10 digits are

0   7   1   7   4   2   0   0   0   1

Using these numbers to simulate sibling groups, we get

Sibling group 1	0 7		girl, boy
Sibling group 2	1		boy
Sibling group 3	7		boy
Sibling group 4	4 2 0 0 0 1		girl, girl, girl, girl, girl, boy

Continuing along row 15 of the random number table,

Sibling group 5	3		boy
Sibling group 6	1		boy
Sibling group 7	2 0 4 7		girl, girl, girl, boy
Sibling group 8	8 4 1		girl, girl, boy

After simulating eight sibling groups, we have 8 boys among 19 children. The proportion of boys is 8/19, which is close to .5. Continuing the simulation to obtain a large number of observations suggests that the long-run proportion of boys in the population would still be .5, which is indeed the case.

EXAMPLE 6.32

Can a close friend read your mind? Try the following experiment. Write the word *blue* on one piece of paper and the word *red* on another, and place the two slips of paper in a box. Select one slip of paper from the box, look at the word written on it, and then try to convey the word by sending a mental message to a friend who is seated in the same room. Ask your friend to select either red or blue, and record whether the response is correct. Repeat this ten times and total up the number of correct responses. How did your friend do? Is your friend receiving your mental messages or just guessing?

Let's investigate this issue by using simulation to get the approximate probabilities of the various possible numbers of correct responses for someone who is guessing. Someone who is guessing should have an equal chance of responding correctly or incorrectly. We can use a random digit to represent a response, with an even digit representing a correct response and an odd digit representing an incorrect response. A sequence of ten digits can be used to simulate one execution of the experiment.

For example, using the last ten digits in row 25 of the random number table gives

5	2	8	3	4	3	0	7	3	5
X	C	C	X	C	X	C	X	X	X

a simulated experiment resulting in four correct responses. We used MINITAB to generate 150 sequences of 10 random digits and obtained the following results:

Sequence Number	Digits	Number Correct
1	3996285890	5
2	1690555784	3
3	9133190550	2
⋮	⋮	⋮
149	3083994450	5
150	9202078546	7

Table 6.2 summarizes the results of our simulation.

Table 6.2 *Estimated probabilities for Example 6.32*

Number Correct	Number of Sequences	Estimated Probability
0	0	.0000
1	1	.0067
2	8	.0533
3	16	.1067
4	30	.2000
5	36	.2400
6	35	.2333
7	17	.1133
8	7	.0467
9	0	.0000
10	0	.0000
Total	150	1.0000

The estimated probabilities in Table 6.2 are based on the assumption that a correct and an incorrect response are equally likely (guessing). Evaluate your friend's performance in light of the information in Table 6.2. Is it likely that someone who is guessing would have been able to get as many correct as your friend did? Do you think your friend was receiving your mental messages? How are the estimated probabilities in Table 6.2 used to support your answer?

Exercises 6.70 – 6.78

6.70 The *Los Angeles Times* (June 14, 1995) reported that the U.S. Postal Service is getting speedier, with higher overnight on-time delivery rates than in the past. Postal Service standards call for overnight delivery within a zone of about 60 miles for any first-class letter deposited by the last collection time posted on a mailbox. Two-day delivery is promised within a 600-mile zone, and three-day delivery is promised for distances over 600 miles. The Price Waterhouse accounting firm conducted an independent audit by "seeding" the mail with letters and recording on-time delivery rates for these letters. Suppose the results of the Price Waterhouse study were as follows. (These numbers are fictitious, but are compatible with summary values given in the article.)

	Number of Letters Mailed	Number Arriving on Time
Los Angeles	500	425
New York	500	415
Washington, D.C.	500	405
Nationwide	6000	5220

Use the given information to estimate the following probabilities.
a. The probability of an on-time delivery in Los Angeles
b. The probability of late delivery in Washington, D.C.
c. The probability that two letters mailed in New York are both delivered on time
d. The probability of on-time delivery nationwide

6.71 Five hundred first-year students at a state university were classified according to both high school GPA and whether they were on academic probation at the end of their first semester.

	High School GPA			
Probation	2.5–<3.0	3.0–<3.5	3.5 and Above	Total
Yes	50	55	30	135
No	45	135	185	365
Total	95	190	215	500

a. Construct a table of the estimated probabilities for each GPA–probation combination by dividing the number of students in each of the six cells of the given table by 500.
b. Use the table constructed in part **a** to approximate the probability that a randomly selected first-year student at this university will be on academic probation at the end of the first semester.
c. What is the estimated probability that a randomly selected first-year student at this university had a high school GPA of 3.5 or above?
d. Are the two outcomes *selected student has a GPA of 3.5 or above* and *selected student is on academic probation at the end of the first semester* independent outcomes? How can you tell?
e. Estimate the proportion of first-year students with high school GPAs between 2.5 and 3.0 who are on academic probation at the end of the first semester.
f. Estimate the proportion of those first-year students with high school GPAs 3.5 and above who are on academic probation at the end of the first semester.

6.72 The table at the top of page 292 describes (approximately) the distribution of students by gender and college at a midsized public university in the West.
If we were to randomly select one student from this university:
a. What is the probability the selected student is a male?

Gender	Education	Engineering	Liberal Arts	College Science and Math	Agriculture	Business	Architecture
Female	300	3200	1500	1500	2100	1500	200
Male	200	800	1500	1500	900	1500	300

Table for Exercise 6.72

b. What is the probability the selected student is in the College of Agriculture?

c. What is the probability the selected student is a male in the College of Agriculture?

d. What is the probability that the selected student is a male who is not from Agriculture?

6.73 On April 1, 2000, the Bureau of the Census in the United States attempted to count every U.S. citizen and every resident. When counting such a large number of people, however, there are likely to be mistakes of various sorts. In the year 2000, many people were employed by the Bureau of the Census who were, in essence, trying to verify some of the information obtained by other Census workers. Suppose that the counts at the bottom of the page are obtained for four counties in one region:

a. If a Census fact-checker selects one person at random from this region, what is the probability that the selected person is from Ventura County?

b. If one person is selected at random from Ventura County, what is the probability that the selected person is Hispanic?

c. If one Hispanic person is selected at random from this region, what is the probability that the selected individual is from Ventura?

d. If one person is selected at random from this region, what is the probability that the selected person is an Asian from San Luis Obispo County?

e. If one person is selected at random from this region, what is the probability that the person is either Asian or from San Luis Obispo County?

f. If one person is selected at random from this region,

what is the probability that the person is Asian or from San Luis Obispo County but not both?

g. If two people are selected at random from this region, what is the probability that both are Caucasians?

h. If two people are selected at random from this region, what is the probability that neither is Caucasian?

i. If two people are selected at random from this region, what is the probability that exactly one is a Caucasian?

j. If two people are selected at random from this region, what is the probability that both are residents of the same county?

k. If two people are selected at random from this region, what is the probability that both are from different racial/ethnic groups?

6.74 A medical research team wishes to evaluate two different treatments for a disease. Subjects are selected two at a time, and then one of the pair is assigned to each of the two treatments. The treatments are applied, and each is either a success (S) or a failure (F). The researchers keep track of the total number of successes for each treatment. They plan to continue the experiment until the number of successes for one treatment exceeds the number of successes for the other treatment by 3. For example, they might observe the results in the table at the top of page 293. The experiment would stop after the sixth pair, because treatment 1 has 3 more successes than treatment 2. The researchers would conclude that treatment 1 is preferable to treatment 2. Suppose that treatment 1 has a success rate of .7 (that is, $P(\text{success}) = .7$ for treatment 1) and that treatment 2 has a success rate of .4. Use simulation to estimate the following probabilities. (See the hint that follows.)

Table for Exercise 6.73

County	Race/Ethnicity				
	Caucasian	Hispanic	Black	Asian	American Indian
Monterey	163,000	139,000	24,000	39,000	4000
San Luis Obispo	180,000	37,000	7000	9000	3000
Santa Barbara	230,000	121,000	12,000	24,000	5000
Ventura	430,000	231,000	18,000	50,000	7000

Table for Exercise 6.74

Pair	Treatment 1	Treatment 2	Total Number of Successes for Treatment 1	Total Number of Successes for Treatment 2
1	S	F	1	0
2	S	S	2	1
3	F	F	2	1
4	S	F	3	1
5	F	F	3	1
6	S	F	4	1

a. The probability that more than 5 pairs must be treated before a conclusion can be reached

b. The probability that the researchers will incorrectly conclude that treatment 2 is the better treatment (*Hint*: Use a pair of random digits to simulate one pair of subjects. Let the first digit represent treatment 1 and use 1–7 as an indication of a success and 8, 9, and 0 to indicate a failure. Let the second digit represent treatment 2, with 1–4 representing a success. For example, if the two digits selected to represent a pair were 8 and 3, you would record failure for treatment 1 and success for treatment 2. Continue to select pairs, keeping track of the total number of successes for each treatment. Stop the trial as soon as the number of successes for one treatment exceeds that for the other by 3. This would complete one trial. Now repeat this whole process until you have results for at least 20 trials (more is better). Finally, use the simulation results to estimate the desired probabilities.)

6.75 Many cities regulate the number of taxi licenses, and there is a great deal of competition for both new and existing licenses. Suppose that a city has decided to sell 10 new licenses for $25,000 each. A lottery will be held to determine who gets the licenses, and no one may request more than 3 licenses. Twenty individuals and taxi companies have entered the lottery. Six of the 20 entries are requests for 3 licenses, 9 are requests for 2 licenses, and the rest are requests for a single license. The city will select requests at random, filling as much of the request as possible. For example, the city might fill requests for 2, 3, 1, and 3 licenses and then select a request for 3. Since there is only one license left, the last request selected would receive a license, but only one.

a. An individual who wishes to be an independent driver has put in a request for a single license. Use simulation to approximate the probability that the request will be granted. Perform *at least* 20 simulated lotteries (more is better!).

b. Do you think that this is a fair way of distributing licenses? Can you propose an alternative procedure for distribution?

6.76 Four students must work together on a group project. They decide that each will take responsibility for a particular part of the project, as follows:

Person	Maria	Alex	Juan	Jacob
Task	Survey design	Data collection	Analysis	Report writing

Because of the way the tasks have been divided, one student must finish before the next student can begin work. To ensure that the project is completed on time, a timeline is established, with a deadline for each team member. If any one of the team members is late, the timely completion of the project is jeopardized. Assume the following probabilities:

1. The probability that Maria completes her part on time is .8.
2. If Maria completes her part on time, the probability that Alex completes on time is .9, but if Maria is late, the probability that Alex completes on time is only .6.
3. If Alex completes his part on time, the probability that Juan completes on time is .8, but if Alex is late, the probability that Juan completes on time is only .5.
4. If Juan completes his part on time, the probability that Jacob completes on time is .9, but if Juan is late, the probability that Jacob completes on time is only .7.

Use simulation (with at least 20 trials) to estimate the probability that the project is completed on time. Think carefully about this one. For example, you might use a random digit to represent each part of the project (four in all). For the first digit (Maria's part), 1–8 could represent *on time* and 9 and 0 could represent *late*. Depending on what happened with Maria (late or on time), you would then look at the digit representing Alex's part. If Maria was on time, 1–9 would represent *on time* for Alex, but if Maria was late, only 1–6 would represent *on time*. The parts for Juan and Jacob could be handled similarly.

6.77 In Exercise 6.76, the probability that Maria completes her part on time was .8. Suppose that this probability is really only .6. Use simulation (with at least 20 trials) to estimate the probability that the project is completed on time.

6.78 Refer to Exercises 6.76 and 6.77. Suppose that the probabilities of timely completion are as in Exercise 6.76

for Maria, Alex, and Juan, but that Jacob has a probability of completing on time of .7 if Juan is on time and .5 if Juan is late.

a. Use simulation (with at least 20 trials) to estimate the probability that the project is completed on time.

b. Compare the probability from part **a** to the one computed in Exercise 6.77. Which decrease in the probability of on-time completion (Maria's or Jacob's) made the biggest change in the probability that the project is completed on time?

Summary of Key Concepts and Formulas

Term or Formula	Comment
Chance experiment	Any experiment for which there is uncertainty concerning the resulting outcome.
Sample space	The collection of all possible outcomes from a chance experiment.
Event	Any collection of possible outcomes from a chance experiment.
Simple event	Any event that consists of a single outcome.
Events 1. *not A*, A^c 2. *A or B*, $A \cup B$ 3. *A and B*, $A \cap B$	1. The event consisting of all outcomes not in A. 2. The event consisting of all outcomes in at least one of the two events. 3. The event consisting of outcomes common to both events.
Disjoint (mutually exclusive) events	Events that have no outcomes in common and so cannot occur at the same time.
Basic properties of probability	1. The probability of any event must be a number between 0 and 1. 2. If S is the sample space for an experiment, $P(S) = 1$. 3. If E and F are disjoint events, $P(E \cup F) = P(E) + P(F)$ 4. $P(E) + P(E^c) = 1$
$P(E) = \dfrac{\text{number of outcomes in } E}{N}$	$P(E)$ *when the outcomes are equally likely* and where N is the number of outcomes in the sample space.
$P(E \cup F) = P(E) + P(F)$ $P(E_1 \cup \cdots \cup E_k) = P(E_1) + \cdots + P(E_k)$	Addition rules when events are disjoint.
$P(E \mid F) = \dfrac{P(E \cap F)}{P(F)}$	The conditional probability of the event E given that the event F has occurred.
Independence of events E and F $P(E \mid F) = P(E)$	Events E and F are independent if the probability that E has occurred given F is the same as the probability that E will occur with no knowledge of F.
$P(E \cap F) = P(E) \cdot P(F)$ $P(E_1 \cap \cdots \cap E_k) = P(E_1)P(E_2) \cdots P(E_k)$	Multiplication rules for *independent* events.
$P(E \cup F) = P(E) + P(F) - P(E \cap F)$	The general addition rule for two events.

$$P(E \cap F) = P(E \mid F)P(F)$$ The general multiplication rule for two events.

$$P(E) = P(E \mid B_1)P(B_1) + \cdots + P(E \mid B_k)P(B_k)$$ The law of total probability, where $B_1, B_2, \ldots, B_k$ are disjoint events with $P(B_1) + \cdots + P(B_k) = 1$.

$$P(B_i \mid E) = \frac{P(E \mid B_i)P(B_i)}{P(E \mid B_1)P(B_1) + \cdots + P(E \mid B_k)P(B_k)}$$ Bayes' rule, where $B_1, B_2, \ldots, B_k$ are disjoint events with $P(B_1) + \cdots + P(B_k) = 1$.

Supplementary Exercises 6.79 – 6.94

6.79 Two individuals, A and B, are finalists for a chess championship. They will play a sequence of games, each of which can result in a win for A, a win for B, or a draw. Suppose that the outcomes of successive games are independent, with $P(\text{A wins game}) = .3$, $P(\text{B wins game}) = .2$, and $P(\text{draw}) = .5$. Each time a player wins a game, he earns one point and his opponent earns no points. The first player to win 5 points wins the championship. For the sake of simplicity, assume that the championship will end in a draw if both players obtain 5 points at the same time.

a. What is the probability that A wins the championship in just five games?

b. What is the probability that it takes just five games to obtain a champion?

c. If a draw earns $\frac{1}{2}$ point for each player, describe how you would perform a simulation experiment to estimate $P(\text{A wins the championship})$.

d. If neither player earns any points from a draw, would the simulation requested in part **c** take longer to perform? Explain your reasoning.

6.80 Consider a clinical trials experiment for comparing two treatments, A and B, and suppose that patients arrive for treatment one by one. If the result of treating each patient, success or failure, is determined before the next patient arrives, a "play the winner" rule is often used to assign treatments. The first patient is equally likely to be assigned A or B. Suppose that A is assigned. Then, if the treatment is a success, A is assigned to the next patient, but if the treatment is a failure, the treatment is switched to B for the next patient. Continuing in this manner, treatment is switched from A to B or from B to A if the previous patient was a failure and remains the same otherwise. If $P(\text{A is a success}) = .7$, $P(\text{B is a success}) = .4$, and the experiment is continued until one treatment has achieved 5 successes, perform a simulation experiment to estimate $P(\text{A is the "winner"})$.

6.81 A single-elimination tournament with four players is to be held. A total of three games will be played. In game 1, the players seeded (rated) first and fourth play. In game 2, the players seeded second and third play. In game 3, the winners of games 1 and 2 play, with the winner of game 3 declared the tournament winner. Suppose that the following probabilities are given:

$P(\text{seed 1 defeats seed 4}) = .8$
$P(\text{seed 1 defeats seed 2}) = .6$
$P(\text{seed 1 defeats seed 3}) = .7$
$P(\text{seed 2 defeats seed 3}) = .6$
$P(\text{seed 2 defeats seed 4}) = .7$
$P(\text{seed 3 defeats seed 4}) = .6$

a. Describe how you would use a selection of random digits to simulate game 1 of this tournament.

b. Describe how you would use a selection of random digits to simulate game 2 of this tournament.

c. How would you use a selection of random digits to simulate the third game in the tournament? (This will depend on the outcomes of games 1 and 2.)

d. Simulate one complete tournament, giving an explanation for each step in the process.

e. Simulate 10 tournaments and use the resulting information to estimate the probability that the first seed wins the tournament.

f. Ask four classmates for their simulation results. Along with your own results, this should give you information on 50 simulated tournaments. Use this information to estimate the probability that the first seed wins the tournament.

g. Why do the estimated probabilities from parts **e** and **f** differ? Which do you think is a better estimate of the true probability? Explain.

6.82 A student has a box containing 25 computer disks, of which 15 are blank and 10 are not. She randomly selects disks one by one and examines each one, terminating the process only when she finds a blank disk. What is the probability that she must examine at least 2 disks? (*Hint*: What must be true of the first disk?)

6.83 In a school machine shop, 60% of all machine breakdowns occur on lathes and 15% occur on drill presses. Let E denote the event that the next machine breakdown is on a lathe and let F denote the event that a drill press is the next machine to break down. With $P(E) = .60$ and $P(F) = 15$, calculate:
   a. $P(E^c)$ **b.** $P(E \cup F)$ **c.** $P(E^c \cap F^c)$

6.84 There are five faculty members in a certain academic department. These individuals have 3, 6, 7, 10, and 14 years of teaching experience, respectively. Two of these individuals are randomly selected to serve on a personnel review committee. What is the probability that the chosen representatives have a total of at least 15 years of teaching experience? (*Hint:* Consider all possible committees.)

6.85 The general addition rule for three events states that

$$P(A \text{ or } B \text{ or } C) = P(A) + P(B) + P(C) - P(A \text{ and } B)$$
$$- P(A \text{ and } C) - P(B \text{ and } C)$$
$$+ P(A \text{ and } B \text{ and } C)$$

A new magazine publishes columns entitled "Art" (A), "Books" (B), and "Cinema" (C). Suppose that 14% of all subscribers read A, 23% read B, 37% read C, 8% read A and B, 9% read A and C, 13% read B and C, and 5% read all three columns. What is the probability that a randomly selected subscriber reads at least one of these three columns?

6.86 A theater complex is currently showing four R-rated movies, three PG-13 movies, two PG movies, and one G movie. The accompanying table gives the number of people at the first showing of each movie on a certain Saturday.

Theater	Rating	No. of Viewers
1	R	600
2	PG-13	420
3	PG-13	323
4	R	196
5	G	254
6	PG	179
7	PG-13	114
8	R	205
9	R	139
10	PG	87

Suppose that a single one of these viewers is randomly selected.
   a. What is the probability that the selected individual saw a PG movie?

b. What is the probability that the selected individual saw a PG or a PG-13 movie?
   c. What is the probability that the selected individual did not see an R movie?

6.87 Refer to Exercise 6.86, and suppose that two viewers are randomly selected (without replacement). Let R_1 and R_2 denote the events that the first and second individuals, respectively, watched an R-rated movie. Are R_1 and R_2 independent events? Explain. From a practical point of view, can these events be regarded as independent? Explain.

6.88 Components of a certain type are shipped to a supplier in batches of ten. Suppose that 50% of all batches contain no defective components, 30% contain one defective component, and 20% contain two defective components. Two components from a batch are randomly selected and tested.
   a. If the batch from which the components were selected actually contains two defectives, what is the probability that neither of these is selected for testing?
   b. What is the probability that the batch contains two defectives and neither of these is selected for testing?
   c. What is the probability that neither component selected for testing is defective? (*Hint:* This could happen with any one of the three types of batches. A tree diagram might help.)

6.89 On Monday morning a loan officer at a credit union is given 15 auto loan applications to process. Three of these are for 36 months, 5 are for 48 months, and the other 7 are for 60 months. Only 2 applications can be processed that morning. Suppose that the 2 to be processed are randomly selected from among the 15.
   a. Let A_1 be the event that the first application processed is for 36 months, and let A_2 be the event that the second application processed is for 36 months. What is $P(A_1)$? What is $P(A_2 \mid A_1)$? Describe the event A_1 *and* A_2 in words, and calculate the probability of this event.
   b. What is the probability that both applications processed that morning are for loans of the same duration?

6.90 A company has just placed an order with a supplier for two different products. Let

   E = event that first product is out of stock

   F = event that second product is out of stock

Suppose that $P(E) = .3$, $P(F) = .2$, and the probability that at least one product is out of stock is .4.
   a. What is the probability that both products are out of stock?
   b. Are E and F independent events?

c. Given that the first product is in stock, what is the probability that the second is also?

6.91   A certain light fixture requires 3 lightbulbs. Suppose that a box contains 25 bulbs, of which 20 are good and the other 5 are defective. Consider randomly selecting 3 bulbs without replacement. Let E denote the event that the first bulb selected is good, F be the event that the second is good, and G represent the event that the third bulb selected is a good one.
   a. What is $P(E)$?
   b. What is $P(F \mid E)$?
   c. What is $P(G \mid E \cap F)$?
   d. What is the probability that all 3 bulbs selected are good?

6.92   Return to Exercise 6.91 and suppose that a fixture requires 4 bulbs. If 4 bulbs are randomly selected from the 25, what is the probability that all 4 are good? What is the probability that at least 1 selected bulb is bad? (*Hint:* Define a fourth event H, and then extend the multiplication rule to the event $E \cap F \cap G \cap H$.)

6.93   A sales representative for a company that manufactures water-treatment systems can make sales presentations to either one or two potential customers on any given day. The probability that there will be a single presentation is .6. If just one presentation is made, it is successful with probability .8. If two presentations are made, the first is successful with probability .7, the second is successful with probability .8, and these two events ("success on 1st" and "success on 2nd") are independent.
   a. What is the probability that just one presentation is made on a particular day and it is successful?
   b. What is the probability that two presentations are made and both are successful?
   c. What is the probability that two presentations are made and neither is successful?
   d. What is the probability that the sales rep has no successful presentations on a given day? (*Hint:* Construct a tree diagram with first-generation branches for one presentation (E_1) and two presentations (E_2). The branches leading out from the E_2 branch should specify all possibilities for each presentation.)

6.94   A transmitter is sending a message by using a binary code, namely, a sequence of 0's and 1's. Each transmitted "bit" (0 or 1) must pass through three relays to reach the receiver. At each relay, the probability is .20 that the bit sent on will be different from the bit received (a reversal). Assume that the relays operate independently of one another.

transmitter $\rightarrow$ relay 1 $\rightarrow$ relay 2 $\rightarrow$ relay 3 $\rightarrow$ receiver

   a. If a 1 is sent from the transmitter, what is the probability that a 1 is sent on by all three relays?
   b. If a 1 is sent from the transmitter, what is the probability that a 1 is received by the receiver? (*Hint:* The eight experimental outcomes can be displayed on a tree diagram with three generations of branches, one generation for each relay.)

References

Devore, Jay L. *Probability and Statistics for Engineering and the Sciences,* 5th ed. Pacific Grove, CA: Brooks/Cole, 2000. (The treatment of probability in this source is more comprehensive and at a somewhat higher mathematical level than is ours.)

Mosteller, Frederick, Robert Rourke, and George Thomas. *Probability with Statistical Applications.* Reading, MA: Addison-Wesley, 1970. (A very good introduction to probability at a modest mathematical level.)

GRAPHING CALCULATOR EXPLORATIONS

6.1 Probabilities as Long-Run Relative Frequencies

One very interesting use of your calculator is for simulating chance experiments. As you know, the long-run behavior of a chance event is stable, with the relative frequency of occurrence approaching a number that is defined as the probability of the event. A very simple chance experiment is flipping a coin. We will use a calculator to "simulate" flipping a coin; that is, the calculator will produce behavior that mimics the behavior of a coin flip. To do this, we need to generate random numbers on the calculator. Recall that in Graphing Calculator Exploration 2.2, we discussed the process of generating a random number between 0 and 1, using a sequence of keystrokes we defined as *rand.* We will use this process to flip some virtual coins. Just in case you want to consider an unfair coin or simulate rolling dice or spinning a spinner, let's define the probability of success, S, for the event of interest to be equal to p.

The calculator generates what are known as uniform random numbers on the interval from 0 to 1. One characteristic of this process is that for any number, p, the probability of generating a number between 0 and p is equal to p. Using this fact, we can perform the chance experiment as follows: Generate a uniform random number between 0 and 1, using your *rand* keystroke sequence. If the number generated is less than or equal to p, count that as a success. If the number generated is greater then p, count that as a failure. For a fair coin, p is equal to .5. Thus, any number generated by *rand* that is less than or equal to .5 will count as a success. The following table shows the beginning of such a simulation:

Toss Number	*Rand*	S/F	Cumulative Number of Successes	Relative Frequency of Successes
1	.9351587791	F	0	0/1 = .0000
2	.1080114624	S	1	1/2 = .5000
3	.0062633066	S	2	2/3 = .6667
4	.5489861799	F	2	2/4 = .5000
5	.8555803143	F	2	2/5 = .4000

To see how the relative frequency of success behaves, you can plot toss number on the horizontal axis and relative frequency of success on the vertical axis. A plot from the table above is shown in the accompanying figure.

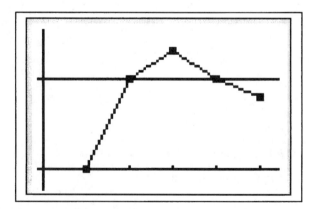

In order to see that the relative frequency settles down and gets close to the true probability of success, we would need to simulate a very large number of tosses. This is a slow process, but you might be able to use your calculator to automate the process. Your calculator may be able to perform this experiment many times and store the results, using some of its list commands. The commands will, of course, be different from calculator to calculator, but the logic that underlies the process will be very similar. If you wish to simulate a large number of tosses, you will definitely want to use the list commands! Remember, since calculators will differ, you will need to look up the correct syntax for the list commands in your calculator manual. (You may wish to refer to Graphing Calculator Exploration 3.1.) Here is a typical sequence of instructions to make this happen:

Step #	Generic Keystrokes	Description
1	Sequence(x, x, 1, 100) $\rightarrow$ List1	Generate a sequence of numbers from 1 to 100, and store this sequence in List1.
2	Sequence(*rand*, x, 1, 100) $\rightarrow$ List2	Create a sequence of random numbers and store them in List2.
3	Int(p − List2) + 1 $\rightarrow$ List3	Convert the random numbers from List2 into 0's and 1's in List3.
4	CumSum(List3) $\rightarrow$ List4	Store the cumulative frequency from List3 in List4.
5	List4/List1 $\rightarrow$ List5	Convert the cumulative frequency to a relative frequency "so far" and store in List5.

Now you can make a scatter plot using the numbers in List1 and the relative frequencies in List5. The accompanying figure shows what a typical screen might look like after performing the procedure. We used $p = .5$ for our coin, and so we added the line *relative frequency = .5* for reference.

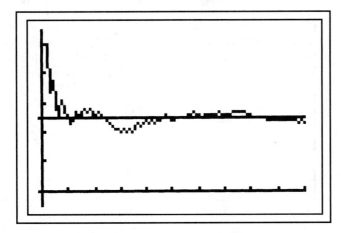

6.2 Simulating Events That Are Independent

As noted earlier, your calculator has a sequence of keystrokes that we have referred to as "*rand.*" The numbers generated by *rand* may be used to construct simulations that involve independent events.

The *rand* function automatically generates "independent" random numbers. This means, for example, that the probability of *rand* returning a number between .5 and 1.0 is unaffected by the value of the previously returned random number. The practical effect of this is that simply pressing the rand button repeatedly generates a sequence of such numbers. Let's see how we can use this capability in a simulation.

We will consider a simplified version of an important problem from meteorology: In a world where a photon of light may be absorbed or deflected in its path, how far will it travel in its original direction? The safety of drivers in fog depends on enough light reaching an oncoming car so that they are aware of the presence of another car.

For purposes of our simulation we will assume that (1) an absorbed photon of light is no longer visible, (2) a photon once deflected does *not* return to its original path, and (3) there are constant probabilities of deflection and absorption. Furthermore, the probability of the fate of the photon one second from now is independent of its status now. It is this real-life independence that requires us to generate "independent" random numbers to faithfully mirror the situation being modeled.

Our conceptual model of a photon's travel will be a point traveling from left to right in the grid below, starting — naturally enough — at square one. (*Note*: The probabilities we will use are for illustrative purposes and do not represent real probabilities in actual fog.)

1	2	3	4	5	6	7	8	9	10

At each position, the photon has three possible fates:

1. From square x it will go on to square $x + 1$ with probability .90.
2. From square x it will be absorbed with probability .05.
3. From square x it may be deflected "up" or "down," each with probability .025.

We arbitrarily define square 10 to be the position at which we would like an oncoming driver to be aware of our presence. We wish to find the probability a photon will get all the way to square 10. To estimate this probability via simulation, we need to link the probabilities of the fates listed above with random numbers generated by *rand*.

Photon Fate	Probability	*Rand* Interval
Continue on path	.9	$0 \leq rand < .9$
Be absorbed	.05	$.9 \leq rand < .95$
Be deflected	.05	$.95 \leq rand < 1.0$

Here are two rows of five random numbers generated by our *rand* procedure. A trial proceeds by reading across the first row, then across the second row.

.812788	.710859	.704518	.146311	.956130
.050035	.126108	.330590	.390195	.114339

In this trial, the photon of light would go straight for 4 intervals and then — according to the interpretation of the random number .956130 — be deflected.

Here is a second trial:

| .576809 | .715678 | .404556 | .635070 | .811570 |
| .062257 | .490527 | .285129 | .758026 | .231308 |

In this second trial, the simulated photon makes it all the way to the oncoming car; all the random numbers are less than .90.

Forty succeeding trials were initiated with the following results, where Success is defined as a photon making it to square 10:

FSSFFFSFFFSSSFFFFSSSFFFFSFFSFFFSFFFFSSFF

Based on these 40 simulated trials, we estimate the probability of a photon arriving uninterrupted at square 10 to be approximately 14/40, or .35.

Here we have considered only whether the photon arrives successfully at square 10. This is not the only simulation we could consider or the most interesting! We could, for example, consider the typical distance a photon travels before it is absorbed or deflected. We might ask the question "At least how far do 50% of the photons travel?" To address this question we would want to know the median of the distribution of photon travel distances. In this situation the outcome of a trial would not be success or failure, but the actual number of squares traveled by the photon before being absorbed or deflected In the long run, the distribution of simulation results should look more and more like the "true" distribution of photon distances traveled, and the median of our simulation results should get closer and closer to the true median of photon travel distance.

6.3 Simulating Events That Are Not Independent

The process of simulating events that are not independent uses the definition of conditional probability. We must consider the outcome of one event before we can simulate the other event. We can simulate the outcome of an event by generating a random number; simulating two dependent events will require the generating of at least two random numbers, depending on the complexity of the simulation.

To illustrate, consider the following example. Burglaries occur about once a month in a neighborhood troubled by crime. The neighbors decided to get a watchdog, whose function would be to bark loudly and scare away burglars. They have rescued an elderly German shepherd from the local animal shelter, and his bark is very much worse than his bite; much of his time is spent sleeping, and there is some concern about his hearing. But he *does* have a relatively ferocious bark, and they got him cheap. From actual trial runs using local neighbors disguised as burglars, it was discovered that Rover actually noticed the simulated burglars 90% of the time, and when he did so, he emitted a ferocious bark about 75% of the time. The question the neighbors have is, When can they expect a payoff from Rover? On average, how many months will go by before a burglar is noticed by Rover and frightened by Rover's ferocious bark?

Of course, basic probability rules could be used to calculate the probability that Rover notices a burglar *and* barks, but it isn't obvious how to determine the number of months that will go by before this event occurs. A simulation can be used to provide some insight in this case.

A particular burglar-event can be modeled using the definition of conditional probability and two random numbers generated by *rand*:

1. If the first random number is less than or equal to .90, this will count as detecting the burglar.
2. If the burglar is detected, we will use the second random number to determine whether Rover barks. A value less than or equal to .75 will indicate a "ferocious bark."
3. The first two steps represent the burglar-event for one month. We will repeat steps 1 and 2 until we get a "ferocious bark," thus completing a trial of the simulation. We can note the number of months required to get an occurrence of "detected *and* ferocious bark."
4. Many repetitions (trials) would generate a distribution of times until a ferocious burglar-scaring bark occurs, and we could examine this distribution to find a "typical" number of months between successful burglar scarings.

For example, here is a record of 5 trials of this experiment, using random numbers generated by the calculator:

First Random #	Detection?	Second Random #	Ferocious Bark?
.8767844333	Yes	.7900592279	No
.9546136582	No	.5215775123	(not detected)
.5329329133	Yes	.2805102664	Yes — a 3-month wait
.1320755632	Yes	.7338039727	Yes — a 1-month wait
.6958584068	Yes	.5688014123	Yes — a 1-month wait
.9961435487	No	.6002686874	(not detected)
.7286825076	Yes	.3887292922	Yes — a 2-month wait
.4319389989	Yes	.0143592629	Yes — a 1-month wait

With only 5 trials, of course, we cannot reach any reliable conclusions, but after many repetitions we would be able to estimate the probability that a burglar would be detected and scared away after 1, 2, 3, or any other number of months.

You should notice that we were led to simulation because our basic probability rules alone were not adequate to solve the problem of interest. Of course, an analytic solution is preferable to one based on simulation because it gives a correct answer quickly; simulation and an appeal to the law of large numbers can only give an approximate answer, and may take considerable time and bookkeeping to execute.

6.4 Larger Simulations

Our examples of simulations have been fairly small so far. Usually, simulations are undertaken with a large number of trials. Just as more data gives us a better idea of a distribution of data, more trials in a simulation will yield a better idea about the distribution of outcomes in a simulation.

If a simulation is not too complicated, it is possible to use an interesting calculator function called a "Sequence" function, or "seq" for short. This function can be used with the list capability of your calculator to perform some interesting simulations and analyze the results graphically. The sequence command will allow a sequence of operations to be performed repetitiously, with results then stored in a list. If 1 random number is desired, you can just use *rand*; if 200 random numbers in a list are desired, the seq command can be used to fill the list. For example, the command

seq(*rand,* x, 1, 200) → List1

stores 200 random numbers in List1. As always, you should check your calculator manual to see what the syntax of the command is on your calculator.

Now we will perform a simulation that illustrates this command. As every algebra student knows, adding apples and apples gives you apples. Here's a question about random numbers generated using *rand*. Will the distribution of the average of random numbers look like the distribution of random numbers used to calculate the average? Let's see what we can find out. We will use the seq command to put 200 numbers generated by *rand,* in each of 5 lists.

Seq(*rand,* x, 1, 200) → List1
Seq(*rand,* x, 1, 200) → List2
Seq(*rand,* x, 1, 200) → List3
Seq(*rand,* x, 1, 200) → List4
Seq(*rand,* x, 1, 200) → List5

Now we'll add the results, store them in List6, then divide their sum by 5 to get the averages, and store in List6:

(List1 + List2 + List3 + List4 + List5)/5 → List6

The settings in the window shown in the accompanying figure were used to generate the histograms on page 304. The first histogram is of the data in List2, and is representative of the histograms in Lists 1–5. The second histogram is a histogram of List6, the sum of the first 5 lists, divided by 5 (the average of five numbers). Do they have the same shape? It doesn't look like it.

```
WINDOW
 Xmin=-.1
 Xmax=1.1
 Xscl=.1
 Ymin=-.1
 Ymax=80
 Yscl=20
 Xres=1▮
```

We can see the effect of averaging numbers. As you continue the study of statistics you will find that this possibly unexpected behavior is due to a result known as the Central Limit Theorem, an important result in statistics. Our point here is that a calculator, together with data analysis skills, can be used to perform nontrivial simulations and analyze the results statistically.

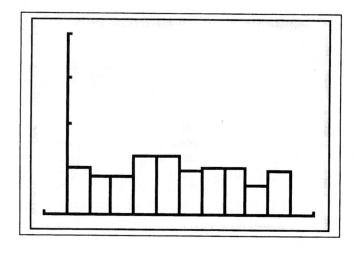

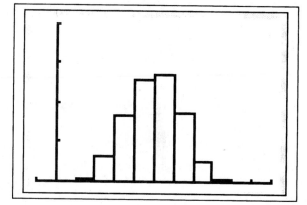

7

Random Variables and Probability Distributions

INTRODUCTION

This chapter is the first of two that together link the basic ideas of probability explored in the previous chapter with the techniques of statistical inference. Chapter 6 used probability to describe the long-run frequency of occurrence of various types of outcomes. This chapter introduces probability models that can be used to describe the distribution of values of a variable. Such models are essential if we are to reach conclusions based on a sample from the population of interest.

In a chance experiment, we often focus on some numerical aspect of the outcome. An environmental scientist who obtains an air sample from a specified location might be especially concerned with the concentration of ozone (a major constituent of smog). A quality control inspector who must decide whether to accept a large shipment of components may base the decision on the number of defectives in a group of 20 components randomly selected from the shipment.

Prior to selection of the air sample, there is uncertainty as to what value of ozone concentration will result. Similarly, the number of defective components among the 20 selected might be any whole number between 0 and 20. Because the value of a variable quantity such as ozone concentration or number of defectives is subject to uncertainty, it is called a *random variable*.

This chapter begins by distinguishing between categorical and numerical variables and between discrete and continuous numerical variables. We show how variation in both discrete and continuous numerical variables can be described by a probability distribution, which can also be used to make probability statements about values of the random variable. Special emphasis is given to three commonly encountered probability distributions: the binomial, geometric, and normal distributions.

7.1   Random Variables

In most chance experiments, an investigator will focus attention on one or more variable quantities. As an example, consider a management consultant who is studying the operation of a supermarket. The chance experiment might involve randomly selecting a customer leaving the store. One interesting numerical variable would be the number of items x purchased by the customer. Possible values of this variable are 0 (a frustrated customer), 1, 2, 3, and so on. Until a customer is selected and the number of items counted, there is uncertainly as to what value of x will result. Another variable of potential interest is the time y (minutes) spent in a checkout line. One possible value of y is 3.0 minutes and another is 4.0 minutes, but *any* other number between 3.0 and 4.0 is also a possibility. Whereas possible values of x are isolated points on the number line, possible y values form an entire interval (a continuum) on the number line.

> **Definition**
>
> A numerical variable whose value depends on the outcome of a chance experiment is called a **random variable.** A random variable associates a numerical value with each outcome of a chance experiment.
>
> A random variable is **discrete** if its set of possible values is a collection of isolated points on the number line. The variable is **continuous** if its set of possible values includes an entire interval on the number line.
>
> We will use lowercase letters, such as x and y, to represent random variables.*

Figure 7.1 shows a set of possible values for each type of random variable. In practice, a discrete random variable almost always arises in connection with counting (for example, the number of items purchased, the number of gas pumps in use, the number of broken eggs in a carton). A continuous random variable is one whose value is typically obtained by measurement (temperature in a freezer compartment, weight of a pineapple, amount of time spent in the store, etc.). Because there is a limit to the accuracy of any measuring instrument, such as a watch or a scale, it may seem that any variable should be regarded as discrete. However, when there are a

FIGURE 7.1 Two different types of random variables

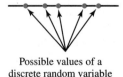

Possible values of a
discrete random variable

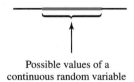

Possible values of a
continuous random variable

---

*In some books, uppercase letters are used to name random variables, with lowercase letters representing a particular value that the variable might assume. We have opted to use a simpler and less formal notation.

large number of very closely spaced values, the variable's behavior is most easily studied by conceptualizing it as continuous. (Doing so allows the use of calculus to solve some types of probability problems.)

EXAMPLE 7.1 Consider an experiment in which the type of car, new (N) or used (U), chosen by each of three successive customers at a discount car dealership is noted. Define a random variable x by

x = number of customers purchasing a new car

The experimental outcome in which the first and third customers purchase a new car and the second customer purchases a used car can be abbreviated NUN. The associated x value is 2, since two of the three customers selected a new car. Similarly, the x value for the outcome NNN (all three purchase a new car) is 3. We display each of the eight possible experimental outcomes and the corresponding value of x in the accompanying table.

Outcome	UUU	NUU	UNU	UUN	NNU	NUN	UNN	NNN
x value	0	1	1	1	2	2	2	3

There are only four possible x values — 0, 1, 2, and 3 — and these are isolated points on the number line. Thus, x is a discrete random variable.

In some situations, the random variable of interest is discrete, but the number of possible values is not finite. This is illustrated in the following example.

EXAMPLE 7.2 Two friends agree to play a game that consists of a sequence of trials. The game will continue until one player wins two trials in a row. One random variable of interest might be

x = number of trials required to complete the game

Letting A denote a win for player 1 and B denote a win for player 2, the simplest possible experimental outcomes are AA (the case where player 1 wins the first two trials and the game ends) and BB (the case where player 2 wins the first two trials). With either of these two outcomes, $x = 2$. There are also two outcomes for which $x = 3$: ABB and BAA. Some other possible outcomes and associated x values are

Outcomes	x value
AA, BB	2
BAA, ABB	3
ABAA, BABB	4
ABABB, BABAA	5
⋮	⋮
ABABABABAA, BABABABABB	10
etc.	

Any positive integer that is at least 2 is a possible value. For example, one possibility that would result in $x = 50$ occurs if the first trial is A, the first 48 trials alternate A and B, and the final two trials are A. Because the values 2, 3, 4, . . . are isolated points on the number line (x is determined by counting), x is a discrete random variable even though there is no upper limit to the number of possible values.

EXAMPLE 7.3

In an engineering stress test, pressure is applied to a thin 1-foot-long bar until the bar snaps. There is uncertainty concerning the precise location at which the bar will snap. Let x be the distance from the left end of the bar to the break. Then $x = .25$ is one possibility, $x = .9$ is another, and in fact any number between 0 and 1 is a possible value of x. (Figure 7.2 shows the case of the outcome $x = .6$.) This set of possible values is an entire interval on the number line, so x is a continuous random variable.

FIGURE 7.2 The bar for Example 7.3 and the outcome $x = .6$

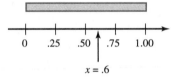

Even though in practice we may only be able to measure the distance to the nearest tenth of an inch or hundredth of an inch, the *actual* distance could be any number between 0 and 1. Even though the recorded values might be rounded due to the accuracy of the measuring instrument, the variable is still continuous.

In data analysis, random variables often arise in the context of summarizing sample data when a sample is selected from some population. This is illustrated in the following example.

EXAMPLE 7.4

Suppose that a counselor plans to select a random sample of 50 seniors at a large high school and ask each student in the sample whether he or she plans to attend college after graduation. The process of sampling is a chance experiment. The sample space for this experiment consists of all the different possible random samples of size 50 that might result (there are a very large number of these), and for simple random sampling, each of these outcomes is equally likely. Let

x = number of successes in the sample

where a success in this instance is defined as a student who plans to attend college. Then x is a random variable, because it associates a numerical value with each of the possible outcomes (random samples) that might occur. One possible sample might result in $x = 18$, whereas other possible samples might result in $x = 12$ or $x = 22$. Possible values of x are 0, 1, 2, . . . , 50, and x is a discrete random variable.

Exercises 7.1 – 7.7

7.1 State whether each of the following random variables is discrete or continuous.
   a. The number of defective tires on a car
   b. The body temperature of a hospital patient
   c. The number of pages in a book
   d. The number of draws (with replacement) from a deck of cards until a heart is selected
   e. The lifetime of a lightbulb

7.2 Classify each of the following random variables as either discrete or continuous.
   a. The fuel efficiency (mpg) of an automobile
   b. The amount of rainfall at a particular location during the next year
   c. The distance that a person throws a baseball
   d. The number of questions asked during a 1-hour lecture
   e. The tension (lb/in^2) at which a tennis racket is strung
   f. The amount of water used by a household during a given month
   g. The number of traffic citations issued by the Highway Patrol in a particular county on a given day

7.3 Starting at a particular time, each car entering an intersection is observed to see whether it turns left (L), right (R), or goes straight ahead (S). The experiment terminates as soon as a car is observed to go straight. Let y denote the number of cars observed. What are possible y values? List five different outcomes and their associated y values.

7.4 A point is randomly selected from the interior of a square, as pictured. Let x denote the distance from the lower left corner A of the square to the selected point.

What are possible values of x? Is x a discrete or continuous variable?

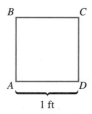

7.5 A point is randomly selected on the surface of a lake that has maximum depth of 100 ft. Let y = the depth of the lake at the randomly chosen point. What are possible values of y? Is y discrete or continuous?

7.6 A person stands at the corner marked A of the square pictured in Exercise 7.4 and tosses a coin. If it lands heads, the person moves one corner clockwise, to B. If the coin lands tails, the person moves one corner counterclockwise, to D. This process is then repeated until the person arrives back at A. Let y denote the number of coin tosses. What are possible values of y? Is y discrete or continuous?

7.7 A box contains four slips of paper marked 1, 2, 3, and 4. Two slips are selected without replacement. List the possible values for each of the following random variables.
   a. x = sum of the two numbers
   b. y = difference between first and second numbers
   c. z = number of slips selected that show an even number
   d. w = number of slips selected that show a 4

7.2 Probability Distributions for Discrete Random Variables

The probability distribution for a random variable is a model that describes the long-run behavior of the variable. For example, suppose that the Department of Animal Regulation in a particular county is interested in studying the variable

x = number of licensed dogs or cats for a household

County regulations prohibit more than five dogs or cats per household. If we consider the chance experiment of randomly selecting a household in this county, then x is a discrete random variable because it associates a numerical value (0, 1, 2, 3, 4, or 5) with each of the possible outcomes (households) in the sample space. Although we know what the possible values for x are, it would also be useful to know how this variable would behave in repeated observation. What would be the most common value? What proportion of the time would $x = 5$ be observed? $x = 3$? A probability distribution provides this type of information about the long-run behavior of a random variable.

Definition

The **probability distribution of a discrete random variable** x gives the probability associated with each possible x value. Each probability is the limiting relative frequency of occurrence of the corresponding x value when the experiment is repeatedly performed.

Common ways to display a probability distribution for a discrete random variable are a table, a probability histogram, or a formula.

If one possible value of x is 2, we often write $p(2)$ in place of $P(x = 2)$. Similarly, $p(5)$ denotes the probability that $x = 5$, and so on.

EXAMPLE 7.5

Suppose that each of four randomly selected customers purchasing a hot tub at a certain store chooses either an electric (E) or a gas (G) model. Assume that these customers make their choices independently of one another and that 40% of all customers select an electric model. This implies that for any particular one of the four customers, $P(E) = .4$ and $P(G) = .6$. One possible experimental outcome is EGGE, where the first and fourth customers select electric models and the other two choose gas models. Because the customers make their choices independently, the multiplication rule for independent events implies that

$P(\text{EGGE}) = P(\text{1st chooses E } and \text{ 2nd chooses G } and \text{ 3rd chooses G } and \text{ 4th chooses E})$
$= P(E) \cdot P(G) \cdot P(G) \cdot P(E)$
$= (.4)(.6)(.6)(.4)$
$= .0576$

Similarly,

$P(\text{EGEG}) = P(E) \cdot P(G) \cdot P(E) \cdot P(G)$
$= (.4)(.6)(.4)(.6)$
$= .0576$ (identical to $P(\text{EGGE})$)

and

$P(\text{GGGE}) = (.6)(.6)(.6)(.4) = .0864$

The number among the four customers who purchase an electric hot tub is a random variable. Let

x = number of electric hot tubs purchased by the four customers

Table 7.1 displays the 16 possible experimental outcomes, the probability of each outcome, and the value of the random variable x that is associated with each outcome.

Table 7.1 *Outcomes and probabilities for Example 7.5*

Outcome	Probability	x value	Outcome	Probability	x value
GGGG	.1296	0	GEEG	.0576	2
EGGG	.0864	1	GEGE	.0576	2
GEGG	.0864	1	GGEE	.0576	2
GGEG	.0864	1	GEEE	.0384	3
GGGE	.0864	1	EGEE	.0384	3
EEGG	.0576	2	EEGE	.0384	3
EGEG	.0576	2	EEEG	.0384	3
EGGE	.0576	2	EEEE	.0256	4

The probability distribution of x is easily obtained from this information. Consider the smallest possible x value, 0. The only outcome for which $x = 0$ is GGGG, so

$$p(0) = P(x = 0) = P(\text{GGGG}) = .1296$$

There are four different outcomes for which $x = 1$, so $p(1)$ results from summing the four corresponding probabilities:

$$
\begin{aligned}
p(1) = P(x = 1) &= P(\text{EGGG } or \text{ GEGG } or \text{ GGEG } or \text{ GGGE}) \\
&= P(\text{EGGG}) + P(\text{GEGG}) + P(\text{GGEG}) + P(\text{GGGE}) \\
&= .0864 + .0864 + .0864 + .0864 \\
&= 4(.0864) \\
&= .3456
\end{aligned}
$$

Similarly,

$$
\begin{aligned}
p(2) &= P(\text{EEGG}) + \cdots + P(\text{GGEE}) = 6(.0576) = .3456 \\
p(3) &= 4(.0384) = .1536 \\
p(4) &= .0256
\end{aligned}
$$

The probability distribution of x is summarized in the accompanying table.

x value	0	1	2	3	4
$p(x)$ = probability of value	.1296	.3456	.3456	.1536	.0256

To interpret $p(3) = .1536$, think of performing the experiment repeatedly, each time with a new group of four customers. In the long run, 15.36% of these groups will have exactly three customers purchasing an electric hot tub.

The probability distribution can be used to determine probabilities of various events involving x. For example, the probability that at least two of the four customers choose electric models is

$$
\begin{aligned}
P(x \geq 2) &= P(x = 2 \ or \ x = 3 \ or \ x = 4) \\
&= p(2) + p(3) + p(4) \\
&= .5248
\end{aligned}
$$

Thus, in the long run, 52.48% of the time a group of four hot tub purchasers will include at least two who select electric models.

When a probability distribution for a discrete variable is given in tabular form, as was done in Example 7.5, the table shows the possible x values and also $p(x)$ for each possible x value. Because $p(x)$ is a probability, it must be a number between 0 and 1, and because the probability distribution lists all possible x values, the sum of all the $p(x)$ values must be equal to 1. These properties of discrete probability distributions are summarized in the accompanying box.

Properties of Discrete Probability Distributions

1. For every possible x value, $0 \leq p(x) \leq 1$.

2. $\displaystyle\sum_{\text{all } x \text{ values}} p(x) = 1$

In tabular form, the probability distribution for a discrete random variable looks exactly like a relative frequency distribution of the sort discussed in Chapter 3. There we introduced a histogram as a pictorial representation of a relative frequency distribution. An analogous picture for a discrete probability distribution is called a *probability histogram*. The picture has a rectangle centered above each possible value of x, and the area of each rectangle is the probability of the corresponding value. Figure 7.3 displays the probability histogram for the probability distribution of Example 7.5.

In Example 7.5, the probability distribution was derived by starting with a simple experimental situation and applying basic probability rules. Often a derivation from

FIGURE 7.3 Probability histogram for the distribution of Example 7.5

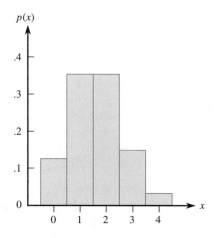

fundamental probabilities is not possible because of the complexity of the experimental situation. In most cases, an investigator conjectures a probability distribution consistent with empirical evidence and prior knowledge. When a probability distribution is constructed to be consistent with empirical evidence, it must also be consistent with rules of probability. Specifically,

1. $p(x) \geq 0$ for every x value

2. $\sum_{\text{all } x \text{ values}} p(x) = 1$

EXAMPLE 7.6

A consumer organization that evaluates new automobiles customarily reports the number of major defects on each car examined. Let x denote the number of major defects on a randomly selected car of a certain type. A large number of automobiles were evaluated, and a probability distribution consistent with these observations is

x	0	1	2	3	4	5	6	7	8	9	10
$p(x)$	.041	.010	.209	.223	.178	.114	.061	.028	.011	.004	.001

The corresponding probability histogram appears in Figure 7.4. The probabilities in this distribution reflect the organization's experience. For example, $p(3) = .223$ indicates that 22.3% of new automobiles had 2 major defects. The probability that the number of major defects is between 2 and 5 inclusive is

$$P(2 \leq x \leq 5) = p(2) + p(3) + p(4) + p(5) = .724$$

If car after car of this type were examined, in the long run, 72.4% would have 2, 3, 4, or 5 major defects.

FIGURE 7.4 Probability histogram for the distribution of number of major defects

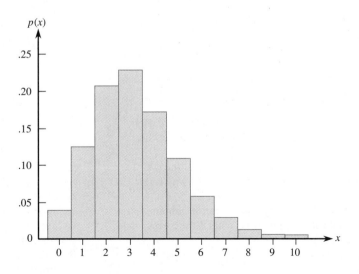

We have seen examples where the probability distribution of a discrete random variable has been given as a table or as a probability (relative frequency) histogram. It is also possible to give a formula that allows calculation of the probability for each of the possible values of the random variable. Examples of this approach are given in Section 7.5.

Exercises 7.8 – 7.19

7.8 Let x be the number of courses for which a randomly selected student at a certain university is registered. The probability distribution of x appears in the accompanying table.

x	1	2	3	4	5	6	7
$p(x)$	.02	.03	.09	.25	.40	.16	.05

a. What is $P(x = 4)$?
b. What is $P(x \leq 4)$?
c. What is the probability that the selected student is taking at most five courses?
d. What is the probability that the selected student is taking at least five courses? More than five courses?
e. Calculate $P(3 \leq x \leq 6)$ and $P(3 < x < 6)$. Explain in words why these two probabilities are different.

7.9 Let y denote the number of broken eggs in a randomly selected carton of one dozen "store brand" eggs at a certain market. Suppose that the probability distribution of y is as follows.

y	0	1	2	3	4
$p(y)$	.65	.20	.10	.04	?

a. Only y values 0, 1, 2, 3, and 4 have positive probabilities. What is $p(4)$?
b. How would you interpret $p(1) = .20$?
c. Calculate $P(y \leq 2)$, the probability that the carton contains at most two broken eggs, and interpret this probability.
d. Calculate $P(y < 2)$, the probability that the carton contains *fewer than* two broken eggs. Why is this smaller than the probability in part **c**?
e. What is the probability that the carton contains exactly ten unbroken eggs? (*Hint*: What is the corresponding value of y?)
f. What is the probability that at least ten eggs are unbroken?

7.10 A restaurant has four bottles of a certain wine in stock. Unbeknownst to the wine steward, two of these bottles (numbers 1 and 2) are bad. Suppose that two bottles are ordered, and let x be the number of good bottles among these two.
a. One possible experimental outcome is (1, 2) [bottles 1 and 2 are the ones selected] and another is (2, 4). List all possible outcomes.
b. Assuming that the two bottles are randomly selected from among the four (equally likely outcomes), what is the probability of each outcome in part **a**?
c. The value of x for the (1, 2) outcome is 0 (neither selected bottle is good), and $x = 1$ for the outcome (2, 4). Determine the x value for each possible outcome. Then use the probabilities in part **b** to determine the probability distribution of x.

7.11 Airlines sometimes overbook flights. Suppose that for a plane with 100 seats, an airline takes 110 reservations. Define the variable x as the number of people who actually show up for a sold-out flight. From past experience, the probability distribution of x is given in the following table.

x	95	96	97	98	99	100	101	102	103
$p(x)$	.05	.10	.12	.14	.24	.17	.06	.04	.03

x	104	105	106	107	108	109	110
$p(x)$	.02	.01	.005	.005	.005	.0037	.0013

a. What is the probability that the airline can accommodate everyone who shows up for the flight?
b. What is the probability that not all passengers can be accommodated?
c. If you are trying to get a seat on such a flight and you are number 1 on the standby list, what is the probability that you will be able to take the flight? What if you are number 3?

7.12 Many manufacturers have quality control programs that include inspection of incoming materials for defects. Suppose that a computer manufacturer receives computer boards in lots of five. Two boards are selected from each lot for inspection. We can represent possible outcomes of the selection process by pairs. For example, the pair (1, 2) represents the selection of boards 1 and 2 for inspection.

a. List the ten different possible outcomes.

b. Suppose that boards 1 and 2 are the only defective boards in a lot of five. Two boards are to be chosen at random. Define x to be the number of defective boards observed among those inspected. Find the probability distribution of x.

7.13 Simulate the experiment described in Exercise 7.12 using five slips of paper, with two marked defective and three marked nondefective. Place the slips in a box, mix them well, and draw out two. Record the number of defectives. Replace the slips and repeat until you have 50 observations on the variable x. Construct a relative frequency distribution for the 50 observations and compare this with the probability distribution obtained in Exercise 7.12.

7.14 Of all airline flight requests received by a certain discount ticket broker, 70% are for domestic travel and 30% are for international flights. Let x be the number of requests among the next three received that are for domestic flights. Assuming independence of successive requests, determine the probability distribution of x. (*Hint*: One possible outcome is DID, with the probability $(.7)(.3)(.7) = .147$.)

7.15 Some parts of California are particularly earthquake-prone. Suppose that in one such area, 20% of all homeowners are insured against earthquake damage. Four homeowners are to be selected at random; let x denote the number among the four who have earthquake insurance.

a. Find the probability distribution of x. (*Hint*: Let S denote a homeowner who has insurance and F one who does not. Then one possible outcome is SFSS, with probability $(.2)(.8)(.2)(.2)$ and associated x value of 3. There are 15 other outcomes.)

b. What is the most likely value for x?

c. What is the probability that at least two of the four selected have earthquake insurance?

7.16 A box contains five slips of paper, marked $1, $1, $1, $10, and $25. The winner of a contest will select two

slips of paper at random and will then get the larger of the dollar amounts on the two slips. Define a random variable w by

w = amount awarded

Determine the probability distribution of w. (*Hint*: Think of the slips as numbered 1, 2, 3, 4, and 5, so that an outcome of the experiment consists of two of these numbers.)

7.17 Components coming off an assembly line are either free of defects (S, for success) or defective (F, for failure). Suppose that 70% of all such components are defect-free. Components are independently selected and tested one by one. Let y denote the number of components that must be tested until a defect-free component is obtained.

a. What is the smallest possible y value, and what experimental outcome gives this y value? What is the second smallest y value, and what outcome gives rise to it?

b. What is the set of all possible y values?

c. Determine the probability of each of the five smallest y values. You should see a pattern that leads to a simple formula for $p(y)$, the probability distribution of y.

7.18 A contractor is required by a county planning department to submit anywhere from one to five forms (depending on the nature of the project) in applying for a building permit. Let y = the number of forms required of the next applicant. The probability that y forms are required is known to be proportional to y; that is, $p(y) = ky$ for $y = 1, \ldots, 5$.

a. What is the value of k? (*Hint*: $\Sigma\, p(y) = 1$.)

b. What is the probability that at most three forms are required?

c. What is the probability that between two and four forms (inclusive) are required?

d. Could $p(y) = y^2/50$ for $y = 1, \ldots, 5$ be the probability distribution of y? Explain.

7.19 A library subscribes to two different weekly news magazines, each of which is supposed to arrive in Wednesday's mail. In actuality, each one could arrive on Wednesday, Thursday, Friday, or Saturday. Suppose that the two arrive independently of one another and that for each one,

$$P(W) = .4 \qquad P(T) = .3 \qquad P(F) = .2 \qquad P(S) = .1$$

Define a random variable y by

y = the number of days beyond Wednesday that it takes for both magazines to arrive

For example, if the first arrives on Friday and the second on Wednesday, then $y = 2$, whereas $y = 1$ if both magazines arrive on Thursday. Obtain the probability distribution of y. (*Hint*: Draw a tree diagram with two generations of branches, the first labeled with arrival days for magazine 1 and the second for magazine 2.)

7.3 Probability Distributions for Continuous Random Variables

A continuous random variable is one that has as its set of possible values an entire interval on the number line. An example is the weight x (lb) of a newborn child. Suppose for the moment that weight is recorded only to the nearest pound. Then possible x values are whole numbers, such as 4 or 9. The probability distribution can be pictured as a probability histogram in which the area of each rectangle is the probability of the corresponding weight value. The total area of all the rectangles is 1, and the probability that a weight (to the nearest pound) is between two values, such as 6 and 8, is the sum of the corresponding rectangular areas. Figure 7.5(a) illustrates this.

Now suppose that weight is measured to the nearest tenth of a pound. There are many more possible weight values than before, such as 5.0, 5.1, 5.7, 7.3, and 8.9. As shown in Figure 7.5(b), the rectangles in the probability histogram are much narrower, and this histogram has a much smoother appearance than the first one. Again, this histogram can be drawn so that the area of each rectangle equals the corresponding probability and the total area of all the rectangles is 1.

Figure 7.5(c) shows what happens as weight is measured to a greater and greater degree of accuracy. The sequence of probability histograms approaches a smooth curve. The curve cannot go below the horizontal measurement scale, and the total area under the curve is 1 (because this is true of each probability histogram). The probability that x falls in an interval such as $6 \leq x \leq 8$ is the area under the curve and above that interval.

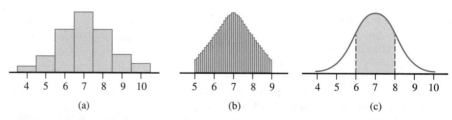

FIGURE 7.5 Probability distribution for birth weight: (a) Weight measured to the nearest pound; (b) Weight measured to the nearest tenth of a pound; (c) Limiting curve as measurement accuracy increases; shaded area $= P(6 \leq \text{weight} \leq 8)$

> **Definition**
>
> A **probability distribution for a continuous random variable** x is specified by a mathematical function denoted by $f(x)$ and called the **density function.** The graph of a density function is a smooth curve (the **density curve**). The following requirements must be met:
>
> 1. $f(x) \geq 0$ (so that the curve cannot dip below the horizontal axis).
> 2. The total area under the density curve is equal to 1.
>
> The probability that x falls in any particular interval is the area under the density curve and above the interval.

Many probability calculations for continuous random variables involve the following three events:

1. $a < x < b$, the event that the random variable x assumes a value between two given numbers, a and b
2. $x < a$, the event that the random variable x assumes a value less than a given number, a
3. $b < x$, the event that the random variable x assumes a value greater than a given number, b

Figure 7.6 illustrates how the probabilities of these events are identified with areas under a density curve.

FIGURE 7.6 Probabilities as areas under a probability density curve

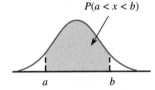
$P(a < x < b)$
a b

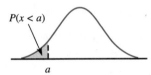
$P(x < a)$
a

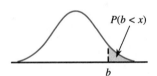
$P(b < x)$
b

EXAMPLE 7.7   Define a continuous random variable x by

> x = amount of time (min) taken by a clerk to process a certain type of application form

Suppose that x has a probability distribution with density function

$$f(x) = \begin{cases} .5 & 4 < x < 6 \\ 0 & \text{otherwise} \end{cases}$$

The graph of $f(x)$, the density curve, is shown in Figure 7.7(a) on page 318. It is especially easy to use, since calculating probabilities requires simply finding the area of rectangles using the formula

> area = (base) · (height)

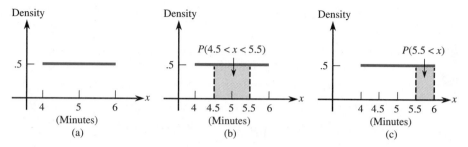

FIGURE 7.7 The uniform distribution for Example 7.7

The curve has positive height, .5, only between $x = 4$ and $x = 6$. The total area under the curve is just the area of the rectangle with base extending from 4 to 6 and with height .5. This gives

$$\text{area} = (6 - 4)(.5) = 1$$

as required.

When the density is constant over an interval (resulting in a flat density curve), the probability distribution is called a *uniform distribution.*

As illustrated in Figure 7.7(b), the probability that x is between 4.5 and 5.5 is

$$
\begin{aligned}
P(4.5 < x < 5.5) &= \text{area of shaded rectangle} \\
&= (\text{base width}) \cdot (\text{height}) \\
&= (5.5 - 4.5)(.5) \\
&= .5
\end{aligned}
$$

Similarly (see Figure 7.7(c)), because $x > 5.5$ is equivalent to $5.5 < x < 6$, we have

$$P(5.5 < x) = (6 - 5.5)(.5) = .25$$

According to this model, in the long run, 25% of all forms that are processed will have processing times that exceed 5.5 min.

▬▬▬▬

The probability that a discrete random variable x lies in the interval between two limits a and b depends on whether either limit is included in the interval. Suppose, for example, that x is the number of major defects on a new automobile. Then

$$P(3 \le x \le 7) = p(3) + p(4) + p(5) + p(6) + p(7)$$

whereas

$$P(3 < x < 7) = p(4) + p(5) + p(6)$$

However, if x is a *continuous* random variable, such as task completion time, then

$$P(3 \le x \le 7) = P(3 < x < 7)$$

because the area under a density curve and above a single value such as 3 or 7 is zero. Geometrically, we can think of finding the area above a single point as finding the area of a rectangle with width = 0. The area above an interval of values therefore does not depend on whether either endpoint is included.

For any two numbers a and b with $a < b$,

$$P(a \leq x \leq b) = P(a < x \leq b) = P(a \leq x < b) = P(a < x < b)$$

when x is a continuous random variable.

Probabilities for continuous random variables are often calculated using cumulative areas. A cumulative area is all the area under the density curve to the left of a particular value. Figure 7.8 illustrates the cumulative area to the left of .5, which is $P(x < .5)$. The probability that x is in any particular interval, $P(a < x < b)$, is the difference between two cumulative areas.

FIGURE 7.8 A cumulative area under a density curve

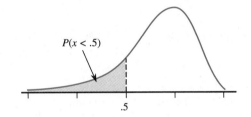

$P(x < .5)$

.5

The probability that a continuous random variable x lies between a lower limit a and an upper limit b is

$$P(a < x < b) = \text{(cumulative area to the left of } b) - \text{(cumulative area to the left of } a)$$
$$= P(x < b) - P(x < a)$$

The foregoing property is illustrated in Figure 7.9 for the case $a = .25$ and $b = .75$. We will use this result extensively in Section 7.6 when we calculate probabilities using the normal distribution.

For some continuous distributions, cumulative areas can be calculated using methods from the branch of mathematics called integral calculus. However, since

FIGURE 7.9 Calculation of $P(a < x < b)$ using cumulative areas

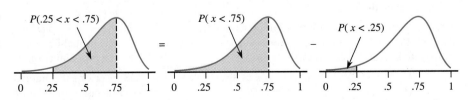

we are not assuming knowledge of calculus, we will rely on tables that have been constructed for the commonly encountered continuous probability distributions.

Exercises 7.20 – 7.26

7.20 Let x denote the lifetime (in thousands of hours) of a certain type of fan used in diesel engines. The density curve of x is as pictured. Shade the area under the curve corresponding to each of the following probabilities (draw a new curve for each part).

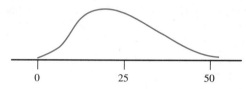

a. $P(10 < x < 25)$
b. $P(10 \leq x \leq 25)$
c. $P(x < 30)$
d. The probability that the lifetime is at least 25,000 hours
e. The probability that the lifetime exceeds 25,000 hours

7.21 A particular professor never dismisses class early. (Do you know anyone like this?) Let x denote the amount of time past the hour (minutes) that elapses before the professor dismisses class. Suppose that x has a uniform distribution on the interval from 0 to 10 min. The density curve is shown in the accompanying figure.

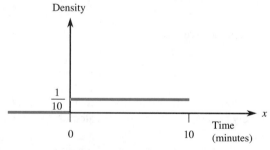

a. What is the probability that at most 5 min elapse before dismissal?
b. What is the probability that between 3 and 5 min elapse before dismissal?

7.22 Refer to the probability distribution given in Exercise 7.21. Put the following probabilities in order, from smallest to largest. Explain your reasoning.

(i) $P(2 < x < 3)$ (ii) $P(2 \leq x \leq 3)$
(iii) $P(x < 2)$ (iv) $P(x > 7)$

7.23 The article "Modeling Sediment and Water Column Interactions for Hydrophobic Pollutants" (*Water Research* (1984): 1169–1174) suggests the uniform distribution on the interval from 7.5 to 20 as a model for x = depth (cm) of the bioturbation layer in sediment for a certain region. (Please don't ask us what the bioturbation layer is!)

a. Draw the density curve for x.
b. What must the height of the density curve be?
c. What is the probability that x is at most 12?
d. What is the probability that x is between 10 and 15? Between 12 and 17? Why are these two probabilities equal?

7.24 Let x denote the amount of gravel sales (tons) during a randomly selected week at a particular sales facility. Suppose that the density curve has height $f(x)$ above the value x, where

$$f(x) = \begin{cases} 2(1-x) & 0 \leq x \leq 1 \\ 0 & \text{otherwise} \end{cases}$$

The density curve (the graph of $f(x)$) is shown in the accompanying figure. Use the fact that the area of a triangle $= \frac{1}{2}(base)(height)$ to calculate each of the following probabilities.

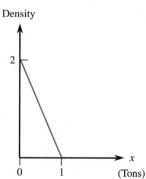

a. $P(x < \frac{1}{2})$
b. $P(x \leq \frac{1}{2})$
c. $P(x < \frac{1}{4})$
d. $P(\frac{1}{4} < x < \frac{1}{2})$ (*Hint*: Use the results of parts **a–c.**)
e. The probability that sales exceeds $\frac{1}{2}$ ton
f. The probability that sales is at least $\frac{1}{4}$ ton

7.25 Let x be the amount of time (min) that a particular San Francisco commuter must wait for a BART train.

Suppose that the density curve is as pictured (a uniform distribution).

Density

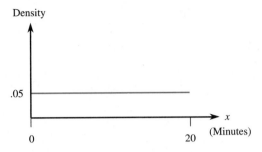

a. What is the probability that x is less than 10 min? More than 15 min?
b. What is the probability that x is between 7 and 12 min?
c. Find the value c for which $P(x < c) = .9$.

7.26 Referring to Exercise 7.25, let x and y be waiting times on two independently selected days. Define a new random variable w by $w = x + y$, the sum of the two waiting times. The set of possible values for w is the interval from 0 to 40 (since both x and y can range from 0 to 20).

It can be shown that the density curve of w is as pictured. (It is called a triangular distribution for obvious reasons!)

Density

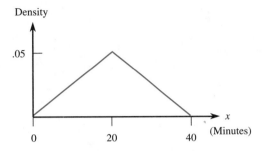

a. Verify that the total area under the density curve is equal to 1. (*Hint*: The area of a triangle is $\frac{1}{2}(base) \cdot (height)$.)
b. What is the probability that w is less than 20? Less than 10? More than 30?
c. What is the probability that w is between 10 and 30? (*Hint*: It might be easier first to find the probability that w is not between 10 and 30.)

7.4 The Mean and Standard Deviation of a Random Variable

We study a random variable x, such as the number of insurance claims made by a homeowner (a discrete variable) or the birth weight of a baby (a continuous variable), to learn something about how its values are distributed along the measurement scale. The sample mean $\bar{x}$ and sample standard deviation s summarize center and spread for the values in a sample. Similarly, the mean value and standard deviation of a random variable describe where the variable's probability distribution is centered and the extent to which it spreads out about the center.

> The **mean value of a random variable x,** denoted by μ_x, describes where the probability distribution of x is centered.
>
> The **standard deviation of a random variable x,** denoted by σ_x, describes variability in the probability distribution. When σ_x is small, observed values of x will tend to be close to the mean value (little variability). When the value of σ_x is large, there will be more variability in observed x values.

Figure 7.10(a) on page 322 shows two discrete probability distributions with the same standard deviation (spread) but different means (center). One distribution has a mean of $\mu_x = 6$, and the other has $\mu_x = 10$. Which is which? Figure 7.10(b) shows two continuous probability distributions that have the same mean but different standard deviations. Which distribution — (i) or (ii) — has the larger standard

FIGURE 7.10(a) Different values of μ_x with the same value of σ_x

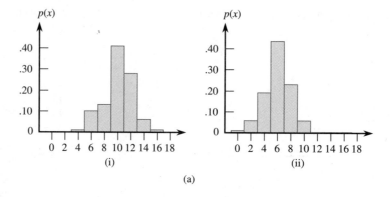

(a)

FIGURE 7.10(b) Different values of σ_x with the same value of μ_x

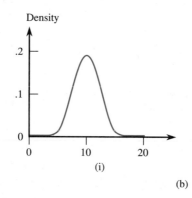
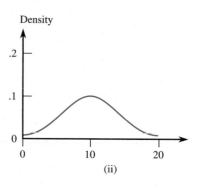

(b)

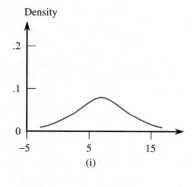
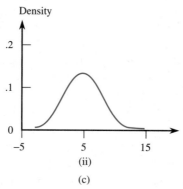
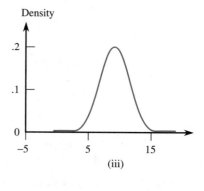

(c)

FIGURE 7.10(c) Different values of μ_x and σ_x

deviation? Finally, Figure 7.10(c) shows three continuous distributions with different means and standard deviations. Which of the three distributions has the largest mean? Which has a mean of about 5? Which distribution has the smallest standard deviation? (Check your answers to the questions above: 7.10(a)(ii) has a mean of 6, and 7.10(a)(i) has a mean of 10; 7.10(b)(ii) has the larger standard deviation; 7.10(c)(iii) has the largest mean, 7.10(c)(ii) has a mean of about 5, and 7.10(c)(iii) has the smallest standard deviation.)

It is customary to use the terms *mean of the random variable x* and *mean of the probability distribution of x* interchangeably. The same is true for the standard deviation: The standard deviation of the random variable x and the standard deviation of the probability distribution of x refer to the same thing. Although the mean and standard deviation are computed differently for discrete and continuous random variables, the interpretation is the same in both cases.

The Mean Value of a Discrete Random Variable

Consider the experiment consisting of the random selection of an automobile licensed in a particular state. Let the discrete random variable x be the number of low-beam headlights on the selected car that need adjustment. Possible x values are 0, 1, and 2, and the probability distribution of x might be as follows:

x value	0	1	2
Probability	.5	.3	.2

The corresponding probability histogram appears in Figure 7.11.

FIGURE 7.11 Probability histogram for the distribution of the number of headlights needing adjustment

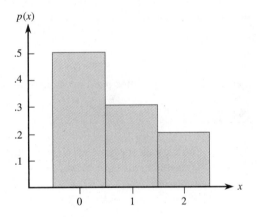

In a sample of 100 such cars, the sample relative frequencies might differ somewhat from these probabilities (which are the limiting relative frequencies). We might see

x value	0	1	2
Frequency	46	33	21

The sample average value of x for these 100 observations is then the sum of 46 zeros, 33 ones, and 21 twos, all divided by 100:

$$\bar{x} = \frac{(46)(0) + (33)(1) + (21)(2)}{100}$$

$$= \left(\frac{46}{100}\right)(0) + \left(\frac{33}{100}\right)(1) + \left(\frac{21}{100}\right)(2)$$

$$= (\text{rel. freq. of } 0)(0) + (\text{rel. freq. of } 1)(1) + (\text{rel. freq. of } 2)(2) = .75$$

As the sample size increases, each relative frequency will approach the corresponding probability. In a very long sequence of experiments, the value of $\bar{x}$ will approach

(probability that $x = 0$)(0) + (probability that $x = 1$)(1) + (probability that $x = 2$)(2)

$$= (.5)(0) + (.3)(1) + (.2)(2)$$
$$= .70$$
$$= \text{mean value of } x$$

Notice that the expression for $\bar{x}$ was a weighted average of possible x values; the weight for each value is the observed relative frequency. Similarly, the mean value of x is a weighted average, but now the weights are the probabilities from the probability distribution, consistent with the definition in the accompanying box.

Definition

The **mean value of a discrete random variable x,** denoted by μ_x, is computed by first multiplying each possible x value by the probability of observing that value and then adding the resulting quantities. Symbolically,

$$\mu_x = \sum_{\substack{\text{all possible} \\ x \text{ values}}} x \cdot p(x)$$

The term **expected value** is sometimes used in place of mean value, and $E(x)$ is alternative notation for μ_x.

EXAMPLE 7.8

Individuals applying for a certain license are allowed up to four attempts to pass the licensing exam. Let x denote the number of attempts made by a randomly selected applicant. The probability distribution of x is as follows:

x	1	2	3	4
$p(x)$	.10	.20	.30	.40

Thus x has mean value

$$\mu_x = \sum_{x=1,2,3,4} x \cdot p(x)$$
$$= (1) \cdot p(1) + (2) \cdot p(2) + (3) \cdot p(3) + (4) \cdot p(4)$$
$$= (1)(.10) + (2)(.20) + (3)(.30) + (4)(.40)$$
$$= .10 + .40 + .90 + 1.60$$
$$= 3.00$$

It is no accident that the symbol μ_x for the mean value is the same symbol used previously for a population mean. When the probability distribution describes how

x values are distributed among the members of a population (so probabilities are population relative frequencies), the mean value of x is exactly the average value of x in the population.

EXAMPLE 7.9

At 1 min after birth and again at 5 min, each newborn child is given a numerical rating called an Apgar score. Possible values of this score are $0, 1, 2, \ldots, 9$, and 10. A child's score is determined by five factors: muscle tone, skin color, respiratory effort, strength of heartbeat, and reflex, with a high score indicating a healthy infant. Let the random variable x denote the Apgar score (at 1 min) of a randomly selected newborn infant at a particular hospital, and suppose that x has the following probability distribution.

x	0	1	2	3	4	5	6	7	8	9	10
$p(x)$	.002	.001	.002	.005	.02	.04	.17	.38	.25	.12	.01

The mean value of x is

$$\mu_x = (0)p(0) + (1)p(1) + \cdots + (9)p(9) + (10)p(10)$$
$$= (0)(.002) + (1)(.001) + \cdots + (9)(.12) + (10)(.01)$$
$$= 7.16$$

The average Apgar score for a *sample* of newborn children born at this hospital may be $\bar{x} = 7.05$, $\bar{x} = 8.30$, or any one of a number of other possible values between 0 and 10. However, as child after child is born and rated, the average score will approach the value 7.16. This value can be interpreted as the mean Apgar score for the population of all babies born at this hospital.

The Standard Deviation of a Discrete Random Variable

The mean value μ_x provides only a partial summary of a probability distribution. Two different distributions may both have the same value of μ_x, yet a long sequence of sample values from one distribution may exhibit considerably more variability than a long sequence of values from the other distribution.

EXAMPLE 7.10

A television manufacturer receives certain components in lots of four from two different suppliers. Let x and y denote the number of defective components in randomly selected lots from the first and second suppliers, respectively. The probability distributions for x and y are as follows:

x	0	1	2	3	4		y	0	1	2	3	4
$p(x)$	.4	.3	.2	.1	0		$p(y)$	.2	.6	.2	0	0

Probability histograms for x and y are given in Figure 7.12.

It is easy to verify that the mean values of both x and y are 1, so for either supplier the long-run average number of defectives per lot is 1. However, the two probability histograms show that the probability distribution for the second supplier is

FIGURE 7.12 Probability distribution for the number of defective components: (a) Supplier 1; (b) Supplier 2

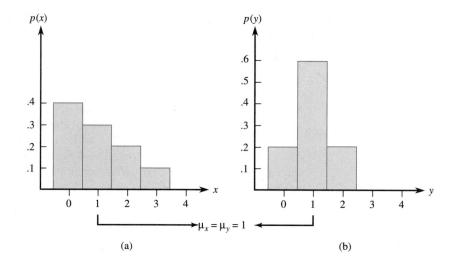

(a) (b)

concentrated nearer the mean value than is the first supplier's distribution. The greater spread of the first distribution implies that there will be more variability in a long sequence of observed x values than in an observed sequence of y values. For example, the y sequence will contain no 3's, whereas in the long run, 10% of the observed x values will be 3.

As with s^2 and s, the variance and standard deviation of x involve squared deviations from the mean. A value far from the mean results in a large squared deviation. However, such a value contributes substantially to variability in x only if the probability associated with that value is not too small. For example, if $\mu_x = 1$ and $x = 25$ is a possible value, the squared deviation is $(25 - 1)^2 = 576$. If, however, $P(x = 25) = .000001$, the value 25 will hardly ever be observed, so it won't contribute much to variability in a long sequence of observations. This is why each squared deviation is multiplied by the probability associated with the value (and thus weighted) to obtain a measure of variability.

Definition

The **variance of a discrete random variable x,** denoted by σ_x^2 is computed by first subtracting the mean from each possible x value to obtain the deviations, then squaring each deviation and multiplying the result by the probability of the corresponding x value, and finally adding these quantities. Symbolically,

$$\sigma_x^2 = \sum_{\substack{\text{all possible} \\ x \text{ values}}} (x - \mu)^2 \cdot p(x)$$

The **standard deviation of x,** denoted by σ_x, is the square root of the variance.

When the probability distribution describes how x values are distributed among members of a population (so probabilities are population relative frequencies), σ_x^2 and σ_x are the population variance and standard deviation (of x), respectively.

EXAMPLE 7.11

(Example 7.10 continued)   For x = number of defectives in a lot from the first supplier,

$$\sigma_x^2 = (0 - 1)^2 \cdot p(0) + (1 - 1)^2 \cdot p(1) + (2 - 1)^2 \cdot p(2) + (3 - 1)^2 \cdot p(3)$$
$$= (1)(.4) + (0)(.3) + (1)(.2) + (4)(.1)$$
$$= 1.0$$

so $\sigma_x = 1.0$. For y = the number of defectives in a lot from the second supplier,

$$\sigma_y^2 = (0 - 1)^2(.2) + (1 - 1)^2(.6) + (2 - 1)^2(.2) = .4$$

Then $\sigma_y = \sqrt{.4} = .632$. The fact that $\sigma_x > \sigma_y$ confirms the impression conveyed by Figure 7.12 concerning the variability of x and y.

EXAMPLE 7.12

(Example 7.9 continued)   Reconsider the distribution of Apgar scores for children born at a certain hospital. What is the probability that a child's score will be within 2 standard deviations of the mean score? As Figure 7.13 shows, values of x within 2 standard deviations of the mean are those for which $\mu - 2\sigma < x < \mu + 2\sigma$.

We already have $\mu_x = 7.16$. The variance is

$$\sigma^2 = \Sigma (x - \mu)^2 \cdot p(x) = \Sigma (x - 7.16)^2 \cdot p(x)$$
$$= (0 - 7.16)^2(.002) + (1 - 7.16)^2(.001) + \cdots + (10 - 7.16)^2(.01)$$
$$= 1.5684$$
$$\sigma = \sqrt{1.5684} = 1.25$$

This gives

$$P(\mu - 2\sigma \leq x \leq \mu + 2\sigma) = P(7.16 - 2.50 \leq x \leq 7.16 + 2.50)$$
$$= P(4.66 \leq x \leq 9.66)$$
$$= P(x = 5 \, or \, x = 6 \, or \, x = 7 \, or \, x = 8 \, or \, x = 9)$$
$$= p(5) + \cdots + p(9)$$
$$= .96$$

FIGURE 7.13 Values within 2 standard deviations of the mean

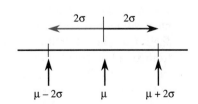

The Mean and Standard Deviation When x Is Continuous

Figure 7.14 illustrates how the density curve for a continuous random variable can be approximated by a probability histogram of a discrete random variable. Computing the mean value and the standard deviation using this discrete distribution gives an approximation to μ_x and σ_x for the continuous random variable x. If an even more accurate approximating probability histogram is used (narrower rectangles), better approximations to μ_x and σ_x result.

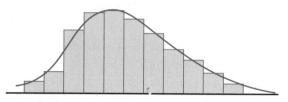

FIGURE 7.14 Approximating a density curve by a probability histogram

In practice, such an approximation method is often unnecessary. Instead, μ_x and σ_x can be defined and computed using methods from calculus. The details need not concern us; what is important is that μ_x and σ_x play exactly the same role here as they did in the discrete case. The mean value μ_x locates the center of the continuous distribution and gives the approximate long-run average of many observed x values. The standard deviation σ_x measures the extent that the continuous distribution (density curve) spreads out about μ_x and gives information about the amount of variability that can be expected in a long sequence of observed x values.

EXAMPLE 7.13

A company receives concrete of a certain type from two different suppliers. Define random variables x and y:

x = compressive strength (lb/in²) of a randomly selected batch from supplier 1

y = compressive strength of a randomly selected batch from supplier 2

Suppose that

$$\mu_x = 4650 \text{ lb/in}^2 \qquad \sigma_x = 200 \text{ lb/in}^2$$
$$\mu_y = 4500 \text{ lb/in}^2 \qquad \sigma_y = 275 \text{ lb/in}^2$$

The long-run average strength per batch for many, many batches from the first supplier will be roughly 4650 lb/in². This is 150 lb/in² greater than the long-run average for batches from the second supplier. In addition, a long sequence of batches from supplier 1 will exhibit substantially less variability in compressive strength values than will a similar sequence from supplier 2. The first supplier is preferred to the second both in terms of average value and variability. Figure 7.15 displays density curves that are consistent with this information.

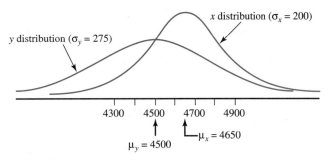

FIGURE 7.15 Density curves for Example 7.13

The Mean and Variance of Linear Functions and Linear Combinations

We have seen how the mean and standard deviation of a random variable provide useful information about its long-run behavior. In some cases, the mean and standard deviation of one or more random variables may be known, but we may also be interested in the behavior of some function of these variables.

For example, consider the experiment consisting of randomly selecting a customer of a propane gas company. The company serves homes heated by propane gas, and customers call when they wish to have their propane tank filled. Suppose that the mean and standard deviation of the random variable

x = number of gallons required to fill the tank

are known to be 318 gallons and 42 gallons, respectively. The company is considering two different pricing models:

Model 1: $2 per gallon
Model 2: service charge of $50 + $1.80 per gallon

and is interested in the variable

y = amount billed

For each of the two models, y can be expressed as a function of the random variable x:

Model 1: $y_{\text{model 1}} = 2x$
Model 2: $y_{\text{model 2}} = 50 + 1.8x$

Both of these are examples of a linear function of x. The mean and standard deviation of a linear function of x can be computed from the mean and standard deviation of x as described in the box on page 330.

The Mean and Variance of a Linear Function

If x is a random variable with mean μ_x and variance σ_x^2 and a and b are numerical constants, the random variable y defined by

$$y = a + bx$$

is called a **linear function of the random variable** x.

The mean of $y = a + bx$ is

$$\mu_y = \mu_{a+bx} = a + b\mu_x$$

The variance of y is

$$\sigma_y^2 = \sigma_{a+bx}^2 = b^2\sigma_x^2$$

from which it follows that the standard deviation of y is

$$\sigma_y = \sigma_{a+bx} = |b|\sigma_x$$

We can use the results in the preceding box to compute the mean and standard deviation of the billing amount variable for the propane gas example as follows: For model 1:

$$\mu_{\text{model 1}} = \mu_{2x} = 2\mu_x = 2(318) = 636$$
$$\sigma_{\text{model 1}}^2 = \sigma_{2x}^2 = 2^2\sigma_x^2 = 4(42)^2 = 7056.00$$
$$\sigma_{\text{model 1}} = \sqrt{7056} = 84.00 = 2(42)$$

For model 2:

$$\mu_{\text{model 2}} = \mu_{50+1.8x} = 50 + 1.8\mu_x = 50 + 1.8(318) = 622.40$$
$$\sigma_{\text{model 2}}^2 = \sigma_{50+1.8x}^2 = 1.8^2\sigma_x^2 = 1.8^2(42)^2 = 5715.36$$
$$\sigma_{\text{model 2}} = \sqrt{5715.36} = 75.60 = 1.8(42)$$

The mean billing amount for pricing model 1 is a bit higher than for pricing model 2, as is the variability in billing amounts. Pricing model 2 would result in slightly more consistency from bill to bill in the amount charged.

Now let's consider a different type of problem. Suppose that you have three tasks that you plan to complete on the way home from school: stop at the public library to return an overdue book for which you must pay a fine, deposit your most recent paycheck at the bank, and stop by the office supply store to purchase paper for your computer printer. Define the following variables:

x_1 = time required to return book and pay fine

x_2 = time required to deposit paycheck

x_3 = time required to buy printer paper

We can then define a new variable, y, to represent the total amount of time to complete these tasks as

$$y = x_1 + x_2 + x_3$$

Defined in this way, y is an example of a linear combination of random variables.

If $x_1, x_2, \ldots, x_n$ are random variables and $a_1, a_2, \ldots, a_n$ are numerical constants, the random variable y defined as

$$y = a_1 x_1 + a_2 x_2 + \cdots + a_n x_n$$

is a **linear combination of the x_i's.**

For example, $y = 10x_1 - 5x_2 + 8x_3$ is a linear combination of $x_1, x_2,$ and x_3, with $a_1 = 10, a_2 = -5,$ and $a_3 = 8$.

It is easy to compute the mean of a linear combination of x_i's if the individual means, $\mu_1, \mu_2, \ldots, \mu_n$, are known. The variance and standard deviation of a linear combination of x_i's is also easily computed *if the x_i's are independent.* Two random variables x_i and x_j are independent if any event defined solely by x_i is independent of any event defined solely by x_j. When the x_i's are not independent, computation of the variance and standard deviation of a linear combination of x_i's is more complicated and is not considered here.

Mean and Variance for Linear Combinations

If $x_1, x_2, \ldots, x_n$ are random variables with means $\mu_1, \mu_2, \ldots, \mu_n$ and variances $\sigma_1^2, \sigma_2^2, \ldots, \sigma_n^2$, respectively, and

$$y = a_1 x_1 + a_2 x_2 + \cdots + a_n x_n$$

then

1. $\mu_y = \mu_{a_1 x_1 + a_2 x_2 + \cdots + a_n x_n} = a_1 \mu_1 + a_2 \mu_2 + \cdots + a_n \mu_n$

   This result is true regardless of whether the x_i's are independent.

2. When $x_1, x_2, \ldots, x_n$ are independent random variables

   $$\sigma_y^2 = \sigma_{a_1 x_1 + a_2 x_2 + \cdots + a_n x_n}^2 = a_1^2 \sigma_1^2 + a_2^2 \sigma_2^2 + \cdots + a_n^2 \sigma_n^2$$

   $$\sigma_y = \sigma_{a_1 x_1 + a_2 x_2 + \cdots + a_n x_n} = \sqrt{a_1^2 \sigma_1^2 + a_2^2 \sigma_2^2 + \cdots + a_n^2 \sigma_n^2}$$

   This result is true only when the x_i's are independent.

The following examples illustrate the use of these rules.

EXAMPLE 7.14

Three different roads feed into a particular freeway entrance. Suppose that during a fixed time period, the number of cars coming from each road onto the freeway is a random variable with mean values as follows:

Road	1	2	3
Mean	800	1000	600

With x_i representing the number of cars entering from road i, we can define $y = x_1 + x_2 + x_3$, the total number of cars entering the freeway. The mean value of y is

$$\mu_y = \mu_{x_1+x_2+x_3}$$
$$= \mu_{x_1} + \mu_{x_2} + \mu_{x_3}$$
$$= 800 + 1000 + 600$$
$$= 2400$$

EXAMPLE 7.15

A nationwide standardized exam consists of a multiple choice section and a free response section. For each section, the mean and standard deviation are reported to be

	Mean	Standard Deviation
Multiple Choice	38	6
Free Response	30	7

Let's define x_1 and x_2 to be the multiple choice and free response score, respectively, of a student selected at random from those taking this exam. We are also interested in the variable y = total score. Suppose that the total score is computed as $y = x_1 + 2x_2$. What are the mean and standard deviation of y?

Since $y = x_1 + 2x_2$ is a linear combination of x_1 and x_2, the mean of y is

$$\mu_y = \mu_{x_1+2x_2}$$
$$= \mu_{x_1} + 2\mu_{x_2}$$
$$= 38 + 2(30)$$
$$= 98$$

What about the variance and standard deviation of y? To use rule 2 in the preceding box, x_1 and x_2 must be independent. It is unlikely that the value of x_1 (a student's multiple choice score) would be unrelated to the value of x_2 (the same student's free response score), since it seems probable that students who score well on one section of the exam will also tend to score well on the other. So, it would not be appropriate to calculate the variance and standard deviation from the information given.

EXAMPLE 7.16

A commuter airline flies small planes between San Luis Obispo and San Francisco. For small planes, the baggage weight is a concern, especially on foggy mornings, because the weight of the plane has an effect on how quickly the plane can ascend. Suppose that it is known that the variable x = weight of baggage checked by a randomly selected passenger has a mean and standard deviation of 42 and 16, respectively. Consider a flight on which ten passengers, all traveling alone, are flying. If we use x_i to denote the baggage weight for passenger i (for i ranging from 1 to 10), the total weight of checked baggage, y, is then

$$y = x_1 + x_2 + \cdots + x_{10}$$

Note that y is a linear combination of the x_i's. The mean value of y is

$$\mu_y = \mu_{x_1} + \mu_{x_2} + \cdots + \mu_{x_n}$$
$$= 42 + 42 + \cdots + 42$$
$$= 420$$

Since the ten passengers are all traveling alone, it is reasonable to think that the ten baggage weights are unrelated and that the x_i's are independent. Then the variance of y is

$$\sigma_y^2 = \sigma_{x_1}^2 + \sigma_{x_2}^2 + \cdots + \sigma_{x_n}^2$$
$$= 16^2 + 16^2 + \cdots + 16^2$$
$$= 2560$$

and

$$\sigma_y = \sqrt{2560} = 50.596$$

Calculation of the variance and standard deviation in this way required that the x_i's be independent. This may not have been a reasonable assumption if the ten passengers had not been traveling alone.

Exercises 7.27 – 7.41

7.27 An express mail service charges a special rate for any package that weighs less than 1 lb. Let x denote the weight of a randomly selected parcel that qualifies for this special rate. The probability distribution of x is specified by the accompanying density curve.

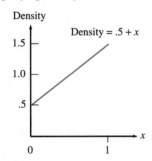

Density

Density = .5 + x

Use the fact that

area of a trapezoid = (base) · (average of two side lengths)

to answer each of the following questions.

a. What is the probability that a randomly selected package of this type is at most $\frac{1}{2}$ lb? Between $\frac{1}{4}$ lb and $\frac{1}{2}$ lb? At least $\frac{3}{4}$ lb?

b. It can be shown that $\mu_x = 7/12$ and $\sigma_x^2 = 11/144$. What is the probability that the value of x is more than 1 standard deviation from the mean value?

7.28 The probability distribution of x, the number of defective tires on a randomly selected automobile checked at a certain inspection station, is given in the accompanying table.

x	0	1	2	3	4
$p(x)$	.54	.16	.06	.04	.20

a. Calculate μ_x, the mean value of x.
b. What is the probability that x exceeds its mean value?

7.29 Exercise 7.9 introduced the accompanying probability distribution for $y =$ the number of broken eggs in a carton.

y	0	1	2	3	4
$p(y)$	.65	.20	.10	.04	.01

a. Calculate and interpret μ_y.
b. In the long run, for what percentage of cartons will the number of broken eggs be less than μ_y? Does this surprise you?
c. Why doesn't $\mu_y = (0 + 1 + 2 + 3 + 4)/5 = 2.0$?

7.30 Referring to Exercise 7.29, use the result of part **a** along with the fact that a carton contains 12 eggs to determine the mean value of $z =$ the number of unbroken eggs. (*Hint*: z can be written as a linear function of x.)

7.31 The mean value of x, the number of defective tires, whose distribution appears in Exercise 7.28, is $\mu_x = 1.2$, from which we obtain the following information:

x	0	1	2	3	4
$x - \mu$	-1.2	-.2	.8	1.8	2.8
$p(x)$	.54	.16	.06	.04	.20

Calculate σ_x^2 and σ_x.

7.32 Exercise 7.8 gave the accompanying probability distribution for x = the number of courses for which a randomly selected student at a certain university is registered. It is easily verified that $\mu = 4.66$ and $\sigma = 1.20$.

x	1	2	3	4	5	6	7
$p(x)$	.02	.03	.09	.25	.40	.16	.05

a. Since $\mu - \sigma = 3.46$, the x values 1, 2, and 3 are more than 1 standard deviation below the mean. What is the probability that x is more than 1 standard deviation below its mean?

b. What x values are more than 2 standard deviations away from the mean value (that is, either less than $\mu - 2\sigma$ or greater than $\mu + 2\sigma$)? What is the probability that x is more than 2 standard deviations away from its mean value?

7.33 A personal computer salesperson working on a commission receives a fixed amount for each system sold. Suppose that for a given month, the probability distribution of x = the number of systems sold is given in the accompanying table.

x	1	2	3	4	5	6	7	8
$p(x)$	.05	.10	.12	.30	.30	.11	.01	.01

a. Find the mean value of x (the mean number of systems sold).

b. Find the variance and standard deviation of x. How would you interpret these values?

c. What is the probability that the number of systems sold is within 1 standard deviation of its mean value?

d. What is the probability that the number sold is more than 2 standard deviations from the mean?

7.34 A local television station sells 15-sec, 30-sec, and 60-sec advertising spots. Let x denote the length of a randomly selected commercial appearing on this station, and suppose that the probability distribution of x is given in the accompanying table.

x	15	30	60
$p(x)$	.1	.3	.6

a. Find the average length for commercials appearing on this station.

b. If a 15-sec spot sells for $500, a 30-sec spot for $800, and a 60-sec spot for $1000, find the average amount paid for commercials appearing on this station. (*Hint*: Consider a new variable, y = cost, and then find the probability distribution and mean value of y.)

7.35 An author has written a book and submitted it to a publisher. The publisher offers to print the book and gives the author the choice between a flat payment of $10,000

and a royalty plan. Under the royalty plan the author would receive $1 for each copy of the book sold. The author thinks that the accompanying table gives the probability distribution of the variable x = the number of books that will be sold. Which payment plan should the author choose? Why?

x	1000	5000	10,000	20,000
$p(x)$	.05	.30	.40	.25

7.36 A grocery store has an express line for customers purchasing at most five items. Let x be the number of items purchased by a randomly selected customer using this line. Give examples of two different assignments of probabilities such that the resulting distributions have the same mean but have standard deviations that are quite different.

7.37 A gas station sells gasoline at the following prices (cents per gallon, depending on the type of gas and service): 115.9, 118.9, 129.9, 139.9, 144.9, and 159.7. Let y denote the price per gallon paid by a randomly selected customer.

a. Is y a discrete random variable? Explain.

b. Suppose that the probability distribution of y is as follows:

y	115.9	118.9	129.9	139.9	144.9	159.7
$p(y)$	.36	.24	.10	.16	.08	.06

What is the probability that a randomly selected customer has paid more than $1.20 per gallon? Less than $1.40 per gallon?

c. Refer to part **b,** and calculate the mean value and standard deviation of y. Interpret these values.

7.38 A chemical supply company currently has in stock 100 lb of a certain chemical, which it sells to customers in 5-lb lots. Let x = the number of lots ordered by a randomly chosen customer. The probability distribution of x is as follows:

x	1	2	3	4
$p(x)$	.2	.4	.3	.1

a. Calculate the mean value of x.

b. Calculate the variance and standard deviation of x.

7.39 Return to Exercise 7.38, and let y denote the amount of material left (lb) after the next customer's order is shipped. Find the mean and variance of y. (*Hint*: y is a linear function of x.)

7.40 An appliance dealer sells three different models of upright freezers having 13.5, 15.9, and 19.1 cubic feet of storage space, respectively. Let x = the amount of storage space purchased by the next customer to buy a freezer. Suppose that x has the following probability distribution:

x	13.5	15.9	19.1
$p(x)$	.2	.5	.3

a. Calculate the mean and variance of x.

b. If the price of the freezer depends on the size of the storage space, x, in the following way, what is the mean value of the variable price paid by the next customer?

Price $= 25x - 8.5$

c. What is the variance of price paid?

7.41 Consider a small ferry that can accommodate cars and buses. The toll for cars is \$3 and the toll for buses is \$10. Let x and y denote the number of cars and buses, respectively, carried on a single trip. Suppose that x and y are independent and have the probability distributions shown below.

x	0	1	2	3	4	5
$p(x)$	.05	.10	.25	.30	.20	.10

y	0	1	2
$p(y)$	.50	.30	.20

a. Compute the mean and variance of x.

b. Compute the mean and variance of y.

c. Compute the mean and variance of the total amount of money collected in tolls from cars.

d. Compute the mean and variance of the total amount of money collected in tolls from buses.

e. Compute the mean and variance of $z =$ total number of vehicles (cars and buses) on the ferry.

f. Compute the mean and variance of $w =$ total amount of money collected in tolls.

7.5　The Binomial and Geometric Distributions

This section introduces two of the more commonly encountered discrete probability distributions: the binomial and geometric distributions. Each of these distributions arise when the experiment of interest consists of making a sequence of dichotomous observations (two possible values for each observation). The process of making a single such observation is called a *trial.* For example, one characteristic of blood type is Rh factor, which can be either positive or negative. We can think of an experiment that consists of noting the Rh factor for each of 25 blood donors as a sequence of 25 dichotomous trials, where each trial consists of observing the Rh factor (positive or negative) of a single donor.

We could also conduct a different experiment that consists of observing the Rh factor of blood donors until a donor who is Rh negative is encountered. This second experiment can also be viewed as a sequence of dichotomous trials, but the total number of trials in this experiment is not predetermined as it was in the previous example where we knew in advance that there would be 25 trials. Experiments of the two types just described are characteristic of those leading to the binomial and geometric probability distributions, respectively.

The Binomial Distribution

Suppose we decide to record the gender of each of the next 25 newborn children at a particular hospital. What is the chance that at least 15 are female? What is the chance that between 10 and 15 are female? How many among the 25 can we expect to be female? These and other similar questions can be answered by studying the *binomial probability distribution.* This distribution arises when the experiment of interest is a *binomial experiment,* one having the following characteristics.

> **Properties of a Binomial Experiment**
>
> 1. It consists of a fixed number of observations called trials.
> 2. Each trial can result in one of only two mutually exclusive outcomes labeled success (S) and failure (F).
> 3. Outcomes of different trials are independent.
> 4. The probability that a trial results in S is the same for each trial.
>
> The **binomial random variable** x is defined as
>
> x = number of successes observed when experiment is performed
>
> The probability distribution of x is called the *binomial probability distribution.*

The term *success* here does not have any of its usual connotations. Which of the two possible outcomes is labeled "success" will be determined by the random variable of interest. For example, if the variable counts the number of female births among the next 25 births at a particular hospital, a female birth would be labeled success (because this is what the variable counts). This labeling is arbitrary: if male births were counted instead, a male birth would be labeled a success and a female birth a failure.

One illustration of a binomial probability distribution was given in Example 7.5. There we considered x = number among four customers who selected an electric (as opposed to gas) hot tub. This is a binomial experiment with

number of trials = 4 $P(\text{success}) = P(\text{E}) = .4$

The 16 possible outcomes, along with their probabilities, were displayed in Table 7.1.

Consider now the case of five customers, a binomial experiment with five trials. Here the binomial distribution will tell us how much probability is associated with each of the possible x values 0, 1, 2, 3, 4, and 5. There are 32 possible outcomes, and five of them yield $x = 1$:

   SFFFF   *FSFFF*   *FFSFF*   *FFFSF*   *FFFFS*

By independence, the first of these has probability

$$
\begin{aligned}
P(SFFFF) &= P(S) \cdot P(F) \cdot P(F) \cdot P(F) \cdot P(F) \\
&= (.4)(.6)(.6)(.6)(.6) \\
&= (.4)(.6)^4 \\
&= .05184
\end{aligned}
$$

The probability calculation will be the same for any outcome with only one success ($x = 1$). It does not matter where in the sequence the single success occurs. Thus

$$
\begin{aligned}
p(1) &= P(x = 1) \\
&= P(SFFFF \text{ or } FSFFF \text{ or } FFSFF \text{ or } FFFSF \text{ or } FFFFS) \\
&= .05184 + .05184 + .05184 + .05184 + .05184 \\
&= (5)(.05184) \\
&= .25920
\end{aligned}
$$

Similarly, there are ten outcomes for which $x = 2$, because there are ten ways to select two from among the five trials to be the S's: *SSFFF, SFSFF,* ... , and *FFFSS.* The probability of each results from multiplying together (.4) two times and (.6) three times. For example,

$$P(SSFFF) = (.4)(.4)(.6)(.6)(.6)$$
$$= (.4)^2(.6)^3$$
$$= .03456$$

and so

$$p(2) = P(x = 2)$$
$$= P(SSFFF) + \cdots + P(FFFSS)$$
$$= (10)(.4)^2(.6)^3$$
$$= .34560$$

The general form of the distribution here is

$$p(x) = P(x\ S\text{'s among the five trials})$$
$$= (\text{no. of outcomes with } x\ S\text{'s}) \cdot (\text{probability of any particular outcome with } x\ S\text{'s})$$
$$= (\text{no. of outcomes with } x\ S\text{'s}) \cdot (.4)^x \cdot (.6)^{5-x}$$

This form was seen previously where $p(2) = 10(.4)^2(.6)^3$.

Let n denote the number of trials in the experiment. Then the number of outcomes with x S's is the number of ways of selecting x from among the n trials to be the success trials. A simple expression for this quantity is

$$\text{number of outcomes with } x \text{ successes} = \frac{n!}{x!(n-x)!}$$

where, for any positive whole number m, the symbol $m!$ (read "m factorial") is defined by

$$m! = m(m-1)(m-2)\cdots(2)(1)$$

and $0! = 1$.

The Binomial Distribution

Let

n = number of independent trials in a binomial experiment

π = constant probability that any particular trial results in a success*

Then

$$p(x) = P(x \text{ successes among } n \text{ trials})$$
$$= \frac{n!}{x!(n-x)!} \cdot \pi^x(1-\pi)^{n-x} \qquad x = 0, 1, 2, \ldots, n$$

(continued)

---

*Some sources use p to represent the probability of success rather than π. We prefer the use of Greek letters for characteristics of a population or probability distribution, thus the use of π.

The expressions $\binom{n}{x}$ or $_nC_x$ are sometimes used in place of $\dfrac{n!}{x!(n-x)!}$. Both are read as "n choose x" and represent the number of ways of choosing x items from a set of n. The binomial probability function can then be written as

$$p(x) = \binom{n}{x}\pi^x(1-\pi)^{n-x} \qquad x = 0, 1, 2, \ldots, n$$

or

$$_nC_x\pi^x(1-\pi)^{n-x} \qquad x = 0, 1, 2, \ldots, n$$

Notice that the probability distribution is being specified by a formula that allows calculation of the various probabilities rather than by giving a table or probability histogram.

EXAMPLE 7.17

Sixty percent of all watches sold by a large discount store have a digital display and 40% have an analog display. The type of watch purchased by each of the next 12 customers will be noted. Define a random variable x by

x = number of watches among these 12 that have a digital display

Since x counts the number of digital watches, we use S to denote the sale of a digital watch. Then x is a binomial random variable with $n = 12$ and $\pi = P(S) = .60$. The probability distribution of x is given by

$$p(x) = \frac{12!}{x!(12-x)!}\cdot(.6)^x(.4)^{12-x} \qquad x = 0, 1, 2, \ldots, 12$$

The probability that exactly four watches are digital is

$$\begin{aligned}
p(4) &= P(x = 4) \\
&= \frac{12!}{4!8!}\cdot(.6)^4(.4)^8 \\
&= (495)(.6)^4(.4)^8 \\
&= .042
\end{aligned}$$

If group after group of 12 purchases is examined, the long-run percentage of those with exactly four digital watches will be 4.2%. According to this calculation, 495 of the possible outcomes (there are $2^{12} = 4096$) have $x = 4$.

The probability that between four and seven (inclusive) are digital is

$$\begin{aligned}
P(4 \le x \le 7) &= P(x = 4 \ or \ x = 5 \ or \ x = 6 \ or \ x = 7) \\
&= p(4) + p(5) + p(6) + p(7) \\
&= \frac{12!}{4!8!}\cdot(.6)^4(.4)^8 + \cdots + \frac{12!}{7!5!}\cdot(.06)^7(.4)^5 \\
&= .042 + .101 + .177 + .227 \\
&= .547
\end{aligned}$$

Notice that

$$P(4 < x < 7) = P(x = 5 \text{ or } x = 6)$$
$$= p(5) + p(6)$$
$$= .278$$

so the probability depends on whether $<$ or $\le$ appears. (This is typical of *discrete random variables*.)

The binomial distribution formula can be tedious to use unless n is very small. Appendix Table X gives binomial probabilities for selected n in combination with various values of π. This should help you practice using the binomial distribution without getting bogged down in arithmetic.

Using Appendix Table X

To find $p(x)$ for any particular value of x,

1. Locate the part of the table corresponding to your value of n (5, 10, 15, 20, or 25).

2. Move down to the row labeled with your value of x.

3. Go across to the column headed by the specified value of π.

The desired probability is at the intersection of the designated x row and π column. For example, when $n = 20$ and $\pi = .8$,

$p(15) = P(x = 15) = $ [entry at intersection of $n = 15$ row and $\pi = .8$ column] $= .175$

Although $p(x)$ is positive for every possible x value, many probabilities are zero to three decimal places, so they appear as .000 in the table. There are much more extensive binomial tables available. Alternatively, most statistics computer packages and graphing calculators are programmed to calculate these probabilities.

Sampling Without Replacement

Suppose that a population consists of N individuals or objects, each one classified as an S or an F. Usually, sampling is carried out without replacement; that is, once an element has been selected into the sample, it is not a candidate for future selection. If the sampling were accomplished by selecting an element from the population, observing whether it is an S or F, and then returning it to the population before the next selection is made, the variable $x = $ number of successes observed in the sample would fit all of the requirements of a binomial random variable. When sampling is done without replacement, the trials (individual selections) are not independent. In this case, the number of successes observed in the sample does not have a binomial distribution, but rather a different type of distribution called a *hypergeometric distribution*. Not only does the name of this distribution sound forbidding, probability calculations for this distribution are even more tedious than for the binomial. Fortunately, when the sample size n is much smaller than N, the population size, probabilities calculated using the binomial distribution and the hypergeometric distribution are very close in value. They are so close, in fact, that statisticians often ignore the difference and use the binomial probabilities in place of the hypergeometric probabilities. Most statisticians recommend the following guideline for determining whether the binomial probability distribution is appropriate when sampling without replacement.

Let x denote the number of S's in a sample of size n selected without replacement from a population consisting of N individuals or objects. If $n/N \le .05$ (that is, if at most 5% of the population is sampled), then the binomial distribution gives a good approximation to the probability distribution of x.

EXAMPLE 7.18   In recent years, homeowners have become increasingly security conscious. A *Los Angeles Times* poll (Nov. 10, 1991) reported that almost 20% of Southern California homeowners questioned had installed a home security system. Suppose that exactly 20% of all such homeowners have a system. Consider a random sample of $n = 20$ homeowners (much less than 5% of the population). Then x, the number of homeowners in the sample who have a security system, has (approximately) a binomial distribution with $n = 20$ and $\pi = .20$. The probability that five of those sampled have a system is

$$p(5) = P(x = 5)$$
$$= (\text{entry in } x = 5 \text{ row and } \pi = .20 \text{ column in Appendix Table X } (n = 20))$$
$$= .175$$

The probability that at least 40% of those in the sample — that is, eight or more — have a system is

$$P(x \ge 8) = P(x = 8, 9, 10, \ldots, 19, \text{ or } 20)$$
$$= p(8) + p(9) + \cdots + p(20)$$
$$= .022 + .007 + .002 + .000 + \cdots + .000$$
$$= .031$$

If, in fact, $\pi = .20$, only about 3% of all samples of size 20 would result in at least eight homeowners having a security system. Because $P(x \ge 8)$ is so small when $\pi = .20$, if $x \ge 8$ were actually observed, we would have to wonder whether the reported value of $\pi = .20$ was correct. Although it is possible that we would observe $x \ge 8$ when $\pi = .20$ (this would happen about 3% of the time in the long run), it might also be the case that π is actually greater than .20. In Chapter 10, we show how hypothesis-testing methods can be used to decide which of two contradictory claims about a population (for example, $\pi = .20$ or $\pi > .20$) is more plausible.

The binomial formula or tables can be used to compute each of the 21 probabilities $p(0), p(1), \ldots, p(20)$. Figure 7.16 shows the probability histogram for the

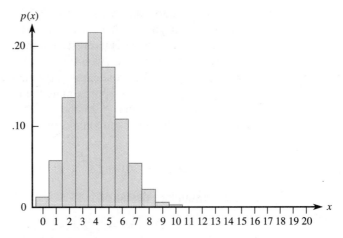

FIGURE 7.16 The binomial probability histogram when *n* = **20** and π = .20

binomial distribution with $n = 20$ and $\pi = .20$. Notice that the distribution is skewed to the right. (The binomial distribution is symmetric only when $\pi = .5$.)

The Mean and Standard Deviation of a Binomial Random Variable

A binomial random variable x based on n trials has possible values 0, 1, 2, . . . , n, so the mean value is

$$\mu_x = \Sigma\, x \cdot p(x) = (0) \cdot p(0) + (1) \cdot p(1) + \cdots + n \cdot p(n)$$

and the variance of x is

$$\sigma^2 = \Sigma\, (x - \mu)^2 \cdot p(x)$$
$$= (0 - \mu)^2 \cdot p(0) + (1 - \mu)^2 \cdot p(1) + \cdots + (n - \mu)^2 \cdot p(n)$$

These expressions appear to be very tedious to evaluate for any particular values of n and π. Fortunately, algebraic manipulation results in considerable simplification, making summation unnecessary.

The mean value and the standard deviation of a binomial random variable are, respectively,

$$\mu_x = n\pi \qquad \sigma_x = \sqrt{n\pi(1 - \pi)}$$

EXAMPLE 7.19 It has been reported (*Newsweek,* Dec. 2, 1991) that one-third of all credit card users pay their bills in full each month. This figure is, of course, an average across different cards and issuers. Suppose that 30% of all individuals holding Visa cards issued by a certain bank pay in full each month. A random sample of $n = 25$ cardholders is to be selected. The bank is interested in the variable x = number in the sample who pay in full each month. Even though sampling is done without replacement, the sample size $n = 25$ is most likely very small compared to the total number of credit card holders, so we can approximate the probability distribution of x using a binomial distribution with $n = 25$ and $\pi = .4$. We have defined "paid in full" as a success because this is the outcome counted by the random variable x. The mean value of x is then

$$\mu_x = n\pi = 25(.30) = 7.5$$

and the standard deviation is

$$\sigma_x = \sqrt{n\pi(1 - \pi)} = \sqrt{25(.30)(.70)} = \sqrt{5.25} = 2.29$$

The probability that x is farther than 1 standard deviation from its mean value is

$$
\begin{aligned}
P(x < \mu_x - \sigma_x \; or \; x > \mu_x + \sigma_x) &= P(x < 5.21 \; or \; x > 9.79) \\
&= P(x \leq 5) + P(x \geq 10) \\
&= p(0) + \cdots + p(5) + p(10) + \cdots + p(25) \\
&= .382 \quad \text{(using Appendix Table X)}
\end{aligned}
$$

The value of σ_x is zero when $\pi = 0$ or $\pi = 1$. In these two cases, there is no uncertainty in x: we are sure to observe $x = 0$ when $\pi = 0$ and $x = n$ when $\pi = 1$. It is also easily verified that $\pi(1 - \pi)$ is largest when $\pi = .5$. Thus the binomial distribution spreads out the most when sampling from a 50–50 population. The farther π is from .5, the less spread out is the distribution.

The Geometric Distribution

A binomial random variable is defined as the number of successes in n independent trials, where each trial can result in either a success or a failure and the probability of success is the same for each trial. Suppose, however, that we are not interested in the number of successes in a fixed number of trials, but rather the number of trials that must be carried out before a success occurs. Two examples are counting the number of boxes of cereal that must be purchased before finding one with a rare toy and counting the number of games that a professional bowler must play before achieving a score over 250.

The variable

$$x = \text{number of trials to first success}$$

is called a *geometric random variable,* and the probability distribution that describes its behavior is called a *geometric probability distribution.*

Suppose an experiment consists of a sequence of trials with the following conditions:

1. The trials are independent.
2. Each trial can result in one of two possible outcomes, success and failure.
3. The probability of success is the same for all trials.

A **geometric random variable** is defined as

  x = number of trials until the first success is observed (including the success trial)

The probability distribution of x is called the *geometric probability distribution.*

For example, suppose that 40% of the students who drive to campus at your university carry jumper cables. Your car has a dead battery and you don't have jumper cables, so you decide to stop students who are headed to the parking lot and ask them if they have a pair of jumper cables. You might be interested in the number of students you would have to stop before finding one who had jumper cables. If we define success as a student with jumper cables, a trial would consist of asking an individual student for help. The random variable

  x = number of students who must be stopped before finding one with
      jumper cables

is an example of a geometric random variable, since it can be viewed as the number of trials to the first success in a sequence of independent trials.

The probability distribution of a geometric random variable is easy to construct. We will use π to denote the probability of success on any given trial. Possible outcomes can be denoted as follows:

Outcome	x = number of trials to first success
S	1
FS	2
FFS	3
$\vdots$	$\vdots$
$FFFFFFS$	7
$\vdots$	$\vdots$

Each possible outcome consists of 0 or more failures followed by a single success. So,

$$p(x) = P(x \text{ trials to first success})$$
$$= P(\underbrace{FF \ldots F}_{x-1 \text{ failures followed by a success on trial } x}S)$$

Since the probability of success is π for each trial, the probability of failure for each trial is $1 - \pi$. Because the trials are independent,

$$p(x) = P(x \text{ trials to first success}) = P(FF \dots FS)$$
$$= P(F)P(F) \cdots P(F)P(S)$$
$$= (1 - \pi)(1 - \pi) \cdots (1 - \pi)\pi$$
$$= (1 - \pi)^{x-1}\pi$$

This leads us to the formula for the geometric probability distribution.

Geometric Probability Distribution

If x is a geometric random variable with probability of success $= \pi$ for each trial, then

$$p(x) = (1 - \pi)^{x-1}\pi \qquad x = 1, 2, 3, \dots$$

EXAMPLE 7.20

Consider the jumper cable problem described previously. For this problem, $\pi = .4$, since 40% of the students who drive to campus carry jumper cables. The probability distribution of

x = number of students who must be stopped before finding a student with jumper cables

is

$$p(x) = (.6)^{x-1}(.4) \qquad x = 1, 2, 3, \dots$$

The probability distribution can now be used to compute various probabilities. For example, the probability that the first student stopped has jumper cables (that is, $x = 1$) is

$$p(1) = (.6)^{1-1}(.4) = (.6)^0(.4) = .4$$

The probability that three or fewer students must be stopped is

$$P(x \le 3) = p(1) + p(2) + p(3)$$
$$= (.6)^0(.4) + (.6)^1(.4) + (.6)^2(.4)$$
$$= .4 + .24 + .144$$
$$= .784$$

Exercises 7.42 – 7.58

7.42 a. In a binomial experiment consisting of six trials, how many outcomes have exactly one S, and what are these outcomes?
b. In a binomial experiment consisting of 20 trials, how many outcomes have exactly ten S's? Exactly fifteen S's? Exactly five S's?

7.43 Suppose that in a certain metropolitan area, nine out of ten households have a VCR. Let x denote the number among four randomly selected households that

have a VCR, so x is a binomial random variable with $n = 4$ and $\pi = .9$.
a. Calculate $p(2) = P(x = 2)$, and interpret this probability.
b. Calculate $p(4)$, the probability that all four selected households have a VCR.
c. Determine $P(x \le 3)$.

7.44 The *Los Angeles Times* (Dec. 13, 1992) reported that what airline passengers like to do most on long flights is rest or sleep; in a survey of 3697 passengers, almost 80% did so. Suppose that for a particular route, the actual percentage is exactly 80%, and consider randomly selecting six passengers. Then x, the number among the selected six who rested or slept, is a binomial random variable with $n = 6$ and $\pi = .8$.

a. Calculate $p(4)$, and interpret this probability.
b. Calculate $p(6)$, the probability that all six selected passengers rested or slept.
c. Determine $P(x \geq 4)$.

7.45 Refer to Exercise 7.44, and suppose that ten rather than six passengers are selected ($n = 10$, $\pi = .8$), so that Appendix Table X can be used.

a. What is $p(8)$?
b. Calculate $P(x \leq 7)$.
c. Calculate the probability that more than half of the selected passengers rested or slept.

7.46 Twenty-five percent of the customers entering a grocery store between 5 P.M. and 7 P.M. use an express checkout. Consider five randomly selected customers, and let x denote the number among the five who use the express checkout.

a. What is $p(2)$, that is, $P(x = 2)$?
b. What is $P(x \leq 1)$?
c. What is $P(2 \leq x)$? (*Hint*: Make use of your computation in part **b**.)
d. What is $P(x \neq 2)$?

7.47 A breeder of show dogs is interested in the number of female puppies in a litter. If a birth is equally likely to result in a male or female puppy, give the probability distribution of the variable

$$x = \text{number of female puppies in a litter of size 5}$$

7.48 Industrial quality control programs often include inspection of incoming materials from suppliers. If parts are purchased in large lots, a typical plan might be to select 20 parts at random from a lot and inspect them. A lot might be judged acceptable if one or fewer defective parts are found among those inspected. Otherwise, the lot is rejected and returned to the supplier. Use Table X to find the probability of accepting lots that have each of the following. (*Hint*: Identify success with a defective part.)

a. 5% defective parts
b. 10% defective parts
c. 20% defective parts

7.49 An experiment was conducted to investigate whether a graphologist (handwriting analyst) could distinguish a normal person's handwriting from that of a psychotic. A well-known expert was given ten files, each containing handwriting samples from a normal person and from a person diagnosed as psychotic. The graphologist was then asked to identify the psychotic's handwriting. The graphologist made correct identifications in six of the ten trials (data taken from *Statistics in the Real World*, Larsen and Stroup, New York: Macmillan, 1976). Does this evidence indicate that the graphologist has an ability to distinguish the handwriting of psychotics? (*Hint*: What is the probability of correctly guessing six or more times out of ten? Your answer should depend on whether this probability is relatively small or relatively large.)

7.50 If the temperature in Florida falls below 32°F during certain periods of the year, there is a chance that the citrus crop will be damaged. Suppose that the probability is .1 that any given tree will show measurable damage when the temperature falls to 30°F. If the temperature does drop to 30°F, what is the expected number of trees showing damage in orchards of 2000 trees? What is the standard deviation of the number of trees that show damage?

7.51 Thirty percent of all automobiles undergoing an emission inspection at a certain inspection station fail the inspection.

a. Among 15 randomly selected cars, what is the probability that at most 5 fail the inspection?
b. Among 15 randomly selected cars, what is the probability that between 5 and 10 (inclusive) fail to pass inspection?
c. Among 25 randomly selected cars, what is the mean value of the number that pass inspection, and what is the standard deviation of the number that pass inspection?
d. What is the probability that among 25 randomly selected cars, the number that pass is within 1 standard deviation of the mean value?

7.52 You are to take a multiple-choice exam consisting of 100 questions with five possible responses to each. Suppose that you have not studied and so must guess (select one of the five answers in a completely random fashion) on each question. Let x represent the number of correct responses on the test.

a. What kind of probability distribution does x have?
b. What is your expected score on the exam? (*Hint*: Your expected score is the mean value of the x distribution.)
c. Compute the variance and standard deviation of x.
d. Based on your answers to parts **b** and **c**, is it likely that you would score over 50 on this exam? Explain the reasoning behind your answer.

7.53 Suppose that 20% of the 10,000 signatures on a certain recall petition are invalid. Would the number of invalid signatures in a sample of size 1000 have (approximately) a binomial distribution? Explain.

7.54 A coin is to be tossed 25 times. Let x = the number of tosses that result in heads (H). Consider the following rule for deciding whether or not the coin is fair:

Judge the coin to be fair if $8 \leq x \leq 17$.
Judge it to be biased if either $x \leq 7$ or $x \geq 18$.

a. What is the probability of judging the coin to be biased when it is actually fair?
b. What is the probability of judging the coin to be fair when $P(H) = .9$, so that there is a substantial bias? Repeat for $P(H) = .1$.
c. What is the probability of judging the coin to be fair when $P(H) = .6$? When $P(H) = .4$? Why are the probabilities so large compared to the probabilities in part **b**?
d. What happens to the "error probabilities" of parts **a** and **b** if the decision rule is changed so that the coin is judged fair if $7 \leq x \leq 18$ and unfair otherwise? Is this a better rule than the one first proposed?

7.55 A city ordinance requires that a smoke detector be installed in all residential housing. There is concern that too many residences are still without detectors, so a costly inspection program is being contemplated. Let π = the proportion of all residences that have a detector. A random sample of 25 residences will be selected. If the sample strongly suggests that $\pi < .80$ (fewer than 80% have detectors), as opposed to $\pi \geq .80$, the program will be implemented. Let x = the number of residences among the 25 that have a detector, and consider the following decision rule:

Reject the claim that $\pi \geq .8$ and implement the program if $x \leq 15$.

a. What is the probability that the program is implemented when $\pi = .80$?
b. What is the probability that the program is not implemented if $\pi = .70$? If $\pi = .60$?
c. How do the "error probabilities" of parts **a** and **b** change if the value 15 in the decision rule is changed to 14?

7.56 Exit polling has been a controversial practice in recent elections, since early release of the resulting information appears to affect whether or not those who have not yet voted will do so. Suppose that 90% of all registered California voters favor banning the release of information from exit polls in presidential elections until after the polls in California close. A random sample of 25 California voters is selected.

a. What is the probability that more than 20 favor the ban?
b. What is the probability that at least 20 favor the ban?
c. What are the mean value and standard deviation of the number who favor the ban?
d. If fewer than 20 in the sample favor the ban, is this at odds with the assertion that (at least) 90% of the populace favors the ban? (*Hint*: Consider $P(x < 20)$ when $\pi = .9$.)

7.57 Sophie is a dog who loves to play catch. Unfortunately, she isn't very good, and the probability that she catches a ball is only .1. Let x = number of tosses required until Sophie catches a ball.

a. Does x have a binomial or a geometric distribution?
b. What is the probability that it will take exactly two tosses for Sophie to catch a ball?
c. What is the probability that more than three tosses will be required?

7.58 Selected boxes of a breakfast cereal contain a prize. Suppose that 5% of the boxes contain the prize and the other 95% contain the message "Sorry, try again." A consumer determined to find a prize decides to continue to buy boxes of cereal until a prize is found. Consider the random variable x, where x = number of boxes purchased until a prize is found.

a. What is the probability that at most 2 boxes must be purchased?
b. What is the probability that exactly four boxes must be purchased?
c. What is the probability that more than four boxes must be purchased?

7.6 The Normal Distribution

Normal distributions formalize the notion of mound-shaped histograms introduced in Chapter 4. Normal distributions are widely used because they provide a reasonable approximation to the distribution of many different variables. They

FIGURE 7.17 A normal distribution

also play a central role in many of the inferential procedures that will be discussed in later chapters. Normal distributions are continuous probability distributions that are bell shaped and symmetric, as shown in Figure 7.17. Normal distributions are sometimes referred to as a *normal curves.*

There are many different normal distributions, and they are distinguished from one another by their mean μ and standard deviation σ. The mean μ of a normal distribution describes where the corresponding curve is centered, and the standard deviation σ describes how much the curve spreads out around the center. As with all continuous probability distributions, the total area under any normal curve is equal to 1. Several normal distributions are shown in Figure 7.18. Notice that the smaller the standard deviation, the taller and narrower the corresponding curve. Recall that areas under a continuous probability distribution curve represent probabilities, so when the standard deviation is small, a larger area is concentrated near the center of the curve, and there is a much greater chance of observing a value near the mean (since μ is at the center).

FIGURE 7.18 Three normal distributions

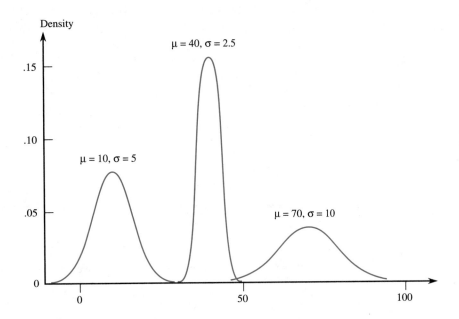

The value of μ is the number on the measurement axis lying directly below the top of the bell. The value of σ can also be ascertained from a picture of the curve. Consider the normal curve in Figure 7.19 (page 348). Starting at the top of the bell

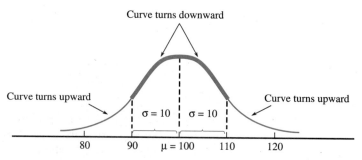

FIGURE 7.19 μ and σ for a normal curve

(above 100) and moving to the right, the curve turns downward until it is above the value 110. After that point, it continues to decrease in height but is turning up rather than down. Similarly, to the left of 100, the curve turns down until it reaches 90 and then begins to turn up. The curve changes from turning down to turning up at a distance of 10 on either side of μ, so $\sigma = 10$. In general, σ is the distance to either side of μ at which a normal curve changes from turning downward to turning upward.

If a particular normal distribution is to be used to describe the behavior of a random variable, a mean and standard deviation must be specified. For example, a normal distribution with mean 7 and standard deviation 1 might be used as a model for the distribution of x = birth weight from Section 7.3. If this model is a reasonable description of the probability distribution, we could use areas under the normal curve with $\mu = 7$ and $\sigma = 1$ to approximate various probabilities related to birth weight. The probability that a birth weight is over 8 lb (expressed symbolically as $P(x > 8)$) corresponds to the shaded area in Figure 7.20(a). The shaded area in Figure 7.20(b) is the (approximate) probability $P(6.5 < x < 8)$ of a birth weight falling between 6.5 and 8 lb.

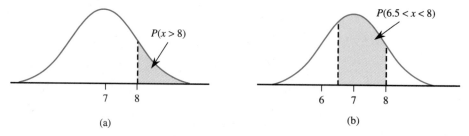

FIGURE 7.20 Normal distribution for birth weight: (a) Shaded area = $P(x > 8)$; (b) Shaded area = $P(6.5 < x < 8)$

Unfortunately, direct computation of such probabilities (areas under a normal curve) is not simple. To overcome this difficulty, we rely on a table of areas for a reference normal distribution called the *standard normal distribution*.

> **Definition**
>
> The **standard normal distribution** is the normal distribution with
>
> $\mu = 0$ and $\sigma = 1$
>
> The corresponding density curve is called the *standard normal curve*. It is customary to use the letter z to represent a variable whose distribution is described by the standard normal curve. The term *z curve* is often used in place of standard normal curve.

Few naturally occurring variables have distributions that are well described by the standard normal distribution, but this distribution is important because it is used in probability calculations for other normal distributions as well. When we are interested in finding a probability based on some other normal curve, we first translate our problem into an "equivalent" problem that involves finding an area under the standard normal curve. A table for the standard normal distribution is then used to find the desired area. To be able to do this, we must first learn to work with the standard normal distribution.

The Standard Normal Distribution

In working with normal distributions, two general skills are required.

1. We must be able to use the normal distribution to compute probabilities, which are areas under a normal curve and above given intervals.

2. We must be able to characterize extreme values in the distribution, such as the largest 5%, the smallest 1%, and the most extreme 5% (which would include the largest 2.5% and the smallest 2.5%).

Let's begin by looking at how to accomplish these tasks when the distribution of interest is the standard normal distribution.

The standard normal or z curve is shown in Figure 7.21(a). It is centered at $\mu = 0$, and the standard deviation, $\sigma = 1$, is a measure of the extent to which it spreads

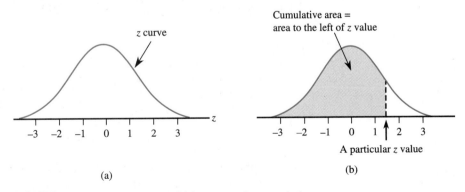

(a)

(b)

FIGURE 7.21 A standard normal (z) curve and a cumulative area

out about its mean (in this case, 0). Note that this picture is consistent with the Empirical Rule of Chapter 4, with about 95% of the area (probability) associated with values that are within 2 standard deviations of the mean (between -2 and 2) and almost all of the area associated with values that are within 3 standard deviations of the mean (between -3 and 3).

Appendix Table II tabulates cumulative z curve areas of the sort shown in Figure 7.21(b) for many different values of z. The smallest value for which the cumulative area is given is -3.89, a value far out in the lower tail of the z curve. The next smallest value for which the area appears is -3.88, then -3.87, then -3.86, and so on in increments of .01, terminating with the cumulative area to the left of 3.89.

Using the Table of Standard Normal Curve Areas

For any number z^* between -3.89 and 3.89 and rounded to two decimal places, Appendix Table II gives

(area under z curve to the left of z^*) $= P(z < z^*) = P(z \leq z^*)$

where the letter z is used to represent a random variable whose distribution is the standard normal distribution.

To find this probability, locate the following:

1. The row labeled with the sign of z^* and the digit to either side of the decimal point (for example, -1.7 or 0.5)
2. The column identified with the second digit to the right of the decimal point in z^* (for example, .06 if $z^* = -1.76$)

The number at the intersection of this row and column is the desired probability, $P(z < z^*)$.

A portion of the table of standard normal curve areas appears in Figure 7.22. To find the area under the z curve to the left of 1.42, look in the row labeled 1.4 and the column labeled .02 (the highlighted row and column in Figure 7.22). From the table, the corresponding cumulative area is .9222. So

z curve area to the left of $1.42 = .9222$

We can also use the table to find the area to the right of a particular value. Since the total area under the z curve is 1, it follows that

z curve area to the right of $1.42 = 1 - (z$ curve area to the left of 1.42)
$$= 1 - .9222$$
$$= .0778$$

These probabilities can be interpreted as meaning that in a long sequence of observations, roughly 92.22% of the observed z values will be smaller than 1.42, and 7.78% will be larger than 1.42.

FIGURE 7.22

z^*	.00	.01	.02	.03	.04	.05
0.0	.5000	.5040	.5080	.5120	.5160	.5199
0.1	.5398	.5438	.5478	.5517	.5557	.5596
0.2	.5793	.5832	.5871	.5910	.5948	.5987
0.3	.6179	.6217	.6255	.6293	.6331	.6368
0.4	.6554	.6591	.6628	.6664	.6700	.6736
0.5	.6915	.6950	.6985	.7019	.7054	.7088
0.6	.7257	.7291	.7324	.7357	.7389	.7422
0.7	.7580	.7611	.7642	.7673	.7704	.7734
0.8	.7881	.7910	.7939	.7967	.7995	.8023
0.9	.8159	.8186	.8212	.8238	.8264	.8289
1.0	.8413	.8438	.8461	.8485	.8508	.8531
1.1	.8643	.8665	.8686	.8708	.8729	.8749
1.2	.8849	.8869	.8888	.8907	.8925	.8944
1.3	.9032	.9049	.9066	.9082	.9099	.9115
1.4	.9192	.9207	.9222	.9236	.9251	.9265
1.5	.9332	.9345	.9357	.9370	.9382	.9394
1.6	.9452	.9463	.9474	.9484	.9495	.9505
1.7	.9554	.9564	.9573	.9582	.9591	.9599
1.8	.9641	.9649	.9656	.9664	.9671	.9678

$P(z < 1.42)$

EXAMPLE 7.21 The probability $P(z < -1.76)$ is found at the intersection of the -1.7 row and the .06 column of the z table. The result is

$$P(z < -1.76) = .0392$$

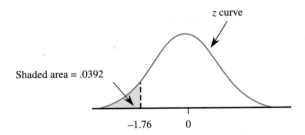

Shaded area = .0392

z curve

-1.76 0

In other words, in a long sequence of observations, roughly 3.9% of the observed z values will be smaller than -1.76. Similarly,

$$P(z \le 0.58) = \text{entry in 0.5 row and .08 column of Table II} = .7190$$

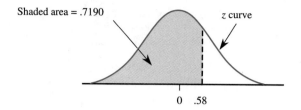

Now consider $P(z < -4.12)$. This probability does not appear in Appendix Table II; there is no -4.1 row. However, it must be less than $P(z < -3.89)$, the smallest z value in the table, because -4.12 is farther out in the lower tail of the z curve. Since $P(z < -3.89) = .0000$ (that is, zero to four decimal places), it follows that

$$P(z < -4.12) \approx 0$$

Similarly,

$$P(z < 4.18) > P(z < 3.89) = 1.0000$$

from which we conclude that

$$P(z < 4.18) \approx 1$$

As illustrated in the preceding examples, we can use the cumulative areas tabulated in Appendix Table II to calculate other probabilities involving z. The probability that z is larger than a value c is

$$P(z > c) = \text{area under the } z \text{ curve to the right of } c = 1 - P(z \leq c)$$

In words, the area to the right of a value (a right-tail area) is 1 minus the corresponding cumulative area. This is illustrated in Figure 7.23.

Similarly, the probability that z falls in the interval between a lower limit a and an upper limit b is

$$P(a < z < b) = \text{area under the } z \text{ curve and above the interval from } a \text{ to } b$$
$$= P(z < b) - P(z < a)$$

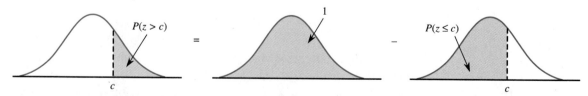

FIGURE 7.23 The relationship between an upper-tail area and a cumulative area

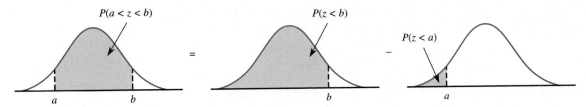

FIGURE 7.24 $P(a < z < b)$ as a difference of two cumulative areas

That is, $P(a < z < b)$ is the difference between two cumulative areas, as illustrated in Figure 7.24.

EXAMPLE 7.22   The probability that z is between -1.76 and 0.58 is

$$P(-1.76 < z < 0.58) = P(z < 0.58) - P(z < -1.76)$$
$$= .7190 - .0392$$
$$= .6798$$

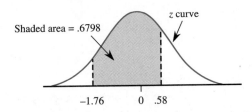

The probability that z is between -2 and $+2$ (within 2 standard deviations of its mean, since $\mu = 0$ and $\sigma = 1$) is

$$P(-2.00 < z < 2.00) = P(z < 2.00) - P(z < -2.00)$$
$$= .9772 - .0228$$
$$= .9544$$
$$\approx .95$$

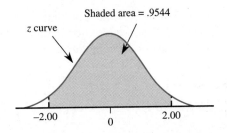

This last probability is the basis for one part of the Empirical Rule, which states that when a histogram is well approximated by a normal curve, roughly 95% of the values are within 2 standard deviations of the mean.

The probability that the value of z exceeds 1.96 is

$$P(z > 1.96) = 1 - P(z \leq 1.96)$$
$$= 1 - .9750$$
$$= .0250$$

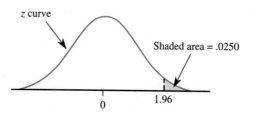

That is, 2.5% of the area under the z curve lies to the right of 1.96 in the upper tail. Similarly,

$$P(z > -1.28) = \text{area to the right of } -1.28$$
$$= 1 - P(z \leq -1.28)$$
$$= 1 - .1003$$
$$= .8997$$
$$\approx .90$$

Identifying Extreme Values

Let's now see how we can identify extreme values in the distribution. For example, we might want to describe the values included in the smallest 2% of the distribution or the values making up the most extreme 5% (which includes the largest 2.5% and the smallest 2.5%).

EXAMPLE 7.23

Suppose we want to describe the values that make up the smallest 2% of the standard normal distribution. Symbolically, we are trying to find a value (call it z^*) such that

$$P(z < z^*) = .02$$

This is illustrated in Figure 7.25. The figure shows that the cumulative area for z^* is .02, so we look for a cumulative area of .0200 in the body of Appendix Table II. The closest cumulative area in the table is .0202, in the -2.0 row and .05 column;

FIGURE 7.25 The smallest 2% of the standard normal distribution

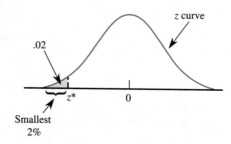

so we will use $z^* = -2.05$, the best approximation from the table. Variable values less than -2.05 make up the smallest 2% of the standard normal distribution.

Suppose that we had been interested in the largest 5% of all z values. We would then be trying to find a value of z^* for which

$$P(z > z^*) = .05$$

as illustrated in Figure 7.26. Since Appendix Table II always works with cumulative area (area to the left), the first step is to determine

area to the left of $z^* = 1 - .05 = .95$

FIGURE 7.26 The largest 5% of the standard normal distribution

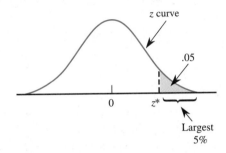

Looking for the cumulative area closest to .95 in Table II, we find that .95 falls exactly halfway between .9495 (corresponding to a z value of 1.64) and .9505 (corresponding to a z value of 1.65). Since .9500 is exactly halfway between the two areas, we will use a z value that is halfway between 1.64 and 1.65. (If one had been closer to .9500 than the other, we would just use the z value corresponding to the closest area). This gives

$$z^* = \frac{1.64 + 1.65}{2} = 1.645$$

Values greater than 1.645 make up the largest 5% of the standard normal distribution. By symmetry, -1.645 separates the smallest 5% of all z values from the others.

EXAMPLE 7.24

Sometimes we are interested in identifying the most extreme (unusually large *or* small) values in a distribution. Consider describing the values that make up the most extreme 5% of the standard normal distribution. That is, we want to separate the middle 95% from the extreme 5%. This is illustrated in Figure 7.27. Since the standard

FIGURE 7.27 The most extreme 5% of the standard normal distribution

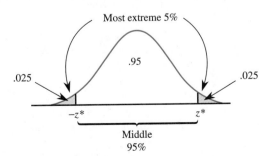

normal distribution is symmetric, the most extreme 5% would be equally divided between the high side and the low side of the distribution, resulting in an area of .025 for each of the tails of the z curve. Symmetry about 0 implies that if $z*$ denotes the value that separates the largest 2.5%, the value that separates the smallest 2.5% will be simply $-z*$.

To find $z*$, first determine the cumulative area for $z*$, which is

area to the left of $z* = .95 + .025 = .975$

The cumulative area .9750 appears in the 1.9 row and .06 column of Table II, so $z* = 1.96$. For the standard normal distribution, 95% of the variable values fall between -1.96 and 1.96; the most extreme 5% are those values that are either greater than 1.96 or less than -1.96.

Other Normal Distributions

We will now show how z curve areas can be used to calculate probabilities and describe values for any normal distribution. Remember that the letter z is reserved for those variables whose distribution is standard normal; the letter x is used more generally for any variable whose distribution is described by a normal curve with mean μ and standard deviation σ.

Suppose that we want to compute $P(a < x < b)$, the probability that the variable x lies in a particular range. This probability corresponds to an area under a normal curve and above the interval from a to b, as shown in Figure 7.28(a).

Our strategy for obtaining this probability is to find an "equivalent" problem involving the standard normal distribution. Finding an equivalent problem means determining an interval $(a*, b*)$ that has the same probability for z (same area under the z curve) as does the interval (a, b) in our original normal distribution. (Figure 7.28(b)). The * notation is used to distinguish a and b from $a*$ and $b*$; $a*$ and $b*$ are values from the z curve, whereas a and b are values from the original normal distribution with mean μ and standard deviation σ. To find $a*$ and $b*$, we simply calculate z scores for the endpoints of the interval for which a probability is desired. This process is called **standardizing** the endpoints. For example, suppose that the

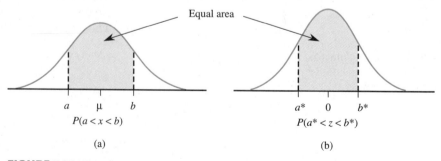

FIGURE 7.28 Equality of nonstandard and standard normal curve areas

variable x has a normal distribution with mean $\mu = 100$ and standard deviation $\sigma = 5$. To calculate

$$P(98 < x < 107)$$

we first translate this problem into an equivalent problem for the standard normal distribution. Recall from Chapter 4 that a z score, or standardized score, tells how many standard deviations away from the mean a value lies; it is calculated by first subtracting the mean and then dividing by the standard deviation. Converting the lower endpoint $a = 98$ to a z score gives

$$a^* = \frac{98 - 100}{5} = \frac{-2}{5} = -.40$$

and converting the upper endpoint yields

$$b^* = \frac{107 - 100}{5} = \frac{7}{5} = 1.40$$

Then

$$P(98 < x < 107) = P(-.40 < z < 1.40)$$

The probability $P(-.40 < z < 1.40)$ can now be evaluated using Appendix Table II.

Finding Probabilities

To calculate probabilities for any normal distribution, standardize the relevant values and then use the table of z curve areas. More specifically, if x is a variable whose behavior is described by a normal distribution with mean μ and standard deviation σ, then

$$P(x < b) = P(z < b^*)$$
$$P(a < x) = P(a^* < z) \quad [\text{Equivalently, } P(x > a) = P(z > a^*)]$$
$$P(a < x < b) = P(a^* < z < b^*)$$

where z is a variable whose distribution is standard normal and

$$a^* = \frac{a - \mu}{\sigma} \qquad b^* = \frac{b - \mu}{\sigma}$$

EXAMPLE 7.25

In poor countries, the growth of children can be an important indicator of general levels of nutrition and health. Data from the article "The Osteological Paradox: Problems in Inferring Prehistoric Health from Skeletal Samples" (*Current Anthropology* (1992): 343–370) suggests that a reasonable model for the probability distribution of the continuous numerical variable x = height of a randomly selected five-year-old child is a normal distribution with a mean of $\mu = 100$ cm and standard deviation $\sigma = 6$ cm. What proportion of the heights is between 94 cm and 112 cm? To answer this question, we must find

$$P(94 < x < 112)$$

Translating to an equivalent problem for the standard normal distribution,

$$a^* = \frac{a - \mu}{\sigma} = \frac{94 - 100}{6} = -1.00$$

$$b^* = \frac{b - \mu}{\sigma} = \frac{112 - 100}{6} = 2.00$$

Then

$$P(94 < x < 112) = P(-1.00 < z < 2.00)$$
$$= (z \text{ curve area to the left of } 2.00) - (z \text{ curve area to the left of } -1.00)$$
$$= .9772 - .1587$$
$$= .8185$$

The probabilities for x and z are shown in Figure 7.29. If height were observed for many children from this population, about 82% of them would fall between 94 and 112 cm.

FIGURE 7.29

$P(94 < x < 112)$ and corresponding z curve area

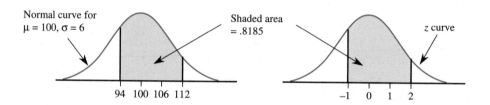

What is the probability that a randomly chosen child will be taller than 110 cm? To evaluate $P(x > 110)$, we first compute

$$a^* = \frac{a - \mu}{\sigma} = \frac{110 - 100}{6} = 1.67$$

Then (see Figure 7.30)

$$P(x > 110) = P(z > 1.67)$$
$$= z \text{ curve area to the right of } 1.67$$
$$= 1 - (z \text{ curve area to the left of } 1.67)$$
$$= 1 - .9525$$
$$= .0475$$

FIGURE 7.30 $P(x > 110)$ and corresponding z curve area

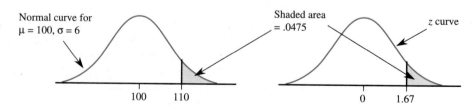

EXAMPLE 7.26   Although there is some controversy regarding the appropriateness of IQ scores as a measure of intelligence, they are commonly used for a variety of purposes. One type of IQ score is scaled so that the mean is 100 and the standard deviation is 15, and scores are approximately normally distributed. (IQ score is actually a discrete variable (since it is based on the number of correct responses on a test), but its population distribution closely resembles a normal curve.) If we define the random variable

x = IQ score of a randomly selected individual

then x has approximately a normal distribution with $\mu = 100$ and $\sigma = 15$.

One way to become eligible for membership in MENSA, an organization purportedly for those of high intelligence, is to have a Stanford–Binet IQ score above 130. What proportion of the population would quality for MENSA membership? An answer to this question requires evaluating $P(x > 130)$. This probability is shown in Figure 7.31. With $a = 130$,

$$a^* = \frac{130 - 100}{15} = 2.00$$

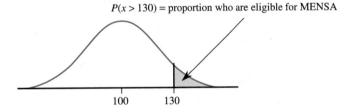

FIGURE 7.31 Normal distribution and desired proportion for Example 7.26

So (see Figure 7.32)

$$
\begin{aligned}
P(x > 130) &= P(z > 2.00) \\
&= z \text{ curve area to the right of } 2.00 \\
&= 1 - (z \text{ curve area to the left of } 2.00) \\
&= 1 - .9772 \\
&= .0228
\end{aligned}
$$

Only 2.28% of the population would qualify for MENSA membership.

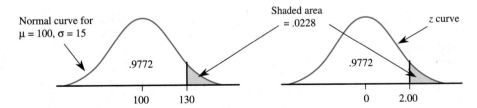

FIGURE 7.32 $P(x > 130)$ and corresponding z curve area

Suppose that we were interested in the proportion of the population with IQ scores below 80 — that is, $P(x < 80)$. With $b = 80$,

$$b* = \frac{80 - 100}{15} = -1.33$$

So

$$\begin{aligned}
P(x < 80) &= P(z < -1.33) \\
&= z \text{ curve area to the left of } -1.33 \\
&= .0918
\end{aligned}$$

as shown in Figure 7.33. This probability (.0918) tells us that just a little over 9% of the population have IQ scores below 80.

FIGURE 7.33 $P(x < 80)$ and corresponding z curve area

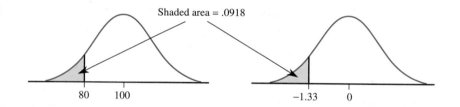

Now consider the proportion of the population with IQs between 75 and 125. Using $a = 75$ and $b = 125$,

$$a* = \frac{75 - 100}{15} = -1.67 \qquad b* = \frac{125 - 100}{15} = 1.67$$

so

$$\begin{aligned}
P(75 < x < 125) &= P(-1.67 < z < 1.67) \\
&= z \text{ curve area between } -1.67 \text{ and } 1.67 \\
&= (z \text{ curve area to the left of } 1.67) - (z \text{ curve area to the left of } -1.67) \\
&= .9525 - .0475 \\
&= .9050
\end{aligned}$$

This is illustrated in Figure 7.34. This tells us that 90.5% of the population have IQ scores between 75 and 125. Of the 9.5% whose IQ score is not between 75 and 125,

FIGURE 7.34
$P(75 < x < 125)$ and corresponding z curve area

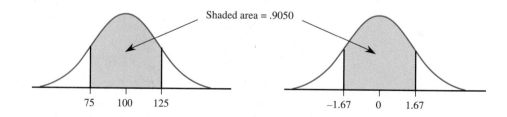

half (4.75%) would have scores over 125, and the other half would have scores below 75.

⸻

When we translate from a problem involving a normal distribution with mean μ and standard deviation σ to one involving the standard normal, we convert to z scores:

$$z = \frac{x - \mu}{\sigma}$$

Since a z score can be interpreted as giving the distance of an x value from the mean in units of the standard deviation, a z score of 1.4 corresponds to an x value that is 1.4 standard deviations above its mean, and a z score of -2.1 corresponds to an x value that is 2.1 standard deviations below its mean.

Suppose that we are trying to evaluate $P(x < 60)$ for a variable whose distribution is normal with $\mu = 50$ and $\sigma = 5$. Converting the endpoint 60 to a z score gives

$$z = \frac{60 - 50}{5} = 2$$

which tells us that the value 60 is 2 standard deviations above its mean. We then have

$$P(x < 60) = P(z < 2)$$

where z is a standard normal variable. Notice that for the standard normal distribution, the value 2 is 2 standard deviations above the mean, since the mean is 0 and the standard deviation is 1. The value $z = 2$ is located the same distance (measured in standard deviations) from the mean of the standard normal as is the value $x = 60$ from the mean in the normal distribution with $\mu = 50$ and $\sigma = 5$. This is why the translation using z scores results in an "equivalent" problem involving the standard normal distribution.

Describing Extreme Values in a Normal Distribution

To describe the extreme values for a normal distribution with mean μ and standard deviation σ, we first solve the corresponding problem for the standard normal distribution and then translate our answer into one for the normal distribution of interest. This process is illustrated in the following example.

EXAMPLE 7.27

Data on the length of time required to complete registration for classes using a telephone registration system suggests that the distribution of the variable

$$x = \text{time to register}$$

for students at a particular university can be well approximated by a normal distribution with mean $\mu = 12$ minutes and standard deviation $\sigma = 2$ minutes. (The normal distribution might not be an appropriate model for $x =$ time to register at another university. Many factors influence the shape, center, and spread of such a distribution.) Because some students do not sign off properly, the university would like to disconnect students automatically after some amount of time has elapsed. It is decided to choose this time such that only 1% of the students will be disconnected while they are still attempting to register. To determine the amount of time that should be allowed before disconnecting a student, we need to describe the largest 1% of the distribution of time to register. These are the individuals who will be mistakenly disconnected. This is illustrated in Figure 7.35(a). To determine the value of x^*, we first solve the analogous problem for the standard normal distribution, as shown in Figure 7.35(b).

FIGURE 7.35 Capturing the largest 1% in a normal distribution

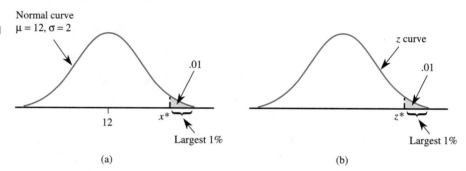

Looking in Appendix Table II for a cumulative area of .99, we find the closest entry (.9901) in the 2.3 row and the .03 column, from which $z^* = 2.33$. For the standard normal distribution, the largest 1% of the distribution is made up of those values greater than 2.33. An equivalent statement is that the largest 1% are those with z scores greater than 2.33. This implies that in the distribution of time to register, x, (or any other normal distribution), the largest 1% will be those values with z scores greater than 2.33 or, equivalently, those x values more than 2.33 standard deviations above the mean. Here the standard deviation is 2, so 2.33 standard deviations is 2.33(2), and it follows that

$$x^* = 12 + 2.33(2) = 12 + 4.66 = 16.66$$

The largest 1% of the *time to register* distribution is made up of values that are greater than 16.66 minutes. If the university system were set to disconnect students after 16.66 minutes, only 1% of the students registering would be disconnected prior to completing their registration.

A general formula for converting a z score back to an x value results from solving $z^* = (x^* - \mu)/\sigma$ for x^*, as shown in the accompanying box.

To convert a z score z^* back to an x value, use

$$x^* = \mu + z^*\sigma$$

EXAMPLE 7.28   The Environmental Protection Agency (EPA) has developed a testing program to monitor vehicle emission levels of several pollutants. The article "Determining Statistical Characteristics of a Vehicle Emissions Audit Procedure" (*Technometrics* (1980): 483–493) describes the program, which measures pollutants from various types of vehicles, all following a fixed driving schedule. Data from the article suggests that the emissions of oxides of nitrogen, which are major constituents of smog, can be plausibly modeled using a normal distribution. Let x denote the amount of this pollutant emitted by a randomly selected vehicle. The distribution of x can be described by a normal distribution with $\mu = 1.6$ and $\sigma = .4$.

Suppose that the EPA wants to offer some sort of incentive to get the worst polluters off the road. What emission levels constitute the worst 10% of the vehicles? The worst 10% would be the 10% with the highest emissions level, as shown in the illustration.

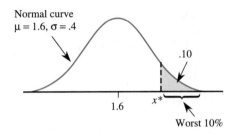

Normal curve
$\mu = 1.6$, $\sigma = .4$

.10

1.6 x^*

Worst 10%

For the standard normal distribution, the largest 10% are those with z values greater than $z^* = 1.28$ (from Table II, based on a cumulative area of .90). Then

$$x^* = \mu + z^*\sigma$$
$$= 1.6 + 1.28(.4)$$
$$= 1.6 + .512$$
$$= 2.112$$

In the population of vehicles of the type considered, about 10% would have oxide emission levels greater than 2.112.

Exercises 7.59 – 7.75

7.59   Determine the following standard normal (z) curve areas.

   a. The area under the z curve to the left of 1.75
   b. The area under the z curve to the left of $-.68$
   c. The area under the z curve to the right of 1.20
   d. The area under the z curve to the right of -2.82
   e. The area under the z curve between -2.22 and .53
   f. The area under the z curve between -1 and 1
   g. The area under the z curve between -4 and 4

7.60   Determine each of the areas under the standard normal (z) curve.

   a. To the left of -1.28 **b.** To the right of 1.28
   c. Between -1 and 2   **d.** To the right of 0
   e. To the right of -5 **f.** Between -1.6 and 2.5
   g. To the left of .23

7.61   Let z denote a random variable having a standard normal distribution. Determine each of the following probabilities.

   a. $P(z < 2.36)$ **b.** $P(z \le 2.36)$
   c. $P(z < -1.23)$ **d.** $P(1.14 < z < 3.35)$
   e. $P(-.77 \le z \le -.55)$ **f.** $P(z > 2)$
   g. $P(z \ge -3.38)$ **h.** $P(z < 4.98)$

7.62 Let z denote a random variable having a normal distribution with $\mu = 0$ and $\sigma = 1$. Determine each of the following probabilities.

a. $P(z < .10)$ **b.** $P(z < -.10)$
c. $P(.40 < z < .85)$ **d.** $P(-.85 < z < -.40)$
e. $P(-.40 < z < .85)$ **f.** $P(z > -1.25)$
g. $P(z < -1.50 \text{ or } z > 2.50)$

7.63 Let z denote a variable that has a standard normal distribution. Determine the value z^* to satisfy the following conditions.

a. $P(z < z^*) = .025$ **b.** $P(z < z^*) = .01$
c. $P(z < z^*) = .05$ **d.** $P(z > z^*) = .02$
e. $P(z > z^*) = .01$ **f.** $P(z > z^* \text{ or } z < -z^*) = .20$

7.64 Determine the value z^* that
a. Separates the largest 3% of all z values from the others
b. Separates the largest 1% of all z values from the others
c. Separates the smallest 4% of all z values from the others
d. Separates the smallest 10% of all z values from the others

7.65 Determine the value of z^* such that
a. z^* and $-z^*$ separate the middle 95% of all z values from the most extreme 5%
b. z^* and $-z^*$ separate the middle 90% of all z values from the most extreme 10%
c. z^* and $-z^*$ separate the middle 98% of all z values from the most extreme 2%
d. z^* and $-z^*$ separate the middle 92% of all z values from the most extreme 8%

7.66 Because $P(z < .44) = .67$, sixty-seven percent of all z values are less than .44, and .44 is the 67th percentile of the standard normal distribution. Determine the value of each of the following percentiles for the standard normal distribution. (If the cumulative area that you must look for does not appear in the z table, use the closest entry.)
a. The 91st percentile (*Hint*: Look for area .9100.)
b. The 77th percentile
c. The 50th percentile
d. The 9th percentile
e. What is the relationship between the 70th z percentile and the 30th z percentile?

7.67 Consider the population of all 1-gallon cans of dusty rose paint manufactured by a particular paint company. Suppose that a normal distribution with mean $\mu = 5$ ml and standard deviation $\sigma = .2$ ml is a reasonable model for the distribution of the variable $x =$ amount of red dye

in the paint mixture. Use the normal distribution model to calculate the following probabilities.

a. $P(x < 5.0)$ **b.** $P(x < 5.4)$
c. $P(x \leq 5.4)$ **d.** $P(4.6 < x < 5.2)$
e. $P(x > 4.5)$ **f.** $P(x > 4.0)$

7.68 Consider babies born in the "normal" range of 37–43 weeks gestational age. Extensive data supports the assumption that for such babies born in the United States, birth weight is normally distributed with mean 3432 g and standard deviation 482 g. (The article "Are Babies Normal" (*The American Statistician* (1999): 298–302) analyzed data from a particular year; for a sensible choice of class intervals, a histogram did not look at all normal but after further investigations is was determined that this was due to some hospitals measuring weight in grams and others measuring to the nearest ounce and then converting to grams. A modified choice of class intervals that allowed for this gave a histogram that was well described by a normal distribution.)
a. What is the probability that the birth weight of a randomly selected baby of this type exceeds 4000 grams? Is between 3000 and 4000 grams?
b. What is the probability that the birth weight of a randomly selected baby of this type is either less than 2000 grams or greater than 5000 grams?
c. What is the probability that the birth weight of a randomly selected baby of this type exceeds 7 lb?
d. How would you characterize the most extreme .1% of all birth weights?
e. If x is a random variable with a normal distribution and a is a numerical constant $(a \neq 0)$, then $y = ax$ also has a normal distribution. Use this to determine the distribution of birth weight expressed in pounds (shape, mean, and standard deviation), and then recalculate the probability from part **c.** How does this compare to your previous answer?

7.69 A machine that cuts corks for wine bottles operates in such a way that the distribution of the diameter of the corks produced is well approximated by a normal distribution with mean 3 cm and standard deviation .1 cm. The specifications call for corks with diameters between 2.9 and 3.1 cm. A cork not meeting the specifications is considered defective. (A cork that is too small leaks and causes the wine to deteriorate; a cork that is too large doesn't fit in the bottle.) What proportion of corks produced by this machine are defective?

7.70 Refer to Exercise 7.69. Suppose that there are two machines available for cutting corks. The one described in the preceding problem produces corks with diameters that are approximately normally distributed with mean 3 cm and standard deviation .1 cm. The second machine produces corks with diameters that are approximately normally distributed with mean 3.05 cm and standard deviation .01 cm. Which machine would you recommend? (*Hint*: Which machine would produce the fewest defective corks?)

7.71 A gasoline tank for a certain car is designed to hold 15 gallons of gas. Suppose that the variable x = actual capacity of a randomly selected tank has a distribution that is well approximated by a normal curve with mean 15.0 gal and standard deviation .1 gal.

a. What is the probability that a randomly selected tank will hold at most 14.8 gal?

b. What is the probability that a randomly selected tank will hold between 14.7 and 15.1 gal?

c. If two such tanks are independently selected, what is the probability that both hold at most 15 gal?

7.72 The time that it takes a randomly selected job applicant to perform a certain task has a distribution that can be approximated by a normal distribution with a mean value of 120 sec and a standard deviation of 20 sec. The fastest 10% are to be given advanced training. What task times qualify individuals for such training?

7.73 A machine that produces ball bearings has initially been set so that the true average diameter of the bearings it produces is .500 in. A bearing is acceptable if its diameter is within .004 in. of this target value. Suppose, however, that the setting has changed during the course of production, so that the distribution of the diameters produced is well approximated by a normal distribution with mean .499 in. and standard deviation .002 in. What percentage of the bearings produced will not be acceptable?

7.74 Suppose that the distribution of net typing rate in words per minute (wpm) for experienced typists can be approximated by a normal curve with mean 60 wpm and standard deviation 15 wpm. (The paper "Effects of Age and Skill in Typing" (*J. of Exper. Psych.* (1984): 345–371) describes how net rate is obtained from gross rate by using a correction for errors.)

a. What is the probability that a randomly selected typist's net rate is at most 60 wpm? Less than 60 wpm?

b. What is the probability that a randomly selected typist's net rate is between 45 and 90 wpm?

c. Would you be surprised to find a typist in this population whose net rate exceeded 105 wpm? (*Note*: The largest net rate in a sample described in the paper cited is 104 wpm.)

d. Suppose that two typists are independently selected. What is the probability that both their typing rates exceed 75 wpm?

e. Suppose that special training is to be made available to the slowest 20% of the typists. What typing speeds would qualify individuals for this training?

7.75 Consider the variable x = time required for a college student to complete a standardized exam. Suppose that for the population of students at a particular university, the distribution of x is well approximated by a normal curve with mean 45 min and standard deviation 5 min.

a. If 50 minutes is allowed for the exam, what proportion of students at this university would be unable to finish in the allotted time?

b. How much time should be allowed for the exam if we wanted 90% of the students taking the test to be able to finish in the allotted time?

c. How much time is required for the fastest 25% of all students to complete the exam?

7.7 Checking for Normality and Normalizing Transformations

Some of the most frequently used statistical methods are valid only when a sample $x_1, x_2, \ldots, x_n$, has come from a population distribution that is at least approximately normal. One way to see whether an assumption of population normality is plausible is to construct a **normal probability plot**. One version of this plot utilizes certain quantities called **normal scores**. The values of the normal scores depend on the sample size n. For example, the normal scores when $n = 10$ are as follows:

$$-1.539 \quad -1.001 \quad -.656 \quad -.376 \quad -.123$$
$$.123 \quad\quad .376 \quad\quad .656 \quad\quad 1.001 \quad\quad 1.539$$

To interpret these numbers, think of selecting sample after sample from a standard normal distribution, each one consisting of $n = 10$ observations. Then -1.539 is the long-run average of the smallest observation from each sample, -1.001 is the long-run average of the second smallest observation from each sample, and so on. In other words, -1.539 is the mean value of the smallest observation in a sample of size 10 from the z distribution, -1.001 is the mean value of the second smallest observation, etc.

Extensive tabulations of normal scores for many different sample sizes are available. Alternatively, many software packages (such as MINITAB and SAS) and some graphing calculators can compute these scores on request and then construct a normal probability plot. Some graphing calculators and software packages use a slightly different algorithm to compute normal scores. This does not change the overall character of a normal probability plot, so either the tabled values or those given by the computer or calculator can be used.

After ordering the sample observations from smallest to largest, the smallest normal score is paired with the smallest observation, the second smallest normal score with the second smallest observation, and so on. The first number in a pair is the normal score, and the second is the observed data value. A normal probability plot is just a scatter plot of the (normal score, observed value) pairs.

If the sample has actually been selected from a *standard* normal distribution, the second number in each pair should be reasonably close to the first number (ordered observation $\approx$ corresponding mean value). Then the n plotted points will fall near a line with slope equal to 1 (a 45° line) passing through $(0, 0)$. When the sample has been obtained from *some* normal population distribution, the plotted points should be close to *some* straight line.

Definition

A **normal probability plot** is a scatter plot of the (normal score, observation) pairs.

A substantial linear pattern in a normal probability plot suggests that population normality is plausible. On the other hand, a systematic departure from a straight-line pattern (such as curvature in the plot) casts doubt on the legitimacy of assuming a normal population distribution.

EXAMPLE 7.29   The following ten observations are widths of contact windows in integrated circuit chips:

| 3.21 | 2.49 | 2.94 | 4.38 | 4.02 | 3.62 | 3.30 | 2.85 | 3.34 | 3.81 |

The ten pairs for the normal probability plot are then

$(-1.539, 2.49)$	$(.123, 3.34)$
$(-1.001, 2.85)$	$(.376, 3.62)$
$(-.656, 2.94)$	$(.656, 3.81)$
$(-.376, 3.21)$	$(1.001, 4.02)$
$(-.123, 3.30)$	$(1.539, 4.38)$

FIGURE 7.36 A normal probability plot for Example 7.29

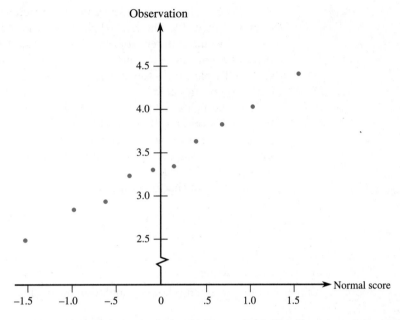

The normal probability plot is shown in Figure 7.36. The linearity of the plot supports the assumption that the window width distribution from which these observations were drawn is normal.

The decision as to whether a plot shows a substantial linear pattern is somewhat subjective. Particularly when n is small, normality should not be ruled out unless the departure from linearity is very clear-cut. Figure 7.37 displays several plots that suggest a nonnormal population distribution.

FIGURE 7.37 Plots suggesting nonnormality: (a) Indication that the population distribution is skewed; (b) Indication that the population distribution has heavier tails than a normal curve; (c) Presence of an outlier

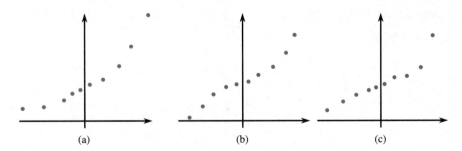

Using the Correlation Coefficient to Check Normality

The correlation coefficient r was introduced in Chapter 5 as a quantitative measure of the extent to which the points in a scatter plot fall close to a straight line. Let's denote the n (normal score, observation) pairs as follows:

(smallest normal score, smallest observation)

$$\vdots$$

(largest normal score, largest observation)

Then the correlation coefficient can be computed using the defining equation for r given in Chapter 5. The normal probability plot always slopes upward (since it is based on values ordered from smallest to largest), so r will be a positive number. A value of r quite close to 1 indicates a very strong linear relationship in the normal probability plot. If r is too much smaller than 1, normality of the underlying distribution is questionable.

How far below 1 does r have to be before we begin to doubt seriously the plausibility of normality? The answer depends on the sample size n. If n is small, an r value somewhat below 1 would not be surprising even when the distribution is normal, but if n is large, only an r value very close to 1 would support the assumption of normality. For selected values of n, Table 7.2 gives critical values to which r can be compared in checking for normality.

If

$r <$ critical r for corresponding n

considerable doubt is cast on the assumption of population normality.

If your sample size is in between two tabled values of n, use the critical value for the larger sample size. (For example, if $n = 46$, use the value of .966 for sample size 50.)

Table 7.2 *Values to which r can be compared to check for normality**

n	5	10	15	20	25	30	40	50	60	75
Critical r	.832	.880	.911	.929	.941	.949	.960	.966	.971	.976

*Source: MINITAB User's Manual.

How were the critical values in Table 7.2 obtained? Consider the critical value .941 for $n = 25$. Suppose that the underlying distribution is actually normal. Consider obtaining a large number of different samples, each one consisting of 25 observations, and computing the value of r for each one. Then it can be shown that only 1% of the samples result in an r value less than the critical value .941. That is, .941 was chosen to guarantee a 1% error rate: In only 1% of all cases will we judge normality implausible when the distribution is really normal. The other critical values are also chosen to yield a 1% error rate for the corresponding sample sizes. It might have occurred to you that another type of error is possible: obtaining a large value of r and concluding that normality is a reasonable assumption when the distribution is actually quite nonnormal. This type of error is more difficult to control than the type mentioned previously, but the procedure we have described generally does a good job in both respects.

EXAMPLE 7.30   (Example 7.29 continued)   The sample size for the contact window width data of Example 7.29 is $n = 10$. The critical r, from Table 7.2 is then .880. The correlation coefficient calculated using the (normal score, observation) pairs is $r = .995$. Since r is larger than the critical r for a sample of size 10, it is plausible that the population distribution of window widths from which this sample was drawn is approximately normal.

Transforming Data to Obtain a Distribution That Is Approximately Normal

Many of the most frequently used statistical methods are valid only when the sample is selected at random from a population whose distribution is at least approximately normal. When the sample histogram shows a distinctly nonnormal shape, it is common to use a transformation or reexpression of the data. By *transforming* data, we mean using some specified mathematical function (such as square root, logarithm, or reciprocal) on each data value to produce a set of transformed data. We can then study and summarize the distribution of these transformed values using methods that require normality. We saw in Chapter 5 that, with bivariate data, one or both of the variables might be transformed in an attempt to find two variables that are linearly related. With univariate data, a transformation is usually chosen to yield a distribution of transformed values that is more symmetric and more closely approximated by a normal curve than was the original distribution.

EXAMPLE 7.31   Data that has been used by several authors to introduce the concept of transformation (for example, "Exploratory Methods for Choosing Power Transformations," *J. Amer. Stat. Assoc.* (1982): 103–108) consists of values of March precipitation for Minneapolis–St. Paul over a period of 30 years. These values are given in Table 7.3 (page 370), along with the square root of each value. Histograms of both the original and transformed data appear in Figure 7.38. The distribution of the original data is clearly skewed, with a long upper tail. The square-root transformation has resulted in a substantially more symmetric distribution, with a typical (that is, central) value near the 1.25 boundary between the third and fourth class intervals.

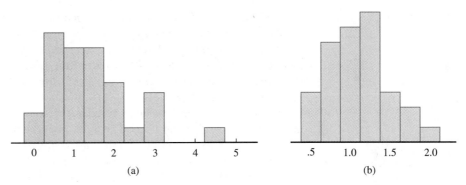

FIGURE 7.38 Histograms of precipitation data: (a) Untransformed data; (b) Square-root transformed data

Table 7.3 *Original and square-root transformed values of March precipitation in Minneapolis–St. Paul over a 30-year period*

Year	Precipitation	$\sqrt{\text{Precipitation}}$	Year	Precipitation	$\sqrt{\text{Precipitation}}$
1	.77	.88	16	1.62	1.27
2	1.74	1.32	17	1.31	1.14
3	.81	.90	18	.32	.57
4	1.20	1.10	19	.59	.77
5	1.95	1.40	20	.81	.90
6	1.20	1.10	21	2.81	1.68
7	.47	.69	22	1.87	1.37
8	1.43	1.20	23	1.18	1.09
9	3.37	1.84	24	1.35	1.16
10	2.20	1.48	25	4.75	2.18
11	3.00	1.73	26	2.48	1.57
12	3.09	1.76	27	.96	.98
13	1.51	1.23	28	1.89	1.37
14	2.10	1.45	29	.90	.95
15	.52	.72	30	2.05	1.43

Logarithmic transformations are also common and, as with bivariate data, either the natural or $\log_{10}$ can be used. A log transformation is usually applied to data that is positively skewed (a long upper tail). This affects values in the upper tail substantially more than values in the lower tail, yielding a more symmetric — and often more normal — distribution.

EXAMPLE 7.32

Exposure to beryllium is known to produce adverse effects on lungs as well as on other tissues and organs in both laboratory animals and humans. The article "Time Lapse Cinematographic Analysis of Beryllium — Lung Fibroblast Interactions" (*Envir. Research* (1983): 34–43) reported the results of experiments designed to study the behavior of certain individual cells that had been exposed to beryllium. An important characteristic of such an individual cell is its interdivision time (IDT). IDTs were determined for a large number of cells both in exposed (treatment) and unexposed (control) conditions. The authors of the article state, "The IDT distributions are seen to be skewed, but the natural logs do have an approximate normal distribution." The same property holds for $\log_{10}$-transformed data. We give representative IDT data in Table 7.4 and the resulting histograms in Figure 7.39, which are in agreement with the authors' statement.

Table 7.4 *Original and log₁₀ (IDT) values*

IDT	$\log_{10}$(IDT)	IDT	$\log_{10}$(IDT)	IDT	$\log_{10}$(IDT)
28.1	1.45	31.2	1.49	13.7	1.14
46.0	1.66	25.8	1.41	16.8	1.23
34.8	1.54	62.3	1.79	28.0	1.45
17.9	1.25	19.5	1.29	21.1	1.32
31.9	1.50	28.9	1.46	60.1	1.78
23.7	1.37	18.6	1.27	21.4	1.33
26.6	1.42	26.2	1.42	32.0	1.51
43.5	1.64	17.4	1.24	38.8	1.59
30.6	1.49	55.6	1.75	25.5	1.41
52.1	1.72	21.0	1.32	22.3	1.35
15.5	1.19	36.3	1.56	19.1	1.28
38.4	1.58	72.8	1.86	48.9	1.69
21.4	1.33	20.7	1.32	57.3	1.76
40.9	1.61				

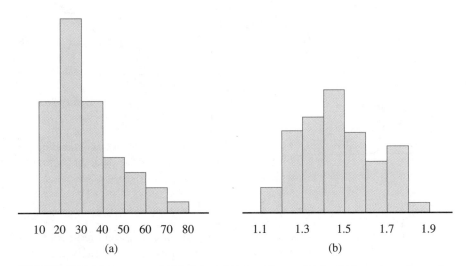

FIGURE 7.39 Histograms of IDT data: (a) Untransformed data; (b) $\log_{10}$-transformed data

The sample size for the IDT data is $n = 40$. The correlation coefficient for the (normal score, original (untransformed) data) pairs is .950, which is less than the critical r for $n = 40$ (critical $r = .960$). The correlation coefficient using the transformed data is .998, which is much larger than the critical r, supporting the assertion that $\log_{10}$(IDT) has approximately a normal distribution. Figure 7.40 (page 372)

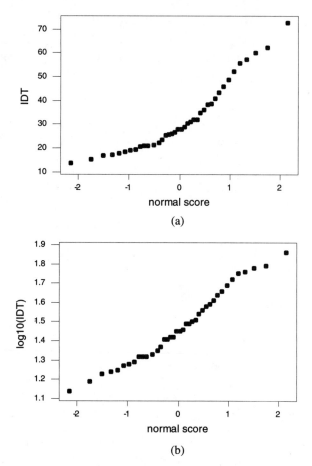

FIGURE 7.40 MINITAB-generated normal probability plots: (a) Original IDT data; (b) Log-transformed IDT data

displays MINITAB normal probability plots for the original data and for the transformed data. The latter plot is clearly more linear in appearance than the former.

Selecting a Transformation

A particular transformation may on occasion be dictated by some theoretical argument, but often this is not the case. In this case, you may wish to try several different transformations to find one that is satisfactory. Figure 7.41, from the article "Distribution of Sperm Counts in Suspected Infertile Men" (*J. of Reproduction*

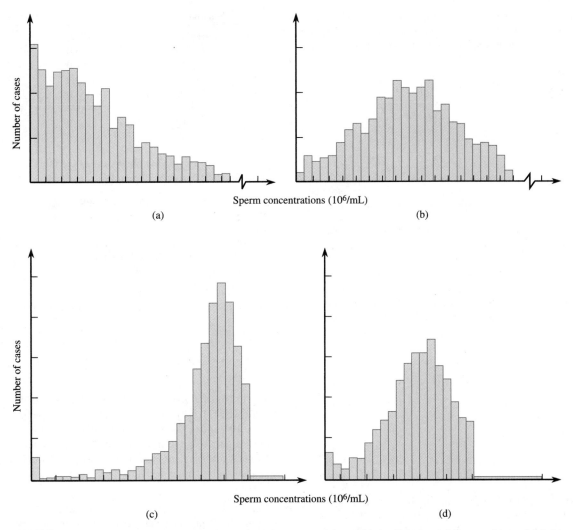

FIGURE 7.41 Histograms of sperm concentrations for 1711 suspected infertile men: (a) Untransformed (highly skewed); (b) Log transformed (reasonably symmetric); (c) Square-root transformed; (d) Cube-root transformed

and Fertility (1983): 91–96), shows what can result from such a search. Other workers in this field had previously used all three of the transformations illustrated.

Exercises 7.76 – 7.87

7.76 Ten measurements of the steam rate (lb/hr) of a distillation tower were used to construct the given normal probability plot ("A Self-Descaling Distillation Tower," *Chem. Eng. Process* (1968): 79–84). Based on the accompanying plot, do you think it is reasonable to assume that the normal distribution provides an adequate description of the steam rate distribution? Explain.

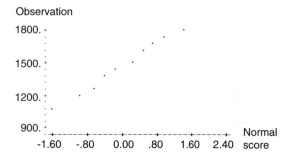

Observation

7.77 The accompanying normal probability plot was constructed using part of the data appearing in the paper "Trace Metals in Sea Scallops" (*Environ. Concentration and Toxicology* 19: 1326–1334). The variable under study was the amount of cadmium in North Atlantic scallops. Does the sample data suggest that the cadmium concentration distribution is not normal? Explain.

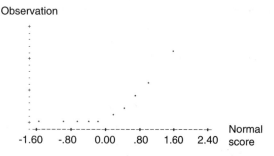

Observation

7.78 Consider the following ten observations on the lifetime (hr) for a certain type of component: 152.7, 172.0, 172.5, 173.3, 193.0, 204.7, 216.5, 234.9, 262.6, 422.6. Construct a normal probability plot, and comment on the plausibility of a normal distribution as a model for component lifetime.

7.79 The paper "The Load-Life Relationship for M50 Bearings with Silicon Nitride Ceramic Balls" (*Lubric. Eng.* (1984): 153–159) reported the accompanying data on bearing load life (million revs); the corresponding normal scores are also given. Construct a normal probability plot. Is normality plausible?

x	Normal Score	x	Normal Score
47.1	−1.867	68.1	−1.131
68.1	−1.408	90.8	−.921

(continued)

x	Normal Score	x	Normal Score
103.6	−.745	278.0	.315
106.0	−.590	278.0	.448
115.0	−.448	289.0	.590
126.0	−.315	289.0	.745
146.6	−.187	367.0	.921
229.0	−.062	385.9	1.131
240.0	.062	392.0	1.408
240.0	.187	395.0	1.867

7.80 The accompanying observations are DDT concentrations in the blood of 20 people.

24	26	30	35	35	38	39	40	40	41
42	52	56	58	61	75	79	88	102	42

Use the normal scores from Exercise 7.79 to construct a normal probability plot, and comment on the appropriateness of a normal probability model.

7.81 Consider the following sample of 25 observations on the diameter x (cm) of a disk used in a certain system.

16.01	16.08	16.13	15.94	16.05	16.27	15.89	15.84
15.95	16.10	15.92	16.04	15.82	16.15	16.06	15.66
15.78	15.99	16.29	16.15	16.19	16.22	16.07	16.13
16.11							

The 13 largest normal scores for a sample of size 25 are 1.965, 1.524, 1.263, 1.067, .905, .764, .637, .519, .409, .303, .200, .100, and 0. The 12 smallest scores result from placing a negative sign in front of each of the given nonzero scores. Construct a normal probability plot. Does it appear plausible that disk diameter is normally distributed? Explain.

7.82 Example 7.31 examined rainfall data for Minneapolis–St. Paul. The square-root transformation was used to obtain a distribution of values that was more symmetric than the distribution of the original data. Another power transformation that has been suggested by meteorologists is the cube root: Transformed value = (original value)$^{1/3}$. The original values and their cube roots (the transformed values) are given in the accompanying table. Construct a histogram of the transformed data. Compare your histogram to those given in Figure 7.38. Which of the cube-root and the square-root transformations appears to result in the more symmetric histogram?

Original	.32	.47	.52	.59	.77
Transformed	.68	.78	.80	.84	.92

(continued)

Original	.81	.81	.90	.96	1.18
Transformed	.93	.93	.97	.99	1.06
Original	1.20	1.20	1.31	1.35	1.43
Transformed	1.06	1.06	1.09	1.11	1.13
Original	1.51	1.62	1.74	1.87	1.89
Transformed	1.15	1.17	1.20	1.23	1.24
Original	1.95	2.05	2.10	2.20	2.48
Transformed	1.25	1.27	1.28	1.30	1.35
Original	2.81	3.00	3.09	3.37	4.75
Transformed	1.41	1.44	1.46	1.50	1.68

7.83 The accompanying table gives a sample of survival times (days from diagnosis) for patients suffering from chronic leukemia of a certain type (*Statistical Methodology for Survival Time Studies,* National Cancer Institute, 1986).

7	47	58	74	177
232	273	285	317	429
440	445	455	468	495
497	532	571	579	581
650	702	715	779	881
900	930	968	1077	1109
1314	1334	1367	1534	1712
1784	1877	1886	2045	2056
2260	2429	2509		

a. Construct a relative frequency distribution for this data set, and draw the corresponding histogram.
b. Would you describe this histogram as having a positive or a negative skew?
c. Would you recommend transforming the data? Explain.

7.84 In a study of warp breakage during the weaving of fabric (*Technometrics* (1982): 63), 100 pieces of yarn were tested. The number of cycles of strain to breakage was recorded for each yarn sample. The resulting data is given in the accompanying table.

86	146	251	653	98	249	400	292	131
176	76	264	15	364	195	262	88	264
42	321	180	198	38	20	61	121	282
180	325	250	196	90	229	166	38	337
341	40	40	135	597	246	211	180	93
571	124	279	81	186	497	182	423	185
338	290	398	71	246	185	188	568	55
244	20	284	93	396	203	829	239	236
277	143	198	264	105	203	124	137	135

(continued)

169	157	224	65	315	229	55	286	350
193	175	220	149	151	353	400	61	194
188								

a. Construct a frequency distribution using the class intervals 0–<100, 100–<200, and so on.
b. Draw the histogram corresponding to the frequency distribution in part **a.** How would you describe the shape of this histogram?
c. Find a transformation for this data that results in a more symmetric histogram than what you obtained in part **b.**

7.85 The article "The Distribution of Buying Frequency Rates" (*J. of Marketing Research* (1980): 210–216) reported the results of a $3\frac{1}{2}$-year study of dentifrice purchases. The authors conducted their research using a national sample of 2071 households and recorded the number of toothpaste purchases for each household participating in the study. The results are given in the accompanying frequency distribution.

Number of Purchases	Number of Households (frequency)
10–<20	904
20–<30	500
30–<40	258
40–<50	167
50–<60	94
60–<70	56
70–<80	26
80–<90	20
90–<100	13
100–<110	9
110–<120	7
120–<130	6
130–<140	6
140–<150	3
150–<160	0
160–<170	2

a. Draw a histogram for this frequency distribution. Would you describe the histogram as positively or negatively skewed?
b. Does the square-root transformation result in a histogram that is more symmetric than that of the original data? (Be careful — this one is a bit tricky, since you don't have the raw data; transforming the endpoints of the class intervals will result in class intervals that are not necessarily of equal widths, so the histogram of the transformed values will have to be drawn with this in mind.)

7.86 Ecologists have long been interested in factors that might explain how far north or south particular animal species are found. As part of one such study, the paper "Temperature and the Northern Distributions of Wintering Birds" (*Ecology* (1991): 2274–2285) gave the accompanying body masses (in grams) for 50 different bird species that had previously been thought to have northern boundaries corresponding to a particular isotherm.

7.7	10.1	21.6	8.6	12.0	11.4	16.6	9.4
11.5	9.0	8.2	20.2	48.5	21.6	26.1	6.2
19.1	21.0	28.1	10.6	31.6	6.7	5.0	68.8
23.9	19.8	20.1	6.0	99.6	19.8	16.5	9.0
448.0	21.3	17.4	36.9	34.0	41.0	15.9	12.5
10.2	31.0	21.5	11.9	32.5	9.8	93.9	10.9
19.6	14.5						

a. Construct a stem-and-leaf display in which 448.0 is shown beside the display as a HI value, the stem of an observation is the tens digit, the leaf is the ones digit, and the tenths digit is suppressed (so, e.g., 21.5 has stem 2 and leaf 1). What do you perceive as the most prominent feature of the display?

b. Draw a histogram based on class intervals 5 –<10, 10 –<15, 15 –<20, 20 –<25, 25 –<30, 30 –<40, 40 –<50, 50 –<100, and 100 –<500. Is a transformation of the data desirable? Explain.

c. Use a calculator or statistical computer package to calculate logarithms of these observations and construct a histogram. Is the log transformation satisfactory?

d. Consider transformed value = $1/\sqrt{\text{original value}}$, and construct a histogram of the transformed data. Does it appear to resemble a normal curve?

7.87 The accompanying figure appeared in the paper "EDTA-Extractable Copper, Zinc and Manganese in Soils of the Canterbury Plains" (*New Zealand J. of Ag. Res.* (1984): 207–217). A large number of topsoil samples were analyzed for manganese (Mn), zinc (Zn), and copper (Cu), and the resulting data was summarized using histograms. The authors transformed each data set using logarithms in an effort to obtain more symmetric distributions of values. Do you think the transformations were successful? Explain.

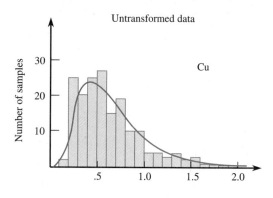

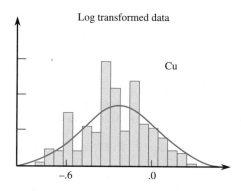

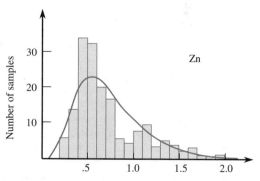

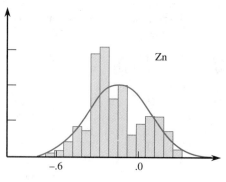

Figure for Exercise 7.87

(*continued*)

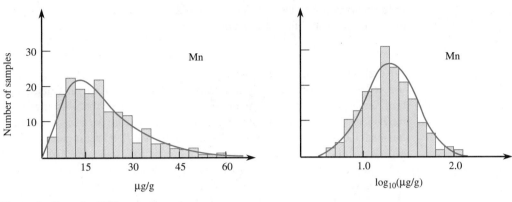

Figure for Exercise 7.87—*continued*

7.8   Using the Normal Distribution to Approximate a Discrete Distribution (Optional)

One useful characteristic of normal curves is that the distribution of many random variables can be approximated by a carefully chosen normal distribution. In this section, we show how probabilities for some discrete random variables can be approximated using a normal curve. The most important case of this concerns the approximation of binomial probabilities.

The Normal Curve and Discrete Variables

The probability distribution of a discrete random variable x is represented pictorially by a probability histogram. The probability of a particular value is the area of the rectangle centered at that value. Possible values of x are isolated points on the number line, usually whole numbers. For example, if $x =$ the IQ of a randomly selected 8-year-old child, then x is a discrete random variable, since an IQ score must be a whole number.

Often a probability histogram can be very well approximated by a normal curve, as illustrated in Figure 7.42. In such cases, it is customary to say that x has approxi-

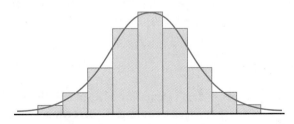

FIGURE 7.42 A normal curve approximation to a probability histogram

mately a normal distribution. The normal distribution can then be used to calculate approximate probabilities of events involving x.

EXAMPLE 7.33   The number of express mail packages mailed at a certain post office on a randomly selected day is approximately normally distributed with mean value 18 and standard deviation 6. Let's first calculate the approximate probability that $x = 20$. Figure 7.43(a) shows a portion of the probability histogram for x with the approximating normal curve superimposed. The area of the shaded rectangle is $P(x = 20)$. The left edge of this rectangle is at 19.5 on the horizontal scale, and the right edge is at 20.5. Therefore, the desired probability is approximately the area under the normal curve between 19.5 and 20.5. Standardizing these limits gives

$$\frac{20.5 - 18}{6} = .42 \qquad \frac{19.5 - 18}{6} = .25$$

from which we get

$$P(x = 20) \approx P(.25 < z < .42) = .6628 - .5987 = .0641$$

In a similar fashion, Figure 7.43(b) shows that $P(x \le 10)$ is approximately the area under the normal curve to the left of 10.5. Then

$$P(x \le 10) \approx P\left(z \le \frac{10.5 - 18}{6} \right) = P(z \le -1.25) = .1056$$

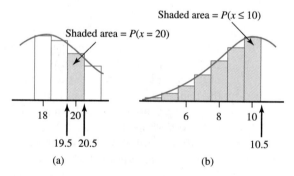

FIGURE 7.43 The normal approximation for Example 7.33

The calculation of probabilities in Example 7.33 illustrates the use of what is known as a **continuity correction.** Because the rectangle for $x = 10$ extends to 10.5 on the right, we use the normal curve area to the left of 10.5 rather than 10. In general, if possible x values are consecutive whole numbers, then $P(a \leq x \leq b)$ will be approximately the normal curve area between limits $a - 1/2$ and $b + 1/2$.

The Normal Approximation to a Binomial Distribution

Figure 7.44 shows the probability histograms for two binomial distributions, one with $n = 25$, $\pi = .4$ and the other with $n = 25$, $\pi = .1$. For each distribution, we computed

$$\mu = n\pi \quad \text{and} \quad \sigma = \sqrt{n\pi(1 - \pi)}$$

and then we superimposed a normal curve with this μ and σ on the corresponding probability histogram. A normal curve fits the probability histogram very well in the first case (Figure 7.44(a)). When this happens, binomial probabilities can be accurately approximated by areas under the normal curve. Because of this, statisticians say that both x (the number of S's) and x/n (the proportion of S's) are approximately normally distributed. In the second case (Figure 7.44(b)), the normal curve does not give a good approximation because the probability histogram is skewed, whereas the normal curve is symmetric.

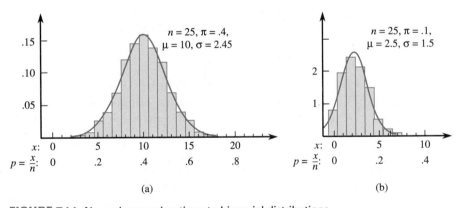

FIGURE 7.44 Normal approximations to binomial distributions

Let x be a binomial random variable based on n trials and success probability π, so that

$$\mu = n\pi \quad \text{and} \quad \sigma = \sqrt{n\pi(1 - \pi)}$$

If n and π are such that

$$n\pi \geq 10 \quad \text{and} \quad n(1 - \pi) \geq 10$$

then x has approximately a normal distribution. Combining this result with the continuity correction implies that

(continued)

$$P(a \le x \le b) \approx P\left(\frac{a - \frac{1}{2} - \mu}{\sigma} \le z \le \frac{b + \frac{1}{2} - \mu}{\sigma} \right)$$

That is, the probability that x is between a and b inclusive is approximately the area under the approximating normal curve between $a - 1/2$ and $b + 1/2$.

Similarly,

$$P(x \le b) \approx P\left(z \le \frac{b + \frac{1}{2} - \mu}{\sigma} \right)$$

$$P(a \le x) \approx P\left(\frac{a - \frac{1}{2} - \mu}{\sigma} \le z \right)$$

When either $n\pi < 10$ or $n(1 - \pi) < 10$, the binomial distribution is too skewed for the normal approximation to give accurate results.

EXAMPLE 7.34

Premature babies are those born more than 3 weeks early. *Newsweek* (May 16, 1988) reports that 10% of the live births in this country are premature. Suppose that 250 live births are randomly selected and the number x of "preemies" is determined. Since

$$n\pi = 250(.10) = 25 \ge 10$$
$$n(1 - \pi) = 250(.90) = 225 \ge 10$$

x has approximately a normal distribution, with

$$\mu = 250(.10) = 25 \qquad \sigma = \sqrt{250(.1)(.9)} = 4.743$$

The probability that x is between 15 and 30 (inclusive) is

$$\begin{aligned}
P(15 \le x \le 30) &= P\left(\frac{14.5 - 25}{4.743} < z < \frac{30.5 - 25}{4.743} \right) \\
&= P(-2.21 < z < 1.16) \\
&= .8770 - .0136 \\
&= .8634
\end{aligned}$$

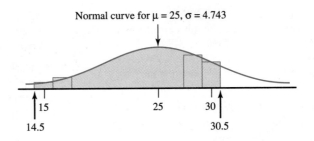

Normal curve for μ = 25, σ = 4.743

The event that fewer than 20 births are premature includes x values 19, 18, 17, . . . , 1, and 0, so

$$P(x < 20) = P(x \leq 19) \approx P\left(z < \frac{19.5 - 25}{4.743}\right)$$
$$= P(z < -1.16)$$
$$= .1230$$

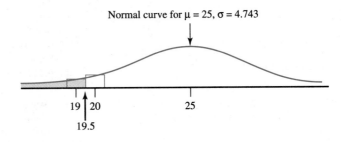

Normal curve for μ = 25, σ = 4.743

Exercises 7.88 – 7.96

7.88 Let x denote the IQ for an individual selected at random from a certain population. The value of x must be a whole number. Suppose that the distribution of x can be approximated by a normal distribution with mean value 100 and standard deviation 15. Approximate the following probabilities:

  a. $P(x = 100)$
  b. $P(x \leq 110)$
  c. $P(x < 110)$ (*Hint:* $x < 110$ is the same as $x \leq 109$.)
  d. $P(75 \leq x \leq 125)$

7.89 Suppose that the distribution of the number of items x produced by an assembly line during an 8-hour shift can be approximated by a normal distribution with mean value 150 and standard deviation 10.

  a. What is the probability that the number of items produced is at most 120?

b. What is the probability that at least 125 items are produced?
c. What is the probability that between 135 and 160 (inclusive) items are produced?

7.90 The number of vehicles leaving a turnpike at a certain exit during a particular time period has approximately a normal distribution with mean value 500 and standard deviation 75. What is the probability that the number of cars exiting during this period is

  a. at least 650?
  b. strictly between 400 and 550? ("Strictly" means that the values 400 and 550 are not included.)
  c. between 400 and 550 (inclusive)?

7.91 Let x have a binomial distribution with $n = 50$ and $\pi = .6$, so that $\mu = n\pi = 30$ and $\sigma = \sqrt{n\pi(1 - \pi)} = $

3.4641. Calculate the following probabilities using the normal approximation with the continuity correction.

a. $P(x = 30)$ **b.** $P(x = 25)$
c. $P(x \leq 25)$ **d.** $P(25 \leq x \leq 40)$
e. $P(25 < x < 40)$ (*Hint:* $25 < x < 40$ is the same as $26 \leq x \leq 39$.)

7.92 Seventy percent of the bicycles sold by a certain store are mountain bikes. Among 100 randomly selected bike purchases, what is the approximate probability that

a. at most 75 are mountain bikes?
b. between 60 and 75 (inclusive) are mountain bikes?
c. more than 80 are mountain bikes?
d. at most 30 are not mountain bikes?

7.93 Suppose that 25% of the fire alarms in a large city are false alarms. Let x denote the number of false alarms in a random sample of 100 alarms. Give approximations to the following probabilities.

a. $P(20 \leq x \leq 30)$
b. $P(20 < x < 30)$
c. $P(35 \leq x)$
d. The probability that x is farther than 2 standard deviations from its mean value

7.94 Suppose that 65% of all registered voters in a certain area favor a 7-day waiting period prior to purchase of a handgun. Among 225 randomly selected voters, what is the probability that

a. at least 150 favor such a waiting period?
b. more than 150 favor such a waiting period?
c. fewer than 125 favor such a waiting period?

7.95 Flash bulbs manufactured by a certain company are sometimes defective.

a. If 5% of all such bulbs are defective, could the techniques of this section be used to approximate the probability that at least 5 of the bulbs in a random sample of size 50 are defective? If so, calculate this probability; if not, explain why not.
b. Reconsider the question posed in part **a** for the probability that at least 20 bulbs in a random sample of size 500 are defective.

7.96 A company that manufactures mufflers for cars offers a lifetime warranty on its products provided that ownership of the car does not change. Suppose that only 20% of its mufflers are replaced under this warranty.

a. In a random sample of 400 purchases, what is the approximate probability that between 75 and 100 (inclusive) are replaced under warranty?
b. Among 400 randomly selected purchases, what is the probability that at most 70 are ultimately replaced under warranty?
c. If you were told that fewer than 50 among 400 randomly selected purchases were ever replaced under warranty, would you question the 20% figure? Explain.

Summary of Key Concepts and Formulas

Term or Formula	Comment
Random variable: discrete or continuous	A numerical variable with a value determined by the outcome of a chance experiment. It is discrete if its possible values are isolated points along the number line and continuous if its possible values form an entire interval on the number line.
Probability distribution $p(x)$ of a discrete random variable x	A formula, table, or graph that gives the probability associated with each x value. Conditions on $p(x)$ are (a) $p(x) \geq 0$, and (b) $\Sigma\, p(x) = 1$, where the sum is over all possible x values.
Probability distribution of a continuous random variable x	Specified by a smooth (density) curve for which the total area under the curve is 1. The probability $P(a < x < b)$ is the area under the curve and above the interval from a to b; this is also $P(a \leq x \leq b)$.
μ_x and σ_x	The mean and standard deviation, respectively, of a random variable x. These quantities describe the center and extent of spread about the center of the variable's probability distribution.

$\mu_x = \Sigma\, xp(x)$

The mean value of a discrete random variable x; it locates the center of the variable's probability distribution.

$\sigma_x^2 = \Sigma\, (x - \mu)^2 p(x)$
$\sigma_x = \sqrt{\sigma_x^2}$

The variance and standard deviation, respectively, of a discrete random variable; these are measures of the extent to which the variable's distribution spreads out about μ.

Binomial probability distribution

$p(x) = \dfrac{n!}{x!(n-x)!}\pi^x(1-\pi)^{n-x}$

This formula gives the probability of observing x successes ($x = 0, 1, \ldots, n$) among n trials of a binomial experiment.

$\mu_x = np$
$\sigma_x = \sqrt{n\pi(1-\pi)}$

The mean and standard deviation of a binomial random variable.

Normal distribution

A continuous probability distribution that has a bell-shaped density curve. A particular normal distribution is determined by specifying values of μ and σ.

Standard normal distribution

This is the normal distribution with $\mu = 0$ and $\sigma = 1$. The density curve is called the z curve, and z is the letter commonly used to denote a variable having this distribution. Areas under the z curve to the left of various values are given in Appendix Table II.

z critical value

A number on the z measurement scale that captures a specified tail area or central area.

$z = \dfrac{x - \mu}{\sigma}$

z is obtained by "standardizing": subtracting the mean and then dividing by the standard deviation. When x has a normal distribution, z has a standard normal distribution. This fact implies that probabilities involving any normal random variable (any μ or σ) can be obtained from z curve areas.

Normal probability plot

A picture used to judge the plausibility of the assumption that a sample has been selected from a normal population distribution. If the plot is reasonably straight, this assumption is reasonable.

Normal approximation to the binomial distribution

When both $n\pi \geq 10$ and $n(1-\pi) \geq 10$, binomial probabilities are well approximated by corresponding areas under a normal curve with $\mu = n\pi$ and $\sigma = \sqrt{n\pi(1-\pi)}$.

Supplementary Exercises 7.97 – 7.119

7.97 An article in the Dec. 8, 1991, *Los Angeles Times* reported that there are 40,000 travel agencies nationwide, of which 11,000 are members of the American Society of Travel Agents (booking a tour through an ASTA member increases the likelihood of a refund in the event of cancellation).

  a. If x is the number of ASTA members among 5000 randomly selected agencies, could you use the methods of Section 7.8 to approximate $P(1200 < x < 1400)$? Why or why not?

  b. In a random sample of 100 agencies, what are the mean value and standard deviation of the number of ASTA members?

  c. If the sample size in part **b** is doubled, does the standard deviation double? Explain.

7.98 A soft-drink machine dispenses only regular Coke and Diet Coke. Sixty percent of all purchases from this machine are diet drinks. The machine currently has ten cans of each type. If 15 customers want to purchase drinks prior to the machine being restocked, what is the probability that each of the 15 is able to purchase the type of drink desired? (*Hint:* Let x denote the number among the 15 who want a diet drink. For which possible values of x is everyone satisfied?)

7.99   A mail-order computer software business has six telephone lines. Let x denote the number of lines in use at a specified time. The probability distribution of x is as follows:

x	0	1	2	3	4	5	6
$p(x)$	.10	.15	.20	.25	.20	.06	.04

Write each of the following events in terms of x, and then calculate the probability of each one.
   a. At most three lines are in use.
   b. Fewer than three lines are in use.
   c. At least three lines are in use.
   d. Between two and five lines (inclusive) are in use.
   e. Between two and four lines (inclusive) are not in use.
   f. At least four lines are not in use.

7.100   Return to Exercise 7.99.
   a. Calculate the mean value and standard deviation of x.
   b. What is the probability that the number of lines in use is farther than 3 standard deviations from the mean value?

7.101   A new battery's voltage may be acceptable (A) or unacceptable (U). A certain flashlight requires two batteries, so batteries will be independently selected and tested until two acceptable ones have been found. Suppose that 80% of all batteries have acceptable voltages, and let y denote the number of batteries that must be tested.
   a. What is $p(2)$, that is, $P(y = 2)$?
   b. What is $p(3)$? (*Hint:* There are two different outcomes that result in $y = 3$.)
   c. In order to have $y = 5$, what must be true of the fifth battery selected? List the four outcomes for which $y = 5$, and then determine $p(5)$.
   d. Use the pattern in your answers for parts **a–c** to obtain a general formula for $p(y)$.

7.102   A pizza company advertises that it puts .5 lb of real mozzarella cheese on its medium pizzas. In fact, the amount of cheese on a randomly selected medium pizza is normally distributed with a mean value of .5 lb and a standard deviation of .025 lb.
   a. What is the probability that the amount of cheese on a medium pizza is between .525 and .550 lb?
   b. What is the probability that the amount of cheese on a medium pizza exceeds the mean value by more than 2 standard deviations?

c. What is the probability that three randomly selected medium pizzas all have at least .475 lb of cheese?

7.103   There are at least two things to keep in mind when interpreting EPA fuel efficiency ratings (mpg) for different automobiles. The first is that their values are determined under experimental conditions that are not necessarily representative of actual driving conditions. Second, each reported value is an average, so fuel efficiency for a randomly selected car may differ (considerably) from the average. Suppose, then, that fuel efficiency for a particular model car under specified conditions is normally distributed with a mean value 30.0 mpg and a standard deviation 1.2 mpg.
   a. What is the probability that the fuel efficiency for a randomly selected car of this type is between 29 and 31 mpg?
   b. Would it surprise you to find that the efficiency of a randomly selected car of this model is less than 25 mpg?
   c. If three cars of this model are randomly selected, what is the probability that all three have efficiencies exceeding 32 mpg?
   d. Find a number $z*$ such that 95% of all cars of this model have efficiencies exceeding c (that is, $P(x > z*) = .95$).

7.104   The amount of time spent by a statistical consultant with a client at their first meeting is a random variable having a normal distribution with a mean value 60 min and a standard deviation 10 min.
   a. What is the probability that more than 45 min is spent at the first meeting?
   b. What amount of time is exceeded by only 10% of all clients at a first meeting?
   c. If the consultant assesses a fixed charge of $10 (for overhead) and then charges $50 per hour, what is the mean revenue from a client's first meeting?

7.105   The lifetime of a certain brand of battery is normally distributed with mean value 6 hr and standard deviation .8 hr when it is used in a particular cassette player. Suppose two new batteries are independently selected and put into the player. The player will cease to function as soon as one of the batteries fails.
   a. What is the probability that the player functions for at least 4 hr?
   b. What is the probability that the cassette player works for at most 7 hr?
   c. Find a number $z*$ such that only 5% of all cassette players will function without battery replacement for more than $z*$ hr.

7.106 A machine producing vitamin E capsules operates so that the actual amount of vitamin E in each capsule is normally distributed with mean 5 mg and standard deviation .05 mg. What is the probability that a randomly selected capsule contains less than 4.9 mg of vitamin E? At least 5.2 mg?

7.107 Accurate labeling of packaged meat is difficult because of weight decrease due to moisture loss (defined as a percentage of the package's original net weight). Suppose that moisture loss for a package of chicken breasts is normally distributed with mean value 4.0% and standard deviation 1.0%. (This model is suggested in the paper "Drained Weight Labeling for Meat and Poultry: An Economic Analysis of a Regulatory Proposal" (*J. of Consumer Affairs* (1980): 307–325).) Let x denote the moisture loss for a randomly selected package.

a. What is the probability that x is between 3.0% and 5.0%?
b. What is the probability that x is at most 4.0%?
c. What is the probability that x is at least 7.0%?
d. Find a number z^* such that 90% of all packages have moisture losses below z^*%.
e. What is the probability that moisture loss differs from the mean value by at least 1%?

7.108 The *Wall Street Journal* (Feb. 15, 1972) reported that General Electric was sued in Texas for sex discrimination over a minimum height requirement of 5 ft 7 in. The suit claimed that this restriction eliminated more than 94% of adult females from consideration. Let x represent the height of a randomly selected adult woman. Suppose that x is approximately normally distributed with mean 66 in (5 ft 6 in) and standard deviation 2 in.

a. Is the claim that 94% of all women are shorter than 5 ft 7 in correct?
b. What proportion of adult women would be excluded from employment due to the height restriction?

7.109 The longest "run" of S's in the sequence *SSFSSSSFFS* has length four, corresponding to the S's on the fourth, fifth, sixth, and seventh trials. Consider a binomial experiment with $n = 4$, and let y be the length (number of trials) in the longest run of S's.

a. When $\pi = .5$, the 16 possible outcomes are equally likely. Determine the probability distribution of y in this case (first list all outcomes and the y value for each one). Then calculate μ_y.
b. Repeat part **a** for the case of $\pi = .6$.
c. Let z denote the longest run of either S's or F's. Determine the probability distribution of z when $\pi = .5$.

7.110 Two sisters, Allison and Teri, have agreed to meet between 1 and 6 P.M. on a particular day. In fact, Allison is equally likely to arrive at exactly 1 P.M., 2 P.M., 3 P.M., 4 P.M., 5 P.M., or 6 P.M. Teri is also equally likely to arrive at each of these six times, and Allison's and Teri's arrival times are independent of one another. There are thus 36 equally likely (Allison, Teri) arrival-time pairs, for example, (2, 3) or (6, 1). Suppose the first person to arrive waits until the second person does also; let w be the amount of time the first person has to wait.

a. What is the probability distribution of w?
b. How much time do you expect to elapse between the two arrivals?

7.111 Four people — a, b, c, and d — are waiting to give blood. Of these four, a and b have type AB blood, whereas c and d do not. An emergency call has just come in for some type AB blood. If blood samples are taken one by one from the four in random order for blood typing, and x is the number of samples taken to obtain an AB individual (so possible x values are 1, 2, and 3), what is the probability distribution of x?

7.112 Bob and Lygia are going to play a series of Trivial Pursuit games. The first person to win four games will be declared the winner. Suppose that outcomes of successive games are independent and that the probability of Lygia winning any particular game is .6. Define a random variable x as the number of games played in the series.

a. What is $p(4)$? (*Hint*: Either Bob or Lygia could win four straight.)
b. What is $p(5)$? (*Hint*: For Lygia to win in exactly five games, what has to happen in the first four games and in game 5?)
c. Determine the probability distribution of x.
d. How many games can you expect the series to last?

7.113 Refer to Exercise 7.112, and let y be the number of games won by the series loser. Determine the probability distribution of y.

7.114 A sporting goods store has a special sale on three brands of tennis balls, call them D, P, and W. Because the sale price is so low, only one can of balls will be sold to each customer. If 40% of all customers buy brand W, 35% buy brand P, 25% buy brand D, and x is the number among three randomly selected customers who buy brand W, what is the probability distribution of x?

7.115 Suppose that your statistics professor tells you that the scores on a midterm exam were approximately normally distributed with a mean of 78 and a standard deviation of 7. The top 15% of all scores have been designated as A's. Your score is 89. Did you receive an A? Explain.

7.116   Suppose that the pH of soil samples taken from a certain geographic region is normally distributed with a mean pH of 6.00 and a standard deviation of .10. If the pH of a randomly selected soil sample from this region is determined, answer the following questions about it.

   a. What is the probability that the resulting pH is between 5.90 and 6.15?

   b. What is the probability that the resulting pH exceeds 6.10?

   c. What is the probability that the resulting pH is at most 5.95?

   d. What value will be exceeded by only 5% of all such pH values?

7.117   The lightbulbs used to provide exterior lighting for a large office building have an average lifetime of 700 hr. If length of life is approximately normally distributed with a standard deviation of 50 hr, how often should all of the bulbs be replaced so that no more than 20% of the bulbs will have already burned out?

7.118   Soaring insurance rates have made it difficult for many people to afford automobile insurance. Suppose that 16% of all those driving in a certain city are uninsured. Consider a random sample of 200 drivers.

   a. What is the mean value of the number who are uninsured, and what is the standard deviation of the number who are uninsured?

   b. What is the (approximate) probability that between 25 and 40 (inclusive) drivers in the sample were uninsured?

   c. If you learned that more than 50 among the 200 were uninsured, would you doubt the 16% figure? Explain.

7.119   Let x denote the duration of a randomly selected pregnancy (the time elapsed between conception and birth). Accepted values for the mean value and standard deviation of x are 266 days and 16 days, respectively. Suppose that the probability distribution of x is (approximately) normal.

   a. What is the probability that the duration of pregnancy is between 250 and 300 days?

   b. What is the probability that the duration of pregnancy is at most 240 days?

   c. What is the probability that the duration of pregnancy is within 16 days of the mean duration?

   d. A "Dear Abby" column dated Jan. 20, 1973, contained a letter from a woman who stated that the duration of her pregnancy was exactly 310 days. (She wrote that the last visit with her husband, who was in the Navy, occurred 310 days prior to birth.) What is the probability that the duration of pregnancy is at least 310 days? Does this probability make you a bit skeptical of the claim?

   e. Some insurance companies will pay the medical expenses associated with childbirth only if the insurance has been in effect for more than 9 months (275 days). This restriction is designed to ensure that the insurance company has to pay benefits only for those pregnancies where conception occurred during coverage. Suppose that conception occurred 2 weeks after coverage began. What is the probability that the insurance company will refuse to pay benefits because of the 275-day insurance requirement?

GRAPHING CALCULATOR EXPLORATIONS

7.1 Discrete Probability Distributions

Discrete random variables will showcase the calculator at its finest — as a handy aid for transforming minutes of mindless calculation into seconds of easy button-pushing. We will capitalize extensively on the list capabilities of your calculator, and discuss not only how to graph a discrete probability distribution, but also how to find the mean and standard deviation of a discrete random variable.

First recall that we have handled a similar problem before when we considered the problem of graphing a relative frequency histogram from a frequency distribution. You may wish to review that procedure now. Recall that frequencies were converted into relative frequencies for plotting. For the discrete random variable calculations we already have those long-run relative frequencies as probabilities. Begin by entering the possible values in your calculator's equivalent of List1 and the corresponding probabilities in List2. For the Apgar scores of Example 7.9, part of the calculator screen will look like the display on the left below. After you enter the data you can, using the calculator's syntax, graph the probability distribution by supplying the proper lists in the histogram command (as was done for the relative frequency histogram). The graph for the probability distribution of Example 7.9 is shown in the display on the right below. The window is set so that the horizontal and vertical axis would show in the screen to give a more informative display. The horizontal axis runs from $-.5$ to 12, and the vertical axis runs from $-.01$ to 0.5.

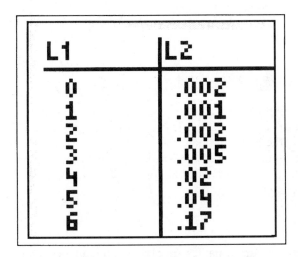
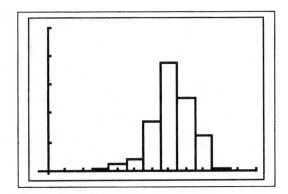

Now we turn our attention to calculating the mean and standard deviation of the random variable. The data is already entered, and we begin by recalling the definition of the mean of a discrete random variable,

$$\mu_x = \sum_{\substack{\text{all possible} \\ x \text{ values}}} xp(x)$$

Because we have stored exactly what is needed in Lists 1 and 2, we can virtually duplicate this definition using the language of lists and list operations for our calculator:

$$\mu_x = \sum_{\substack{\text{all possible} \\ x \text{ values}}} \text{List1} * \text{List2}$$

The strategy for finding the mean of the Apgar random variable is to calculate the products of numbers in our Lists 1 and 2, store the results in List3, and then find the sum of all the numbers in List3. Multiplying to obtain the product is fairly easy when we appeal again to the language of lists:

List1 * List2 → List3

Now we need to find the sum of the numbers, and how this is done will vary from calculator to calculator. The most likely scenario would be for you to calculate the "1-variable statistics" for List3. Calculators will usually report the sum — look for this symbol: ($\Sigma\, x$). Be careful! Don't be misled by the symbol for the mean; we want the *sum*. You should get the value 7.16, which agrees with the text example.

We use a similar strategy to compute the value of the standard deviation. We will find the variance first, then take the square root. The formula for the variance,

$$\sigma_x^2 = \sum_{\substack{\text{all possible} \\ x \text{ values}}} (x - \mu_x)^2\, p(x)$$

also easily translates into the language of lists:

$$\sigma_x^2 = \sum_{\substack{\text{all possible} \\ x \text{ values}}} (\text{List1} - 7.16)^2 * \text{List2}$$

The list language is only slightly more complicated than for the mean:

$$(\text{List1} - 7.16)^2 * \text{List2} \rightarrow \text{List3}$$

Then the sum of the numbers in List3 is the variance. Taking the square root gives the standard deviation. Performing these calculations, you should get a variance of 1.5684 (from ($\Sigma\, x$) for List3; again, don't be misled and write down the σ_x or s_x!). The standard deviation is then found by taking the square root of 1.5684, resulting in 1.2524, which again agrees with the text.

7.2 Binomial Probability Calculations

Most calculations having to do with random variables are of one of three types. They are finding (i) the probability the variable will assume a value between two given numbers; (ii) the probability the variable will assume a value less than a given number; or (iii) the probability the random variable will assume a value greater than a given number. Because these calculations are so common in statistics, your calculator may have a built-in capability for finding these probabilities. In the case of a discrete random variable such as the binomial distribution, there is also a special case of (i) above, the probability that the random variable will actually assume a particular value.

For the binomial distribution, we will illustrate these calculations using Example 7.17. Recall that in Example 7.17 the probability exactly four watches were digital was

$$p(x) = \binom{n}{x}\pi^x(1 - \pi)^{n-x} = \binom{12}{4}(.6)^4(1 - .6)^{12-4} = .042$$

Even if your calculator does not have special binomial functions, it is likely to have a key for the combinations ($_nC_r$), possibly cleverly hidden under the "math" or "probability" menu. If you use this key, don't forget that to perform the calculation above you will have to press n, then the $_nC_r$ key, and then r. Calculator keystrokes might look like this:

$$12\ _nC_r 4 * .6\char`^4 * .4\char`^8$$

If your calculator has built-in binomial capabilities you will have fewer keystrokes, but you will have to remember the "right" function to call, and the order the information is entered into the calculator. Let's take these problems one at a time. First, the function names. If your calculator has a built-in function for binomial calculations, it probably has two: a function for finding the probability that "x is equal to a given value," and a function for finding the probability that "x is less than or equal to a given value." The first function is known to statisticians as a "density" function, and is commonly abbreviated "pdf" for "probability density function." The second is known as a "cumulative distribution function" and is commonly labeled "cdf." These two functions will in all likelihood have very similar names on your calculator—be alert!

The second problem you will face is that the binomial probabilities are more complicated than the square-root button on your calculator. The calculator needs more than just one number to find the binomial probabilities, and the order you enter the numbers *does* make a difference! Look in your calculator manual for something that looks like "binomial," especially with a "pdf" somewhere (it could be easy, like "binompdf," or possibly a little more cryptic, like "binpdf"), and turn to that page. Your manual will be very careful to specify both what the needed function parameters are, and what order you should enter them. As an example, one type of calculator has the following:

binompdf(*numtrials, p,* [x])

The manual informs that "*numtrials*" is the number of trials, "p" is the probability of success, and that "x" can be either an integer or a list of integers. This information collectively explains what is known as the "syntax" of the function. It is your responsibility to get the numbers right, and get them in the right order! The square brackets, "[ ]," are very common notation in the calculator and computer world, and they usually indicate that the bracketed quantity is either optional or defaults to a preselected option if you do not enter a number in that space. Should you not enter a number, the function would look like: **binompdf**(*numtrials, p*). For our example, the number of trials is 12, and the probability of success is .6. Since the probability of exactly 4 digital watches is desired, we enter

binompdf(12, .6, 4)

(Notice that when the number is actually entered, the brackets are not keyed in.) The number .042042 appears, which we know from Example 7.17 is the correct answer. This is a good sign! Now try this on your own calculator. Remember, you must navigate to the function in the manner presented in your manual, and you have to pay attention to the syntax. While you are learning how to use this function (or any calculator function), use examples with known answers and check the results.

Now suppose you wish to find the probability of getting 4 *or fewer* digital watches out of 12. The appropriate function here is the cumulative density function, or "cdf":

binomcdf(*numtrials, p,* [x])

Does this look disturbingly familiar? Except for the "c" instead of the "p," they look exactly alike! The good news is that we already understand the syntax; the bad news is that if we aren't careful we might get the wrong function in haste. Be careful! The function **binomcdf**(12, .6, 4) gives the answer, .0573099213. This calculation was not performed in Example 7.17, but we can use binompdf to check the result:

$$\textbf{binompdf}(12, .6, 0) + \textbf{binompdf}(12, .6, 1) + \cdots + \textbf{binompdf}(12, .6, 4)$$
$$= .000016777 + .0003019898 + \cdots + .042042$$
$$= .05731$$

Now let's go on to another of the common calculations with random variables: What is the probability that the random variable will assume a value between 4 and 7? The word *between* is disturbingly ambiguous. Do we mean to include 4 and 7, or do we mean only the probability of getting a 5 or 6? In Example 7.17 this was made clear by using the word *inclusive,* but it is always wise to realize the chameleon-like nature of the word *between*. To evaluate the probability desired, we will use the *cumulative* distribution function for the binomial, binomcdf. (Remember, your calculator function may have a different name!) The logic is elementary, as Sherlock Holmes would say. The probability that the binomial random variable will assume a value between 4 and 7 (inclusive) is equal to the probability of assuming a value less than or equal to 7, minus the probability of assuming a value less than or equal to 3 (*not* 4). In symbols,

$$P(4 \le x \le 7) = P(x \le 7) - P(x \le 3)$$

which is found as follows:

$$\textbf{binomcdf}(12, .6, 7) - \textbf{binomcdf}(12, .6, 3)$$

which gives us .5465545, in agreement with the answer in the text.

The last type of binomial random variable calculation problem is finding the probability of assuming a value greater than a given value. What, for instance, is the probability of more than 7 watches being digital? Using a fundamental property of probability, we know

$$P(x > 7) = 1 - P(x \le 7)$$

which we translate into

$$1 - \textbf{binomcdf}(12, .6, 7)$$

which gives .438178222.

We have gone into some detail to explain how these binomial probability problems can be solved using the calculator. This detail is justified not only because of the importance of the binomial distribution, but also because these same calculator procedures will be used for finding probabilities involving the geometric and normal random variables, yet to come. Because the discussions in this exploration have been detailed, the discussions in those cases will be briefer.

In Graphing Calculator Exploration 7.1, we discussed how to graph a discrete distribution. When we graphed the probability density function for the Apgar scores, we manually entered the outcomes and their associated probabilities. Anticipating that you may wish to consider binomial chance experiments with many potential successes, we will streamline the data entry process using some commands and functions we have already discussed in previous calculator explorations.

Graphing a binomial distribution will involve three steps:

1. Construct the list of possible values in List1 using the seq command (or your calculator's equivalent).
2. Construct the probabilities in List2 using the binompdf function (or your calculator's equivalent).
3. Draw the graph (in the form of a histogram) of the probability distribution.

Figure 7.16 gave a graph of the binomial probability distribution when $n = 20$ and $\pi = .20$. Carrying out the steps below puts the integers 0 to 20 in List1, and $p(x)$ for x values from 0 to 20 in List2.

1. seq($x, x, 0, 20$) → List1 puts a sequence of 21 integers into List1. (Remember to verify your calculator syntax and the order of the information to be entered for your calculator!)
2. binompdf(20, .2) → List2. (Remember to verify . . . .)
3. Now graph the probability distribution, where List1 contains the possible data values and List2 contains the probabilities.

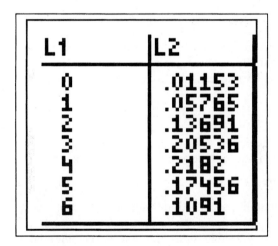

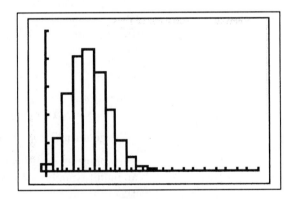

7.3 Geometric Probability Calculations

The calculator exploration of geometric random variables will be an echo of the binomial we have already discussed. We again consider (i) the probability the variable will assume a value between two given numbers; (ii) the probability the variable will assume a value less than a given number; and (iii) the probability the random variable will assume a value greater than a given number.

We will illustrate these calculations using Example 7.20, the jumper cable problem. The probability the first student stopped has jumper cables is

$$p(1) = (1 - \pi)^{1-1}\pi = (1 - .4)^{1-1}(.4) = .4$$

The corresponding keystrokes will be something like

$(1 - .4)^0 * .4$

If your calculator has the density function and cumulative density function for the geometric distribution, they will probably be named something like geompdf and geomcdf. The calculator syntax for the probability density function will probably look something like

geompdf(p, x)

where "p" is the probability of success and "x" is the number of students you would ask until success. For our example, we want the probability of jumper cables on the very first stop. We enter **geompdf**(.4, 1), and the function returns .4.

Now suppose you wish to find the probability of jumper cables after 4 *or fewer* stops. Using the cumulative density function, "geomcdf"(which has the same parameters as the geompdf function), we enter **geomcdf**(.4, 4), which returns .8704. As with the binomial, we can check this by summing:

geompdf(.4, 1) + **geompdf**(.4, 2) + **geompdf**(.4, 4) + **geompdf**(.4, 4)

$= .4 + .24 + .144 + .0864$

$= .87041$

The probability that a geometric random variable will assume a value between 4 and 7 (inclusive) is equal to the probability of observing a value less than or equal to 7, minus the probability of observing a value less than or equal to 3 (*not* 4). In symbols,

$P(4 \leq x \leq 7) = P(x \leq 7) - P(x \leq 3)$

which is found using **geomcdf**(.4, 7) $-$ **geomcdf**(.4, 3), giving .1880064.

What is the probability of more than 7 stops before we get jumper cables?

$1 -$ **geomcdf**(.4, 7),   which gives us .0279936.

Graphing an entire geometric probability distribution is not possible, since there is an infinite number of possible values — 1, 2, 3, ... Nevertheless, we can graph parts of the distribution. The method for graphing is just like that for the binomial random variable. Use the seq function to create a list of integers in List1; (ii) use the geompdf function to find the corresponding probabilities and store them in List2; and (iii) plot the distribution as you would a histogram, and as we have previously done with the binomial. These steps for the geometric distribution of Example 7.20 are summarized below:

1. seq($x, x, 1, 20$) $\rightarrow$ List1

2. geompdf(.4, List1) $\rightarrow$ List2

3. Graph a histogram with the domain List1, and probabilities in List2.

The data editing window and graph of this geometric distribution appear in the following figures.

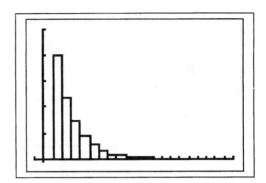

Remember that, for the geometric distribution, we are really calculating probabilities for only part of the distribution. The graph should tail to the right in a gradual manner, not suddenly drop out of sight. If you notice a sudden plummeting, it could be that there are more significant probabilities to the right. For example, suppose we consider the chance experiment of flipping a coin until a head appears. The distribution of x = number of tosses is geometric with success probability .5. If the distribution is plotted using the previous steps but using only a sequence of integers from 1 to 4, the picture below on the left results. Clearly, there are values with probabilities far from zero that are not represented in the graph. The solution is to construct the sequence of integers over a larger range of values, say 1 to 16. At some point, of course, the geometric probabilities become very close to zero, as in the figure below on the right. If your graph looks similar to the one on the right, you can be fairly certain you have captured the essential behavior of the particular geometric distribution.

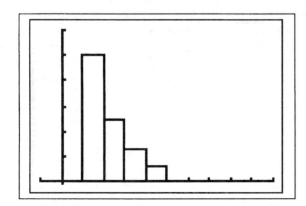

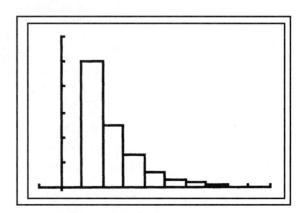

7.4 Normal Curves and the Normal Probability Distribution

The normal distribution is arguably the most famous distribution in all of statistics. As we have learned, "the" normal distribution is really a family of distributions with the same shape, but different means and standard deviations. The standard normal distribution is the normal probability distribution with $\mu = 0$ and $\sigma = 1.0$. From the calculator perspective, working with the normal distribution is slightly different from the binomial and geometric distributions because the normal distribution is continuous. As a consequence, it will not be graphed as a histogram. Normal curves are graphed just as a straight line or any other function is graphed. However, the normal curve is not a simple function. Because of this, we will graph normal curves by appealing to built-in functions rather than keying in a complex function. The built-in function will have "normal" in it somewhere, and might be something like this:

normalpdf(x, [μ, σ]).

If you are a glutton for punishment or your calculator does not have a built-in normalpdf function, here are the keystrokes:

y1 = $1/(\mathrm{sqr}(2 * \pi) * \sigma) * \exp(-(x - \mu)^2/(2 * \sigma^2))$

Assuming you are smiling because your calculator has a built-in normalpdf function, let's put it to good use. The syntax above for the normalpdf function might seem complicated, but actual use is simple once you get used to it. You should check your calculator manual for two *very important* pieces of information. First, make sure you know the required order for the information you must provide. Second, look closely at the sigma, wherever it is in your calculator's syntax. Make sure you discern whether you must enter σ or σ^2. One symbol (σ) would mean you are to enter the standard deviation of the normal curve you want to graph; the other (σ^2) requests the variance. Now let's tackle the notation. First of all, if your calculator's syntax has those square brackets $-[\mu, \sigma]-$ remember that they indicate numbers that are optional. If you leave them out, the normalpdf function will simply default to the standard normal curve, with mean 0 and standard deviation 1.

Let's graph the three normal curves from Figure 7.18. The first has a mean and standard deviation of 10 and 5, respectively. The second has a mean and standard deviation of 40 and 2.5, and the third a mean and standard deviation of 70 and 10.

You must once again navigate your calculator's menu system to find the normal curve function, and paste this function into the function definition window where you usually define simpler functions. Using the syntax above, you should see your calculator's equivalent of the following:

y1 = normalpdf(x, 10, 5)

y2 = normalpdf(x, 40, 2.5)

y3 = normalpdf(x, 70, 10)

Graphing these functions using the window setting shown at the left, we see the graphs in the figure at the right. Notice that this faithfully reproduces Figure 7.18.

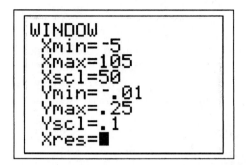

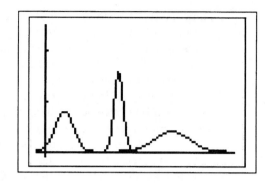

Now let's graph the standard normal distribution. If your calculator syntax indicates that it defaults to a standard normal, you will only have to enter your calculator equivalent of

y1 = normalpdf(x).

It is also possible that your calculator does *not* default to standard normal, in which case you would have to specify the mean and standard deviation as 0 and 1, something like

y1 = normalpdf(x, 0, 1).

Set your graphing window to be consistent with Figure 7.21, with x values running from about -3.5 to 3.5 and the y values from $-.05$ to .40. These values should be fine for the standard normal distribution. If you don't see a distribution filling the screen like the accompanying figure, something is amiss and you need to verify your keystrokes and check your calculator's manual.

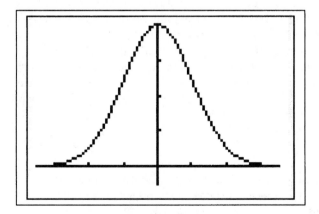

Since the normal probability distribution is a continuous distribution, the probability that x would be equal to a specific value is, of course, 0. For continuous distributions we are usually interested in finding (i) the area under the curve between two specific values; and (ii) the area in the extremes, or "tails" of the distribution. The function that we will use to find these values will be symbolized by the notation "normalcdf," which stands for the "normal cumulative distribution function." This actually is a misnomer, because the functions calling themselves "cdf" functions on many calculators actually calculate the probability that the standard normal variable is *between* two values. Calculators seem to get it right for the discrete probability density functions, but for some reason have elected to use similar names for very different kinds of calculation for continuous probability density functions — don't let this minor inconvenience confuse you!

Two strategies are used by calculator manufacturers for evaluating the probability that z is between two values, a and b. It is possible your calculator has a table for you to fill in the values as in the accompanying figure. For a calculator utilizing this strategy, you would have to fill in the lower bound, upper bound, and standard deviation (1.0) and mean (0.0).

Other calculators ask for the mean and standard deviation as parameters of the function. If your calculator uses this strategy, your built-in cumulative distribution function will have a syntax something like this:

normalcdf(lower bound, upper bound $[, \mu, \sigma]$)

For a calculator using this syntax, you would fill in the lower bound and upper bound with the appropriate values for z, and ignore the optional parameters, since z will have a standard normal distribution. You will specify other values for μ and σ when performing calculations that are not already in terms of z scores. After navigating your calculator's menus, you will enter something like this:

normalcdf(z-lower, z-upper)

Let's find a probability we already know, from Example 7.22. The probability that z is between -1.76 and .58 is .6798. We would enter the function as normalcdf(-1.76, .58) and the calculator would return a value of .6798388789. We, of course, would not suggest writing all those digits; rounding off to .6798 is perfectly fine (as you may have surmised from considering Appendix Table II).

As you might guess from its name, the normalcdf function can also be used for calculation of the cumulative distribution function — that is, finding the probability that z will be below a specific value. Once again, we will use a problem we already know the answer to, from Example 7.21. We want to find the probability that z is less than -1.76. Remembering that the set of possible values for a standard normal random variable is the entire real line, you might think to enter the following:

normalcdf($-\infty$, -1.76)

If so, your thinking is right on target, except for one thing — there is no "$-\infty$" on your calculator. Some calculators will have a special symbol for "$-\infty$", which the calculator translates internally to its equivalent of a "very small number." You should check your manual for this number and how to find it. The representation will probably be something like "$-1E99$" or "$-1e999$," which is calculator-speak for -1 times 10 raised to the highest power the calculator can handle. In the case of the standard normal curve, it may be just as easy to enter a different but still very small number in

---

Normal C.D.

Lower :0
Upper :0
σ :0
μ :0
Execute

place of the "$-\infty$," perhaps normalcdf (-10, -1.76). On our calculator, .0392038577 is returned, which agrees with the tabled answer. (If you are squeamish about -10, use -50; with -50 we get .0392038577 also!)

Finding the probability that a z is greater than a particular value is also easy. For example, using another calculation from the text, we find the area to the right of $z = 1.42$ as follows:

$$1 - \text{normalcdf}("-\infty", 1.42)$$

Using -10 for the lower bound, we get .0778038883, agreeing with the text once again.

The last type of problem examined will be the identification of extreme values. The easiest way to do this is with a built-in function, typically called "InvNormal," which stands for "inverse normal." The "InvNormal" function — or whatever it is named on your calculator — will be the reverse of finding the probability that z is less than a specified value. Earlier in our discussion, we found the probability that z is less than -1.76 to be .0392038577. The InvNormal function returns a z value when given the probability. Thus, InvNormal(.0392038577) equals -1.76. Except for the difference in function name, the syntax for this function should be the same as for normalcdf:

$$\text{InvNormal(cumulative probability, } [, \mu, \sigma])$$

On our calculator, InvNormal(.0392038577) returns -1.760000538.

7.5 The Normal Approximation to the Binomial

In earlier Graphing Calculator Explorations, you saw how to calculate probabilities associated with the binomial and normal distributions, using built-in calculator functions. We used the generic terms "binompdf," "binomcdf," "normalpdf," and "normalcdf" to refer to these functions. In this exploration we will focus on the normal approximation to the binomial distribution. Whenever a continuous distribution is used to approximate a discrete distribution, the question naturally occurs, "How good is the approximation?" The answer usually given by statistics instructors is, quite rightly, "it depends." In the case of the approximation to the binomial, the goodness of fit depends on the two quantities that define the binomial distribution: n and π. Most statisticians have a simple "rule of thumb" they apply for approximating the binomial with a normal distribution, such as:

When either $n\pi < 10$ or $n(1 - \pi) < 10$, the binomial distribution is too skewed for the normal approximation to give accurate results.

Different statisticians have different rules of thumb, some feeling comfortable with using 5 rather than 10 in the rule above. In days of yore — that is, in precalculator days — students would have to accept the rule of thumb as one of the mysteries of statistics. Today a statistics student, armed with his or her calculator, can not only understand what the rules of thumb are all about, but investigate the various rules of thumb.

It might be argued that using the normal distribution to approximate a binomial seems a little foolish, because the calculator gives us exact values. There is something to this argument, but remember: We will not always be able to find exact probabilities in other situations in statistics, and must rely on approximations. Even in the case of the binomial, the built-in function will have limits imposed on the numbers it can successfully evaluate because of round-off error or because intermediate results will be too small or too large for the calculator's microprocessor. When the calculator can evaluate "exact" probabilities for the binomial, using an approximation involves a fundamental tradeoff between ease of calculation and exactness of answer. An understanding of this with the normal approximation to the binomial will give us a better feel for the issues involved when we encounter similar tradeoffs in statistics courses yet to come.

Let's reacquaint ourselves with some syntax using Example 7.33 about express mail packages. Following that example, we will use the normalpdf function to approximate the probability that $x = 20$. Remembering the syntax from earlier discussions,

normalcdf(lower bound, upper bound $[, \mu, \sigma]$)

we will enter normalcdf(19.5, 20.5, 18, 6). The calculator returns .062832569, which is not quite as close to the text's value .0641 as we would like. However, remember that the x values in the text were transformed to standard z's. Recalculating using the z values from the text, normalcdf(0.25, 0.42) returns a value of .0640509632. It appears the discrepancy was due to rounding in the z values, and it looks like we have the right keystrokes.

Now consider Example 7.34 on premature babies, and compare the binomial calculations with the normal approximations. To find the binomial probability, we recall the built-in function we called binomcdf. Remember that this function includes the rightmost value indicated; therefore we subtract the probability of getting x less than or equal to 14 from the probability of getting x less than or equal to 30 to get the desired probability:

$$P(15 \le x \le 30) = \text{binomcdf}(250, .1, 30) - \text{binomcdf}(250, .1, 14)$$
$$= .8753286537 - .00931244487$$
$$= .8660162088$$

For the normal curve approximation we will key in the mean and standard deviation $\mu = 25$ and $\sigma = 4.74341649$:

normalcdf(lower bound, upper bound $[, \mu, \sigma]$)
$$= \text{normalcdf}(14.5, 30.5, 25, 4.74341649)$$
$$= .8634457937$$

The difference between the two calculated probabilities is .0025704151. This does not seem to be a large difference, but it *is* a difference. According to our rule of thumb, it is appropriate to use the normal approximation in this instance, but the investigator in the context of his or her situation must evaluate the practical importance of the error.

Now let's redo the calculations not with a sample size of 250, but only 50. Keeping the results proportionally the same by dividing by 5, we will consider approximating the probability of getting between 3 and 6 premature babies (inclusive) from a random sample of 50 babies. In this case,

$$n\pi = 50(.10) = 5 < 10$$
$$n(1 - \pi) = 50(1 - .10) = 45 \geq 10$$

Because $n\pi < 10$, our rule of thumb would regard this binomial distribution as too skewed for the normal curve approximation to give accurate results. Let's see what happens:

$$P(3 \leq x \leq 6) = \text{binomcdf}(50, .1, 6) - \text{binomcdf}(50, .1, 2)$$
$$= .7702268435 - .1117287563$$
$$= .6584980872$$

To evaluate this probability using the normal curve approximation, we use

$$\mu = 50(.1) = 5 \quad \text{and} \quad \sigma = \sqrt{50(.1)(.9)} = 2.121320344$$

and then

normalcdf(lower bound, upper bound $[, \mu, \sigma]$)
$$= \text{normalcdf}(2.5, 6.5, 5, 2.121320344)$$
$$= .6409535402$$

The difference between the binomial and the normal approximation in this case is .017544547. It is interesting to note that using a rule of thumb with 5 instead of 10 would call this difference "acceptable." Once again, the individual judgment by the investigator must be used in evaluating the goodness of the approximation for any particular study.

Finally, we will superimpose the appropriate normal distribution over the binomial distribution to get a visual sense of the approximation. It is entirely possible that a given approximation will do a better job for different choices of values of the endpoints of the interval, and the graphs may give us an overall sense of when a normal approximation might be acceptable.

Graphing a binomial distribution and a normal distribution at the same time involves skills we have seen in previous explorations. (You may want to refer back to Graphing Calculator Explorations 7.2 and 7.4 about the binomial and normal distributions to refresh your memory.)

We will graph the binomial and normal distributions for four situations, each with sample size 20, but with probabilities of success of .05, .1, .25, and .5. We will change the windows to make the graphs fill the windows, but this should not affect any interpretations of the goodness of fit to the binomial by the normal distribution. As a reminder, binomial preparations for the first graph are:

1. Seq($x, x, 0, 20$) → List1
2. Binompdf(20, .05) → List2
3. Specify that we want a histogram with the values in List1, and the corresponding binomial probabilities in List2.

For the normal curve plot, define the graphing function by supplying the mean and standard deviation of the binomial as parameters for the normalpdf function:

$$Y1 = \text{normalpdf}(x, 1, 0.97468)$$

The four plots appear below.

$n = 20; \pi = .05$

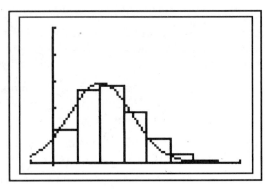

$n = 20; \pi = .10$

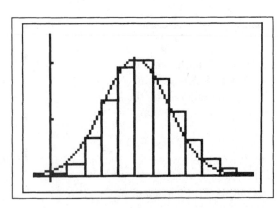

$n = 20; \pi = .25$

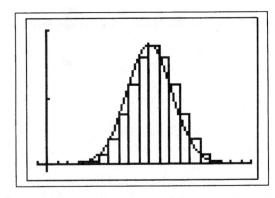

$n = 20; \pi = .50$

As can be seen from a comparison of the plots, the normal approximation gets "closer and closer" as π gets closer and closer to .5. For $\pi = .25$, the rule of thumb is not satisfied for $n = 20$, and for $\pi = .5$, the rule of thumb is satisfied using $n = 20$. It is a bit difficult to judge whether the normal approximation to the binomial is "adequate" for a particular situation by just looking at the plots.

Modern technology makes it possible to do binomial calculations quickly, so the normal approximation to the binomial is not as widely used as it once was. However, there are other distributions in statistics that are "approximately" normal as long as certain conditions are satisfied. Working with the approximation to the binomial should give you an appreciation for the uses of the normal distribution to approximate these other distributions.

8

Sampling Variability and Sampling Distributions

INTRODUCTION

The inferential methods presented in subsequent chapters will use information contained in a sample to reach conclusions about one or more characteristics of the population from which the sample was selected. For example, let μ denote the true mean fat content of quarter-pound hamburgers marketed by a national fast-food chain. To learn something about μ, a sample of $n = 50$ hamburgers might be obtained and the fat content determined for each burger in the sample. The sample data might produce a mean of $\bar{x} = 28.4$ grams. How close is this sample mean to the population mean, μ? If another sample of 50 were selected and $\bar{x}$ computed, would this second $\bar{x}$ value be near 28.4, or might it be quite different?

These questions can be addressed by studying what is called the *sampling distribution* of $\bar{x}$. Just as the distribution of a numerical variable describes its long-run behavior, the sampling distribution of $\bar{x}$ provides information about the long-run behavior of $\bar{x}$.

Later in the text, properties of the sampling distribution of $\bar{x}$ will be used to develop inferential procedures for drawing conclusions about μ. In a similar manner, we will study the sampling distribution of a sample proportion (the fraction of individuals or objects in a sample who have some characteristic of interest). This will enable us to describe procedures for making inferences about the corresponding population proportion, π.

8.1   Statistics and Sampling Variability

Many investigations and research projects try to draw conclusions about how the values of some variable x are distributed in a population. Often, attention is focused on a single characteristic of that distribution. Examples include the following:

1. x = fat content (grams) of an Oscar Mayer beef hot dog, with interest centered on the mean fat content μ of all such hot dogs

2. x = fuel efficiency (miles per gallon) for a 2000 Honda Accord, with interest focused on the variability in fuel efficiency as described by σ, the standard deviation for the fuel efficiency population distribution

3. x = time to first recurrence of skin cancer for a patient treated using a particular therapy, with attention focused on the proportion π of such individuals whose first recurrence is within 5 years of treatment

4. x = marital status (a categorical variable) of a student at a particular university, with interest centered on π, the proportion of all students at the university who are married

The usual way of obtaining information regarding the value of a population characteristic is by selecting a sample from the population. For example, to gain insight about the mean value of the fat content distribution for the population of all Oscar Mayer beef hot dogs, we might select a sample of 50 hot dogs for analysis. Each item in the sample could be tested to yield a value of x = fat content. As in Chapter 7, we could construct a histogram of the 50 sample x values, and we could view this histogram as a rough approximation of the population distribution of x. In a similar way, we could view the sample mean, $\bar{x}$ (the mean of the sample of n values) as an approximation of μ, the mean of the population distribution. It would be nice if the value of $\bar{x}$ were equal to the value of μ. It could happen that $\bar{x} = \mu$, but this would be an unusual occurrence; $\bar{x}$ is only an *estimate* of μ. An important question addressed later in this chapter is, How good is this estimate?

Not only does the value of $\bar{x}$ for a particular sample from a population usually differ from μ, but the $\bar{x}$ values from different samples also typically differ from one another. (For example, two different samples of 50 hot dogs will usually result in different $\bar{x}$ values.) This sample-to-sample variability makes it challenging to generalize from a sample to the population from which it was selected. To meet this challenge, we must understand sample-to-sample variability.

EXAMPLE 8.1

When we sample from a population, many different samples might result. Consider a small population consisting of the board of directors of a day care center. (Also listed are the values of x = number of children for each board member.)

Board member	Jay	Carol	Allison	Teresa	Lygia	Bob	Roxy	Kyle
Number of children	2	2	0	0	2	2	0	3

The average number of children for the entire group of eight (the population) is

$$\mu = \frac{2 + 2 + 0 + 0 + 2 + 2 + 0 + 3}{8} = \frac{11}{8} = 1.38$$

There are 28 different possible outcomes when a sample of size 2 is selected (without replacement) from the board; these are shown in Table 8.1. If sample 1 (Jay, Carol) is selected, the associated sample mean is $\bar{x}_1 = (2 + 2)/2 = 2$; if sample 14 (Allison, Teresa) is selected, the associated sample mean is $\bar{x}_{14} = (0 + 0)/2 = 0$. Each of the different possible samples has an $\bar{x}$ value associated with it. The $\bar{x}$ values not only differ from μ (neither $\bar{x}_1$ nor $\bar{x}_{14}$ is equal to 1.38), but they may also differ from sample to sample (for example, $\bar{x}_1 \neq \bar{x}_{14}$).

Table 8.1 *All possible samples of size 2 and their respective means*

Sample	$\bar{x}$	Sample	$\bar{x}$	Sample	$\bar{x}$	Sample	$\bar{x}$
1. Jay, Carol	2	8. Carol, Allison	1	15. Allison, Lygia	1	22. Teresa, Kyle	1.5
2. Jay, Allison	1	9. Carol, Teresa	1	16. Allison, Bob	1	23. Lygia, Bob	2
3. Jay, Teresa	1	10. Carol, Lygia	2	17. Allison, Roxy	0	24. Lygia, Roxy	1
4. Jay, Lygia	2	11. Carol, Bob	2	18. Allison, Kyle	1.5	25. Lygia, Kyle	2.5
5. Jay, Bob	2	12. Carol, Roxy	1	19. Teresa, Lygia	1	26. Bob, Roxy	1
6. Jay, Roxy	1	13. Carol, Kyle	2.5	20. Teresa, Bob	1	27. Bob, Kyle	2.5
7. Jay, Kyle	2.5	14. Allison, Teresa	0	21. Teresa, Roxy	0	28. Roxy, Kyle	1.5

Table 8.1 also gives the value of $\bar{x}$ for each of the 28 different possible samples, and a dotplot of $\bar{x}$ values in the table is shown in Figure 8.1.

Figure 8.1 provides some insight into the behavior of the mean of a sample of size 2 from this population. We see that the value $\bar{x} = 1$ is more common than any other value. However, $\bar{x}$ values can differ greatly not only from one sample to another, but also from the value of the population mean μ. If we were to select a random sample of size 2 (so that each different sample has the same chance of being selected), we could not count on the resulting value of $\bar{x}$ being very close to the population mean $\mu = 1.38$.

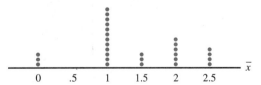

FIGURE 8.1 A dotplot of the $\bar{x}$ values in Example 8.1

To formalize these ideas, we must introduce some terminology.

> **Definition**
>
> Any quantity computed from values in a sample is called a **statistic.**
>
> The observed value of a statistic depends on the particular sample selected from the population; typically, it varies from sample to sample. This variability is called **sampling variability.**

The statistic considered in Example 8.1 was $\bar{x}$, the sample mean. Values of statistics such as $\bar{x}$, s, the sample median, the sample interquartile range, and so on, will be our primary source of information regarding various population characteristics.

Sampling Distributions

It is possible to view a statistic as a random variable and to use a distribution to describe its behavior. Before examining how this is done, let's review how a population distribution is used to describe the distribution of the values of some variable in a particular population. For example, suppose that we are studying the population of all mail carriers in Los Angeles County. Each carrier has an assigned route, and so associated with each carrier is a value of the variable

x = length of route (in miles)

One carrier might have a 3.8-mile route ($x = 3.8$), whereas for another carrier, we might find that $x = 2.4$. If we knew the x value for every individual in the population, we could construct a density histogram that would represent the population distribution for the variable x. The distribution might look like the histogram in Figure 8.2. The population distribution tells us about the distribution of individual x values in the population. If a single carrier were randomly selected from the population, the population distribution could be used to evaluate probabilities such as $P(x > 4)$ or $P(2 < x < 3)$.

FIGURE 8.2 Histogram of x = length of route

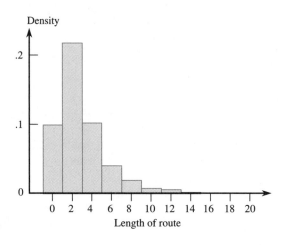

Now suppose that we are interested in μ, the mean value of x (the average route length) for this population. If the population distribution is known, the value of μ can be determined. For the distribution in Figure 8.2, $\mu \approx 3$.

What if the population distribution is not known? Then we must estimate μ using information in a sample. Suppose that we plan to take a random sample of size $n = 100$ from the population of all mail carriers in Los Angeles County. There are many, many different possible samples that might result (many more than the 28 from the day care example). We now define a hypothetical "population," which consists of all the different possible samples of a given size n. That is, we consider a *population of samples*. (This is a bit abstract, but it is a very important idea!) The population of samples is viewed as a population because it consists of *every* different sample; it is the complete collection of all possible samples.

Just as a variable associates a value with every individual or object in a population and can be described by its distribution, a statistic associates a value with each individual sample in the population of samples. Thus, a statistic can also be described by a distribution.

Definition

The distribution of a statistic is called its **sampling distribution.**

EXAMPLE 8.2   (Example 8.1 continued)   In Example 8.1, we considered a small population consisting of the board of directors of a day care center. The variable of interest was x = number of children, and the statistic of interest was the sample mean $\bar{x}$. Figure 8.1 showed a plot of the $\bar{x}$ values for the population of all 28 samples of size $n = 2$. The corresponding density histogram appears in Figure 8.3; this is the sampling

FIGURE 8.3 Sampling distribution of $\bar{x}$ for Example 8.2

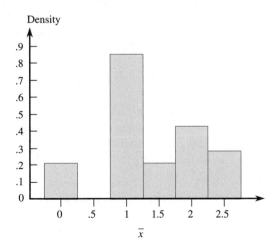

distribution of the statistic $\bar{x}$ when a random sample of size 2 is selected from the original population of eight people.

Looking at the sampling distribution of $\bar{x}$ depicted in Figure 8.3 helps us to understand what happens when a random sample of size 2 is selected from this population. For example, we can compute the probability of observing a sample mean that is within .5 of $\mu = 1.38$, the mean number of children in the population. This probability is

$$P(\bar{x} \text{ is within .5 of } \mu) = P(\bar{x} = 1 \text{ or } \bar{x} = 1.5)$$
$$= \frac{12}{28} + \frac{3}{28}$$
$$= .4286 + .1071$$
$$= .5357$$

The probabilities $P(\bar{x} = 1) = 12/28$ and $P(\bar{x} = 1.5) = 3/28$ were computed using the information in Figure 8.1. Note that they are also the areas of the corresponding rectangles in the density histogram of Figure 8.3.

The sampling distribution of $\bar{x}$ can also be displayed in table form in this case (because there are only a few different possible $\bar{x}$ values), as shown below.

$\bar{x}$	0	1	1.5	2	2.5
$p(\bar{x})$	$\frac{3}{28}$	$\frac{12}{28}$	$\frac{3}{28}$	$\frac{6}{28}$	$\frac{4}{28}$

The sampling distribution of $\bar{x}$ provides important information about the behavior of the statistic $\bar{x}$ and how it relates to μ. However, you may have noticed that obtaining the sampling distribution of $\bar{x}$, even for a population of just eight individuals and a sample size of 2, was a lot of work. For realistic examples with larger population and sample sizes, the situation becomes worse, since there are so many possible samples. Fortunately, as we look at more examples, patterns emerge that enable us to describe some important aspects of the sampling distribution of the statistic $\bar{x}$ without actually having to look at all possible samples.

EXAMPLE 8.3

Consider a small population consisting of the 20 students enrolled in an upper-division class. The students are numbered from 1 to 20, and the amount of money (dollars) each student spent on textbooks for the current semester is shown in the accompanying table.

Student	Amount Spent on Books	Student	Amount Spent on Books
1	267	3	342
2	258	4	261
			(*continued*)

Student	Amount Spent on Books	Student	Amount Spent on Books
5	275	13	265
6	295	14	262
7	222	15	333
8	270	16	184
9	278	17	231
10	168	18	159
11	319	19	230
12	263	20	323

For this population,

$$\mu = \frac{267 + 258 + \cdots + 323}{20} = 260.25$$

Suppose that the value of the population mean is unknown to us and that we want to estimate μ by taking a random sample of five students and computing the sample mean amount spent on textbooks, $\bar{x}$. Is this a reasonable thing to do? Will the estimate that we produce be close to the true population mean? To answer these questions, we can perform a simple experiment that allows us to look at the behavior of the statistic $\bar{x}$ when random samples of size five are taken from the population. (Note that this is not a realistic example. If a population consisted of only 20 individuals, we would probably conduct a census rather than select a sample. However, this small population size makes it easier to work with as we develop the ideas of sampling distributions.)

Let's select a random sample of size 5 from this population. This can be done by writing the numbers from 1 to 20 on slips of paper, mixing them well, and then selecting five slips without replacement. The numbers on the slips selected identify which of the 20 students will be included in our sample. Alternatively, either a table of random digits or a random number generator can be used to determine which five should be selected. We used MINITAB to obtain five numbers between 1 and 20, resulting in

    17   20   7   11   9

Selecting students numbered 17, 20, 7, 11, and 9 gives the following sample of amounts spent on books:

    231   323   222   319   278

For this sample,

$$\bar{x} = \frac{1373}{5} = 274.60$$

The sample mean is larger than the population mean of 260.25 by about \$15. Is this difference typical, or is this sample mean unusually far away from μ? Taking additional samples could provide some insight.

Four more random samples (samples 2–5) from this same population are shown here.

Sample 2		Sample 3		Sample 4		Sample 5	
Student	x	Student	x	Student	x	Student	x
4	261	15	333	20	323	18	159
15	333	12	263	16	184	8	270
12	263	3	342	19	230	9	278
1	267	7	222	1	267	7	222
18	159	18	159	8	270	14	262

The value of $\bar{x}$ can be computed for each sample:

Sample	2	3	4	5
$\bar{x}$	256.60	263.80	254.80	238.20

Since $\mu = 260.25$, we can see the following:

1. The value of $\bar{x}$ differs from one random sample to another (sampling variability).
2. Some samples produced $\bar{x}$ values larger than μ (samples 1 and 3), whereas others produced $\bar{x}$ values smaller than μ (samples 2, 4, and 5).
3. Samples 2, 3, and 4 produced $\bar{x}$ values that were fairly close to the population mean, but sample 5 resulted in an $\bar{x}$ value that was $22 below the population mean.

Continuing with the experiment, we selected 45 additional random samples (each of size 5). The resulting sample means are as follows:

Sample	$\bar{x}$	Sample	$\bar{x}$	Sample	$\bar{x}$
1	274.6	16	255.0	31	253.4
2	256.6	17	307.2	32	279.6
3	263.8	18	280.0	33	252.6
4	254.8	19	277.4	34	242.2
5	238.2	20	241.2	35	262.6
6	275.6	21	216.0	36	215.4
7	279.2	22	270.0	37	266.2
8	241.6	23	301.0	38	261.4
9	255.4	24	247.0	39	275.0
10	263.8	25	273.8	40	301.4
11	239.6	26	282.8	41	237.0
12	248.2	27	220.4	42	287.4
13	330.8	28	213.6	43	249.2
14	288.8	29	287.6	44	236.8
15	252.8	30	214.8	45	264.6

Figure 8.4, a MINITAB density histogram of the 45 sample means, provides us with a lot of information about the behavior of the statistic $\bar{x}$. Most samples resulted in sample means that were reasonably near $\mu = 260.25$, falling between 235 and 295. A few samples, however, produced $\bar{x}$ values that were very far from μ. If we were to take a sample of size 5 from this population and use $\bar{x}$ as an estimate of the population mean μ, we should not expect $\bar{x}$ to be really close to μ.

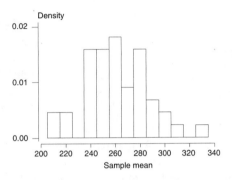

FIGURE 8.4 $\bar{x}$ values from 45 random samples

If we had continued sampling indefinitely, the resulting histogram of $\bar{x}$ values would represent the sampling distribution of $\bar{x}$ for samples of size 5 from this population. The information provided by this sampling distribution enables us to evaluate the behavior of the statistic $\bar{x}$ and is critical for developing inferential procedures based on $\bar{x}$. You may have noticed some general patterns emerging as we did this type of sampling experiment with different sample sizes and different populations, so we do not actually have to carry out such experiments on a regular basis! These general results are described in the next section.

Exercises 8.1 – 8.13

8.1 Explain the difference between a population characteristic and a statistic.

8.2 What is the difference between $\bar{x}$ and μ? between σ and s?

8.3 For each of the following statements, identify the number that appears in boldface type as the value of either a population characteristic or a statistic.

a. A department store reports that **84%** of all customers who use the store's credit plan pay their bills on time.

b. A sample of 100 students at a large university had a mean age of **24.1** years.

c. The Department of Motor Vehicles reports that **22%** of all vehicles registered in a particular state are imports.

d. A hospital reports that based on the ten most recent cases, the mean length of stay for surgical patients is **6.4** days.

e. A consumer group, after testing 100 batteries of a certain brand, reported an average life of **63** hours of use.

8.4 Consider the following "population": {1, 2, 3, 4}. Note that the population mean is

$$\mu = \frac{1 + 2 + 3 + 4}{4} = 2.5$$

a. Suppose that a random sample of size 2 is to be selected without replacement from this population. There are 12 possible samples (provided that the order in which observations are selected is taken into account):

1, 2	1, 3	1, 4	2, 1	2, 3	2, 4
3, 1	3, 2	3, 4	4, 1	4, 2	4, 3

Compute the sample mean for each of the 12 possible samples. Use this information to construct the sampling distribution of $\bar{x}$. (Display it in table form.)

b. Suppose that a random sample of size 2 is to be selected, but this time sampling will be done with replacement. Using a method similar to that of part **a,** construct the sampling distribution of $\bar{x}$. (*Hint:* There are 16 different possible samples in this case.)

c. In what ways are the two sampling distributions of parts **a** and **b** similar? In what ways are they different?

8.5 Simulate sampling from the population of Exercise 8.4 by using four slips of paper individually marked 1, 2, 3, and 4. Select a sample of size 2 without replacement, and compute $\bar{x}$. Repeat this process 50 times, and construct a relative frequency distribution of the 50 $\bar{x}$ values. How does the relative frequency distribution compare to the sampling distribution of $\bar{x}$ derived in Exercise 8.4, part **a**?

8.6 Use the method of Exercise 8.4 to find the sampling distribution of the sample range when a random sample of size 2 is to be selected with replacement from the population {1, 2, 3, 4}. (Recall that sample range = largest observation − smallest observation.)

8.7 On four different occasions, you have borrowed money from a friend in the amounts $1, $5, $10, and $20. The friend has four IOU slips with each of these amounts written on a different slip. He will randomly select two slips from among the four and ask that those IOUs be repaid immediately. Let *t* denote the amount that you must repay immediately.

a. List all possible outcomes and the value of *t* for each, and then determine the sampling distribution of *t*.

b. Calculate μ_t, the mean value of *t*. How is μ_t related to the "population" mean μ?

8.8 Refer to Exercise 8.7. Suppose that instead of asking that the total amount on the two selected slips be repaid immediately, the maximum *m* of the two selected amounts is to be repaid immediately (e.g., *m* = 10 if the $5 and $10 slips are drawn). Obtain the sampling distribution of *m* if the two slips are selected

a. without replacement
b. with replacement

8.9 Consider the following population: {2, 3, 3, 4, 4}. The value of μ is 3.2, but suppose that this is not known to an investigator, who therefore wants to estimate μ from sample data. Three possible statistics for estimating μ are

> Statistic no. 1: the sample mean, $\bar{x}$
> Statistic no. 2: the sample median
> Statistic no. 3: the average of the largest and the
> smallest values in the sample

A random sample of size 3 will be selected without replacement. Provided that we disregard the order in which observations are selected, there are ten possible samples that might result (writing 3 and 3*, 4 and 4* to distinguish the two 3's and the two 4's in the population):

2, 3, 3*	2, 3, 4	2, 3, 4*	2, 3*, 4	2, 3*, 4*
2, 4, 4*	3, 3*, 4	3, 3*, 4*	3, 4, 4*	3*, 4, 4*

For each of these ten samples, compute statistics 1, 2, and 3. Construct the sampling distribution of each of these statistics. Which statistic would you recommend for estimating μ? Explain the reasons for your choice.

8.10 On your shelf, you have four books that you are planning to read in the near future. Two are fictional works, containing 212 and 379 pages, respectively, and the other two are nonfiction, with 350 and 575 pages, respectively.

a. Suppose that you randomly select two books from among these four to take on a one-week ski trip (in case you injure yourself). Let $\bar{x}$ denote the sample average number of pages for the two books selected. Obtain the sampling distribution of $\bar{x}$ for this sampling method, and determine the mean of the $\bar{x}$ distribution.

b. Suppose that you randomly select one of the two fiction books and also randomly select one of the two nonfiction books. Determine the sampling distribution of $\bar{x}$, and then calculate the mean of the $\bar{x}$ distribution.

8.11 Consider a population consisting of the following five values, which represent the number of video rentals during the academic year for each of five housemates.

8	14	16	10	11

a. Compute the mean of this population.

b. Select a random sample of size 2 by writing the numbers on slips of paper and selecting two. Compute the mean of your sample.

c. Repeatedly select samples of size 2, and compute the associated $\bar{x}$ values until you have looked at 25 samples.

d. Construct a histogram using the 25 $\bar{x}$ values. Are most of the sample $\bar{x}$ values near the population mean? Do the $\bar{x}$ values differ a lot from sample to sample, or do they tend to be similar?

8.12 Select 10 additional random samples of size 5 from the population of 20 students given in Example 8.3, and compute the mean amount spent on books for each of the 10 samples. Are the $\bar{x}$ values consistent with the results of the sampling experiment summarized in Figure 8.4?

8.13 Suppose the sampling experiment described in Example 8.3 had used samples of size 10 rather than size 5. If 45 samples of size 10 were selected, $\bar{x}$ computed, and a histogram constructed, how do you think this histogram would differ from that for samples of size 5 (Figure 8.4)? In what way would it be similar?

8.2   The Sampling Distribution of a Sample Mean

When the objective of a statistical investigation is to make an inference about the population mean, μ, it is natural to consider the sample mean, $\bar{x}$. To understand how inferential procedures based on $\bar{x}$ work, we must first study how sampling variability causes $\bar{x}$ to vary in value from one sample to another. The behavior of $\bar{x}$ is described by its sampling distribution. The sample size n and characteristics of the population — its shape, mean value μ, and standard deviation σ — are important in determining properties of the sampling distribution of $\bar{x}$.

It is helpful first to consider the results of some sampling experiments. In the two examples that follow, we start with a specified x population distribution, fix a sample size n, and select 500 different random samples of this size. We then compute $\bar{x}$ for each sample and construct a sample histogram of these 500 $\bar{x}$ values. Because 500 is reasonably large (a reasonably long sequence of samples), the sample histogram should rather closely resemble the true sampling distribution of $\bar{x}$ (which would be obtained from an unending sequence of $\bar{x}$ values). We repeat the experiment for several different values of n to see how the choice of sample size affects the sampling distribution. Careful examination of these sample histograms will aid in understanding the general properties to be stated shortly.

EXAMPLE 8.4   The article "Platelet Size in Myocardial Infarction" (*Brit. Med. J.* (1983): 449–451) presented evidence suggesting that the distribution of platelet volume was approximately normal in shape both for patients after acute myocardial infarction (a heart attack) and for control subjects who had no history of serious illness. The suggested values of μ and σ for the control-subject distribution were $\mu = 8.25$ and $\sigma = .75$. Figure 8.5 (page 412) shows the corresponding normal curve centered at 8.25, the mean value of platelet volume. The value of the population standard deviation, .75, determines the extent to which the x distribution spreads out about its mean value.

We first used MINITAB to select 500 random samples from this normal distribution, with each sample consisting of $n = 5$ observations. A histogram of the resulting 500 $\bar{x}$ values appears in Figure 8.6(a). This procedure was repeated for samples of size $n = 10$, again for $n = 20$, and finally for $n = 30$. The resulting sample histograms of $\bar{x}$ values are displayed in Figures 8.6(b)–(d).

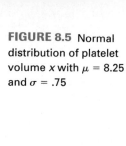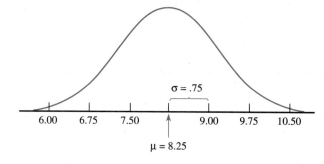

FIGURE 8.5 Normal distribution of platelet volume x with $\mu = 8.25$ and $\sigma = .75$

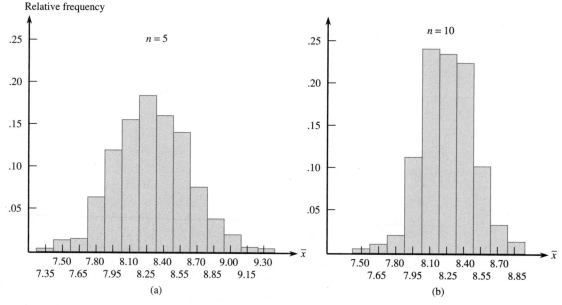

(a)

(b)

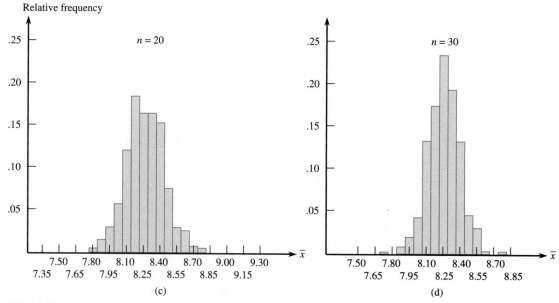

(c)

(d)

FIGURE 8.6 Sample histograms for $\bar{x}$ based on 500 samples, each consisting of n observations: (a) $n = 5$; (b) $n = 10$; (c) $n = 20$; (d) $n = 30$

The first thing to notice about the histograms is their shape. To a reasonable approximation, each of the four looks like a normal curve. The resemblance would be even more striking if each histogram had been based on many more than 500 $\bar{x}$ values. Second, each histogram is centered approximately at 8.25, the mean of the population being sampled. Had the histograms been based on an unending sequence of $\bar{x}$ values, their centers would have been exactly the population mean, 8.25.

The final aspect of the histograms to note is their spread relative to one another. The smaller the value of n, the greater the extent to which the sampling distribution spreads out about the population mean value. This is why the histograms for $n = 20$ and $n = 30$ are based on narrower class intervals than those for the two smaller sample sizes. For the larger sample sizes, most of the $\bar{x}$ values are quite close to 8.25. This is the effect of averaging. When n is small, a single, unusual x value can result in an $\bar{x}$ value far from the center. With a larger sample size, any unusual x values, when averaged with the other sample values, still tend to yield an $\bar{x}$ value close to μ. Combining these insights yields a result that should appeal to your intuition: **$\bar{x}$ based on a large n tends to be closer to μ than does $\bar{x}$ based on a small n.**

EXAMPLE 8.5   Now consider properties of the $\bar{x}$ distribution when the population is quite skewed (and thus very unlike a normal distribution). The Winter 1995 issue of *Chance* magazine gave data on the length of the overtime period for all 251 National Hockey League play-off games between 1970 and 1993 that went into overtime. In hockey, the overtime period ends as soon as either team scores a goal. Figure 8.7 displays a histogram of the data. The histogram has a very long upper tail, indicating that although most overtimes lasted less than 20 minutes, there were a few games with long overtime periods.

If we think of the 251 values as a population, the histogram in Figure 8.7 shows the distribution of values in that population. The skewed shape makes identification of the mean value from the picture more difficult than for a normal distribution.

FIGURE 8.7 The population distribution for Example 8.5 ($\mu = 9.841$)

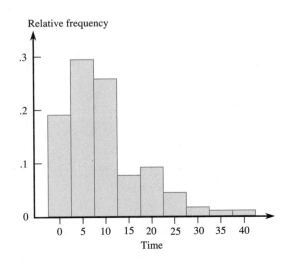

We found the average of these 251 values to be $\mu = 9.841$, so that is the balance point for the population histogram. The median value for the population, 8.000, is less than μ as a result of the distribution's positively skewed nature.

For each of the sample sizes $n = 5, 10, 20,$ and 30, we selected 500 random samples of size n. This was done with replacement to approximate more nearly the usual situation in which the sample size n is only a small fraction of the population size. We then constructed a histogram of the 500 $\bar{x}$ values for each of the four sample sizes. These histograms are displayed in Figure 8.8.

As with samples from a normal population, the averages of the 500 $\bar{x}$ values for the four different sample sizes are all close to the population mean $\mu = 9.841$. If each histogram had been based on an unending sequence of sample means, rather than just 500 of them, each one would have been centered at exactly 9.841. Comparison of the four $\bar{x}$ histograms in Figure 8.8 also shows that as n increases, the

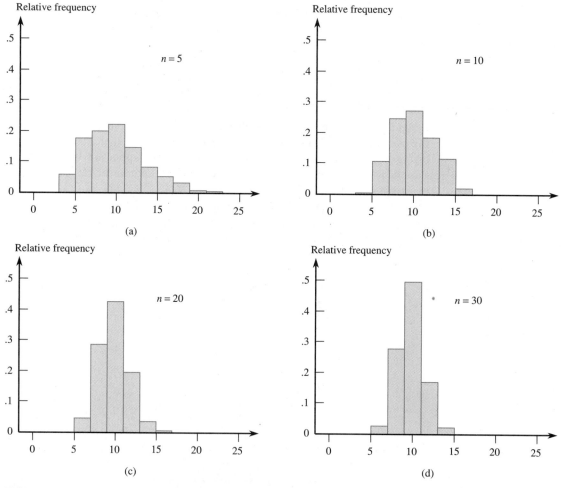

FIGURE 8.8 Four histograms of 500 $\bar{x}$ values for Example 8.5

histogram's spread about its center decreases. This was also true of increasing sample sizes from a normal population: $\bar{x}$ is less variable for a large sample size than it is for a small sample size.

Another aspect of these histograms distinguishes them from the distribution of $\bar{x}$ based on a sample from a normal population. They are skewed and differ in shape more, but they become progressively more symmetric as the sample size increases. The four histograms in Figure 8.8 were drawn using the same horizontal scale, to make it easy to see the decrease in spread as n increases. However, this makes it harder to see how the histograms' shapes change with sample size. Figure 8.9 is a histogram based on narrower class intervals for the $\bar{x}$ values from samples of size 30. This figure shows that for $n = 30$, the histogram has a shape much like a normal curve. Again this is the effect of averaging. Even when n is large, one of the few large x values in the population appears infrequently in the sample. When one does appear, its contribution to $\bar{x}$ is swamped by the contributions of more typical sample values. The normal-curve shape of the histogram for $n = 30$ is what is predicted by the Central Limit Theorem, which will be introduced shortly. According to this theorem, even if the population distribution bears no resemblance whatsoever to a normal curve, the $\bar{x}$ sampling distribution is approximately normal when the sample size n is reasonably large.

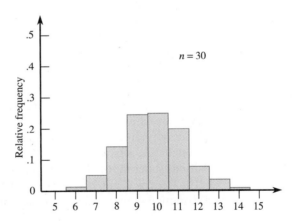

FIGURE 8.9 Histogram of 500 $\bar{x}$ values

General Properties of the Sampling Distribution of $\bar{x}$

Examples 8.4 and 8.5 suggest that for any n, the center of the $\bar{x}$ distribution (the mean value of $\bar{x}$) coincides with the mean of the population being sampled, and the spread of the $\bar{x}$ distribution decreases as n increases, indicating that the standard deviation of $\bar{x}$ is smaller for large n than for small n. The sample histograms also suggest that in some cases, the $\bar{x}$ distribution is approximately normal in shape. These observations are stated more formally in the following general rules.

General Properties of the Sampling Distribution of $\bar{x}$

Let $\bar{x}$ denote the mean of the observations in a random sample of size n from a population having mean μ and standard deviation σ. Denote the mean value of the $\bar{x}$ distribution by $\mu_{\bar{x}}$ and the standard deviation of the $\bar{x}$ distribution by $\sigma_{\bar{x}}$. Then the following rules hold.

Rule 1. $\mu_{\bar{x}} = \mu$

Rule 2. $\sigma_{\bar{x}} = \dfrac{\sigma}{\sqrt{n}}$ This rule is approximately correct as long as no more than 5% of the population is included in the sample.

Rule 3. When the population distribution is normal, the sampling distribution of $\bar{x}$ is also normal for any sample size n.

Rule 4. **(Central Limit Theorem)** When n is sufficiently large, the sampling distribution of $\bar{x}$ is well approximated by a normal curve, even when the population distribution is not itself normal.

Rule 1, $\mu_{\bar{x}} = \mu$, states that the sampling distribution of $\bar{x}$ is always centered at the mean of the population sampled. Rule 2, $\sigma_{\bar{x}} = \sigma/\sqrt{n}$, not only states that the spread of the sampling distribution of $\bar{x}$ decreases as n increases, but also gives a precise relationship between the standard deviation of the $\bar{x}$ distribution and the population standard deviation. When $n = 4$, for example,

$$\sigma_{\bar{x}} = \frac{\sigma}{\sqrt{n}} = \frac{\sigma}{\sqrt{4}} = \frac{\sigma}{2}$$

so the $\bar{x}$ distribution has a standard deviation only half as large as the population standard deviation. Rules 3 and 4 specify circumstances under which the $\bar{x}$ distribution is normal or approximately normal: when the population is normal or when the sample size is large. Figure 8.10 illustrates these rules by showing several $\bar{x}$ distributions superimposed over a graph of the population distribution.

The Central Limit Theorem of rule 4 states that when n is sufficiently large, the $\bar{x}$ distribution is approximately normal, no matter what the population distribution

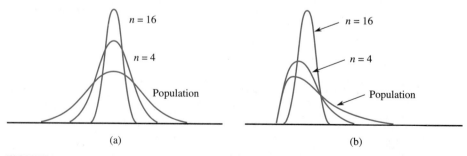

FIGURE 8.10 Population distribution and sampling distributions of $\bar{x}$: (a) A symmetric population; (b) A skewed population

looks like. This result has enabled statisticians to develop large-sample procedures for making inferences about a population mean μ even when the shape of the population distribution is unknown.

Recall that a variable is standardized by subtracting the mean value and then dividing by its standard deviation. Using rules 1 and 2 to standardize $\bar{x}$ gives an important consequence of the last two rules.

If n is large or the population distribution is normal, the standardized variable

$$z = \frac{\bar{x} - \mu_{\bar{x}}}{\sigma_{\bar{x}}} = \frac{\bar{x} - \mu}{\sigma / \sqrt{n}}$$

has (approximately) a standard normal (z) distribution.

Application of the Central Limit Theorem in specific situations requires a rule of thumb for deciding whether n is indeed sufficiently large. Such a rule is not as easy to come by as you might think. Look back at Figure 8.8, which shows the approximate sampling distribution of $\bar{x}$ for $n = 5, 10, 20$, and 30 when the population distribution is quite skewed. Certainly the histogram for $n = 5$ is not well described by a normal curve, and this is still true of the histogram for $n = 10$, particularly in the tails of the histogram (far away from the mean value). Among the four histograms, only the one for $n = 30$ has a reasonably normal shape.

On the other hand, when the population distribution is normal, the sampling distribution of $\bar{x}$ is normal for any n. If the population distribution is somewhat skewed but not to the extent of Figure 8.7, we might expect the $\bar{x}$ sampling distribution to be a bit skewed for $n = 5$ but quite well fit by a normal curve for n as small as 10 or 15. How large an n is needed for the $\bar{x}$ distribution to approximate a normal curve will depend on how much the population distribution differs from a normal distribution. The closer the population distribution is to being normal, the smaller will be the value of n necessary for the Central Limit Theorem approximation to be accurate.

The following conservative rule is recommended by many statisticians.

The Central Limit Theorem can safely be applied if n exceeds 30.

If the population distribution is believed to be reasonably close to a normal distribution, an n of 15 or 20 is often large enough for $\bar{x}$ to have approximately a normal distribution. At the other extreme, we can imagine a distribution with a much longer tail than that of Figure 8.7, in which case even $n = 40$ or 50 would not suffice for approximate normality of $\bar{x}$. In practice, however, very few population distributions are likely to be this badly behaved.

EXAMPLE 8.6

Let x denote the duration of a randomly selected song (in minutes) for a certain type of songbird. Suppose that the mean value of song duration is $\mu = 1.5$ min and that the standard deviation of song duration is $\sigma = .9$ min. (These values are suggested by the results of a large random sample of house finch songs, as reported in "Song Dialects and Colonization in the House Finch" (*Condor* (1975): 407–422).) The sampling distribution of $\bar{x}$ based on a random sample of $n = 25$ song durations then also has mean value

$$\mu_{\bar{x}} = 1.5 \text{ min}$$

That is, the sampling distribution of $\bar{x}$ is centered at 1.5. The standard deviation of $\bar{x}$ is

$$\sigma_{\bar{x}} = \frac{\sigma}{\sqrt{n}} = \frac{.9}{\sqrt{25}} = .18$$

which is only one-fifth as large as the population standard deviation, σ.

EXAMPLE 8.7

A soft-drink bottler claims that, on average, cans contain 12 oz of soda. Let x denote the actual volume of soda in a randomly selected can. Suppose that x is normally distributed with $\sigma = .16$ oz. Sixteen cans are to be selected and the soda volume determined for each one; let $\bar{x}$ denote the resulting sample average soda volume. Because the x distribution is normal, the sampling distribution of $\bar{x}$ is also normal. *If the bottler's claim is correct,* the sampling distribution of $\bar{x}$ has mean value

$$\mu_{\bar{x}} = \mu = 12$$

and standard deviation

$$\sigma_{\bar{x}} = \frac{\sigma}{\sqrt{n}} = \frac{.16}{\sqrt{16}} = .04$$

To calculate a probability involving $\bar{x}$, we simply standardize by subtracting the mean value, 12, and dividing by the standard deviation (of $\bar{x}$), which is .04. For example, the probability that the sample average soda volume is between 11.96% and 12.08% is calculated by first standardizing the interval limits:

Lower limit: $\quad a^* = \dfrac{11.06 - 12}{.04} = -1.0$

Upper limit: $\quad b^* = \dfrac{12.08 - 12}{.04} = 2.0$

Then

$$P(11.96 \leq \bar{x} \leq 12.08) = \text{area under the } z \text{ curve between } -1.0 \text{ and } 2.0$$
$$= (\text{area to the left of } 2.0) - (\text{area to the left of } -1.0)$$
$$= .9772 - .1587$$
$$= .8185$$

The probability that the sample average soda volume is at least 12.1 oz is

$$P(\bar{x} \geq 12.1) = P\left(z \geq \frac{12.1 - 12}{.04} = 2.5 \right) = (\text{area under the } z \text{ curve to the right of } 2.5) = .0062$$

If the x distribution is as described and the claim is correct, a sample average soda volume based on 16 observations exceeds 12.1 oz for less than 1% of all such samples. Thus, observation of an $\bar{x}$ value that exceeds 12.1 oz casts doubt on the bottler's claim that the average soda volume is 12 oz.

EXAMPLE 8.8

A hot dog manufacturer asserts that one of its brands of hot dogs has an average fat content of $\mu = 18$ grams per hot dog. Consumers of this brand would probably not be disturbed if the mean was less than 18, but would be unhappy if it exceeded 18. Let x denote the fat content of a randomly selected hot dog, and suppose that σ, the standard deviation of the x distribution, is 1.

An independent testing organization is asked to analyze a random sample of 36 hot dogs. Let $\bar{x}$ be the average fat content for this sample. The sample size, $n = 36$, is large enough to invoke the Central Limit Theorem and regard the $\bar{x}$ distribution as being approximately normal. The standard deviation of the $\bar{x}$ distribution is

$$\sigma_{\bar{x}} = \frac{\sigma}{\sqrt{n}} = \frac{1}{\sqrt{36}} = .1667$$

If the manufacturer's claim is correct, we know that

$$\mu_{\bar{x}} = \mu = 18 \text{ grams}$$

Suppose the sample resulted in a mean of $\bar{x} = 18.4$. Does this result indicate that the manufacturer's claim is incorrect?

We can answer this question by looking at the sampling distribution of $\bar{x}$. Because of sampling variability, even if $\mu = 18$, we know that $\bar{x}$ will typically deviate somewhat from this value. Is it likely that we would see a sample mean at least as large as 18.4 when the population mean is really 18? If the company's claim is correct,

$$P(\bar{x} \geq 18.4) \approx P\left(z \geq \frac{18.4 - 18}{.1667} \right)$$
$$= P(z \geq 2.4)$$
$$= \text{area under the } z \text{ curve to the right of } 2.4$$
$$= .0082$$

The value $\bar{x} = 18.4$ exceeds 18 by enough to cast substantial doubt on the manufacturer's claim. Values of $\bar{x}$ at least as large as 18.4 will be observed only about .82% of the time when a random sample of size 36 is taken from a population with mean 18 and standard deviation 1.

▬▬▬▬▬▬

Other Cases

We now know a great deal about the sampling distribution of $\bar{x}$ in two cases, that of a normal population distribution and that of a large sample size. What happens when the population distribution is not normal and n is small? Although it is still true that $\mu_{\bar{x}} = \mu$ and $\sigma_{\bar{x}} = \sigma/\sqrt{n}$, unfortunately there is no general result about the shape of the $\bar{x}$ distribution. When the objective is to make an inference about the center of such a population, one way to proceed is to replace the normality assumption with some other distributional model for the population. Statisticians have proposed and studied a number of such models. Theoretical methods or simulation can be used to describe the $\bar{x}$ distribution corresponding to the assumed model. One alternative strategy is to use one of the transformations presented in Chapter 7 to create a data set that more closely resembles a sample from a normal population and then base inferences on the transformed data. Yet another path is to use an inferential procedure based on statistics other than $\bar{x}$.

Exercises 8.14 – 8.26

▬▬▬▬

8.14 A random sample is to be selected from a population with mean $\mu = 100$ and standard deviation $\sigma = 10$. Determine the mean and standard deviation of the $\bar{x}$ sampling distribution for each of the following sample sizes.

   a. $n = 9$ **b.** $n = 15$
   c. $n = 36$ **d.** $n = 50$
   e. $n = 100$ **f.** $n = 400$

8.15 For which of the sample sizes given in Exercise 8.14 would it be reasonable to think that the $\bar{x}$ sampling distribution will be approximately normal in shape?

8.16 Explain the difference between σ and $\sigma_{\bar{x}}$. Between μ and $\mu_{\bar{x}}$.

8.17 Suppose that a random sample of size 64 is to be selected from a population with mean 40 and standard deviation 5.

   a. What are the mean and standard deviation of the $\bar{x}$ sampling distribution? Describe the shape of the $\bar{x}$ sampling distribution.

   b. What is the approximate probability that $\bar{x}$ will be within .5 of the population mean μ?

   c. What is the approximate probability that $\bar{x}$ will differ from μ by more than .7?

8.18 The time that a randomly selected individual waits for an elevator in an office building has a uniform distribution over the interval from 0 to 1 min. It can be shown that for this distribution, $\mu = .5$ and $\sigma = .289$.

   a. Let $\bar{x}$ be the sample average waiting time for a random sample of 16 individuals. What are the mean value and standard deviation of the sampling distribution of $\bar{x}$?

   b. Answer part **a** for a random sample of 50 individuals. In this case, sketch a picture of a good approximation to the actual $\bar{x}$ distribution.

8.19 Let x denote the time (min) that it takes a fifth-grade student to read a certain passage. Suppose that the mean value and standard deviation of x are $\mu = 2$ min and $\sigma = .8$ min, respectively.

   a. If $\bar{x}$ is the sample average time for a random sample of $n = 9$ students, where is the $\bar{x}$ distribution centered, and how much does it spread out about the center (as described by its standard deviation)?

   b. Repeat part **a** for a sample of size of $n = 20$ and again for a sample of size $n = 100$. How do the centers and spreads of the three $\bar{x}$ distributions compare to one another? Which sample size would be most likely to result in an $\bar{x}$ value close to μ, and why?

8.20 In the library on a university campus, there is a sign in the elevator that indicates a limit of 16 persons. Furthermore, there is a weight limit of 2500 pounds. Assume the average weight of students, faculty, and staff on campus is 150 pounds, the standard deviation is 27 pounds, and that the distribution of weights of individuals on campus is approximately normal. If a random sample of 16 persons from the campus is taken:
  a. What is the expected value of the sample mean of their weights?
  b. What is the standard deviation of the sampling distribution?
  c. What average weight for these 16 people will result in the total weight exceeding the weight limit of 2500 pounds?
  d. What is the chance that a random sample of 16 persons on the elevator will exceed the weight limit?

8.21 Suppose that the mean value of interpupillary distance for all adult males is 65 mm, and the population standard deviation is 5 mm.
  a. If the distribution of interpupillary distance is normal and a sample of $n = 25$ adult males is selected, what is the probability that the sample average distance $\bar{x}$ for these 25 will be between 64 and 67 mm? At least 68 mm?
  b. Suppose that a sample of 100 adult males is obtained. Without assuming that interpupillary distance is normally distributed, what is the approximate probability that the sample average distance is between 64 and 67 mm? At least 68 mm?

8.22 Suppose that a sample of size 100 is to be drawn from a population with standard deviation 10.
  a. What is the probability that the sample mean will be within 2 of the value of μ?
  b. For this example ($n = 100$, $\sigma = 10$), complete each statement by computing the appropriate value.
    i. Approximately 95% of the time, $\bar{x}$ is within _____ of μ.
    ii. Approximately .3% of the time, $\bar{x}$ is farther than _____ from μ.

8.23 A manufacturing process is designed to produce bolts with a .5-in diameter. Once each day, a random sample of 36 bolts is selected and the diameters recorded. If the resulting sample mean is less than .49 in or greater than .51 in, the process is shut down for adjustment. The standard deviation for diameter is .02 in. What is the probability that the manufacturing line will be shut down unnecessarily? (*Hint*: Find the probability of observing an $\bar{x}$

in the shut-down range when the true process mean really is .5 in.)

8.24 College students with a checking account typically write relatively few checks in any given month, whereas full-time residents typically write many more checks during a month. Suppose that 50% of a bank's accounts are held by students and 50% by full-time residents. Let x denote the number of checks written in a given month by a randomly selected bank customer.
  a. Give a sketch of what the probability distribution of x might look like.
  b. Suppose the mean value of x is 22.0 and the standard deviation is 16.5. If a random sample of $n = 100$ customers is selected and $\bar{x}$ denotes the sample average number of checks written during a particular month, where is the sampling distribution of $\bar{x}$ centered, and what is the standard deviation of the $\bar{x}$ distribution? Sketch a rough picture of the sampling distribution.
  c. Referring to part **b**, what is the approximate probability that $\bar{x}$ is at most 20? At least 25? What result are you using to justify your computations?

8.25 An airplane with room for 100 passengers has a total baggage limit of 6000 lb. Suppose that the total weight of the baggage checked by an individual passenger is a random variable x with mean value 50 lb and standard deviation 20 lb. If 100 passengers board a flight, what is the approximate probability that the total weight of their baggage will exceed the limit? (*Hint*: With $n = 100$, the total weight exceeds the limit when the average weight $\bar{x}$ exceeds 6000/100.)

8.26 The thickness (mm) of the coating applied to disk drives is a characteristic that determines the usefulness of the product. When no unusual circumstances are present, the thickness (x) has a normal distribution with mean 3 mm and standard deviation .05 mm. Suppose that the process will be monitored by selecting a random sample of 16 drives from each shift's production and determining $\bar{x}$, the mean coating thickness for the sample.
  a. Describe the sampling distribution of $\bar{x}$ (for a sample size 16).
  b. When no unusual circumstances are present, we expect $\bar{x}$ to be within $3\sigma_{\bar{x}}$ of 3 mm, the desired value. An $\bar{x}$ value farther from 3 than $3\sigma_{\bar{x}}$ is interpreted as an indication of a problem that needs attention. Compute $3 \pm 3\sigma_{\bar{x}}$. (A plot over time of $\bar{x}$ values with horizontal lines drawn at the limits $\mu \pm 3\sigma_{\bar{x}}$ is called a process control chart.)
  c. Referring to part **b**, what is the probability that a

sample mean will be outside $3 \pm 3\sigma_{\bar{x}}$ just by chance (that is, when there are no unusual circumstances)?

d. Suppose that a machine used to apply the coating is out of adjustment, resulting in a mean coating thick-

ness of 3.05 mm. What is the probability that a problem will be detected when the next sample is taken? (*Hint:* This will occur if $\bar{x} > 3 + 3\sigma_{\bar{x}}$ or $\bar{x} < 3 - 3\sigma_{\bar{x}}$ when $\mu = 3.05$.)

8.3   The Sampling Distribution of a Sample Proportion

The objective of many statistical investigations is to draw a conclusion about the proportion of individuals or objects in a population that possess a specified property; for example, Maytag washers that don't require service during the warranty period, Europeans who favor the deployment of a certain type of missile, or coffee drinkers who regularly drink decaffeinated coffee. Traditionally, any individual or object that possesses the property of interest is labeled a success (S), and one that does not possess the property is termed a failure (F). The letter π denotes the proportion of S's in the population. The value of π is a number between 0 and 1, and 100π is the percentage of S's in the population. If $\pi = .75$, 75% of the population members are S's, and if $\pi = .01$, this means that a population contains only 1% S's and 99% F's.

The value of π is usually unknown to an investigator. When a random sample of size n is selected from this type of population, some of the individuals in the sample are S's, and the remaining individuals in the sample are F's. The statistic that will provide a basis for making inferences about π is p, the **sample proportion of S's**:

$$p = \frac{\text{number of } S\text{'s in the sample}}{n}$$

For example, if $n = 5$ and three S's result, then $p = 3/5 = .6$.

Just as making inferences about μ requires knowing something about the sampling distribution of the statistic $\bar{x}$, making inferences about π requires first learning about properties of the sampling distribution of the statistic p. For example, when $n = 5$, the six possible values of p are 0, .2 (from 1/5), .4, .6, .8, and 1. The sampling distribution of p gives the probability of each of these six possible values, the long-run proportion of the time that each value would occur if samples with $n = 5$ were selected over and over again. Similarly, when $n = 100$, the 101 possible values of p are 0, .01, .02, . . . , .98, .99, and 1. The sampling distribution of p can then be used to calculate probabilities such as $P(.3 \le p \le .7)$ and $P(.75 \le p)$.

As we did for the distribution of the sample mean, we will look at some simulation experiments to develop an intuitive understanding of the distribution of the sample proportion before stating general rules. In each example, 500 random samples (each of size n) are selected from a population having a specified value of π. We compute p for each sample and then construct a sample histogram of the 500 values.

EXAMPLE 8.9   In the fall of 1999, there were 16,250 students enrolled at Cal Poly, San Luis Obispo. Of these students, 6825 (42%) were female. To illustrate properties of the sampling distribution of a sample proportion, we decided to simulate sampling from Cal

Poly's student population. With S denoting a female student and F a male student, the proportion of S's in the population is $\pi = .42$. A statistical software package was used to select 500 samples of size $n = 5$, then 500 samples of size $n = 10$, then 500 samples with $n = 25$, and finally 500 samples with $n = 50$. The sample histograms of the 500 values of p for the four sample sizes are displayed in Figure 8.11.

The most noticeable feature of the histogram shapes is the progression toward the shape of a normal curve as n increases. The histogram for $n = 10$ is more bell shaped than the histogram for samples of size 5, although it still has a slight positive skew. The histograms for $n = 25$ and $n = 50$ look much like a normal curve.

Although the skewness of the first two histograms makes it a bit difficult to locate their centers, all four histograms appear to be centered at roughly .42, the value of π for the population sampled. Had the histograms been based on an unending sequence of samples, each histogram would have been centered at exactly .42. Finally, as was the case with the sampling distribution of $\bar{x}$, the histograms spread out

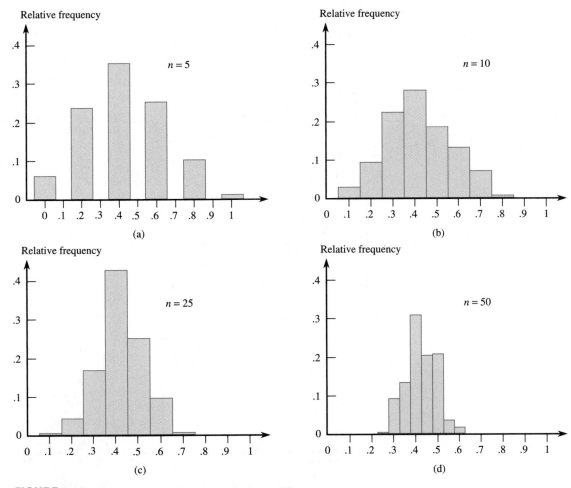

FIGURE 8.11 Histograms for 500 values of p ($\pi = .42$)

more for small n than for large n. The value of p based on a large sample size tends to be closer to π, the population proportion of S's, than does p from a small sample.

EXAMPLE 8.10

The development of viral hepatitis subsequent to a blood transfusion can cause serious complications for a patient. The article "Hepatitis in Patients with Acute Non-lymphatic Leukemia" (*Amer. J. of Med.* (1983): 413–421) reported that in spite of careful screening for those having a hepatitis antigen, viral hepatitis occurs in 7% of blood recipients. Here we simulate sampling from the population of blood re-

FIGURE 8.12 Histograms of 500 values of p ($\pi = .07$)

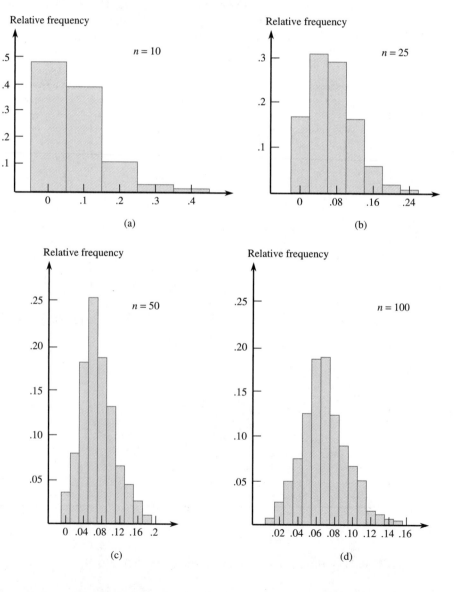

cipients, with S denoting a recipient who contracts hepatitis (not the sort of characteristic one usually thinks of as a success, but the $S-F$ labeling is arbitrary), so $\pi = .07$. Figure 8.12 displays histograms of 500 values of p for the four sample sizes $n = 10, 25, 50,$ and 100.

As was the case in Example 8.9, all four histograms are centered at approximately the value of π for the population being sampled. (The average values of p for these simulations are .0690, .0677, .0707, and .0694.) If the histograms had been based on an unending sequence of samples, they would all have been centered at exactly $\pi = .07$. Again, the spread of a histogram based on a large n is smaller than the spread of a histogram resulting from a small sample size. The larger the value of n, the closer the sample proportion, p, tends to be to the value of the population proportion, π.

Furthermore, there is a progression toward the shape of a normal curve as n increases. However, the progression is much slower here than in the previous example, because the value of π is so extreme. (The same thing would happen for $\pi = .93$, except that the histograms would be negatively rather than positively skewed.) The histograms for $n = 10$ and $n = 25$ exhibit substantial skew, and the skew of the histogram for $n = 50$ is still moderate (compare Figure 8.12(c) to Figure 8.11(d)). Only the histogram for $n = 100$ is reasonably well fit by a normal curve. It appears that whether a normal curve provides a good approximation to the sampling distribution of p depends on the values of both n and π. Knowing only that $n = 50$ is not enough to guarantee that the shape of the histogram is approximately normal.

General Properties of the Sampling Distribution of p

The previous two examples suggest that the sampling distribution of p depends both on n, the sample size, and on π, the proportion of S's in the population. These results are stated more formally in the following general rules.

General Properties of the Sampling Distribution of p

Let p be the proportion of successes in a random sample of size n from a population whose proportion of S's is π. Denote the mean value of p by μ_p and the standard deviation by σ_p. Then the following rules hold.

Rule 1. $\mu_p = \pi$

Rule 2. $\sigma_p = \sqrt{\dfrac{\pi(1 - \pi)}{n}}$

Rule 3. When n is large and π is not too near 0 or 1, the sampling distribution of p is approximately normal.

Thus, the sampling distribution of p is always centered at the value of the population success proportion π, and the extent to which the distribution spreads out about π decreases as the sample size n increases.

Examples 8.9 and 8.10 indicate that both π and n must be considered in judging whether the sampling distribution of p is approximately normal.

> The farther the value of π is from .5, the larger n must be for a normal approximation to the sampling distribution of p to be accurate. A conservative rule of thumb is that if both $n\pi \geq 10$ and $n(1 - \pi) \geq 10$, then it is safe to use a normal approximation.

A sample size of $n = 100$ is not by itself sufficient to justify the use of a normal approximation. If $\pi = .01$, the distribution of p will be very positively skewed, so a bell-shaped curve will not give a good approximation. Similarly, if $n = 100$ and $\pi = .99$ (so $n(1 - \pi) = 1 < 10$), the distribution of p will have a substantial negative skew. The conditions $n\pi \geq 10$ and $n(1 - \pi) \geq 10$ ensure that the sampling distribution of p is not too skewed. If $\pi = .5$, the normal approximation can be used for n as small as 20, whereas for $\pi = .05$ or .95, n should be at least 200.

EXAMPLE 8.11

The proportion of all blood recipients stricken with viral hepatitis was given as .07 in the article referenced in Example 8.10. Suppose that a new treatment is believed to reduce the incidence rate of viral hepatitis. This treatment is given to $n = 200$ blood recipients. Only 6 of the 200 patients contract hepatitis. This appears to be a favorable result, since $p = 6/200 = .03$. The question of interest to medical researchers is: Does this result indicate that the true (long-run) proportion of patients who contract hepatitis after the experimental treatment is less than .07, or could this result be plausibly attributed to sampling variability (i.e., the fact that p will typically deviate from its mean value, π)? If the treatment is ineffective,

$$\mu_p = \pi = .07 \qquad \sigma_p = \sqrt{\frac{\pi(1 - \pi)}{200}} = \sqrt{\frac{(.07)(.93)}{200}} = .018$$

Furthermore, since

$$n\pi = 200(.07) = 14 \geq 10$$

and

$$n(1 - \pi) = 200(.93) = 186 \geq 10$$

the sampling distribution of p is approximately normal. Then

$$P(p < .03) \approx P\left(z < \frac{.03 - .07}{.018}\right)$$
$$= P(z < -2.22)$$
$$= .0132$$

Thus, it is unlikely that a sample proportion .03 or smaller would be observed if the treatment really were ineffective. The new treatment appears to yield a smaller incidence rate for the disease than occurs without any treatment.

Exercises 8.27 – 8.33

8.27 A random sample is to be selected from a population that has a proportion of successes $\pi = .65$. Determine the mean and standard deviation of the sampling distribution of p for each of the following sample sizes.
 a. $n = 10$ **b.** $n = 20$ **c.** $n = 30$
 d. $n = 50$ **e.** $n = 100$ **f.** $n = 200$

8.28 For which of the sample sizes given in Exercise 8.27 would the sampling distribution of p be approximately normal if $\pi = .65$? If $\pi = .2$?

8.29 A certain chromosome defect occurs in only one out of 200 Caucasian adult males. A random sample of $n = 100$ adult Caucasian males is obtained.
 a. What is the mean value of the sample proportion p, and what is the standard deviation of the sample proportion?
 b. Does p have approximately a normal distribution in this case? Explain.
 c. What is the smallest value of n for which the sampling distribution of p is approximately normal?

8.30 The article "Should Pregnant Women Move? Linking Risks for Birth Defects with Proximity to Toxic Waste Sites" (*Chance* (1992): 40–45) reported that in a large study carried out in the state of New York, approximately 30% of the study subjects lived within 1 mile of a hazardous waste site. Let π denote the proportion of all New York residents who live within 1 mile of such a site, and suppose that $\pi = .3$.
 a. Would p based on a random sample of only 10 residents have approximately a normal distribution? Explain why or why not.
 b. What are the mean value and standard deviation of p based on a random sample of size 400?
 c. When $n = 400$, what is $P(.25 \le p \le .35)$?
 d. Is the probability calculated in part **c** larger or

smaller than would be the case if $n = 500$? Answer without actually calculating this latter probability.

8.31 The article "Thrillers" (*Newsweek,* Apr. 22, 1985) states, "Surveys tell us that more than half of America's college graduates are avid readers of mystery novels." Let π denote the actual proportion of college graduates who are avid readers of mystery novels. Consider p to be based on a random sample of 225 college graduates.
 a. If $\pi = .5$, what are the mean value and standard deviation of p? Answer this question for $\pi = .6$. Does p have approximately a normal distribution in both cases? Explain.
 b. Calculate $P(p \ge .6)$ both for $\pi = .5$ and for $\pi = .6$.
 c. Without doing any calculations, how do you think the probabilities in part **b** would change if n were 400 rather than 225?

8.32 Suppose that a particular candidate for public office is in fact favored by 48% of all registered voters in the district. A polling organization takes a random sample of 500 voters and will use p, the sample proportion, to estimate π. What is the approximate probability that p will be greater than .5, causing the polling organization to incorrectly predict the result of the upcoming election?

8.33 A manufacturer of computer printers purchases plastic ink cartridges from a vendor. When a large shipment is received, a random sample of 200 cartridges is selected, and each is inspected. If the sample proportion of defectives is more than .02, the entire shipment will be returned to the vendor.
 a. What is the approximate probability that the shipment will be returned if the true proportion of defectives in the shipment is .05?
 b. What is the approximate probability that the shipment will not be returned when the true proportion of defectives in the shipment is .10?

Summary of Key Concepts and Formulas

Term or Formula	Comment
Statistic	Any quantity whose value is computed from sample data.
Sampling distribution	The probability distribution of a statistic: The sampling distribution describes the long-run behavior of the statistic.
Sampling distribution of $\bar{x}$	The probability distribution of the sample mean $\bar{x}$, based on a random sample of size n. Properties of the $\bar{x}$ sampling distribution: $\mu_{\bar{x}} = \mu$ and $\sigma_{\bar{x}} = \sigma/\sqrt{n}$

Term or Formula	Comment
	(where μ and σ are the population mean and standard deviation, respectively). In addition, when the population distribution is normal or the sample size is large, the sampling distribution of $\bar{x}$ is (approximately) normal.
Central Limit Theorem	This important theorem states that when n is sufficiently large, the $\bar{x}$ distribution will be approximately normal. The standard rule of thumb is that the theorem can safely be applied when n exceeds 30.
Sampling distribution of p	The probability distribution of the sample proportion p, based on a random sample of size n. When the sample size is sufficiently large, the sampling distribution of p is approximately normal, with $\mu_p = \pi$ and $\sigma_p = \sqrt{\pi(1-\pi)/n}$ (where π is the true population proportion).

Supplementary Exercises 8.34 – 8.40

8.34 The nicotine content in a single cigarette of a particular brand has a distribution with mean .8 mg and standard deviation .1 mg. If 100 of these cigarettes are analyzed, what is the probability that the resulting sample mean nicotine content will be less than .79? Less than .77?

8.35 Let $x_1, x_2, \ldots, x_{100}$ denote the actual net weights (in lb) of 100 randomly selected bags of fertilizer. Suppose that the weight of a randomly selected bag has a distribution with mean 50 lb and variance 1 lb^2. Let $\bar{x}$ be the sample mean weight ($n = 100$).
  a. Describe the sampling distribution of $\bar{x}$.
  b. What is the probability that the sample mean is between 49.75 lb and 50.25 lb?
  c. What is the probability that the sample mean is less than 50 lb?

8.36 Suppose that 20% of the subscribers of a cable television company watch the shopping channel at least once a week. The cable company is trying to decide whether to replace this channel with a new local station. A survey of 100 subscribers will be undertaken. The cable company has decided to keep the shopping channel if the sample proportion is greater than .25. What is the approximate probability that the cable company will keep the shopping channel, even though the true proportion who watch it is only .20?

8.37 Although a lecture period at a certain university lasts exactly 50 minutes, the actual lecture time of a statistics instructor on any particular day has a distribution with mean value 52 min and standard deviation 2 min. Suppose that times of different lectures are independent of one another. Let $\bar{x}$ represent the mean of 36 randomly selected lecture times.

  a. What are the mean value and standard deviation of the sampling distribution of $\bar{x}$?
  b. What is the probability that the sample mean exceeds 50 min? 55 min?

8.38 Water permeability of concrete is an important characteristic in assessing suitability for various applications. Permeability can be measured by letting water flow across the surface and determining the amount lost (in/hr). Suppose that the permeability index x for a randomly selected concrete specimen of a particular type is normally distributed with mean value 1000 and standard deviation 150.
  a. How likely is it that a single specimen will have a permeability index between 850 and 1300?
  b. If the permeability index is determined for each specimen in a random sample of size 10, how likely is it that the sample average permeability index will be between 950 and 1100? Between 850 and 1300?

8.39 *Newsweek* (Nov. 23, 1992) reported that 40% of all U.S. employees participate in "self-insurance" health plans ($\pi = .40$).
  a. In a random sample of 100 employees, what is the approximate probability that at least half of those in the sample participate in such a plan?
  b. Suppose you were told that at least 60 of the 100 employees in a sample from your state participated in such a plan. Would you think $\pi = .40$ for your state?

8.40 The amount of money spent by a customer at a discount store has a mean of $100 and a standard deviation of $30. What is the probability that a randomly selected group of 50 shoppers will spend a total of more than $5300? (*Hint*: The total will be more than $5300 when the sample average exceeds what value?)

References

Freedman, David, Robert Pisani, Roger Purves. *Statistics,* 3rd ed. New York: W. W. Norton, 1997. (This book gives an excellent informal discussion of sampling distributions.)

GRAPHING CALCULATOR EXPLORATIONS

8.1 The Sampling Distribution of the Mean

The characteristics of the sampling distribution of a statistic are very important for drawing inferences in statistics. Because the sample mean is widely used, its sampling distribution is of particular interest. The Central Limit Theorem guarantees that as long as n is large enough, the sampling distribution of the sample mean is approximately normal even though the parent population may not be normal. We relied on the results of simulations to illustrate this property of the sampling distribution of $\bar{x}$. The graphing calculator can be used to simulate the process of sampling from a very large population. We have already discussed most of the calculator commands needed to perform such a simulation.

Although simulations are generally done on a very large scale, we will operate on a very small scale in this calculator exploration. Our methods will be suitable for performing a small number of trials of a simulation and then combining the results over a whole statistics class. To go further into simulation of sampling distributions would require programming your calculator (more on this later) or using a computer with appropriate software.

The process of selecting a random sample from a very large population is generally modeled in the following manner:

1. Construct a mathematical description — a probability distribution — of the population distribution of interest.
2. Generate a random sample from the specified distribution.
3. Calculate the statistic of interest (here, the sample mean) for the simulated random sample.
4. Repeat steps 2 and 3 a large number of times.
5. Construct a distribution of the resulting values of the sample statistic.

We will illustrate how this might be done using a graphing calculator. First, we need a population to sample from. A convenient population, though not a very common one, is the uniform distribution from 0 to 1. (This distribution is convenient for the calculator, as it is very easy to use the *rand* function to generate the random sample.)

Now we need to sample from this population — that is, generate random selections from a uniform population distribution. We will do this as discussed in previous calculator explorations, using the *seq* command and storing the results in a list.

If you are following along with calculator in hand, you should note a couple things. First, the number of random numbers you will be able to generate will depend on both your calculator's capabilities and the amount of memory your calculator has left to use. For this example we will use $n = 500$. If your calculator gives you an error message, try a number smaller than 500 to check that you have the calculator syntax correct. If your calculator works for a small value of n but not a large value, the large value may exceed a maximum list size or you may have already stored many programs, pictures, and/or data in your calculator and you may not have enough memory left to handle the requested computation. Second, this command *can* take awhile to complete. Do not be worried if your calculator's attention seems to be wandering.

Here is a generic statement to generate 500 random numbers and store them in a list:

$$seq(rand, \text{x}, 1, 500) \rightarrow \text{List1}$$

Now that we have generated a sample of size 500, let's check our results by making a histogram of the results. If we have been successful, we should see a distribution that mirrors the population — that is, a fairly uniform distribution of results. **It is very important to check this by displaying your simulated sampling results!** Verifying these results is the equivalent of checking to see that you have a "right answer so far." Although you should expect some fluctuation from a perfect match — we are, after all, sampling — you should not get seriously distorted results with an n as large as 500. Our edit window and histogram results are displayed in the accompanying figures. Pay particular attention to the Xmin and XScl values in the window. (Recall from an earlier calculator exploration that these values will determine the class size for a histogram.) We want to partition the interval from 0 to 1 into 10 equal intervals, and other choices for these values will give you a graph that is difficult to interpret. From the graph, the sample appears to be representative of the population from which we are sampling.

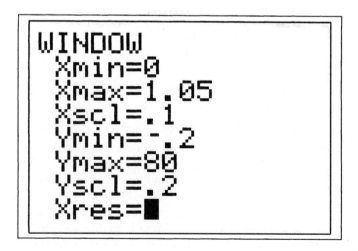

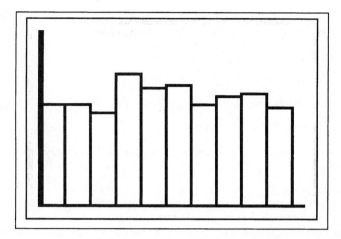

Now we will calculate the sample mean for the data in List1. Our result: $\bar{x} =$ .50223. This sample mean represents *one trial* of our simulation. In order to do a respectable simulation, we would have to repeat the preceding commands a large number of times. This single mean from our one trial is what we meant earlier by a "very small scale" simulation. Even on this small scale, however, combining your results with similar results from your classmates can generate very good simulations. Obtaining 10 trials from each of 25 students is easily accomplished in a class period, and 250 trials is nothing to sneeze at!

We should note in passing that this small-scale restriction of generating sample means one at a time is easily overcome *if* you have someone in your class who knows how to do a little programming on the calculator. It is even possible that if you have *no* experience programming a calculator you could do a simulation. Having spent many frustrating hours at programming, we cannot in good conscience suggest that you learn to program your calculator just for simulations. However, if you want to explore simulations on a medium-sized scale, and have someone to help you set up the program below, you could do some simulations very quickly. As written, it will perform 500 trials of the simulation we have been discussing, using a sample size of 50. Remember, syntax will vary among calculators, but something very close to this program will perform steps 2, 3, and 4 outlined above, and store the results in List2. Consult your friendly local calculator gurus for details.

```
For (X, 1, 500)
seq(rand, X, 1, 50) → List1
mean(List1) → List2(X)
End
```

In the example above, we generated random numbers from a population uniformly distributed from 0 to 1. Sampling distributions of the mean from other populations may also be considered; the populations available are limited only by your random number generation capability. Some populations might be more appropriately modeled as normal or binomial. Some calculators have built-in functions to generate the random values you would get if sampling from a normal or binomial population. Here are some other possible sampling simulations:

From a population of integers, 1 to 10:

$seq(int(rand \times 10) + 1, x, 1, 500) \rightarrow List1$

or, if there is a built-in function (keystrokes denoted here by *randInt*):

$seq(randInt, x, 1, 500) \rightarrow List1$

From a normal population (assuming a built-in function, keystrokes denoted by *randNormal*):

$seq(randNormal, x, 1, 500) \rightarrow List1$

Your calculator may have different built-in functions or different syntax but your exploration of the sampling distribution of the mean can begin as soon as you check that manual!

8.2 The Sampling Distribution of the Sample Proportion

We saw in the previous exploration how to simulate sampling from a population in order to study the sampling distribution of the sample mean. In this exploration we will consider a similar process for studying the sampling distribution of the sample proportion. To review a bit, the process of random sampling from a very large population is modeled as follows:

1. Construct a mathematical description — a probability distribution — of the population.
2. Generate a random sample from this population density function.
3. Calculate the statistic of interest (here, the sample proportion) for the simulated random sample.
4. Repeat steps 2 and 3 a large number of times.
5. Construct a distribution of the resulting values of the sample proportion.

In the case of modeling the sampling distribution of a sample proportion, we conceive of a population consisting of a set of 0's and 1's, where the proportion of 1's in the population is equal to the given population proportion, π. The process of simulating a random sample consists of generating a fixed number n of 0's and 1's, and finding the proportion of 1's in the sample. This value would be the simulated sample proportion. For example, suppose we are simulating sampling from a population with a population proportion $\pi = .75$ using a sample size of $n = 12$. One simulated sample might look like this:

111010011110

We note that 8 out of the 12 are 1's, giving a sample proportion of $p = .67$. How might we accomplish this simulation on our calculator? The easiest method is to capitalize on a built-in function to generate outcomes that are from a binomial distribution. If your calculator has such a function it is probably somewhere in a "math" or "prob" menu — look in your manual's index for something like "RandBin," or "RandBinom." To save keystrokes, we will use the generic name *RandBin* to refer to this built-in calculator function. This function will need two numbers in order to generate a binomial random variable: your sample size and the population proportion. For our example above, we might enter *RandBin*(12, .75) or *RandBin*(.75, 12), depending on the function's syntax.

On your calculator, find the built-in function and enter the appropriate syntax with the parameters, $n = 12$ and $\pi = .75$. When you do this, the output might be 8.

We will simulate the process of sampling and calculating a sample proportion by using the *seq* function as we did earlier, but now we will convert the outcomes into proportions:

$$seq(RandBin(12, .75)/12, x, 1, 500) \rightarrow List1$$

The distribution of our results, together with summary values, are shown in the accompanying figures. Note that the simulated results are close to the theoretical values for the mean (.75) and standard deviation (.125) for the sampling distribution of *p*.

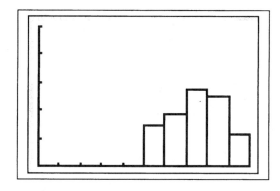
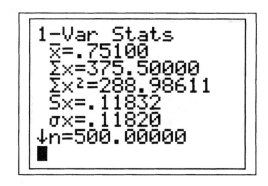

If your calculator does not have a *RandBin* built-in function, you can still simulate random sampling from a population by using a slightly modified program from the one used for the sampling distribution of the mean. Be forewarned, this slight modification involves some mathematics, but is fairly easy to implement.

The use of the same program as before capitalizes on the fact that the sample proportion is actually the mean of the 0's and 1's in the sample. The programming goal is to generate 12 random numbers between 0 and 1, and then convert them to 0's and 1's before calculating the mean in the program below. That's where the "slight modification" rears its ugly head! The hitch in the plan is to find a simple (for the calculator) function that will translate random numbers on the interval from 0 to 1 into 0's and 1's with the right probability. We will use two common functions to make this happen: the absolute value function, usually called "abs" on calculators, and the "greatest integer" function, usually named "Int" or possibly "floor." The mathematical function that will perform the translation is

$$Y = abs(int(rand - \pi)), \text{ where } \pi \text{ is the population proportion}$$

For our example, this would be

$Y = abs(int(rand - .75))$

To generate a random sample, substitute the expression for *rand* appropriate to your calculator in the *seq* function below:

For (X, 1, 500)
$seq(abs(int(rand - .75)), X, 1, 12) \rightarrow List1$
$mean(List1) \rightarrow List2(X)$
End

Before you use the program to study the sampling distribution of the sample proportion, be sure to make a couple trial runs and check the results with the theoretical expectations!

9

Estimation Using a Single Sample

INTRODUCTION

Despite protests from civil libertarians and gay rights activists, many people favor mandatory AIDS testing of certain at-risk groups, and some even believe that all citizens should be tested. What proportion of the adults in the United States favor mandatory testing for all citizens? For doctors, dentists, and other health care workers? For those convicted of a crime? To assess public opinion on this issue, researchers conducted a survey of 1014 adult U.S. citizens ("Large Majorities Continue to Back AIDS Testing," *Gallup Poll Monthly* (1991): 25–28). The researchers wanted to use the resulting information to estimate the true proportion in favor of testing various groups. The methods introduced in this chapter are used to produce the desired estimates. Since the estimates are based only on a sample rather than on a census of all U.S. adults, it is important that these estimates are constructed in a way that also conveys information about the anticipated accuracy.

The objective of inferential statistics is to use sample data to decrease our uncertainty about the corresponding population. Often, data is collected to obtain information that allows the investigator to estimate the value of some population characteristic, such as a population mean, μ, or a population proportion, π. One way to accomplish this uses the sample data to arrive at a single number that represents a plausible value for the characteristic of interest. Alternatively, an entire range of plausible values for the characteristic can be reported. These two estimation techniques, *point estimation* and *interval estimation,* are introduced in this chapter.

9.1 Point Estimation

The simplest approach to estimating a population characteristic involves using sample data to compute a single number that can be regarded as a plausible value of the characteristic. For example, sample data might suggest that 1000 hours is a plausible value for μ, the true mean lifetime for lightbulbs of a particular brand. In a different setting, a sample survey of students at a particular university might lead to the statement that .41 is a plausible value for π, the true proportion of students who favor a fee for recreational activities.

Definition

A **point estimate** of a population characteristic is a single number that is based on sample data and represents a plausible value of the characteristic.

In the examples just given, 1000 is a point estimate of μ, and .41 is a point estimate of π. The adjective *point* reflects the fact that the estimate corresponds to a single point on the number line.

A point estimate is obtained by first selecting an appropriate statistic. The estimate is then the value of the statistic for the given sample. For example, the computed value of the sample mean $\bar{x}$ provides a point estimate of a population mean μ.

EXAMPLE 9.1

One of the purposes of the survey on AIDS testing described in the chapter introduction was to estimate the proportion of the U.S. population who believe that every citizen should be tested for AIDS. The article reported that 466 of the 1014 people surveyed believed that all citizens should be tested. Let's use this information to estimate π, where

π = true proportion of all U.S. adults who favor AIDS testing of all citizens

With "success" identified as a person who favors testing, π is then just the population proportion of successes. The statistic

$$p = \frac{\text{number of successes in the sample}}{n}$$

which is the sample proportion of successes, is an obvious choice for obtaining a point estimate of π. Based on the reported information, the point estimate of π is

$$p = \frac{466}{1014} = .460$$

That is, based on the sample, we estimate that 46% of the adults in the United States favor AIDS testing of all citizens.

For purposes of estimating a population proportion π, it would be difficult to suggest a statistic other than p that could reasonably be used. In other situations, such as the one illustrated in Example 9.2, there may be several statistics that can be used to obtain an estimate.

EXAMPLE 9.2

The article "Effects of Roadside Conditions on Plants and Insects" (*J. Appl. Ecology* (1988): 709–715) reported the results of an experiment to evaluate the effect of nitrous oxide from automobile exhaust on roadside soil. The concentration of soluble nitrogen (in mg/g) was recorded at 20 roadside locations. Suppose that the observed readings are as follows (data is compatible with summary values given in the article):

| 1.7 | 1.9 | 2.0 | 1.7 | 1.6 | 1.7 | 1.8 | 1.7 | 1.8 | 1.8 |
| 1.6 | 1.7 | 2.1 | 1.7 | 1.8 | 1.7 | 1.8 | 1.4 | 1.9 | 1.7 |

A dotplot of the data is shown here:

Suppose further that a point estimate of μ, the true mean soluble nitrogen concentration, is desired. An obvious choice of a statistic for estimating μ is the sample mean, $\bar{x}$. However, there are other possibilities. We might consider using a trimmed mean or even the sample median, since the data set exhibits some symmetry. (If the corresponding population distribution is symmetric, μ is also the population median.)

The three statistics and the resulting estimates of μ calculated from the data are

$$\text{sample mean} = \bar{x} = \frac{\Sigma x}{n} = \frac{35.1}{20} = 1.76$$

$$\text{sample median} = \frac{1.7 + 1.7}{2} = 1.70$$

$$10\% \text{ trimmed mean} = \left(\begin{array}{c} \text{average of middle} \\ \text{16 observations} \end{array} \right) = \frac{28.0}{16} = 1.75$$

The estimates differ somewhat from one another. The choice from among them should depend on which statistic tends, on average, to produce an estimate closest to the true value. The following subsection discusses criteria for choosing among competing statistics.

Choosing a Statistic for Computing an Estimate

The point of the previous example is that there may be more than one statistic that can reasonably be used to obtain a point estimate of a specified population characteristic. Loosely speaking, the statistic used should be one that tends to yield an accurate estimate — that is, an estimate close to the value of the population characteristic. Information about the accuracy of estimation for a particular statistic is provided by the statistic's sampling distribution. Figure 9.1 displays the sampling distributions of three different statistics. The value of the population characteristic, which we refer to as the *true value,* is marked on the measurement axis.

The distribution pictured in Figure 9.1(a) is that of a statistic unlikely to yield an estimate close to the true value. The distribution is centered to the right of the true value, making it very likely that an estimate (a value of the statistic for a particular sample) will be larger than the true value. If this statistic is used to compute an estimate based on a first sample, then another estimate based on a second sample, and another estimate based on a third sample, and so on, the long-run average value of these estimates will exceed the true value.

The sampling distribution of Figure 9.1(b) is centered at the true value. Thus, while one estimate may be smaller than the true value and another may be larger, when this statistic is used many times over with different samples, there will be no long-run tendency to over- or underestimate the true value. Note that even though the sampling distribution is correctly centered, it spreads out quite a bit about the true value. Because of this, some estimates resulting from the use of this statistic will be far above or below the true value — even though there is no systematic tendency to underestimate or overestimate the true value. In contrast, the mean value of the statistic with the distribution shown in Figure 9.1(c) is exactly the true value of the population characteristic (implying no systematic error in estimation), and the statistic's standard deviation is relatively small. Estimates using this third statistic will almost always be quite close to the true value — certainly more often than estimates resulting from the statistic with the sampling distribution shown in Figure 9.1(b).

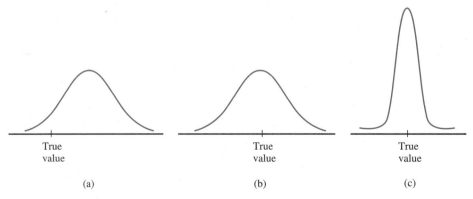

True value   True value   True value

(a)   (b)   (c)

FIGURE 9.1 Sampling distributions of three different statistics for estimating a population characteristic

> **Definition**
>
> A statistic with mean value equal to the value of the population characteristic being estimated is said to be an **unbiased statistic.** A statistic that is not unbiased is said to be **biased.**

Let $x_1, x_2, \ldots, x_n$ represent the values in a random sample. One of the general results concerning the sampling distribution of $\bar{x}$, the sample mean, is that $\mu_{\bar{x}} = \mu$. This result says that the $\bar{x}$ values from many different random samples of size n center around μ, the population mean. For example, if $\mu = 100$, the $\bar{x}$ distribution is centered at 100, whereas if $\mu = 5200$, then the $\bar{x}$ distribution is centered at 5200. Therefore, $\bar{x}$ is an unbiased statistic for estimating μ. Similarly, since $\mu_p = \pi$, the sampling distribution of p is centered at π, and it follows that p is an unbiased statistic for estimating a population proportion.

Using an unbiased statistic that has a small standard deviation guarantees that there will be no systematic tendency to under- or overestimate the value of the population characteristic *and* that estimates will almost always be relatively close to the true value.

> Given a choice between several unbiased statistics that could be used for estimating a population characteristic, the best statistic to use is the one with the smallest standard deviation.

Consider the problem of estimating a population mean, μ. The obvious choice of statistic for obtaining a point estimate of μ is the sample mean, $\bar{x}$, an unbiased statistic for this purpose. However, when the population distribution is symmetric, $\bar{x}$ is not the only choice. Other unbiased statistics for estimating μ in this case include the sample median and any trimmed mean (with the same number of observations trimmed from each end of the ordered sample). Which statistic should be used? The following facts may be helpful in making a choice.

1. If the population distribution is normal, then $\bar{x}$ has a smaller standard deviation than any other unbiased statistic for estimating μ. However, in this case, a trimmed mean with a small trimming percentage (such as 10%) performs almost as well as $\bar{x}$.

2. When the population distribution is symmetric with heavy tails compared to the normal curve, a trimmed mean is a better statistic than $\bar{x}$ for estimating μ.

When the population distribution is unquestionably normal, the choice is clear: Use $\bar{x}$ to estimate μ. But with a heavy-tailed distribution, a trimmed mean gives protection against one or two outliers in the sample that might otherwise have a large effect on the value of the estimate.

Now consider estimating another population characteristic, the population variance σ^2. The sample variance

$$s^2 = \frac{\Sigma\,(x - \bar{x})^2}{n - 1}$$

is a good choice for obtaining a point estimate of the population variance σ^2. It can be shown that s^2 is an unbiased statistic for estimating σ^2; that is, whatever the value of σ^2, the sampling distribution of s^2 is centered at that value. It is precisely for this reason — to obtain an unbiased statistic — that the divisor $n - 1$ is used. An alternative statistic is the average squared deviation

$$\frac{\Sigma\,(x - \bar{x})^2}{n}$$

which one might think has a more natural divisor than s^2. However, the average squared deviation is biased, with its values tending to be smaller, on average, than σ^2.

EXAMPLE 9.3

The article "Sensory and Mechanical Assessment of the Quality of Frankfurters" (*J. of Texture Studies* (1990): 395–409) reported the following salt content (percentage by weight) for ten frankfurters:

| 2.26 | 2.11 | 1.64 | 1.17 | 1.64 | 2.36 | 1.70 | 2.10 | 2.19 | 2.40 |

Then

$$\Sigma\,x = 19.57 \qquad \Sigma\,x^2 = 39.7335 \qquad n = 10$$

and

$$\Sigma\,(x - \bar{x})^2 = \Sigma\,x^2 - \frac{(\Sigma\,x)^2}{n}$$
$$= 39.7335 - \frac{(19.57)^2}{10}$$
$$= 1.435$$

Let σ^2 denote the true variance in salt content for frankfurters. Using the sample variance s^2 to provide a point estimate of σ^2 yields

$$s^2 = \frac{\Sigma\,(x - \bar{x})^2}{n - 1} = \frac{1.435}{9} = .1594$$

Using the average squared deviation (with divisor $n = 10$), the resulting point estimate is

$$\frac{\Sigma\,(x - \bar{x})^2}{n} = \frac{1.435}{10} = .1435$$

Because s^2 is an unbiased statistic for estimating σ^2, most statisticians would recommend using the point estimate .1594.

An obvious choice of a statistic for estimating the population standard deviation σ is the sample standard deviation s. For the data given in Example 9.3,

$$s = \sqrt{.1594} = .3992$$

Unfortunately, the fact that s^2 is an unbiased statistic for estimating σ^2 does not imply that s is an unbiased statistic for estimating σ. The sample standard deviation tends to underestimate slightly the true value of σ. However, unbiasedness is not the only criterion by which a statistic can be judged, and there are other good reasons for using s to estimate σ. In what follows, whenever we need to estimate σ based on a single random sample, we use the statistic s to obtain a point estimate.

Exercises 9.1 – 9.9

9.1 Three different statistics are being considered for estimating a population characteristic. The sampling distributions of the three statistics are shown in the illustration at the bottom of the page. Which statistic would you recommend? Explain your choice.

9.2 Why is an unbiased statistic generally preferred over a biased statistic for estimating a population characteristic? Does unbiasedness alone guarantee that the estimate will be close to the true value? Explain. Under what circumstances might you choose a biased statistic over an unbiased statistic if two statistics are available for estimating a population characteristic?

9.3 Radon, an odorless gas that can cause lung cancer, is released in the decay of radium, which is naturally present in soil and rocks. The Environmental Protection Agency (EPA) considers radon to be the second leading cause of lung cancer (behind cigarette smoking). One radon hot spot has been identified on the Spokane Indian

reservation in Washington. The Associated Press (June 10, 1991) reported that of 270 randomly selected homes on the reservation, 68 showed radon readings above 4 pCi (the level that the EPA considers hazardous). Estimate π, the proportion of all homes on the reservation with a radon level above 4 pCi.

9.4 The article "Diallel Analysis of Pod Length and Shelling Percent of Winged Beans" (Field Crops Research (1987): 209–216) reported the following lengths for nine winged bean pods:

12.1   18.5   17.2   27.7   17.9   12.9   17.0   15.0   14.5

a. Use the given data to produce a point estimate of μ, the true mean length for winged bean pods. (*Hint*: $\Sigma x = 152.8$)
b. Use the given data to produce a point estimate of σ^2, the variance of pod length for all winged bean pods. (*Hint*: $\Sigma x^2 = 2762.86$)

Figure for Exercise 9.1

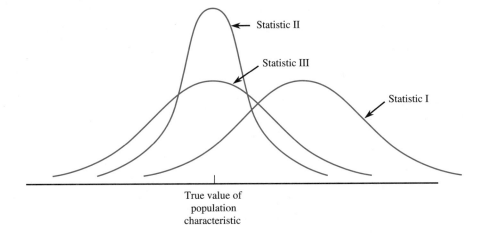

Statistic II

Statistic III

Statistic I

True value of
population
characteristic

c. Use the given data to produce an estimate of σ, the standard deviation of pod length. Is the statistic you used to produce your estimate unbiased?

9.5 A study reported in *Newsweek* (Dec. 23, 1991) involved a sample of 935 smokers. Each individual received a nicotine patch, which delivers nicotine to the bloodstream but at a much slower rate than cigarettes. Dosage was decreased to zero over a 12-week period. Suppose that 245 of the subjects were still not smoking 6 months after treatment (this figure is consistent with information given in the article). Estimate the percentage of all smokers who, when given this treatment, would refrain from smoking for at least 6 months.

9.6 The following data on gross efficiency (ratio of work accomplished per minute to calorie expenditure per minute) for trained endurance cyclists was given in the article "Cycling Efficiency Is Related to the Percentage of Type I Muscle Fibers" (*Medicine and Science in Sports and Exercise* (1992): 782–788).

18.3	18.9	19.0	20.9	21.4	20.5	20.1
20.1	20.8	20.5	19.9	20.5	20.6	22.1
21.9	21.2	20.5	22.6	22.6		

a. Assuming that the distribution of gross energy in the population of all endurance cyclists is normal, give a point estimate of μ, the population mean gross efficiency. (*Hint*: $\Sigma x = 392.4$)

b. Making no assumptions about the shape of the population distribution, estimate the proportion of all such cyclists whose gross efficiency is at most 20.

9.7 A random sample of $n = 12$ 4-year-old red pine trees was selected, and the diameter (in) of each tree's main stem was measured. The resulting observations are as follows:

11.3	10.7	12.4	15.2	10.1	12.1
16.2	10.5	11.4	11.0	10.7	12.0

a. Compute a point estimate of σ, the population standard deviation of main stem diameter. What statistic did you use to obtain your estimate?

b. Making no assumptions whatsoever about the shape of the population distribution of diameter, give a point estimate for the population median diameter (that is, for the middle diameter value in the entire population

of 4-year-old red pine trees). What statistic did you use to obtain the estimate?

c. Suppose that the population distribution of diameter is symmetric but with heavier tails than the normal distribution. Give a point estimate of the population mean diameter based on a statistic that gives some protection against the presence of outliers in the sample. What statistic did you use?

d. Suppose that the diameter distribution is normal. Then the 90th percentile of the diameter distribution is $\mu + 1.28\sigma$ (so 90% of all trees have diameters less than this value). Compute a point estimate for this percentile. (*Hint*: First compute an estimate of μ in this case; then use it along with your estimate of σ from part **a.**)

9.8 Each person in a random sample of 20 students at a particular university was asked whether he or she was registered to vote. The responses (R = registered, N = not registered) are given here:

R	R	N	R	N	N	R	R	R	N
R	R	R	R	R	N	R	R	R	N

Use this data to estimate π, the true proportion of all students at the university who were registered to vote.

9.9 a. A random sample of ten houses in a particular area, each of which is heated with natural gas, is selected, and the amount of gas (therms) used during the month of January is determined for each house. The resulting observations are as follows:

103	156	118	89	125	147	122	109	138	99

Let μ_J denote the average gas usage during January by all houses in this area. Compute a point estimate of μ_J.

b. Suppose that there are 10,000 houses in this area that use natural gas for heating. Let τ denote the total amount of gas used by all of these houses during January. Estimate τ using the data of part **a.** What statistic (sample-based quantity) did you use in computing your estimate?

c. Use the data in part **a** to estimate π, the proportion of all houses that used at least 100 therms.

d. Give a point estimate of the population median usage (the middle value in the population of all houses) based on the sample of part **a.** Which statistic did you use?

9.2 A Large-Sample Confidence Interval for a Population Proportion

In Section 9.1, we saw how to use a statistic to produce a point estimate of a population characteristic. The value of a point estimate depends on which sample, out of all the possible samples, happens to be selected. Different samples will usually yield different estimates due to chance differences from one sample to another. Because of sampling variability, rarely is the point estimate from a sample exactly equal to the true value of the population characteristic. We must hope that the chosen statistic produces an estimate that is close, on average, to the true value. Although a point estimate may represent our best single-number guess for the value of the population characteristic, it is not the only plausible value. These considerations suggest the need to indicate in some way how precisely the population characteristic has been estimated. A point estimate by itself does not provide this information.

As an alternative to a point estimate, suppose that we report not just a single credible value for the population characteristic, but an entire interval of reasonable values based on the sample data. For example, we might be confident that for all calls made from AT&T pay phones, the proportion, π, of calls that are billed to a credit card is in the interval from .53 to .57. The narrowness of this interval implies that we have rather precise information about the value of π. If, with the same high degree of confidence, we could state only that π was between .32 and .74, it would be clear that we had relatively imprecise knowledge of the value of π.

Definition

A **confidence interval** for a population characteristic is an interval of plausible values for the characteristic. It is constructed so that, with a chosen degree of confidence, the value of the characteristic will be captured inside the interval.

Associated with each confidence interval is a *confidence level*. The confidence level provides information on how much "confidence" we can have in the *method* used to construct the interval estimate, *not* our confidence in any one particular interval. Usual choices for confidence levels are 90%, 95%, and 99%, although other levels are also possible. If we were to construct a 95% confidence interval using the technique to be described shortly, we would be using a method that is "successful" 95% of the time. That is, if this method were used to generate an interval estimate over and over again with different samples, in the long run, 95% of the resulting intervals would capture the true value of the characteristic being estimated. Similarly, a 99% confidence interval is one that is constructed using a method that is, in the long run, successful in capturing the true value of the population characteristic 99% of the time.

Definition

The **confidence level** associated with a confidence interval estimate is the success rate of the *method* used to construct the interval.

Many factors influence the choice of confidence level. These factors will be discussed after we develop the method for constructing confidence intervals. We will first consider a large-sample confidence interval for a population proportion, π.

Often an investigator wishes to make an inference about the proportion of individuals or objects in a population that possesses a particular property of interest. For example, a university administrator might be interested in the proportion of students who prefer a new web-based computer registration system to the previous registration method. In a different setting, a quality control engineer might be concerned about the proportion of defective parts manufactured using a particular process.

Let

π = proportion of the population that possesses the property of interest

In this section, we consider the problem of estimating π using information in a random sample of size n from the population. Previously, we used the sample proportion

$$p = \frac{\text{number in sample that possesses property of interest}}{n}$$

to calculate a point estimate of π. We can also use p to form a confidence interval for π.

Although a small-sample confidence interval for π can be obtained, our focus will be on the large-sample case. The justification for the large-sample interval rests on properties of the sampling distribution of the statistic p:

1. The sampling distribution of p is centered at π; that is, $\mu_p = \pi$. Therefore, p is an unbiased statistic for estimating π.
2. The standard deviation of p is $\sigma_p = \sqrt{\pi(1 - \pi)/n}$.
3. As long as n is large ($n\pi \geq 10$ and $n(1 - \pi) \geq 10$), the sampling distribution of p is well approximated by a normal curve.

The accompanying box summarizes these properties.

When n is large, the statistic p has a sampling distribution that is approximately normal with mean π and standard deviation $\sqrt{\pi(1 - \pi)/n}$.

The development of a confidence interval for π is easier to understand if we select a particular confidence level (say, 95%). Appendix Table II, the table of standard normal (z) curve areas, can be used to determine a value z^* such that a central area of .95 falls between $-z^*$ and z^*. In this case, the remaining area of .05 is divided equally between the two tails, as shown in Figure 9.2. The total area to the left of the desired z^* is .975 (.95 central area + .025 area below $-z^*$). Locating .9750 in the body of Appendix Table II, we find that the corresponding z critical value is 1.96.

FIGURE 9.2 Capturing a central area of .95 under the z curve

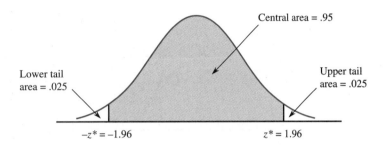

Central area = .95

Lower tail area = .025

Upper tail area = .025

$-z^* = -1.96$ $z^* = 1.96$

Generalizing this result to normal distributions other than the standard normal tells us that for *any* normal distribution, about 95% of the values are within 1.96 standard deviations of the mean. Since (for large samples) the sampling distribution of p is approximately normal with mean $\mu_p = \pi$ and standard deviation $\sigma_p = \sqrt{\pi(1 - \pi)/n}$, we get the following result.

Approximately 95% of all large samples will result in a value of p that is within $1.96\sigma_p = 1.96\sqrt{\pi(1 - \pi)/n}$ of the true population proportion π.

If p is within $1.96\sqrt{\pi(1 - \pi)/n}$ of π, the interval

$$p - 1.96\sqrt{\frac{\pi(1 - \pi)}{n}} \quad \text{to} \quad p + 1.96\sqrt{\frac{\pi(1 - \pi)}{n}}$$

will capture π (and this will happen for 95% of all possible samples). However, if p is farther away from π than $1.96\sqrt{\pi(1 - \pi)/n}$ (which will happen for about 5% of all possible samples), the interval will not include the true value of π. This is shown in Figure 9.3 on page 446.

Since p is within $1.96\sigma_p$ of π 95% of the time, this implies that in repeated sampling, 95% of the time the interval

$$p - 1.96\sqrt{\frac{\pi(1 - \pi)}{n}} \quad \text{to} \quad p + 1.96\sqrt{\frac{\pi(1 - \pi)}{n}}$$

FIGURE 9.3 The population proportion π is captured in the interval from $p - 1.96\sqrt{\pi(1 - \pi)/n}$ to $p + 1.96\sqrt{\pi(1 - \pi)/n}$ when p is within $1.96\sqrt{\pi(1 - \pi)/n}$ of π.

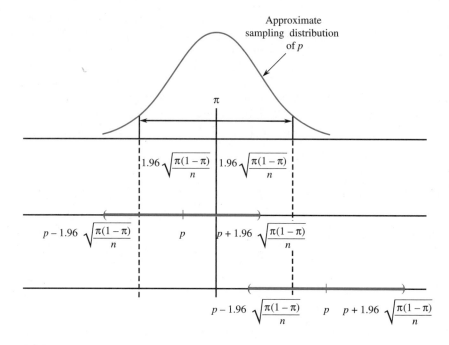

will contain π. Since π is unknown, $\sqrt{\pi(1 - \pi)/n}$ must be estimated. As long as the sample size is large, the value of $\sqrt{p(1 - p)/n}$ should be close to $\sqrt{\pi(1 - \pi)/n}$ and can be used in its place.

When n is large, a 95% confidence interval for π is

$$\left(p - 1.96\sqrt{\frac{p(1 - p)}{n}}, \quad p + 1.96\sqrt{\frac{p(1 - p)}{n}} \right)$$

An abbreviated formula for the interval is

$$p \pm 1.96\sqrt{\frac{p(1 - p)}{n}}$$

where $+$ gives the upper endpoint of the interval and $-$ gives the lower endpoint of the interval. The interval can be used as long as $np \geq 10$ and $n(1 - p) \geq 10$.

EXAMPLE 9.4   Let's return to the information gathered by the Gallup Organization in the survey on attitudes towards AIDS testing (see the chapter introduction and Example 9.1):

Total number of people surveyed:	1014
Number favoring testing of every citizen:	466
Number favoring testing of doctors and dentists:	882

A point estimate of π = true proportion who favor AIDS testing of all citizens is

$$p = \frac{466}{1014} = .460$$

Since

$$np = (1014)(.46) = 466 \geq 10 \quad \text{and} \quad n(1 - p) = (1014)(.54) = 548 \geq 10$$

the large-sample interval can be used. For a 95% confidence level, the appropriate z critical value is 1.96, and a 95% confidence interval for π is

$$
\begin{aligned}
p \pm 1.96\sqrt{\frac{p(1 - p)}{n}} &= .460 \pm 1.96\sqrt{\frac{(.460)(.540)}{1014}} \\
&= .460 \pm (1.96)(.016) \\
&= .460 \pm .031 \\
&= (.429, .491)
\end{aligned}
$$

Based on what was seen in this sample, we can be 95% confident that π, the true proportion favoring AIDS testing of all citizens, is between .429 and .491. That is, we used a *method* to construct this estimate that is successful in capturing the true value of π 95% of the time.

The calculations for a 95% confidence interval for the proportion who favor testing of doctors and dentists are

$$p = \frac{882}{1014} = .870$$

Since $np = 1014(.870) = 882 \geq 10$ and $n(1 - p) = 1014(.130) = 132 \geq 10$, the large-sample interval can be used. Then

$$
\begin{aligned}
p \pm 1.96\sqrt{\frac{p(1 - p)}{n}} &= .870 \pm 1.96\sqrt{\frac{(.870)(.130)}{1014}} \\
&= .870 \pm .021 \\
&= (.849, .891)
\end{aligned}
$$

Based on this interval, we conclude that between 84.9% and 89.1% of the population favors AIDS testing of doctors and dentists. A comparison of these intervals indicates that the sentiment for testing doctors and dentists is much stronger than that for testing the general public.

The second 95% confidence interval for π calculated in Example 9.4 is (.849, .891). It is tempting to say that there is a probability of .95 that π is between .849 and .891. **Do not yield to this temptation!** The 95% refers to the percentage of *all* possible samples resulting in an interval that includes π. In other words, if we take sample after sample from the population and use each one separately to compute a 95% confidence interval, in the long run roughly 95% of these intervals will capture π. Figure 9.4 on page 448 illustrates this for intervals generated from

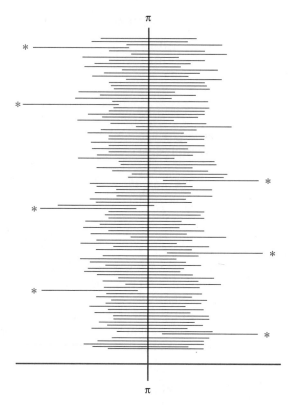

FIGURE 9.4 One hundred 95% confidence intervals for π computed from 100 different random samples (Asterisks identify intervals that do not include π.)

100 different random samples; 93 of the intervals include π, whereas 7 do not. Any specific interval, and our interval (.849, .891) in particular, either includes π or it does not (remember, the value of π is fixed but not known to us). We cannot make a chance (probability) statement concerning this particular interval. *The confidence level 95% refers to the **method** used to construct the interval rather than to any particular interval, such as the one we obtained.*

The formula given for a 95% confidence interval can easily be adapted for other confidence levels. The choice of a 95% confidence level led to the use of the z value 1.96 (chosen to achieve a central area of .95 under the standard normal curve) in the formula. Any other confidence level can be obtained by using an appropriate z critical value in place of 1.96. For example, suppose we wanted to achieve a confidence level of 99%. To obtain a central area of .99, the approximate z critical value would have a cumulative area (area to the left) of .995, as illustrated in Figure 9.5. From Appendix Table II, we find that the corresponding z critical value is 2.58. A 99% confidence interval for π is then obtained by using 2.58 in place of 1.96 in the formula for the 95% confidence interval.

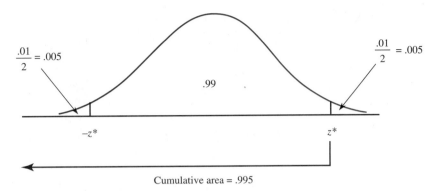

FIGURE 9.5 Finding the z critical value for a 99% confidence level

---

The Large-Sample Confidence Interval for π

The general formula for a confidence interval for a population proportion π when

1. p is the sample proportion from a **random sample,** and
2. the sample size **n is large** ($np \geq 10$ and $n(1 - p) \geq 10$)

is

$$p \pm (z \text{ critical value})\sqrt{\frac{p(1 - p)}{n}}$$

The desired confidence level determines which z critical value is used. The three most commonly used confidence levels, 90%, 95%, and 99%, use z critical values 1.645, 1.96, and 2.58, respectively.

Note: This interval is not appropriate for small samples. It is possible to construct a confidence interval in the small-sample case (based on the binomial distribution), but this is beyond the scope of this text.

Why settle for 95% confidence when 99% confidence is possible? Because the higher confidence level comes with a price tag. The resulting interval is wider than the 95% interval. The width of the 95% interval is $2(1.96\sqrt{p(1 - p)/n})$, whereas the 99% interval has width $2(2.58\sqrt{p(1 - p)/n})$. The higher *reliability* of the 99% interval (where "reliability" is specified by the confidence level) entails a loss in precision (as indicated by the wider interval). In the opinion of many investigators, a 95% interval gives a reasonable compromise between reliability and precision.

EXAMPLE 9.5   An Associated Press article on potential violent behavior reported the results of a survey of 750 workers who were employed full time (*San Luis Obispo Tribune,* Sept. 7, 1999). Of those surveyed, 125 indicated that they were so angered by a coworker during the past year that he or she felt like hitting the person (but didn't). Assuming that it is reasonable to regard this sample of 750 as a random sample from the population of full-time workers, we can use this information to construct an estimate π, the true proportion of full-time workers so angered in the last year that they wanted to hit a colleague.

For this sample

$$p = \frac{125}{750} = .167$$

Since $np = 125$ and $n(1 - p) = 625$ are both greater than or equal to 10, the sample size is large enough to use the formula for a large-sample confidence interval. A 90% confidence interval for π is then

$$p \pm (z \text{ critical value})\sqrt{\frac{p(1 - p)}{n}} = .167 \pm 1.645\sqrt{\frac{(.167)(.833)}{750}}$$
$$= .167 \pm (1.645)(.014)$$
$$= .167 \pm .022$$
$$= (.145, .189)$$

Based on this sample data, we can be 90% confident that the true proportion of full-time workers who have been angry enough in the last year to consider hitting a coworker is between .145 and .189. We have used a *method* to construct this interval estimate that has a 10% error rate.

The General Form of a Confidence Interval

Many confidence intervals have the same general form as the large-sample intervals for π just considered. We started with a statistic p, from which a point estimate for π was obtained. The standard deviation of this statistic is $\sqrt{\pi(1 - \pi)/n}$. This resulted in a confidence interval of the form

$$\begin{pmatrix} \text{point estimate using} \\ \text{a specified statistic} \end{pmatrix} \pm \begin{pmatrix} \text{critical} \\ \text{value} \end{pmatrix} \cdot \begin{pmatrix} \text{standard deviation} \\ \text{of the statistic} \end{pmatrix}$$

Since π was unknown, we estimated the standard deviation of the statistic by $\sqrt{p(1 - p)/n}$, which yielded the interval

$$\begin{pmatrix} \text{point estimate using} \\ \text{a specified statistic} \end{pmatrix} \pm \begin{pmatrix} \text{critical} \\ \text{value} \end{pmatrix} \cdot \begin{pmatrix} \text{estimated standard} \\ \text{deviation of the} \\ \text{statistic} \end{pmatrix}$$

For a population characteristic other than π, a statistic for estimating the characteristic will be selected. Then (drawing on statistical theory) a formula for the standard deviation of the statistic will be given. In practice, it will almost always be necessary to estimate this standard deviation (using something analogous to $\sqrt{p(1 - p)/n}$ rather than $\sqrt{\pi(1 - \pi)/n}$), so the second interval will be the prototype confidence interval. The estimated standard deviation of the statistic is usually referred to as the *standard error* in the statistical literature.

Definition

The **standard error** of a statistic is the estimated standard deviation of the statistic.

Choosing the Sample Size

The 95% confidence interval for π is based on the fact that, for approximately 95% of all random samples, p will be within $1.96\sqrt{\pi(1-\pi)/n}$ of π. The quantity $1.96\sqrt{\pi(1-\pi)/n}$ is sometimes called the *bound on the error of estimation* associated with a 95% confidence level — we have 95% confidence that the point estimate p will be no farther than this from π.

Definition

The **bound on error of estimation, B,** associated with a 95% confidence interval is
$(1.96) \cdot$ (standard error of the statistic).

Before collecting any data, an investigator may wish to determine a sample size for which a particular value of the bound is achieved. For example, with π representing the true proportion of students at a university who purchase textbooks over the Internet, the objective of an investigation may be to estimate π to within .05 with 95% confidence. The value of n necessary to achieve this is obtained by equating .05 to $1.96\sqrt{\pi(1-\pi)/n}$ and solving for n.

In general, suppose that we wish to estimate π to within an amount B (the specified bound on error of estimation) with 95% confidence. Finding the necessary sample size requires solving the equation

$$B = 1.96\sqrt{\frac{\pi(1-\pi)}{n}}$$

Solving this equation for n results in

$$n = \pi(1-\pi)\left(\frac{1.96}{B}\right)^2$$

Unfortunately, the use of this formula requires the value of π, which is unknown. One possible way to proceed is to carry out a preliminary study and use the resulting data to get a rough estimate of π. In other cases, prior knowledge may suggest a reasonable estimate of π. If there is no reasonable basis for estimating π and a preliminary study is not feasible, a conservative solution follows from the observation that $\pi(1 - \pi)$ is never larger than .25 (its value when $\pi = .5$). Replacing $\pi(1 - \pi)$ with .25, the maximum value, yields

$$n = .25\left(\frac{1.96}{B}\right)^2$$

Using this formula to obtain n gives us a sample size for which we can be 95% confident that p will be within B of π, no matter what the value of π.

The sample size required to estimate a population proportion π to within an amount B with 95% confidence is

$$n = \pi(1 - \pi)\left(\frac{1.96}{B}\right)^2$$

The value of π may be estimated using prior information. In the absence of any such information, using $\pi = .5$ in this formula gives a conservatively large value for the required sample size (this value of π gives a larger n than would any other value).

EXAMPLE 9.6

The 1991 publication of the book *Final Exit*, which includes chapters on doctor-assisted suicide, caused a great deal of controversy in the medical community. The Society for the Right to Die and the American Medical Association quoted very different figures regarding the proportion of primary-care physicians who have participated in some form of doctor-assisted suicide for terminally ill patients (*USA Today*, July 1991). Suppose that a survey of physicians is to be designed to estimate this proportion to within .05 with 95% confidence. How many primary-care physicians should be included in a random sample?

Using a conservative value of $\pi = .5$ in the formula for required sample size gives

$$n = \pi(1 - \pi)\left(\frac{1.96}{B}\right)^2 = .25\left(\frac{1.96}{.05}\right)^2 = 384.16$$

Thus, a sample of at least 385 doctors should be used.

Exercises 9.10 – 9.28

9.10 For each of the following, indicate which would result in a wider large-sample confidence interval for π.
  a. 90% confidence level or 95% confidence level
  b. $n = 100$ or $n = 400$

9.11 The formula used to compute a confidence interval for π is

$$p \pm (z \text{ critical value}) \cdot \sqrt{\frac{p(1 - p)}{n}}$$

What is the appropriate z critical value for each of the following confidence levels?
  a. 95%   **b.** 90%   **c.** 99%   **d.** 80%   **e.** 85%

9.12 The use of the interval

$$p \pm (z \text{ critical value}) \cdot \sqrt{\frac{p(1 - p)}{n}}$$

requires a large sample. For each of the following combinations of n and p, indicate whether the given interval would be appropriate.

a. $n = 50$ and $p = .30$ **b.** $n = 50$ and $p = .05$
c. $n = 15$ and $p = .45$ **d.** $n = 100$ and $p = .01$
e. $n = 100$ and $p = .70$ **f.** $n = 40$ and $p = .25$
g. $n = 60$ and $p = .25$ **h.** $n = 80$ and $p = .10$

9.13   Discuss how each of the following factors affects the width of the confidence interval for π.

   a. The confidence level
   b. The sample size
   c. The value of p

9.14   Retailers report that the use of cents-off coupons is increasing. The Scripps Howard News Service (July 9, 1991) reported the proportion of all households that use coupons as .77. Suppose that this estimate was based on a random sample of 800 households (that is, $n = 800$ and $p = .77$). Construct a 95% confidence interval for π, the true proportion of all households that use coupons.

9.15   The article "Ban on Smoking Shown to Improve Lung Health" (*Los Angeles Times,* Dec. 9, 1998) reported the results of a study of San Francisco bartenders. Of 39 bartenders who reported respiratory problems before a ban on smoking in bars, 21 were symptom-free 2 months after the ban began. Suppose that it is reasonable to regard this group of 39 bartenders as a random sample of bartenders with reported respiratory problems. Use this information to construct a 95% confidence interval for the proportion of bartenders with respiratory problems who were symptom-free 2 months after the smoking ban.

9.16   The *Princeton Metro Times* (Sept. 25, 1999) reported that 48% of a random sample of 369 students at The College of New Jersey indicated that they were "binge drinkers." Binge drinking was defined to be consuming five to six drinks in one sitting for men and four to five drinks for women. Construct and interpret a 90% confidence interval for π, the proportion of students at The College of New Jersey who are binge drinkers.

9.17   An article in the *Chicago Tribune* (Aug. 29, 1999) reported that in a poll of residents of the Chicago suburbs, 43% felt that their financial situation had improved during the last year. The following statement is from the article:

> The findings of this Tribune poll are based on interviews with 930 randomly selected suburban residents. The sample included suburban Cook County plus DuPage, Kane, Lake, McHenry and Will Counties. In a sample of this size, one can say with 95% certainty that results will differ by no more than 3 percent from results obtained if all residents had been included in the poll.

Comment on this statement. Give a statistical argument to justify the claim that the estimate of 43% is within 3% of the true proportion of residents who feel that their financial situation has improved.

9.18   According to a 1998 survey of 4000 randomly selected teachers, many full-time public school teachers feel inadequately prepared for various classroom situations. Of those surveyed, 21% indicated that they felt unprepared to address the needs of students with disabilities, and 20% stated that they felt unprepared to integrate technology into the grade or subject they taught.

   a. Construct a 99% confidence interval for the true proportion of teachers who feel unprepared to integrate technology into the classroom. Is your interval relatively narrow or wide? What aspect of this problem explains the interval width?
   b. Construct and interpret a 95% confidence interval for the proportion of teachers who feel unprepared to address the needs of students with disabilities.
   c. Since a higher percentage of the sample reported that they felt unprepared to address the needs of students with disabilities (21%) than reported that they felt unprepared to integrate technology (20%), is it reasonable to conclude that this proportion is also higher for the entire population of teachers? Use the results of parts **a** and **b** to justify your answer.

9.19   In an article titled "Fluoridation Brushed off by Utah" (Associated Press, Aug. 24, 1998), it was reported that a small but vocal minority in Utah has been successful in keeping fluoride out of Utah water supplies in spite of evidence that fluoridation reduces tooth decay and the fact that a clear majority of Utah residents favor fluoridation. To support this statement, the article included the result of a survey of Utah residents that found 65% to be in favor of fluoridation. Suppose that this result was based on a random sample of 150 Utah residents. Construct and interpret a 90% confidence interval for π, the true proportion of Utah residents who favor fluoridation. Is this interval consistent with the statement that fluoridation is favored by a clear majority of residents?

9.20   In a study of 1710 school children in Australia (*The Herald Sun,* Oct. 27, 1994), 1060 indicated that they normally watch TV before school in the morning. (Interestingly, only 35% of the parents said their children watched TV before school!). Construct a 95% confidence interval for the true proportion of Australian children who say they watch TV before school. What assumption about the sample must be true for the method used to construct the interval to be valid?

9.21 The *Chronicle of Higher Education* (Jan. 13, 1993) reported that 72.1% of those responding to a national survey of college freshmen were attending the college of their first choice. Suppose that $n = 500$ students responded to the survey (the actual sample size was much larger).

    a. Using the sample size $n = 500$, calculate a 99% confidence interval for the proportion of college students who are attending their first choice of college.

    b. Compute and interpret a 95% confidence interval for the proportion of students who are *not* attending their first choice of college.

    c. The actual sample size for this survey was much larger than 500. Would a confidence interval based on the actual sample size have been narrower or wider than the one computed in part **a**?

9.22 The Gallup Organization conducted a telephone survey on attitudes towards AIDS (*Gallup Monthly*, 1991). A total of 1014 individuals were contacted. Each was asked whether they agreed with the following statement: "Landlords should have the right to evict a tenant from an apartment because that person has AIDS." One hundred one individuals in the sample agreed with this statement. Use this data to construct a 90% confidence interval for the proportion who are in agreement with this statement. Give an interpretation of your interval.

9.23 Many Americans think it does not matter which political party controls Congress. In an Associated Press article (*San Luis Obispo Telegram-Tribune*, Sept. 6, 1995), it was reported that 442 individuals in a random sample of 1005 U.S. adults said it wouldn't make a difference which party is in power.

    a. Construct a 90% confidence interval for the proportion of U.S. adults who believe that it wouldn't make a difference which party is in power.

    b. Based on your interval in part **a**, is it plausible that a majority of U.S. adults feel that it makes no difference which party is in control? Explain your reasoning.

9.24 Recent high-profile legal cases have many people reevaluating the jury system. Many believe that juries in criminal trials should be able to convict on less than a unanimous vote. To assess support for this idea, each individual in a random sample of Californians was asked whether they favored allowing conviction by a 10–2 verdict in criminal cases not involving the death penalty. The

Associated Press (*San Luis Obispo Telegram-Tribune*, Sept. 13, 1995) reported that 71% supported the 10–2 verdict. Suppose that the sample size for this survey was $n = 900$. Compute and interpret a 99% confidence interval for the proportion of Californians who favor the 10–2 verdict.

9.25 The article "What Kinds of People Do Not Use Seat Belts?" (*Amer. J. of Public Health* (1977): 1043–1049) reported on a survey with the objective of studying characteristics of drivers and seat belt usage. Let π denote the proportion of all drivers of cars with seat belts who use them. Suppose that at the outset of the study, the investigators wished to estimate π to within an amount .02 with 95% confidence. What is the required sample size? Why is the recommended sample size so large?

9.26 The article "Television and Human Values: A Case for Cooperation" (*J. of Home Econ.* (Summer 1982): 18–23) reported that 48% of U.S. homes had more than one television set.

    a. Suppose that the statistic reported in this article was based on a random sample of 200 homes. Construct a 95% confidence interval for π, the proportion of all U.S. homes that have more than one television set.

    b. The same article reported that the average viewing time per household was about 6 hours per day. If you wanted to construct an interval estimate for the average television viewing time for all U.S. households, would you use the following interval? Explain your answer.

$$p \pm (z \text{ critical value}) \cdot \sqrt{\frac{p(1-p)}{n}}$$

9.27 A consumer group is interested in estimating the proportion of packages of ground beef sold at a particular store that have an actual fat content exceeding the fat content stated on the label. How many packages of ground beef should be tested to estimate this proportion to within .05 with 95% confidence?

9.28 A manufacturer of small appliances purchases plastic handles for coffeepots from an outside vendor. If a handle is cracked, it is considered defective and must be discarded. A very large shipment of plastic handles is received. The proportion of defective handles, π, is of interest. How many handles from the shipment should be inspected to estimate π to within .1 with 95% confidence?

9.3   A Confidence Interval for a Population Mean

The purpose of an investigation is often to estimate the population mean, μ. In this section, we consider how to use information from a random sample to construct a confidence interval estimate of a population mean.

We begin by considering the case where (1) σ, **the population standard deviation, is known** (not realistic, but we will see shortly how to handle the more common situation where σ is unknown) and (2) **the sample size, n, is large** enough for the Central Limit Theorem to apply. In this case, the following three properties about the sampling distribution of $\bar{x}$ hold:

1. The sampling distribution of $\bar{x}$ is centered at μ, so $\bar{x}$ is an unbiased statistic for estimating μ.
2. The standard deviation of $\bar{x}$ is $\sigma_{\bar{x}} = \sigma/\sqrt{n}$.
3. As long as n is large (generally $n \geq 30$), the sampling distribution of $\bar{x}$ is approximately normal, even when the population distribution itself is not normal.

The same reasoning that was used to develop the large-sample confidence interval for a population proportion π can be used to obtain a confidence interval estimate for μ.

The One-Sample z Confidence Interval for μ

The general formula for a confidence interval for a population mean μ when

1. $\bar{x}$ is the sample mean from a **random sample,**
2. the **sample size n is large** (generally $n \geq 30$), and
3. σ, **the population standard deviation, is known**

is

$$\bar{x} \pm (z \text{ critical value})\left(\frac{\sigma}{\sqrt{n}}\right)$$

EXAMPLE 9.7   Public Citizens, a consumer group, ranks the 111 U.S. nuclear reactors according to employee exposure to radiation. Based on a sample of workers at Diablo Canyon Nuclear Power Plant, Public Citizens rated Diablo's Unit 2 reactor as the 28th worst in the country. They reported (*San Luis Obispo Telegram-Tribune,* April 11, 1990) a mean annual radiation exposure of .481 rems for a sample of Unit 2 workers. Suppose that this mean was based on a random sample of 100 workers.

Let μ denote the true mean radiation exposure for Unit 2 workers at Diablo Canyon. Although σ, the true population standard deviation, is not usually known, suppose for illustrative purposes that $\sigma = .35$. Then a 95% confidence interval for μ is

$$\bar{x} \pm (z \text{ critical value})\left(\frac{\sigma}{\sqrt{n}}\right) = .481 \pm (1.96)\left(\frac{.35}{\sqrt{100}}\right)$$
$$= .481 \pm .069$$
$$= (.412, .550)$$

Based on this sample, *plausible* values of μ, the true mean annual radiation exposure for Diablo's Unit 2 workers, are between .412 and .550 rems. A 95% confidence level is associated with the method used to produce this interval estimate.

The confidence interval just introduced is appropriate when σ is known and n is large, and it can be used irrespective of the shape of the population distribution. This is because it is based on the Central Limit Theorem, which says that when n is sufficiently large, the sampling distribution of $\bar{x}$ is approximately normal for any population distribution. When n is small, the Central Limit Theorem cannot be used to justify the normality of the sampling distribution of $\bar{x}$, therefore the interval we developed previously should not be used. One way to proceed in the small-sample case is to make a specific assumption about the shape of the population distribution and then to use a method of estimation that is valid only under this assumption.

The one instance where this is easy to do is when it is reasonable to think that the population distribution is normal in shape. Recall that in the case of a normal population distribution, the sampling distribution of $\bar{x}$ is normal even for small sample sizes. So, if n is small but the population distribution is normal, the same confidence interval formula just introduced can still be used.

If n is small (generally $n < 30$) but it is reasonable to believe that the distribution of values in the population is normal, a confidence interval for μ (when σ is known) is

$$\bar{x} \pm (z \text{ critical value})\left(\frac{\sigma}{\sqrt{n}}\right)$$

Sample data can be used in several ways to assess the plausibility of normality. The two most common ways are to look at a normal probability plot of the sample data (looking for a plot that is reasonably straight) and to construct a boxplot of the data (looking for approximate symmetry and no outliers).

A Confidence Interval for μ When σ Is Unknown

The confidence intervals just developed have one major drawback: To compute the interval endpoints, σ must be known. Unfortunately, this is rarely the case. We now turn our attention to the situation when σ is unknown. The development of the confidence interval in this instance depends on the assumption that the population distribution is normal. Although this assumption is not critical if the sample size is large, it is important when the sample size is small.

To understand the derivation of this confidence interval, it is instructive to begin by taking another look at the 95% confidence interval. When σ is known, we know that $\mu_{\bar{x}} = \mu$ and $\sigma_{\bar{x}} = \sigma/\sqrt{n}$; also, when the population distribution is normal, the $\bar{x}$ distribution is normal. These facts imply that the standardized variable

$$z = \frac{\bar{x} - \mu}{\dfrac{\sigma}{\sqrt{n}}}$$

has approximately a standard normal distribution. Since the interval from -1.96 to 1.96 captures an area of .95 under the z curve, approximately 95% of all samples result in an $\bar{x}$ value that satisfies

$$-1.96 < \frac{\bar{x} - \mu}{\dfrac{\sigma}{\sqrt{n}}} < 1.96$$

Manipulating these inequalities to isolate μ in the middle results in the equivalent inequalities

$$\bar{x} - 1.96\left(\frac{\sigma}{\sqrt{n}}\right) < \mu < \bar{x} + 1.96\left(\frac{\sigma}{\sqrt{n}}\right)$$

The term $\bar{x} - 1.96(\sigma/\sqrt{n})$ is the lower endpoint of the 95% large-sample confidence interval for μ, and $\bar{x} + 1.96(\sigma/\sqrt{n})$ is the upper endpoint.

However, if σ is unknown, we must use the sample data to estimate σ. If we use the sample standard deviation as our estimate, the result is a different standardized variable denoted by t:

$$t = \frac{\bar{x} - \mu}{\dfrac{s}{\sqrt{n}}}$$

The value of s may not be all that close to σ, especially when n is small. As a consequence, the use of s in place of σ introduces extra variability, so the distribution of t is more spread out than the standard normal (z) curve. (The value of z will vary from sample to sample, because different samples generally result in different $\bar{x}$ values. There is even more variability in t, because different samples may result in different values of both $\bar{x}$ and s.)

To develop an appropriate confidence interval, we must know about the probability distribution of the standardized variable t for a sample from a normal population. This requires that we first learn about probability distributions called *t distributions*.

t Distributions

Just as there are many different normal distributions, there are also many different *t* distributions. Although normal distributions are distinguished from one another by their mean μ and standard deviation σ, the t distributions are distinguished by a positive whole number called the number of *degrees of freedom* (df). There is a t distribution with 1 df, another with 2 df, and so on.

Important Properties of t Distributions

1. The t curve corresponding to any fixed number of degrees of freedom is bell shaped and centered at zero (just like the standard normal (z) curve).

2. Each t curve is more spread out than the z curve.

3. As the number of degrees of freedom increases, the spread of the corresponding t curve decreases.

4. As the number of degrees of freedom increases, the corresponding sequence of t curves approaches the z curve.

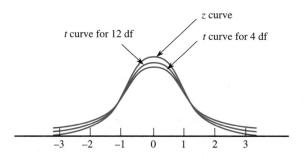

FIGURE 9.6 Comparison of the z curve and t curves for 12 df and 4 df

These properties are illustrated in Figure 9.6, which shows two t curves along with the z curve.

Appendix Table III gives selected critical values for various t distributions. The central areas for which values are tabulated are .80, .90, .95, .98, .99, .998, and .999. To find a particular critical value, go down the left margin of the table to the row labeled with the desired number of degrees of freedom. Then move over in that row to the column headed by the desired central area. For example, the value in the 12-df row under the column corresponding to central area .95 is 2.18, so 95% of the area under the t curve with 12 df lies between -2.18 and 2.18. Moving over two columns, we find the critical value for central area .99 (still with 12 df) to be 3.06 (see Figure 9.7). Moving down the .99 column to the 20-df row, we see the critical value is 2.85, so the area between -2.85 and 2.85 under the t curve with 20 df is .99. Notice that the critical values increase as you move to the right in each row. This is necessary to capture a larger central area and a smaller tail area. In each column, the critical values decrease as you move downward, reflecting decreasing spread for t distributions with larger degrees of freedom.

The larger the number of degrees of freedom, the more closely the t curve re-

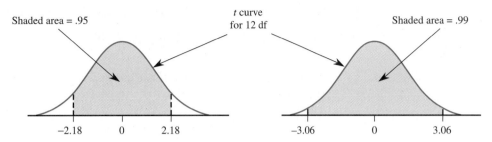

FIGURE 9.7 *t* critical values illustrated

sembles the z curve. To emphasize this, we have included the z critical values as the last row of the t table. Furthermore, once the number of degrees of freedom exceeds 30, the critical values change very little as the number of degrees of freedom increases. For this reason, Table III jumps from 30 df to 40 df, then to 60 df, then to 120 df, and finally to the row of z critical values. If we need a critical value for a number of degrees of freedom between those tabulated, we just use the critical value for the closest df. For df $>$ 120, we use the z critical values. Many graphing calculators calculate t critical values for any df; so, if you are using such a calculator, it will not be necessary to approximate the t critical values as described.

The One-Sample t Confidence Interval

The fact that the sampling distribution of $(\bar{x} - \mu)/(\sigma/\sqrt{n})$ is approximately the z (standard normal) distribution when n is large is what led to the z interval when σ is known. In the same way, the following proposition provides the key to obtaining a confidence interval when the population distribution is normal and σ is unknown.

Let $x_1, x_2, \ldots, x_n$ constitute a random sample from a normal population distribution. Then the probability distribution of the standardized variable

$$t = \frac{\bar{x} - \mu}{\dfrac{s}{\sqrt{n}}}$$

is the t distribution with $n - 1$ df.

To see how this result leads to the desired confidence interval, consider the case $n = 25$, so we use the t distribution with df $= 24$. From Table III, the interval between -2.06 and 2.06 captures a central area of .95 under the t curve with 24 df. Then 95% of all samples (with $n = 25$) from a normal population result in values of $\bar{x}$ and s for which

$$-2.06 < \frac{\bar{x} - \mu}{\dfrac{s}{\sqrt{n}}} < 2.06$$

Algebraically manipulating these inequalities to isolate μ yields

$$\bar{x} - 2.06\left(\frac{s}{\sqrt{25}}\right) < \mu < \bar{x} + 2.06\left(\frac{s}{\sqrt{25}}\right)$$

The 95% confidence interval for μ in this situation extends from the lower endpoint $\bar{x} - 2.06(s/\sqrt{25})$ to the upper endpoint $\bar{x} + 2.06(s/\sqrt{25})$. This interval can be written

$$\bar{x} \pm 2.06\left(\frac{s}{\sqrt{25}}\right)$$

The major difference between this interval and the interval when σ is known is the use of the t critical value 2.06 rather than the z critical value 1.96. The extra uncertainty that results from estimating σ causes the t interval to be wider than the z interval.

If the sample size is something other than 25 or if the desired confidence level is something other than 95%, a different t critical value (obtained from Table III) is used in place of 2.06.

The One-Sample t Confidence Interval for μ

The general formula for a confidence interval for a population mean μ based on a sample of size n when

1. $\bar{x}$ is the sample mean from a **random sample,**
2. the **population distribution is normal,** OR the **sample size n is large** (generally $n \geq 30$), and
3. σ, **the population standard deviation, is unknown**

is

$$\bar{x} \pm (t \text{ critical value})\left(\frac{s}{\sqrt{n}}\right)$$

where the t critical value is based on $n - 1$ df. Appendix Table III gives critical values appropriate for each of the confidence levels 90%, 95%, and 99%, as well as several other less frequently used confidence levels.

If n is large (generally $n \geq 30$), the normality of the population distribution is not critical. *However, this confidence interval is appropriate for small n only when the population distribution is (at least approximately) normal.* If this is not the case, as might be suggested by a normal probability plot or boxplot, another method should be used.

EXAMPLE 9.8   "Executives whose wives stay at home earn more" was the headline for an article that appeared in the *San Luis Obispo Telegram-Tribune* (Oct. 14, 1994). To support this claim, the article presented data from a random sample of 231 married men who received MBA degrees in the late 1970s. For this sample, the mean salary in 1993 for married men in two-income families was $95,140, whereas the mean for the married men who were the sole source of family income was $124,510. The article did not give the sample standard deviations, but suppose for purposes of this example that they were as given in the accompanying table.

	n	Mean Salary	Standard Deviation
Two-income family	140	95,140	15,000
Sole source of income	91	124,510	18,000

If we had access to the raw data (the $140 + 91 = 231$ salary observations), we might begin by looking at boxplots. Data consistent with the given summary quantities was used to generate the boxplots of Figure 9.8. The boxplots show that salaries for the one-income group tend to be higher than those for the two-income group. There is some overlap, and the smallest salary in each group is similar. The outliers in the boxplots indicate that the income distributions may not be normal, but because the sample sizes are large, we can still use the t confidence interval.

FIGURE 9.8 Boxplots for Example 9.8

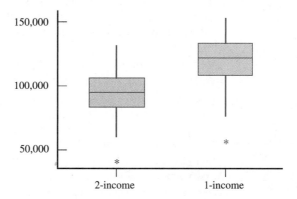

As a next step, we can use the confidence interval of this section to estimate the true mean salary of each group. Let's first focus on the sample of men from two-income families. For this group,

sample size $= n = 140$

sample mean salary $= \bar{x} = 95,140$

sample standard deviation $= s = 15,000$

The newspaper article generalized from this sample to the population of "executives whose wives work." Is this appropriate? Remember that everyone in the sample received an MBA degree in the late 1970s. A more cautious approach would be to consider this group as a sample from the population of married men who received MBAs in the late 1970s and whose wives also work. Then, with μ denoting the mean salary for this population, we can estimate μ using a 90% confidence interval.

From Appendix Table III, we will use t critical value = 1.645 (from the z critical value row since df = 139 > 120, the largest tabled df). The 90% confidence interval for μ is

$$\bar{x} \pm (t \text{ critical value})\left(\frac{s}{\sqrt{n}}\right) = 95{,}140 \pm (1.645)\left(\frac{15{,}000}{\sqrt{140}}\right)$$

$$= 95{,}140 \pm 2085.42$$

$$= (\$93{,}054.58, \$97{,}225.42)$$

Based on this sample, we are confident that μ is between \$93,054.58 and \$97,225.42. This interval is fairly narrow given the context of this problem; the interval spans a range of only \$4200. This indicates that our information about the value of μ is relatively precise.

A 90% confidence interval estimate for the mean of the population of married men who received MBAs in the late 1970s and who are the sole source of family income is (\$121,406.03, \$127,613.97). This interval is wider than the first interval for two reasons: The sample size is smaller (91 rather than 140), and the sample standard deviation is larger (18,000 rather than 15,000).

Based on these two interval estimates, it does appear that the mean salary for the two-income family group is lower than the mean for the one-income group. However, this analysis does not prove the existence of a cause-and-effect relationship between wife's work status and increased salary. The cited article falls into this trap by speculating that "the upper ranks of corporate America favor men whose wives stay at home" and that stay-at-home wives "actually further their husbands careers." Can you see the flaw in this argument?

EXAMPLE 9.9

Misconceptions about everyday scientific phenomena, such as the seasons, gravity, and light, are common in the general population. To assess the degree to which university students about to receive a middle school teaching credential shared these misconceptions, a science quiz consisting of 10 multiple-choice questions was given to a random sample of 52 such students ("Assessing Selected Physical Science and Earth Science Misconceptions of Middle School Through University Preservice Teachers," *J. College Science Teaching* (1994): 38–42). Let x denote the number of correct responses (out of 10), and let μ denote the mean value of x in the population of students about to receive a middle school teaching credential. For this sample of $n = 52$ students,

sample mean number of correct responses = $\bar{x}$ = 5.14

sample standard deviation = s = 1.30

(For the purpose of this example, we approximated the standard deviation, but it is consistent with information given in the article.) A histogram for sample data consistent with the given summary quantities is shown in Figure 9.9. Even though the histogram suggests that the corresponding population distribution may be positively skewed, we can use the t confidence interval formula, because the sample size is large.

FIGURE 9.9 Histogram for Example 9.9

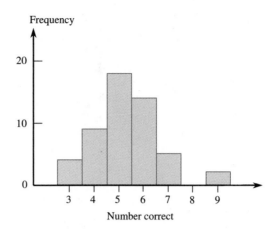

Since df = 51, we will use the *t* critical value for df = 60 (the closest df value in Appendix Table III). A 95% confidence interval for μ is then

$$\bar{x} \pm (t \text{ critical value})\left(\frac{s}{\sqrt{n}}\right) = 5.14 \pm (2.00)\left(\frac{1.30}{\sqrt{52}}\right)$$
$$= 5.14 \pm .36$$
$$= (4.78, 5.50)$$

Based on this sample, we believe that the mean number of correct responses for preservice middle school teachers is somewhere between 4.78 and 5.50.

EXAMPLE 9.10

A study of the ability of individuals to walk in a straight line ("Can We Really Walk Straight?" *Amer. J. of Physical Anthropology* (1992): 19–27) reported the accompanying data on *cadence* (strides per second) for a sample of *n* = 20 randomly selected healthy men.

.95	.85	.92	.95	.93	.86	1.00	.92	.85	.81
.78	.93	.93	1.05	.93	1.06	1.06	.96	.81	.96

Figure 9.10 (page 464) is a normal probability plot of this data. The plot is reasonably straight, so it seems plausible that the population distribution is approximately normal. Calculation of a confidence interval for the population mean cadence requires $\bar{x}$ and s:

$$\bar{x} = .926 \qquad s^2 = .006552 \qquad s = \sqrt{.006552} = .0809$$

The *t* critical value for a 99% confidence interval based on 19 df is 2.86. The interval is

$$\bar{x} \pm (t \text{ critical value})\left(\frac{s}{\sqrt{n}}\right) = .926 \pm (2.86)\left(\frac{.0809}{\sqrt{20}}\right)$$
$$= .926 \pm .052$$
$$= (.874, .978)$$

FIGURE 9.10 Normal probability plot for Example 9.10

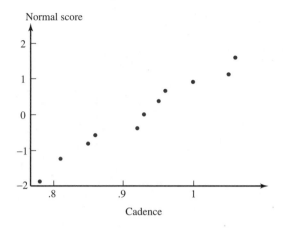

With 99% confidence, we estimate the population mean cadence to be between .874 and .978 strides per second. Remember that the 99% confidence level implies that if the same formula is used to calculate intervals for sample after sample randomly selected from the population, in the long run 99% of these intervals will capture μ between the lower and upper confidence limits.

Any of the commercially available statistical computing packages will produce a one-sample t confidence interval upon request. The following is an example of output from MINITAB. The slight discrepancy between the previous interval and MINITAB's interval occurs because MINITAB uses more decimal accuracy in $\bar{x}$, s, and t critical values.

	N	MEAN	STDEV	SE MEAN	99.0 PERCENT C.I.
CADENCE	20	0.9255	0.0809	0.0181	(0.8737, 0.9773)

EXAMPLE 9.11

How much time do school-age children spend helping with housework? The article "The Three Corners of Domestic Labor: Mothers', Fathers', and Children's Week-day and Weekend Housework" (*J. Marriage and the Family* (1994): 657–668) gave information on the number of minutes per weekday spent on housework. The mean and standard deviation in the accompanying table are for a random sample of girls in two-parent families where both parents work full-time.

Sample Size	Mean Time Spent per Weekday (minutes)	Standard Deviation
26	14.0	8.6

The authors of the article analyzed this data using methods designed for population distributions that are approximately normal. This assumption appears a bit questionable based on the reported mean and standard deviation (it is impossible to spend less than 0 minutes per day on housework, so the smallest possible value, 0, is only 1.63 standard deviations below the mean). However, since the authors reported that there were no outliers in the data and since n is relatively close to 30, we will use the t confidence interval formula of this section to compute a 95% confidence interval.

Since $n = 26$, df $= 25$, and the appropriate t critical value is 2.06. The confidence interval is then

$$\bar{x} \pm (t \text{ critical value})\left(\frac{s}{\sqrt{n}}\right) = 14.0 \pm (2.06)\left(\frac{8.6}{\sqrt{26}}\right)$$
$$= 14.0 \pm 3.5$$
$$= (10.5, 17.5)$$

Based on the sample data, we believe that the true mean time per weekday spent on housework is between 10.5 minutes and 17.5 minutes for girls in two-parent families where both parents work. We used a method that has a 5% error rate to construct this interval. We should be somewhat cautious in interpreting this confidence interval because of the concern expressed about the normality of the population distribution.

Choosing the Sample Size

When estimating μ using a large sample or a small sample from a normal population, the bound on error of estimation, B, associated with a 95% confidence interval is

$$B = 1.96\left(\frac{\sigma}{\sqrt{n}}\right)$$

Before collecting any data, an investigator may wish to determine a sample size for which a particular value of the bound is achieved. For example, with μ representing the average fuel efficiency (mpg) for all cars of a certain type, the objective of an investigation may be to estimate μ to within 1 mpg with 95% confidence. The value of n necessary to achieve this is obtained by equating 1 to $1.96(\sigma/\sqrt{n})$ and solving for n.

In general, suppose that we wish to estimate μ to within an amount B (the specified bound on error of estimation) with 95% confidence. Finding the necessary sample size requires solving the equation $B = 1.96(\sigma/\sqrt{n})$ for n. The result is

$$n = \left[\frac{1.96\sigma}{B}\right]^2$$

Notice that a large value of σ forces n to be large, as does a small value of B. Use of the sample-size formula requires that σ be known, but this is rarely the case. One possibility is to carry out a preliminary study and use the resulting sample standard deviation (or a somewhat larger value, to be conservative) to determine n for the main part of the study. Another possibility is simply to make an educated guess about the value of σ and use it in calculating n. For a population distribution that is not too skewed, dividing the range (the difference between the largest and smallest values) by 4 often gives a rough idea of the value of the standard deviation.

The sample size required to estimate a population mean μ to within an amount B with 95% confidence is

$$n = \left[\frac{1.96\sigma}{B}\right]^2$$

If σ is unknown, it may be estimated based on previous information or, for a population that is not too skewed, by using (range)/4.

If the desired confidence level is something other than 95%, 1.96 is replaced by the appropriate z critical value (e.g., 2.58 for 99% confidence).

EXAMPLE 9.12

The financial aid office wishes to estimate the mean cost of textbooks per quarter for students at a particular university. For the estimate to be useful, it should be within $20 of the true population mean. How large a sample should be used to be 95% confident of achieving this level of accuracy?

To determine the required sample size, we must have a value for σ. The financial aid office is pretty sure that the amount spent on books varies widely, with most values between $50 and $450. A reasonable estimate of σ is then

$$\frac{\text{range}}{4} = \frac{450 - 50}{4} = \frac{400}{4} = 100$$

The required sample size is

$$n = \left[\frac{196\sigma}{B}\right]^2 = \left[\frac{(1.96)(100)}{20}\right]^2 = [9.8]^2 = 96.04$$

Rounding up, a sample size of 97 or larger is recommended.

Exercises 9.29 – 9.49

9.29 Given a variable that has a t distribution with the specified degrees of freedom, what percentage of the time will its value fall in the indicated region?
  a. 10 df, between -1.81 and 1.81
  b. 10 df, between -2.23 and 2.23
  c. 24 df, between -2.06 and 2.06
  d. 24 df, between -2.80 and 2.80
  e. 24 df, outside the interval from -2.80 to 2.80
  f. 24 df, to the right of 2.80
  g. 10 df, to the left of -1.81

9.30 The formula used to compute a confidence interval for the mean of a normal population when n is small is

$$\bar{x} \pm (t \text{ critical value}) \cdot \frac{s}{\sqrt{n}}$$

What is the appropriate t critical value for each of the following confidence levels and sample sizes?
  a. 95% confidence, $n = 17$
  b. 90% confidence, $n = 12$
  c. 99% confidence, $n = 24$
  d. 90% confidence, $n = 25$
  e. 90% confidence, $n = 13$
  f. 95% confidence, $n = 10$

9.31 Each of the following is a confidence interval for μ = true average resonance frequency (in Hz) for all tennis rackets of a certain type: (114.4, 115.6) and (114.1, 115.9).
  a. What is the value of the sample mean resonance frequency?

b. The confidence level for one of these intervals is 90% and for the other it is 99%. Which is which, and why?

9.32 Samples of two different types of automobiles were selected, and the actual speed for each car was determined when the speedometer registered 50 mph. The resulting 95% confidence intervals for true average actual speed were (51.3, 52.7) and (49.4, 50.6). Assuming that the two sample standard deviations were identical, which confidence interval is based on the larger sample size? Explain your reasoning.

9.33 Suppose that a random sample of 50 bottles of a particular brand of cough medicine is selected and the alcohol content of each bottle is determined. Let μ denote the average alcohol content for the population of all bottles of the brand under study. Suppose that the sample of 50 results in a 95% confidence interval for μ of (7.8, 9.4).

a. Would a 90% confidence interval have been narrower or wider than the given interval? Explain your answer.

b. Consider the following statement: There is a 95% chance that μ is between 7.8 and 9.4. Is this statement correct? Why or why not?

c. Consider the following statement: If the process of selecting a sample of size 50 and then computing the corresponding 95% confidence interval is repeated 100 times, 95 of the resulting intervals will include μ. Is this statement correct? Why or why not?

9.34 The McClatchy News Service (*San Luis Obispo Telegram-Tribune,* June 13, 1991) reported on a sample of prime-time television hours. The following table summarizes the information reported for four networks.

Network	Mean Number of Violent Acts per Hour
ABC	15.6
CBS	11.9
FOX	11.7
NBC	11.0

Suppose that each of these sample means was computed on the basis of viewing $n = 50$ randomly selected prime-time hours and that the population standard deviation for each of the four networks is known to be $\sigma = 5$.

a. Compute a 95% confidence interval for μ_{ABC}, the true mean number of violent acts per prime-time hour for ABC.

b. Compute 95% confidence intervals for the mean number of violent acts per prime-time hour for each of the other three networks.

c. The National Coalition on Television Violence claims that shows on ABC are more violent than on the other networks. Based on the confidence intervals from parts **a** and **b,** do you agree with this conclusion? Explain.

9.35 Seventy-seven students at the University of Virginia were asked to keep a diary of a conversation with their mothers, recording any lies they told during these conversations (*San Luis Obispo Telegram-Tribune,* August 16, 1995). It was reported that the mean number of lies per conversation was .5. Suppose that the standard deviation (which was not reported) was .4.

a. Suppose that this group of 77 is a random sample from the population of students at this university. Construct a 95% confidence interval for the mean number of lies per conversation for this population.

b. The interval in part **a** does not include 0. Does this imply that all students lie to their mothers? Explain.

9.36 The article "Selected Characteristics of High-Risk Students and their Enrollment Persistence" (*J. of College Student Development* (1994): 54–60) examined factors that affect whether students stay in college. The accompanying summary statistics are based on data from a sample of 44 students who did not return to college after the first quarter (the nonpersisters) and a sample of 257 students who did return (the persisters).

	Number of Hours Worked per Week During the First Quarter	
	Mean	**Standard Deviation**
Nonpersisters	25.62	14.41
Persisters	18.10	15.31

a. Consider the 44 nonpersisters as a random sample from the population of all nonpersisters at the university where the data was collected. Compute a 98% confidence interval for the mean number of hours worked per week for nonpersisters.

b. Consider the 257 persisters as a random sample from the population of all persisters at the university where the data was collected. Compute a 98% confidence interval for the mean number of hours worked per week for persisters.

c. The 98% confidence interval for persisters is narrower than the corresponding interval for nonpersisters even though the standard deviation for persisters is larger than that for nonpersisters. Explain why this happened.

d. Based on the interval in part **a,** do you think that the mean number of hours worked per week for nonpersisters is greater than 20? Explain.

9.37 The article "Financial Characteristics of Leveraged Buyouts" (*J. of Business Research* (1992): 241–252) reported summary data covering the period 1986–1988 for random samples of two different types of companies. Consider the accompanying data on net income per share of common stock. (*Note:* Standard error = $s/\sqrt{n}$)

Type of Firm	Sample Size	Sample Mean	Standard Error of Mean
Involved in leveraged buyout	45	.6214	.4000
Not involved	297	.9504	.2076

a. Calculate confidence intervals at the 95% confidence level for true average earnings per share both for leveraged buyout (LBO) firms and for non-LBO firms.
b. Why is the non-LBO interval narrower than the LBO interval?

9.38 Increases in worker injuries and disability claims have prompted renewed interest in workplace design and regulation. As one particular aspect of this, employees required to do regular lifting should not have to handle unsafe loads. The article "Anthropometric, Muscle Strength, and Spinal Mobility Characteristics as Predictors of the Rating of Acceptable Loads in Parcel Sorting" (*Ergonomics* (1992): 1033–1044) reported on a study involving a random sample of $n = 18$ male postal workers. The sample mean rating of acceptable load attained with a work-simulating test was found to be $\bar{x} = 9.7$ kg, and the sample standard deviation was $s = 4.3$ kg. Suppose that in the population of all male postal workers, the distribution of rating of acceptable load can be modeled approximately using a normal distribution with mean value μ. Construct and interpret a 95% confidence interval for μ.

9.39 A number of cities have implemented needle exchange programs in an attempt to slow the spread of AIDS among intravenous drug users. These programs are designed to remove infected needles from circulation. One such program in New Haven kept information on the length of time a needle remained in circulation. It was reported that the mean circulation time (in days) was 2.2 ("What Happened to HIV Transmission Among Drug Injectors in New Haven?" *Chance* (Spring 1993)). The sample size and standard deviation for circulation time were not given in the article, but suppose that they were $n = 25$ needles and $s = 1.2$ days. Compute a 90% confidence interval for μ, the mean circulation time for needles recovered by this exchange program.

9.40 The eating habits of 12 bats were examined in the article "Foraging Behavior of the Indian False Vampire Bat" (*Biotropica* (1991): 63–67). These bats consume insects and frogs. For these 12 bats, the mean time to consume a frog was $\bar{x} = 21.9$ min. Suppose the standard deviation was $s = 7.7$ min. Construct and interpret a 90% confidence interval for the mean suppertime of a vampire bat whose meal consists of a frog. What assumptions must be reasonable for the one-sample t interval to be appropriate?

9.41 Fat content (in percentage) for ten randomly selected hot dogs was given in the article "Sensory and Mechanical Assessment of the Quality of Frankfurters" (*J. of Texture Studies* (1990): 395–409). Use the given data to construct a 90% confidence interval for the true mean fat percentage of hot dogs.

25.2 21.3 22.8 17.0 29.8 21.0 25.5 16.0 20.9 19.5

9.42 A triathlon consisting of swimming, cycling, and running is one of the more strenuous amateur sporting events. The article "Cardiovascular and Thermal Response of Triathlon Performance" (*Medicine and Science in Sports and Exercise* (1988): 385–389) reported on a research study involving nine randomly selected male triathletes. Maximum heart rate (beats/min) was recorded while performing each of the three events.

	$\bar{x}$	s
Swimming	188	7.2
Biking	186	8.5
Running	194	7.8

a. Assuming that the heart-rate distribution for each event is approximately normal, construct 95% confidence intervals for the true mean heart rate of triathletes for each event.
b. Do the intervals in part **a** overlap? Based on the computed intervals, do you think there is evidence that the mean maximum heart rate is higher for running than for the other two events? Explain.

9.43 Five students visiting the student health center for a free dental examination during National Dental Hygiene Month were asked how many months had passed since their last visit to a dentist. Their responses were as follows:

6   17   11   22   29

Assuming that these five students can be considered to be a random sample of all students participating in the free checkup program, construct a 95% confidence interval for the mean number of months elapsed since the last visit to a dentist for the population of students participating in the program.

9.44 The article "First Year Academic Success: A Prediction Combining Cognitive and Psychosocial Variables for Caucasian and African American Students" (*J. College Student Devel.* (1999): 599) reported that the sample mean and standard deviation for high school grade point average for students enrolled at a large research university were 3.73 and .45, respectively. Suppose that the mean and standard deviaition were based on a random sample of 900 students at the university.

a. Construct a 95% confidence interval for the mean high school GPA for students at this university.

b. Suppose you wanted to make a statement about the range of GPAs for students at this university. Is it reasonable to say that 95% of the students at the university have GPAs in the interval you computed in part **a**? Explain.

9.45 The following data are the calories per $\frac{1}{2}$-cup serving for 17 popular chocolate ice cream brands reviewed by *Consumer Reports* (July 1999).

| 270 | 150 | 170 | 140 | 160 | 160 | 160 | 290 |
| 190 | 190 | 160 | 170 | 150 | 110 | 180 | 170 |

Is it reasonable to use the *t* confidence interval of this section to compute a confidence interval for μ, the true mean calories per $\frac{1}{2}$-cup serving of chocolate ice cream? Explain why or why not.

9.46 The Bureau of Alcohol, Tobacco, and Firearms (BATF) has been concerned about lead levels in California wines. In a previous testing of wine specimens, lead levels ranging from 50 to 700 parts per billion were recorded (*San Luis Obispo Telegram-Tribune,* June 11, 1991). How many wine specimens should be tested if BATF wishes to estimate the true mean lead level for California wines to within 10 parts per billion with 95% confidence?

9.47 A manufacturer of college textbooks is interested in estimating the strength of the bindings produced by a particular binding machine. Strength can be measured by recording the force required to pull the pages from the binding. If this force is measured in pounds, how many books should be tested to estimate with 95% confidence to within .1 lb the average force required to break the binding? Assume that σ is known to be .8 lb.

9.48 The article "*National Geographic,* the Doomsday Machine," which appeared in the *Journal of Irreproducible Results* (yes, there really is a journal by that name — it's a spoof of technical journals!) predicted dire consequences resulting from a nationwide buildup of *National Geographic.* The author's predictions are based on the observation that the number of subscriptions for *National Geographic* is on the rise and that no one ever throws away a copy of *National Geographic.* A key to the analysis presented in the article is the weight of an issue of the magazine. Suppose that you were assigned the task of estimating the average weight of an issue of *National Geographic.* How many issues should you sample to estimate the average weight to within .1 oz with 95% confidence? Assume that σ is known to be 1 oz.

9.49 The formula described in this section for determining sample size corresponds to a confidence level of 95%. What would be the appropriate formula for determining sample size when the desired confidence level is 90%? 98%?

9.4   Interpreting the Results of Statistical Analyses

The purpose of most surveys and many research studies is to produce estimates of population characteristics. Unfortunately, there is no customary way of reporting the resulting estimates. Possibilities include

confidence interval

estimate $\pm$ bound on error

estimate $\pm$ standard error

and, if the population characteristic is a population mean,

mean $\pm$ sample standard deviation

If the interval reported is described as a confidence interval, it should be accompanied by a confidence level. These intervals can be interpreted just as we have interpreted the confidence intervals in this chapter, and the confidence level specifies the long-run error rate associated with the method used to construct the interval (for example, a 95% confidence level specifies a 5% long-run error rate).

A form particularly common in news articles is estimate ± bound on error, where bound on error is also sometimes called **margin of error.** The bound on error reported is usually two times the standard deviation of the estimate. This method of reporting is a little more informal than a confidence interval and, if the sample size is reasonably large, is roughly equivalent to reporting a 95% confidence interval. You can interpret these intervals like a confidence interval with approximate confidence level of 95%.

You must use care in interpreting intervals reported in the form

estimate ± standard error

Recall from Section 9.2 that the general form of a confidence interval is

estimate ± (critical value)(standard deviation of the estimate)

In journal articles, the estimated standard deviation of the estimate is usually referred to as the *standard error.* The critical value in the confidence interval formula was determined by the form of the sampling distribution of the estimate and by the confidence level. Note that the reported form, estimate ± standard error, is equivalent to a confidence interval with the critical value set equal to 1. For an estimate whose sampling distribution is (approximately) normal (such as the mean of a large sample or a large-sample proportion), a critical value of 1 corresponds to an approximate confidence level of 68%. Since a confidence level of 68% is rather low, you may want to use the given information and the confidence interval formula to convert to an interval with a higher confidence level.

When researchers are trying to estimate a population mean, they sometimes report

sample mean ± sample standard deviation

Be particularly careful here. To convert this information into a useful interval estimate of the population mean, you must first convert the sample standard deviation to the standard error of the sample mean (by dividing by $\sqrt{n}$) and then use the standard error and an appropriate critical value to construct a confidence interval.

For example, suppose that a random sample of size 100 is to be used to estimate the population mean. If the sample resulted in a sample mean of 500 and a sample standard deviation of 20, you might find the published results summarized in any of the following ways:

95% confidence interval for the population mean: (496.08, 503.92)

mean ± bound on error: 500 ± 4

mean ± standard error: 500 ± 2

mean ± standard deviation: 500 ± 20

What to Look for in Published Data

Here are some questions to ask when you encounter interval estimates in research reports.

- Is the reported interval a confidence interval, mean $\pm$ bound on error, mean $\pm$ standard error, or mean $\pm$ standard deviation? If the reported interval is not a confidence interval, you may want to construct a confidence interval from the given information.

- What confidence level is associated with the given interval? Is the choice of confidence level reasonable? What does the confidence level say about the long-run error rate of the method used to construct the interval?

- Is the reported interval relatively narrow or relatively wide? Has the population characteristic been estimated precisely?

For example, the article "Increased Vital and Total Lung Capacity in Tibetan Compared to Han Residents of Lhasa" (*Amer. J. of Physical Anthro.* (1991): 341–351) compared various physical characteristics of people who live at high altitudes to those of people who live a sea level. The article includes the following statements:

> We studied 38 Tibetan and 43 Han residents. . . . The Tibetan compared with the Han subjects had a larger total lung capacity [6.80 $\pm$ 0.19 (mean $\pm$ SEM) vs. 6.24 $\pm$ 0.18 liters].

The reported intervals are of the form estimate $\pm$ standard error. We can use this information to construct a confidence interval for the mean total lung capacity for residents of each of the two locations. Since the sample sizes are both large, we can use the t confidence interval formula

mean $\pm$ (t critical value)(standard deviation of the mean)

The Tibetan sample has df = 38 − 1 = 37 and the Han sample has df = 43 − 1 = 42. For both of these df's, the closest df value in Appendix Table III is for df = 40, and the corresponding t critical value for a 95% confidence level is 2.02. The corresponding intervals are

Tibetan residents: $6.80 \pm 2.02(0.19) = 6.80 \pm .38 = (6.42, 7.18)$
Han residents: $6.24 \pm 2.02(0.18) = 6.24 \pm .36 = (5.88, 6.60)$

The chosen confidence level of 95% implies that the method used to construct each of the intervals has a 5% long-run error rate. Assuming it is reasonable to view these samples as random samples, we can interpret these intervals as follows:

- Based on the information provided in the sample of Tibetan residents, we can be 95% confident that the mean total lung capacity of Tibetan residents is between 6.42 and 7.18 liters.

- Based on the information provided in the sample of Han residents, we can be 95% confident that the mean total lung capacity of Han residents is between 5.88 and 6.60 liters.

These intervals are relatively wide, indicating that the value of the population mean has not been estimated precisely in either case. This is not surprising given the reported sample sizes. Note that the two intervals overlap. This may cause us to be skeptical of the statement that Tibetans have a higher total lung capacity than Han residents. Formal methods for directly comparing two groups, covered in Chapter 11, could be used to further investigate this issue.

Summary of Key Concepts and Formulas

Term or Formula	Comment
Point estimate	A single number, based on sample data, that represents a plausible value of a population characteristic.
Unbiased statistic	A statistic that has a sampling distribution with a mean equal to the value of the population characteristic to be estimated.
Confidence interval	An interval that is computed from sample data and provides a range of plausible values for a population characteristic.
Confidence level	A number that provides information on how much "confidence" we can have in the method used to construct a confidence interval estimate. The confidence level specifies the percentage of all possible samples that will produce an interval containing the true value of the population characteristic.
$p \pm (z \text{ critical value})\sqrt{\dfrac{p(1-p)}{n}}$	A formula used to construct a confidence interval for π when the sample size is large.
$n = \pi(1 - \pi)\left[\dfrac{1.96}{B}\right]^2$	A formula used to compute the sample size necessary for estimating π to within an amount B with 95% confidence. (For other confidence levels, replace 1.96 with an appropriate z critical value.)
$\bar{x} \pm (z \text{ critical value})\dfrac{\sigma}{\sqrt{n}}$	A formula used to construct a confidence interval for μ when σ is known and either the sample size is large or the population distribution is normal.
$\bar{x} \pm (t \text{ critical value})\dfrac{s}{\sqrt{n}}$	A formula used to construct a confidence interval for μ when σ is unknown and either the sample size is large or the population distribution is normal.
$n = \left[\dfrac{1.96\sigma}{B}\right]^2$	A formula used to compute the sample size necessary for estimating μ to within an amount B with 95% confidence. (For other confidence levels, replace 1.96 with an appropriate z critical value.)

Supplementary Exercises 9.50 – 9.58

9.50 Scripps News Service (Sept. 14, 1991) reported that 4% of the members of the American Bar Association (ABA) are African American. Suppose that this figure is based on a random sample of 400 ABA members.

a. Is the sample size large enough to justify the use of the large-sample confidence interval for a population proportion?

b. Construct and interpret a 90% confidence interval

for π, the true proportion of all ABA members who are African American.

9.51 The Center for Urban Transportation Research released a report stating that the average commuting distance in the United States is 10.9 mi (*USA Today,* Aug. 13, 1991). Suppose that this average is actually the mean of a random sample of 300 commuters and that the sample standard deviation was 6.2 mi. Estimate the true mean commuting distance using a 99% confidence interval.

9.52 In 1991, California imposed a "snack tax" (a sales tax on snack food) in an attempt to help balance the state budget. A proposed alternative tax was a 12¢-per-pack increase in the cigarette tax. In a poll of 602 randomly selected California registered voters, 445 responded that they would have preferred the cigarette tax increase to the snack tax (*Reno Gazette-Journal,* Aug. 26, 1991). Estimate the true proportion of California registered voters who preferred the cigarette tax increase; use a 95% confidence interval.

9.53 Systolic blood pressure (mm Hg) was determined for each individual in a random sample of nine athletes who had used anabolic steroids ("The Blood Pressure Response to Exercise in Anabolic Steroid Users," *Medicine and Science in Sports and Exercise* (1992): 633–637), resulting in a sample mean and sample standard deviation of 207.8 and 10.9, respectively.

   a. Calculate and interpret a confidence interval for mean blood pressure in the population of all athletes who used anabolic steroids; use a confidence level of 95%.

   b. Does the validity of the interval calculated in part **a** require any assumptions about the population distribution? Explain.

9.54 The confidence intervals presented in this chapter give both lower and upper bounds on plausible values for the population characteristic being estimated. In some instances, only an upper bound or only a lower bound is appropriate. Using the same reasoning that gave the large-sample interval in Section 9.3, we can say that when n is large, 99% of all samples have

$$\mu < \bar{x} + 2.33\frac{s}{\sqrt{n}}$$

(because the area under the z curve to the left of 2.33 is .99). Thus, $\bar{x} + 2.33(s/\sqrt{n})$ is a 99% upper confidence bound for μ. Use the data of Example 9.8 to calculate the 99% upper confidence bound for the true average salary for married men in two-income families.

9.55 The Associated Press (Dec. 16, 1991) reported that in a random sample of 507 people, only 142 correctly described the Bill of Rights as the first ten amendments to the U.S. Constitution. Calculate a 95% confidence interval for the proportion of the entire population that could give a correct description.

9.56 When n is large, the statistic s is approximately unbiased for estimating σ and has approximately a normal distribution. The standard deviation of this statistic when the population distribution is normal is $\sigma_s \approx \sigma/\sqrt{2n}$, which can be estimated by $s/\sqrt{2n}$. A large-sample confidence interval for the population standard deviation σ is then

$$s \pm (z \text{ critical value})\frac{s}{\sqrt{2n}}$$

Use the data of Exercise 9.51 to obtain a 95% confidence interval for the true standard deviation of commuting distance.

9.57 The interval from -2.33 to 1.75 captures an area of .95 under the z curve. This implies that another large-sample 95% confidence interval for μ has lower limit $\bar{x} - 2.33(\sigma/\sqrt{n})$ and upper limit $\bar{x} + 1.75(\sigma/\sqrt{n})$. Would you recommend using this 95% interval over the 95% interval $\bar{x} \pm 1.96(\sigma/\sqrt{n})$ discussed in the text? Explain. (*Hint:* Look at the width of each interval.)

9.58 Suppose that an individual's waiting time for a morning bus is known to have a uniform distribution on the interval from 0 min to an unknown upper limit θ min. A 95% confidence interval for θ based on a random sample of n waiting times can be shown to have lower limit $\max(x)$ and upper limit $\max(x)/(.05)^{1/n}$. Assuming that $n = 5$ and that the resulting waiting times are 4.2, 3.5, 1.7, 1.2, and 2.4, obtain the confidence interval. (*Hint:* $(.05)^{1/5} = .5493$) Notice that the confidence interval here is not of the form (estimate) $\pm$ (critical value)(standard deviation).

References

Devore, Jay L. *Probability and Statistics for Engineering and the Sciences,* 5th ed. Pacific Grove, CA: Brooks/Cole, 2000. (This book gives a somewhat general introduction to confidence intervals.)

Freedman, David, Robert Pisani, and Roger Purves, *Statistics,* 3rd ed. New York: W. W. Norton, 1997. (This book contains a very informal discussion of confidence intervals.)

GRAPHING CALCULATOR EXPLORATIONS

9.1 The Confidence Interval for a Population Proportion

Since confidence intervals are widely used, it will come as no surprise to you that your calculator may have a built-in capability for accomplishing the task of determining the confidence interval for a population proportion.

Using this capability is so simple that your hardest task will be to identify which confidence interval you want. (Your calculator will list many confidence interval choices.) Once again you must navigate your calculator's menu system, looking for key words such as INTR for "interval," or possibly TESTS. (Confidence intervals are frequently associated with "hypothesis tests," a topic to come later in this book.) Once you find the right menu, look for words like "1" and "prop." and "z" — together these key words should indicate a "1 sample z confidence interval for a proportion."

Once you select the correct choice, you will be presented with a screen for providing information. The information you must provide will, of course, be the same information you would need to calculate the confidence interval: the number of successes, the sample size, and the confidence level. Here are two representative screens:

```
1-Prop Zinterval
C-Level   :
x   :
n   :
Execute
```

```
1-Prop ZInt
x:
n:
C-Level:
Calculate
```

From Example 9.4, supply the following information: $x = 466$, $n = 1014$, and the C-Level = .95. Move your cursor down to Execute or Calculate, and press the Enter, Execute, or Calculate button, depending on your calculator. The confidence interval should appear immediately. Again, here are two representative screens:

```
1-Prop Zinterval
  Left=0.42889
  Right=0.49024
  p̂=0.46
  n=1014
```

```
1-Prop ZInt
(.42889, .49024)
p̂=.459566075
n=1014
```

Notice that the formatting in these screens is slightly different, and this brings up the reasonable question, Which of these is the "right" format? The answer is, probably neither. Recall a previous graphing calculator exploration in Chapter 3, where we discussed the differences between a calculator's presentation of information and the appropriate way to communicate this information. **Check with your instructor to ascertain her or his preferences about what information is required.** You will almost certainly want to report more information than what is presented on these calculator screens. Specifically, you will notice that the calculator does not verify the appropriateness of the large-sample assumption! Some confidence intervals may stray outside the interval from 0 to 1, which makes sense to the calculator, but not to the statistician. Thus, you may have to modify the interval returned by your calculator. Also, don't forget that your calculator does not know the context of the problem; you will have to provide that contextual information irrespective of the form of the calculator presentation. Remember: The calculator only does the calculation; you must do the thinking.

9.2 A Confidence Interval for a Population Mean

Finding the confidence interval for a single mean on your calculator will require you to navigate the menu system much as you did to find the confidence interval for a proportion. The confidence interval for the mean will have some added challenges, however:

1. You will have to decide whether to base the interval on the z or t distribution, and

2. You may use previous calculations of the sample mean and standard deviation, OR the calculator will evaluate these statistics from data contained in a List.

We will follow Example 9.10, and use the t distribution for our confidence interval. We have entered the data in List1 and are ready to proceed. Since we have chosen to use the t confidence interval, the normality of the population becomes an issue. Since the original data is at hand, we can assess the plausibility of a normal population via a normal probability plot. The accompanying figure shows the resulting normal probability plot.

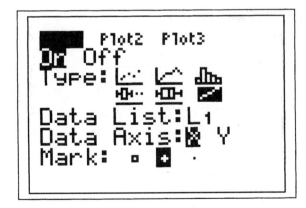
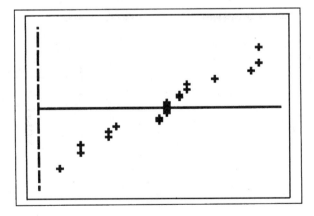

After verifying the plausibility of the normality of the population, we will construct the confidence interval. In this case, we have a choice of entering the sample calculations or letting the calculator evaluate the sample mean and standard deviation. Based on that choice, we see one of the following screens.

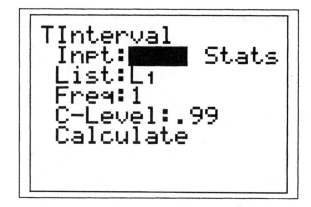

The accompanying figure shows one calculator's version of the confidence interval. Once again, we caution you that the calculator only calculates — you must still do the thinking and present the solution in the context of the particular problem at hand.

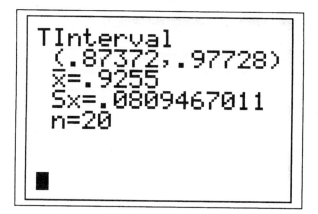

10

Hypothesis Testing Using a Single Sample

INTRODUCTION

In the previous chapter, we considered situations in which the primary goal was to estimate the unknown value of some population characteristic. Sample data may also be used to decide whether some claim or *hypothesis* about a population characteristic is plausible. For example, the lead story in the August 27, 1995, *San Francisco Examiner* focused on health care problems resulting from an increase in the number of Americans without health insurance. Serious difficulties for state and local agencies arise if the coverage rate for the entire population is less than 90%. The same day, the *Los Angeles Times* reported that in its survey of 3297 California adults, 2750, or 83.4%, had health insurance coverage. Let π denote the proportion of all California adults who have health insurance. The hypothesis-testing methods presented in this chapter could be used to decide whether the sample data provides strong support for the hypothesis $\pi < .90$.

As another example, a report released by the National Association of Colleges and Employers stated that the average starting salary for students graduating in 1999 with a degree in marketing was $31,542 (*San Luis Obispo Tribune,* Sept. 8, 1999). Suppose that you are interested in determining whether the mean starting salary for students graduating with a marketing degree from your university is greater than the national average of $31,542. You select a random sample of $n = 10$ marketing graduates from the 1999 graduating class of your university and determine the starting salary of each one. If this sample produced a mean starting salary of $31,758 and a standard deviation of $1214, is it reasonable to conclude that μ, the true mean starting salary for marketing graduates at your university, is greater than $31,542 (that is, $\mu > 31,542$)? We will see in this chapter how this sample data might be analyzed to decide whether $\mu > 31,542$ is an appropriate conclusion.

10.1 Hypotheses and Test Procedures

A hypothesis is a claim or statement either about the value of a single population characteristic or about the values of several population characteristics. The following are examples of legitimate hypotheses:

$\mu = 1000$ where μ is the mean number of characters in an e-mail message

$\pi < .01$ where π is the proportion of e-mail messages that are undeliverable

In contrast, the statements $\bar{x} > 1000$ and $p = .01$ are *not* hypotheses, because $\bar{x}$ and p are *sample* characteristics.

A **test of hypotheses** or **test procedure** is a method for using sample data to decide between two competing claims (hypotheses) about a population characteristic. One hypothesis might be $\mu = 1000$ and the other $\mu \neq 1000$, or one might be $\pi = .01$ and the other $\pi < .01$. If it were possible to carry out a census of the entire population, we would know which of the two hypotheses is correct, but usually we must decide between them using information from a sample.

A criminal trial is a familiar situation in which a choice between two contradictory claims must be made. The person accused of the crime must be judged either guilty or not guilty. Under the U.S. system of justice, the individual on trial is initially presumed not guilty. Only strong evidence to the contrary will cause the not guilty claim to be rejected in favor of a guilty verdict. The burden is thus put on the prosecution to prove the guilty claim. The French perspective in criminal proceedings is the opposite of ours. There, once enough evidence has been presented to justify bringing an individual to trial, the initial assumption is that the accused is guilty. The burden of proof then falls on the accused to establish otherwise.

As in a judicial proceeding, we shall initially assume that a particular hypothesis, called the *null hypothesis,* is the correct one. We then consider the evidence (the sample data) and reject the null hypothesis in favor of the competing hypothesis, called the *alternative hypothesis,* only if there is *convincing* evidence against the null hypothesis.

Definition

The **null hypothesis,** denoted by H_0, is a claim about a population characteristic that is initially assumed to be true.

The **alternative hypothesis,** denoted by H_a, is the competing claim.

In carrying out a test of H_0 versus H_a, the hypothesis H_0 will be rejected in favor of H_a only if sample evidence strongly suggests that H_0 is false. If the sample does not contain such evidence, H_0 will not be rejected. The two possible conclusions are then *reject H_0* or *fail to reject H_0*.

EXAMPLE 10.1   Because of variation in the manufacturing process, tennis balls produced by a particular machine do not have identical diameters. Let μ denote the true average diameter for tennis balls currently being produced. Suppose that the machine was initially calibrated to achieve the design specification $\mu = 3$ in However, the manufacturer is now concerned that the diameters no longer conform to this specification. That is, $\mu \neq 3$ in must now be considered a possibility. If sample evidence suggests that $\mu \neq 3$ in, the production process will have to be halted while recalibration takes place. Because this is costly, the manufacturer wants to be quite sure that $\mu \neq 3$ in before undertaking recalibration. Under these circumstances, a sensible choice of hypotheses is

H_0: $\mu = 3$ (the specification is being met, so recalibration is unnecessary)

H_a: $\mu \neq 3$

Only compelling sample evidence would then result in H_0 being rejected in favor of H_a.

EXAMPLE 10.2

K mart-brand 60-watt lightbulbs state on the package "Avg. Life 1000 Hr." Let μ denote the true mean life of K mart 60-watt lightbulbs. Then the advertised claim is $\mu = 1000$ hr. People who purchase this brand would be unhappy if μ is actually less than the advertised value. Suppose that a sample of K mart lightbulbs is selected and the lifetime for each bulb in the sample is recorded. The sample results can then be used to test the hypothesis $\mu = 1000$ hr against the hypothesis $\mu < 1000$ hr. The accusation that the company is overstating the mean lifetime is a serious one, and it is reasonable to require compelling evidence from the sample before concluding that $\mu < 1000$. This suggests that the claim $\mu = 1000$ should be selected as the null hypothesis and $\mu < 1000$ selected as the alternative hypothesis. Then

H_0: $\mu = 1000$

would be rejected in favor of

H_a: $\mu < 1000$

only when sample evidence strongly suggests that the initial assumption $\mu = 1000$ hr is no longer tenable.

Since the alternative hypothesis in Example 10.2 asserted that $\mu < 1000$ (true average lifetime is less than the advertised value), it might have seemed sensible to state H_0 as the inequality $\mu \geq 1000$. This assertion is in fact the *implicit* null hypothesis, but we will state H_0 explicitly as a claim of equality. There are several reasons for this. First of all, the development of a decision rule is most easily understood if there is only a single value of μ (or π, or whatever other population characteristic is under consideration) when H_0 is true. Secondly, suppose that the sample data provided compelling evidence that H_0: $\mu = 1000$ should be rejected in favor of H_a: $\mu < 1000$. This means that we were convinced by the sample data that the true mean was smaller than 1000. It follows that we would have also been convinced that the true mean could not have been 1001 or 1010 or any other value that was larger than 1000. As a consequence, the conclusion when testing $H_0: \mu = 1000$ versus $H_a: \mu < 1000$ will always be the same as the conclusion for a test where the null hypothesis is $H_0: \mu \geq 1000$. For the sake of simplicity, we will state the null hypothesis, H_0, as a claim of equality.

The form of a null hypothesis is

H_0:   population characteristic = hypothesized value

where the hypothesized value is a specific number determined by the problem context.

The alternative hypothesis will have one of the following three forms:

H_a: population characteristic > hypothesized value

H_a: population characteristic < hypothesized value

H_a: population characteristic ≠ hypothesized value

Thus, we might test $H_0: \pi = .1$ versus $H_a: \pi < .1$; but we won't test $H_0: \mu = 50$ versus $H_a: \mu > 100$. The number appearing in the alternative hypothesis must be identical to the hypothesized value in H_0.

The following example illustrates how the selection of H_0 (the claim initially believed true) and H_a depend on the objectives of a study.

EXAMPLE 10.3   A medical research team has been given the task of evaluating a new laser treatment for certain types of tumors. Consider the following two scenarios:

Scenario 1:   The current standard treatment is considered reasonable and safe by the medical community, has no major side effects, and has a known success rate of .85 (85%).

Scenario 2:   The current standard treatment sometimes has serious side effects, is very costly, and has a known success rate of .30 (30%).

In the first scenario, research efforts would probably be directed toward determining whether the new treatment has a higher success rate than the standard treatment. Unless convincing evidence of this is presented, it is unlikely that current medical practice would be changed. With π representing the true proportion of successes for the laser treatment, the following hypotheses would be tested:

$$H_0: \quad \pi = .85 \qquad \text{versus} \qquad H_a: \quad \pi > .85$$

In this case, rejection of the null hypothesis would be indicative of compelling evidence that the success rate is higher for the new treatment.

In the second scenario, the current standard treatment does not have much to recommend it. The new laser treatment may be considered preferable because of cost or because it has fewer or less serious side effects, as long as the success rate for the new procedure is no worse than that of the standard treatment. Here, researchers might decide to test the hypothesis

$$H_0: \quad \pi = .30 \qquad \text{versus} \qquad H_a: \quad \pi < .30$$

If the null hypothesis is rejected, the new treatment will not be put forward as an alternative to the standard treatment, because there is strong evidence that the laser method has a lower success rate.

If the null hypothesis is not rejected, we are able to conclude only that there is not convincing evidence that the success rate for the laser treatment is lower than that for the standard. This is *not* the same as saying we have evidence that the laser treatment is as good as the standard. If medical practice were to embrace the new procedure, it would not be because it has a higher success rate but rather because it costs less or has fewer side effects, and there is not strong evidence that it has a lower success rate than the standard.

You should be careful in setting up the hypotheses for a test. Remember that a statistical hypothesis test is only capable of demonstrating strong support for the alternative hypothesis (by rejection of the null hypothesis). When the null hypothesis is not rejected, it does not mean strong support for H_0—only lack of strong evidence against it. In the lightbulb scenario of Example 10.2, if $H_0: \mu = 1000$ is rejected in favor of $H_a: \mu < 1000$, we have strong evidence for believing that true average lifetime is less than the advertised value. However, nonrejection of H_0 does not necessarily provide strong support for the advertised claim. If the objective is to demonstrate that the average lifetime is greater than 1000 hours, the hypothesis tested should be $H_0: \mu = 1000$ versus $H_a: \mu > 1000$. Now rejection of H_0 indicates strong evidence that $\mu > 1000$. *Keep the research objectives in mind when selecting the hypotheses to be tested.*

Exercises 10.1 – 10.9

10.1 Explain why the statement $\bar{x} = 50$ is not a legitimate hypothesis.

10.2 For the following pairs, indicate which do not comply with our rules for setting up hypotheses, and explain why.
  a. $H_0: \mu = 15, H_a: \mu = 15$
  b. $H_0: \pi = .4, H_a: \pi > .6$
  c. $H_0: \mu = 123, H_a: \mu < 123$
  d. $H_0: \mu = 123, H_a: \mu = 125$
  e. $H_0: p = .1, H_a: p \neq .1$

10.3 To determine whether the pipe welds in a nuclear power plant meet specifications, a random sample of welds is selected, and tests are conducted on each weld in the sample. Weld strength is measured as the force required to break the weld. Suppose that the specifications state that the mean strength of welds should exceed 100 lb/in². The inspection team decides to test $H_0: \mu = 100$ versus $H_a: \mu > 100$. Explain why this alternative was chosen rather than $\mu < 100$.

10.4 The mean length of long-distance telephone calls placed with a particular phone company was known to be 7.3 minutes under an old rate structure. In an attempt to be more competitive with other long-distance carriers, the phone company lowered long-distance rates, thinking that its customers would be encouraged to make longer calls, so there would not be a big loss in revenue. Let μ

denote the true mean length of long-distance calls after the rate reduction. What hypotheses should the phone company test to determine whether the mean length of long-distance calls increased with the lower rates?

10.5   A certain university has decided to introduce the use of + and − with letter grades, as long as there is evidence that more than 60% of the faculty favor the change. A random sample of faculty will be selected and the resulting data used to test the relevant hypotheses. If π represents the true proportion of all faculty that favor a change to +/− grading, which of the following pair of hypotheses should the administration test? Explain your choice.

$$H_0: \quad \pi = .6 \quad \text{versus} \quad H_a: \quad \pi < .6$$

or

$$H_0: \quad \pi = .6 \quad \text{versus} \quad H_a: \quad \pi > .6$$

10.6   A certain television station has been providing live coverage of a particularly sensational criminal trial. The station's program director wishes to know whether more than half the potential viewers prefer a return to regular daytime programming. A survey of randomly selected viewers is to be conducted. Let π represent the true proportion of viewers who prefer regular daytime programming. What hypotheses should the program director test to answer the question of interest?

10.7   Researchers have postulated that because of differences in diet, Japanese children have a lower mean blood cholesterol level than U.S. children. Suppose that the mean level for U.S. children is known to be 170. Let μ represent the true mean blood cholesterol level for Japanese children. What hypotheses should the researchers test?

10.8   A county commissioner must vote on a resolution that would commit substantial resources to the construction of a sewer in an outlying residential area. Her fiscal decisions have been criticized in the past, so she decides to take a survey of constituents to find out whether they favor spending money for a sewer system. She will vote to appropriate funds only if she can be fairly certain that a majority of the people in her district favor the measure. What hypotheses should she test?

10.9   Many older homes have electrical systems that use fuses rather than circuit breakers. A manufacturer of 40-amp fuses wants to make sure that the mean amperage at which its fuses burn out is in fact 40. If the mean amperage is lower than 40, customers will complain because the fuses require replacement too often. If the mean amperage is higher than 40, the manufacturer might be liable for damage to an electrical system as a result of fuse malfunction. To verify the mean amperage of the fuses, a sample of fuses is to be selected and inspected. If a hypothesis test is to be performed using the resulting data, what null and alternative hypotheses would be of interest to the manufacturer?

10.2   Errors in Hypothesis Testing

Once hypotheses have been formulated, we need a method for using sample data to determine whether H_0 should be rejected. The method that we use for this purpose is called a **test procedure.** Just as a jury may reach the wrong verdict in a trial, there is some chance that the use of a test procedure with sample data may lead us to the wrong conclusion about a population characteristic. This section discusses the kinds of errors that may occur and considers how the choice of a test procedure influences the chances of these errors.

One erroneous conclusion in a criminal trial is for a jury to convict an innocent person, and another is for a guilty person to be set free. Similarly, there are two different types of errors that might be made when making a decision in a hypothesis-testing problem. One type of error involves rejecting H_0 even though H_0 is true. The second type of error results from failing to reject H_0 when it is false.

> **Definition**
>
> **type I error**   the error of rejecting H_0 when H_0 is true
>
> **type II error**   the error of failing to reject H_0 when H_0 is false

The only way to guarantee that neither type of error will occur is to make such decisions on the basis of a census of the entire population. The risk of error is introduced when we try to make an inference about a population using incomplete information. This is the price paid for basing an inference on a sample. With any reasonable procedure, there is some chance that a type I error will be made and some chance that a type II error will result.

EXAMPLE 10.4   The U.S. Department of Transportation reported that during a recent period, 77% of all domestic passenger flights arrived "on time" (meaning within 15 minutes of the scheduled arrival). Suppose that an airline with a poor on-time record decides to offer its employees a bonus if, in an upcoming month, the airline's proportion of on-time flights exceeds the overall industry rate of .77. Let π be the true proportion of the airline's flights that are on time during the month of interest. A random sample of flights might be selected and used as a basis for choosing between

$$H_0: \quad \pi = .77 \quad \text{and} \quad H_a: \quad \pi > .77$$

In this context, a type I error (rejecting a true H_0) would result in the airline's rewarding its employees when in fact their true proportion of on-time flights did not exceed .77. A type II error (not rejecting a false H_0) would result in the airline employees *not* receiving a reward that in fact they deserved.

EXAMPLE 10.5   Researchers at the National Cancer Institute announced plans to begin studies of a cancer treatment thought to slow the growth of tumors (Associated Press, March 30, 1993). Having tested the treatment on only 13 patients, the researchers had very little information, but they stated that the treatment appears to be less toxic than standard chemotherapy treatments. An experiment to study the treatment more extensively was reported as being in the planning stages.

Let μ denote the true mean growth rate of tumors for patients receiving the new treatment. Data resulting from the planned experiments could be used to test

$H_0: \quad \mu =$ mean growth rate of tumors without treatment

$H_a: \quad \mu <$ mean growth rate of tumors without treatment

The null hypothesis states that the new treatment is not effective — that the mean growth rate of tumors for patients receiving the new treatment is the same as for patients who are not treated. The alternative hypothesis states that the new treatment is effective in reducing the mean growth rate of tumors. In this context, a type I error consists of incorrectly concluding that the new treatment is effective in slowing the growth rate of tumors. A type II error consists of concluding that the new treatment is ineffective when, in fact, the mean growth rate of tumors is reduced.

Examples 10.4 and 10.5 illustrate the two different types of error that might occur when testing hypotheses. Type I and type II errors, and the associated consequences of making such errors, are quite different. In choosing a test procedure (a method for deciding whether to reject H_0), we would like a small probability of drawing an incorrect conclusion. The accompanying box introduces the terminology and notation used to describe error probabilities.

Definition

The probability of a type I error is denoted by α and is called the **level of significance** of the test. Thus, a test with $\alpha = .01$ is said to have a level of significance of .01 or to be a level .01 test.

The probability of a type II error is denoted by β.

EXAMPLE 10.6   A type of lie detector that measures brain waves was developed by a professor of neurobiology at Northwestern University (Associated Press, July 7, 1988). He said, "It would probably not falsely accuse any innocent people and it would probably pick up 70% to 90% of guilty people." Suppose that the result of this lie detector test is allowed as evidence in a criminal trial as the sole basis of a decision between two rival hypotheses:

H_0:   accused is innocent

H_a: accused is guilty

[*Note*: Although these are not "statistical hypotheses" (statements about a population characteristic), the possible decision errors are analogous to type I and type II errors.]

In this situation, a type I error would be finding an innocent person guilty — rejecting the null hypothesis of innocence when it was, in fact, true. A type II error would be finding a guilty person innocent — not rejecting the null hypothesis of innocence when it was, in fact, false. If the developer of the lie detector is correct in his statements, the probability of a type I error, α, is approximately 0, and β, the probability of type II error, is somewhere between .1 and .3.

The ideal test procedure has both $\alpha = 0$ and $\beta = 0$. However, if we must base our decision on incomplete information — a sample rather than a census — it is impossible to achieve this ideal. The standard test procedures do allow the user to control α, but they provide no direct control over β. Since α represents the probability of rejecting a true null hypothesis, selecting a significance level $\alpha = .05$ results in a test procedure that, used over and over with different samples, rejects a *true* H_0 about five times in a hundred. Selecting $\alpha = .01$ results in a test procedure with a type I error rate of 1% in long-term repeated use. Choosing a small value for α implies that the user wants to employ a procedure for which the risk of a type I error is quite small.

One question arises naturally at this point: If the user can select α, the probability of making a type I error, why would anyone ever select $\alpha = .05$ rather than $\alpha = .01$? Why not always select a very small value for α? To achieve a small probability of making a type I error, the corresponding test procedure will require the evidence against H_0 to be very strong before the null hypothesis can be rejected. Although this makes a type I error unlikely, it increases the risk of a type II error (*not* rejecting H_0 when it should have been rejected). Frequently the investigator must balance the consequences of type I and type II errors. If a type II error has serious consequences, it may be a good idea to select a somewhat larger value for α.

In general, there is a compromise between small α and small β, leading to the following widely accepted principle for specifying a test procedure.

After assessing the consequences of type I and type II errors, identify the largest α that is tolerable for the problem. Then employ a test procedure that uses this maximum acceptable value — rather than anything smaller — as the level of significance (because using a smaller α increases β). In other words, don't make α smaller than it needs to be.

Thus, if you decide that $\alpha = .05$ is tolerable, you should not use a test with $\alpha = .01$, because the smaller α inevitably results in a larger β. The values of α most commonly used in practical problems are .05 and .01 (a 1-in-20 or 1-in-100 chance of rejecting H_0 when it is actually true), but the choice in any given problem depends on the seriousness of a type I error relative to a type II error in that context.

EXAMPLE 10.7 The Associated Press (May 13, 1993) reported that the Environmental Protection Agency (EPA) had warned 819 communities that their tap water contained too much lead. Drinking water is considered unsafe if the mean concentration of lead is 15 parts per billion or greater. The EPA requires the cited communities to take corrective actions and to monitor lead levels. With μ denoting the mean concentration of lead, a cited community could test

$$H_0: \ \mu = 15 \quad \text{versus} \quad H_a: \ \mu < 15$$

The null hypothesis (which is equivalent to the assertion $\mu \geq 15$) states that the mean lead concentration is excessive by EPA standards. The alternative hypothesis states that the mean lead concentration is at an acceptable level and that the water system meets EPA standards for lead.

In this context, a type I error would lead to the conclusion that a water source meets EPA standards for lead when, in fact, it does not. Possible consequences of this type of error include health risks associated with excessive lead consumption (for example, increased blood pressure, hearing loss, and, in severe cases, anemia and kidney damage). A type II error would be to conclude that the water does not meet EPA standards for lead when, in fact, it actually does. Possible consequences of a type II error include elimination of a community water source. Since a type I error might result in potentially serious public health risks, a small value of α (type I error probability), such as $\alpha = .01$, could be selected. Of course, selecting a small value for α increases the risk of a type II error. If the community has only one water source, a type II error could also have very serious consequences for the community, and we might want to rethink our choice of α.

Exercises 10.10 – 10.19

10.10 Researchers at the University of Washington and Harvard University analyzed records of breast cancer screening and diagnostic evaluations ("Mammogram Cancer Scares More Frequent Than Thought," *USA Today,* April 16, 1998). Discussing the benefits and downsides of the screening process, the article states that, although the rate of false positives is higher than previously thought, if radiologists were less aggressive in following up on suspicious tests, the rate of false positives would fall but the rate of missed cancers would rise. Suppose that such a screening test is to be used to decide between a null hypothesis of H_0: no cancer is present and an alternative hypothesis of H_a: cancer is present.

a. Would a false positive (thinking that cancer is present when in fact it is not) be a type I error or a type II error?

b. Describe a type I error in the context of this problem, and discuss the consequences of making a type I error.

c. Describe a type II error in the context of this problem, and discuss the consequences of making a type II error.

d. What aspect of the relationship between the probability of type I and II errors is being described by the statement in the article that if radiologists were less aggressive in following up on suspicious tests, the rate of false positives would fall but the rate of missed cancers would rise?

10.11 The National Cancer Institute conducted a 2-year study to determine whether cancer death rates for areas near nuclear power plants were higher than for areas without nuclear facilities (*San Luis Obispo Telegram-Tribune,* Sept. 17, 1990). A spokesperson for the Cancer Institute said, "From the data at hand, there was no convincing evidence of any increased risk of death from any of the cancers surveyed due to living near nuclear facilities. However, no study can prove the absence of an effect."

a. Let π denote the true proportion of the population in areas near nuclear power plants who die of cancer during a given year. The researchers at the Cancer Institute might have considered the two rival hypotheses of the form

H_0: π = value for areas without nuclear facilities

H_a: π > value for areas without nuclear facilities

Did the Cancer Institute reject H_0 or fail to reject H_0?

b. If the Cancer Institute was incorrect in its conclusion that there is no increased cancer risk associated with living near a nuclear power plant, would they be making a type I or a type II error? Explain.

c. Comment on the spokesperson's last statement that no study can *prove* the absence of an effect. Do you agree with this statement?

10.12 Ann Landers, in her advice column of Oct. 24, 1994 (*San Luis Obispo Telegram-Tribune*), described the reliability of DNA paternity testing as follows: "To get a completely accurate result, you would have to be tested, and so would (the man) and your mother. The test is 100 percent accurate if the man is NOT the father and 99.9 percent accurate if he is."

a. Consider using the results of DNA paternity testing to decide between the following two hypotheses:

H_0: a particular man is the father

H_a: a particular man is not the father

In the context of this problem, describe type I and II errors. (Although these are not hypotheses about a population characteristic, this exercise illustrates the definitions of type I and type II errors.)

b. Based on the information given, what are the values of α, the probability of type I error, and β, the probability of type II error?

c. Ann Landers also states, "If the mother is not tested, there is a 0.8 percent chance of a false positive." For the hypotheses given in part **a,** what are the values of α and β if the decision will be based on DNA testing where the mother is not tested?

10.13   Pizza Hut, after test-marketing a new product called the Bigfoot Pizza, concluded that introduction of the Bigfoot nationwide would increase its sales by more than 14% (*USA Today,* April 2, 1993). This conclusion was based on recording sales information for a random sample of Pizza Hut restaurants selected for the market-ing trial. With μ denoting the mean percentage increase in sales for all Pizza Hut restaurants, consider using the sample data to decide between

$$H_0: \quad \mu = 14 \quad \text{and} \quad H_a: \quad \mu > 14$$

a. Is Pizza Hut's conclusion consistent with a decision to reject H_0 or to fail to reject H_0?

b. If Pizza Hut is incorrect in its conclusion, is the com-pany making a type I or a type II error?

10.14   A television manufacturer claims that (at least) 90% of its sets will need no service during the first 3 years of operation. A consumer agency wishes to check this claim, so it obtains a random sample of $n = 100$ purchasers and asks each whether the set purchased needed repair during the first 3 years. Let p be the sample proportion of responses indicating no repair (so that no repair is iden-tified with a success). Let π denote the true proportion of successes. The agency does not want to claim false ad-vertising unless sample evidence strongly suggests that $\pi < .9$. The appropriate hypotheses are then

$$H_0: \quad \pi = .9 \quad \text{versus} \quad H_a: \quad \pi < .9$$

a. In the context of this problem, describe type I and type II errors, and discuss the possible consequences of each.

b. Would you recommend a test procedure that uses $\alpha = .10$ or one that uses $\alpha = .01$? Explain.

10.15   A manufacturer of handheld calculators receives very large shipments of printed circuits from a supplier. It is too costly and time-consuming to inspect all incom-ing circuits, so when each shipment arrives, a sample is se-lected for inspection. Information from the sample is then used to test $H_0: \pi = .05$ versus $H_a: \pi > .05$, where π is the true proportion of defectives in the shipment. If the null hypothesis is not rejected, the shipment is accepted, and the circuits are used in the production of calculators. If the null hypothesis is rejected, the entire shipment is re-turned to the supplier because of inferior quality. (A ship-ment is defined to be of inferior qualify if it contains more than 5% defectives.)

a. In this context, define type I and type II errors.

b. From the calculator manufacturer's point of view, which type of error would be considered more serious?

c. From the printed circuit supplier's point of view, which type of error would be considered more serious?

10.16   Water samples are taken from water used for cooling as it is being discharged from a power plant into a river. It has been determined that as long as the mean temperature of the discharged water is at most 150°F, there will be no negative effects on the river's ecosystem. To investigate whether the plant is in compliance with regulations that prohibit a mean discharge water temper-ature above 150°, 50 water samples will be taken at ran-domly selected times and the temperature of each sample recorded. The resulting data will be used to test the hy-potheses

$$H_0: \quad \mu = 150° \quad \text{versus} \quad H_a: \quad \mu > 150°$$

In the context of this example, describe type I and type II errors. Which type of error would you consider to be more serious? Explain.

10.17   Occasionally, warning flares of the type contained in most automobile emergency kits fail to ignite. A con-sumer advocacy group is to investigate a claim against a manufacturer of flares brought by a person who claims that the proportion of defectives is much higher than the value of .1 claimed by the manufacturer. A large number of flares will be tested and the results used to decide be-tween $H_0: \pi = .1$ and $H_a: \pi > .1$, where π represents the true proportion of defectives for flares made by this man-ufacturer. If H_0 is rejected, charges of false advertising will be filed against the manufacturer.

a. Explain why the alternative hypothesis was chosen to be $H_a: \pi > .1$.

b. In this context, describe type I and type II errors, and discuss the consequences of each.

10.18   Suppose that you are an inspector for the Fish and Game Department, and you are given the task of deter-mining whether to prohibit fishing along part of the Ore-gon coast. You will close an area to fishing if it is deter-mined that fish in that region have an unacceptably high mercury content.

a. Assuming that a mercury concentration of 5 ppm is considered to be the maximum safe concentration, which of the following pairs of hypotheses would you test? Give the reasons for your choice.

$$H_0: \quad \mu = 5 \qquad \text{versus} \qquad H_a: \quad \mu > 5$$

or

$$H_0: \quad \mu = 5 \qquad \text{versus} \qquad H_a: \quad \mu < 5$$

b. Would you prefer a significance level of .1 or .01 for your test? Explain.

10.19 An automobile manufacturer is considering using robots for part of its assembly process. Converting to robots is an expensive process, so it will be undertaken only if there is strong evidence that the proportion of defective installations is lower for the robots than for human assemblers. Let π denote the true proportion of defective installations for the robots. It is known that human assemblers have a defect proportion of .02.

a. Which of the following pairs of hypotheses should the manufacturer test? Explain your answer.

$$H_0: \quad \pi = .02 \qquad \text{versus} \qquad H_a: \quad \pi < .02$$

or

$$H_0: \quad \pi = .02 \qquad \text{versus} \qquad H_a: \quad \pi > .02$$

b. In the context of this example, describe type I and type II errors.

c. Would you prefer a test with $\alpha = .01$ or $\alpha = .1$? Explain your reasoning.

10.3 Large-Sample Hypothesis Tests for a Population Proportion

Now that some general concepts of hypothesis testing have been introduced, we are ready to turn our attention to the development of procedures for using sample information to decide between the null and alternative hypotheses. There are two possibilities: We will either reject H_0 or else fail to reject H_0. The fundamental idea behind hypothesis-testing procedures is this: *We reject the null hypothesis if the observed sample is very unlikely to have occurred when H_0 is true.* In this section, we consider testing hypotheses about a population proportion when the sample size n is large.

Let π denote the proportion of individuals or objects in a specified population that possess a certain property. A random sample of n individuals or objects is to be selected from the population. The sample proportion

$$p = \frac{\text{number in the sample that possess property}}{n}$$

is the natural statistic for making inferences about π.

The large-sample test procedure is based on the same properties of the sampling distribution of p that were used previously to obtain a confidence interval for π, namely,

1. $\mu_p = \pi$

2. $\sigma_p = \sqrt{\dfrac{\pi(1 - \pi)}{n}}$

3. When n is large, the sampling distribution of p is approximately normal.

These three results imply that the standardized variable

$$z = \frac{p - \pi}{\sqrt{\dfrac{\pi(1 - \pi)}{n}}}$$

has approximately a standard normal distribution when n is large. With this background, let's look at an example.

EXAMPLE 10.8   The authors of the article "Perceived Risks of Heart Disease and Cancer Among Cigarette Smokers" (*J. Amer. Medical Assn.* (1999): 1019–1021) expressed the concern that a majority of smokers do not view themselves as being at increased risk of heart disease or cancer. Because of this, the authors call for a public health campaign to educate smokers about the associated risks. In support of this recommendation, the authors offered the results of a study of 737 current smokers selected at random from U.S. households with telephones. Of the 737 smokers surveyed, 295 indicated that they believed they have a higher than average risk of cancer. Does this data suggest that π, the true proportion of smokers who view themselves as being at increased risk of cancer is, in fact, less than .5 as claimed by the authors of the paper?

This question can be answered by testing hypotheses, where

π = true proportion of smokers who believe that they are at increased risk
       of cancer

H_0: $\pi = .5$

H_a: $\pi < .5$ (The proportion of smokers who believe that they are at
                        increased risk of cancer is less than .5. That is, less than
                        half of all smokers believe that they are at increased risk
                        of cancer.)

Recall that in a hypothesis test, the null hypothesis will be rejected only if there is convincing evidence against it — in this case, convincing evidence that $\pi < .5$. If H_0 is rejected, the authors would have strong support for the claim made in the article and for their recommended action.

For this sample,

$$p = \frac{295}{737} = .400$$

The observed sample proportion is certainly less than .5, but this could just be due to sampling variability. That is, when $\pi = .5$ (meaning H_0 is true), the sample proportion p will usually differ somewhat from .5 simply because of chance variation from one sample to another. Is it plausible that a sample proportion of $p = .40$ occurred as a result of this chance variation, or is it unusual to observe a sample proportion this small when $\pi = .5$?

To answer this question, we form a *test statistic,* the quantity used as a basis for making a decision between H_0 and H_a, by replacing π by the hypothesized value in the foregoing z variable:

$$z = \frac{p - .5}{\sqrt{\dfrac{(.5)(.5)}{n}}}$$

If the null hypothesis is true, this statistic should have approximately a standard normal distribution, because when the sample size is large and H_0 is true,

1. $\mu_p = .5$

2. $\sigma_p = \sqrt{\dfrac{(.5)(.5)}{n}}$

3. p has approximately a normal distribution.

The calculated value of z expresses the distance between p and the hypothesized value as a number of standard deviations. If, for example, $z = -3$, then the value of p that came from the sample is 3 standard deviations (of p) less than what we would have expected if the null hypothesis were true. How likely is it that a z value at least this contradictory to H_0 would be observed if in fact H_0 is true? Because the test statistic z is constructed using the hypothesized value from the null hypothesis, when H_0 is true it has (approximately) a standard normal distribution, so

$$P(z \leq -3 \text{ when } H_0 \text{ is true}) = \text{area under the } z \text{ curve to the left of } -3.00$$
$$= .0013$$

That is, if H_0 were true, very few samples (much less than 1% of all samples) would produce a value of z at least as contradictory to H_0 as $z = -3$. Since this z value is in the most extreme 1%, it would be sensible to reject H_0.

For our data,

$$z = \frac{p - .5}{\sqrt{\dfrac{(.5)(.5)}{n}}} = \frac{.40 - .5}{\sqrt{\dfrac{(.5)(.5)}{737}}} = \frac{-.10}{.018} = -5.56$$

That is, $p = .40$ is more than 5.5 standard deviations less than what we would expect it to be if H_0: $\pi = .5$ were true. The sample data appears to be much more consistent with H_a: $\pi < .5$. In particular,

$$P(\text{value of } z \text{ is at least as contradictory to } H_0 \text{ as } -5.56 \text{ when } H_0 \text{ is true})$$
$$= P(z \leq -5.56 \text{ when } H_0 \text{ is true})$$
$$= \text{area under the } z \text{ curve to the left of } -5.56$$
$$\approx 0$$

There is virtually no chance of seeing a sample proportion and hence a z value this extreme as a result of chance variation alone when H_0 is true. If p is more than 5.5 standard deviations away from .5, how can we possibly believe that $\pi = .5$? The evidence for rejecting H_0 in favor of H_a is very compelling.

The preceding example illustrates the rationale behind large-sample procedures for testing hypotheses about π (and other test procedures as well). We begin by assuming that the null hypothesis is correct. The sample is then examined in light of this assumption. If the observed sample proportion would not be unusual when H_0

is true, then chance variability from one sample to another is a plausible explanation for what has been observed, and H_0 should not be rejected. On the other hand, if the observed sample proportion would have been quite unlikely were H_0 true, we take the sample as convincing evidence against the null hypothesis and we should reject H_0. We base a decision to reject or to fail to reject the null hypothesis on an assessment of how extreme or unlikely the observed sample is when H_0 is true.

The assessment of how contradictory the observed data is to H_0 is based on first computing the value of the test statistic

$$z = \frac{p - \text{hypothesized value}}{\sqrt{\dfrac{(\text{hypothesized value})(1 - \text{hypothesized value})}{n}}}$$

We then calculate the *P-value,* the probability, assuming that H_0 is true, of obtaining a z value at least as contradictory to H_0 as what was actually observed.

Definition

A **test statistic** is the function of sample data on which a conclusion to reject or fail to reject H_0 is based.

The **P-value** (also sometimes called the **observed significance level**) is a measure of inconsistency between the hypothesized value for a population characteristic and the observed sample. It is the probability, assuming that H_0 is true, of obtaining a test statistic value at least as inconsistent with H_0 as what actually resulted.

EXAMPLE 10.9

A number of initiatives on the topic of legalized gambling have appeared on state ballots in recent years. Suppose that a political candidate has decided to support legalization of casino gambling if he is convinced that more than 2/3 of U.S. adults approve of casino gambling. *USA Today* (June 17, 1999) reported the results of a Gallup poll in which 1523 adults (selected at random from households with telephones) were asked if they approved of casino gambling. The number in the sample who approved was 1035. Does the sample provide convincing evidence that more than 2/3 approve?

With

$$\pi = \text{true proportion of U.S. adults who approve of casino gambling}$$

the relevant hypotheses are

$$H_0: \quad \pi = 2/3 = .667$$
$$H_a: \quad \pi > 2/3 = .667$$

The sample proportion is

$$p = \frac{1035}{1523} = .680$$

Does the value of p exceed 2/3 by enough to cast substantial doubt on H_0?

Since the sample size is large, the statistic

$$z = \frac{p - .667}{\sqrt{\dfrac{(.667)(1 - .667)}{n}}}$$

has approximately a standard normal distribution when H_0 is true. The calculated value of the test statistic is

$$z = \frac{.680 - .667}{\sqrt{\dfrac{(.667)(1 - .667)}{1523}}} = \frac{.013}{.012} = 1.08$$

The probability that a z value at least this inconsistent with H_0 would be observed if in fact H_0 is true is

$$P\text{-value} = P(z \geq 1.08 \text{ when } H_0 \text{ is true})$$
$$= \text{area under the } z \text{ curve to the right of } 1.08$$
$$= 1 - .8599$$
$$= .1401$$

This probability indicates that when $\pi = .667$, it would not be all that unusual to observe a sample proportion as large as .680. When H_0 is true, roughly 14% of all samples would have a sample proportion larger than .680, so a sample proportion of .680 is reasonably consistent with the null hypothesis. Although .680 is larger than the hypothesized value of $\pi = .667$, chance variation from sample to sample is a plausible explanation for what was observed. There is not strong evidence that the proportion who favor casino gambling is greater than 2/3.

As illustrated by Examples 10.8 and 10.9, small P-values indicate that sample results are inconsistent with H_0, whereas larger P-values are interpreted as meaning that the data is consistent with H_0 and that sampling variability alone is a plausible explanation for what was observed in the sample. As you probably noticed, the two cases examined (P-value ≈ 0 and P-value $= .1401$) were such that a decision between rejecting or not rejecting H_0 was clear-cut. A decision in other cases might not be so obvious. For example, what if the sample had resulted in P-value $= .04$? Is this unusual enough to warrant rejection of H_0? How small must the P-value be before H_0 should be rejected?

The answer depends on the significance level α (the probability of a type I error) selected for the test. For example, suppose that $\alpha = .05$. This means that the probability of rejecting a true null hypothesis is to be .05. To obtain a test procedure with this probability of type I error, the null hypothesis should be rejected if the sample result is among the most unusual 5% of all samples when H_0 is true. That is, H_0 will be rejected if the computed P-value $\leq .05$. If $\alpha = .01$ had been selected, H_0 would be rejected only if we observed a sample result so extreme that it would be among the most unusual 1% if H_0 is true (that is, if P-value $\leq .01$).

A decision as to whether H_0 should be rejected results from comparing the P-value to the chosen α:

H_0 should be rejected if P-value $\leq \alpha$.

H_0 should not be rejected if P-value $> \alpha$.

Suppose, for example, that P-value $= .0352$ and that a significance level of $.05$ is chosen. Then, since

$$P\text{-value} = .0352 \leq .05 = \alpha$$

H_0 would be rejected. This would not be the case, though, for $\alpha = .01$, since then P-value $> \alpha$.

Computing a P-value for a Large-Sample Test Concerning π

The computation of the P-value depends on the form of the inequality in H_a. Suppose, for example, that we wish to test:

$$H_0: \quad \pi = .6 \qquad \text{versus} \qquad H_a: \quad \pi > .6$$

based on a large sample. The appropriate test statistic is

$$z = \frac{p - .6}{\sqrt{\dfrac{(.6)(1 - .6)}{n}}}$$

Values of p contradictory to H_0 and much more consistent with H_a are those much *larger* than .6 (since $\pi = .6$ when H_0 is true and $\pi > .6$ when H_0 is false and H_a is true). Such values of p correspond to z values considerably greater than zero. If

$$n = 400 \qquad p = .679$$

then

$$z = \frac{.679 - .6}{\sqrt{\dfrac{(.6)(.4)}{400}}} = \frac{.079}{.025} = 3.16$$

The value $p = .687$ is a bit more than 3 standard deviations larger than what we would have expected were H_0 true. Thus,

$$
\begin{aligned}
P\text{-value} &= P(z \text{ at least as contradictory to } H_0 \text{ as } 3.16 \text{ when } H_0 \text{ is true}) \\
&= P(z \geq 3.16 \text{ when } H_0 \text{ is true}) \\
&= \text{area under the } z \text{ curve to the right of } 3.16 \\
&= 1 - .9992 \\
&= .0008
\end{aligned}
$$

This P-value is illustrated in Figure 10.1 (page 494). If H_0 were true, in the long run only 8 out of 10,000 samples would result in a z value more extreme than what

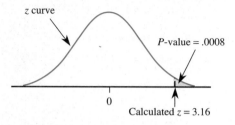

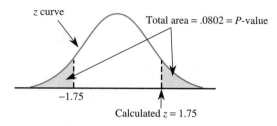

FIGURE 10.1 Calculating a *P*-value

FIGURE 10.2 *P*-value as the sum of two tail areas

actually resulted; most of us would consider such a z quite unusual. Using a significance level of .01, we would reject H_0 because *P*-value = .0008 ≤ .01 = α.

Now consider testing H_0: $\pi = .3$ versus H_a: $\pi \neq .3$. A value of p *either* much greater than .3 *or* much less than .3 is inconsistent with H_0 and provides support for H_a. Such a p corresponds to a z value far out in *either* tail of the z curve. If

$$z = \frac{p - .3}{\sqrt{\dfrac{(.3)(1 - .3)}{n}}} = 1.75$$

then (as shown in Figure 10.2)

$$
\begin{aligned}
\textit{P-value} &= P(z \text{ value at least as inconsistent with } H_0 \text{ as } 1.75 \text{ when } H_0 \text{ is true}) \\
&= P(z \geq 1.75 \text{ or } z \leq -1.75 \text{ when } H_0 \text{ is true}) \\
&= (z \text{ curve area to the right of } 1.75) + (z \text{ curve area to the left of } -1.75) \\
&= (1 - .9599) + .0401 \\
&= .0802
\end{aligned}
$$

The *P*-value in this situation would also be .0802 if $z = -1.75$, since 1.75 and -1.75 are equally inconsistent with H_0 and consistent with H_a: $\pi \neq .3$.

Determination of the *P*-value When the Test Statistic Is *z*

1. **Upper-tailed test**
   H_a: $\pi >$ hypothesized value

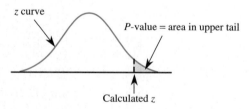

2. **Lower-tailed test**
   H_a: $\pi <$ hypothesized value

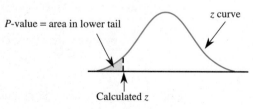

(continued)

3. **Two-tailed test**
   H_a: $\pi \neq$ hypothesized value

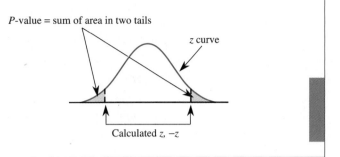

P-value = sum of area in two tails

z curve

Calculated z, $-z$

The symmetry of the z curve implies that when the test is two-tailed (the $\neq$ alternative), it is not necessary to add two curve areas. Instead,

If z is positive, P-value $= 2 \cdot$ (area to the right of z).

If z is negative, P-value $= 2 \cdot$ (area to the left of z).

EXAMPLE 10.10   In December 1999, a countywide water conservation campaign was conducted in a particular county. In January 2000, a random sample of 500 homes was selected, and water usage was recorded for each home in the sample. The county supervisors would like to know whether the data supports the claim that fewer than half of the households in the county reduced water consumption. The relevant hypotheses are

$$H_0: \quad \pi = .5 \qquad \text{versus} \qquad H_a: \quad \pi < .5$$

where π = true proportion of households in the county with reduced water usage. Suppose that the sample results were

$$n = 500 \qquad p = .440$$

Since the sample size is large and this is a lower-tailed test, we can compute the P-value by first calculating the value of the z test statistic

$$z = \frac{p - .5}{\sqrt{\dfrac{(.5)(1 - .5)}{n}}}$$

and then finding the area under the z curve to the left of this z.
Based on the observed sample data,

$$z = \frac{.440 - .5}{\sqrt{\dfrac{(.5)(1 - .5)}{500}}} = \frac{-.060}{.0224} = -2.68$$

The P-value is then equal to the area under the z curve and to the left of -2.68. From the entry in the -2.6 row and .08 column of Appendix Table II, we find

$$P\text{-value} = .0037$$

Using a .01 significance level, we would reject H_0 (since .0037 < .01), suggesting that the proportion with reduced water usage was less than .5. Notice that rejection of H_0 would not be justified if a *very* small significance level, such as .001, had been selected.

─────

Example 10.10 illustrates the calculation of a *P*-value for a lower-tailed test. The use of *P*-values in upper-tailed and two-tailed tests is illustrated in Examples 10.11 and 10.12. But first we summarize large-sample tests of hypotheses about a population proportion and introduce a step-by-step procedure for carrying out a hypothesis test.

Summary of Large-Sample z Test for π

Null hypothesis: H_0: π = hypothesized value

Test statistic: $z = \dfrac{p - \text{hypothesized value}}{\sqrt{\dfrac{(\text{hypothesized value})(1 - \text{hypothesized value})}{n}}}$

Alternative hypothesis:

H_a: π > hypothesized value

H_a: π < hypothesized value

H_a: π ≠ hypothesized value

P-value:

Area under z curve to right of calculated z

Area under z curve to left of calculated z

(i) $2 \cdot$ (area to right of z) if z is positive

(ii) $2 \cdot$ (area to left of z) if z is negative

Assumptions: 1. p is the sample proportion from a *random sample*.

2. The *sample size is large*. This test can be used if n satisfies both $n(\text{hypothesized value}) \geq 10$ and $n(1 - \text{hypothesized value}) \geq 10$.

We recommend that the following sequence of steps be used when carrying out a hypothesis test.

Steps in a Hypothesis-Testing Analysis

1. Describe the population characteristic about which hypotheses are to be tested.

2. State the null hypothesis, H_0.

3. State the alternative hypothesis, H_a.

4. Select the significance level α for the test.

5. Display the test statistic to be used, with substitution of the hypothesized value identified in step 2 but without any computation at this point.

6. Check to make sure that any assumptions required for the test are reasonable.

(continued)

7. Compute all quantities appearing in the test statistic and then the value of the test statistic itself.

8. Determine the *P*-value associated with the observed value of the test statistic.

9. State the conclusion (which will be to reject H_0 if *P*-value $\leq \alpha$ and not to reject H_0 otherwise). The conclusion should then be stated in the context of the problem, and the level of significance should be included.

Steps 1–4 constitute a statement of the problem, steps 5–8 give the analysis that will lead to a decision, and step 9 provides the conclusion.

EXAMPLE 10.11

The article "Credit Cards and College Students: Who Pays, Who Benefits?" (*J. College Student Development* (1998): 50–56) described a study of credit card payment practices of college students. According to the authors of the article, the credit card industry asserts that at most 50% of college students carry a credit card balance from month to month. However, the authors of the article report that, in a random sample of 310 college students, 217 carried a balance each month. Does this sample provide sufficient evidence to reject the industry claim? We will answer this question by carrying out a hypothesis test using a .05 significance level.

1. *Population characteristic of interest*:

π = true proportion of college students who carry a balance from month to month

2. *Null hypothesis*: H_0: $\pi = .5$

3. *Alternative hypothesis*: H_a: $\pi > .5$ (the proportion of students who carry a balance from month to month is greater than .5)

4. *Significance level*: $\alpha = .05$

5. *Test statistic*:

$$z = \frac{p - \text{hypothesized value}}{\sqrt{\dfrac{(\text{hypothesized value})(1 - \text{hypothesized value})}{n}}} = \frac{p - .5}{\sqrt{\dfrac{(.5)(.5)}{n}}}$$

6. *Assumptions*:   This test requires a random sample and a large sample size. The given sample was a random sample with a sample size of $n = 310$. Since $310(.5) \geq 10$ and $310(1 - .5) \geq 10$, the large-sample test is appropriate.

7. *Computations*: $n = 310$ and $p = 217/310 = .700$, so

$$z = \frac{.700 - .5}{\sqrt{\dfrac{(.5)(.5)}{310}}} = \frac{.200}{.028} = 7.14$$

8. *P-value*:   This is an upper-tailed test (the inequality in H_a is >), so the *P*-value is the area to the right of the computed *z* value. Since $z = 7.14$ is so far out in

the upper tail of the standard normal distribution, the area to its right is negligible. Thus,

$$P\text{-value} \approx 0$$

9. *Conclusion*: Since P-value $\leq \alpha$ ($0 \leq .05$), H_0 is rejected at the .05 level of significance. We conclude that the proportion of students who carry a credit card balance from month to month is greater than .5. That is, the sample provides convincing evidence that the industry claim is not correct.

EXAMPLE 10.12

The Associated Press (Feb. 27, 1995) reported that 71% of Americans age 25 and older are overweight, a substantial increase over the 58% figure from 1983. Although this information came from a Harris Poll (a survey) rather than a census of the population, let's assume for purposes of this example that the nationwide population proportion is exactly .71. Suppose that an investigator wishes to know whether the proportion of such individuals in her state that are overweight differs from the national proportion. A random sample of size $n = 600$ results in 450 who are classified as overweight. What can the investigator conclude? We will answer this question by carrying out a hypothesis test with $\alpha = .01$.

1. π = proportion of all residents of a particular state age 25 and over who are overweight
2. $H_0: \pi = .71$
3. $H_a: \pi \neq .71$ (differs from the national proportion)
4. *Significance level*: $\alpha = .01$
5. *Test statistic*:

$$z = \frac{p - \text{hypothesized value}}{\sqrt{\dfrac{(\text{hypothesized value})(1 - \text{hypothesized value})}{n}}} = \frac{p - .71}{\sqrt{\dfrac{(.71)(.29)}{n}}}$$

6. *Assumptions*:   This test requires a random sample and a large sample size. The given sample was a random sample and had sample size $n = 600$. Since $600(.71) \geq 10$ and $600(.29) \geq 10$, the large-sample test is appropriate.
7. *Computations*: $p = 450/600 = .750$, from which

$$z = \frac{.750 - .71}{\sqrt{\dfrac{(.71)(.29)}{600}}} = \frac{.040}{.01852} = 2.16$$

8. *P-value*:   The area under the z curve to the right of 2.16 is $1 - .9846 = .0154$, so P-value $= 2(.0154) = .0308$.

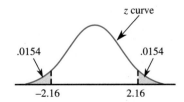

9. *Conclusion*: At significance level .01, H_0 cannot be rejected, because *P*-value = .0308 > .01 = α. The data does not convince us that the proportion in this state differs from the nationwide proportion. (Notice that the opposite conclusion would be reached had we used α = .05.)

Most statistical computer packages and graphing calculators will calculate and report *P*-values for a variety of hypothesis-testing situations, including the large-sample test for a proportion. MINITAB was used to carry out the test of Example 10.10, and the resulting computer output follows. (MINITAB uses p instead of π to denote the population proportion.)

Test and Confidence Interval for One Proportion

Test of p = 0.5 vs p < 0.5

Sample	X	N	Sample p	95.0 % CI	Z-Value	P-Value
1	220	500	0.440000	(0.396491, 0.483509)	-2.68	0.004

From the MINITAB output, $z = -2.68$ and the associated *P*-value is .004. The small difference in the value of the test statistic is the result of rounding in the hand computations of Example 10.10.

It is also possible to compute the value of the z test statistic and then use a statistical computer package or graphing calculator to determine the corresponding *P*-value as an area under the standard normal curve. For example, the user can specify a value, and MINITAB will determine the area to the left of this value for any particular normal distribution. Because of this, the computer can be used in place of Appendix Table II. In Example 10.12 (a two-tailed test), the computed z was 2.16. Using MINITAB gives

z	area to left
-2.16	0.015386
2.16	0.984614

from which

area to the left of -2.16 = .015386

area to the right of 2.16 = 1 − .984614 = .015386

and

P-value = .015386 + .015386 = .030772

which agrees with the value obtained by using the tables.

Exercises 10.20 – 10.37

10.20 a. Use the definition of the *P*-value to explain why H_0 would certainly be rejected if *P*-value = .0003.
b. Use the definition of the *P*-value to explain why H_0 would definitely not be rejected if *P*-value = .350.

10.21 For which of the given *P*-values will the null hypothesis be rejected when performing a level .05 test?
a. .001 **b.** .021 **c.** .078 **d.** .047 **e.** .148

10.22 Pairs of P-values and significance levels, α, are given. For each pair, state whether the observed P-value will lead to rejection of H_0 at the given significance level.

a. P-value = .084, α = .05
b. P-value = .003, α = .001
c. P-value = .498, α = .05
d. P-value = .084, α = .10
e. P-value = .039, α = .01
f. P-value = .218, α = .10

10.23 Let π denote the proportion of grocery store customers that use the store's "club card." For a large-sample z test of $H_0: \pi = .5$ versus $H_a: \pi > .5$, find the P-value associated with each of the given values of the test statistic.

a. 1.40 **b.** .93 **c.** 1.96 **d.** 2.45 **e.** −.17

10.24 For which of the following null hypotheses H_0 and sample sizes n is the large-sample z test appropriate?

a. $H_0: \pi = .2, n = 25$
b. $H_0: \pi = .6, n = 10$
c. $H_0: \pi = .9, n = 100$
d. $H_0: \pi = .05, n = 75$

10.25 Although arsenic is known to be a poison, it also has some beneficial medicinal uses. In one study of the use of arsenic to treat a rare type of blood cell cancer denoted by APL (for acute promyelocytic leukemia), APL patients were given an arsenic compound as part of their treatment. Of those receiving arsenic, 42% were in remission and showed no signs of leukemia in a subsequent examination (*The Washington Post,* Nov. 5, 1998). It is known that 15% of APL patients are in remission after the conventional treatment. Suppose that the study had included 40 randomly selected patients (the actual number in the study was much smaller). Is there sufficient evidence to conclude that the proportion in remission for the arsenic treatment is greater than .15, the remission proportion for the conventional treatment? Test the relevant hypotheses using a .01 significance level.

10.26 According to the article "Which Adults Do Underage Youth Ask for Cigarettes?" (*American Journal of Public Health* (1999): 1561–1564), 43.6% of the 149 18- to 19-year-olds in a random sample have been asked to buy cigarettes for an underage smoker.

a. Is there convincing evidence that fewer than half of 18- to 19-year-olds have been approached to buy cigarettes by an underage smoker? Be sure to calculate the P-value for your hypothesis test.
b. The article went on to state that of the 110 non-

smoking 18- to 19-year-olds, only 38.2% had been approached to buy cigarettes for an underage smoker. Is there evidence that less than half of nonsmoking 18- to 19-year-olds have been approached to buy cigarettes?

10.27 Many people have misconceptions about how profitable small, consistent investments can be. In a survey of 1010 randomly selected U.S. adults (Associated Press, Oct. 29, 1999), only 374 responded that they thought that an investment of $25 per week over 40 years with a 7% annual return would result in a sum of over $100,000 (the correct amount is $286,640). Is there sufficient evidence to conclude that fewer than half of U.S. adults are aware that such an investment would result in a sum of over $100,000? Test the relevant hypotheses using α = .05.

10.28 The same survey described in Exercise 10.27 also asked the individuals in the sample what they thought was their best chance to obtain more than $500,000 in their lifetime. Twenty eight percent responded "win a lottery or sweepstakes." Does this provide convincing evidence that more than a fourth of U.S. adults see a lottery or sweepstakes win as their best chance at accumulating $500,000? Test using a significance level of .01.

10.29 "Clerk Tosses Out Two Recall Petitions — Random Check Fails to Confirm Enough Valid Signatures" was the headline of an article that appeared in the *San Luis Obispo Tribune* (Sept. 9, 1999). The article states that a random check of the petitions showed that there were not enough valid signatures to force a recall election. For one of the petitions, 11,800 signatures were submitted and 10,384 valid signatures were required to force a recall election. There would be enough valid signatures to force a recall if 88% or more of the signatures could be verified. The county clerk selected a random sample of 592 signatures and verified the signature, address, and registration status against voter registration records. It was determined that 411 of the 592 were valid.

a. Is there sufficient evidence to conclude that the proportion of valid signatures on the petition is less than .88? Test the relevant hypotheses using α = .01.
b. For the hypotheses in part **a**, describe a type I error and discuss the consequence of making such an error.
c. Describe a type II error, and discuss the consequence of making such an error.
d. If the conclusion to the test in part **a** is in error, which type of error (I or II) would you be making?
e. Is the result of your hypothesis test consistent with the county clerk's decision to "toss out" the recall petition? Explain.

10.30 Duck hunting in populated areas faces opposition on the basis of safety and environmental issues. The *San Luis Obispo Telegram-Tribune* (June 18, 1991) reported the results of a survey to assess public opinion regarding duck hunting on Morro Bay (located along the central coast of California). A random sample of 750 local residents included 560 who strongly opposed hunting on the bay. Does this sample provide sufficient evidence to conclude that the majority of local residents oppose hunting on Morro Bay? Test the relevant hypotheses using $\alpha = .01$.

10.31 Drug testing of job applicants is becoming increasingly common. The Associated Press (May 24, 1990) reported that 12.1% of those tested in California tested positive. Suppose that this figure had been based on a random sample of size 600, with 73 testing positive. Does this sample support a claim that more than 10% of job applicants in California test positive for drug use?

10.32 In an experiment, ("Effect of Preparation Methods on Total Fat Content, Moisture Content, and Sensory Characteristics of Breaded Chicken Nuggets and Beef Steak Fingers," *Family and Consumer Sciences Research Journal* (1999): 18–27), 25 of 37 panelists preferred the flavor of deep-fat fried chicken nuggets to baked chicken nuggets.
a. Is there evidence to show that a clear majority of consumers prefer either baked or deep-fat fried chicken nuggets? Assume that the 37 panelists can be viewed as a random sample of consumers.
b. Describe a type I error.
c. Which sort of error, type I or type II, could not have been committed in this test.
d. In another part of this experiment, it was noted that 23 of 37 panelists preferred the flavor of baked beef steak fingers to hot-air fried beef steak fingers. Is there evidence to show that the majority of consumers prefer the flavor of baked beef steak fingers to hot-air fried beef steak fingers?

10.33 The state of Georgia's HOPE scholarship program guarantees fully paid tuition to Georgia public universities for Georgia high school seniors who have a B average in academic requirements as long as they maintain a B average in college. Of 137 randomly selected students enrolling in the Ivan Allen College at Georgia Tech (social science and humanities majors) in 1996 who had a B average going into college ("Who Loses HOPE? Attrition from Georgia's College Scholarship Program,"

Southern Economic Journal (1999): 379-390) 53.2% had a GPA below 3.0 at the end of their first year. Does this data provide convincing evidence that a majority of students at Ivan Allen College who enroll with a HOPE scholarship will lose their scholarship?

10.34 Seat belts help prevent injuries in automobile accidents, but they certainly don't offer complete protection in extreme situations. A random sample of 319 front-seat occupants involved in head-on collisions in a certain region resulted in 95 who sustained no injuries ("Influencing Factors on the Injury Severity of Restrained Front Seat Occupants in Car-to-Car Head-on Collisions," *Accid. Anal. and Prev.* (1995): 143–150). Does this suggest that the true (population) proportion of uninjured occupants exceeds .25? State and test the relevant hypotheses using a significance level of .05.

10.35 Academic dishonesty continues to be a problem on college and university campuses. One study ("Cheating at Small Colleges: An Examination of Student and Faculty Behaviors," *J. of College Student Devel.* (1994): 255–260) reported that in a random sample of 480 students, 124 said that they had looked at notes during an exam. Assuming that the survey was designed and carried out in a manner resulting in truthful responses, does it appear that more than 20% of all students in the population sampled engaged in this particular form of cheating?

10.36 White remains the most popular car color in the United States, but its popularity appears to be slipping relative to other colors. According to an annual survey by DuPont (*Los Angeles Times,* Feb. 22, 1994), white was the color of 20% of the vehicles purchased during 1993, a decline of 4% from the previous year. (According to a DuPont spokesperson, white represents "innocence, purity, honesty, and cleanliness.") A random sample of 400 cars purchased during this period in a certain metropolitan area resulted in 100 that were white. Does the proportion of all cars purchased in this area that were white appear to differ from the national percentage? Test the relevant hypotheses using $\alpha = .05$. Does your conclusion change if $\alpha = .01$ is used?

10.37 A random sample of 2700 California lawyers revealed only 1107 who felt that the ethical standards of most lawyers are high (Associated Press, Nov. 12, 1994). Does this provide strong evidence for concluding that fewer than 50% of all California lawyers feel this way? Carry out a test of appropriate hypotheses.

10.4   Hypothesis Tests for a Population Mean

We will now turn our attention to developing a method for testing hypotheses about a population mean. The test procedures in this case are based on the same two results that led to the z and t confidence intervals in Chapter 9. These results follow:

1. When either n is large or the population distribution is approximately normal, then

$$z = \frac{\bar{x} - \mu}{\dfrac{\sigma}{\sqrt{n}}}$$

   has approximately a standard normal distribution.

2. When either n is large or the population distribution is approximately normal, then

$$t = \frac{\bar{x} - \mu}{\dfrac{s}{\sqrt{n}}}$$

   has approximately a t distribution with df $= n - 1$.

A consequence of these two results is that if we are interested in testing a null hypothesis of the form

H_0: $\mu =$ hypothesized value

then, depending on whether σ is known or unknown, we can use (as long as n is large or the population distribution is approximately normal) either the accompanying z or t test statistic:

Case 1 σ known

Test statistic: $z = \dfrac{\bar{x} - \text{hypothesized value}}{\dfrac{\sigma}{\sqrt{n}}}$

P-value: Computed as an area under the z curve

Case 2 σ unknown

Test statistic: $t = \dfrac{\bar{x} - \text{hypothesized value}}{\dfrac{s}{\sqrt{n}}}$

P-value: Computed as an area under the t curve with df $= n - 1$

Since it is rarely the case that σ, the population standard deviation, is known, we will focus our attention on the test procedure for the σ unknown case.

When testing a hypothesis about a population mean, the null hypothesis will specify a particular hypothesized value for μ:

H_0: μ = hypothesized value

The alternative hypothesis will have one of the following three forms

H_a: $\mu >$ hypothesized value
H_a: $\mu <$ hypothesized value
H_a: $\mu \neq$ hypothesized value

depending on the research question being addressed. If n is large or the population distribution is approximately normal, the test statistic

$$t = \frac{\bar{x} - \text{hypothesized value}}{\dfrac{s}{\sqrt{n}}}$$

can be used. For example, if the null hypothesis to be tested is H_0: $\mu = 100$, the test statistic becomes

$$t = \frac{\bar{x} - 100}{\dfrac{s}{\sqrt{n}}}$$

Consider the alternative hypothesis H_a: $\mu > 100$, and suppose that a sample of size $n = 24$ gives $\bar{x} = 104.20$ and $s = 8.23$. The resulting test statistic value is

$$t = \frac{104.20 - 100}{\dfrac{8.23}{\sqrt{24}}} = \frac{4.20}{1.6799} = 2.50$$

Since this is an upper-tailed test, if the test statistic had been z rather than t, the P-value would be the area under the z *curve* to the right of 2.50. With a t statistic, the P-value is the area under an appropriate t *curve* (here with df = 24 − 1 = 23) to the right of 2.50. The 23-df row of the t table (Appendix Table III) used previously for calculating confidence intervals shows that 2.50 is the critical value for a central t curve area of .98. Thus, the area to the right of 2.50 under this t curve is .01 (see Figure 10.3), and this area is the P-value. That is, if H_0 were true, only 1% of all

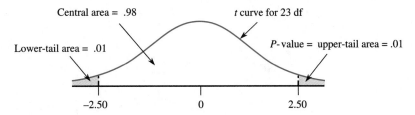

FIGURE 10.3 *P*-value for an upper-tailed *t* test

samples of size 24 would result in a value of t at least as extreme as the one we obtained. Using a significance level of .05, H_0 would be rejected, because

P-value $= .01 < .05 = \alpha$

Suppose that we had calculated $t = 2.7$ for an upper-tailed test based on 23 df. The 23-df row of Table III contains 2.50 (upper-tail area .01) and 2.81 (upper-tail area .005), but not 2.7. We can therefore not determine exactly the P-value (captured tail area); at best, we can say that it is between .005 and .01. More generally, because any particular row of Table III contains only seven critical values, for most calculated values of t the captured tail area cannot be determined exactly. To remedy this deficiency, we have included as Appendix Table IV a tabulation of t curve tail areas. Each column of the table is for a different number of degrees of freedom: 1, 2, 3, . . . , 30, 35, 40, 60, 120, and a last column for df $= \infty$, which is the same as for the z curve. The table gives the area under each t curve to the right of values ranging from 0.0 to 4.0 in increments of .1. Part of this table appears in Figure 10.4. For example,

area under the 23-df t curve to the right of 2.7 $= .006$

$= P$-value for an upper-tailed t test

Suppose that $t = -2.7$ for a lower-tailed test based on 23 df. Then, because any t curve is symmetric about 0,

P-value $=$ area to the left of $-2.7 =$ area to the right of 2.7 $= .006$

For a two-tailed z test, we had to double the captured tail area to obtain the P-value. This is also the case for a two-tailed t test. Thus, if $t = 2.7$ or if $t = -2.7$ for a 23-df two-tailed test,

P-value $= 2 \cdot (.006) = .012$

df $\diagdown$ t	1	2	...	22	23	24	...	60	120
0.0									
0.1									
⋮				⋮	⋮	⋮			
2.6			...	.008	.008	.008	...		
2.7			...	.007	.006	.006	...		
2.8			...	.005	.005	.005	...		
⋮				⋮	⋮	⋮			
4.0									

Area under 23-df t curve to right of 2.7

FIGURE 10.4 Part of Appendix Table IV: t curve tail areas

Once past 30 df, the tail areas change very little, so the last (z) column provides a very good approximation.

The following two boxes give a general description of the test procedure and show how the *P*-value is obtained as a *t* curve area or sum of areas.

Finding *P*-values for a *t* Test

1. **Upper-tailed test**
   H_a: $\mu >$ hypothesized value

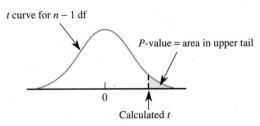

2. **Lower-tailed test**
   H_a: $\mu <$ hypothesized value

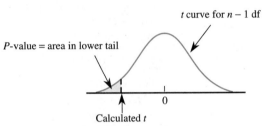

3. **Two-tailed test**
   H_a: $\mu \neq$ hypothesized value

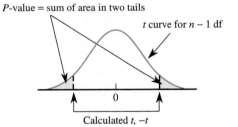

Appendix Table IV gives upper-tail *t* curve areas to the right of values 0.0, 0.1, . . . , 4.0. These areas are *P*-values for upper-tailed tests and, by symmetry, also for lower-tailed tests. Doubling an area gives the *P*-value for a two-tailed test.

The One–Sample *t* Test for a Population Mean

Null hypothesis: H_0: $\mu =$ hypothesized value

Test statistic: $t = \dfrac{\bar{x} - \text{hypothesized value}}{\dfrac{s}{\sqrt{n}}}$

(continued)

Alternative hypothesis:

P-value:

H_a: μ > hypothesized value   Area under t curve with df = $n - 1$ to right of calculated t

H_a: μ < hypothesized value   Area under t curve with df = $n - 1$ to left of calculated t

H_a: $\mu \neq$ hypothesized value

    (i) 2 · (area to right of t) if t is positive

    (ii) 2 · (area to left of t) if t is negative

Assumptions: 1. $\bar{x}$ and s are the sample mean and sample standard deviation from a *random sample.*

          2. The *sample size is large* (generally $n \geq 30$) or *the population distribution is at least approximately normal.*

EXAMPLE 10.13

The recommended daily dietary allowance (RDA) for zinc among males older than 50 years is 15 mg/day (*World Almanac*, 1992). The article "Nutrient Intakes and Dietary Patterns of Older Americans: A National Study" (*J. of Gerontology* (1992): M145–M150) reported the following data on zinc intake for a random sample of males age 65–74 years:

$$n = 115 \qquad \bar{x} = 11.3 \qquad s = 6.43$$

The researcher is concerned that males in this age group may not be getting the RDA for zinc in their diets. Is there evidence to support this concern? That is, does this data suggest that μ, the average daily zinc intake for the population of all males age 65–74, is less than the RDA? We will test

$$H_0: \quad \mu = 15 \qquad \text{versus} \qquad H_a: \quad \mu < 15$$

The null hypothesis will be rejected only if there is convincing evidence that $\mu < 15$ (that is, strong evidence against H_0).

Figure 10.5 shows a boxplot of data consistent with the given summary quantities. Roughly 75% of the sample observations are smaller than 15 (the top edge of the box is at the upper quartile). Furthermore, the observed $\bar{x}$ value, 11.3, is certainly smaller than 15, but this could be simply the result of sampling variability. That is, even when $\mu = 15$, meaning H_0 is true, the resulting $\bar{x}$ value will usually differ somewhat from 15 simply because of chance variation from one sample to another. Is it plausible that a sample mean of 11.3 occurred as a result of this chance variation, or is it unusual to observe a sample mean as small as 11.3 when $\mu = 15$?

FIGURE 10.5 Boxplot for zinc intake data

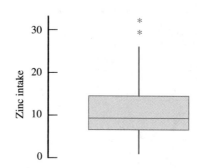

To answer this question, we will carry out a hypothesis test with a significance level of .05 using the same step-by-step procedure described in the previous section.

1. *Population characteristic of interest*:

   μ = true mean zinc intake for males age 65–74

2. *Null hypothesis*: H_0: $\mu = 15$
3. *Alternative hypothesis*: H_a: $\mu < 15$ (the mean is less than the RDA of 15)
4. *Significance level*: $\alpha = .05$
5. *Test statistic*: $t = \dfrac{\bar{x} - \text{hypothesized value}}{\dfrac{s}{\sqrt{n}}} = \dfrac{\bar{x} - 15}{\dfrac{s}{\sqrt{n}}}$

6. *Assumptions*:   This test requires a random sample and either a large sample size or approximately a normal population distribution. Since the given sample was a random sample and the sample size was $n = 115$, the t test is appropriate.
7. *Computations*: $n = 115$, $\bar{x} = 11.3$, and $s = 6.43$, so

   $$t = \frac{11.3 - 15}{\dfrac{6.43}{\sqrt{15}}} = \frac{-3.7}{.600} = -6.17$$

8. *P-value*:   This is a lower-tailed test (the inequality in H_a is $<$), so the P-value is the area to the left of the computed t value. Since df $= 115 - 1 = 114$, we can use the df $= 120$ column (closest to df $= 114$) of Appendix Table IV to find the P-value. With $t = -6.17$,

   P-value $=$ area to the left of $-6.17 \approx 0$

   (since 6.17 is greater than 4.0, the largest tabled value).

9. *Conclusion*:   Since P-value $\leq \alpha$, we reject H_0 at the .05 level of significance. There is virtually no chance of seeing a sample mean, and hence a t value, this extreme as a result of just chance variation when H_0 is true. There is convincing evidence that the mean zinc intake for males age 65–74 is less than the recommended daily allowance.

EXAMPLE 10.14

Speed, size, and strength are thought to be important factors in football performance. The article "Physical and Performance Characteristics of NCAA Division I Football Players" (*Research Quarterly for Exercise and Sport* (1990): 395–401) reported on physical characteristics of Division I starting football players in the 1988 football season. Information for teams ranked in the top 20 was easily obtained, and it was reported that the mean weight of starters on top-20 teams was 105 kg. A random sample of 33 starting players (various positions were represented) from Division I teams that were not ranked in the top 20 resulted in a sample mean weight of 103.3 kg and a sample standard deviation of 16.3 kg. Is there sufficient evidence to conclude that the mean weight for non–top-20 starters is less than 105, the known value for top-20 teams? Since the sample size is large, the $\bar{x}$ distribution will be approximately normal even though the distribution of weights of college football players is probably not normal in shape. We will carry out a test with significance level .01.

1. μ = true mean weight of starters on teams that are not ranked in the top 20
2. H_0: $\mu = 105$
3. H_a: $\mu < 105$
4. *Significance level*: $\alpha = .01$
5. *Test statistic*: $t = \dfrac{\bar{x} - \text{hypothesized value}}{\dfrac{s}{\sqrt{n}}} = \dfrac{\bar{x} - 105}{\dfrac{s}{\sqrt{n}}}$

6. *Assumptions*: This test requires a random sample and either a large sample size or a normal population distribution. The given sample was a random sample and even though the population distribution may not be normal, the sample size was $n = 33$ so use of the t test is reasonable.

7. *Computations*: $n = 33, \bar{x} = 103.3$, and $s = 16.3$, so

$$t = \dfrac{103.3 - 105}{\dfrac{16.3}{\sqrt{33}}} = \dfrac{-1.7}{2.838} = -.60$$

8. *P-value*: This is a lower-tailed test, so the *P*-value is the area under the t curve with $33 - 1 = 32$ df and to the left of the computed t value:

   P-value = area to the left of $-.60 = .277$

   (using the df = 30 column of Appendix Table IV).

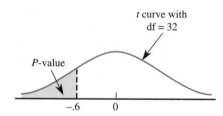

9. *Conclusion*: Since *P*-value $> .01$, we fail to reject H_0 at the .01 level of significance. There is not sufficient evidence to conclude that the mean weight of non–top-20 starters is below that of top-20 teams.

EXAMPLE 10.15   A well-designed and safe workplace can contribute greatly to increasing productivity. It is especially important that workers not be asked to perform tasks, such as lifting, that exceed their capabilities. The accompanying data on maximum weight of lift (MWAL, in kg) for a frequency of four lifts per minute was reported in the article "The Effects of Speed, Frequency, and Load on Measured Hand Forces for a Floor-to-Knuckle Lifting Task" (*Ergonomics* (1992): 833–843):

25.8   36.6   26.3   21.8   27.2

Subjects were randomly selected from the population of healthy males age 18–30. Does the data suggest that the population mean MWAL exceeds 25? Let's carry out a test of the relevant hypotheses using a .05 significance level.

1. μ = population mean MWAL
2. H_0: $\mu = 25$
3. H_a: $\mu > 25$
4. $\alpha = .05$
5. $t = \dfrac{\bar{x} - \text{hypothesized value}}{\dfrac{s}{\sqrt{n}}} = \dfrac{\bar{x} - 25}{\dfrac{s}{\sqrt{n}}}$

6. This test requires a random sample and either a large sample size or a normal population distribution. The given sample was a random sample. Since the sample size is only 5, for the t test to be appropriate, we must be able to assume that the population distribution of MWAL values is approximately normal. Is this reasonable? The accompanying graph gives a boxplot of the data.

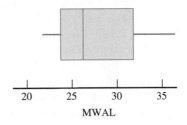

MWAL

Although the boxplot is not perfectly symmetrical, there are no outliers. It is very difficult to assess the reasonableness of normality with only five observations. Based on their understanding of MWAL values, the authors of the article thought it was reasonable to assume that the population distribution was approximately normal. So, based on this judgment, we will proceed with the t test.

7. $\Sigma x = 137.7$ and $\Sigma x^2 = 3911.97$, from which $\bar{x} = 27.54$, $s = 5.47$, and

$$t = \frac{27.54 - 25}{\dfrac{5.47}{\sqrt{5}}} = 1.0$$

8. The test is based on $n - 1 = 4$ df. Appendix Table IV shows that the area under the 4-df curve to the right of 1.0 is .187. Because the test is upper-tailed, *P*-value = .187.

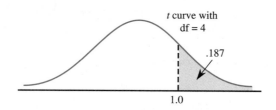

9. Since P-value $= .187 > .05 = \alpha$, H_0 cannot be rejected at this significance level. The data does not provide strong evidence for concluding that the population mean MWAL exceeds 25.

   Output from MINITAB follows. The P-value differs slightly from ours because MINITAB uses more decimal accuracy in computing t. Thus, if H_0 were true, about 18% of all samples would yield a value of t more extreme than what we obtained. Since 1.04 is not in the most extreme 5% of all values, H_0 is not rejected.

TEST OF MU = 25.000 VS MU G.T. 25.000

N	MEAN	STDEV	SE MEAN	T	P VALUE
5	27.540	5.471	2.447	1.04	0.18

EXAMPLE 10.16

One concern employers have about the use of technology is the amount of time that employees spend each day making personal use of company technology, such as personal phone, e-mail, Internet, and computer games. The Associated Press (Sept. 7, 1999) reported that a management consultant believes that, on average, workers spend 75 minutes a day making personal use of company technology. Suppose that the CEO of a large corporation wanted to determine whether the average amount of time spent in personal use of company technology for her employees was greater than the reported value of 75 minutes. Each person in a random sample of 10 employees was contacted and asked about daily personal use of company technology. (Participants would probably have to be guaranteed anonymity to obtain truthful responses.) The resulting data is given along with some summary statistics.

Employee	1	2	3	4	5	6	7	8	9	10
Time	66	70	75	88	69	89	71	71	63	86

$n = 10$ $\bar{x} = 74.80$ $s = 9.45$

Does this data provide evidence that the mean for this company is greater than 75 minutes? To answer this question, let's carry out a hypothesis test with $\alpha = .05$.

1. μ = mean daily time spent in personal use of company technology for this company

2. $H_0: \mu = 75$

3. $H_a: \mu > 75$

4. $\alpha = .05$

5. $t = \dfrac{\bar{x} - \text{hypothesized value}}{\dfrac{s}{\sqrt{n}}} = \dfrac{\bar{x} - 75}{\dfrac{s}{\sqrt{n}}}$

6. This test requires a random sample and either a large sample or a normal population distribution. The given sample was a random sample of employees. Since the sample size is small, we must be willing to assume that the population distribution of times is approximately normal. The accompanying normal probability plot appears to be reasonably straight, and although a boxplot reveals some skewness in the sample, it does not reveal any outliers. Also, the correlation between the expected normal scores and the observed data for this sample is .943, which is well above the critical r value for $n = 10$ of .880. All of these observations support the assumption of a normal population distribution, so we will proceed with the t test.

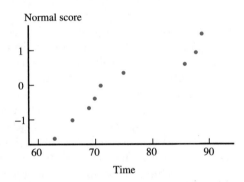

Correlations (Pearson)
Correlation of Time and Normal Score = 0.943

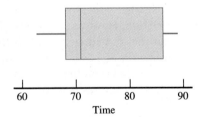

7. $t = \dfrac{74.80 - 75}{\dfrac{9.45}{\sqrt{10}}} = -.07$

8. From the df = 9 column of Appendix Table IV and rounding the test statistic value to $-.1$, we get

P-value = area to the right of $-.1 = 1 -$ area to the left of $-.1$
$= 1 - .461 = .539$

9. Because the P-value $> \alpha$, we fail to reject H_0. There is not sufficient evidence to conclude that the mean time spent in personal use of company technology is greater than 75 minutes per day for this company.

MINITAB could also have been used to carry out the test, as shown in the accompanying output.

T-Test of the Mean
Test of mu = 75.00 vs mu > 75.00

Variable	n	Mean	StDev	SE Mean	t	P
Time	10	74.80	9.45	2.99	-0.07	0.53

Although we had to round the computed t value to $-.1$ to use Appendix Table IV, MINITAB was able to compute the P-value corresponding to the actual value of the test statistic.

Statistical Versus Practical Significance

Carrying out a test amounts to deciding whether the value obtained for the test statistic could plausibly have resulted when H_0 is true. If the value does not deviate too much from what is expected when H_0 is true, there is no compelling reason for rejecting H_0 in favor of H_a. But suppose that the observed value is quite far out in the appropriate tail of the test statistic's sampling distribution when H_0 is true (for example, a large positive value of t when H_a contains the inequality $>$). One could continue to believe that H_0 is true and that such a value arose just through chance variation (a very unusual and "unrepresentative" sample). However, in this case, a more plausible explanation for what was observed is that H_0 is false and H_a is true.

When the value of the test statistic leads to rejection of H_0, it is customary to say that the result is **statistically significant** at the chosen level α. The finding of statistical significance means that, in the investigator's opinion, the observed deviation from what was expected under H_0 cannot plausibly be attributed only to chance variation. However, statistical significance cannot be equated with the conclusion that the true situation differs from what H_0 states in any practical sense. That is, even after H_0 has been rejected, the data may suggest that there is no *practical* difference between the true value of the population characteristic and what the null hypothesis states that value to be.

EXAMPLE 10.17

Let μ denote the true average score on a standardized test for children in a certain region of the United States. The average score for all children in the United States is 100. Regional education authorities are interested in testing

$$H_0: \quad \mu = 100 \qquad \text{versus} \qquad H_a: \quad \mu > 100$$

using a significance level of .001. A sample of 2500 children resulted in the values

$$n = 2500 \qquad \bar{x} = 101.0 \qquad s = 15.0$$

Then

$$t = \frac{101.0 - 100}{\frac{15}{\sqrt{2500}}} = 3.3$$

This is an upper-tailed test, so (using the z column of Appendix Table IV since df $= 2499$)

P-value $=$ area to the right of $3.33 = .000$

Because P-value $< .001$, we reject H_0. The true mean score for this region does appear to exceed 100.

However, with $n = 2500$, the point estimate $\bar{x} = 101.0$ is almost surely very close to the true value of μ. So it looks as though H_0 was rejected because $\mu \approx 101$ rather than 100. And, from a practical point of view, a 1-point difference will most likely have no significance. A statistically significant result does not necessarily mean that there are any practical consequences.

Exercises 10.38 – 10.55

10.38 Newly purchased automobile tires of a certain type are supposed to be filled to a pressure of 30 lb/in^2. Let μ denote the true average pressure. Find the P-value associated with each given z statistic value for testing H_0: $\mu = 30$ versus H_a: $\mu \neq 30$ when σ is known.
   a. 2.10   **b.** -1.75 **c.** $-.58$ **d.** 1.44   **e.** -5.00

10.39 A soda manufacturer is interested in determining whether its bottling machine tends to overfill. Each bottle is supposed to contain 12 oz of fluid. A random sample of size 36 is taken from bottles coming off the production line, and the contents of each bottle are carefully measured. It is found that the mean amount of soda for the sample of bottles is 12.1 oz. Suppose it is known that $\sigma = .2$ oz. The manufacturer will use this information to test H_0: $\mu = 12$ versus H_a: $\mu > 12$. Use the accompanying MINITAB output to answer the following questions.

TEST OF MU = 12.0000 VS MU G.T. 12.0000

	N	MEAN	STDEV	SE MEAN	Z	P VALUE
soda	36	12.1001	0.1999	0.0333	3.00	0.0014

   a. What value does the z test statistic take for this data?
   b. What is the P-value associated with the computed value of z?
   c. If the manufacturer decides on a level .05 test, should H_0 be rejected in favor of the conclusion that the machine is overfilling?

10.40 The desired percentage of silicon dioxide in a certain type of cement is 5.0. A random sample of $n = 36$ specimens gave a sample average percentage of $\bar{x} = 5.21$. Let $\mu =$ the true average percentage of silicon dioxide in this type of cement, and suppose σ is known to be .38. Test H_0: $\mu = 5$ versus H_a: $\mu \neq 5$ using a significance level of .01.

10.41 Give as much information as you can about the P-value of a t test in each of the following situations:
   a. Upper-tailed test, df $= 8$, $t = 2.0$
   b. Upper-tailed test, $n = 14$, $t = 3.2$
   c. Lower-tailed test, df $= 10$, $t = -2.4$
   d. Lower-tailed test, $n = 22$, $t = -4.2$
   e. Two-tailed test, df $= 15$, $t = -1.6$
   f. Two-tailed test, $n = 16$, $t = 1.6$
   g. Two-tailed test, $n = 16$, $t = 6.3$

10.42 Give as much information as you can about the P-value of a t test in each of the following situations:
   a. Two-tailed test, df $= 9$, $t = .73$
   b. Upper-tailed test, df $= 10$, $t = -.5$
   c. Lower-tailed test, $n = 20$, $t = -2.1$
   d. Lower-tailed test, $n = 20$, $t = -5.1$
   e. Two-tailed test, $n = 40$, $t = 1.7$

10.43 Paint used to paint lines on roads must reflect enough light to be clearly visible at night. Let μ denote the true average reflectometer reading for a new type of paint under consideration. A test of H_0: $\mu = 20$ versus H_a: $\mu > 20$ based on a sample of 15 observations gave $t = 3.2$. What conclusion is appropriate at each of the following significance levels?

   a. $\alpha = .05$ **b.** $\alpha = .01$ **c.** $\alpha = .001$

10.44 A certain pen has been designed so that true average writing lifetime under controlled conditions (involving the use of a writing machine) is at least 10 hours. A random sample of 18 pens is selected, the writing lifetime of each is determined, and a normal probability plot of the resulting data supports the use of a one-sample t test. The relevant hypotheses are H_0: $\mu = 10$ versus H_a: $\mu < 10$.

   a. If $t = -2.3$ and $\alpha = .05$ is selected, what conclusion is appropriate?

   b. If $t = -1.83$ and $\alpha = .01$ is selected, what conclusion is appropriate?

   c. If $t = .47$, what conclusion is appropriate?

10.45 The true average diameter of ball bearings of a certain type is supposed to be .5 in. What conclusion is appropriate when testing H_0: $\mu = .5$ versus H_a: $\mu \neq .5$ in each of the following situations?

   a. $n = 13, t = 1.6, \alpha = .05$

   b. $n = 13, t = -1.6, \alpha = .05$

   c. $n = 25, t = -2.6, \alpha = .01$

   d. $n = 25, t = -3.6$

10.46 There is now broad evidence that stressful work conditions and critical personal characteristics contribute to the development of coronary dysfunction and disease. The article "Job Stressors and Coping Characteristics in Work-Related Disease" (*Work and Stress* (1994): 130–140) reported that, for a random sample of 42 blue-collar workers suffering from ischemic heart disease (IHD), the sample mean systolic blood pressure and sample standard deviation were 145.3 and 17.6, respectively. Suppose that for the population of all blue-collar non-IHD workers, the mean is 138 (a value consistent with information given in the article). Does it appear that true average blue-collar IHD pressure exceeds the non-IHD value? State and test the relevant hypotheses at significance level .01 using the suggested sequence of steps.

10.47 Typically, only very brave students are willing to speak out in a college classroom. Student participation may be especially difficult if the individual is from a different culture or country. The article "An Assessment of Class Participation by International Graduate Students" (*J. of College Student Development* (1995): 132–140) considered a numerical "speaking-up" scale, with possible values from 3 to 15 (a low value means that a student rarely speaks). For a random sample of 64 males from Asian countries where English is not the official language, the sample mean and sample standard deviation were 8.75 and 2.57, respectively. Suppose that the mean for the population of all males having English as their native language is 10.0 (suggested by data in the article). Does it appear that the population mean for males from non–English-speaking Asian countries is smaller than 10.0? Employ the sequence of nine steps to analyze the data and reach a conclusion.

10.48 Are young women delaying marriage and marrying at a later age? This question was addressed in a report issued by the Census Bureau (Associated Press, June 8, 1991). The report stated that in 1970 (based on census results) the mean age of brides marrying for the first time was 20.8 years. In 1990 (based on a sample, since census results were not yet available), the mean was 23.9. Suppose that the 1990 sample mean had been based on a random sample of size 100 and that the sample standard deviation was 6.4. Is there sufficient evidence to support the claim that women are now marrying later in life? Test the relevant hypotheses using $\alpha = .01$. (*Note*: It is probably not reasonable to think that the distribution of age at first marriage is normal in shape.)

10.49 In the article "Religious Outlook, Culture War Politics, and Antipathy Toward Christian Fundamentalists" (*Public Opinion Quarterly* (1999): 29–61), it is stated that nonfundamentalists have negative feelings overall toward Christian fundamentalists. The data offered to support this claim is that the mean "temperature" rating given to Christian fundamentalists for a random sample of 960 nonfundamentalists is 47 degrees (higher temperatures mean more approval). The sample standard deviation is 21. Test the hypothesis that the mean rating given by all nonfundamentalists to Christian fundamentalists is below 57 (the average rating given across all religious, racial, and political groups).

10.50 In recent years, female athletes have been identified as a population particularly at risk for developing eating disorders. The article "Disordered Eating in Female Collegiate Gymnasts" (*J. of Sport and Exercise Psychology* (1993): 424–436) reported that for a sample of 17 gymnasts who were classified as binge eaters, the sample mean difference between current weight and ideal weight was 6.65 lb. The accompanying boxplot is based on data consistent with this information; the sample standard deviation is 2.97 lb. Suppose that the true average

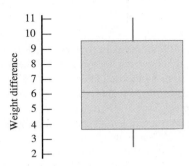

weight difference for female gymnasts with normal eating habits is 4.5 lb (a value suggested in the article).

a. Does the boxplot indicate that μ, the true average weight difference for bingers, exceeds the value for normal eaters?

b. Carry out a test of the hypotheses suggested in part **a**.

10.51 An article titled "Teen Boys Forget Whatever it Was" appeared in the Australian newspaper *The Mercury* (April 21, 1997). It describes a study of academic performance and attention span, and reports that the mean time to distraction for teenage boys working on an independent task was 4 minutes. Although the sample size was not given in the article, suppose that this mean was based on a random sample of 50 teenage Australian boys and that the sample standard deviation was 1.4 min. Does this data provide convincing evidence that the average attention span for teenage boys is less than 5 minutes? Test the relevant hypotheses using $\alpha = .01$.

10.52 According to the article "Workaholism in Organizations: Gender Differences" (*Sex Roles* (1999): 333–346), the following data was reported on 1996 income for random samples of male and female MBA graduates from a certain Canadian business school:

	n	$\bar{x}$	*s*
Males	258	$133,442	$131,090
Females	233	$105,156	$98,525

(*Note*: These salary figures are in Canadian dollars.)

a. Test the hypothesis that the mean salary of a male MBA grad from this school was in excess of $100,000 in 1996.

b. Is there convincing evidence that the mean salary for all female MBA grads is above $100,000? Test using $\alpha = .10$.

c. If a significance level of .05 or .01 were used instead of .05 in the test of part **b**, would you still reach the same conclusion? Explain.

10.53 Many consumers pay careful attention to stated nutritional contents on packaged foods when making purchases. It is therefore important that the information on packages be accurate. A random sample of $n = 12$ frozen dinners of a certain type was selected from production during a particular period, and the calorie content of each one was determined. (This determination entails destroying the product, so a census would certainly not be desirable!) Here are the resulting observations, along with a boxplot and normal probability plot.

255	244	239	242	265	245
259	248	225	226	251	233

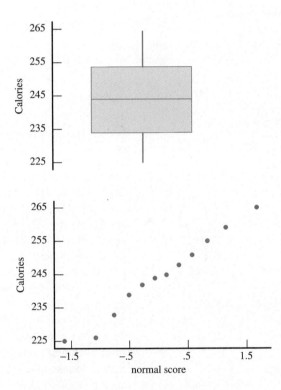

a. Is it reasonable to test hypotheses about true average calorie content μ by using a *t* test?

b. The stated calorie content is 240. Does the boxplot suggest that true average content differs from the stated value? Explain your reasoning.

c. Carry out a formal test of the hypotheses suggested in part **b**.

10.54 Much concern has been expressed in recent years regarding the practice of using nitrates as meat preservatives. In one study involving possible effects of these chemicals, bacteria cultures were grown in a medium containing nitrates. The rate of uptake of radio-labeled amino acid was then determined for each one, yielding the accompanying observations:

7251 6871 9632 6866 9094 5849 8957 7978
7064 7494 7883 8178 7523 8724 7468

Suppose that it is known that true average uptake for cultures without nitrates is 8000. Does the data suggest that the addition of nitrates results in a decrease in true average uptake? Test the appropriate hypotheses using a significance level of .10. Be sure to state any assumptions that are necessary to validate the use of your test.

10.55 The times of first sprinkler activation (sec) for a series of tests of fire-prevention sprinkler systems that use an aqueous film-forming foam were as follows:

27 41 22 27 23 35 30
33 24 27 28 22 24

("Use of AFFF in Sprinkler Systems," *Fire Tech.* (1976): 5). The system has been designed so that the true average activation time is supposed to be at most 25 sec. Does the data strongly indicate that the design specifications have not been met?

a. Test the relevant hypotheses using a significance level of .05. What assumptions, if any, are you making about the distribution of activation times?

b. MINITAB was used to test H_0: $\mu = 25$ versus H_a: $\mu > 25$. Is the output consistent with your calculations in part **a**?

TEST OF MU = 25.000 VS MU G.T. 25.000

	N	MEAN	STDEV	SE MEAN	T	P VALUE
time	13	27.923	5.619	1.559	1.88	0.043

10.5   Power and Probability of Type II Error

In this chapter, we have introduced test procedures for testing hypotheses about a population characteristic. What characterizes a "good" test procedure? It makes sense to think that a good test procedure is one that has both a small probability of rejecting H_0 when it is true (a type I error) and a high probability of rejecting H_0 when it is false. The test procedures presented in this chapter allow us to directly control the probability of rejecting a true H_0 by our choice of the significance level, α. But what about the probability of rejecting H_0 when it is false? As we will see, several factors influence this probability. Let's begin by considering an example.

Suppose that the student body president at a university is interested in studying the amount of money that students spend on textbooks each semester. The director of the Financial Aid Office believes that the average amount spent on books is $300 per semester and uses this figure in determining the amount of financial aid for which a student is eligible. The student body president plans to ask each individual in a random sample of students how much he or she spent on books this semester and has decided to use the resulting data to test

$$H_0: \quad \mu = 300 \qquad \text{versus} \qquad H_a: \quad \mu > 300$$

using a significance level of .05. If the true mean is 300 (or less than 300), the correct decision would be to fail to reject the null hypothesis (incorrectly rejecting the null hypothesis would be a type I error). On the other hand, if the true mean is 325 or 310 or even 301, the correct decision would be to reject the null hypothesis (not rejecting would be a type II error). How likely is it that the null hypothesis will, in fact, be rejected? If the true mean is 301, the probability that we will reject H_0: $\mu = 300$ is not very great. This is because when we carry out the test, we are essentially looking at the sample mean and asking, Does this look like what we would have

FIGURE 10.6 Sampling distribution of $\bar{x}$ when $\mu =$ 300, 305, and 325

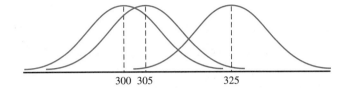

expected to see if the population mean were 300? As illustrated in Figure 10.6, if the true mean is greater than, but very close to, 300, chances are that the sample mean will look pretty much like what we would have expected to see if the population mean had been 300, and we will be unconvinced that the null hypothesis should be rejected. If the true mean is 325, it is less likely that the sample will be mistaken for a sample from a population with mean 300; sample means will tend to cluster around 325, and so it is more likely that we will correctly reject H_0. If the true mean is 350, rejection of H_0 is even more likely.

When we consider the probability of rejecting the null hypothesis, we are looking at what statisticians refer to as the **power** of the test.

The **power of a test** is the probability of rejecting the null hypothesis.

From the previous discussion, it should be apparent that the power of the test when a hypothesis about a population mean is being tested depends on the true value of the population mean, μ. Because the true value of μ is unknown (if we knew the value of μ, we wouldn't be doing the hypothesis test!), we cannot know what the power is for the actual true value of μ. It is possible, however, to gain some insight into the power of a test by looking at a number of "what if" scenarios. For example, we might ask, What is the power if the true mean is 325?, What is the power if the true mean is 310?, and so on. That is, we can calculate the power at $\mu = 325$, the power at $\mu = 310$, and at any other value of interest.

Although it is technically possible to consider power when the null hypothesis is true, an investigator is usually concerned about the power only at values for which the null hypothesis is false. In general, when testing a hypothesis about a population characteristic, there are three factors that influence the power of the test:

1. The size of the difference between the true value of the population characteristic and the hypothesized value (the value that appears in the null hypothesis)

2. The choice of significance level, α, for the test

3. The sample size

The effect of each of these factors on the power of a test is described in the accompanying box.

Effect of Various Factors on the Power of a Test

1. The larger the size of the discrepancy between the hypothesized value and the true value of the population characteristic, the higher the power.
2. The larger the significance level, α, the higher the power of the test.
3. The larger the sample size, the higher the power of the test.

Let's consider each of these three statements. The first has already been discussed in the context of the textbook example. Since power is the probability of rejecting the null hypothesis, it makes sense that power will be higher when the true value of a population characteristic is very different from the hypothesized value than when it is close to it.

The effect of significance level on power is not quite as obvious. To understand the relationship between power and significance level, it helps to see the relationship between power and β, the probability of a type II error.

When H_0 is false, power $= 1 - \beta$.

This relationship follows from the definitions of power and type II error. A type II error results from *not* rejecting a false H_0. Since power is the probability of rejecting H_0, it follows that *when H_0 is false*

$$\begin{aligned} \text{power} &= \text{probability of rejecting a false } H_0 \\ &= 1 - \text{probability of not rejecting a false } H_0 \\ &= 1 - \beta \end{aligned}$$

Recall from Section 10.2 that the choice of α, the type I error probability, affects the value of β, the type II error probability. Choosing a larger value for α results in a smaller value for β (and thus a larger value for $1 - \beta$). In terms of power, this means that choosing a larger value for α results in a larger value for the power of the test. That is, the larger the type I error probability we are willing to tolerate, the more likely it is that the test will be able to detect any particular departure from H_0.

The third factor that affects the power of a test is the sample size. When H_0 is false, the power of a test is the probability that we will "detect" that it is false and, based on the observed sample, reject H_0. Intuition suggests that we will be more likely to detect a departure from H_0 with a large sample than with a small sample. This is in fact the case — the larger the sample size, the higher the power.

Consider testing the hypotheses presented previously:

$$H_0: \quad \mu = 300 \qquad \text{versus} \qquad H_a: \quad \mu > 300$$

These observations about power imply the following, for example:

1. The power of a test based on a sample of size 100 will be higher than the power of a test based on a sample of size 75 (assuming the same significance level).
2. The power of a test using a significance level of .05 will be higher than the power of a test using a significance level of .01 (assuming the same sample size).
3. The power of the test will be greater if the true mean is 350 than if the true mean is 325 (assuming the same sample size and significance level).

As was mentioned previously in this section, it is impossible to calculate the *actual* power of a test because in practice we do not know the true value of population characteristics. However, we can evaluate the power at a selected alternative value if we want to know whether the power would be high or low if this were in fact the true value.

The following optional subsection shows how type II error probabilities and power can be evaluated for selected tests.

Calculating Power and Type II Error Probabilities for Selected Tests (Optional)

The test procedures presented in this chapter are designed to control the probability of a type I error (rejecting H_0 when H_0 is true) at the desired level α. However, little has been said so far about calculating the value of β, the probability of a type II error (not rejecting H_0 when H_0 is false). Here we consider the determination of β and power for the hypothesis tests previously introduced.

When we carry out a hypothesis test, we specify the desired value of α, the probability of a type I error. The probability of a type II error, β, is the probability of not rejecting H_0 even though it is false. Suppose we are testing

$$H_0: \quad \mu = 1.5 \qquad \text{versus} \qquad H_a: \quad \mu > 1.5$$

Since we do not know the true value of μ, we cannot calculate the actual value of β. However, the vulnerability of the test to type II error can be investigated by calculating β for several different values of μ, such as $\mu = 1.55$, $\mu = 1.6$, and $\mu = 1.7$. Once the value of β has been determined, the power of the test at the corresponding alternative value is just $1 - \beta$.

EXAMPLE 10.18

A cigarette manufacturer claims that the mean nicotine content of its cigarettes is 1.5 mg. We might investigate this claim by testing

$$H_0: \quad \mu = 1.5 \qquad \text{versus} \qquad H_a: \quad \mu > 1.5$$

where μ = true mean nicotine content. A random sample of $n = 36$ cigarettes is to be selected and the resulting data used to reach a conclusion. Suppose that the standard deviation of nicotine content (σ) is known to be .20 mg and that a significance level of .01 is to be used. Our test statistic is (since $\sigma = .20$)

$$z = \frac{\bar{x} - 1.5}{\dfrac{.20}{\sqrt{n}}} = \frac{\bar{x} - 1.5}{\dfrac{.20}{\sqrt{36}}} = \frac{\bar{x} - 1.5}{.0333}$$

The inequality in H_a implies that

P-value = area under z curve to the right of calculated z

From Appendix Table II, it is easily verified that the z critical value 2.33 captures an upper-tail z curve area of .01. Thus, rejecting H_0 if P-value $\leq$ .01 is equivalent to the decision rule

reject H_0 if calculated $z \geq 2.33$

This becomes

reject H_0 if $\dfrac{\bar{x} - 1.5}{.0333} \geq 2.33$

Solving this inequality for $\bar{x}$, we get

$\bar{x} \geq 1.5 + 2.33(.0333)$

or

$\bar{x} \geq 1.578$

So if $\bar{x} \geq 1.578$, we will reject H_0, and if $\bar{x} < 1.578$, we will fail to reject H_0. This decision rule corresponds to $\alpha = .01$.

Suppose now that $\mu = 1.6$ (H_0 is false). A type II error will then occur if $\bar{x} < 1.578$. What is the probability that this occurs? If $\mu = 1.6$, the sampling distribution of $\bar{x}$ is approximately normal, centered at 1.6, and has a standard deviation of .0333. The probability of observing an $\bar{x}$ value less than 1.578 can then be determined by finding an area under a normal curve with mean 1.6 and standard deviation .0333, as illustrated in Figure 10.7.

Since this is not the standard normal (z) curve, we must first convert to a z score before using Appendix Table II to find the area. Here,

$$z \text{ score for } 1.578 = \frac{1.578 - \mu_{\bar{x}}}{\sigma_{\bar{x}}} = \frac{1.578 - 1.6}{.0333} = -.66$$

and

area under z curve to left of $-.66 = .2546$

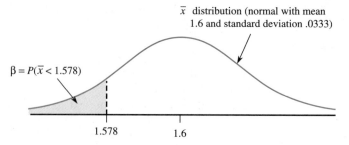

FIGURE 10.7 β when $\mu = 1.6$ in Example 10.18

So, if $\mu = 1.6$, $\beta = .2546$. This means that if μ is 1.6, about 25% of all samples would result in $\bar{x}$ values less than 1.578 and failure to reject H_0.

The power of the test at $\mu = 1.6$ is then

$$
\begin{aligned}
(\text{power at } \mu = 1.6) &= 1 - (\beta \text{ when } \mu \text{ is } 1.6) \\
&= 1 - .2546 \\
&= .7454
\end{aligned}
$$

Thus, if the true mean is 1.6, the probability of rejecting H_0: $\mu = 1.5$ in favor of H_a: $\mu > 1.5$ is .7454. That is, if μ is 1.6 and the test is used repeatedly with random samples selected from the population, in the long run about 75% of the samples will result in the correct conclusion to reject H_0.

Now consider β and power when $\mu = 1.65$. The normal curve in Figure 10.7 would then be centered at 1.65. Since β is the area to the left of 1.578 and the curve has shifted to the right, β decreases. Converting 1.578 to a z score and using Appendix Table II gives $\beta = .0154$. Also

$$
(\text{power at } \mu = 1.65) = 1 - .0154 = .9846
$$

As expected, the power at $\mu = 1.65$ is higher than the power at $\mu = 1.6$ because 1.65 is farther from the hypothesized value of 1.5.

MINITAB will calculate power for specified values of σ, α, n, and the difference between the true and hypothesized value of μ. The accompanying output shows power calculations corresponding to those in Example 10.18.

1-Sample Z Test

Testing mean = null (versus > null)
Alpha = 0.01   Sigma = 0.2   Sample Size = 36

Difference	Power
0.10	0.7497
0.15	0.9851

The probability of a type II error and power for z tests concerning a population proportion is calculated in an analogous manner.

EXAMPLE 10.19

A package delivery service advertises that at least 90% of all packages brought to its office by 9 A.M. for delivery in the same city are delivered by noon that day. Let π denote the proportion of all such packages actually delivered by noon. The hypotheses of interest are

$$
H_0: \quad \pi = .9 \qquad \text{versus} \qquad H_a: \quad \pi < .9
$$

where the alternative hypothesis states that the company's claim is untrue. The value $\pi = .8$ represents a substantial departure from the company's claim. If the hypotheses are tested at level .01 using a sample of $n = 225$ packages, what is the probability that the departure from H_0 represented by this alternative value will go undetected?

At significance level .01, H_0 is rejected if P-value $\leq .01$. For the case of a lower-tailed test, this is the same as rejecting H_0 if

$$z = \frac{p - \mu_p}{\sigma_p} = \frac{p - .9}{\sqrt{\frac{(.9)(.1)}{225}}} = \frac{p - .9}{.02} \leq -2.33$$

(Since -2.33 captures a lower-tail z curve area of .01, the smallest 1% of all z values satisfy $z \leq -2.33$.) This inequality is equivalent to $p \leq .853$, so H_0 will *not* be rejected if $p > .853$. When $\pi = .8$, p has approximately a normal distribution with

$$\mu_p = .8 \qquad \sigma_p = \sqrt{\frac{(.8)(.2)}{225}} = .0267$$

Then β is the probability of obtaining a sample proportion greater than .853, as illustrated in Figure 10.8.

FIGURE 10.8 β when $\pi = .8$ in Example 10.19

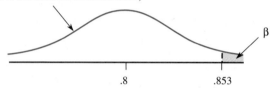

Sampling distribution of p (normal with mean .8 and standard deviation .0267)

Converting to a z score results in

$$z = \frac{.853 - .8}{.0267} = 1.99$$

and Appendix Table II gives

$$\beta = 1 - .9767 = .0233$$

When $\pi = .8$ and a level .01 test is used, less than 3% of all samples of size $n = 225$ will result in a type II error. The power of the test at $\pi = .8$ is $1 - .0233 = .9767$. This means that the probability of rejecting H_0: $\pi = .9$ in favor of H_a: $\pi < .9$ when π is really .8 is .9767, which is quite high.

β and Power for the t Test (Optional)

Power and β values for t tests can be determined by using a set of graphs specially constructed for this purpose. As with the z test, the value of β depends not only on the true value of μ, but also on the selected significance level α; β increases as α is made smaller. In addition, β depends on the number of degrees of freedom, $n - 1$. For any fixed level α, it should be easier for the test to detect a specific departure from H_0 when n is large than when n is small. This is indeed the case; for a fixed alternative value, β decreases as $n - 1$ increases.

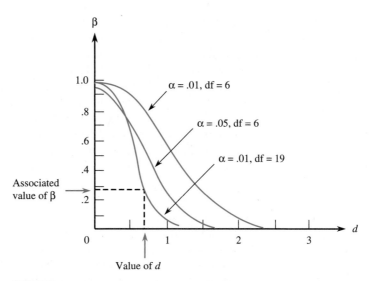

FIGURE 10.9 β curves for the one-tailed t test

Unfortunately, there is one other quantity on which β depends: the population standard deviation σ. As σ increases, so does $\sigma_{\bar{x}}$. This in turn makes it more likely that an $\bar{x}$ value far from μ will be observed, resulting in an erroneous conclusion. Once α is specified and n is fixed, the determination of β at a particular alternative value of μ requires that a value of σ be chosen, since each different value of σ yields a different value of β. (This did not present a problem with the z test because when using a z test the value of σ is known.) If the investigator can specify a range of plausible values for σ, then using the largest such value will give a pessimistic β (one on the high side) and a pessimistic value of power (one on the low side).

Figure 10.9 shows three different β curves for a one-tailed t test (appropriate for $H_a: \mu >$ hypothesized value or for $H_a: \mu <$ hypothesized value). A more complete set of curves for both one- and two-tailed tests when $\alpha = .05$ and when $\alpha = .01$ appears in Appendix Table V. To determine β, first compute the quantity

$$d = \frac{|\text{alternative value} - \text{hypothesized value}|}{\sigma}$$

Then locate d on the horizontal axis, move directly up to the curve for $n - 1$ df, and move over to the vertical axis to read β.

EXAMPLE 10.20

Consider testing

$$H_0: \quad \mu = 100 \qquad \text{versus} \qquad H_a: \quad \mu > 100$$

and focus on the alternative value $\mu = 110$. Suppose that $\sigma = 10$, the sample size is $n = 7$, and a significance level of .01 has been selected. For $\sigma = 10$,

$$d = \frac{|110 - 100|}{10} = \frac{10}{10} = 1$$

Figure 10.9 (using df $= 7 - 1 = 6$) gives $\beta \approx .6$. The interpretation is that if $\sigma = 10$ and a level .01 test based on $n = 7$ is used when $\mu = 110$ (and thus H_0 is false), roughly 60% of all samples will result in erroneously not rejecting H_0! Equivalently, the power of the test at $\mu = 110$ is only $1 - .6 = .4$. The probability of rejecting H_0 when $\mu = 110$ is not very large. If a level .05 test is used instead, then $\beta \approx .3$, which is still rather large. Using a level .01 test with $n = 20$ (df $= 19$) yields, from Figure 10.9, $\beta \approx .05$. At the alternative value $\mu = 110$, for $\sigma = 10$ the level .01 test based on $n = 20$ has smaller β than the level .05 test with $n = 7$. Substantially increasing n counterbalances using the smaller α.

Now consider the alternative $\mu = 105$, again with $\sigma = 10$, so that

$$d = \frac{|105 - 100|}{10} = \frac{5}{10} = .5$$

Then from Figure 10.9, $\beta \approx .95$ when $\alpha = .01, n = 7$; $\beta \approx .7$ when $\alpha = .05, n = 7$; and $\beta \approx .65$ when $\alpha = .01, n = 20$. These values of β are all quite large; with $\sigma = 10, \mu = 105$ is too close to the hypothesized value of 100 for any of these three tests to have a good chance of detecting such a departure from H_0. A substantial decrease in β necessitates using a much larger sample size. For example, from Table V, $\beta \approx .08$ when $\alpha = .05$ and $n = 40$.

The curves in Figure 10.9 also give β when testing H_0: $\mu = 100$ versus H_a: $\mu < 100$. If the alternative value $\mu = 90$ is of interest and $\sigma = 10$,

$$d = \frac{|90 - 100|}{10} = \frac{10}{10} = 1$$

and values of β are the same as those given in the first paragraph of this example.

Since curves for only selected degrees of freedom appear in Table V, other degrees of freedom require a visual approximation. For example, the 27-df curve (for $n = 28$) would lie between the 19- and 29-df curves, which do appear, and it would be closer to the latter. This type of approximation is adequate because it is the general magnitude of β — large, small, or moderate — that is of primary concern.

It is also possible to use MINITAB to evaluate power for the t test. For example, the accompanying output shows MINITAB calculations for power at $\mu = 110$ for samples sized 7 and 20 when $\alpha = .01$. The corresponding approximate values from Table V found in Example 10.20 are fairly close to the MINITAB values.

1-Sample t Test

Testing mean = null (versus > null)
Calculating power for mean = null + 10
Alpha = 0.01   Sigma = 10

Sample Size	Power
7	0.3968
20	0.9653

The β curves in Table V are those for t tests. When the alternative corresponds to a value of d relatively close to zero, β for a t test may be rather large. One might ask whether there is another type of test that has the same level of significance α as does the t test and smaller values of β. The following result provides the answer to this question.

> When the population distribution is normal, the t test for testing hypotheses about μ has smaller β than does any other test procedure that has the same level of significance α.

Stated another way, among all tests with level of significance α, the t test makes β as small as it can possibly be. In this sense, the t test is a best test. Statisticians have also shown that when the population distribution is not too far from a normal distribution, no test procedure can improve on the t test (have the same α and substantially smaller β) by very much. But when the population distribution is believed to be very nonnormal (heavy-tailed, highly skewed, or multimodal), the t test should not be used. Then it's time to consult your friendly neighborhood statistician, who could provide you with alternative methods of analysis.

Exercises 10.56 – 10.62

10.56 The power of a test is influenced by the sample size and the choice of significance level.
a. Explain how increasing the sample size affects the power (when significance level is held fixed).
b. Explain how increasing the significance level affects the power (when sample size is held fixed).

10.57 Water samples are taken from water used for cooling as it is being discharged from a power plant into a river. It has been determined that as long as the mean temperature of the discharged water is at most 150°F, there will be no negative effects on the river ecosystem. To investigate whether the plant is in compliance with regulations that prohibit a mean discharge water temperature above 150°, 50 water samples will be taken at randomly selected times, and the water temperature of each sample will be recorded. A z statistic

$$z = \frac{\bar{x} - 150}{\frac{\sigma}{\sqrt{n}}}$$

will then be used to decide between the hypotheses

$$H_0: \quad \mu = 150 \quad \text{and} \quad H_a: \quad \mu > 150$$

where μ is the true mean temperature of discharged water. Assume that σ is known to be 10.
a. Explain why use of the z statistic would be appropriate in this setting.
b. Describe type I and type II errors in this context.
c. The rejection of H_0 when $z \geq 1.8$ corresponds to what value of α? (That is, what is the area under the z curve to the right of 1.8?)
d. Suppose that the true value for μ is 153 and that H_0 is to be rejected if $z \geq 1.8$. Draw a sketch (similar to that of Figure 10.7) of the sampling distribution of $\bar{x}$, and shade the region that would represent β, the probability of making a type II error.
e. For the hypotheses and test procedure described, compute the value of β when $\mu = 153$.
f. For the hypotheses and test procedure described, what value would β take if $\mu = 160$?
g. If H_0 is rejected when $z \geq 1.8$ and $\bar{x} = 152.4$, what is the appropriate conclusion? What type of error might have been made in reaching this conclusion?

10.58 Let μ denote the true average lifetime for a certain type of pen under controlled laboratory conditions.

A test of H_0: $\mu = 10$ versus H_a: $\mu < 10$ will be based on a sample of size 36. Suppose that σ is known to be .6, from which $\sigma_{\bar{x}} = .1$. The appropriate test statistic is then

$$z = \frac{\bar{x} - 10}{.1}$$

a. What is α for the test procedure that rejects H_0 if $z \leq -1.28$?

b. If the test procedure of part **a** is used, calculate β when $\mu = 9.8$, and interpret this error probability.

c. Without doing any calculation, explain how β when $\mu = 9.5$ compares to β when $\mu = 9.8$. Then verify your assertion by computing β when $\mu = 9.5$.

d. What is the power of the test when $\mu = 9.8$? When $\mu = 9.5$?

10.59 The city council in a large city has become concerned about the trend toward exclusion of renters with children in apartments within the city. The housing coordinator has decided to select a random sample of 125 apartments and determine for each whether children would be permitted. Let π be the true proportion of apartments that prohibit children. If π exceeds .75, the city council will consider appropriate legislation.

a. If 102 of the 125 sampled exclude renters with children, would a level .05 test lead you to the conclusion that more than 75% of all apartments exclude children?

b. What is the power of the test when $\pi = .8$ and $\alpha = .05$?

10.60 The amount of shaft wear after a fixed mileage was determined for each of seven randomly selected internal combustion engines, resulting in a mean of .0372 in and a standard deviation of .0125 in.

a. Assuming that the distribution of shaft wear is normal, test at level .05 the hypotheses H_0: $\mu = .035$ versus

H_a: $\mu > .035$.

b. Using $\sigma = .0125$, $\alpha = .05$, and Table V, what is the value of β, the probability of a type II error, when $\mu = .04$?

c. What is the power of the test when $\mu = .04$ and $\alpha = .05$?

10.61 Optical fibers are used in telecommunications to transmit light. Current technology allows production of fibers that will transmit light about 50 km (*Research at Rensselaer,* 1984). Researchers are trying to develop a new type of glass fiber that will increase this distance. In evaluating a new fiber, it would be of interest to test H_0: $\mu = 50$ versus H_a: $\mu > 50$, with μ denoting the true average transmission distance for the new optical fiber.

a. Assuming $\sigma = 10$ and $n = 10$, use Table V to find β, the probability of a type II error, for each of the given alternative values of μ when a level .05 test is employed.

   i. 52   **ii.** 55   **iii.** 60   **iv.** 70

b. What happens to β in each of the cases in part **a** if σ is actually larger than 10? Explain your reasoning.

10.62 Let μ denote the true average diameter for bearings of a certain type. A test of H_0: $\mu = .5$ versus H_a: $\mu \neq .5$ will be based on a sample of n bearings. The diameter distribution is believed to be normal. Determine the value of β in each of the following cases:

a. $n = 15$, $\alpha = .05$, $\sigma = .02$, $\mu = .52$

b. $n = 15$, $\alpha = .05$, $\sigma = .02$, $\mu = .48$

c. $n = 15$, $\alpha = .01$, $\sigma = .02$, $\mu = .52$

d. $n = 15$, $\alpha = .05$, $\sigma = .02$, $\mu = .54$

e. $n = 15$, $\alpha = .05$, $\sigma = .04$, $\mu = .54$

f. $n = 20$, $\alpha = .05$, $\sigma = .04$, $\mu = .54$

g. Is the way in which β changes as n, α, σ, and μ vary consistent with your intuition? Explain.

10.6   Interpreting the Results of Statistical Analyses

The step-by-step procedure that we have proposed for testing hypotheses provides a systematic approach for carrying out a complete test. However, when the results of such a test are reported in a journal article, it is common to find only the value of the test statistic and the associated *P*-value accompanying the discussion of conclusions drawn from the data analyzed. When a statistical computer package is used to perform a hypothesis test, the output generally includes the value of the test statistic and the *P*-value, leaving the rest of the analysis and interpretation to the user.

For example, the article "Forgeries Found on Jackson Petition" (*Austin American Statesman,* Jan. 23, 1988) indicates that forgeries were discovered on nominating petitions submitted to qualify Jesse Jackson for the 1988 Texas presidential primary. Five thousand valid signatures were required to qualify for the ballot, and 8500 signatures were submitted. However, because of the reported forgeries, it was unclear whether enough valid signatures were submitted. Jackson would qualify for the ballot as long as the proportion of valid signatures was greater than 5000/8500 = .588. A sample of size 162 was randomly selected from the names that appeared on the petition, and 46 of the 162 individuals selected denied signing the petition. The article states that Jackson's spokesperson was sure that "the petition contained at least 5000 valid signatures." Is this a reasonable conclusion based on the sample data?

The z test for a population proportion would be appropriate for testing

$$H_0: \quad \pi = .588 \qquad \text{versus} \qquad H_a: \quad \pi > .588$$

where π represents the true proportion of valid signatures on the petition. Typical computer output for this test follows.

Test for One Proportion
Test of p = 0.588 vs p > 0.588

Sample	X	N	Sample p	Z-Value	P-Value
1	116	162	0.716049	3.31	0.0005

The stated P-value would lead to the rejection of H_0 for any $\alpha > .0005$, and so the data supports the statement by Jackson's spokesperson.

What to Look for in Published Data

Here are some questions to consider when you are reading a report that contains the results of a hypothesis test.

■ What hypotheses are being tested? Are the hypotheses about a population mean, a population proportion, or some other population characteristic?

■ Was the appropriate test used? Does the validity of the test depend on any assumptions about the population sampled? If so, are the assumptions reasonable?

■ What is the P-value associated with the test? What is the significance level selected for the test? Is the chosen significance level reasonable?

■ Are the conclusions drawn consistent with the results of the hypothesis test?

For example, consider the following statement (based on information in the article "Serum Transferrin Receptor for the Detection of Iron Deficiency in Pregnancy," *Amer. J. Clinical Nutrition* (1991): 1077–1081):

In a total sample of 176 pregnant women, mean serum receptor concentration did not differ significantly from 5.63, the mean for women who are not pregnant ($P > 0.10$).

The statement does not indicate what test was performed or what the value of the test statistics was. It appears that the hypotheses of interest are

$$H_0: \quad \mu = 5.63 \qquad \text{versus} \qquad H_a: \quad \mu \neq 5.63$$

where μ represents the true mean serum receptor concentration for pregnant women. Since the sample size is large, if the sample can be considered a random sample the one-sample t test would be appropriate. With the large sample size, no assumptions about the shape of the population distribution of serum receptor concentration values is necessary. Because the reported P-value is so large (P-value $>$ .10), there is no reason to reject H_0. We cannot conclude that the mean for pregnant women differs from the known mean of 5.63 for women who are not pregnant.

Summary of Key Concepts and Formulas

Term or Formula	Comment
Hypothesis	A claim about the value of a population characteristic.
Null hypothesis, H_0	The hypothesis initially assumed to be true. It has the form H_0: population characteristic = hypothesized value.
Alternative hypothesis, H_a	A hypothesis that specifies a claim that is contradictory to H_0 and is judged the more plausible claim when H_0 is rejected.
Type I error	Rejection of H_0 when H_0 is true; the probability of a type I error is denoted by α and is referred to as the significance level for the test.
Type II error	Nonrejection of H_0 when H_0 is false; the probability of a type II error is denoted by β.
Test statistic	The quantity computed from sample data and used to make a decision between H_0 and H_a.
P-value	The probability, computed assuming H_0 to be true, of obtaining a value of the test statistic at least as contradictory to H_0 as what actually resulted. H_0 is rejected if P-value $\leq \alpha$ and not rejected if P-value $> \alpha$, where α is the chosen significance level.
$z = \dfrac{p - \text{hypothesized value}}{\sqrt{\dfrac{(\text{hyp. val})(1 - \text{hyp. val})}{n}}}$	A test statistic for testing H_0: π = hypothesized value when the sample size is large. The P-value is determined from the z curve.
$z = \dfrac{\bar{x} - \text{hypothesized value}}{\dfrac{\sigma}{\sqrt{n}}}$	A test statistic for testing H_0: μ = hypothesized value when σ is known and either the population distribution is normal or the sample size is large. The P-value is determined from the z curve.
$t = \dfrac{\bar{x} - \text{hypothesized value}}{\dfrac{s}{\sqrt{n}}}$	A test statistic for testing H_0: μ = hypothesized value when σ is unknown and either the population distribution is normal or the sample size is large. The P-value is determined from the t curve with df = $(n - 1)$.
Power	The power of a test is the probability of rejecting the null hypothesis. Power is affected by the size of the difference between the hypothesized value and the true value, the sample size, and the significance level.

Supplementary Exercises 10.63 – 10.75

10.63 When a published article reports the results of many hypothesis tests, the *P*-values are not usually given. Instead, the following type of coding scheme is frequently used ("Hostility in Depression," *Psychological Reports* (1994): 1391–1401): $*p < .05, ^\dagger p < .01, ^\ddagger p < .001,$ $^a p < .0001$. Which of the symbols would be used to code for each of the following *P*-values?

  a. .037   **b.** .0026   **c.** .072   **d.** .0003

10.64 A random sample of $n = 44$ individuals with a B.S. degree in accounting who had started with a "Big Eight" accounting firm and subsequently changed jobs resulted in a sample mean time to change of 35.02 months and a sample standard deviation of 18.94 months ("The Debate Over Post-Baccalaureate Education: One University's Experience," *Issues in Accounting Educ.* (1992): 18–36). Can it be concluded that the true average time to change exceeds 2 years? Test the appropriate hypotheses using a significance level of .01.

10.65 What motivates companies to offer stock ownership plans to their employees? In a random sample of 87 companies having such plans, 54 said that the primary rationale was tax-related ("The Advantages and Disadvantages of ESOPs: A Long-Range Analysis," *J. of Small Business Management* (1991): 15–21). Does this information provide strong support for concluding that more than half of all such firms feel this way?

10.66 Sixty-five randomly selected clerical workers at a large financial service organization participated in a health risk analysis, and the resulting data was summarized in the article "Workplace Stress and Indicators of Coronary-Disease Risk" (*Academy of Mgmt. J.* (1988): 686–698). For systolic blood pressure, the sample mean and standard deviation were 111.63 and 11.94, respectively. Is there sufficient evidence to conclude that the mean systolic blood pressure for all clerical workers at this business exceeds 110? Test the relevant hypotheses using a .01 significance level.

10.67 The article "Caffeine Knowledge, Attitudes, and Consumption in Adult Women" (*J. of Nutrition Educ.* (1992): 179–184) reported the following summary data on daily caffeine consumption for a random sample of adult women:

$n = 47$ $\bar{x} = 215$ mg $s = 235$ mg   range: 5 to 1176

  a. Does it appear plausible that the population distribution of daily caffeine consumption is normal? Is it

necessary to assume a normal population distribution to test hypotheses about the value of the population mean consumption? Explain your reasoning.
  b. Suppose that it had previously been believed that mean consumption was at most 200 mg. Does the given data contradict this prior belief? Test the appropriate hypotheses at significance level .10.

10.68 A standard method for recovering minerals and metals from biological materials results in a mean copper recovery of 63 ppm when used to treat oyster tissue. A new treatment method was described in the article "Simple Sample Digestion of Sewage and Sludge for Multi-Element Analysis" (*J. Environ. Sci. and Health* (1984): 959–972). Suppose that this new treatment is used to treat $n = 40$ bits of oyster tissue, resulting in a sample mean copper recovery and a sample standard deviation of 62.6 ppm and 3.7 ppm, respectively. Is there evidence to suggest that the mean copper recovery is lower for the new method than for the standard one? Use a .01 significance level.

10.69 Past experience has indicated that the true response rate is 40% when individuals are approached with a request to fill out and return a particular questionnaire in a stamped and addressed envelope. An investigator believes that if the person distributing the questionnaire is stigmatized in some obvious way, potential respondents would feel sorry for the distributor and thus tend to respond at a rate higher than 40%. To investigate this theory, a distributor is fitted with an eyepatch. Of the 200 questionnaires distributed by this individual, 109 were returned. Does this strongly suggest that the response rate in this situation does exceed the rate in the past? State and test the appropriate hypotheses at significance level .05.

10.70 An automobile manufacturer who wishes to advertise that one of its models achieves 30 miles per gallon (mpg) decides to carry out a fuel efficiency test. Six nonprofessional drivers are selected, and each one drives a car from Phoenix to Los Angeles. The resulting figures are, in mpg,

  27.2   29.3   31.2   28.4   30.3   29.6

Assuming that fuel efficiency (mpg) is normally distributed under these circumstances, does the data contradict the claim that true average fuel efficiency is (at least) 30? Use of MINITAB gave the output at the top of the next page.

TEST OF MU = 30.0 VS MU L.T. 30.0

N	MEAN	STDEV	SE MEAN	T	P VALUE
6	29.33	1.41	0.57	-1.16	0.15

What would you conclude, and why?

10.71 The effect of discharging wastewater from a dairy processing plant into groundwater used to irrigate kidney beans was examined in the article "Effect of Industrial Dairy Processing Effluent on Soil and Crop Plants" (*Environ. Pollution* (1984): 97–106). The wastewater was rich in bicarbonates and calcium, so it was thought that irrigating with a 50% solution of wastewater might promote growth. Suppose that 40 kidney bean plants were irrigated with this mixture, resulting in a sample mean root length of 5.46 cm and a sample standard deviation of .55 cm. The mean root length for kidney bean plants irrigated with uncontaminated water is known to be 5.20 cm. Does this data support the hypothesis that irrigation with the 50% wastewater solution results in a mean root length that is greater than 5.20? Use a .05 significance level.

10.72 The article referenced in Exercise 10.71 also gave information on root length for pearl millet. When irrigated with uncontaminated water, the mean root length is 6.40 cm. A sample of 40 plants irrigated with a 50% wastewater solution resulted in a sample mean root length of 4.76 cm and a sample standard deviation of .48 cm. Does the data strongly suggest that irrigation with the wastewater mixture results in a mean root length that differs from 6.40? Use a .05 level test.

10.73 A student organization uses the proceeds from a particular soft-drink dispensing machine to finance its activities. The price per can had been $.50 for a long time, and the average daily revenue during that period had been $50.00. The price was recently increased to $.60 per can. A random sample of $n = 20$ days subsequent to the price increase yielded a sample average revenue and sample standard deviation of $47.30 and $4.20, respectively. Does this data suggest that the true average daily revenue has decreased from its value prior to the price increase? Test the appropriate hypotheses using $\alpha = .05$.

10.74 A hot-tub manufacturer advertises that with its heating equipment, a temperature of 100°F can be achieved in at most 15 min. A random sample of 25 tubs is selected, and the time necessary to achieve a 100°F temperature is determined for each tub. The sample average time and sample standard deviation are 17.5 min and 2.2 min, respectively. Does this data cast doubt on the company's claim? Carry out a test of hypotheses using significance level .05.

10.75 Let π denote the proportion of voters in a certain state who favor a particular proposed constitutional amendment. Consider testing H_0: $\pi = .5$ versus H_a: $\pi > .5$ at significance level .05 based on a sample of size $n = 50$.

   a. Suppose that H_0 is in fact true. Use Appendix Table I (our table of random numbers) to simulate selecting a sample, and use the resulting data to carry out the test.

   b. If you repeated part **a** a total of 100 times (a simulation consisting of 100 replications), how many times would you expect H_0 to be rejected?

   c. Now suppose that H_0 is false because $\pi = .6$. Again use Appendix Table I to simulate selecting a sample, and carry out the test. If you repeated this a total of 100 times, would you expect H_0 to be rejected more frequently than when H_0 is true?

References

The books by Freedman et al. and by Moore listed in previous chapter references are excellent sources. Their orientation is primarily conceptual, with a minimum of mathematical development, and both sources offer many valuable insights.

GRAPHING CALCULATOR EXPLORATIONS

10.1 The Hypothesis Test for a Population Proportion

Using your calculator's hypothesis-testing capability begins, as usual, by navigating your calculator's menu system, this time looking for key words such as "hypothesis" and "tests." As before with confidence intervals, look for things like "1" and "prop." and "z."

Once you select the correct choice, you will be presented with a screen for providing information. The information you must provide will be exactly what you would need to test the hypothesis on paper: the sample number of successes, the sample size, the level of significance, etc. Here are two representative screens, with information from Example 10.10 in the text. (The screen on the right has the "<" alternative hypothesis selected, although the shading doesn't show up in the figure.)

<table>
<tr><td>

1-Prop Ztest
Prop   : $<p_0$
p_0 :.5
x :220
n :500
Execute

</td><td>

1-Prop ZTest
p_0:.5
x:200
n:500
Prop $\neq p_0$ $<p_0$ $>p_0$
Calculate Draw

</td></tr>
</table>

Move your cursor down to Execute or Calculate, and press the Enter, Execute, and Calculate button, depending on your calculator. The results should appear immediately. Again, here are two representative screens:

<table>
<tr><td>

1-Prop Ztest
Prop<0.5
$Z = -2.683281573$
$p = .003645226$
$\hat{p} = .44$
$n = 500$

</td><td>

1-Prop ZInt
prop$<.5$
$z = -2.683281573$
$p = .003645226$
$\hat{p} = .44$
$n = 500$

</td></tr>
</table>

Notice the slight difference between the text and the calculator answers due to rounding in the hand calculations.

The calculator does the work for only a few of the steps in a hypothesis test. Remember, there is still work for you to do in completing the other steps!

10.2 The Hypothesis Test for a Population Mean

Testing a hypothesis for a single mean on your calculator will require you to navigate the menu system once again. As was true with the confidence interval for the mean,

1. You must decide whether to base the test on the z or a t distribution, and

2. You may use previous calculations of the sample mean and standard deviation, or the calculator will evaluate these statistics from data contained in a List.

We will follow Example 10.16, and use the *t* test after entering the data in List1. The normality of the population must be assessed as before. A calculator boxplot and normal probability plot are shown here.

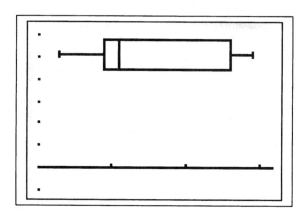

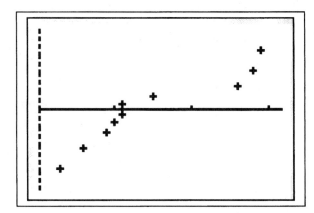

After assessing the plausibility of the normality of the population, we will test the hypotheses. Again we have the choice of entering the sample calculations or providing data in a list and letting the calculator evaluate the sample mean and standard deviation. Based on the choice, we see one of the following screens.

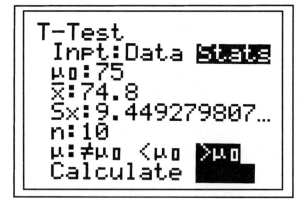

The accompanying figure shows calculator output for the hypothesis test.

```
T-Test
  μ>75
  t=-.0669316122
  P=.5259502751
  x̄=74.8
  Sx=9.449279808
  n=10
█
```

Remember that you must complete some steps in the hypothesis-testing analysis yourself.

11

Comparing Two Populations or Treatments

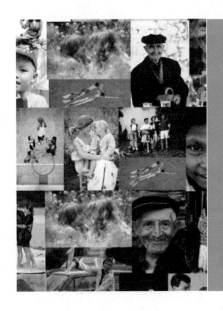

INTRODUCTION

Many investigations are carried out for the purpose of comparing two populations or treatments. For example, the article "Learn More, Earn More?" (ETS Policy Notes (1999)) described how the job market treats high school graduates who do not go to college. By comparing data from a random sample of students who had high grades in high school with data from a random sample of students who had low grades, the authors of the article were able to investigate whether the proportion employed for those with high grades was higher than the proportion employed for those with low grades. The authors were also interested in whether the mean monthly salary was higher for those with good grades than for those who did not earn good grades. To answer these questions, hypothesis tests that compare the proportions or means for two different populations were used. This chapter introduces hypothesis tests and confidence intervals that can be used when comparing two populations or treatments.

11.1 Inferences Concerning the Difference Between Two Population or Treatment Means Using Independent Samples

In this section, we consider using sample data to compare two population means or two treatment means. An investigator may wish to estimate the difference between two population means or to test hypotheses about this difference. For example, a university financial aid director may want to determine whether the mean cost of textbooks is different for students enrolled in the Engineering College than for students enrolled in the Liberal Arts College. Here, two populations (one consisting of all students enrolled in the Engineering College and the other consisting of all students enrolled in the Liberal Arts College) are to be compared on the basis of their respective means. Information from a random sample from each population could be the basis for making such a comparison.

In other cases, an experiment might be carried out to compare two different treatments or to compare the effect of a treatment with the effect of no treatment (treatment versus control). For example, an agricultural experimenter might wish to compare weight gains for animals placed on two different diets (each diet is a treatment), or an educational researcher might wish to compare on-line instruction to traditional classroom instruction by studying the difference in mean scores on a common final exam (each type of instruction is a treatment).

In previous chapters, the symbol μ was used to denote the mean of a single population under study. When comparing two populations or treatments, we must use notation that distinguishes between the characteristics of the first and those of the second. This is accomplished by using subscripts on quantities such as μ and σ^2. Similarly, subscripts on sample statistics, such as $\bar{x}$, indicate to which sample these quantities refer.

Notation

	Mean Value	Variance	Standard Deviation
Population or treatment 1	μ_1	σ_1^2	σ_1
Population or treatment 2	μ_2	σ_2^2	σ_2

	Sample Size	Mean	Variance	Standard Deviation
Sample from population or treatment 1	n_1	$\bar{x}_1$	s_1^2	s_1
Sample from population or treatment 2	n_2	$\bar{x}_2$	s_2^2	s_2

Comparison of means focuses on the difference, $\mu_1 - \mu_2$. When $\mu_1 - \mu_2 = 0$, the two population or treatment means are identical. That is,

$\mu_1 - \mu_2 = 0$ is equivalent to $\mu_1 = \mu_2$

Similarly,

$\mu_1 - \mu_2 > 0$ is equivalent to $\mu_1 > \mu_2$

and

$\mu_1 - \mu_2 < 0$ is equivalent to $\mu_1 < \mu_2$

We will first consider the problem of comparing two population means.

Before developing inferential procedures concerning $\mu_1 - \mu_2$, we must consider how the two samples, one from each population, are selected. Two samples are said to be **independent** if the selection of the individuals or objects that make up one sample does not influence the selection of those in the other sample. However, when observations from the first sample are paired in some meaningful way with observations in the second sample, the data is said to be **paired.** For example, to study the effectiveness of a speed-reading course, the reading speed of subjects could be measured before they take the class and again after they complete the course. This gives rise to two related samples — one from the population of individuals who have not taken this particular course (the "before" measurements) and one from the population of individuals who have had such a course (the "after" measurements). These samples are paired. The two samples are not independently chosen, since the selection of individuals from the first (before) population completely determines which individuals make up the sample from the second (after) population. In this section, we consider procedures based on independent samples. Methods for analyzing data resulting from paired samples are presented in Section 11.2.

Because $\bar{x}_1$ provides an estimate of μ_1 and $\bar{x}_2$ gives an estimate of μ_2, it is natural to use $\bar{x}_1 - \bar{x}_2$ as a point estimate of $\mu_1 - \mu_2$. The value of $\bar{x}_1$ varies from sample to sample (it is a *statistic*), as does the value of $\bar{x}_2$. Since the difference $\bar{x}_1 - \bar{x}_2$ is calculated from sample values, it is also a statistic and therefore has a sampling distribution. Our inferential methods will be based on information about the sampling distribution of $\bar{x}_1 - \bar{x}_2$.

Properties of the Sampling Distribution of $\bar{x}_1 - \bar{x}_2$

If the random samples on which $\bar{x}_1$ and $\bar{x}_2$ are based are selected independently of one another, then

1. $\mu_{\bar{x}_1 - \bar{x}_2} = \left(\begin{array}{c} \text{the mean value} \\ \text{of } \bar{x}_1 - \bar{x}_2 \end{array} \right) = \mu_{\bar{x}_1} - \mu_{\bar{x}_2} = \mu_1 - \mu_2$

   Thus, the sampling distribution of $\bar{x}_1 - \bar{x}_2$ is always centered at the value of $\mu_1 - \mu_2$, so $\bar{x}_1 - \bar{x}_2$ is an unbiased statistic for estimating $\mu_1 - \mu_2$.

2. $\sigma^2_{\bar{x}_1 - \bar{x}_2} = \left(\begin{array}{c} \text{the variance of} \\ \bar{x}_1 - \bar{x}_2 \end{array} \right) = \sigma^2_{\bar{x}_1} + \sigma^2_{\bar{x}_2} = \dfrac{\sigma_1^2}{n_1} + \dfrac{\sigma_2^2}{n_2}$

   and

   $\sigma_{\bar{x}_1 - \bar{x}_2} = \left(\begin{array}{c} \text{standard deviation of} \\ \bar{x}_1 - \bar{x}_2 \end{array} \right) = \sqrt{\dfrac{\sigma_1^2}{n_1} + \dfrac{\sigma_2^2}{n_2}}$

3. If n_1 and n_2 are both large or the population distributions are (at least approximately) normal, $\bar{x}_1$ and $\bar{x}_2$ each have (at least approximately) a normal distribution. This implies that the sampling distribution of $\bar{x}_1 - \bar{x}_2$ is also normal or approximately normal.

Properties 1 and 2 follow from the following general results:

1. The mean value of a difference is the difference of the two individual mean values.

2. The variance of a difference of *independent* quantities is the *sum* of the two individual variances.

When the sample sizes are large or the population distributions are (at least approximately) normal, the properties of the sampling distribution of $\bar{x}_1 - \bar{x}_2$ imply that $\bar{x}_1 - \bar{x}_2$ can be standardized to obtain a variable with a sampling distribution that is (at least approximately) the standard normal (z) curve. This gives the following result.

When n_1 and n_2 are both large or the population distributions are (at least approximately) normal, the distribution of

$$z = \frac{\bar{x}_1 - \bar{x}_2 - (\mu_1 - \mu_2)}{\sqrt{\dfrac{\sigma_1^2}{n_1} + \dfrac{\sigma_2^2}{n_2}}}$$

is described (at least approximately) by the standard normal (z) distribution.

Although it is possible to base a test procedure and confidence interval on this result, the values of σ_1^2 and σ_2^2 are rarely known. As a result, the applicability of z is limited. When σ_1^2 and σ_2^2 are unknown, we must estimate them using the corresponding sample variances, s_1^2 and s_2^2. The result on which a test procedure and confidence interval are based is given in the accompanying box.

If n_1 and n_2 are both large or if the population distributions are normal and when the two random samples are independently selected, the standardized variable

$$t = \frac{\bar{x}_1 - \bar{x}_2 - (\mu_1 - \mu_2)}{\sqrt{\dfrac{s_1^2}{n_1} + \dfrac{s_2^2}{n_2}}}$$

has approximately a t distribution with

$$\text{df} = \frac{(V_1 + V_2)^2}{\dfrac{V_1^2}{n_1 - 1} + \dfrac{V_2^2}{n_2 - 1}} \qquad \text{where } V_1 = \frac{s_1^2}{n_1} \text{ and } V_2 = \frac{s_2^2}{n_2}$$

df should be truncated (rounded down) to an integer.

If one or both sample sizes are small, normal probability plots or boxplots can be used to evaluate whether it is reasonable to assume that the population distributions are normal.

Test Procedures

In a test designed to compare two population means, the null hypothesis will be of the form

$$H_0: \quad \mu_1 - \mu_2 = \text{hypothesized value}$$

Often the hypothesized value will be zero, indicating that there is no difference between the population means. The alternative hypothesis will involve the same hypothesized value but will use one of the three inequalities $>$, $<$, or $\neq$. As an example, let μ_1 and μ_2 denote the average fuel efficiencies (mpg) for two models of a certain type of car, equipped with 4-cylinder and 6-cylinder engines, respectively. The hypotheses under consideration might be

$$H_0: \quad \mu_1 - \mu_2 = 5 \qquad \text{versus} \qquad H_a: \quad \mu_1 - \mu_2 > 5$$

With this null hypothesis, the researcher claims that average efficiency for the 4-cylinder engine exceeds the average efficiency for the 6-cylinder engine by exactly 5 mpg. The alternative hypothesis states that the difference between the true average efficiencies is more than 5 mpg.

A test statistic is obtained by replacing $\mu_1 - \mu_2$ in the standardized t variable (given in the previous box) by the hypothesized value which appears in H_0. Thus, the t statistic for testing $H_0: \mu_1 - \mu_2 = 5$ is

$$t = \frac{\bar{x}_1 - \bar{x}_2 - 5}{\sqrt{\dfrac{s_1^2}{n_1} + \dfrac{s_2^2}{n_2}}}$$

When H_0 is true and the sample sizes are large or the population distributions are normal, the sampling distribution of the test statistic is approximately a t distribution. The P-value for the test is obtained by first computing the appropriate df and then using Appendix Table IV. The next box gives a general description of the test procedure.

Summary of the Two-Sample t Test for Comparing Two Population Means

Null hypothesis: $H_0: \mu_1 - \mu_2 = \text{hypothesized value}$

Test statistic: $t = \dfrac{\bar{x}_1 - \bar{x}_2 - (\mu_1 - \mu_2)}{\sqrt{\dfrac{s_1^2}{n_1} + \dfrac{s_2^2}{n_2}}}$

The appropriate df for the two-sample t test is

$$df = \frac{(V_1 + V_2)^2}{\dfrac{V_1^2}{n_1 - 1} + \dfrac{V_2^2}{n_2 - 1}} \qquad \text{where } V_1 = \frac{s_1^2}{n_1} \text{ and } V_2 = \frac{s_2^2}{n_2}$$

df should be truncated (rounded down) to an integer.

(continued)

Alternative hypothesis:

$H_a: \mu_1 - \mu_2 >$ hypothesized value

P-value:

Area under appropriate t curve to the right of the computed t

$H_a: \mu_1 - \mu_2 <$ hypothesized value

Area under appropriate t curve to the left of the computed t

$H_a: \mu_1 - \mu_2 \neq$ hypothesized value

(i) $2 \cdot$ (area to the right of the computed t) if t is positive

or

(ii) $2 \cdot$ (area to the left of the computed t) if t is negative

Assumptions: 1. The two samples are *independently selected random samples*.
2. The *sample sizes are large* (generally 30 or larger)
    OR
    the population distributions are (at least approximately) normal.

EXAMPLE 11.1

The article "Affective Variables Related to Mathematics Achievement Among High-Risk College Freshmen" (*Psych. Reports* (1991): 399–403) examines the relationship between attitudes toward mathematics and success at college-level mathematics. Twenty men and thirty-eight women selected at random from those identified as being at high risk of failure (because they did not meet the usual admission requirements for the university) participated in the study. Each student was asked to respond to a series of questions, and the answers were combined to obtain a math anxiety score. For this particular scale, the higher the score, the lower the level of anxiety toward mathematics. Summary values appear in the accompanying table.

	n	$\bar{x}$	s
Males	20	35.9	11.9
Females	38	36.6	12.3

Does this data provide evidence that, as many researchers have hypothesized, the mean anxiety score for women is different than that for men? Using the nine-step procedure introduced in Chapter 10, we will test the relevant hypotheses using a .05 level of significance.

1. $\mu_1 =$ true mean math anxiety score for male at-risk students
      $\mu_2 =$ true mean math anxiety score for female at-risk students
   $\mu_1 - \mu_2 =$ difference in mean math anxiety scores

2. $H_0: \mu_1 - \mu_2 = 0$

3. $H_a: \mu_1 - \mu_2 \neq 0$

4. *Significance level*: $\alpha = .05$

5. *Test statistic*: $t = \dfrac{(\bar{x}_1 - \bar{x}_2) - \text{hypothesized value}}{\sqrt{\dfrac{s_1^2}{n_1} + \dfrac{s_2^2}{n_2}}} = \dfrac{(\bar{x}_1 - \bar{x}_2) - 0}{\sqrt{\dfrac{s_1^2}{n_1} + \dfrac{s_2^2}{n_2}}}$

6. *Assumptions*:   The samples were selected independently (the selection of men did not influence the selection of women), and we are told that the selec-

tion was random. Since the sample sizes are not both large, we must be willing to assume that the population distributions are approximately normal to justify using the t statistic. To decide whether this is reasonable, we would have to examine the actual data in each sample. The article gave only summary statistics (no raw data), but if the data from the two samples were available, we could construct boxplots or normal probability plots. For purposes of this example, let's suppose that the boxplots looked like those in the accompanying picture. Since the boxplots are reasonably symmetric and there are no outliers, the assumption of normality would be plausible.

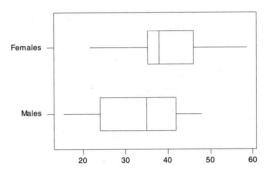

7. *Calculation:* $t = \dfrac{(35.9 - 36.6) - 0}{\sqrt{\dfrac{(11.9)^2}{20} + \dfrac{(12.3)^2}{38}}} = \dfrac{-.70}{\sqrt{7.081 + 3.981}} = \dfrac{-.70}{3.326} = -.21$

8. *P-value:*   We first compute the df for the two-sample t test:

$$V_1 = \frac{s_1^2}{n_1} = 7.081 \qquad V_2 = \frac{s_2^2}{n_2} = 3.981$$

$$\text{df} = \frac{(V_1 + V_2)^2}{\dfrac{V_1^2}{n_1 - 1} + \dfrac{V_2^2}{n_2 - 1}} = \frac{(7.081 + 3.981)^2}{\dfrac{(7.081)^2}{19} + \dfrac{(3.981)^2}{37}} = \frac{122.3678}{3.0673} = 39.89$$

We truncate df to 39. Appendix Table IV shows that the area under the t curve with 40 df (as close to 39 as possible in the table) to the right of .2 is .421. By symmetry, the area to the left of $-.2$ is also .421. Because this is a two-tailed test,

$$P\text{-value} \approx 2(.421) = .842$$

9. *Conclusion:*   Since $.842 > .05$, H_0 is not rejected. The data does not provide strong evidence that the mean math anxiety score of at-risk males differs from that of at-risk females.

EXAMPLE 11.2

Tennis elbow is thought to be aggravated by the impact experienced when hitting the ball. The article "Forces on the Hand in the Tennis One-Handed Backhand" (*Int. J. of Sport Biomechanics* (1991): 282–292) reported the force (N) on the hand just after impact on a one-handed backhand drive for six advanced players and for eight intermediate players. Summary statistics from the article, as well as data consistent with these summary quantities, appear in the accompanying table.

	Data								n	$\bar{x}$	s
Advanced	44.7	26.31	55.75	28.54	46.99	39.46			6	40.3	11.3
Intermediate	15.58	19.16	24.13	10.56	32.88	21.47	14.32	33.09	8	21.4	8.3

The authors of the article assumed in their analysis of the data that both force distributions (advanced and intermediate) were normal. We will use the given information to determine whether the mean force after impact is greater for advanced tennis players than it is for intermediate players.

1. μ_1 = mean force after impact for advanced players
      μ_2 = mean force after impact for intermediate players
   $\mu_1 - \mu_2$ = difference in mean force

2. H_0: $\mu_1 - \mu_2 = 0$

3. H_a: $\mu_1 - \mu_2 > 0$

4. *Significance level*: $\alpha = .01$

5. *Test statistic*: $t = \dfrac{(\bar{x}_1 - \bar{x}_2) - \text{hypothesized value}}{\sqrt{\dfrac{s_1^2}{n_1} + \dfrac{s_2^2}{n_2}}} = \dfrac{(\bar{x}_1 - \bar{x}_2) - 0}{\sqrt{\dfrac{s_1^2}{n_1} + \dfrac{s_2^2}{n_2}}}$

6. *Assumptions*:   For the two-sample t test to be appropriate, we must be willing to assume that the two samples can be viewed as independently selected random samples from the populations of advanced and intermediate tennis players. Since the sample sizes are both small, it is necessary to assume that the force distribution is approximately normal for each of these two populations. Boxplots constructed using the sample data are shown here, and they are consistent with the authors assumption of normal population distributions.

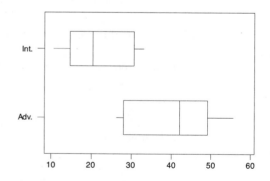

7. *Calculation*: $t = \dfrac{(40.3 - 21.4) - 0}{\sqrt{\dfrac{(11.3)^2}{6} + \dfrac{(8.3)^2}{8}}} = \dfrac{18.9}{5.467} = 3.46$

8. *P-value*: We first compute df for the two-sample *t* test:

$$V_1 = \frac{s_1^2}{n_1} = 21.282 \qquad V_2 = \frac{s_2^2}{n_2} = 8.611$$

$$df = \frac{(V_1 + V_2)^2}{\dfrac{V_1^2}{n_1 - 1} + \dfrac{V_2^2}{n_2 - 1}} = \frac{(21.282 + 8.611)^2}{\dfrac{(21.282)^2}{5} + \dfrac{(8.611)^2}{7}} = \frac{893.5914}{101.1775} = 8.8321$$

Truncating to an integer gives df = 8. Appendix Table IV shows that the area under the *t* curve with 8 df to the right of 3.5 is .004. Because this is an upper-tailed test,

P-value ≈ .004

9. *Conclusion*:   Since *P*-value ≤ α (.004 ≤ .01), we reject H_0. There is sufficient evidence to conclude that the mean force after impact is greater for advanced players than it is for intermediate players.

Many statistical computer packages will perform the calculations for the two-sample *t* test. The accompanying MINITAB output shows the df, test statistic value, and *P*-value for the hypothesis test of Example 11.2.

Two sample T for Adv. vs Int.

	N	Mean	StDev	SE Mean
Adv.	6	40.3	11.3	4.6
Int.	8	21.40	8.30	2.9

T-Test mu Adv. = mu Int. (vs >): T = 3.46 P = 0.0043 DF = 8

Comparing Treatments

When an experiment is carried out to compare two treatments (or a single treatment with a control), the investigator is interested in the effect of the treatments on some response variable. The treatments are "applied" to individuals (as in an experiment to compare two different medications for decreasing blood pressure) or objects (as in an experiment to compare two different baking temperatures on the density of bread), and the value of some response variable (e.g., blood pressure, density) is recorded. Based on the resulting data, the investigator might wish to determine whether there is a difference in the mean response for the two treatments.

In many actual experimental situations, the individuals or objects to which the treatments will be applied are not selected at random from some larger population. A consequence of this is that it is not possible to generalize the results of the experiment to some larger population. However, *if the experimental design provides for random assignment of treatments to the individuals or objects used in the experiment (or random assignment of the individuals or objects to treatments), it is possible to test hypotheses about treatment differences.*

It is common practice to use the two-sample *t* test statistic previously described if the experiment employs random assignment and if either the samples sizes are large or it is reasonable to think that the treatment response distributions (the distributions of response values that would result if the treatments were applied to a *very* large number of individuals or objects) are approximately normal.

Two-Sample *t* Test for Comparing Two Treatments

When

1. *treatments are randomly assigned* to individuals or objects (or vice versa; that is, individuals or objects are randomly assigned to treatments), and

2. the *sample sizes are large* (generally 30 or larger) OR the *treatment response distributions are approximately normal,*

the two-sample *t* test can be used to test H_0: $\mu_1 - \mu_2 = $ *hypothesized value,* where μ_1 and μ_2 represent the mean response for treatments 1 and 2, respectively.

In this case, these two conditions replace the assumptions previously stated for comparing two population means. The validity of the assumption of normality of the treatment response distributions can be assessed by constructing a normal probability plot or a boxplot of the values in each sample.

Although the two-sample *t* test to compare two treatments when the individuals or objects used in the experiment are not randomly selected from some population is only an approximate test (the reported *P*-values are only approximate), this is still the most common way to analyze such data.

EXAMPLE 11.3

The article "Effects of Learning-Style Intervention on College Student's Achievement, Anxiety, Anger, and Curiosity" (*J. of College Student Development* (1994): 461–466) reported on an experiment designed to compare a treatment group with a control group. First-year nursing students enrolled in a science course were randomly assigned to one of two groups. Students in the control group were provided with conventional study-skill guidelines. Each student in the treatment group was evaluated to determine his or her preferred learning style. Students in the treatment group were then provided with homework prescriptions based on their identified learning style, as well as the conventional study-skill guidelines. Does the treatment (the homework prescriptions based on learning style) improve science grades? We will test the relevant hypotheses using a significance level of .01.

Group	Sample Size	Sample Mean	Sample SD
Control	100	2.75	.85
Treatment	103	3.28	.70

1. Let μ_1 denote the mean science score for the treatment *conventional study-skill guidelines* and define μ_2 analogously for the treatment *homework prescriptions based on learning style in addition to conventional study-skill guide-*

lines. Then $\mu_1 - \mu_2$ is the difference between the mean science scores for the two treatments.

2. $H_0: \mu_1 - \mu_2 = 0$

3. $H_a: \mu_1 - \mu_2 < 0$

4. *Significance level*: $\alpha = .01$

5. *Test statistic*: $t = \dfrac{(\bar{x}_1 - \bar{x}_2) - \text{hypothesized value}}{\sqrt{\dfrac{s_1^2}{n_1} + \dfrac{s_2^2}{n_2}}} = \dfrac{(\bar{x}_1 - \bar{x}_2) - 0}{\sqrt{\dfrac{s_1^2}{n_1} + \dfrac{s_2^2}{n_2}}}$

6. *Assumptions*:   Subjects were randomly assigned to the treatment groups, and both sample sizes are large, so the two-sample t test is appropriate.

7. *Computation*: $t = \dfrac{(2.75 - 3.28) - 0}{\sqrt{\dfrac{(.85)^2}{100} + \dfrac{(.70)^2}{103}}} = \dfrac{-.53}{.1095} = -4.84$

8. *P-value*:   We first compute the df for the two-sample t test:

$$V_1 = \frac{s_1^2}{n_1} = .0072 \qquad V_2 = \frac{s_2^2}{n_2} = .0048$$

$$\text{df} = \frac{(V_1 + V_2)^2}{\dfrac{V_1^2}{n_1 - 1} + \dfrac{V_2^2}{n_2 - 1}} = \frac{(.0072 + .0048)^2}{\dfrac{(.0072)^2}{99} + \dfrac{(.0048)^2}{102}} = \frac{.000144}{.0000075} = 192.69$$

This is a lower-tailed test, so the *P*-value is the area under the t curve with df $= 192$ and to the left of -4.84. Since -4.84 is so far out in the lower tail of this t curve, *P*-value ≈ 0.

9. *Conclusion*:   Since *P*-value $\leq \alpha$, H_0 is rejected. There is evidence that the mean science score for the treatment group is higher. The data supports the conclusion that homework prescriptions based on learning style are effective in raising the mean science score.

You have probably noticed that evaluating the formula for degrees of freedom for the two-sample t test involves quite a bit of arithmetic. An alternative approach is to compute a conservative estimate of the *P*-value — one that is close to but larger than the actual *P*-value. If H_0 were rejected using this conservative estimate, then it would also be rejected if the actual *P*-value had been used. **A conservative estimate of the *P*-value for the two-sample t test can be found by using the t curve with df equal to the smaller of $n_1 - 1$ and $n_2 - 1$.**

The Pooled t Test

The two-sample t test procedure just described is appropriate when it is reasonable to assume that the population distributions are approximately normal. If it is also known that the variances of the two populations are equal ($\sigma_1^2 = \sigma_2^2$), an alternative procedure known as the *pooled t test* can be used. This test procedure combines information from both samples to obtain a "pooled" estimate of the common variance and then uses this pooled estimate of the variance in place of s_1^2 and s_2^2 in the t test statistic. This test procedure was widely used in the past, but it has fallen into some disfavor because it is quite sensitive to departures from the assumption of equal population variances. If the population variances are equal, the pooled t procedure has a slightly better chance of detecting departures from H_0 than does the two-sample t test of this section. However, P-values based on the pooled t procedure can be seriously in error if the population variances are not equal, so in general the two-sample t procedure is a better choice.

Comparisons and Causation

If the assignment of treatments to the individuals or objects used in a comparison of treatments is not made by the investigators, the study is said to be **observational.**

As an example, the article "Lead and Cadmium Absorption among Children near a Nonferrous Metal Plant" (*Environ. Research* (1978): 290–308) reported data on blood-lead concentrations for two different samples of children. The first sample was drawn from a population residing within 1 km of a lead smelter, whereas those in the second sample were selected from a rural area much farther from the smelter. It was the parents of children, rather than the investigators, who determined whether the children would be in the close-to-smelter group or the far-from-smelter group. As a second example, a letter in the *Journal of the American Medical Association* (May 19, 1978) reported on a comparison of doctors' longevity after medical school graduation for those with an academic affiliation and those in private practice. (The letter writer's stated objective was to see whether "publish or perish" really meant "publish *and* perish.") Here again, an investigator did not start out with a group of doctors and assign some to academic and others to nonacademic careers. The doctors themselves selected their groups.

The difficulty with drawing conclusions based on an observational study is that a statistically significant difference may be due to some underlying factors that have not been controlled rather than to conditions that define the groups. Does the type of medical practice itself have an effect on longevity, or is the observed difference in lifetimes caused by other factors, which themselves led graduates to choose academic or nonacademic careers? Similarly, is the observed difference in blood-lead concentration levels due to proximity to the smelter? Perhaps there are other physical and socioeconomic factors related both to choice of living area and to concentration.

In general, rejection of H_0: $\mu_1 - \mu_2 = 0$ in favor of H_a: $\mu_1 - \mu_2 > 0$ suggests that, on the average, higher values of the variable are *associated* with individuals in the first population or receiving the first treatment than with those in the second population or receiving the second treatment. But *association does not imply causation*. Strong statistical evidence for a causal relationship can be built up over time through many different comparative studies that point to the same conclusions (as in the many investigations linking smoking to lung cancer). A **randomized controlled experiment,** in which investigators assign subjects in some prescribed random fashion to the treatments or conditions being compared, is particularly effective in suggesting causality. With such random assignment, the investigator and other interested parties will have more confidence in the conclusion that an observed difference was caused by the difference in treatments or conditions. (Recall the discussion in Chapter 2 of the role of randomization in designing an experiment.)

A Confidence Interval

A confidence interval for $\mu_1 - \mu_2$ is easily obtained from the basic t variable of this section. Both the derivation of and formula for the interval are very similar to that of the one-sample t interval discussed in Chapter 9.

The Two-Sample t Confidence Interval for the Difference Between Two Population or Treatment Means

The general formula for a confidence interval for $\mu_1 - \mu_2$ when

1. the two samples are *independently chosen random samples,* and
2. the *sample sizes are both large* (generally $n_1 \geq 30$ and $n_2 \geq 30$) OR the *population distributions are approximately normal*

is

$$\bar{x}_1 - \bar{x}_2 \pm (t \text{ critical value})\sqrt{\frac{s_1^2}{n_1} + \frac{s_2^2}{n_2}}$$

The t critical value is based on

$$df = \frac{(V_1 + V_2)^2}{\dfrac{V_1^2}{n_1 - 1} + \dfrac{V_2^2}{n_2 - 1}} \qquad \text{where } V_1 = \frac{s_1^2}{n_1} \text{ and } V_2 = \frac{s_2^2}{n_2}$$

df should be truncated (rounded down) to an integer. The t critical values for the usual confidence levels are given in Appendix Table III.

For a comparison of two treatments, when

1. *treatments are randomly assigned* to individuals or objects (or vice versa), and
2. the *sample sizes are large* (generally 30 or larger) OR the *treatment response distributions are approximately normal,*

the two-sample t confidence interval formula can be used to estimate $\mu_1 - \mu_2$.

EXAMPLE 11.4 Much research effort has been expended in studying possible causes of the pharmacological and behavioral effects resulting from smoking marijuana. The article "Intravenous Injection in Man of Δ^9THC and 11-OH-Δ^9THC" (*Science* (1982): 633) reported on a study of two chemical substances thought to be instrumental in marijuana's effects. Subjects were randomly assigned to one of two treatment groups. Individuals in each group were given one of the two substances in increasing amounts and asked to say when the effect was first perceived. The accompanying data values are necessary dose to perception per kilogram of body weight.

Δ^9THC	19.54	14.47	16.00	24.83	26.39	11.49	($\bar{x}_1 = 18.79$, $s_1 = 5.91$)
11-OH-Δ^9THC	15.95	25.89	20.53	15.52	14.18	16.00	($\bar{x}_2 = 18.01$, $s_2 = 4.42$)

Subjects were randomly assigned to treatments. Since both the sample sizes are small, we must first investigate whether it is reasonable to assume that the distribution of dose to perception is approximately normal for the two chemical substances. There are no outliers in either data set, and normal probability plots are reasonably straight, suggesting that the assumption of approximate normality is reasonable.

To estimate $\mu_1 - \mu_2$, the difference in mean dose to perception for the two substances, we will calculate a 95% confidence interval.

$$V_1 = \frac{s_1^2}{n_1} = \frac{(5.91)^2}{6} = 5.821 \qquad V_2 = \frac{s_2^2}{n_2} = \frac{(4.42)^2}{6} = 3.256$$

$$df = \frac{(V_1 + V_2)^2}{\dfrac{V_1^2}{n_1 - 1} + \dfrac{V_2^2}{n_2 - 1}} = \frac{(5.821 + 3.256)^2}{\dfrac{(5.821)^2}{5} + \dfrac{(3.256)^2}{5}} = \frac{82.392}{8.897} = 9.26$$

Truncating to an integer gives df = 9. In the 9-df row of Appendix Table III, the t critical value for a 95% confidence level is 2.26. The interval is then

$$\bar{x}_1 - \bar{x}_2 \pm (t \text{ critical value})\sqrt{\frac{s_1^2}{n_1} + \frac{s_2^2}{n_2}} = (18.79 - 18.01) \pm (2.26)\sqrt{\frac{(5.91)^2}{6} + \frac{(4.42)^2}{6}}$$

$$= .78 \pm (2.26)(3.01)$$
$$= .78 \pm 6.80$$
$$= (-6.02, 7.58)$$

This interval is rather wide, because the two sample variances are large and the sample sizes are small. Notice that the interval includes 0, so 0 is one of the plausible values for $\mu_1 - \mu_2$. That is, it is plausible that there is no difference in the mean dose to perception for the two substances.

Most statistical computer packages will compute the two-sample t confidence interval. MINITAB was used to construct a 95% confidence interval using the data of this example; the resulting output is shown here.

Two Sample T-Test and Confidence Interval

Two sample T for THC vs 11-OH-THC

	N	Mean	StDev	SE Mean
THC	6	18.79	5.91	2.4
11-OH-TH	6	18.01	4.42	1.8

95% CI for mu THC - mu 11-OH-TH: (-6.0, 7.6)

Exercises 11.1 – 11.31

11.1 Consider two populations for which $\mu_1 = 30$, $\sigma_1 = 2$, $\mu_2 = 25$, and $\sigma_2 = 3$. Suppose that two independent random samples of sizes $n_1 = 40$ and $n_2 = 50$ are selected. Describe the approximate sampling distribution of $\bar{x}_1 - \bar{x}_2$ (center, spread, and shape).

11.2 An article in the November 1983 *Consumer Reports* compared various types of batteries. The average lifetimes of Duracell Alkaline AA batteries and Eveready Energizer Alkaline AA batteries were given as 4.1 hr and 4.5 hr, respectively. Suppose that these are the population average lifetimes.

a. Let $\bar{x}_1$ be the sample average lifetime of 100 Duracell batteries and $\bar{x}_2$ be the sample average lifetime of 100 Eveready batteries. What is the mean value of $\bar{x}_1 - \bar{x}_2$; that is, where is the sampling distribution of $\bar{x}_1 - \bar{x}_2$ centered? How does your answer depend on the specified sample sizes?

b. Suppose that population standard deviations of lifetime are 1.8 hr for Duracell batteries and 2.0 hr for Eveready batteries. With the sample sizes as given in part **a**, what is the variance of the statistic $\bar{x}_1 - \bar{x}_2$, and what is its standard deviation?

c. For the sample sizes as given in part **a**, draw a picture of the approximate sampling distribution curve of $\bar{x}_1 - \bar{x}_2$ (include a measurement scale on the horizontal axis). Would the shape of the curve necessarily be the same for sample sizes of 10 batteries of each type? Explain.

11.3 An acquaintance of ours can take either a scenic route to work or a nonscenic route. She decides that use of the nonscenic route can be justified only if it reduces true average travel time by more than 10 min.

a. If μ_1 refers to the scenic route and μ_2 to the nonscenic route, which hypotheses should be tested?

b. If μ_1 refers to the nonscenic route and μ_2 to the scenic route, which hypotheses should be tested?

11.4 When a surgeon repairs injuries he uses sutures (stitched knots) to hold together and stabilize the injured area. If these knots elongate and loosen through use, the injury may not heal properly because the tissues would not be optimally positioned. Researchers at the University of California, San Francisco, tied a series of different types of knots with two types of suture material, Maxon and Ticron. Suppose that 112 tissue specimens were available and that for each specimen the type of knot and suture material were randomly assigned. The investigators tested the knots to see how much the loops elongated; the elongations were measured and the resulting data is summarized here.

Maxon

	n	$\bar{x}$ (mm)	sd (mm)
Square (control)	10	10.0	.1
Duncan loop	15	11.0	.3
Overhand	15	11.0	.9
Roeder	10	13.5	.1
Snyder	10	13.5	2.0

Ticron

	n	$\bar{x}$ (mm)	sd (mm)
Square (control)	10	2.5	.06
Duncan loop	11	10.9	.4
Overhand	11	8.1	1.0
Roeder	10	5.0	.04
Snyder	10	8.1	.06

a. Is there a significant difference in elongation between the square knot and the Duncan loop for Maxon thread?

b. Is there a significant difference between the square knot and the Duncan loop for Ticron thread?

c. For the Duncan loop data, is there a significant difference between the elongation of Maxon versus Ticron threads?

11.5 Snake experts believe that venomous snakes inject different amounts of venom when killing their prey. Researchers at the University of Wyoming tested this hypothesis to determine whether young prairie rattlesnakes use more venom to kill larger mice and less venom for smaller mice ("Venom Metering by Juvenile Prairie Rattlesnakes, Crotalus v. Viridis: Effects of Prey Size and Experience," *Animal Behavior* (1995): 33–40). In the first trial, the researchers used three groups of seven randomly selected snakes that were inexperienced hunters. In the second trial, three different groups of seven randomly selected "experienced" snakes were used. The amount (mg) of venom each snake injected was recorded and categorized according to the size of the mouse.

	Small Mouse		Medium Mouse		Large Mouse	
	$\bar{x}$	s	$\bar{x}$	s	$\bar{x}$	s
Inexperienced	3.1	1.0	3.4	.4	1.8	.3
Experienced	2.6	.3	2.9	.6	4.7	.3

For the small prey, is there a significant difference between the inexperienced and experienced snakes? For the medium prey? For the large prey?

11.6 The coloration of male guppies may affect the mating preference of the female guppy. To test this hypothesis, scientists first identified two types of guppies, Yarra and Paria, that display different colorations ("Evolutionary Mismatch of Mating Preferences and Male Colour Patterns in Guppies," *Animal Behaviour* (1997): 343–51). The relative area of orange was calculated for fish of each type. A random sample of 30 Yarra guppies resulted in a mean relative area of .106 and a standard deviation of .055. A random sample of 30 Paria guppies resulted in a mean relative area of .178 and a standard deviation .058. Is there evidence of a difference in coloration? Test the relevant hypotheses to determine whether the mean area of orange is different for the two types of guppies.

11.7 The article "Trial Lawyers and Testosterone: Blue-Collar Talent in a White-Collar World" (*J. Applied Social Psychology* (1998): 84–94) compared trial lawyers and nontrial lawyers on the basis of mean testosterone level. Random samples of 35 male trial lawyers, 31 male non-trial lawyers, 13 female trial lawyers, and 18 female nontrial lawyers were selected for study. The article includes the following statement:

> Trial lawyers had higher testosterone levels than did nontrial lawyers. This was true for men, $t(64) = 3.75$, $p < .001$, and for women, $t(29) = 2.26$, $p < .05$.

a. Based on the information given, is there a significant difference in the mean testosterone level for male trial and nontrial lawyers?
b. Based on the information given, is there a significant difference in the mean testosterone level for female trial and nontrial lawyers?
c. Do you have enough information to carry out a test to determine whether there is a significant difference in the mean testosterone levels of male and female trial lawyers? If so, carry out such a test. If not, what additional information would you need to be able to conduct the test?

11.8 The article "The Relevance of Sexual Orientation to Substance Abuse and Psychological Distress Among College Students" (*J. College Student Development* ((1998): 157–165) examined substance abuse patterns for individuals in a random sample of gay, lesbian, and bisexual students and in a random sample of heterosexual students at a particular university. One of the variables considered was frequency of marijuana use in the past year. It was reported that the mean and standard devia-

tion for this variable were 15.68 and 43.69, respectively for the sample of 37 gay, lesbian, and bisexual students, and 12.28 and 44.20 for the sample of 156 heterosexual students.

a. Based on the information given, is it reasonable to assume that the population distributions of the variable *frequency of marijuana use* are approximately normal. Explain why or why not.
b. Taking your answer to part **a** into account, do you think it is still reasonable to use the two-sample t test to determine whether there is sufficient evidence to conclude that the mean frequency of marijuana use differs for the two groups? Explain why or why not.
c. If you concluded in part **b** that the independent samples t test is appropriate, use it to carry out a test of the relevant hypotheses using $\alpha = .01$.

11.9 Do faculty and students have similar perceptions of what types of behavior are inappropriate in the classroom? This question was examined by the author of the article "Faculty and Student Perceptions of Classroom Etiquette" (*J. College Student Development* (1998): 515–516). Each individual in a random sample of 173 students in general education classes at a large public university was asked to judge various behaviors on a scale from 1 (totally inappropriate) to 5 (totally appropriate). Individuals in a random sample of 98 faculty members also rated the same behaviors. The mean rating for two of the behaviors studied are shown here (the means are consistent with data provided by the author of the article). The sample standard deviations were not given, but for purposes of this exercise, assume that they are all equal to 1.0.

Student Behavior	Student Mean Rating	Faculty Mean Rating
Wearing hats in the classroom	2.80	3.63
Addressing instructor by first name	2.90	2.11
Talking on a cell phone	1.11	1.10

a. Is there sufficient evidence to conclude that the mean "appropriateness" score assigned to wearing a hat in class differs for students and faculty?
b. Is there sufficient evidence to conclude that the mean "appropriateness" score assigned to addressing an instructor by his or her first name is higher for students than for faculty?
c. Is there sufficient evidence to conclude that the mean "appropriateness" score assigned to talking on a cell phone differs for students and faculty? Does the result of your test imply that students and faculty consider it acceptable to talk on a cell phone during class?

11.10 The article "So Close, Yet So Far: Predictors of Attrition in College Seniors" (*J. College Student Development* (1999): 343–354) attempts to describe differences between college seniors who disenroll before graduating and those who do graduate. Researchers randomly selected 42 nonreturning and 48 returning seniors, none of whom were transfer students. These 90 students rated themselves on personal contact and campus involvement. The resulting data is summarized here:

	Returning (n = 48)		Nonreturning (n = 42)	
	Mean	Standard Deviation	Mean	Standard Deviation
Personal contact	3.22	.93	2.41	1.03
Campus involvement	3.21	1.01	3.31	1.03

a. Construct and interpret a 95% confidence interval for the difference in mean campus involvement rating for returning and nonreturning students. Does your interval support the statement that students who do not return are less involved, on average, than those who do? Explain.

b. Do students who don't return have a lower mean personal contact rating than those who do return? Test the relevant hypotheses using a significance level of .01.

11.11 Does clear-cutting of trees in an area cause local extinction of the tailed frog? We don't really know, but an article that begins to address that issue attempts to quantify aspects of the habitat and microhabitat of streams with and without tailed frogs. The following data is from "Distribution and Habitat of *Ascaphus Truei* in Streams on Managed, Young Growth Forests in North Coastal California" (*Journal of Herpetology* (1999): 71–79).

Assume that the habitats and microhabitats examined were selected independently and at random from those that have tailed frogs and those that do not have tailed frogs. Use a significance level of .01 for any hypothesis tests required to answer the following questions.

	Sites With Tailed Frogs		
	n	$\bar{x}$	s
Habitat Characteristics			
Stream gradient (%)	18	9.1	6.00
Water temp. (°C)	18	12.2	1.71
Microhabitat Characteristics			
Depth (cm)	82	5.32	2.27

	Sites Without Tailed Frogs		
	n	$\bar{x}$	s
Habitat Characteristics			
Stream gradient (%)	31	5.9	6.29
Water temp. (°C)	31	12.8	1.33
Microhabitat Characteristics			
Depth (cm)	267	8.46	5.95

a. Is there evidence of a difference in mean stream gradient between streams with and streams without tailed frogs? (*Note:* Assume that it is reasonable to think that the distribution of stream gradients is approximately normal for both types of streams. It is possible for gradient % to be either positive or negative, so the fact that the mean $-$ 2 standard deviations is negative is not, by itself, an indication of nonnormality in this case.)

b. Is there evidence of a difference in the mean water temperature between streams with and streams without tailed frogs?

c. The article reported that the two depth distributions are both quite skewed. Does this imply that it would be unreasonable to use the independent samples *t* test to compare the mean depth for the two types of streams? If not, carry out a test to determine whether the sample data supports the claim that the mean depth of streams without tailed frogs is greater than the mean depth of streams with tailed frogs.

11.12 Are girls less inclined to enroll in science courses than boys? One recent study ("Intentions of Young Students to Enroll in Science Courses in the Future: An Examination of Gender Differences" (*Science Education* (1999): 55–76) asked randomly selected fourth, fifth, and sixth graders how many science courses they intend to take. The following data was obtained:

	n	Mean	Standard Deviation
Males	203	3.42	1.49
Females	224	2.42	1.35

Calculate a 99% confidence interval for the difference between males and females in mean number of science courses planned. Interpret your interval. Based on your interval, how would you answer the question posed at the beginning of the exercise?

11.13 The article "Workaholism in Organizations: Gender Differences" (*Sex Roles* (1999): 333–346) gave the following data on 1996 income for random samples of male and female MBA graduates from a particular Canadian business school:

	n	$\bar{x}$	s
Males	258	$133,442	$131,090
Females	233	$105,156	$98,525

(*Note*: These salary figures are in Canadian dollars.)

a. For what significance levels would you conclude that the mean salary of female MBA graduates of this business school is above $100,000?

b. Is there convincing evidence that the mean salary for female MBA graduates of this business school is lower than the mean salary for the male graduates?

11.14 The article "Movement and Habitat Use by Lake Whitefish During Spawning in a Boreal Lake: Integrating Acoustic Telemetry and Geographic Information Systems" (*Transactions of the American Fisheries Society,* (1999): 939–952) included the accompanying data on weights of 18 fish caught in 1995 and 1996:

1995	776	580	539	648	538	891	673	783	571	627
1996	571	627	727	727	867	1042	804	832	764	727

Is it reasonable to use the independent samples *t* test to compare the mean weight of fish for the 2 years? Explain why or why not.

11.15 The article "The Relationship of Task and Ego Orientation to Sportsmanship Attitudes and the Perceived Legitimacy of Injurious Acts" (*Research Quarterly for Exercise and Sport* (1991): 79–87) examined the extent of approval of unsporting play and cheating. High school basketball players completed a questionnaire that was used to arrive at an approval score, with higher scores indicating greater approval. A random sample of 56 male players resulted in a mean approval rating for unsportsmanlike play of 2.76, whereas the mean for a random sample of 67 female players was 2.02. Suppose that the two sample standard deviations were .44 for males and .41 for females. Is it reasonable to conclude that the mean approval rating is higher for male players than for female players by more than .5? Use $\alpha = .05$.

11.16 The article "Conflict Resolution Styles in Gay, Lesbian, Heterosexual Nonparent and Heterosexual Parent Couples" (*J. Marriage and Family* (1994): 705–722) reported the results of administering the Ineffective Arguing Inventory (IAI) to both partners of 75 gay, 51 lesbian, 108 married nonparent, and 99 married parent couples. The IAI is a measure of dysfunctionality in couple conflict resolution. The mean and standard deviation of the IAI scores for each group are given in the accompanying table.

	Type of Couple			
	Gay	Lesbian	Hetero-sexual Nonparent	Hetero-sexual Parent
Sample size	75	51	108	99
Mean	17.16	15.92	16.79	18.09
Standard deviation	6.19	5.93	5.81	6.55

a. Is there evidence to suggest that the mean IAI score is higher for heterosexual parent couples than for heterosexual nonparent couples? Test the relevant hypotheses using $\alpha = .01$.

b. Is there evidence to suggest that the mean IAI score is higher for gay couples than for lesbian couples? Test the relevant hypotheses using $\alpha = .05$.

c. Construct a 95% confidence interval for the difference in mean IAI score for gay couples and heterosexual nonparent couples. Interpret this interval.

d. Construct a 95% confidence interval for the difference in mean IAI score for heterosexual parent couples and heterosexual nonparent couples. Interpret this interval.

11.17 Techniques for processing poultry were examined in the article "Texture Profiles of Canned Boned Chicken as Affected by Chilling-Aging Times" (*Poultry Science* (1994): 1475–1478). Whole chickens were chilled 0, 2, 8, or 24 hours before being cooked and canned. To determine whether the chilling time affected the texture of the canned chicken, samples were evaluated by trained tasters. One characteristic of interest was hardness. The accompanying summary quantities were obtained. Each mean is based on 36 ratings.

	Chilling Time			
	0 hr	2 hr	8 hr	24 hr
Mean hardness	7.52	6.55	5.70	5.65
Standard deviation	.96	1.74	1.32	1.50

a. Does the data suggest that there is a difference in mean hardness for chicken chilled 0 hours before cooking and chicken chilled 2 hours before cooking? Use $\alpha = .05$.

b. Does the data suggest that there is a difference in mean hardness for chicken chilled 8 hours before cooking and chicken chilled 24 hours before cooking? Use $\alpha = .05$.

c. Use a 90% confidence interval to estimate the difference in mean hardness for chicken chilled 2 hours before cooking and chicken chilled 8 hours before cooking.

11.18 The Rape Myth Acceptance Scale (RMA) was administered to 333 male students at a public university. Of these students, 155 were randomly selected from students who did not belong to a fraternity, and 178 were randomly selected from students who belonged to a fraternity. The higher the score on the RMA, the greater the acceptance of rape myths, so lower scores are considered desirable. The accompanying data appeared in the article "Rape Supportive Attitudes Among Greek Students Before and After a Date Rape Prevention Program" (*J. of College Student Development* (1994): 450–455). Does the data support the researchers' claim that the mean RMA score is lower for fraternity members? Use $\alpha = .05$.

		Mean RMA	
Sample	**n**	**Score**	**sd**
Fraternity members	178	25.63	6.16
Not fraternity members	155	27.40	5.51

11.19 The discharge of industrial wastewater into rivers affects water quality. To assess the effect of a particular power plant on water quality, 24 water specimens were taken 16 km upstream and 4 km downstream of the plant. Alkalinity (mg/L) was determined for each specimen, resulting in the summary quantities in the accompanying table. Does the data suggest that the true mean alkalinity is higher downstream than upstream by more than 50 mg/L? Use a .05 significance level.

			Standard
Location	**n**	**Mean**	**Deviation**
Upstream	24	75.9	1.83
Downstream	24	183.6	1.70

11.20 According to the Associated Press (*San Luis Obispo Telegram-Tribune,* June 23, 1995), a study by Italian researchers indicated that low cholesterol and depression were linked. The researchers found that among 331 randomly selected patients hospitalized because they had attempted suicide, the mean cholesterol level was 198. The mean cholesterol level of 331 randomly selected patients admitted to the hospital for other reasons was 217. The sample standard deviations were not reported, but suppose that they were 20 for the group who had attempted suicide and 24 for the other group. Does this data provide sufficient evidence to conclude that the mean cholesterol level is lower for those who have attempted suicide? Test the relevant hypotheses using $\alpha = .05$.

11.21 The article "The Sorority Rush Process: Self-Selection, Acceptance Criteria, and the Effect of Rejection" (*J. of College Student Development* (1994): 346–353) reported on a study of factors associated with the decision to rush a sorority. Fifty-four women who rushed a sorority and 51 women who did not were asked how often they drank alcoholic beverages. For the sorority rush group, the mean was 2.72 drinks per week and the standard deviation .86. For the group who did not rush, the mean was 2.11 and the standard deviation 1.02. Is there evidence to support the claim that those who rush a sorority drink more than those who do not rush? Test the relevant hypotheses using $\alpha = .01$. What assumptions are required in order for the two-sample t test to be appropriate?

11.22 The authors of the article "Dynamics of Canopy Structure and Light Interception in *Pinus elliottii,* North Florida" (*Ecological Monographs* (1991): 33–51) planned an experiment to determine the effect of fertilizer on a measure of leaf area. A number of plots were available for the study, and half were selected at random to be fertilized. To ensure that the plots to be treated with fertilizer and the control plots were similar, prior to beginning the experiment tree density (the number of trees per hectare) was recorded for eight plots to be fertilized and eight control plots, resulting in the given data.

Fertilizer plots	1024	1216	1312	1280	1216	1312	992	1120
Control plots	1104	1072	1088	1328	1376	1280	1120	1200

Based on the accompanying MINITAB output, would you conclude that there is a significant difference in the mean tree density for fertilizer and control plots? Use $\alpha = .05$.

Twosample T for fert. vs control

	N	Mean	StDev	SE Mean
Fert.	8	1184	126	44
Control	8	1196	118	42

95% C.I. for mu fert. - mu control: (-144, 120)

T-Test mu fert. = mu control (vs not =): T = -0.20 P = 0.85 DF = 13

11.23 British health officials have expressed concern about problems associated with vitamin D deficiency among certain immigrants. Doctors have conjectured that such a deficiency could be related to the amount of fiber in a person's diet. An experiment was designed to compare the vitamin D plasma half-life for two groups of healthy individuals. One group was placed on a normal diet, whereas the second group was placed on a high-fiber diet. The accompanying table gives the resulting data (from "Reduced Plasma Half-Lives of Radio-Labeled 25(OH)D3 in Subjects Receiving a High-Fibre Diet," *Brit. J. of Nutrit.* (1993): 213–216).

Normal diet	19.1	24.0	28.6	29.7	30.0	34.8	
High-fiber diet	12.0	13.0	13.6	20.5	22.7	23.7	24.8

Use the following MINITAB output to determine whether the data indicates that the mean half-life is higher for those on a normal diet than those on a high-fiber diet. Assume that treatments were assigned at random and the two plasma half-life distributions are normal. Test the appropriate hypotheses using $\alpha = .01$.

Twosample T for normal vs high

	N	Mean	StDev	SE Mean
Normal	6	27.70	5.44	2.2
High	7	18.61	5.55	2.1

95% C.I. for mu normal - mu high: (2.3, 15.9)
T-Test mu normal = mu high(vs >):T = 2.97 P = 0.0070 DF = 10

11.24 A researcher at the Medical College of Virginia conducted a study of 60 randomly selected male soccer players and concluded that frequently "heading" the ball in soccer lowers players' IQs (*USA Today,* Aug. 14, 1995). The soccer players were divided into two groups, based on whether they averaged 10 or more headers per game. Mean IQs were reported in the article, but the sample sizes and standard deviations were not given. Suppose that these values were as given in the accompanying table.

	n	Sample Mean	Sample sd
Fewer than 10 headers	35	112	10
10 or more headers	25	103	8

Does this data support the researcher's conclusion? Test the relevant hypotheses using $\alpha = .05$. Can you conclude that heading the ball *causes* lower IQ?

11.25 The effect of loneliness among college students was examined in the article "The Importance of Perceived Duration: Loneliness and Its Relationship to Self-Esteem and Academic Performance" (*J. College Student Development* (1994): 456–460). Based on reported frequency and duration of loneliness, subjects were divided into two groups. The first group ($n_1 = 72$) was the short-duration loneliness group, and the second ($n_2 = 17$) was the long-duration loneliness group. A self-esteem inventory was administered to students in both groups. For the short-duration group, the reported mean self-esteem score was 76.78 and the standard deviation was 17.80. For the long-duration group, the mean and standard deviation were 64.00 and 15.68, respectively. Does the data support the researcher's claim that mean self-esteem is lower for students classified as having long-duration loneliness? Test the relevant hypotheses using $\alpha = .01$. Be sure to state any assumptions that are necessary for your test to be valid.

11.26 The article referenced in Exercise 11.25 also gave information on college grade point average. Summary values appear in the accompanying table. Is there a significant difference in mean grade point average between students who experience short-duration loneliness and those who experience long-duration loneliness? Use $\alpha = .05$.

Group	n	Mean	sd
Short duration	72	2.58	.75
Long duration	17	2.80	.79

11.27 Wayne Gretzky was one of ice hockey's most prolific scorers when he played for the Edmonton Oilers. During his last season with the Oilers, Gretzky played in 41 games and missed 17 games due to injury. The article "The Great Gretzky" (*Chance* (1991): 16–21) looked at the number of goals scored by the Oilers in games with and without Gretzky, as shown in the accompanying table. If we view the 41 games with Gretzky as a random sample of all Oiler games in which Gretzky played and the 17 games without Gretzky as a random sample of all Oiler games in which Gretzky did not play, is there evidence that the mean number of goals scored by the Oilers is higher for games in which Gretzky played? Use $\alpha = .01$.

	n	Sample Mean	Sample sd
Games with Gretzky	41	4.73	1.29
Games without Gretzky	17	3.88	1.18

11.28 Here's one to sink your teeth into: The authors of the article "Analysis of Food Crushing Sounds During Mastication: Total Sound Level Studies" (*J. of Texture Studies* (1990): 165–178) studied the nature of sounds generated during eating. Peak loudness (in decibels at 20 cm away) was measured for both open-mouth and closed-mouth chewing of potato chips and of tortilla chips. Forty subjects participated, with ten assigned at random to each combination of conditions (such as closed-mouth, potato chip, and so on). We are not making this up! Summary values taken from plots given in the article appear in the accompanying table.

	n	$\bar{x}$	s
Potato chip			
Open mouth	10	63	13
Closed mouth	10	54	16
Tortilla chip			
Open mouth	10	60	15
Closed mouth	10	53	16

a. Construct a 95% confidence interval for the difference in mean peak loudness between open-mouth and closed-mouth chewing of potato chips. Interpret the resulting interval.

b. For closed-mouth chewing (the recommended

method!), is there sufficient evidence to indicate that there is a difference between potato chips and tortilla chips with respect to mean peak loudness? Test the relevant hypotheses using $\alpha = .01$.

c. The means and standard deviations given here were actually for stale chips. When ten measurements of peak loudness were recorded for closed-mouth chewing of fresh tortilla chips, the resulting mean and standard deviation were 56 and 14, respectively. Is there sufficient evidence to conclude that fresh tortilla chips are louder than stale chips? Use $\alpha = .05$.

11.29   Thirteen males and fourteen females participated in a study of grip and leg strength ("Sex Differences in Static Strength and Fatigability in Three Different Muscle Groups," *Research Quarterly for Exercise and Sport* (1990): 238–242). Right-leg strength (in newtons) was recorded for each participant, resulting in the summary statistics given in the accompanying table. Estimate the difference in mean right-leg strength between males and females, using a 95% confidence interval. Interpret the resulting interval.

	n	$\bar{x}$	s
Males	13	2127	513
Females	14	1843	446

11.30   Are very young infants more likely to imitate actions that are modeled by a person or simulated by an object? This question was the basis of a research study summarized in the article "The Role of Person and Object in Eliciting Early Imitation" (*J. Experimental Child Psych.* (1991): 423–433). One action examined was mouth opening. This action was modeled repeatedly by either a person or a doll, and the number of times that the infant imitated the behavior was recorded. Twenty-seven infants participated, with 12 exposed to a human model and 15 exposed to the doll. Summary values are given here. Is there sufficient evidence to conclude that the mean number of imitations is higher for infants who watch a human model than for infants who watch a doll? Test the relevant hypotheses using a .01 significance level.

	Person Model	Doll Model
$\bar{x}$	5.14	3.46
s	1.6	1.3

11.31   Do certain behaviors result in a severe drain on energy resources because a great deal of energy is expended in comparison to energy intake? The article "The Energetic Cost of Courtship and Aggression in a Plethodontid Salamander" (*Ecology* (1983): 979–983) reported on one of the few studies concerned with behavior and energy expenditure. The accompanying table gives oxygen consumption (mL/g/hr) for male-female salamander pairs. (The determination of consumption values was rather complicated. It is partly for this reason that so few studies of this type have been carried out.)

Behavior	Sample Size	Sample Mean	Sample sd
Noncourting	11	.072	.0066
Courting	15	.099	.0071

a. The pooled t test is a test procedure for testing H_0: $\mu_1 - \mu_2 = hypothesized\ value$ when it is reasonable to assume that the two population distributions are normal with equal standard deviations ($\sigma_1 = \sigma_2$). The test statistic for the pooled t test is obtained by replacing both s_1 and s_2 in the two-sample t test statistic with s_p, where

$$s_p = \sqrt{\frac{(n_1 - 1)s_1^2 + (n_2 - 1)s_2^2}{n_1 + n_2 - 2}}$$

When the population distributions are normal with equal standard deviations and H_0 is true, the resulting pooled t statistic has a t distribution with df $= n_1 + n_2 - 2$. For the reported data, the two sample standard deviations are similar. Use the pooled t test with $\alpha = .05$ to determine whether the mean oxygen consumption for courting pairs is higher than the mean oxygen consumption for noncourting pairs.

b. Would the conclusion in part **a** have been different if the two-sample t test had been used rather than the pooled t test?

11.2   Inferences Concerning the Difference Between Two Population or Treatment Means Using Paired Samples

Two samples are said to be *independent* if the selection of the individuals or objects that make up one of the samples has no bearing on the selection of those in the other sample. In some situations, an experiment with independent samples is not the best way to obtain information concerning a possible difference between the populations. For example, suppose that an investigator wants to determine whether regular aerobic exercise affects blood pressure. A random sample of people who jog regularly and a second random sample of people who do not exercise regularly are selected independently of one another. The researcher then uses the two-sample *t* test to conclude that a significant difference exists between the average blood pressures for joggers and nonjoggers. Is it reasonable to think that the difference in mean blood pressure is attributable to jogging? It is known that blood pressure is related to both diet and body weight. Might it not be the case that joggers in the sample tend to be leaner and adhere to a healthier diet than the nonjoggers and that *this* might account for the observed difference? On the basis of this study, the researcher wouldn't be able to rule out the possibility that the observed difference in blood pressure is explained by weight differences between the people in the two samples and that aerobic exercise itself has no effect.

One way to avoid this difficulty would be to match subjects by weight. The researcher would find pairs of subjects so that the jogger and nonjogger in each pair were similar in weight (although weights for different pairs might vary widely). The factor *weight* could then be ruled out as a possible explanation for an observed difference in average blood pressure between the two groups. Matching the subjects by weight results in two samples for which each observation in the first sample is coupled in a meaningful way with a particular observation in the second sample. Such samples are said to be **paired.**

Experiments can be designed to yield paired data in a number of different ways. Some studies involve using the same group of individuals with measurements recorded both before and after some intervening treatment. Others use naturally occurring pairs, such as twins or husbands and wives, and some construct pairs by matching on factors with effects that might otherwise obscure differences (or the lack of them) between the two populations of interest (as might weight in the jogging example). Paired samples often provide more information than would independent samples, because extraneous effects are screened out.

EXAMPLE 11.5   It has been hypothesized that strenuous physical activity affects hormone levels. The article "Growth Hormone Increase During Sleep After Daytime Exercise" (*J. of Endocrinology* (1974): 473–478) reported the results of an experiment involving six healthy male subjects. For each participant, blood samples were taken during sleep on two different nights. The first blood sample (the control) was drawn after a day that included no strenuous activities, and the second was drawn after a day when the subject had engaged in strenuous exercise. The resulting data on growth hormone level (in mg/ml) follows. The samples are paired rather than independent, because both samples are composed of measurements on the same men.

	Subject					
	1	**2**	**3**	**4**	**5**	**6**
Postexercise	13.6	14.7	42.8	20.0	19.2	17.3
Control	8.5	12.6	21.6	19.4	14.7	13.6

Let μ_1 denote the mean nocturnal growth hormone level for the population of all healthy males who participated in strenuous activity on the previous day. Similarly, let μ_2 denote the mean nocturnal hormone level for the population consisting of all healthy males whose activities on the previous day did not include any strenuous physical exercise. The hypotheses of interest are then

$$H_0: \quad \mu_1 - \mu_2 = 0 \quad \text{versus} \quad H_a: \quad \mu_1 - \mu_2 \neq 0$$

Notice that in each of the six data pairs, the postexercise hormone level is higher than the corresponding control level. Intuitively, this suggests that there may be a difference between the population means.

Disregarding the paired nature of the samples results in a loss of information. Nocturnal growth hormone levels vary substantially from one individual to another. It is this variability that may obscure the difference in hormone level associated with strenuous exercise when the two-sample t test is used. If we were to (incorrectly) use the two-sample t test for independent samples on the given data, the resulting t test statistic value would be 1.28. This value would not allow for rejection of the hypothesis $\mu_1 - \mu_2 = 0$, even at level of significance .10. This result might surprise you at first, but remember that this test procedure ignores the fact that the samples are paired. Two plots of the data are given in Figure 11.1. The first one ignores the pairing, and the two samples look quite similar. The plot in which pairs are identified does suggest a difference, since for each pair the exercise observation exceeds the no-exercise observation.

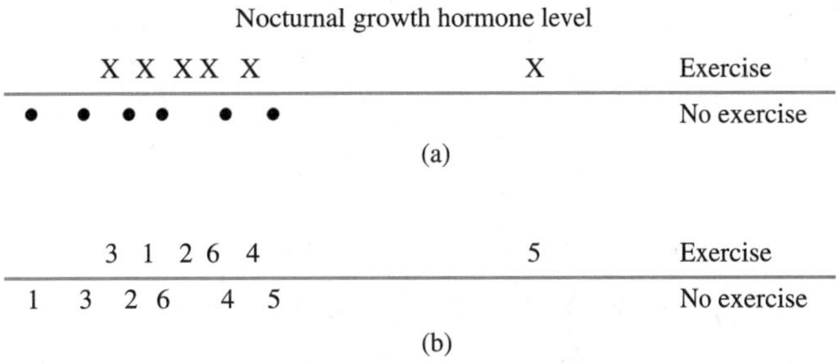

FIGURE 11.1 Two plots of the paired data from Example 11.5: (a) Pairing ignored; (b) Pairs identified

Example 11.5 suggests that the methods of inference developed for independent samples are not adequate for dealing with paired samples. When sample observations from the first population are paired in some meaningful way with sample observations from the second population, inferences can be based on the differences between the two observations within each sample pair. The n sample differences can then be regarded as having been selected from a large population of differences. Thus, in Example 11.5, we can think of the six (postexercise − control) differences as having been selected from an entire population of differences. Let

μ_d = mean value of the difference population

and

σ_d = standard deviation of the difference population

The relationship between μ_d and the two individual population means is

$\mu_d = \mu_1 - \mu_2$

Therefore, when the samples are paired, inferences about $\mu_1 - \mu_2$ are equivalent to inferences about μ_d. Since inferences about μ_d can be based on the n observed sample differences, the original two-sample problem becomes a familiar one-sample problem.

The Paired t Test

To compare two population or treatment means when the samples are paired, we first translate the hypothesis of interest from one about the value of $\mu_1 - \mu_2$ to an equivalent one involving μ_d.

Hypothesis	Equivalent Hypothesis When Samples Are Paired
H_0: $\mu_1 - \mu_2$ = hypothesized value	H_0: μ_d = hypothesized value
H_a: $\mu_1 - \mu_2 >$ hypothesized value	H_a: $\mu_d >$ hypothesized value
H_a: $\mu_1 - \mu_2 <$ hypothesized value	H_a: $\mu_d <$ hypothesized value
H_a: $\mu_1 - \mu_2 \neq$ hypothesized value	H_a: $\mu_d \neq$ hypothesized value

Sample differences (sample 1 value − sample 2 value) are then computed and used as the basis for testing hypotheses about μ_d. When the sample size is large or it is reasonable to assume that the population of differences is approximately normal, the one-sample t test based on the differences is the recommended test procedure. Generally speaking, the difference population will be normal if each of the two individual populations is normal. A normal probability plot or boxplot of the differences can be used to validate this assumption.

Summary of the Paired t Test for Comparing Two Population or Treatment Means

Null hypothesis: H_0: μ_d = hypothesized value

Test statistic: $t = \dfrac{\bar{x}_d - \text{hypothesized value}}{\dfrac{s_d}{\sqrt{n}}}$

where n is the number of sample differences and $\bar{x}_d$ and s_d are the sample mean and standard deviation of the differences. This test is based on df = $n - 1$.

Alternative hypothesis:	*P-value*:
H_a: $\mu_d >$ hypothesized value	Area under the appropriate t curve to the right of the calculated t
H_a: $\mu_d <$ hypothesized value	Area under the appropriate t curve to the left of the calculated t
H_a: $\mu_d \neq$ hypothesized value	(i) $2 \cdot$ (area to the right of t) if t is positive
	or
	(ii) $2 \cdot$ (area to the left of t) if t is negative

(continued)

Assumptions:	1. The samples are *paired*.
	2. The *n* sample differences can be viewed as a *random sample* from a population of differences.
	3. The *number of sample differences is large* (generally at least 30) OR the *population distribution of differences is approximately normal*.

EXAMPLE 11.6　Children with Down syndrome generally show a pattern of retarded mental development, although some achieve higher intellectual levels than others. The intellectual achievements of Down syndrome children have been studied by numerous investigators, and several different types of chromosomal abnormalities associated with the syndrome have been identified. Two such abnormalities are called trisomy 21 and mosaicism. The levels of intellectual functioning of trisomy 21 and mosaic Down syndrome children were compared in the article "Mental Development in Down Syndrome Mosaicism" (*Amer. Journal on Mental Retardation* (1991): 345–351).

Thirty children with mosaic Down syndrome who were treated at USC Medical Center were selected to participate in the research project. The investigators then chose 30 children with trisomy 21 Down syndrome from among 350 who had been seen at the medical center. The 30 chosen were selected (with the help of a computer) to achieve the best possible matches for children in the mosaic group, using age, gender, and parental socioeconomic status as the criteria for matching. The result was 30 matched pairs of children. IQ levels for all of the children were determined.

The researchers proposed using this data to test the theory that children with mosaic Down syndrome generally achieve higher intellectual levels. These samples were not independently chosen, since the 30 trisomy 21 children were specifically selected to match the children in the mosaic sample. Since the data is paired, our analysis will focus on the mosaic–trisomy 21 IQ differences. The data and computed differences are shown in the accompanying table.

Pair	Mosaic	Trisomy 21	Difference	Pair	Mosaic	Trisomy 21	Difference
1	73	71	2	11	58	63	−5
2	43	53	−10	12	71	47	24
3	69	58	11	13	92	44	48
4	89	71	18	14	57	28	29
5	53	50	3	15	76	75	1
6	81	70	11	16	55	60	−5
7	59	55	4	17	61	48	13
8	71	18	53	18	63	55	8
9	65	31	34	19	87	59	28
10	53	57	−4	20	64	78	−14

(*continued*)

Pair	Mosaic	Trisomy 21	Difference		Pair	Mosaic	Trisomy 21	Difference
21	63	55	8		26	61	50	11
22	58	51	7		27	91	63	28
23	50	55	−5		28	43	46	−3
24	59	53	6		29	55	54	1
25	75	47	28		30	88	48	40

Let

$$\mu_1 = \text{mean IQ for children with mosaic Down syndrome}$$
$$\mu_2 = \text{mean IQ for children with trisomy 21 Down syndrome}$$

and

$$\mu_d = \mu_1 - \mu_2 = \text{mean IQ difference between mosaic and trisomy 21 Down syndrome children}$$

The question of interest can be answered by testing the hypothesis

$$H_0: \quad \mu_d = 0 \qquad \text{versus} \qquad H_a: \quad \mu_d > 0$$

Using the 30 differences, we compute

$$\Sigma \text{ diff} = 370 \qquad \Sigma (\text{diff})^2 = 13{,}134$$

$$\bar{x}_d = \frac{\Sigma \text{ diff}}{n} = \frac{370}{30} = 12.33$$

$$s_d^2 = \frac{\Sigma (\text{diff})^2 - \dfrac{(\Sigma \text{ diff})^2}{n}}{n-1} = \frac{13{,}134 - \dfrac{(370)^2}{30}}{29} = 295.54$$

$$s_d = \sqrt{s_d^2} = 17.19$$

We now use the paired t test with a significance level of .05 to carry out the hypothesis test.

1. $\mu_d = \mu_1 - \mu_2 = $ mean IQ difference between mosaic and trisomy 21 Down syndrome children

2. $H_0: \mu_d = 0$

3. $H_a: \mu_d > 0$

4. *Significance level*: $\alpha = .05$

5. *Test statistic*: $t = \dfrac{\bar{x}_d - \text{hypothesized value}}{\dfrac{s_d}{\sqrt{n}}} = \dfrac{\bar{x}_d - 0}{\dfrac{s_d}{\sqrt{n}}}$

6. *Assumptions*: The sample size is large ($n = 30$). The researchers believed that it was reasonable to view the 30 sample differences as a random sample of all such differences, and so we will proceed with the paired t test.

7. *Calculation*: $t = \dfrac{12.33 - 0}{\dfrac{17.19}{\sqrt{30}}} = 3.93$

8. *P-value*:   This is an upper-tailed test, so the *P*-value is the area to the right of the computed *t* value. The appropriate df for this test is df = 30 − 1 = 29. From the 29-df column and the 3.9 row of Appendix Table IV, we find that

   P-value = .000

   The area to the right of 3.9·under the *t* curve with 29 df is so small that when it is rounded to 3 decimal places, the value is .000.

9. *Conclusion*:   Since *P*-value ≤ α (.000 ≤ .05), we reject H_0. The data supports the theory that the mean IQ for mosaic Down syndrome children is higher than the mean IQ for trisomy 21 Down syndrome children.

To use the two-sample *t* test (for independent samples) for the data in Example 11.6 would have been incorrect, because the samples are not independent. Inappropriate use of the two-sample *t* test would have resulted in a computed test statistic value of 3.55. The conclusion would still be to reject the hypothesis of equal mean IQs in this particular example, but this is not always the case.

EXAMPLE 11.7

It has been estimated that between 1945 and 1971, as many as 2 million children were born to mothers treated with diethylstilbestrol (DES), a nonsteroidal estrogen. The FDA banned this drug in 1971 because research indicated a link with the incidence of cervical cancer. The article "Effects of Prenatal Exposure to Diethylstilbestrol (DES) on Hemispheric Laterality and Spatial Ability in Human Males" (*Hormones and Behavior* (1992): 62–75) discussed a study in which 10 males exposed to DES and their unexposed brothers underwent various tests. Here is the summary data on the results of a spatial ability test:

$$n = 10 \qquad \bar{x}_1 = \text{exposed mean} = 12.6 \qquad \bar{x}_2 = \text{unexposed mean} = 13.8$$

$$\bar{x}_d = -1.2 \qquad \text{standard error of difference} = \frac{s_d}{\sqrt{n}} = .5$$

The investigators used a one-tailed test to see whether DES exposure was associated with reduced spatial ability.

1. μ_d = difference between true average score for exposed males and true average score for unexposed males

2. $H_0: \mu_d = 0$

3. $H_a: \mu_d < 0$

4. *Significance level*:   α = .05

5. *Test statistic*: $t = \dfrac{\bar{x}_d - \text{hypothesized value}}{\dfrac{s_d}{\sqrt{n}}} = \dfrac{\bar{x}_d - 0}{\dfrac{s_d}{\sqrt{n}}}$

6. *Assumptions*:   Without the raw data, it is difficult to assess the reasonableness of the assumptions. The author of the article judged the assumptions of the paired *t* test to be reasonable, so we will proceed.

7. *Calculation:* $t = \dfrac{-1.2 - 0}{.5} = -2.4$

8. *P-value:*   This is a lower-tailed test. There are 10 sample differences, so df = $10 - 1 = 9$. From the 9-df column and the 2.4 row of Appendix Table IV, we find that the area to the right of 2.4 is .020. Since the t curve is symmetric around 0, the area to the left of -2.4 is also .020, so

P-value = .020

9. *Conclusion:*   Since P-value $\leq \alpha$, H_0 is rejected. There is evidence that the mean spatial ability score is lower for those exposed to DES than for those who were unexposed.

The numerators $\bar{x}_d$ and $\bar{x}_1 - \bar{x}_2$ of the paired t and two-sample t test statistics are always equal. The difference lies in the denominator. The variability in differences is usually much smaller than the variability in each sample separately (because measurements in a pair tend to be similar). As a result, the value of the paired t statistic is usually larger in magnitude than the value of the two-sample t statistic. Pairing typically reduces variability that might otherwise obscure small but nevertheless significant differences.

A Confidence Interval

The one-sample t confidence interval for μ given in Chapter 9 is easily adapted to obtain an interval estimate for μ_d.

Paired t Confidence Interval for μ_d

When

1. the samples are *paired,*

2. the n sample differences can be viewed as a *random sample* from a population of differences, and

3. the *number of sample differences is large* (generally at least 30) OR the *population distribution of differences is approximately normal,*

the paired t confidence interval for μ_d is

$$\bar{x}_d \pm (t \text{ critical value}) \cdot \frac{s_d}{\sqrt{n}}$$

For a specified confidence level, the $(n - 1)$ df row of Appendix Table III gives the appropriate t critical value.

EXAMPLE 11.8   The effect of exercise on the amount of lactic acid in the blood was examined in the article "A Descriptive Analysis of Elite-Level Racquetball" (*Research Quarterly for Exercise and Sport* (1991): 109–114). Eight males were selected at random from those attending a week-long training camp. Blood lactate levels were measured before and after playing three games of racquetball, as shown in the accompanying table. We will use this data to estimate the mean change in blood lactate level using a 95% confidence interval.

Player	Before	After	Difference
1	13	18	−5
2	20	37	−17
3	17	40	−23
4	13	35	−22
5	13	30	−17
6	16	20	−4
7	15	33	−18
8	16	19	−3

$$\bar{x}_d = -13.63 \qquad s_d = 8.28 \qquad n = 8$$

The eight men were selected at random from training camp participants. The accompanying boxplot of the eight sample differences is consistent with a difference population that is approximately normal, so the paired t confidence interval is appropriate.

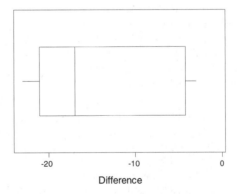

Difference

The t critical value for df = 7 and a 95% confidence level is 2.36, and so the confidence interval is

$$\bar{x}_d \pm (t \text{ critical value}) \cdot \frac{s_d}{\sqrt{n}} = -13.63 \pm (2.37) \cdot \frac{8.28}{\sqrt{8}}$$
$$= -13.63 \pm 6.938$$
$$= (-20.568, -6.692)$$

Based on the sample data, we can be 95% confident that the difference in mean blood lactate level is between −20.568 and −6.692. That is, we are 95% confident that the mean decrease in blood lactate level is somewhere between 6.692 and 20.568 after three games of racquetball.

When two populations must be compared to draw a conclusion on the basis of sample data, a researcher might choose to use independent samples or paired samples. In many situations, paired data provides a more effective comparison by screening out the effects of extraneous variables that might obscure differences between the two populations or that might suggest a difference when none exists.

Exercises 11.32 – 11.43

11.32 Suppose that you were interested in investigating the effect of a drug that is to be used in the treatment of patients who have glaucoma in both eyes. A comparison between the mean reduction in eye pressure for this drug and for a standard treatment is desired. Both treatments are applied directly to the eye.
  a. Describe how you would go about collecting data for your investigation.
  b. Does your method result in paired data?
  c. Can you think of a reasonable method of collecting data that would result in independent samples? Would such an experiment be as informative as a paired experiment? Comment.

11.33 Two different underground pipe coatings for preventing corrosion are to be compared. The effect of a coating (as measured by maximum depth of corrosion penetration on a piece of pipe) may vary with depth, orientation, soil type, pipe composition, etc. Describe how an experiment that filters out the effects of these extraneous factors could be carried out.

11.34 In a study of memory recall, eight students from a large psychology class were selected at random and given 10 min to memorize a list of 20 nonsense words. Each was asked to list as many of the words as he or she could remember both 1 hr and 24 hr later, as shown in the accompanying table. Is there evidence to suggest that the mean number of words recalled after 1 hr exceeds the mean recall after 24 hr by more than 3? Use a level .01 test.

Subject	1	2	3	4	5	6	7	8
1 hr later	14	12	18	7	11	9	16	15
24 hr later	10	4	14	6	9	6	12	12

11.35 Do girls think they don't need to take as many science classes as boys? The article "Intentions of Young Students to Enroll in Science Courses in the Future: An Examination of Gender Differences" (*Science Education* (1999): 55–76) gives information from a survey of children in grades 4, 5, and 6. The 224 girls participating in the survey each indicated the number of science courses they intended to take in the future, and they also indicated the number of science courses they thought boys their age should take in the future. For each girl, the authors calculated the difference between the number of science classes she intends to take and the number she thinks boys should take.
  a. Explain why this data is paired.
  b. The mean of the differences was −.83 (indicating girls intended, on average, to take fewer classes than they thought boys should take), and the standard deviation was 1.51. Construct and interpret a 95% confidence interval for the mean difference.

11.36 As part of a study to determine the effects of allowing the use of credit cards for alcohol purchases in Canada (see "Changes in Alcohol Consumption Patterns Following the Introduction of Credit Cards in Ontario Liquor Stores," *Journal of Studies on Alcohol* (1999): 378–382), randomly selected individuals were given a questionnaire asking them (among other things) how many drinks they had consumed during the previous week. A year later (after liquor stores started accepting credit cards for purchases), these same individuals were again asked how many drinks they had consumed in the previous week. The data shown is consistent with summary statistics presented in the article.

	n	1994 Mean	1995 Mean	$\bar{d}$	s_d
Credit-Card Shoppers	96	6.72	6.34	.38	5.52
Non–Credit-Card Shoppers	850	4.09	3.97	.12	4.58

  a. The standard deviation of the difference was quite large. Explain how this could be the case.
  b. Calculate a 95% confidence interval for the mean difference in drink consumption for credit card shoppers between 1994 and 1995. Did the mean number of drinks decrease?
  c. Test the hypothesis that there was no change in the mean number of drinks between 1994 and 1995 for the non–credit-card shoppers. Be sure to calculate and interpret the P-value for this test.

11.37 The article "Agronomic Performance of Winter Versus Spring Wheat" (*Agronomy J.* (1991): 527–531) described the results of an experiment to compare the yield (kg/ha) of Sundance winter wheat and Manitou spring wheat. Data for nine test plots is given in the accompanying table. Is there sufficient evidence to conclude that the mean yield for the Sundance winter wheat is higher than that for the Manitou spring wheat? Use $\alpha = .01$.

	Location				
	1	**2**	**3**	**4**	**5**
Sundance	3201	3095	3297	3644	3604
Manitou	2386	2011	2616	3094	3069

	Location			
	6	**7**	**8**	**9**
Sundance	2860	3470	2042	3689
Manitou	2074	2308	1525	2779

11.38 The article referenced in Exercise 11.37 also gave information on the protein concentration of Sundance winter wheat and Manitou spring wheat grown at nine locations. Protein concentration was measured in grams of protein per kilogram of wheat.

a. Use the accompanying data to determine whether there is a significant difference in protein concentration for the two varieties of wheat.

b. Construct a 90% confidence interval for the true difference in mean protein concentration for the two varieties of wheat.

	Location				
	1	**2**	**3**	**4**	**5**
Sundance	77	84	116	145	100
Manitou	125	111	144	171	133

	Location			
	6	**7**	**8**	**9**
Sundance	154	144	149	137
Manitou	166	143	176	170

11.39 Several methods of estimating the number of seeds in soil samples have been developed by ecologists. An article in the *Journal of Ecology* ("A Comparison of Methods for Estimating Seed Numbers in the Soil" (1990): 1079–1093) considered three such methods. The accompanying data gives number of seeds detected by the direct method and by the stratified method for 27 soil specimens. Does the data provide sufficient evidence to conclude that the mean number of seeds detected differs for the two methods? Test the relevant hypotheses using $\alpha = .05$.

Sample	Direct	Stratified
1	24	8
2	32	36
3	0	8
4	60	56
5	20	52
6	64	64
7	40	28
8	8	8
9	12	8
10	92	100
11	4	0
12	68	56
13	76	68
14	24	52
15	32	28
16	0	0
17	36	36
18	16	12
19	92	92
20	4	12
21	40	48
22	24	24
23	0	0
24	8	12
25	12	40
26	16	12
27	40	76

11.40 The effect of exercise on the amount of lactic acid in the blood was examined in the article "A Descriptive Analysis of Elite-Level Racquetball" (*Research Quarterly for Exercise and Sport* (1991): 109–114). Eight men and seven women who were attending a week-long training camp participated in the experiment. Blood lactate levels were measured before and after playing three games of racquetball, as shown in the accompanying table.

	Men			Women		
Player	**Before**	**After**		**Player**	**Before**	**After**
1	13	18		1	11	21
2	20	37		2	16	26
3	17	40		3	13	19
4	13	35		4	18	21
5	13	30		5	14	14
6	16	20		6	11	31
7	15	33		7	13	20
8	16	19				

a. Estimate the mean change in blood lactate level for male racquetball players using a 90% confidence interval. Why is this interval wider than the one in Example 11.8?

b. Estimate the mean change for female players using a 90% confidence interval.

c. Based on the intervals from parts **a** and **b**, do you think the mean change in blood lactate level is the same for men as it is for women? Explain.

11.41 Dentists make many people nervous (even more so than statisticians!). To see whether such nervousness elevates blood pressure, the blood pressure and pulse rates of 60 subjects were measured in a dental setting and in a medical setting ("The Effect of the Dental Setting on Blood Pressure Measurement," *Am. J. of Public Health* (1983): 1210–1214). For each subject, the difference (dental-setting blood pressure minus medical-setting blood pressure) was calculated. The analogous differences were also calculated for pulse rates. Summary data follows.

	Mean Difference	Standard Deviation of Differences
Systolic blood pressure	4.47	8.77
Pulse (beats/min)	−1.33	8.84

a. Does the data strongly suggest that true mean blood pressure is higher in a dental setting than in a medical setting? Use a level .01 test.

b. Is there sufficient evidence to indicate that true mean pulse rate in a dental setting differs from the true mean pulse rate in a medical setting? Use a significance level of .05.

11.42 Many people who quit smoking complain of weight gain. The results of an investigation of the relationship between smoking cessation and weight gain are given in the article "Does Smoking Cessation Lead to Weight Gain?" (*Amer. J. of Public Health* (1983): 1303–1305). Three hundred twenty-two subjects, selected at random from those who successfully participated in a program to quit smoking, were weighed at the beginning of the program and again 1 year later. The mean change in weight was 5.15 lb, and the standard deviation of the weight changes was 11.45 lb. Is there sufficient evidence to conclude that the true mean change in weight is positive? Use $\alpha = .05$.

11.43 The article "A Supplementary Behavioral Program to Improve Deficient Reading Performance" (*J. Abnormal Child Psych.* (1973): 390–399) reported the results of an experiment in which seven pairs of children reading below grade level were obtained by matching, so that within each pair the two children were equally deficient in reading ability. Then one child (selected at random) from each pair received experimental training, whereas the other received standard training. Based on the accompanying improvement scores, does the experimental training appear to be superior to the standard training? Use a .1 significance level.

	Pair						
	1	**2**	**3**	**4**	**5**	**6**	**7**
Experimental	.5	1.0	.6	.1	1.3	.1	1.0
Control	.8	1.1	−.1	.2	.2	1.5	.8

11.3 Large-Sample Inferences Concerning a Difference Between Two Population or Treatment Proportions

Large-sample methods for estimating and testing hypotheses about a single population proportion were presented in Chapters 9 and 10. The symbol π was used to represent the true proportion of individuals in the population who possess some characteristic (the "successes"). Inferences about the value of π were based on p, the corresponding sample proportion of successes.

Many investigations are carried out to compare the proportion of successes in one population (or resulting from one treatment) to the proportion of successes in a second population (or from a second treatment). As was the case for means, the subscripts 1 and 2 are used to distinguish between the two population proportions, sample sizes, and sample proportions.

Notation

Population or Treatment 1:   Proportion of "successes" $= \pi_1$

Population or Treatment 2:   Proportion of "successes" $= \pi_2$

	Sample Size	Proportion of Successes
Sample from population or treatment 1	n_1	p_1
Sample from population or treatment 2	n_2	p_2

When comparing two populations or treatments on the basis of "success" proportions, it is common to focus on the quantity $\pi_1 - \pi_2$, the difference between the two proportions. Since p_1 provides an estimate of π_1 and p_2 provides an estimate of π_2, the obvious choice for an estimate of $\pi_1 - \pi_2$ is $p_1 - p_2$.

Because p_1 and p_2 each vary in value from sample to sample, so will the difference $p_1 - p_2$. For example, a first sample from each of two populations might yield

$$p_1 = .69 \qquad p_2 = .70 \qquad p_1 - p_2 = -.01$$

A second sample from each might result in

$$p_1 = .79 \qquad p_2 = .67 \qquad p_1 - p_2 = .12$$

and so on. Since the statistic $p_1 - p_2$ will be the basis for drawing inferences about $\pi_1 - \pi_2$, we need to know something about its behavior.

Properties of the Sampling Distribution of $p_1 - p_2$

If two random samples are selected independently of one another, the following properties hold:

1. $\mu_{p_1-p_2} = \pi_1 - \pi_2$

   This says that the sampling distribution of $p_1 - p_2$ is centered at $\pi_1 - \pi_2$, so $p_1 - p_2$ is an unbiased statistic for estimating $\pi_1 - \pi_2$.

2. $\sigma^2_{p_1-p_2} = \sigma^2_{p_1} + \sigma^2_{p_2} = \dfrac{\pi_1(1 - \pi_1)}{n_1} + \dfrac{\pi_2(1 - \pi_2)}{n_2}$

   and

   $\sigma_{p_1-p_2} = \sqrt{\dfrac{\pi_1(1 - \pi_1)}{n_1} + \dfrac{\pi_2(1 - \pi_2)}{n_2}}$

3. If both n_1 and n_2 are large (that is, if $n_1\pi_1 \geq 10$, $n_1(1 - \pi_1) \geq 10$, $n_2\pi_2 \geq 10$, and $n_2(1 - \pi_2) \geq 10$), then p_1 and p_2 each have a sampling distribution that is approximately normal, and their difference, $p_1 - p_2$, also has a sampling distribution that is approximately normal.

The properties in the box imply that when the samples are independently selected and both sample sizes are large, the distribution of the standardized variable

$$z = \frac{p_1 - p_2 - (\pi_1 - \pi_2)}{\sqrt{\dfrac{\pi_1(1 - \pi_1)}{n_1} + \dfrac{\pi_2(1 - \pi_2)}{n_2}}}$$

is described approximately by the standard normal (z) curve.

A Large-Sample Test Procedure

Comparisons of π_1 and π_2 are often based on large, independently selected samples, and we restrict ourselves to this case. The most general null hypothesis of interest has the form

$$H_0: \quad \pi_1 - \pi_2 = \text{hypothesized value}$$

However, when the hypothesized value is something other than zero, the appropriate test statistic differs somewhat from the test statistic used for $H_0: \pi_1 - \pi_2 = 0$. Since this latter H_0 is almost always the relevant one in applied problems, we will focus exclusively on it.

Our basic testing principle has been to use a procedure that controls the probability of a type I error at the desired level α. This requires using a test statistic with a sampling distribution that is known when H_0 is true. That is, the test statistic should be developed under the assumption that $\pi_1 = \pi_2$ (as specified by the null hypothesis $\pi_1 - \pi_2 = 0$). In this case, π can be used to denote the common value of the two population proportions. The z variable obtained by standardizing $p_1 - p_2$ then simplifies to

$$z = \frac{p_1 - p_2}{\sqrt{\dfrac{\pi(1 - \pi)}{n_1} + \dfrac{\pi(1 - \pi)}{n_2}}}$$

Unfortunately, this cannot serve as a test statistic, because the denominator cannot be computed: H_0 says that there is a common value π but it does not specify the value. A test statistic can be obtained, though, by first *estimating* π from the sample data and then using this estimate in the denominator of z.

When $\pi_1 = \pi_2$, either p_1 or p_2 separately gives an estimate of the common proportion π. However, a better estimate than either of these is a weighted average of the two, in which more weight is given to the sample proportion based on the larger sample.

Definition

The **combined estimate of the common population proportion** is

$$p_c = \frac{n_1 p_1 + n_2 p_2}{n_1 + n_2} = \frac{\text{total number of } S\text{'s in two samples}}{\text{total sample size}}$$

The test statistic for testing $H_0: \pi_1 - \pi_2 = 0$ results from using p_c in place of π in the standardized variable z given previously. This z statistic has approximately a standard normal distribution when H_0 is true, so a test that has the desired significance level α can be obtained by calculating a P-value using the z table.

Summary of Large-Sample z Tests for $\pi_1 - \pi_2 = 0$

Null hypothesis: $H_0: \pi_1 - \pi_2 = 0$

Test statistic: $z = \dfrac{p_1 - p_2}{\sqrt{\dfrac{p_c(1 - p_c)}{n_1} + \dfrac{p_c(1 - p_c)}{n_2}}}$

Alternative hypothesis:	*P-value*:
$H_a: \pi_1 - \pi_2 > 0$	Area under the z curve to the right of the computed z
$H_a: \pi_1 - \pi_2 < 0$	Area under the z curve to the left of the computed z
$H_a: \pi_1 - \pi_2 \neq 0$	(i) $2 \cdot$ (area to the right of z) if z is positive
	or
	(ii) $2 \cdot$ (area to the left of z) is z is negative

Assumptions: 1. The samples are *independently chosen random samples* OR *treatments were assigned at random to individuals or objects* (or vice versa).
2. Both *sample sizes are large*:

$$n_1 p_1 \geq 10 \qquad n_1(1 - p_1) \geq 10 \qquad n_2 p_2 \geq 10 \qquad n_2(1 - p_2) \geq 10$$

EXAMPLE 11.9

Many investigators have studied the effect of the wording of questions on survey responses. Consider the following two versions of a question concerning gun control:

1. Would you favor or oppose a law that would require a person to obtain a police permit before purchasing a gun?

2. Would you favor or oppose a law that would require a person to obtain a police permit before purchasing a gun, or do you think that such a law would interfere too much with the right of citizens to own guns?

Let π_1 denote the proportion of all adults who would respond "favor" when asked question 1, and define π_2 similarly for question 2. The article "Attitude Measurement and the Gun Control Paradox" (*Public Opinion Quarterly* (1977–1978): 427–438) reported the accompanying sample data from two independently chosen random samples:

Sample size	$n_1 = 615$	$n_2 = 585$
Number who favor	463	403
Sample proportion	$p_1 = 463/615 = .753$	$p_2 = 403/585 = .689$

The extra phrase in question 2 that reminds individuals of the right to bear arms might elicit a smaller proportion of favorable responses than would the first question without the phrase. Does the data suggest that this is the case? We will test the relevant hypotheses using $\alpha = .01$. Suppose that $\pi_1 = \pi_2$; let π denote the common value. Then the combined estimate of π is

$$p_c = \frac{n_1 p_1 + n_2 p_2}{n_1 + n_2} = \frac{615(.753) + 585(.689)}{615 + 585} = .722$$

The nine-step procedure can now be used to perform the hypothesis test:

1. $\pi_1 - \pi_2$ is the difference between the true proportions of favorable responses to questions 1 and 2.

2. $H_0: \pi_1 - \pi_2 = 0$ $(\pi_1 = \pi_2)$

3. $H_a: \pi_1 - \pi_2 > 0$ $(\pi_1 > \pi_2$, in which case the extra phrase *does* result in proportionately fewer favorable responses.)

4. *Significance level*: $\alpha = .01$

5. *Test statistic*: $z = \dfrac{p_1 - p_2}{\sqrt{\dfrac{p_c(1 - p_c)}{n_1} + \dfrac{p_c(1 - p_c)}{n_2}}}$

6. *Assumptions*:   The two samples are independently chosen random samples. Checking to make sure that the sample sizes are large enough, we have

   $n_1 p_1 = 615(.753) = 463.10 \geq 10$ $n_1(1 - p_1) = 615(.247) = 151.91 \geq 10$

   $n_2 p_2 = 585(.689) = 403.07 \geq 10$ $n_2(1 - p_2) = 585(.311) = 181.94 \geq 10$

7. *Calculation*: $n_1 = 615, n_2 = 585, p_1 = .753, p_2 = .689$, and $p_c = .722$, so

   $$z = \frac{.753 - .689}{\sqrt{\dfrac{(.722)(.278)}{615} + \dfrac{(.722)(.278)}{585}}} = \frac{.064}{.0259} = 2.47$$

8. *P-value*:   This is an upper-tailed test, so the P-value is the area under the z curve and to the right of the computed $z = 2.47$. From Appendix Table II,

   P-value $= 1 - .9932 = .0068$

9. *Conclusion*:   Since P-value $\leq \alpha$, the hypothesis H_0 is rejected at level .01. Inclusion of the extra phrase about the right to bear arms *does* seem to result in fewer favorable responses than would be elicited without the phrase. This example points out the need to be particularly careful in wording survey questions.

EXAMPLE 11.10   The authors of the article "Accommodating Persons with AIDS: Acceptance and Rejection in Rental Situations" (*J. Applied Social Psychology* (1999): 261–270) state that even though landlords participating in a telephone survey indicated that they would generally be willing to rent to persons with AIDS, they wondered whether this was true in actual practice. To investigate, two random samples of 80 advertisements for rooms for rent were independently selected from newspaper advertisements in three large cities. An adult male caller responded to each ad in the first sample of 80 and inquired about the availability of the room and was told that the room was still available in 61 of these calls. The same caller also responded to each ad in the second sample. In these calls, the caller indicated that he was currently receiving some treatment for AIDS and was about to be released from the hospital and would require a place to live. The caller was told that a room was available in 32 of these calls. Based on this information, the authors concluded that "reference to AIDS substantially decreased the likelihood of a room being described as available." Does the data support this conclusion? Let's carry out a hypothesis test with $\alpha = .01$.

No AIDS Reference	$n_1 = 80$	$p_1 = 61/80 = .763$
AIDS Reference	$n_2 = 80$	$p_2 = 32/80 = .400$

1. $\pi_1 = $ proportion of rooms available when there is no AIDS reference

   $\pi_2 = $ proportion of rooms available with AIDS reference

2. $H_0: \pi_1 - \pi_2 = 0$

3. $H_a: \pi_1 - \pi_2 > 0$

4. *Significance level:* $\alpha = .01$

5. *Test statistic:* $z = \dfrac{p_1 - p_2}{\sqrt{\dfrac{p_c(1 - p_c)}{n_1} + \dfrac{p_c(1 - p_c)}{n_2}}}$

6. *Assumptions:* The two samples are independently chosen random samples. Checking to make sure that the sample sizes are large enough, we have

   $n_1 = 80 \qquad n_2 = 80 \qquad p_1 = .763 \qquad p_2 = .400$

   $p_c = \dfrac{n_1 p_1 + n_2 p_2}{n_1 + n_2} = \dfrac{80(.763) + 80(.400)}{80 + 80} = .582$

   $n_1 p_1 = 61.04 \geq 10 \qquad n_1(1 - p_1) = 18.96 \geq 10$

   $n_2 p_2 = 32.00 \geq 10 \qquad n_2(1 - p_2) = 48.00 \geq 10$

7. *Calculations:* $z = \dfrac{.763 - .400}{\sqrt{\dfrac{(.582)(.418)}{80} + \dfrac{(.582)(.418)}{80}}} = \dfrac{.363}{.078} = 4.65$

8. *P-value:* The P-value for this test is the area under the z curve and to the right of the computed $z = 4.65$. Since 4.65 is so far out in the upper tail of the z curve,

   $P\text{-value} \approx 0$

9. *Conclusion:* Since $P\text{-value} \leq \alpha$, the hypothesis H_0 is rejected at level .01. There is strong evidence that the proportion of rooms reported as available is smaller with the AIDS reference than without it. This supports the claim made by the authors of the article.

MINITAB can also be used to carry out a two-sample z test to compare two population proportions, as shown in the accompanying output.

Test and Confidence Interval for Two Proportions

Sample	X	N	Sample p
1	61	80	0.762500
2	32	80	0.400000

Estimate for p(1) - p(2): 0.3625
95% CI for p(1) - p(2): (0.220302, 0.504698)
Test for p(1) - p(2) = 0 (vs > 0): z = 4.65 P-Value = 0.000

A Confidence Interval

A large-sample confidence interval for $\pi_1 - \pi_2$ is a special case of the general z interval formula

point estimate $\pm$ (z critical value)(estimated standard deviation)

The statistic $p_1 - p_2$ gives a point estimate of $\pi_1 - \pi_2$, and the standard deviation of this statistic is

$$\sigma_{p_1-p_2} = \sqrt{\frac{\pi_1(1 - \pi_1)}{n_1} + \frac{\pi_2(1 - \pi_2)}{n_2}}$$

An estimated standard deviation is obtained by using the sample proportions p_1 and p_2 in place of π_1 and π_2, respectively, under the square root symbol. Notice that this estimated standard deviation differs from the one used previously in the test statistic. Here, there isn't a null hypothesis that claims $\pi_1 = \pi_2$, so there is no common value of π to estimate.

A Large-Sample Confidence Interval for $\pi_1 - \pi_2$

When

1. the samples are *independently selected random samples* OR *treatments were assigned at random to individuals or objects* (or vice versa), and

2. both *sample sizes are large*:

$$n_1 p_1 \geq 10 \qquad n_1(1 - p_1) \geq 10 \qquad n_2 p_2 \geq 10 \qquad n_2(1 - p_2) \geq 10$$

a large-sample confidence interval for $\pi_1 - \pi_2$ is

$$(p_1 - p_2) \pm (z \text{ critical value})\sqrt{\frac{p_1(1 - p_1)}{n_1} + \frac{p_2(1 - p_2)}{n_2}}$$

EXAMPLE 11.11

Researchers at the National Cancer Institute released the results of a study that examined the effect of weed-killing herbicides on house pets (Associated Press, Sept. 4, 1991). The following data is compatible with summary values given in the report. Dogs, some of whom were from homes where the herbicide was used on a regular basis, were examined for the presence of malignant lymphoma. We will use the given data to estimate the difference between the proportion of exposed dogs that develop lymphoma and the proportion of unexposed dogs that develop lymphoma.

Group	Sample Size	Number with Lymphoma	p
Exposed	827	473	.572
Unexposed	130	19	.146

Let π_1 denote the proportion of exposed dogs that develop lymphoma, and define π_2 similarly for unexposed dogs. The sample sizes are large enough for the large-sample interval to be valid ($n_1 p_1 = 827(.572) \geq 10$, $n_1(1 - p_1) = 827(.428) \geq 10$, and so on). A 90% confidence interval for $\pi_1 - \pi_2$ is

$$(p_1 - p_2) \pm (z \text{ critical value})\sqrt{\frac{p_1(1 - p_1)}{n_1} + \frac{p_2(1 - p_2)}{n_2}}$$

$$= (.572 - .146) \pm (1.645)\sqrt{\frac{(.572)(.428)}{827} + \frac{(.146)(.854)}{130}}$$

$$= .426 \pm (1.645)(.0354)$$

$$= .426 \pm .058$$

$$= (.368, .484)$$

Based on the observed sample, we believe that the proportion of exposed dogs that develop lymphoma exceeds that for unexposed dogs by somewhere between .36 and .48.

Exercises 11.44 – 11.58

11.44 Let π_1 and π_2 denote the proportions of all male and all female shoppers, respectively, who buy only name-brand grocery products (as opposed to generic or store brands).
  a. Test $H_0: \pi_1 - \pi_2 = 0$ versus $H_a: \pi_1 - \pi_2 \neq 0$ at level .05, using the following data: $n_1 = 200$, number of successes (only name-brand purchases) = 87, $n_2 = 300$, number of successes = 96.
  b. Use the data of part **a** to compute a 95% confidence interval for $\pi_1 - \pi_2$.

11.45 A university is interested in evaluating registration processes. Students can register for classes by using either a telephone registration system or an on-line system that is accessed through the university's Web site. Independent random samples of 80 students who registered by phone and 60 students who registered on-line were selected. Of those who registered by phone, 57 reported that they were satisfied with the registration process. Of those who registered on-line, 50 reported that they were satisfied.

Based on this data, is it reasonable to conclude that the proportion who are satisfied is higher for those who register on-line? Test the appropriate hypotheses using $\alpha = .05$.

11.46 Do teachers find their work rewarding and satisfying? The article "Work-Related Attitudes" (*Psychological Reports* (1991): 443–450) reported the results of a survey of random samples of 395 elementary school teachers and 266 high school teachers. Of the elementary school teachers, 224 said they were very satisfied with their jobs, whereas 126 of the high school teachers were very satisfied with their work. Based on this data, is it reasonable to conclude that the proportion very satisfied is different for elementary school teachers than it is for high school teachers? Test the appropriate hypotheses using a .05 significance level.

11.47 The article "Foraging Behavior of the Indian False Vampire Bat" (*Biotropica* (1991): 63–67) reported that 36 of 193 female bats in flight spent more than 5 min in the air before locating food. For male bats, 64 of 168 spent more than 5 min in the air. Is there sufficient evidence to conclude that the proportion of flights longer than 5 min in length differs for males and females? Test the relevant hypotheses using $\alpha = .01$.

11.48 The article "Regrets on Early Sex" appeared in the Australian newspaper *The Herald* (Jan. 2, 1998). The article stated that "While 54% of women thought that they should have waited longer before having sex, only 16% of the men felt that way." Suppose that this statement had been based on interviews with independently chosen random samples of 50 women and 50 men from a particular city in New Zealand. (The article summarizes a survey conducted in New Zealand.) Construct and interpret a 90% confidence interval for the difference in the true proportions of men and women who think they should have waited longer before having sex.

11.49 Gender differences in student needs and fears were examined in the article "A Survey of Counseling Needs of Male and Female College Students" (*J. College Student Development* (1998): 205–208). Random samples of male and female students were selected from those attending a particular university. Of 234 males surveyed, 27.5% said that they were concerned about the possibility of getting AIDS. Of 568 female students surveyed, 42.7% reported being concerned about the possibility of getting AIDS. Is there sufficient evidence to conclude that the proportion of female students concerned about the possibility of getting AIDS is greater than the corresponding proportion for males?

11.50 In a study that was described by a writer for the *Washington Post* (*San Luis Obispo Tribune*, Feb. 2, 2000), 500 patients undergoing abdominal surgery were randomly assigned to breathe one of two oxygen mixtures during surgery and for 2 hours afterward. One group received a mixture containing 30% oxygen, a standard generally used in surgery. The other group was given 80% oxygen. Wound infections developed in 28 of the 250 patients who received 30% oxygen and in 13 of the 250 patients who received 80% oxygen. Is there sufficient evidence to conclude that the proportion of patients who develop wound infections is lower for the 80% oxygen treatment than for the 30% oxygen treatment? Test the relevant hypotheses using a significance level of .05.

11.51 The state of Georgia's HOPE scholarship program guarantees fully paid tuition to Georgia public universities for Georgia high school seniors who have a B average in academic requirements as long as they maintain a B average in college. (See "Who Loses HOPE? Attrition

from Georgia's College Scholarship Program" (*Southern Economic Journal* (1999): 379–390).) It was reported that 53.2% of a random sample of 137 students entering Ivan Allen College at Georgia Tech (social science and humanities) with a HOPE scholarship lost the scholarship at the end of the first year because they had a GPA of less than 3.0. It was also reported that 72 of a random sample of 111 students entering the College of Computing with a B average had lost their HOPE scholarship by the end of the first year. Is there evidence that the proportion who lose HOPE scholarships is different for the College of Computing than for the Ivan Allen College?

11.52 The article "Softball Sliding Injuries" (*Amer. J. of Diseases of Children* (1988): 715–716) provided a comparison of breakaway bases (designed to reduce injuries) and stationary bases. Consider the accompanying data (which agrees with summary values given in the paper).

	Number of Games Played	Number of Games Where a Player Suffered a Sliding Injury
Stationary bases	1250	90
Breakaway bases	1250	20

Does the use of breakaway bases reduce the proportion of games with a player suffering a sliding injury? Answer by performing a level .01 test. What assumptions are necessary in order for your test to be valid? Do you think they are reasonable in this case?

11.53 An Associated Press article (*San Luis Obispo Telegram-Tribune*, Sept. 23, 1995) examined the changing attitudes of Catholic priests. National surveys of priests aged 26 to 35 were conducted in 1985 and again in 1993. The priests surveyed were asked whether they agreed with the following statement: Celibacy should be a matter of personal choice for priests. In 1985, 69% of those surveyed agreed; in 1993, 38% agreed. Suppose that the samples were randomly selected and that the sample sizes were both 200. Is there evidence that the proportion of priests who agreed that celibacy should be a matter of personal choice declined from 1985 to 1993? Use $\alpha = .05$.

11.54 A *New York Times* article (Oct. 6, 1993) reported on an alternative sentencing program for those convicted of a crime. Those in the alternative program took a year-long course at the Dartmouth campus of the University of Massachusetts. Of 32 men who took the class, 6 were subsequently convicted of new crimes. Over the same time period, 18 of 40 other men with similar backgrounds were convicted of new crimes. Is there evidence that the pro-portion convicted of new crimes is lower for those taking the class? Test the appropriate hypotheses using $\alpha = .01$.

11.55 Are college students who take a freshman orien-tation course more or less likely to stay in college than those who do not take such a course? The article "A Lon-gitudinal Study of the Retention and Academic Perfor-mance of Participants in Freshmen Orientation Courses" (*J. College Student Development* (1994): 444–449) re-ported that 50 of 94 randomly selected students who did not participate in an orientation course returned for a second year. Of 94 randomly selected students who did take the orientation course, 56 returned for a second year. Construct a 95% confidence interval for $\pi_1 - \pi_2$, the difference in the proportion returning for students who do not take an orientation course and those who do. Give an interpretation of this interval.

11.56 The article "Truth and DARE: Tracking Drug Education to Graduation" (*Social Problems* (1994): 448–456) compared the drug use of 238 randomly selected high school seniors exposed to a drug education program (DARE) and 335 randomly selected high school seniors who were not exposed to such a program. Data for mari-juana use is given in the accompanying table. Is there evidence that the proportion using marijuana is lower for students exposed to the DARE program? Use $\alpha = .05$.

	n	Number Who Use Marijuana
Exposed to DARE	288	141
Not exposed to DARE	335	181

11.57 The positive effect of water fluoridation on dental health is well documented. One study that validates this is described in the article "Impact of Water Fluoridation on Children's Dental Health: A Controlled Study of Two Pennsylvania Communities" (*Amer. Stat. Assoc. Proc. of the Social Statistics Section* (1981): 262–265). Two com-munities were compared. One had adopted fluoridation in 1966, whereas the other had no such program. Of 143 ran-domly selected children from the town without fluoridated water, 106 had decayed teeth, and 67 of 119 randomly se-lected children from the town with fluoridated water had decayed teeth. Let π_1 denote the true proportion of chil-dren drinking fluoridated water who have decayed teeth, and let π_2 denote the analogous proportion for children drinking unfluoridated water. Estimate $\pi_1 - \pi_2$ using a 90% confidence interval. Does the interval contain 0? In-terpret the interval.

11.58 A person released from prison before completing the original sentence is placed under the supervision of a parole board. If that person violates specified conditions of good behavior during the parole period, the board can order a return to prison. The article "Impulsive and Pre-meditated Homicide: An Analysis of the Subsequent Parole Risk of the Murderer" (*J. of Criminal Law and Criminology* (1978): 108–114) reported the accompany-ing data on parole behavior. One random sample of indi-viduals had served time in prison for impulsive murder, and the other random sample had served time for pre-meditated murder. Construct a 98% confidence interval for the difference in the proportion who successfully com-pleted parole for impulsive and premeditated murders.

	Crime	
	Impulsive	**Premeditated**
Sample size	$n_1 = 42$	$n_2 = 40$
Number with no violation	13	22
Sample proportion	$p_1 = 3.10$	$p_2 = .550$

11.4 Distribution-Free Procedures for Inferences Concerning a Difference Between Two Population or Treatment Means Using Independent Samples (Optional)

One approach to making inferences about $\mu_1 - \mu_2$ when n_1 and n_2 are small is to assume that the two population or treatment response distributions are normal and then use the two-sample t test or confidence interval presented in Section 11.1. In some situations, however, the normality assumption may not be reasonable. The validity of the procedures developed in this section does not depend on the nor-mality of the population distributions. The procedures can be used to compare two populations or treatments when it is reasonable to assume that the population or treatment response distributions have the same shape and spread.

Basic Assumptions in This Section

The two population or treatment response distributions have the same shape and spread. The only possible difference between the distributions is that one may be shifted to one side or the other.

Distributions consistent with these assumptions are shown in Figure 11.2(a). The distributions shown in Figure 11.2(b) have different shapes and spreads, therefore the methods of this section would not be appropriate in this case. Inferences that involve comparing distributions with different shapes and spreads can be quite complicated. Good advice from a statistician is particularly important in this case.

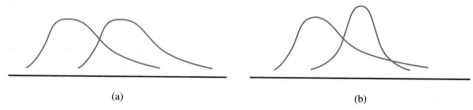

(a)   (b)

FIGURE 11.2 Two possible population distribution pairs: (a) Same shape and spread, differing only in location; (b) Very different shape and spread

Procedures that do not require any overly specific assumptions about the population distributions are said to be **distribution-free** (some texts use the term *nonparametric* instead of distribution-free). The two-sample *t* test of Section 11.1 is not distribution-free, because its use is predicated on the specific assumption of (at least approximate) normality.

Inferences about $\mu_1 - \mu_2$ will be made from information in two independent random samples, one consisting of n_1 observations from the first population and the other consisting of n_2 observations from the second population. Suppose that the two population distributions are in fact identical (so that $\mu_1 = \mu_2$). In this case, each of the $n_1 + n_2$ observations is actually drawn from the same population distribution. The distribution-free procedure presented here is based on regarding the $n_1 + n_2$ observations as a single data set and assigning ranks to the ordered values. The assignment is easiest when there are no ties among the $n_1 + n_2$ values (each observation is different from every one of the others), so assume for the moment that this is the case. Then the smallest among the $n_1 + n_2$ values receives rank 1, the second smallest rank 2, and so on, until finally the largest value is assigned rank $n_1 + n_2$.

EXAMPLE 11.12

An experiment to compare fuel efficiencies for two types of subcompact automobile was carried out by first randomly selecting $n_1 = 5$ cars of type 1 and $n_2 = 5$ cars of type 2. Each car was then driven from Phoenix to Los Angeles by a nonprofessional driver, after which the fuel efficiency (in mpg) was determined. The resulting data, with observations in each sample ordered from smallest to largest, is given here:

Type 1	39.3	41.1	42.4	43.0	44.4
Type 2	37.8	39.0	39.8	40.7	42.1

The data and the associated ranks are shown in the accompanying dotplot.

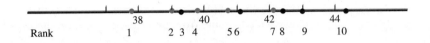

- Sample 1
- Sample 2

The ranks of the five observations in the first sample are 3, 6, 8, 9, 10. If these five observations had all been larger than every value in the second sample, the corresponding ranks would have been 6, 7, 8, 9, and 10. On the other hand, if all five sample 1 observations had been less than each value in the second sample, the ranks would have been 1, 2, 3, 4, and 5. The ranks of the five observations in the first sample might be any set of five numbers from among 1, 2, 3, . . . , 9, 10 — there are actually 252 possibilities.

Testing Hypotheses

Let's first consider testing

$$H_0: \quad \mu_1 - \mu_2 = 0 \quad (\mu_1 = \mu_2)$$

versus

$$H_a: \quad \mu_1 - \mu_2 > 0 \quad (\mu_1 > \mu_2)$$

When H_0 is true, all $n_1 + n_2$ observations in the two samples are actually drawn from identical population distributions. We would then expect that the observations in the first sample would be intermingled with those of the second sample when plotted along the number line. In this case, the ranks of the observations should also be intermingled. For example, with $n_1 = 5$ and $n_2 = 5$, the set of sample 1 ranks 2, 3, 5, 8, 10 would be consistent with $\mu_1 = \mu_2$, as would the set 1, 4, 7, 8, 9. However, when $\mu_1 = \mu_2$, it would be quite unusual for all five values from sample 1 to be larger than every value in sample 2, resulting in the set 6, 7, 8, 9, 10 of sample 1 ranks.

A convenient measure of the extent to which the ranks are intermingled is the sum of the sample 1 ranks. These ranks in Example 11.12 were 3, 6, 8, 9, and 10, so

rank-sum = 3 + 6 + 8 + 9 + 10 = 36

The largest possible rank-sum when $n_1 = n_2 = 5$ is $6 + 7 + 8 + 9 + 10 = 40$. If μ_1 is much larger than μ_2, we would expect the rank-sum to be near its largest possible value. This suggests that we reject H_0 for unusually large values of the rank-sum.

Developing a test procedure requires information about the sampling distribution of the rank-sum statistic when H_0 is true. To illustrate this, consider again the case $n_1 = n_2 = 5$. There are 252 different sets of 5 from among the 10 ranks 1, 2, 3, ..., 9, 10. The key point is that, when H_0 is true, any one of these 252 sets has the same chance of being the sample 1 ranks as does any other set, because all 10 observations come from the same population distribution. The chance under H_0 that any particular set occurs is 1/252 (because the possibilities are equally likely).

Table 11.1 displays the 12 sets of sample 1 ranks that yield the largest rank-sum values. Each of the other 240 possible rank sets has a rank-sum value less than 36.

Table 11.1 *The 12 rank sets that have the largest rank-sums when $n_1 = 5$ and $n_2 = 5$.*

Sample 1 Ranks	Rank-Sum
6 7 8 9 10	40
5 7 8 9 10	39
4 7 8 9 10	38
5 6 8 9 10	38
3 7 8 9 10	37
4 6 8 9 10	37
5 6 7 9 10	37
2 7 8 9 10	36
3 6 8 9 10	36
4 5 8 9 10	36
4 6 7 9 10	36
5 6 7 8 10	36

If we observe rank-sum = 36, we can compute

P(rank-sum $\geq$ 36 when H_0 is true) = 12/252 = .0476

That is, when H_0 is true, a rank-sum at least as large as 36 would be observed only about 4.76% of the time. Thus, a test of H_0: $\mu_1 - \mu_2 = 0$ versus H_a: $\mu_1 - \mu_2 > 0$, based on $n_1 = 5$ and $n_2 = 5$ and a rank-sum of 36, would have an associated P-value of .0476. It is this type of reasoning that allows us to reach a conclusion about whether to reject H_0.

The process of looking at different rank-sum sets to carry out a test can be quite tedious. Fortunately, information about both one- and two-tailed P-values associated with values of the rank-sum statistic has been tabled. Appendix Table VI provides information on P-values for selected values of n_1 and n_2. For example, when

n_1 and n_2 are both 5, Appendix Table VI tells us that, for an upper-tailed test with a rank-sum statistic value of 36, P-value $< .05$. This is consistent with the value of .0476 computed previously. Had the rank-sum been 40, we would have concluded that the P-value was less than .01 (using the table). The use of this table is further illustrated in the examples that follow.

Summary of the Rank Sum Test*

Null hypothesis: $H_0: \mu_1 - \mu_2 = 0$

Test statistic: rank-sum = sum of ranks assigned to the observations in the first sample

Alternative hypothesis:	*Type of test*:
$H_a: \mu_1 - \mu_2 > 0$	Upper-tailed
$H_a: \mu_1 - \mu_2 < 0$	Lower-tailed
$H_a: \mu_1 - \mu_2 \neq 0$	Two-tailed

Information about the P-value associated with this test is given in Appendix Table VI.

Assumptions: 1. *The samples are independent random samples* OR *treatments are randomly assigned to individuals or objects* (or vice versa).
2. The two *population or treatment response distributions have the same shape and spread.*

EXAMPLE 11.13

The extent to which an infant's health is affected by parents' smoking is an important public health concern. The article "Measuring the Exposure of Infants to Tobacco Smoke" (*New Engl. J. of Med.* (1984): 1075–1078) reported on a study in which various measurements were taken both from a random sample of infants who had been exposed to household smoke and from a sample of unexposed infants. The accompanying data consists of observations on urinary concentration of cotanine, a major metabolite of nicotine. (The values constitute a subset of the original data and were read from a plot that appeared in the article.)

Unexposed ($n_1 = 7$)	8	11	12	14	20	43	111	
Rank	1	2	3	4	5	7	11	
Exposed ($n_2 = 8$)	35	56	83	92	128	150	176	208
Rank	6	8	9	10	12	13	14	15

Does the data suggest that the true average cotanine level is higher for exposed than for unexposed infants? The investigators used the rank-sum test to analyze the data.

1. $\mu_1 - \mu_2$ is the difference between true average cotanine concentration for unexposed and exposed infants.

2. $H_0: \mu_1 - \mu_2 = 0$

---

*This procedure is often called the *Wilcoxon rank-sum test* or the *Mann–Whitney test,* after the statisticians who developed it. Some sources use a slightly different (but equivalent) test statistic formula.

3. $H_a: \mu_1 - \mu_2 < 0$ (unexposed average is less than exposed average)
4. *Significance level*: $\alpha = .01$
5. *Test statistic*:   rank-sum = sum of sample 1 ranks
6. *Assumptions*:   The authors of the article indicated that they believe the assumptions of the rank-sum test were reasonable.
7. *Calculation*:   rank-sum $= 1 + 2 + 3 + 4 + 5 + 7 + 11 = 33$
8. *P-value*:   This is a lower-tailed test. With $n_1 = 7$ and $n_2 = 8$, Appendix Table VI tells us that P-value $< .05$ if rank-sum ≤ 41 and P-value $< .01$ if rank-sum ≤ 36. Since rank-sum $= 33$, we conclude that

   P-value $< .01$

9. *Conclusion*:   Since the P-value is less than α (.01), we reject H_0 and conclude that infants exposed to cigarette smoke do have a higher average cotanine level than do unexposed infants.

Many statistical computer packages will perform the rank-sum test and give exact P-values. Partial MINITAB output for the data of Example 11.13 follows. MINITAB uses the symbol **W** to denote the rank-sum statistic and uses the terms **ETA1** and **ETA2** in place of μ_1 and μ_2, but the test statistic value and the associated P-values are the same as for the test presented here. The output indicates that W = 33 and that the test "is significant at 0.0046." These two values, 33 and .0046, are the rank-sum statistic and the P-value, respectively. In Example 11.13, we used Appendix Table VI to determine that P-value $< .01$. This statement is consistent with the actual P-value given in the MINITAB output.

```
Unexpose    N = 7     Median =  14.00
Exposed     N = 8     Median = 110.00
Point estimate for ETA1-ETA2 is     -79.00
95.7 Percent CI for ETA1-ETA2 is (-156.00,-23.99)
W = 33.0
Test of ETA1 = ETA2 vs ETA1 < ETA2 is significant at 0.0046
```

The test procedure just described can be easily modified to handle a hypothesized value other than zero. Consider as an example testing $H_0: \mu_1 - \mu_2 = 5$. This hypothesis is equivalent to $H_0: (\mu_1 - 5) - \mu_2 = 0$. That is, if 5 is subtracted from each population 1 value, then according to H_0, the distribution of the resulting values coincides with the population 2 distribution. This suggests that, if the hypothesized value of 5 is first subtracted from each sample 1 observation, the test can then be carried out as before.

To test $H_0: \mu_1 - \mu_2 = $ *hypothesized value*, subtract the hypothesized value from each observation in the first sample and then determine the ranks of these when combined with the n_2 observations from the second sample.

EXAMPLE 11.14   (Example 11.13 continued)   Reconsider the cotanine concentration data introduced in Example 11.13. Suppose that a researcher wished to know whether average concentration for exposed children exceeds that for unexposed children by more than 25. Recall that μ_1 is the true average concentration for unexposed children. The exposed average exceeds the unexposed average by 25 when $\mu_1 - \mu_2 = -25$ and by more than 25 when $\mu_1 - \mu_2 < -25$. The hypotheses of interest are, therefore,

$$H_0: \quad \mu_1 - \mu_2 = -25 \qquad H_a: \mu_1 - \mu_2 < -25$$

These hypotheses can be tested by first subtracting -25 (or, equivalently, adding 25) to each sample 1 observation.

Sample 1

Unexposed	8	11	12	14	20	43	111
Unexposed $- (-25)$	33	36	37	39	45	68	136
Rank	1	3	4	5	6	8	12

Sample 2

Exposed	35	56	83	92	128	150	176	208
Rank	2	7	9	10	11	13	14	15

1. $\mu_1 - \mu_2 =$ difference in true mean cotanine concentration for unexposed and exposed infants

2. $H_0: \mu_1 - \mu_2 = -25$

3. $H_a: \mu_1 - \mu_2 < -25$

4. *Significance level*: $\alpha = .01$

5. *Test statistic*:   rank-sum = sum of sample 1 ranks

6. *Assumptions*:   Subtracting 25 does not change the shape or spread of a distribution, so if the assumptions were reasonable in Example 11.13, they are also reasonable here.

7. *Calculation*:   rank-sum = $1 + 3 + 4 + 5 + 6 + 8 + 12 = 39$

8. *P-value*:   This is a lower-tailed test. With $n_1 = 7$ and $n_2 = 8$, Appendix Table VI tells us that P-value $< .05$ if rank-sum ≤ 41 and P-value $< .01$ if rank-sum ≤ 36. Since rank-sum = 39, we conclude that

   $.01 < P\text{-value} < .05$

9. *Conclusion*:   Since P-value $> .01$, we fail to reject H_0. Sample evidence does not suggest that the mean concentration level for exposed infants is more than 25 higher than the mean for unexposed infants.

Frequently the $n_1 + n_2$ observations in the two samples are not all different from one another. When this occurs, the rank assigned to each observation in a tied group is the average of the ranks that would be assigned if the values in the group all differed slightly from one another. Consider, for example, the 10 ordered values

| 5.6 | 6.0 | 6.0 | 6.3 | 6.8 | 7.1 | 7.1 | 7.1 | 7.9 | 8.2 |

If the two 6.0 values differed slightly from each other, they would be assigned ranks 2 and 3. Therefore, each one is assigned rank $(2 + 3)/2 = 2.5$. If the three 7.1 observations were all slightly different, they would receive ranks 6, 7, and 8, so each of the three is assigned rank $(6 + 7 + 8)/3 = 7$. The ranks for the above 10 observations are then

| 1 | 2.5 | 2.5 | 4 | 5 | 7 | 7 | 7 | 9 | 10 |

If the proportion of tied values is quite large, it is recommended that the rank-sum statistic be multiplied by a *correction factor*. Several of the chapter references contain additional information.

Appendix Table VI contains information about P-values for the rank-sum test when $n_1 \le 8$ and $n_2 \le 8$. More extensive tables exist for other combinations of sample size. There is also a test procedure based on using a normal distribution to approximate the sampling distribution of the rank-sum statistic. This alternative procedure is often used when the two sample sizes are larger than 8. One of the chapter references can be consulted for details.

A Confidence Interval for $\mu_1 - \mu_2$

A confidence interval based on the rank-sum statistic is not nearly as familiar to users of statistical methods as is the hypothesis-testing procedure. This is unfortunate, because the confidence interval has the same virtues to recommend it as does the rank-sum test. It does not require the assumption that the population distributions are normal, as does the two-sample t interval.

The actual derivation of the rank-sum confidence interval is quite involved, and computing these intervals by hand can be very tedious, so we will rely on computer software.

The **rank sum confidence interval for $\mu_1 - \mu_2$** is the interval consisting of all hypothesized values for which

$$H_0: \mu_1 - \mu_2 = \text{hypothesized value}$$

cannot be rejected when using a two-tailed test.

A 95% confidence interval consists of those hypothesized values for which the previous null hypothesis is not rejected by a test with significance level $\alpha = .05$.

A 99% confidence interval is associated with a level $\alpha = .01$ test, and a 90% confidence interval is associated with a level $\alpha = .10$ test.

Thus, if $H_0: \mu_1 - \mu_2 = 100$ cannot be rejected at level .05, then 100 is included in the 95% confidence interval for $\mu_1 - \mu_2$.

EXAMPLE 11.15   The article "Some Mechanical Properties of Impregnated Bark Board" (*Forest Products J.* (1977): 31–38) reported the accompanying observations on crushing strength for epoxy-impregnated bark board (sample 1) and bark board impregnated with another polymer (sample 2). The sample values are displayed in the accompanying table:

Sample 1	10,860	11,120	11,340	12,130	13,070	14,380
Sample 2	4590	4850	5640	6390	6510	

A 95% confidence interval for $\mu_1 - \mu_2$ was requested using MINITAB. The resulting output follows:

```
Sample 1    N = 6    Median =  11735
Sample 2    N = 5    Median =   5640
Point estimate for ETA1-ETA2 is     6490
96.4 percent CI for ETA1-ETA2 is (4730, 8480)
```

Remember that MINITAB uses **ETA1 − ETA2** in place of $\mu_1 - \mu_2$. Also note that it was not possible to construct an interval for an exact 95% confidence level. MINITAB has calculated a 96.4% confidence interval. The reported interval is (4730, 8480). Based on the sample information, we estimate that the difference in mean crushing strength for epoxy-impregnated bark board and board impregnated with a different polymer is between 4730 and 8480 pounds per square inch.

There are many other distribution-free procedures, including some that allow inferences based on paired samples. The books mentioned in the chapter references can be consulted for further information.

Exercises 11.59 – 11.65

11.59 Urinary fluoride concentration (ppm) was measured both for a sample of livestock that had been grazing in an area previously exposed to fluoride pollution and for a similar sample that had grazed in an unpolluted region. Does the accompanying data indicate strongly that the true average fluoride concentration for livestock grazing in the polluted region is larger than for the unpolluted region? Assume that the distributions of urinary fluoride concentration for both grazing areas have the same shape and spread, and use a level .05 rank-sum test.

Polluted	21.3	18.7	23.0	17.1	16.8	20.9	19.7
Unpolluted	14.2	18.3	17.2	18.4	20.0		

11.60 A modification has been made to the process for producing a certain type of time-zero film (film that begins to develop as soon as a picture is taken). Because the modification involves extra cost, it will be incorporated only if sample data strongly indicates that the modification has decreased true average developing time by more than 1 sec. Assuming that the developing-time distributions differ only with respect to location, if at all, use the rank-sum test at level .05 on the accompanying data to test the appropriate hypotheses.

Original process	8.6	5.1	4.5	5.4	6.3	6.6	5.7	8.5
Modified process	5.5	4.0	3.8	6.0	5.8	4.9	7.0	5.7

11.61 The study reported in "Gait Patterns During Free Choice Ladder Ascents" (*Human Movement Sci.* (1983): 187–195) was motivated by publicity concerning the increased accident rate for individuals climbing ladders. A number of different gait patterns were used by subjects climbing a portable straight ladder according to specified instructions. The accompanying data consist of ascent times for seven subjects who used a lateral gait and six subjects who used a four-beat diagonal gait.

Lateral   .86   1.31   1.64   1.51   1.53   1.39   1.09
Diagonal   1.27   1.82   1.66   .85   1.45   1.24

a. Use the rank-sum test to see whether the data suggests any difference in the true average ascent times for the two gaits.

b. Interpret the 95% confidence interval for the difference between the true average gait times, as indicated by the accompanying MINITAB output.

lateral   N = 7   Median = 1.3900
diagonal   N = 6   Median = 1.3600
Point estimate for ETA1 - ETA2 is   -0.0400
96.2 percent C.I. for ETA1 - ETA2 is (-0.4300, 0.3697)

11.62 A blood-lead level of 70 mg/mL has been commonly accepted as safe. However, researchers have noted that some neurophysiological symptoms of lead poisoning appear in people whose blood-lead levels are below 70 mg/mL. The article "Subclinical Neuropathy at Safe Levels of Lead Exposure" (*Arch. Environ. Health* (1975): 180) gives the following nerve-conduction velocities for a group of workers who were exposed to lead in the workplace but whose blood-lead levels were below 70 mg/mL and for a group of controls who had no exposure to lead:

Exposed to lead   46   46   43   41   38   36   31
Control   54   50.5   46   45   44   42   41

Use a level .05 rank-sum test to determine whether there is a significant difference in mean conduction velocity between workers exposed to lead and those not exposed to lead.

11.63 The effectiveness of antidepressants in treating the eating disorder bulimia was examined in the article "Bulimia Treated with Imipramine: A Placebo-Controlled Double-Blind Study" (*Amer. J. Psych.* (1983): 554–558). A group of patients diagnosed as bulimic were randomly assigned to one of two treatment groups, one receiving imipramine and the other a placebo. One of the variables recorded was binge frequency. The authors chose to analyze the data using a rank-sum test because it makes no

assumption of normality. They state, "Because of the wide range of some measures, such as frequency of binges, the rank sum is more appropriate and somewhat more conservative." Data consistent with the findings of this article is given in the accompanying table.

Number of Binges During 1 Week

Placebo	8	3	15	3	4	10	6	4
Imipramine	2	1	2	7	3	12	1	5

Does this data strongly suggest that imipramine is effective in reducing the mean number of binges per week? Use a level .05 rank-sum test.

11.64 In an experiment to compare the bond strength of two different adhesives, each adhesive was used in five bondings of two surfaces, and the force necessary to separate the surfaces was determined for each bonding. For adhesive 1, the resulting values were 229, 286, 245, 299, and 259, whereas the adhesive 2 observations were 213, 179, 163, 247, and 225. Let μ_1 and μ_2 denote the true average bond strengths of adhesives 1 and 2, respectively. Interpret the 90% distribution-free confidence interval estimate of $\mu_1 - \mu_2$ given in the MINITAB output shown here:

adhes. 1   N = 5   Median = 259.00
adhes. 2   N = 5   Median = 213.00
Point estimate for ETA1 - ETA2 is   61.00
90.5 Percent C.I. for ETA1 - ETA2 is (16.00, 95.98)

11.65 The article "A Study of Wood Stove Particulate Emissions" (*J. Air Poll. Control Assoc.* (1979): 724–728) reported the accompanying data on burn time (hr) for samples of oak and pine. An estimate of the difference between mean burn time for oak and mean burn time for pine is desired.

Oak   1.72   .67   1.55   1.56   1.42   1.23   1.77   .48
Pine   .98   1.40   1.33   1.52   .73   1.20

Interpret the interval given in the MINITAB output.

Oak   N = 8   Median = 1.4850
Pine   N = 6   Median = 1.2650
Point estimate for ETA1 - ETA2 is   0.2100
95.5 Percent C.I. for ETA1 - ETA2 is (-0.4998, 0.5699)

11.5   Interpreting the Results of Statistical Analyses

Many different types of research involve comparing two populations or treatments. It is easy to find examples of the two-sample hypothesis tests introduced in this chapter in published sources in a wide variety of disciplines. As with the one-sample

tests, it is common to find only the value of the test statistic and the associated
P-value (or sometimes only the P-value) in reports.

What to Look for in Published Data

Here are some questions to consider when you are reading a report that contains
the result of a two-sample hypothesis test:

- Are only two groups being compared? If more than two groups are being
  compared two at a time, a different type of analysis probably would be preferable (see Chapter 15).
- Were the samples selected independently, or are the samples paired? If the
  samples are paired, was the analysis performed appropriate for paired
  samples?
- What hypotheses are being tested? Is the test one- or two-sided?
- Does the validity of the test performed depend on any assumptions about the
  populations sampled (such as normality)? If so, do the assumptions appear to
  be reasonable?
- What is the P-value associated with the test? Does the P-value lead to rejection of the null hypothesis?
- Are the conclusions drawn consistent with the results of the hypothesis test?
  In particular, if H_0 was rejected, does this indicate practical significance or
  only statistical significance?

As an example, we consider a study reported in the article "The Relationship
Between Distress and Delight in Males' and Females' Reactions to Frightening
Films" (*Human Communication Research* (1991): 625–637). The investigators measured emotional responses of 50 males and 60 females after viewing a segment from
a horror film. The article included the following statement:

> Females were much more likely to express distress than were males. While
> males did express higher levels of delight than females, the difference was
> not statistically significant.

The accompanying summary data was also contained in the article.

	Distress Index		Delight Index	
Gender	**Mean**	**SD**	**Mean**	**SD**
Males	31.2	10.0	12.02	3.65
Females	40.4	9.1	9.09	5.55
	P-value $< .001$		Not significant (P-value $> .05$)	

The P-values are the only evidence of the hypothesis tests that support the conclusions given. The P-value $< .001$ for the distress index means that the hypothesis

$$H_0: \quad \mu_F - \mu_M = 0$$

was rejected in favor of

$$H_a: \quad \mu_F - \mu_M > 0$$

where μ_F and μ_M are the true mean distress indexes for females and males, respectively.

The nonsignificant P-value (P-value $> .05$) reported for the delight index means that the hypothesis

$$H_0: \quad \mu_F - \mu_M = 0$$

(where μ_F and μ_M now refer to mean delight index for females and males, respectively) could not be rejected. Chance sample-to-sample variability is a plausible explanation for the observed difference in sample means ($12.02 - 9.09$). Thus we would not want to put much emphasis on the author's statement that males express higher levels of delight than females, which is based only on the fact that $12.02 > 9.09$ and could plausibly be due entirely to chance.

The article describes the samples as consisting of undergraduates selected from the student body of a large midwestern university. The authors extrapolate their results to American men and women in general. If it were considered unreasonable to make this type of generalization, we could be more conservative and view the sampled populations to be male and female university students, or male and female midwestern university students, or even male and female students at this particular university.

The comparison of males and females was based on two independently selected groups (not paired). Since the sample sizes were large, the two-sample t test for means could reasonably have been employed, and this would have required no specific assumptions about the two underlying populations.

Summary of Key Concepts and Formulas

Term or Formula	Comment
Independent samples	Two samples where the individuals or objects in the first sample are selected independently from those in the second sample.
Paired samples	Two samples for which each observation in the first sample is paired in a meaningful way with a particular observaton in the second sample.
$t = \dfrac{(\bar{x}_1 - \bar{x}_2) - \text{hyp. value}}{\sqrt{\dfrac{s_1^2}{n_1} + \dfrac{s_2^2}{n_2}}}$	The test statistic for testing $H_0: \mu_1 - \mu_2 =$ hypothesized value when the samples are independently selected and the sample sizes are large or it is reasonable to assume that both population distributions are normal.
$(\bar{x}_1 - \bar{x}_2) \pm (t \text{ crit. value})\sqrt{\dfrac{s_1^2}{n_1} + \dfrac{s_2^2}{n_2}}$	A formula for constructing a confidence interval for $\mu_1 - \mu_2$ when the samples are independently selected and the sample sizes are large or it is reasonable to assume that the population distributions are normal.

Term or Formula	Comment
$\text{df} = \dfrac{(V_1 + V_2)^2}{\dfrac{V_1^2}{n_1 - 1} + \dfrac{V_2^2}{n_2 - 1}}$ where $V_1 = \dfrac{s_1^2}{n_1}$ and $V_2 = \dfrac{s_2^2}{n_2}$	The formula for determining df for the two-sample t test and confidence interval.
$\bar{x}_d$	The sample mean difference.
s_d	The standard deviation of the sample differences.
μ_d	The mean value for the population of differences.
σ_d	The standard deviation for the population of differences.
$t = \dfrac{\bar{x}_d - \text{hypothesized value}}{\dfrac{s_d}{\sqrt{n}}}$	The paired t test statistic for testing $H_0: \mu_d =$ hypothesized value.
$\bar{x}_d \pm (t \text{ critical value})\dfrac{s_d}{\sqrt{n}}$	The paired t confidence interval formula.
$p_c = \dfrac{n_1 p_1 + n_2 p_2}{n_1 + n_2}$	p_c is the statistic for estimating the common population proportion when $\pi_1 = \pi_2$.
$z = \dfrac{p_1 - p_2}{\sqrt{\dfrac{p_c(1 - p_c)}{n_1} + \dfrac{p_c(1 - p_c)}{n_2}}}$	The test statistic for testing $H_0: \pi_1 - \pi_2 = 0$ when both sample sizes are large.
$(p_1 - p_2) \pm (z \text{ crit. value})\sqrt{\dfrac{p_1(1 - p_1)}{n_1} + \dfrac{p_2(1 - p_2)}{n_2}}$	A formula for constructing a confidence interval for $\pi_1 - \pi_2$ when both sample sizes are large.
Rank-sum = sum of sample 1 ranks	The distribution-free test statistic for testing $H_0: \mu_1 - \mu_2 =$ hypothesized value when it is reasonable to assume that the two populations have the same shape and spread.

Supplementary Exercises 11.66 – 11.89

11.66 An eating attitudes test (EAT) was administered both to a sample of female models and to a control group of females, resulting in the accompanying summary data ("Gender Differences in Eating Attitudes, Body Concept, and Self-Esteem Among Models," *Sex Roles* (1992): 413–437).

Group	Sample Size	Sample Mean	Sample sd
Models	30	8.63	7.1
Controls	30	10.97	8.9

Let μ_1 denote the true average EAT score for models and μ_2 the true average EAT score for controls.
   a. Verify that the value of the test statistic for testing $H_0: \mu_1 - \mu_2 = 0$ versus $H_a: \mu_1 - \mu_2 < 0$ is $t = -1.13$.
   b. Calculate the P-value, and then use it to reach a conclusion at significance level .05.
   c. If μ_1 had been the true average score for controls and μ_2 the true average score for models, how would the hypotheses, value of t, and P-value have changed?

11.67 The authors of the article "Hostility, Aggression, and Anxiety Levels of Divorce and Nondivorce Children as Manifested in Their Responses to Projective Tests" (*J. of Personality Assessment* (1991): 438–452) speculated that children of divorced parents would show higher levels of hostility than children of married parents. To test this theory, Rorschach tests were used to compute a measure of hostility for 54 children of divorced parents and 54 children of married parents. All children who participated were between 10 and 12 years of age and lived in Sweden. Does the accompanying data support the researchers' theory?

Group	$\bar{x}$	s
Children of divorced parents	5.38	3.96
Children of married parents	1.94	3.10

11.68 Do teenage boys worry more than teenage girls? This is one of the questions addressed by the authors of the article "The Relationship of Self-Esteem and Attributional Style to Young People's Worries" (*J. of Psych.* (1987): 207–215). A scale called the Worries Scale was administered to a group of teenagers, and the results are summarized in the accompanying table.

Gender	n	Sample Mean Score	Sample sd
Girls	108	62.05	9.5
Boys	78	67.59	9.7

Is there sufficient evidence to conclude that teenage boys score higher on the Worries Scale than teenage girls? Use a significance level of $\alpha = .05$.

11.69 Some astrologers have speculated that people born under certain sun-signs are more extroverted than people born under other signs. The accompanying data was taken from the article "Self-Attribution Theory and the Sun-Sign" (*J. of Social Psych.* (1984): 121–126). The Eysenck Personality Inventory (EPI) was used to measure extroversion and neuroticism.

	Sample Size	Extroversion		Neuroticism	
Sun-sign	n	$\bar{x}$	s	$\bar{x}$	s
Water signs	59	11.71	3.69	12.32	4.15
Other signs	186	12.53	4.14	12.23	4.11
Winter signs	73	11.49	4.28	11.96	4.22
Summer signs	49	13.57	3.71	13.27	4.04

a. Is there sufficient evidence to indicate that those born under water signs have a lower mean extroversion score than those born under other (nonwater) signs? Use a level .01 test.

b. Does the data strongly suggest that those born under winter signs have a lower mean extroversion score than those born under summer signs? Use a level .05 test.

c. Does the data indicate that those born under water signs differ significantly from those born under other signs with respect to mean neuroticism score? Use $\alpha = .05$.

d. Do those born under winter signs differ from those born under summer signs with respect to mean neuroticism score? Carry out a test of appropriate hypotheses using a significance level of .01.

11.70 Recorded speech can be compressed and played back at a faster rate. The article "Comprehension by College Students of Time-Compressed Lectures" (*J. of Exper. Educ.* (Fall 1975): 53–56) gave the results of a study designed to test comprehension of time-compressed speech. Fifty students listened to a 60-min lecture and took a comprehension test. Another 50 students heard the same lecture time-compressed to 40 min. The sample mean and standard deviation of comprehension scores for the normal-speed group were 9.18 and 4.59, respectively, and those for the time-compressed group were 6.34 and 4.93, respectively.

a. Use a level .01 test to determine whether the true mean comprehension score for students hearing a time-compressed lecture is significantly lower than the true mean score for students who hear a lecture at normal speed. What assumption must be reasonable in order for your test to be valid?

b. Estimate the difference in the true mean comprehension scores for normal and time-compressed lectures using a 95% confidence interval.

11.71 Should quizzes be given at the beginning or the end of a lecture period? The article "On Positioning the Quiz: An Empirical Analysis" (*Accounting Review* (1980): 664–670) provides some insight. Two sections of an introductory accounting class were given identical instructions to read and study assigned text materials. Three quizzes and a final exam were given during the term, with one section taking quizzes at the beginning of the lecture and the other section taking quizzes at the end. Final exam scores for the two groups are summarized. Does the accompanying data indicate that there is a significant difference in the true mean final exam scores for students who take quizzes at the beginning of class and those who take quizzes at the end? Do you think that the assumptions required for the two-sample *t* test are reasonable?

	Quiz at Beginning	Quiz at End
Sample size	40	40
Mean	143.7	131.7
Standard deviation	21.2	20.9

11.72 Data on self-esteem, leadership ability, and GPA was used to compare college students who were hired as resident assistants at Mississippi State University and those who were not hired ("Wellness as a Factor in University Housing," *J. of College Student Development* (1994): 248–254), as shown in the accompanying table.

	Self-esteem		Leadership		GPA	
	$\bar{x}$	s	$\bar{x}$	s	$\bar{x}$	s
Hired ($n = 69$)	83.28	12.21	62.51	3.05	2.94	.61
Not hired ($n = 47$)	81.96	12.78	62.43	3.36	2.60	.79

a. Does the data suggest that the true mean self-esteem score for students hired by the university as resident assistants differs from the mean for students who are not hired? Test the relevant hypotheses using $\alpha = .05$.
b. Does the data suggest that the true mean leadership score for students hired by the university as resident assistants differs from the mean for students who are not hired? Test the relevant hypotheses using $\alpha = .05$.
c. Does the data suggest that the true mean GPA for students hired by the university as resident assistants differs from the mean for students who are not hired? Test the relevant hypotheses using $\alpha = .05$.

11.73 The article "Religion and Well-Being Among Canadian University Students: The Role of Faith Groups on Campus" (*J. for the Scientific Study of Religion* (1994): 62–73) compared the self-esteem of students who belonged to Christian clubs and students who did not belong to such groups. Each student in a random sample of $n = 169$ members of Christian groups (the affiliated group) completed a questionnaire designed to measure self-esteem. The same questionnaire was also completed by each student in a random sample of $n = 124$ students who did not belong to a religious club (the unaffiliated group). The mean self-esteem score for the affiliated group was 25.08, and the mean for the unaffiliated group was 24.55.

The sample standard deviations weren't given in the article, but suppose that they were 10 for the affiliated group and 8 for the unaffiliated group. Is there evidence that the true mean self-esteem score differs for affiliated and unaffiliated students? Test using a significance level of .01.

11.74 The article "Social Science Research on Lay Definitions of Sexual Harassment" (*J. of Social Issues* (1995): 21–37) reported on a survey of randomly selected undergraduate and graduate students. Each student was asked whether he or she thought that unwanted pressure for a date from a faculty member was an example of sexual harassment. (See the accompanying table.)

	n	Number Who Perceived Pressure as Sexual Harassment
Undergraduates		
Males	1336	882
Females	1346	1104
Graduate students		
Males	565	469
Females	575	529

a. For undergraduate students, is there evidence that the proportion who perceive pressure for a date as sexual harassment is lower for males than for females? Test using $\alpha = .05$.
b. For graduate students, is there evidence that the proportion who perceive pressure for a date as sexual harassment is lower for males than for females? Test using $\alpha = .05$.
c. Estimate the difference between the proportion of male graduate students who perceive pressure for a date as sexual harassment and the corresponding proportion for male undergraduate students, using a 90% confidence interval. Interpret the resulting interval.

11.75 Meteorologists classify storms as either single-peak or multiple-peak. The total number of lightning flashes was recorded for seven single-peak and four multiple-peak storms, resulting in the accompanying data ("Lightning Phenomenology in the Tampa Bay Area," *J. of Geophys. Research* (1984): 11,789–11,805).

Single-peak	117	56	19	40	82	69	80
Multiple-peak	229	197	242	430			

a. Does the data suggest that the true mean number of lightning flashes differs for the two types of storms? Use a .05 significance level.

b. What assumptions about the distribution of number of flashes for each of the two types of storms are necessary if your test in part **a** is to be valid?

11.76 Nine observations of surface-soil pH were made at each of two different locations at the Central Soil Salinity Research Institute experimental farm, and the resulting data appeared in the article "Sodium-Calcium Exchange Equilibria in Soils as Affected by Calcium Carbonate and Organic Matter" (*Soil Sci.* (1984): 109). Does the accompanying data suggest that the true mean soil pH values differ for the two locations? Test the appropriate hypotheses using a .05 significance level.

Site	pH				
Location A	8.53	8.52	8.01	7.99	7.93
	7.89	7.85	7.82	7.80	
Location B	7.85	7.73	7.58	7.40	7.35
	7.30	7.27	7.27	7.23	

11.77 The accompanying summary data on the ratio of strength to cross-sectional area for knee extensors is taken from the article "Knee Extensor and Knee Flexor Strength: Cross-Sectional Area Ratios in Young and Elderly Men" (*J. of Gerontology* (1992): M204–M210).

Group	n	$\bar{x}$	Standard Error, $\dfrac{s}{\sqrt{n}}$
Young men	13	7.47	.22
Elderly men	12	6.71	.28

Does this data suggest that the true average ratio for young men exceeds that for elderly men? Carry out a test of appropriate hypotheses using $\alpha = .05$.

11.78 The authors of the article "The Color-a-Person Body Dissatisfaction Test" (*J. of Personality Assessment* (1991): 395–413) used a nonverbal measure of body dissatisfaction to compare students at two different universities. A random sample of 47 women at the University of Cincinnati and a random sample of 24 women at the University of New Mexico participated in the study. Mean and standard deviations for a measure of self-esteem are given in the accompanying table. Is it reasonable to conclude that women at the two universities differ with respect to mean self-esteem score? Use $\alpha = .05$.

	Self-esteem Score	
	$\bar{x}$	s
University of Cincinnati	32.55	4.41
University of New Mexico	31.25	4.92

11.79 An experiment to assess the effects of automobile pollution was described in the article "Effects of Roadside Conditions on Plants and Insects" (*J. of Appl. Ecology* (1988): 709–715). Twenty soil specimens taken .6 m from the roadside resulted in a mean nitrogen concentration of 1.70 mg/g. Twenty soil specimens taken 6 m from the roadside resulted in a mean nitrogen concentration of 1.35 mg/g. Suppose that the sample standard deviations were .4 (at .6 m) and .3 (at 6 m). Is there sufficient evidence to conclude that the mean nitrogen concentration in soil .6 m from the roadside is higher than that at 6 m? Test the relevant hypotheses using $\alpha = .01$.

11.80 The article "Effect of Carbohydrate and Vitamin B6 on Fuel Substrates during Exercise in Women" (*Medicine and Science in Sports and Exercise* (1988): 223–239) compared a group of women who participated in regular aerobic exercise with a group who did not. Data on percentage of body fat was obtained by skinfolds measured at seven sites, as summarized in the accompanying table.

	Sample Size	Mean Percent Body Fat	Sample sd
Aerobic exercise	5	20.1	4.2
No aerobic exercise	5	20.4	3.4

a. Estimate the true mean difference in percent body fat between the two groups, using a 90% confidence interval.
b. Does your interval in part **a** contain zero? How would you interpret the interval?
c. What assumptions are necessary for the interval of part **a** to be valid?

11.81 In an effort to assess the risk of brain damage in infants, the authors of the article "Neuropathologic Documentation of Prenatal Brain Damage" (*Amer. J. of Diseases of Children* (1988): 858–866) studied both infants born with brain damage and infants born without brain damage. They also looked at premature babies as well as those carried to term.

	Sample Size	Mean Birth Weight (g)	Sample sd
Term			
Brain damage	12	2998	707
No brain damage	13	2704	627
Premature			
Brain damage	10	1541	448
No brain damage	54	1204	656

a. Does the data indicate that the true mean birth weight for premature infants with brain damage differs from that of premature infants without brain damage? Test the relevant hypotheses using a significance level of .05.

b. Repeat part **a** for infants carried to term.

c. Use a 90% confidence interval to estimate the difference in mean weight between term and premature infants without brain damage.

d. Repeat part **c** for infants with brain damage.

11.82 The accompanying table gives data that appeared in the article "Development and Clinical Trial of a Minimal-Contact, Cognitive-Behavioral Treatment for Tension Headache" (*Cognitive Therapy and Research* (1988): 325–339). Two different treatments for headache (a relaxation therapy and a cognitive-behavioral therapy) were compared. Is there sufficient evidence to support the researchers' claim that the cognitive-behavioral therapy is more effective than relaxation in increasing the mean number of headache-free days (for a 1-week period)? Test the relevant hypotheses using a level .05 test.

Therapy	Sample Size	Mean Number of Headache-free Days	Sample sd
Relaxation	24	3.82	1.75
Cognitive-behavioral	24	5.71	1.43

11.83 Key terms in survey questions too often are not well understood, and such ambiguity can affect responses. As an example, the article "How Unclear Terms Affect Survey Data" (*Public Opinion Quarterly* (1992): 218–231) described a survey in which each individual in a sample was asked, "Do you exercise or play sports regularly?" But what constitutes exercise? The following revised question was then asked of each individual in the same sample: "Do you do any sports or hobbies involving physical activities, or any exercise, including walking, on a regular basis?" The resulting data is shown in the accompanying table.

	Yes	No
Initial question	48	52
Revised question	60	40

Is there any difference between the true proportions of yes responses to these questions? Can a procedure from this chapter be used to answer the question posed? If yes, use it; if not, explain why not.

11.84 An electronic implant that stimulates the auditory nerve has been used to restore partial hearing to a number of deaf people. In a study of implant acceptability (*Los Angeles Times,* Jan. 29, 1985), 250 adults born deaf and 250 adults who went deaf after learning to speak were followed for a period of time after receiving an implant. Of those deaf from birth, 75 had removed the implant, whereas only 25 of those who went deaf after learning to speak had done so. Does this suggest that the true proportion who remove the implants differs for those that were born deaf and those that went deaf after learning to speak? Test the relevant hypotheses using a .01 significance level.

11.85 Samples of both surface soil and subsoil were taken from eight randomly selected agricultural locations in a particular county. The soil samples were analyzed to determine both surface pH and subsoil pH, with the results shown in the accompanying table.

	Location			
	1	**2**	**3**	**4**
Surface pH	6.55	5.98	5.59	6.17
Subsoil pH	6.78	6.14	5.80	5.91

	Location			
	5	**6**	**7**	**8**
Surface pH	5.92	6.18	6.43	5.68
Subsoil pH	6.10	6.01	6.18	5.88

a. Compute a 90% confidence interval for the true average difference between surface and subsoil pH for agricultural land in this county.

b. What assumptions are necessary to validate the interval in part **a**?

11.86 The article "An Evaluation of Football Helmets Under Impact Conditions" (*Amer. J. of Sports Med.* (1984): 233–237) reported that when 44 padded football helmets and 37 suspension-type helmets were subjected to an impact test (a drop of 1.5 m onto a hard surface), 5 of the padded and 24 of the suspension-type helmets showed damage. Using a .01 significance level, test appropriate hypotheses to determine whether there is a difference between the two helmet types with respect to the true proportion of each type that would be damaged by a 1.5-m drop onto a hard surface.

11.87 The article "Post-Mortem Analysis of Neuropeptides in Brains from Sudden Infant Death Victims" (*Brain Research* (1984): 279–285) reported age (in days) at death for infants who died of sudden infant death syndrome (SIDS). Assuming that age at death for SIDS victims is normally distributed, use the accompanying data to construct a 95% confidence interval for the difference in the true mean age at death for female and male SIDS victims. Interpret the resulting interval. How does the interpretation depend on whether zero is included in the interval?

Age at Death (days)

Females	55	120	135	154	54		
Males	56	60	60	60	105	140	147

11.88 The accompanying 1982 and 1983 net earnings (in millions of dollars) of ten food and beverage firms appeared in an article "Capital Expenditures Report" (*Food Engr.* (1984): 93–101). Is there sufficient evidence to indicate that mean net earnings increased from 1982 to 1983? Assume that the ten companies represent a random sample of all food and beverage firms. Perform the appropriate hypothesis test using a .01 significance level.

Firm	1983	1982
Coors	89.0	33.0
ConAgra	28.7	20.6
ADM	110.2	155.0
Heinz	237.5	214.3

(*continued*)

Firm	1983	1982
General Mills	130.7	134.1
Nestlé	113.4	93.7
Beatrice	292.1	320.8
Carnation	155.0	137.6
Hershey	100.2	94.2
Procter & Gamble	58.5	66.5

11.89 Many researchers have investigated the relationship between stress and reproductive efficiency. One such study is described in the article "Stress or Acute Adrenocorticotrophin Treatment Suppresses LHRH-Induced LH Release in the Ram" (*J. of Reprod. and Fertility* (1984): 385–393). Seven rams were used in the study, and LH (luteinizing hormone) release (ng/min) was recorded before and after treatment with ACTH (adrenocorticotrophin, a drug that results in stimulation of the adrenal gland). Use the accompanying data and $\alpha = .01$ to determine whether there is a significant reduction in mean LH release following treatment with ACTH.

Ram	1	2	3	4	5	6	7
Before	2400	1400	1375	1325	1200	1150	850
After	2250	1425	1100	800	850	925	700

References

Daniel, Wayne. *Applied Nonparametric Statistics,* 2nd ed. Boston: PWS-Kent, 1990. (An elementary presentation of distribution-free methods, including the rank-sum test discussed in the last section of this chapter.)

Devore, Jay. *Probability and Statistics for Engineering and the Sciences,* 5th ed. Belmont, Calif.: Duxbury Press, 2000. (Contains a somewhat more comprehensive treatment of the inferential material presented in this and the previous two chapters, although the notation is a bit more mathematical than that of the present text.)

Mosteller, Frederick, and Richard Rourke. *Sturdy Statistics.* Reading, Mass.: Addison-Wesley, 1973. (A very readable intuitive development of distribution-free methods, including those based on ranks.)

GRAPHING CALCULATOR EXPLORATIONS

11.1 Inference About Differences in Means Using Independent Samples

Testing Hypotheses

Constructing confidence intervals and testing a hypothesis about a difference between two population means using your calculator is similar to working with a single mean. This requires that you navigate the menu system of your calculator to the point where these procedures can be selected. Since you are by now an old hand at navigation, we will use Example 11.2 to illustrate a hypothesis test.

Enter the data in your calculator, using the lists of your choice. We will check the assumption of approximate normality by displaying boxplots of the two data sets, as

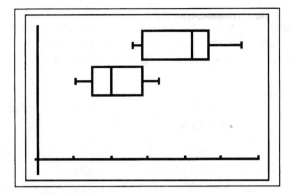

shown in the accompanying figure. (If your calculator does both types of boxplots, be sure to choose the option that shows outliers in your data!)

After verifying the plausibility of the normality of the populations, we navigate our calculator's menu system to choose a procedure. For the two-sample t test, the common calculator screens are not large enough to show all relevant data, so we will again show "larger-than-life" representations rather than capture the actual screens.

2-Sample t Test	2-SampTTest
Data :List	Inpt: Data Stats
$\mu 1$:$>\mu 2$	List1: L1
List1 :List 1	List2: L2
List2 :List 2	Freq1: 1
Freq1 :1	Freq2: 1
Freq2 :1	$\mu 1:\neq\mu 2 <\mu 2 >\mu 2$
Pooled :Off	Pooled:No Yes
Execute	Calculate Draw

Notice that you have choices similar to the hypothesis test for a single population mean, with one exception: pooling. As discussed in the text, pooling has fallen into disfavor among statisticians, so your choice here should be "Off" or "No." The output also requires more than one calculator screen, and we will therefore show a big screen representation.

2-Sample tTest	2-SampleTTest
$\mu 1$ $>\mu 2$	$\mu 1 > \mu 2$
t $=3.457080368$	t$=3.457080368$
p $=.0036966205$	p$=.0036966205$
df $=8.837999538$	df$=8.837999538$
$\bar{x}_1$ $=40.29166667$	$\bar{x}_1=40.29166667$
$\bar{x}_2$ $=21.39875$	$\bar{x}_2=21.39875$
x1$\sigma n-1$ $=11.2919572$	Sx1$=11.2919572$
x2$\sigma n-1$ $=8.30168905$	Sx2$=8.30168905$
n1 $=6$	n1$=6$
n2 $=8$	n2$=8$

Once again, we remind you that, although this information is "all there" in the sense that the results of the hypothesis test are completely displayed, it is *not* in a form that should be used for reporting your results to others. You should still complete and report all of the appropriate steps for a hypothesis test.

Confidence Intervals

We will now turn our attention to confidence intervals for the difference between two population means, using data from Example 11.4. Enter the data in your calculator, using the lists of your choice.

After verifying the plausibility of the normality of the populations, we again navigate the calculator's menu system to choose a procedure. For the two-sample *t* confidence interval, the data entry will be the same as for a hypothesis test, and there should be only one additional piece of information needed: the confidence level. Here are two potential displays. Again, your calculator will display only some of this information, and you will have to scroll to see it all.

2-Sample tInterval
Data :List
C-Level :.95
List1 :List1
List2 :List2
Freq1 :1
Freq2 :1
Pooled:Off
Execute

2-SampTInt
Inpt:Data Stats
List1:L1
List2:L2
Freq1:1
Freq2:1
C-Level:.95
Pooled:No Yes
Calculate

We can see from the screen representations that the data is contained in a list rather than user-supplied statistics, and that the confidence levels are 95%. Once again, remember that we should choose *not* to pool. The results from the calculations are displayed here. These represent two common ways that calculators report confidence intervals.

2-Sample tInterval
Left =-6.009
Right=7.5593
df =9.2598
$\bar{x}_1$ =18.7867
$\bar{x}_2$ =18.0117
x1$\sigma n-1$ =5.9083
x2$\sigma n-1$ =4.4181
n1 =6
n2 =6

2-SampTInt
(-6.009,7.5593)
df=9.259779656
$\bar{x}_1$=18.78666667
$\bar{x}_2$=18.01166667
Sx1=5.90830827
Sx2=4.41808744
n1=6
n2=6

11.2 Inferences About Differences in Means Using Paired Samples

Procedures for making inferences about the difference between two population means using paired samples is similar to using a single sample. The reason, of course, is that we base these inferences on a single set of difference values. To test hypotheses or to construct confidence intervals using these differences, you must enter the observations in your calculator, create the differences by subtraction, and then analyze them. We will use Example 11.8 to illustrate how to do this on the calculator. Recall that the data are blood lactate levels before and after exercise.

If you are following along with your calculator, enter the preexercise data into List1 and the postexercise data into List2. (You may put the data in the lists of your choice as long as you remember where the data is and the order of subtraction.) After entering the data, use list commands to subtract the postexercise score from the preexercise score (i.e., preexercise minus postexercise) and store the result in List3:

List1 − List2 → List3

L1	L2	L3	1
13.000	18.000	-5.000	
20.000	37.000	-17.00	
17.000	40.000	-23.00	
13.000	35.000	-22.00	
13.000	30.000	-17.00	
16.000	20.000	-4.000	
15.000	33.000	-18.00	

At this point, you should see something like this figure in your data edit screen. The list of interest for the paired *t* procedures is List3. We will proceed from here as we did with inference for a single sample. Since the sample size in this example is small, the approximate normality of the differences is an important consideration. We must assess the plausibility of a normal population *of differences.*

The boxplot for the differences (List3) shown in this figure does not indicate any difficulties with the assumption of approximate normality, so we will proceed to the

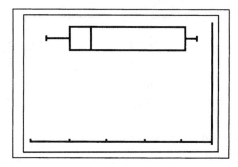

hypothesis test and construction of a confidence interval, as discussed in previous Graphing Calculator Explorations. The results are shown here. The alternative hypothesis was that the mean difference would be less than 0, a one-tailed test. In the case of the confidence interval, we used a 95% confidence level. You should compare the results shown in these figures with your calculator results to check that you have correctly implemented the paired t calculator process. Remember, the calculator will typically give answers with more decimal accuracy than is shown in the text, so slight differences should not be cause for worry.

T-Test
$\mu < 0.00000$
t=-4.65443
p=.00116
x̄=-13.62500
Sx=8.27971
n=8.00000

Tinterval
(-20.55, -6.703)
x̄=-13.62500
Sx=8.27971
n=8.00000

11.3 Inference About a Difference in Proportions

Using your calculator for making inferences about the difference in population proportions is similar to the procedure for making inferences about the difference in population means using independent samples. Once again, navigate your calculator's menu system, and look for something like hypothesis tests and confidence intervals for "2 sample proportions."

Testing Hypotheses

Data entry for a test of the null hypothesis of equal population proportions is very simple. Once you correctly identify the number of "successes" in the problem at hand, the most difficult part is remembering to check the large sample assumptions! We will use Example 11.9 for illustrative purposes. Recall that this example considered wording on a survey question about gun control. The data on "number who favor" is summarized in the accompanying table.

	Question 1	Question 2
Sample size	$n_1 = 615$	$n_2 = 585$
Number in favor	463	403
Sample proportion	$p_1 = 463/615 = .753$	$p_2 = 403/585 = .689$

After transferring this data to the calculator screen, you should have something similar to the accompanying figure.

```
2-Prop ZTest
x1:463
n1:615
x2:403
n2:585

p1: ≠ p2 < p2 > p2
Calculate Draw
```

Notice that for this calculator the notation for the hypothesized population proportions is p rather than π — don't be confused and believe your hypothesis is about *sample* proportions! Notice that you must choose your alternative hypothesis from among the three listed. In Example 11.9, the alternative hypothesis is $H_a: \pi_1 - \pi_2 > 0$, or equivalently, $\pi_1 > \pi_2$. (On this calculator, the coresponding alternative hypothesis would be selected by moving the cursor to ">p2" and pressing Enter.)

When we press Enter, we are presented again with more information than will fit on one screen, so we will present another "larger than life" screen. This calculator screen presents enough information to reconstruct the hypothesis test of the example.

```
2-Prop ZTest
p1 >p2
z=2.470925045
p=.0067382115
p̂1 = .7528355285
p̂2 = .6888888889
p̂ = .7216666667
n1=615
n2=585
```

Confidence Intervals

We will now turn our attention to confidence intervals for the difference between two population proportions, using data from Example 11.11. Recall that this example looked at an experiment that examined the effect of herbicides on house pets. Here is a summary of the data:

Group	Sample Size	Number of Lymphoma	p
Exposed	827	473	.57
Unexposed	130	19	.15

After navigating your calculator's menu system, you should find a screen for data entry. The entry of the data is similar to that for the hypothesis test, but here you must specify the confidence level, which was unnecessary for the hypothesis test.

```
2-Prop ZInt
x1:473
n1:827
x2:19
n2:130
C-Level: .90
Calculate
```

After entering the appropriate values and pressing Enter, you should see something like the accompanying screen.

```
2-Prop ZInt
(.3675, .4841)
p̂1 =.5719
p̂2 =.1462
n1=827.0000
n2=130.0000
```

Notice that there is no point estimate of the difference in population proportions on this screen. If you want to express your confidence interval in point estimate ± error bound form, you must find the point estimate by subtracting the sample proportions. The values to add and subtract from the point estimate can be found by computing one-half the interval length given in the screen. These calculations are illustrated here:

point estimate = .5719 − .1462 = .4257

error bound = (.4841 − .3675)/2 = .0583

The 90% confidence interval is then .4257 ± .0583. Except for the effects of rounding, these values agree with the example.

12

The Analysis of Categorical Data and Goodness-of-Fit Tests

INTRODUCTION

Most of the techniques presented in previous chapters are designed for numerical data. It is often the case, however, that information is collected on categorical variables, such as political affiliation, gender, or college major. As with numerical data, categorical data sets can be univariate (consisting of observations on a single categorical variable), bivariate (observations on two categorical variables), or even multivariate. In this chapter, we first consider inferential methods for analyzing univariate categorical data sets and then turn to techniques appropriate for use with bivariate data.

12.1   Chi-Squared Tests for Univariate Categorical Data

Univariate categorical data sets arise in a variety of settings. If each student in a sample of 100 is classified according to whether he or she is enrolled full-time or part-time, data on a categorical variable with two categories results. Each airline passenger in a sample of 50 might be classified into three categories based on the type of ticket — coach, business class, or first class. Each registered voter in a sample of 100 selected from those registered in a particular city might be asked which of the five city council members he or she favors for mayor. This would yield observations on a categorical variable with five categories.

Univariate categorical data is most conveniently summarized in a **one-way frequency table.** Suppose, for example, that each item returned to a department store is classified according to disposition — cash refund, credit to charge account, merchandise exchange, or return refused. Records of 100 randomly selected returns are examined and each disposition recorded. The first few observations might be

Cash refund Return refused

Exchange Cash refund

Exchange Credit to account

Counting the number of observations of each type might then result in the following one-way table:

	Disposition			
	Cash	**Credit**	**Exchange**	**Refused**
Frequency	34	18	31	17

For a categorical variable with k possible values (k different levels or categories), sample data is summarized in a one-way frequency table consisting of k cells, displayed horizontally or vertically.

In this section, we consider testing hypotheses about the proportion of the population that fall into each of the possible categories. For example, the customer relations manager for a department store might be interested in determining whether the four possible dispositions for a return request occur with equal frequency. If this is indeed the case, the long-run proportion of returns falling into each of the four categories is 1/4, or .25. The test procedure to be presented shortly would allow the manager to decide whether this hypothesis is plausible.

Notation

k = number of categories of a categorical variable

π_1 = true proportion for category 1

π_2 = true proportion for category 2

$\vdots$

π_k = true proportion for category k

 (*Note:* $\pi_1 + \pi_2 + \cdots + \pi_k = 1$)

The hypotheses to be tested have the form

H_0: π_1 = hypothesized proportion for category 1

  π_2 = hypothesized proportion for category 2

  $\vdots$

  π_k = hypothesized proportion for category k

H_a: H_0 is not true, so at least one of the true category proportions differs from the corresponding hypothesized value.

For the example involving department store returns,

π_1 = proportion of all returns that result in a cash refund

π_2 = proportion of all returns that result in an account credit

π_3 = proportion of all returns that result in an exchange

π_4 = proportion of all returns that are refused

The null hypothesis of interest is then

H_0: $\pi_1 = .25, \pi_2 = .25, \pi_3 = .25, \pi_4 = .25$

A null hypothesis of the type just described will be tested by first selecting a random sample of size n and then classifying each sample response into one of the k possible categories. To decide whether the sample data is compatible with the null hypothesis, we must compare the observed cell counts (frequencies) to the cell counts that would have been expected when the null hypothesis is true. In general, the expected cell counts are

expected cell count for category $1 = n\pi_1$

expected cell count for category $2 = n\pi_2$

and so on. The expected cell counts when H_0 is true result from substituting the corresponding hypothesized proportion for each π.

EXAMPLE 12.1 A number of psychological studies have considered the relationship between various deviant behaviors and other variables such as the lunar phase. The article "Psychiatric and Alcoholic Admissions Do Not Occur Disproportionately Close to Patients' Birthdays" (*Psychological Reports* (1992): 944–946) focused on the existence of any relationship between date of patient admission for specified treatment and patient's birthday. Admission date was partitioned into four categories according to how close it was to the patient's birthday:

1. Within 7 days of birthday
2. Between 8 and 30 days, inclusive, from the birthday
3. Between 31 and 90 days, inclusive, from the birthday
4. More than 90 days from the birthday

Let π_1, π_2, π_3, and π_4 denote the true proportions in categories 1, 2, 3, and 4, respectively. If there is no relationship between admission date and birthday, then since there are 15 days included in the first category (from 7 days prior to the patient's birthday to 7 days after, including, of course, the birthday itself),

$$\pi_1 = \frac{15}{365} = .041$$

(For simplicity, we are ignoring Feb. 29 of leap years.) Similarly, in the absence of any relationship,

$$\pi_2 = \frac{46}{365} = .126$$

$$\pi_3 = .329$$

$$\pi_4 = .504$$

The hypotheses of interest are then

H_0: $\pi_1 = .041, \pi_2 = .126, \pi_3 = .329, \pi_4 = .504$

H_a: H_0 is not true.

The cited article gave data for $n = 200$ patients admitted for alcoholism treatment. If H_0 is true, the expected counts are

$$\begin{pmatrix} \text{expected count} \\ \text{for category 1} \end{pmatrix} = n \begin{pmatrix} \text{hypothesized proportion} \\ \text{for category 1} \end{pmatrix} = 200(.041) = 8.2$$

$$\begin{pmatrix} \text{expected count} \\ \text{for category 2} \end{pmatrix} = n \begin{pmatrix} \text{hypothesized proportion} \\ \text{for category 2} \end{pmatrix} = 200(.126) = 25.2$$

$$\begin{pmatrix} \text{expected count} \\ \text{for category 3} \end{pmatrix} = n \begin{pmatrix} \text{hypothesized proportion} \\ \text{for category 3} \end{pmatrix} = 200(.329) = 65.8$$

$$\begin{pmatrix} \text{expected count} \\ \text{for category 4} \end{pmatrix} = n \begin{pmatrix} \text{hypothesized proportion} \\ \text{for category 4} \end{pmatrix} = 200(.504) = 100.8$$

Observed and expected cell counts are given in the following table.

	Category 1	2	3	4
Observed count	11	24	69	96
Expected count	8.2	25.2	65.8	100.8

Since the observed counts are based on a *sample* of admissions, it would be somewhat surprising to see *exactly* 4.1% of the sample falling in the first category, exactly 12.6% in the second, and so on, even when H_0 is true. If the differences between the observed and expected cell counts can reasonably be attributed to sampling variation, the data would be considered compatible with H_0. On the other hand, if the discrepancy between the observed and expected cell counts is too large to be attributed solely to chance differences from one sample to another, H_0 should be rejected in favor of H_a. Thus, we need an assessment of how different the observed and expected counts are.

The goodness-of-fit statistic, denoted by X^2, is a quantitative measure of the extent to which the observed counts differ from those expected when H_0 is true.*

*The Greek letter χ is often used in place of X. The symbol X^2 is referred to as the chi-squared (χ^2) statistic. In using X^2 rather than χ^2, we are adhering to the convention of denoting sample quantities by Roman letters.

The **goodness-of-fit statistic, X^2,** results from first computing the quantity

$$\frac{(\text{observed cell count } - \text{ expected cell count})^2}{\text{expected cell count}}$$

for each cell, where, for a sample of size n,

$$\left(\begin{array}{c}\text{expected cell}\\\text{count}\end{array}\right) = n\left(\begin{array}{c}\text{hypothesized value of corresponding}\\\text{population proportion}\end{array}\right)$$

The X^2 is the sum of these quantities for all k cells:

$$X^2 = \sum_{\text{all cells}} \frac{(\text{observed cell count } - \text{ ecpected cell count})^2}{\text{expected cell count}}$$

The value of the X^2 statistic reflects the magnitude of the discrepancies between observed and expected cell counts. When the differences are sizable, the value of X^2 tends to be large. Therefore, large values of X^2 suggest rejection of H_0. A small value of X^2 (it can never be negative) occurs when the observed cell counts are quite similar to those expected when H_0 is true and so would lend support to H_0.

As with previous test procedures, a conclusion will be reached by comparing a P-value to the significance level for the test. The P-value is computed as the probability of observing a value of X^2 at least as large as the observed value when H_0 is true. This requires information concerning the sampling distribution of X^2 when H_0 is true. A key result is that when the null hypothesis is correct and the sample size is sufficiently large, the behavior of X^2 is described approximately by a **chi-squared distribution.** A chi-squared curve has no area associated with negative values and is asymmetric, with a longer tail on the right. There are actually many chi-squared distributions, each one identified with a different number of degrees of freedom. Curves corresponding to several chi-squared distributions are shown in Figure 12.1.

FIGURE 12.1 Chi-squared curves

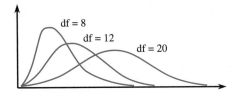

For a test procedure based on the X^2 statistic, the associated P-value is the area under the appropriate chi-squared curve and to the right of the computed X^2 value. Appendix Table IX gives upper-tail areas for chi-squared distributions with up to 20 df. Our chi-squared table has a different appearance from the t table used in previous chapters. In the t table, there is a single "value" column on the far left and then a column of P-values (tail areas) for each different df. A single column of t values works here because all t curves are centered at zero, and the t curves approach the z curve as the df increases. However, because the chi-squared curves move farther and farther to the right and spread out more as df increases, a single "value" column is impractical in this situation.

To find the area to the right for a particular X^2 value, locate the appropriate df column in Appendix Table IX. Determine which listed value is closest to the X^2 value of interest, and read the right-tail area corresponding to this value from the left-hand column of the table. For example, for a chi-squared distribution with df = 2, the area to the right of $X^2 = 4.93$ is .085, as shown in Figure 12.2. For this same chi-squared distribution (df = 2), the area to the right of 6.25 is approximately .045 (the area to the right of 6.20, the closest entry in the table for df = 2).

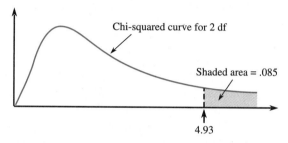

FIGURE 12.2 A chi-squared upper-tail area

Goodness-of-Fit Tests

As long as none of the expected cell counts are too small, when H_0 is true, the X^2 goodness-of-fit statistic has approximately a chi-squared distribution with df = $(k - 1)$. *It is generally agreed that use of the chi-squared distribution is appropriate when the sample size is large enough for every expected cell count to be at least 5.* If any of the expected cell frequencies are less than 5, categories may be combined in a sensible way to create acceptable expected cell counts. Just remember to compute df based on the reduced number of categories.

Goodness-of-Fit Test Procedure

Hypotheses: H_0: π_1 = hypothesized proportion for category 1

$\vdots$

π_k = hypothesized proportion for category k

H_a: H_0 is not true.

Test statistic: $X^2 = \displaystyle\sum_{\text{all cells}} \frac{(\text{observed cell count} - \text{expected cell count})^2}{\text{expected cell count}}$

P-values: When H_0 is true and all expected counts are at least 5, X^2 has approximately a chi-squared distribution with df = $k - 1$. Therefore, the P-value associated with the computed test statistic value is the area to the right of X^2 under the df = $k - 1$ chi-squared curve. Upper-tail areas for chi-squared distribtuions are found in Appendix Table IX.

(continued)

Assumptions: 1. Observed cell counts are based on a *random sample*.
2. The *sample size is large*. The sample size is large enough for the chi-squared test to be appropriate as long as every expected count is at least 5.

EXAMPLE 12.2

(Example 12.1 continued)   We will use the admissions data of Example 12.1 to test the hypothesis that admission date is unrelated to birthday. Let's use a .05 level of significance and the nine-step hypothesis-testing procedure illustrated in previous chapters.

1. Let π_1, π_2, π_3, and π_4 denote the proportions of all admissions for treatment of alcoholism falling in the four categories.

2. H_0: $\pi_1 = .041$, $\pi_2 = .126$, $\pi_3 = .329$, $\pi_4 = .504$

3. H_a: H_0 is not true.

4. *Significance level*: $\alpha = .05$

5. *Test Statistic*: $X^2 = \sum_{\text{all cells}} \dfrac{(\text{observed cell count} - \text{expected cell count})^2}{\text{expected cell count}}$

6. *Assumptions*:   The expected cell counts (from Example 12.1) are 8.2, 25.2, 65.8, and 100.8, all of which are greater than 5. The article did not indicate how the patients were selected. We can proceed with the chi-squared test if it is reasonable to assume that the 200 patients in the sample can be regarded as a random sample of patients admitted for treatment of alcoholism.

7. *Calculation*:
$$X^2 = \frac{(11 - 8.2)^2}{8.2} + \frac{(24 - 25.2)^2}{25.5} + \frac{(69 - 65.8)^2}{65.8} + \frac{(96 - 100.8)^2}{100.8}$$
$$= .96 + .06 + .16 + .23$$
$$= 1.41$$

8. *P-value*:   The P-value is based on a chi-squared distribution with df = $4 - 1 = 3$. The computed value of X^2 is smaller than 6.25 (the smallest entry in the df = 3 column of Appendix Table IX), so P-value > .10.

9. *Conclusion*:   Since P-value > α, H_0 cannot be rejected. There is not sufficient evidence to conclude that a relationship exists between birthday and date admitted for treatment.

EXAMPLE 12.3

Does the color of a car influence the chance that it will be stolen? The Associated Press (*San Luis Obispo Telegram-Tribune*, Sept. 2, 1995) reported the following information for a random sample of 830 stolen vehicles: 140 were white, 100 were blue, 270 were red, 230 were black, and 90 were other colors. We will use the X^2 goodness-of-fit test and a significance level of $\alpha = .01$ to test the hypothesis that proportions stolen are identical to population color proportions. Suppose it is known that 15% of all cars are white, 15% are blue, 35% are red, 30% are black, and 5% are other colors. If these same population color proportions hold for stolen cars, the expected counts are

$$\text{expected count for white} = 830(.15) = 124.5$$
$$\text{expected count for blue} = 830(.15) = 124.5$$
$$\text{expected count for red} = 830(.35) = 290.5$$
$$\text{expected count for black} = 830(.30) = 249.0$$
$$\text{expected count for other colors} = 830(.05) = 41.5$$

These expected counts have been entered in Table 12.1.

Table 12.1

Category	Color	Observed Count	Expected Count
1	White	140	124.5
2	Blue	100	124.5
3	Red	270	290.5
4	Black	230	249.0
5	Other colors	90	41.5

1. Let $\pi_1, \pi_2, \ldots, \pi_5$ denote the true proportion of stolen cars that fall into the five color categories.

2. $H_0: \pi_1 = .15, \pi_2 = .15, \pi_3 = .35, \pi_4 = .30, \pi_5 = .05$

3. $H_a: H_0$ is not true.

4. *Significance level*: $\alpha = .01$

5. *Test statistic*: $X^2 = \sum_{\text{all cells}} \dfrac{(\text{observed cell count} - \text{expected cell count})^2}{\text{expected cell count}}$

6. *Assumptions*: The sample was a random sample of stolen vehicles. All expected counts are greater than 5, so the sample size is large enough to use the chi-squared test.

7. *Calculation*:
$$X^2 = \frac{(140 - 124.5)^2}{124.5} + \frac{(100 - 124.5)^2}{124.5} + \frac{(270 - 290.5)^2}{290.5} + \frac{(230 - 249.0)^2}{249.0} + \frac{(90 - 41.5)^2}{41.5}$$
$$= 1.93 + 4.82 + 1.45 + 1.45 + 56.68$$
$$= 66.33$$

8. *P-value*: All expected counts exceed 5, so the *P*-value can be based on a chi-squared distribution with df $= 5 - 1 = 4$. The computed X^2 value is larger than 18.46, the largest value in the df $= 4$ column of Appendix Table IX, so

$$P\text{-value} < .001$$

9. *Conclusion*: Since *P*-value $\le \alpha$, H_0 is rejected. There is convincing evidence that at least one of the color proportions for stolen cars differs from the corresponding proportion for all cars.

Exercises 12.1 – 12.14

12.1 State what you know about the *P*-value for a chi-squared test in each of the following situations, and state the conclusion for a significance level of $\alpha = .01$.
  a. $X^2 = 7.5$, df $= 2$ **b.** $X^2 = 13.0$, df $= 6$
  c. $X^2 = 18.0$, df $= 9$ **d.** $X^2 = 21.3$, df $= 4$
  e. $X^2 = 5.0$, df $= 3$

12.2 A particular paperback book is published in a choice of four different covers. A certain bookstore keeps copies of each cover on its racks. To test the hypothesis that sales are equally divided among the four choices, a random sample of 100 purchases is identified.
  a. If the resulting X^2 value is 6.4, what conclusion would you reach when using a test with significance level .05?
  b. What conclusion would be appropriate at significance level .01 if $X^2 = 15.3$?
  c. If there were six different covers rather than just four, what would you conclude if $X^2 = 13.7$ and a test with $\alpha = .05$ was used?

12.3 Packages of mixed nuts made by a certain company contain four types of nuts. The percentages of nuts of types 1, 2, 3, and 4 are supposed to be 40%, 30%, 20%, and 10%, respectively. A random sample of nuts is selected, and each one is categorized by type.
  a. If the sample size is 200 and the resulting test statistic value is $X^2 = 19.0$, what conclusion would be appropriate for a significance level of .001?
  b. If the random sample had consisted of only 40 nuts, would you use the chi-squared test here? Explain your reasoning.

12.4 The article "Family Planning: Football Style. The Relative Age Effect in Football" (*Int. Rev. for Soc. of Sport* (1992): 77–88) investigated the relationship between month of birth and achievement in sports. Birth

dates were collected for players on teams competing in the 1990 World Cup soccer games, and they are summarized in the accompanying table.

Birthday	Frequency
Quarter 1 (Aug.–Oct.)	150
Quarter 2 (Nov.–Jan.)	138
Quarter 3 (Feb.–Apr.)	140
Quarter 4 (May–July)	100

The authors of the article used a chi-squared goodness-of-fit test to determine whether the data was consistent with the hypothesis that the birthdays of world-class soccer players are evenly distributed across the four quarters.
  a. Perform a test of this hypothesis, using a .05 significance level.
  b. The article states that this analysis demonstrates that the distribution of player's birthdays is not random and that the number of players is related to the "quarters of the football year." Does your analysis support this conclusion? Explain.

12.5 The color vision of birds plays a role in their foraging behavior: Birds use color to select and avoid certain types of food. The authors of the article "Colour Avoidance in Northern Bobwhites: Effects of Age, Sex, and Previous Experience" (*Animal Behaviour* (1995): 519–526) studied the pecking behavior of 1-day-old bobwhites. In an area painted white, they inserted four pins with different colored heads. The color of the pin chosen on the birds first peck was noted for each of 33 bobwhites, resulting in the accompanying table. Does the data provide evidence of a color preference? Test using $\alpha = .01$.

Color	First Peck Frequency
Blue	16
Green	8
Yellow	6
Red	3

12.6   An article about the California lottery that appeared in the *San Luis Obispo Tribune* (Dec. 15, 1999) gave the following information on the age distribution of adults in California: 35% are between 18 and 34 years old, 51% are between 35 and 64 years old, and 14% are 65 years old or older. The article also gave information on the age distribution of those who purchase lottery tickets. The accompanying table is consistent with the values given in the article. Suppose that the data resulted from a random sample of 200 lottery ticket purchasers.

Age of Purchaser	Frequency
18–34	36
35–64	130
65 and over	34

Based on this sample data, is it reasonable to conclude that one or more of these three age groups buys a disproportionate share of lottery tickets? Use a chi-square goodness-of-fit test with $\alpha = .05$.

12.7   Drivers born under the astrological sign of Capricorn are the worst drivers in Australia, according to an article that appeared in the Australian newspaper *The Mercury* (Oct. 26, 1998). This statement was based on a study of insurance claims that resulted in the following data for male policyholders of a large insurance company.

Astrological Sign	Number of Policyholders
Aquarius	35,666
Aries	37,926
Cancer	38,126
Capricorn	54,906
Gemini	37,179
Leo	37,354
Libra	37,910
Pisces	36,677
Sagittarius	34,175
Scorpio	35,352
Taurus	37,179
Virgo	37,718

a. Assuming that it is reasonable to treat the policyholders of this particular insurance company as a random sample of insured drivers in Australia, is the observed data consistent with the hypothesis that the proportion of policyholders is the same for each of the 12 astrological signs?

b. Why do you think that the proportion of Capricorn policyholders is so much higher than would be expected if the proportions are the same for all astrological signs?

c. Suppose that a random sample of 1000 accident claims submitted to this insurance company is selected and each claim classified according to the astrological sign of the driver. (The accompanying table is consistent with accident rates given in the article.)

Astrological Sign	Observed Number in Sample
Aquarius	85
Aries	83
Cancer	82
Capricorn	88
Gemini	83
Leo	83
Libra	83
Pisces	82
Sagittarius	81
Scorpio	85
Taurus	84
Virgo	81

Test the null hypothesis that the proportion of accident claims submitted by drivers of each astrological sign is consistent with the proportion of policyholders of each sign. Use the information on the distribution of policyholders from part **a** to compute estimated expected frequencies and then carry out an appropriate test.

12.8   According to Census Bureau data, in 1998 the California population consisted of 50.7% whites, 6.6% blacks, 30.6% Hispanics, 10.8% Asians, and 1.3% other ethnic groups. Suppose that a random sample of 1000 students graduating from California colleges and universities in 1998 resulted in the accompanying data on ethnic group. This data is consistent with summary statistics contained in the article titled "Crumbling Public School System a Threat to California's Future (*Investor's Business Daily,* Nov. 12, 1999).

Ethnic Group	Number in Sample
White	679
Black	51
Hispanic	77
Asian	190
Other	3

Does this data provide evidence that the proportion of students graduating from colleges and universities in California for these ethnic group categories differs from the respective proportions in the population for California? Test the appropriate hypotheses using $\alpha = .01$.

12.9   Each person in a 1989 random sample of prison inmates was classified according to the type of offense committed, resulting in the following one-way frequency table. (These values are based on information in the article "Profile of Jail Inmates," *USA Today*, April 25, 1991). In 1983, it was reported that 30.7% of inmates had been convicted of violent crimes, 38.6% of crimes against property, 9.3% of drug-related crimes, 20.6% of public-order offenses, and .8% of other types of crimes. Does this data provide sufficient evidence to conclude that the true 1989 proportions falling in the various offense categories are not all the same as in 1983? Test the relevant hypotheses using $\alpha = .05$.

Type of Offense

	Violent Crime	Crimes Against Property	Drug-related Crimes	Public-order Offense	Other
Frequency	225	300	230	228	16

12.10 Criminologists have long debated whether there is a relationship between weather and violent crime. The author of the article "Is There a Season for Homicide?" (*Criminology* (1988): 287–296) classified 1361 homicides according to season, resulting in the accompanying data. Does this data support the theory that the homicide rate is not the same over the four seasons? Test the relevant hypotheses using a significance level of .05.

Season

Winter	Spring	Summer	Fall
328	334	372	327

12.11 When public opinion surveys are conducted by mail, a cover letter explaining the purpose of the survey is usually included. To determine whether the wording of the cover letter influences the response rate, three different cover letters were used in a survey of students at a Midwestern university ("The Effectiveness of Cover-Letter Appeals," *J. of Soc. Psych.* (1984): 85–91). Suppose that each of the three cover letters accompanied questionnaires sent to an equal number of randomly selected students. Returned questionnaires were then classified according to the type of cover letter (I, II, or III). Use the accompanying data to test the hypothesis that $\pi_1 = 1/3$, $\pi_2 = 1/3$, and $\pi_3 = 1/3$, where π_1, π_2, and π_3 are the true proportions of all returned questionnaires accompanied by cover letters I, II, and III, respectively. Use a .05 significance level.

Cover-letter Type

	I	II	III
Frequency	48	44	39

12.12 A certain genetic characteristic of a particular plant can appear in one of three forms (phenotypes). A researcher has developed a theory, according to which the hypothesized proportions are $\pi_1 = .25$, $\pi_2 = .50$, and $\pi_3 = .25$. A random sample of 200 plants yields $X^2 = 4.63$.
a. Carry out a test of the null hypothesis that the theory is correct, using level of significance $\alpha = .05$.
b. Suppose that a random sample of 300 plants had resulted in the same value of X^2. How would your analysis and conclusion differ from those in part **a**?

12.13 The article "Linkage Studies of the Tomato" (*Trans. Royal Canad. Inst.* (1931): 1–19) reported the accompanying data on phenotypes resulting from crossing tall cut-leaf tomatoes with dwarf potato-leaf tomatoes. There are four possible phenotypes: (1) tall cut-leaf, (2) tall potato-leaf, (3) dwarf cut-leaf, and (4) dwarf potato-leaf. Mendel's laws of inheritance imply that $\pi_1 = 9/16$, $\pi_2 = 3/16$, $\pi_3 = 3/16$, and $\pi_4 = 1/16$. Is the data from this experiment consistent with Mendel's laws? Use a .01 significance level.

Phenotype

	1	2	3	4
Frequency	926	288	293	104

12.14 It is hypothesized that when homing pigeons are disoriented in a certain manner, they will exhibit no preference for any direction of flight after takeoff. To test this, 120 pigeons are disoriented and let loose, and the direction of flight of each is recorded. The resulting data is given in the accompanying table. Use the goodness-of-fit test with significance level .10 to determine whether the data supports the hypothesis.

Direction	Frequency
0°–45°	12
45°–90°	16
90°–135°	17
135°–180°	15
180°–225°	13
225°–270°	20
270°–315°	17
315°–360°	10

12.2   Tests for Homogeneity and Independence in a Two-Way Table

Data resulting from observations made on two different categorical variables can also be summarized using a tabular format. As an example, suppose that residents of a particular city can watch national news on affiliate stations of ABC, CBS, NBC, or PBS (the public television network). A researcher wishes to know whether there is any relationship between political philosophy (liberal, moderate, or conservative) and preferred news program among those residents who regularly watch the national news. Let x denote the variable *political philosophy* and y the variable *preferred network*. A random sample of 300 regular watchers is to be selected, and each one will be asked for his or her x and y values. The data set is bivariate and might initially be displayed as follows:

Observation	x value	y value
1	Liberal	CBS
2	Conservative	ABC
3	Conservative	PBS
⋮	⋮	⋮
299	Moderate	NBC
300	Liberal	PBS

Bivariate categorical data of this sort can most easily be summarized by constructing a **two-way frequency table,** or **contingency table.** This is a rectangular table that consists of a row for each possible value of x (each category specified by this variable) and a column for each possible value of y. There is then a cell in the table for each possible (x, y) combination. Once such a table has been constructed, the number of times each particular (x, y) combination occurs in the data set is determined, and these numbers (frequencies) are entered in the corresponding cells of the table. The resulting numbers are called **observed cell counts.** The table for the example relating political philosophy to preferred network discussed previously contains 3 rows and 4 columns (because x and y have 3 and 4 possible values, respectively). Table 12.2 is one possible table, with the rows and columns labeled with the possible x and y "values." These are often referred to as the *row* and *column categories.*

Table 12.2 *An example of a 3 × 4 frequency table*

	ABC	CBS	NBC	PBS	Row Marginal Total
Liberal	20	20	25	15	80
Moderate	45	35	50	20	150
Conservative	15	40	10	5	70
Column Marginal Total	80	95	85	40	300

Marginal totals are obtained by adding the observed cell counts in each row and also in each column of the table. The row and column marginal totals, along with the total of all observed cell counts in the table — the **grand total** — have been included in Table 12.2. The marginal totals provide information on the distribution of observed values for each variable separately. In this example, the row marginal totals reveal that the sample consisted of 80 liberals, 150 moderates, and 70 conservatives. Similarly, column marginal totals indicated how often each of the preferred program categories occurred: 80 preferred ABC news, 95 preferred CBS, and so on. The grand total, 300, is the number of observations in the bivariate data set — in this case, the sample size (although occasionally such a table results from a census of the entire population).

Two-way frequency tables are often characterized by the number of rows and columns in the table (specified in that order: rows first, then columns). Table 12.2 is called a 3×4 table. The smallest two-way frequency table is a 2×2 table, which has only two rows and two columns and thus four cells.

Two-way tables arise naturally in two different types of investigations. A researcher may be interested in comparing two or more groups on the basis of a single categorical variable and so may obtain a sample separately from each group. For example, data could be collected at a university to compare students, faculty, and staff on the basis of primary mode of transportation to campus (car, bicycle, motorcycle, bus, or on foot). One random sample of 200 students, another of 100 faculty, and a third of 150 staff might be chosen, and the selected individuals could be interviewed to obtain the necessary transportation information. Data from such a study could easily be summarized in a 3×5 two-way frequency table with row categories of student, faculty, and staff and column categories corresponding to the five possible modes of transportation. The observed cell counts could then be used to gain insight into differences and similarities among the three groups with respect to the means of transportation. This type of bivariate categorical data is characterized by having one set of marginal totals fixed (the sample sizes from the different groups), whereas each total in the other set is random. In the 3×5 situation just discussed, the row totals would be fixed at 200, 100, and 150, respectively.

A two-way table also arises when the values of two different categorical variables are observed for all individuals or items in a single sample. For example, a sample of 500 registered voters might be selected. Each voter could then be asked both if he or she favored a particular property tax initiative and if he or she was a registered Democrat, Republican, or independent. This would result in a bivariate data set with x representing the variable *political affiliation* (with categories Democrat, Republican, and independent) and y representing the variable *response* (favor initiative or oppose initiative). The corresponding 3×2 frequency table could then be used to investigate any association between position on the tax initiative and political affiliation. This type of bivariate categorical data is characterized by having both sets of marginal totals random and only the grand total fixed.

Comparing Two or More Populations: A Test of Homogeneity

When the value of a categorical variable is recorded for members of separate random samples obtained from each population under study, the central issue is whether the category proportions are the same for all of the populations. As in

Section 12.1, the test procedure uses a chi-squared statistic that compares the observed counts to those that would be expected if there were no differences between the populations.

EXAMPLE 12.4

In the past, a number of professions were prohibited from advertising. In 1977, the U.S. Supreme Court ruled that prohibiting doctors and lawyers from advertising violated their right to free speech. The article "Should Dentists Advertise?" (*J. of Ad. Research* (June 1982): 33–38) compared the attitudes of consumers and dentists toward the advertising of dental services. Separate random samples of 101 consumers and 124 dentists were asked to respond to the following statement: "I favor the use of advertising by dentists to attract new patients." Possible responses were strongly agree, agree, neutral, disagree, and strongly disagree. The data presented in the article appears in Table 12.3, a 2 × 5 frequency table. The authors were interested in determining whether the two groups — consumers and dentists — differed in their attitudes toward advertising.

Table 12.3 *Observed counts for Example 12.4*

	Response					
	Strongly Agree	Agree	Neutral	Disagree	Strongly Disagree	Row Marg. Total
Consumers	34	49	9	4	5	101
Dentists	9	18	23	28	46	124
Col. Marg. Total	43	67	32	32	51	225
Col. % of Total	19.11	29.78	14.22	14.22	22.67	

Estimates of expected cell counts can be thought of in the following manner: There were 225 responses, of which 43 were "strongly agree." The proportion of the total responding "strongly agree" is then

$$\frac{43}{225} = .19111$$

If there were no difference in response for consumers and dentists, we would then expect about 19.111% of the consumers and 19.111% of the dentists to have strongly agreed. Therefore, the expected cell counts* for the two cells in the "strongly agree" column are

expected count for consumer–strongly agree cell = .19111(101) = 19.30

expected count for dentist–strongly agree cell = .19111(124) = 23.70

---

*In this section, all expected cell counts will be estimated from sample data. All references to "expected counts" should be interpreted as *estimated* expected counts.

Note that the expected cell counts need not be whole numbers. The expected cell counts for the remaining cells can be computed in a similar manner. For example,

$$\frac{67}{225} = .29778$$

of all responses were in the "agree" category, so

expected count for consumer–agree cell = .29778(101) = 30.08

expected count for dentist–agree cell = .29778(124) = 36.92

It is common practice to display the observed cell counts and the corresponding estimated expected cell counts in the same table, with the estimated expected cell counts enclosed in parentheses. Expected cell counts for the remaining six cells have been computed and entered into Table 12.4. Except for small differences due to rounding, each marginal total for expected cell counts is identical to that of the corresponding observed counts.

A quick comparison of the observed and expected cell counts reveals large discrepancies. It appears that most consumers are more favorable toward advertising since the observed counts in the consumer–strongly agree and consumer–agree cells are substantially higher than would be expected if no difference exists, whereas the observed cell counts in the consumer–disagree and consumer–strongly disagree cells are lower than would be expected when no difference exists. The opposite relationship between observed and expected counts is exhibited by the dentists.

In Example 12.4, the expected count for a cell corresponding to a particular group–response combination was computed in two steps. First, the response *marginal proportion* was computed (for example, 43/225 for the "strongly agree" response). Then this proportion was multiplied by a marginal group total (for example, 101(43/225) for the consumer group). This is equivalent to first multiplying the row and column marginal totals and then dividing by the grand total:

$$\frac{(101)(43)}{225}$$

Table 12.4 *Observed and expected counts for Example 12.4*

| Group | Observed and Expected Cell Counts | | | | | |
	Strongly Agree	Agree	Neutral	Disagree	Strongly Disagree	Row Marg. Total
Consumers	34 (19.30)	49 (30.08)	9 (14.36)	4 (14.36)	5 (22.89)	101
Dentists	9 (23.70)	18 (36.92)	23 (17.64)	28 (17.64)	46 (28.11)	124
Col. Marg. Total	43	67	32	32	51	225

To compare two or more groups on the basis of a categorical variable, calculate an **expected cell count** for each cell by selecting the corresponding row and column marginal totals and then computing

$$\text{expected cell count} = \frac{(\text{row marginal total})(\text{column marginal total})}{\text{grand total}}$$

These quantities represent what would be expected when there is no difference between the groups under study.

The X^2 statistic, introduced in Section 12.1, can now be used to compare the observed cell counts to the expected cell counts. A large value of X^2 results when there are substantial discrepancies between the observed and expected counts, and suggests that the hypothesis of no differences between the populations should be rejected. A formal test procedure is described in the accompanying box.

Comparing Two or More Populations Using the X^2 Statistic

Null hypothesis: H_0: The true category proportions are the same for all the populations (homogeneity of populations).

Alternative hypothesis: H_a: The true category proportions are not all the same for all of the populations.

Test statistic: $X^2 = \sum_{\text{all cells}} \dfrac{(\text{observed cell count} - \text{expected cell count})^2}{\text{expected cell count}}$

The expected cell counts are estimated from the sample data (assuming that H_0 is true) using the formula

$$\text{expected cell count} = \frac{(\text{row marginal total})(\text{column marginal total})}{\text{grand total}}$$

P-values: When H_0 is true, X^2 has approximately a chi-squared distribution with

$$\text{df} = (\text{number of rows} - 1)(\text{number of columns} - 1)$$

The *P*-value associated with the computed test statistic value is the area to the right of X^2 under the chi-squared curve with the appropriate df. Upper-tail areas for chi-squared distributions are found in Appendix Table IX.

Assumptions: 1. The data consists of *independently chosen random samples.*
2. *The sample size is large*: all expected counts are at least 5. If some expected counts are less than 5, rows or columns of the table may be combined to achieve a table with satisfactory expected counts.

EXAMPLE 12.5

(Example 12.4 continued)   The following table of observed and expected cell counts appeared in Example 12.4.

Group	Strongly Agree	Agree	Neutral	Disagree	Strongly Disagree
Consumers	34 (19.30)	49 (30.08)	9 (14.36)	4 (14.36)	5 (22.89)
Dentists	9 (23.70)	18 (36.92)	23 (17.64)	28 (17.64)	46 (28.11)

Hypotheses:

   H_0:   Proportions in each response category are the same for consumers and dentists.

   H_a: H_0 is not true.

Significance level: A significance level of $\alpha = .05$ will be used.

Test statistic: $X^2 = \sum_{\text{all cells}} \dfrac{(\text{observed cell count} - \text{expected cell count})^2}{\text{expected cell count}}$

Assumptions: All expected cell counts are at least 5, and the random samples were independently chosen, so use of the X^2 test is appropriate.

Calculation: $X^2 = \dfrac{(34 - 19.30)^2}{19.30} + \cdots + \dfrac{(46 - 28.11)^2}{28.11} = 84.47$

P-value: The two-way table for this example has 2 rows and 5 columns, so the appropriate df is $(2 - 1)(5 - 1) = 4$. Since 84.47 is greater than 18.46, the largest entry in the 4-df column of Appendix Table IX,

   P-value $< .001$

Conclusion: P-value $\leq \alpha$, so H_0 is rejected. There is strong evidence to support the claim that the proportions in the response categories are not the same for dentists and consumers.

EXAMPLE 12.6

The data on drinking behavior for independently chosen random samples of male and female students (Table 12.5, page 616) is similar to data that appeared in the article "Relationship of Health Behaviors to Alcohol and Cigarette Use by College Students" (*J. of College Student Development* (1992): 163–170). Does there appear to be a gender difference with respect to drinking behavior? (*Note*: Low = 1–7 drinks/week, moderate = 8–24 drinks/week, high = 25 or more drinks/week.)

Table 12.5 *Observed and expected counts for Example 12.6*

Drinking Level	Gender		Row Marginal Total
	Men	**Women**	
None	140 (158.6)	186 (167.4)	326
Low	478 (554.0)	661 (585.0)	1139
Moderate	300 (230.1)	173 (242.9)	473
High	63 (38.4)	16 (40.6)	79
Column Marginal Total	981	1036	2017

The relevant hypotheses are

H_0: True proportions for the four drinking levels are the same for males and females.

H_a: H_0 is not true.

Significance level: $\alpha = .01$

Test statistic: $X^2 = \displaystyle\sum_{\text{all cells}} \dfrac{(\text{observed cell count} - \text{expected cell count})^2}{\text{expected cell count}}$

Assumptions: Table 12.5 contains the computed expected counts, all of which are greater than 5. The data consists of independently chosen random samples.

Calculation: $X^2 = \dfrac{(140 - 158.6)^2}{158.6} + \cdots + \dfrac{(16 - 40.6)^2}{40.6} = 96.6$

P-value: The table has 4 rows and 2 columns, so df $= (4 - 1)(2 - 1) = 3$. The computed value of X^2 is greater than the largest entry of the 3-df column of Appendix Table IX, so

P-value $< .001$

Conclusion: Since P-value $\leq \alpha$, H_0 is rejected. The data indicates that males and females differ with respect to drinking level.

────

Testing for Independence of Two Categorical Variables

The X^2 test statistic and test procedure can also be used to investigate association between two categorical variables in a single population. Suppose that each population member has a value of a first categorical variable and also of a second such variable. As an example, television viewers in a particular city might be categorized both with respect to preferred network (ABC, CBS, NBC, or PBS) and with respect to favorite type of programming (comedy, drama, or information-news). The question of interest is often whether knowledge of one variable's value provides any information about the value of the other variable — that is, are the two variables independent? Continuing the example, suppose that those who favor ABC prefer the three types of programming in proportions .4, .5, and .1 and that these proportions are also correct for individuals favoring any of the other three networks. Then learning an individual's preferred network provides no extra information concerning that individual's favorite type of programming. The categorical variables *preferred network* and *favorite program type* are independent.

To see how expected counts are obtained in this situation, recall from Chapter 6 that if two outcomes A and B are independent,

$$P(A \text{ and } B) = P(A) \cdot P(B)$$

so the proportion of time that they occur together in the long run is the product of the two individual long-run relative frequencies. Similarly, two categorical variables are independent in a population if, for each particular category of the first variable and each particular category of the second variable,

$$\begin{pmatrix} \text{proportion} \\ \text{of individuals} \\ \text{in a particular} \\ \text{category combination} \end{pmatrix} = \begin{pmatrix} \text{proportion in} \\ \text{specified category} \\ \text{of 1st variable} \end{pmatrix} \cdot \begin{pmatrix} \text{proportion in} \\ \text{specified category} \\ \text{of 2nd variable} \end{pmatrix}$$

Thus, if 30% of all viewers prefer ABC and the proportions of program type preferences are as previously given, assuming the two variables are independent, the proportion of individuals who both favor ABC and prefer comedy is $(.3)(.4) = .12$ (or 12%).

Multiplying the right-hand side of this expression by the sample size gives us the expected number of individuals in the sample who are in both specified categories of the two variables when the variables are independent. However, these expected counts cannot be calculated, because the individual population proportions are not known. The resolution of this dilemma is to estimate each population proportion using the corresponding sample proportion:

$$\begin{pmatrix} \text{estimated expected number} \\ \text{in specified categories} \\ \text{of the two variables} \end{pmatrix} = \begin{pmatrix} \text{sample} \\ \text{size} \end{pmatrix} \cdot \frac{\begin{pmatrix} \text{obs. number in} \\ \text{categ. of 1st var.} \end{pmatrix}}{\text{sample size}} \cdot \frac{\begin{pmatrix} \text{obs. number in} \\ \text{categ. of 2nd var.} \end{pmatrix}}{\text{sample size}}$$

$$= \frac{\begin{pmatrix} \text{obs. number in} \\ \text{categ. of 1st var.} \end{pmatrix} \cdot \begin{pmatrix} \text{obs. number in} \\ \text{categ. of 2nd var.} \end{pmatrix}}{\text{sample size}}$$

Suppose that the observed counts are displayed in a rectangular table in which rows correspond to the different categories of the first variable and columns to the categories of the second variable. Then the numerator in the preceding expression for estimated expected counts is just the product of the row and column marginal totals. This is exactly how estimated expected counts were computed in the test for homogeneity of several populations.

X^2 Test for Independence

Null hypothesis: H_0: The two variables are independent.

Alternative hypothesis: H_a: The two variables are not independent.

Test statistic: $X^2 = \sum_{\text{all cells}} \dfrac{(\text{observed cell count} - \text{expected cell count})^2}{\text{expected cell count}}$

The expected cell counts are estimated (assuming H_0 is true) by the formula

$$\text{expected cell count} = \frac{(\text{row marginal total})(\text{column marginal total})}{\text{grand total}}$$

P-values: When H_0 is true and the assumptions of the X^2 test are satisfied, X^2 has approximately a chi-squared distribution with

   df = (number of rows − 1)(number of columns − 1)

The *P*-value associated with the computed test statistic value is the area to the right of X^2 under the chi-squared curve with the appropriate df. Upper-tail areas for chi-squared distributions are found in Appendix Table IX.

Assumptions: 1. The observed counts are from a *random sample*.
   2. The *sample size is large*: All expected counts are at least 5. If some expected counts are less than 5, rows or columns of the table should be combined to achieve a table with satisfactory expected counts.

EXAMPLE 12.7

The article "Factors Associated with Sexual Risk-Taking Behaviors Among Adolescents" (*J. Marriage and Family* (1994): 622–632) examined the relationship between gender and contraceptive use by sexually active teens. Each person in a random sample of sexually active teens was classified according to gender and contraceptive use (with three categories: rarely or never use, use sometimes or most of the time, and always use), resulting in a 3 × 2 table. Data consistent with percentages given in the article is in Table 12.6.

Table 12.6 *Observed counts for Example 12.7*

	Gender		
Contraceptive Use	**Female**	**Male**	**Row Marg. Total**
Rarely/Never	210	350	560
Sometimes/Most Times	190	320	510
Always	400	530	930
Col. Marg. Total	800	1200	2000

The authors were interested in determining whether there is an association between gender and contraceptive use. Using a .05 significance level, we will test

H_0: Gender and contraceptive use are independent.

H_a: Gender and contraceptive use are not independent.

Significance level: $\alpha = .05$

Test statistic: $X^2 = \sum_{\text{all cells}} \dfrac{(\text{observed cell count} - \text{expected cell count})^2}{\text{expected cell count}}$

Assumptions: Before we can check the assumptions we must first compute the expected cell counts.

Row	Cell Column	Expected Cell Count
1	1	$\dfrac{(560)(800)}{2000} = 224.0$
1	2	$\dfrac{(560)(1200)}{2000} = 336.0$
2	1	$\dfrac{(510)(800)}{2000} = 204.0$
2	2	$\dfrac{(510)(1200)}{2000} = 306.0$
3	1	$\dfrac{(930)(800)}{2000} = 372.0$
3	2	$\dfrac{(930)(1200)}{2000} = 558.0$

All expected cell counts are greater than 5 and the observed cell counts are based on a random sample, so the X^2 test can be used. The observed and expected counts are given together in Table 12.7.

Table 12.7 *Observed and expected counts for Example 12.7*

Contraceptive Use	Gender		Row Marg. Total
	Female	Male	
Rarely/Never	210 (224)	350 (336)	560
Sometimes/Most Times	190 (204)	320 (306)	510
Always	400 (372)	530 (558)	930
Col. Marg. Total	800	1200	2000

Calculation: $X^2 = \dfrac{(210 - 224)^2}{224} + \cdots + \dfrac{(530 - 558)^2}{558} = 6.572$

P-value: The table has 3 rows and 2 columns, so df $= (3 - 1)(2 - 1) = 2$.
The entry closest to 6.572 in the 2-df column of Appendix Table IX is 6.70, so
the approximate *P*-value for this test is

   P-value $\approx .035$

Conclusion: Since *P*-value $\leq \alpha$, we reject H_0 and conclude that there is an
association between gender and contraceptive use.

EXAMPLE 12.8

The accompanying two-way table (Table 12.8) was constructed using data in the
article "Television Viewing and Physical Fitness in Adults" (*Research Quarterly for
Exercise and Sport* (1990): 315–320). The author hoped to determine whether time
spent watching television is associated with cardiovascular fitness. Subjects were
asked about their television-viewing time (per day, rounded to the nearest hour) and
were classified as physically fit if they scored in the excellent or very good category
on a step test. Expected cell counts (computed under the assumption of no association between TV viewing and fitness) appear in parentheses in the table.

Table 12.8 *Observed and expected counts for Example 12.8*

TV Group	Fitness		
	Physically Fit	**Not Physically Fit**	
0	35 (25.5)	147 (156.5)	182
1–2	101 (102.2)	629 (627.8)	730
3–4	28 (35.0)	222 (215.0)	250
5 or More	4 (5.3)	34 (32.7)	38
	168	1032	1200

The X^2 test with a significance level of .01 will be used to test the relevant
hypotheses:

   H_0:   Fitness and TV viewing are independent.

   H_a: Fitness and TV viewing are not independent.

Significance level: $\alpha = .01$

Test statistic: $X^2 = \displaystyle\sum_{\text{all cells}} \dfrac{(\text{observed cell count} - \text{expected cell count})^2}{\text{expected cell count}}$

Assumptions: All expected cell counts are at least 5. Assuming that the data can be viewed as a random sample of adults, the X^2 test can be used.

Calculation: $X^2 = \dfrac{(35 - 25.5)^2}{25.5} + \cdots + \dfrac{(34 - 32.7)^2}{32.7} = 6.13$

P-value: The table has 4 rows and 2 columns, so df = $(4 - 1)(2 - 1) = 3$. The computed value of $X^2 = 6.13$ is smaller than 6.25, the smallest entry in the 3-df column of Appendix Table IX. Therefore,

$\qquad$ *P*-value $> .10$

Conclusion: Since *P*-value $> \alpha$, H_0 is not rejected. There is not sufficient evidence to conclude that an association exists between TV viewing and fitness.

Most statistical computer packages will calculate expected cell counts, the value of the X^2 statistic, and the associated *P*-value. MINITAB output for the data of Example 12.8 is given in Figure 12.3. The discrepancy between $X^2 = 6.13$ and $X^2 = 6.161$ is due to rounding in the hand calculations. Note also that MINITAB gives *P*-value $= .104$. This is consistent with the statement *P*-value $> .10$ from Example 12.8.

FIGURE 12.3 MINITAB output for Example 12.8

Chi-Square Test
Expected counts are printed below observed counts

	Fit	Not fit	Total
1	35	147	182
	25.48	156.52	
2	101	629	730
	102.20	627.80	
3	28	222	250
	35.00	215.00	
4	4	34	38
	5.32	32.68	
Total	168	1032	1200

Chi-Sq = 3.557 + 0.579 +
$\qquad$ 0.014 + 0.002 +
$\qquad$ 1.400 + 0.228 +
$\qquad$ 0.328 + 0.053 = 6.161
DF = 3, P-Value = 0.104

In some investigations, values of more than two categorical variables are recorded for each individual in the sample. For example, in addition to the variables *television-viewing time* and *fitness level,* the researchers in the study referenced in Example 12.8 might also have recorded age group (with categories under 21, 21–30, 31–40, and so on) for each subject. A number of interesting questions could then be explored: Are all three variables independent of one another? Is it possible that age and fitness level are dependent but that the relationship between them does not depend on TV-viewing time? For a particular age group, are TV-viewing time and fitness independent?

The X^2 test procedure described in this section for analysis of bivariate categorical data can be extended for use with *multivariate categorical data*. Appropriate hypothesis tests can then be used to provide insight into the relationships between variables. However, the computations required to calculate estimated expected cell counts and to compute the value of X^2 are quite tedious, so they are seldom done without the aid of a computer. Several statistical computer packages (including BMDP and SAS) can perform this type of analysis. Consult the chapter references for further information on the analysis of categorical data.

Exercises 12.15 – 12.33

12.15 A particular state university system has six campuses. On each campus, a random sample of students will be selected, and each student will be categorized with respect to political philosophy as liberal, moderate, or conservative. The null hypothesis of interest is that the proportion of students falling in these three categories is the same at all six campuses.

a. On how many degrees of freedom will the resulting X^2 test be based?

b. How does your answer in part **a** change if there are seven campuses rather than six?

c. How does your answer in part **a** change if there are four rather than three categories for political philosophy?

12.16 A random sample of 1000 registered voters in a certain county is selected, and each voter is categorized with respect to both educational level (four categories) and preferred candidate in an upcoming supervisorial election (five possibilities). The hypothesis of interest is that educational level and preferred candidate are independent factors.

a. If $X^2 = 7.2$, what would you conclude at significance level .10?

b. If there were only four candidates vying for election, what would you conclude if $X^2 = 14.5$ and $\alpha = .05$?

12.17 What factors influence people to respond to mail surveys? Researchers have found that the appearance of the survey and the nature of the first question have an influence on the overall response rate, but do they affect the speed of response? The article "The Impact of Cover Design and First Questions on Responses for a Mail Survey of Skydivers" (*Leisure Science* (1991): 67–76) presents the results of an experiment to investigate rate of response. Surveys with a plain cover were mailed to a random sample of 427 skydivers. Surveys with a cover graphic of a skydiver were mailed to a separate random sample of 414 skydivers. The postmark on the returned survey was used to classify each person who received a survey in one of the following response categories: returned in 1–7 days, returned in 8–14 days, returned in 15–31 days, returned in 32–60 days, and not returned. The resulting data is summarized in the two-way table below. Does the data support the theory that the proportions falling into the various response categories are not the same for the two cover designs? Use a significance level of .05 to test the relevant hypotheses.

12.18 The article "Heavy Drinking and Problems Among Wine Drinkers" (*Journal of Studies on Alcohol* (1999): 467–471) analyzed drinking problems among Canadians. The article gave the percentage of drinkers

Data for Exercise 12.17

Cover Design	Response Category					
	1–7	8–14	15–31	32–60	Not Returned	
Graphic	70	76	51	19	198	414
Plain	84	53	51	32	207	427
	154	129	102	51	405	841

who never drank to intoxication for seven different types of drinkers. From the information given in the article, we can obtain the following table:

	Never Intoxicated	Sometimes Intoxicated
Beer Only	515	741
Wine Only	875	232
Spirits Only	387	372
Beer and Wine	774	560
Beer and Spirits	312	727
Wine and Spirits	708	349
Beer, Wine, and Spirits	1032	1119

Assume that each of the seven samples studied can be viewed as a random sample from the respective group.

a. Is there sufficient evidence to conclude that the proportion who never become intoxicated is not the same for the seven groups?

b. Is the test that you performed in part **a** a test of independence or a test of homogeneity?

12.19 Do women have different patterns of work behavior than men? The article "Workaholism in Organizations: Gender Differences" (*Sex Roles: A Journal of Research* (1999): 333–346) attempts to answer this question. Each person in a random sample of 423 graduates of a business school in Canada were polled and classified by gender and workaholism type, resulting in the accompanying table:

	Gender	
Workaholism Types	Female	Male
Work Enthusiasts	20	41
Workaholics	32	37
Enthusiastic Workaholics	34	46
Unengaged Workers	43	52
Relaxed Workers	24	27
Disenchanted Workers	37	30

a. Test the hypothesis that gender and workaholism type are independent.

b. The author writes "women and men fell into each of the six workaholism types to a similar degree." Does the outcome of the test you performed in part **a** support this conclusion? Explain.

12.20 The following table is based on data reported in "Light-to-moderate Alcohol Consumption and Risk of Stroke Among U.S. Male Physicians" (*New England J. of Medicine* (1999): 1557–1564). The table is based on 21,870 male physicians age 40–84 who are participating in the Physicians Health Study.

	Alcohol Consumption (No. of Drinks)				
	<1/wk	1/wk	2–4/wk	5–6/wk	1/day
Never Smoked	3577	1711	2430	1211	1910
Smoked in the Past	1595	1068	1999	1264	2683
Currently Smokes	524	289	470	296	849

a. Calculate row percentages by dividing each observed count by the corresponding row total. Are the proportions falling in each of the alcohol consumption categories similar for the three smoking categories?

b. Test the hypothesis that smoking status and alcohol consumption are independent.

c. Is the result of your test in part **b** consistent with what you expected based on your answer to part **a**? If not, explain why your initial impression based on the percentages may not have been accurate? What aspect of the data set factors into your explanation?

12.21 The article "Motion Sickness in Public Road Transport: The Effect of Driver, Route and Vehicle" (*Ergonomics* (1999): 1646–1664) reported that seat position within a bus may have some effect on whether one experiences motion sickness. The accompanying table classifies each person in a random sample of bus riders by the location of his or her seat and whether nausea was reported.

	Nausea	No Nausea
Front	58	870
Middle	166	1163
Rear	193	806

Based on this data, can you conclude that there is an association between seat location and nausea? Test the relevant hypotheses using $\alpha = .05$.

12.22 The article "Mortality Among Recent Purchasers of Handguns" (*New England J. of Medicine* (1999): 1583–1589) examined the relationship between handgun purchases and cause of death. Suppose that a random sample of 4000 death records was examined and cause of death was noted for each individual in the sample. Gun registration records were also examined to determine which of these 4000 individuals had purchased a handgun within the previous year. Data consistent with summary values given in the article is shown in the accompanying table.

	Suicide	Not Suicide
Handgun Purchased	4	12
No Handgun Purchased	63	3921

Is it reasonable to use the chi-squared test to determine whether there is an association between handgun purchase within the year prior to death and whether the death was a suicide? Justify your answer.

12.23 When charities solicit donations, they must know what approach will encourage people to donate. Would a picture of a needy child tug at the heartstrings and convince someone to contribute to the charity? As reported in the article "The Effect of Photographs of the Handicapped on Donation" (*J. of Applied Social Psychology* (1979): 426–431), experimenters went door-to-door with four types of displays: a picture of a smiling child, a picture of an unsmiling child, a verbal message, and an identification of the charity only. At each door, they would randomly select which display to present. Suppose 18 of 30 subjects contributed when shown the picture of the smiling child, 14 of 30 contributed when shown the picture of an unsmiling child, 16 of 26 contributed when shown the verbal message, and 18 of 22 contributed when shown only the identification of the charity. What conclusions can be drawn from this data? Explain your procedure and reasoning.

12.24 The accompanying table is based on data given in the article "The Association Between Smoking and Unhealthy Behaviors Among a National Sample of Mexican-American Adolescents" (*J. of School Health* (1998): 376–379). Each boy in a sample of Mexican-American males age 10–18 was classified according to smoking status and response to a question asking whether he liked to do risky things. Assume that it is reasonable to regard the sample as a random sample of Mexican-American male adolescents.

	Smoking Status	
	Smoker	**Nonsmoker**
Likes Risky Things	45	46
Doesn't Like Risky Things	36	153

a. Is there sufficient evidence to conclude that there is an association between smoking status and desire to do risky things? Test the relevant hypotheses using $\alpha = .05$.
b. Based on your conclusion in part **a,** is it reasonable to conclude that smoking *causes* an increase in the desire to do risky things? Explain.

12.25 The article "Cooperative Hunting in Lions: The Role of the Individual" (*Behavioral Ecology and Sociobiology* (1992): 445–454) discusses the different roles taken by lionesses as they attack and capture prey. The authors were interested in the effect of the position in line as stalking occurs; an individual lioness may be in the center of the line or on the wing (end of the line) as they advance toward their prey. In addition to position, the role of the lioness was also considered. A lioness could initiate a chase (be the first one to charge the prey), or she could participate and join the chase after it has been initiated. Data from the article is summarized in the accompanying table.

	Role	
Position	**Initiate Chase**	**Participate in Chase**
Center	28	48
Wing	66	41

Is there evidence of an association between position and role? Test the relevant hypotheses using $\alpha = .01$. What assumptions about how the data was collected must be true for the chi-squared test to be an appropriate way to analyze this data?

12.26 The authors of the article "A Survey of Parent Attitudes and Practices Regarding Under Age Drinking" (*J. Youth and Adolescence* (1995): 315–334) conducted a telephone survey of parents with preteen and teenage children. One of the questions asked was "How effective do you think you are in talking to your children about drinking?" Responses are summarized in the accompanying 3×2 table. Using a significance level of .05, carry out a test to determine whether there is an association between age of children and parental response.

	Age of Children	
Response	**Preteen**	**Teen**
Very Effective	126	149
Somewhat Effective	44	41
Not at All Effective or Don't Know	51	26

12.27 The article "Regional Differences in Attitudes Toward Corporal Punishment" (*J. Marriage and Family* (1994): 314–324) presents data resulting from a random sample of 978 adults. Each individual in the sample was asked whether he or she agreed with the following statement: "Sometimes it is necessary to discipline a child with a good, hard spanking." Respondents were also classified according to the region of the United States in which they lived. The resulting data is summarized in the accompanying table. Is there an association between response (agree, disagree) and region of residence? Use $\alpha = .01$.

Response

Region	Agree	Disagree
Northeast	130	59
West	146	42
Midwest	211	52
South	291	47

12.28 A story describing a date rape was read by 352 high school students. To investigate the effect of the victim's clothing on subject's judgment of the situation described, the story was accompanied by either a photograph of the victim dressed provocatively, a photo of the victim dressed conservatively, or no picture. Each student was asked whether the situation described in the story was one of rape. Data from the article "The Influence of Victim's Attire on Adolescent Judgements of Date Rape" (*Adolescence* (1995): 319–323) is given in the accompanying table. Is there evidence that the proportion who believe that the story described a rape differs for the three different photo groups? Test the relevant hypotheses using $\alpha = .01$.

Picture

Response	Provocative	Conservative	No Picture
Rape	80	104	92
Not Rape	47	12	17

12.29 Can people tell the difference between a female nose and a male nose? This important (?) research question was examined in the article "You Can Tell by the Nose: Judging Sex from an Isolated Facial Feature" (*Perception* (1995): 969–973). Eight Caucasian males and eight Caucasian females posed for nose photos. The article states that none of the volunteers wore nose studs or had prominent nasal hair. Each person placed a black Lycra tube over his or her head in such a way that only the nose protruded through a hole in the material. Photos were then taken from three different angles: front view, three-quarter view, and profile. These photos were shown to a sample of undergraduate students. Each student in the sample was shown one of the nose photos and asked whether it was a photo of a male or a female; and the response was classified as either correct or incorrect. The accompanying table was constructed using summary val-

ues reported in the article. Is there evidence that the proportion of correct sex identifications differs for the three different nose views?

View

Sex ID	Front	Profile	Three-quarter
Correct	23	26	29
Incorrect	17	14	11

12.30 The accompanying table appeared in the article "Europe's Receptivity to New Religious Movements: Round Two" (*J. for the Scientific Study of Religion* (1993): 389–397). Suppose that the percentages in the table were based on random samples of size 200 from each of the six countries.

Country	Percent Who Believe in Fortune-Tellers
Great Britain	42
West Germany	32
East Germany	22
Slovenia	55
Ireland	27
Northern Ireland	33

a. Use the given percentages to construct a two-way table with rows corresponding to the six countries and columns corresponding to the categories "believe in fortune-tellers" and "don't believe in fortune-tellers." Enter the observed counts in the table. (*Hint:* The observed counts for Great Britain are $(.42)(200) = 84$ for the "believe" column and $(.58)(200) = 116$ for the "don't believe" column.)

b. Is there evidence that the proportion who believe in fortune-tellers is not the same for all six countries? Test the relevant hypotheses using $\alpha = .01$.

12.31 The two-way frequency table at the top of page 626 appeared in the article "Marijuana Use in College" (*Youth and Society* (1979): 323–334). Four hundred forty-five college students were classified according to both frequency of marijuana use and parental use of alcohol and psychoactive drugs. Use the accompanying SAS computer output to carry out a X^2 test with a significance level of .01 to determine whether there is an association between marijuana use and parental use of drugs and alcohol.

Data for Exercise 12.31

Parental Use of Alcohol and Drugs	Student Level of Marijuana Use			Row Marg. Total
	Never	Occasional	Regular	
Neither	141	54	40	235
One	68	44	51	163
Both	17	11	19	47
Col. Marg. Total	226	109	110	445

TABLE OF PRNT_USE BY STUD_USE

PRINT_USE   STUD_USE

Frequency
Percent
Row Pct

Col Pct	never	occasion	regular	Total
neither	141	54	40	235
	31.69	12.13	8.99	52.81
	60.00	22.98	17.02	
	62.39	49.54	36.36	
one	68	44	51	163
	15.28	9.89	11.46	36.63
	41.72	26.99	31.29	
	30.09	40.37	46.36	
both	17	11	19	47
	3.82	2.47	4.27	10.56
	36.17	23.40	40.43	
	7.52	10.09	17.27	
Total	226	109	110	446
	50.79	24.49	24.72	100.00

STATISTICS FOR TABLE OF PRINT_USE BY STUD_USE

Statistic	DF	Value	Prob
Chi-Square	4	22.373	0.000

12.32 Jail inmates can be classified into one of the following four categories according to the type of crime committed: violent crime, crime against property, drug offenses, and public-order offenses. Suppose that random samples of 500 male inmates and 500 female inmates are selected, and each inmate is classified according to type of offense. The data in the accompanying table is based on summary values given in the article "Profile of Jail Inmates" (*USA Today*, April 25, 1991). We would like to know whether male and female inmates differ with respect to type of offense.

Type of Crime	Gender	
	Male	Female
Violent	117	66
Property	150	160
Drug	109	168
Public-Order	124	106

a. Is this a test of homogeneity or a test of independence?

b. Test the relevant hypotheses using a significance level of .05.

12.33 Job satisfaction of professionals was examined in the article "Psychology of the Scientist: Work Related Attitudes of U.S. Scientists" (*Psychological Reports* (1991): 443–450). Each person in a random sample of 778 teachers was classified according to a job satisfaction variable and also by teaching level, resulting in the accompanying two-way table. Can we conclude that there is an association between job satisfaction and teaching level? Test the relevant hypotheses using $\alpha = .05$.

Teaching Level	Job Satisfaction	
	Satisfied	Unsatisfied
College	74	43
High School	224	171
Elementary	126	140

12.3   Interpreting the Results of Statistical Analyses

Many studies, particularly in the social sciences, result in categorical data. The questions of interest in such studies often lead to an analysis that involves using a chi-squared test. As with the other hypothesis tests considered, it is common to find the result of a chi-squared test summarized by giving the value of the chi-squared test statistic and an associated *P*-value. Because categorical data can be summarized compactly in frequency tables, the data often is given in the article (unlike data for numerical variables, which is rarely given).

What to Look for in Published Data

Here are some questions to consider when you are reading an article that contains the results of a chi-squared test:

- Are the variables of interest categorical rather than numerical?
- Is the data given in the article in the form of a frequency table?
- If a two-way frequency table is involved, is the question of interest one of homogeneity or one of independence?
- What null hypothesis is being tested? Are the results of the analysis reported in the correct context (homogeneity, etc.)?
- Is the sample size large enough to make use of a chi-squared test value? (Are all expected counts at least 5?)
- What is the value of the test statistic? Is the associated *P*-value given? Should the null hypothesis tested be rejected?
- Are the conclusions drawn by the authors consistent with the results of the test?
- How different are the observed and expected counts? Does the result have practical as well as statistical significance?

The authors of the article "Predicting Professional Sports Game Outcomes from Intermediate Game Scores" (*Chance* (1992): 18–22) used a chi-squared test to determine whether there was any merit to the idea that basketball games are not settled until the last quarter, whereas baseball games are over by the seventh inning. They also considered football and hockey. Data was collected for 189 basketball games, 92 baseball games, 80 hockey games, and 93 football games. The games analyzed were sampled randomly from all games played during the 1990 season for baseball and football and for the 1990–1991 season for basketball and hockey. For each game, the late-game leader was determined, and then it was noted whether the late-game leader actually ended up winning the game. The resulting data is summarized in the accompanying table.

Sport	Late-game Leader Wins	Late-game Leader Loses
Basketball	150	39
Baseball	86	6
Hockey	65	15
Football	72	21

The authors state

> *Late-game leader* is defined as the team that is ahead after three quarters in basketball and football, two periods in hockey, and seven innings in baseball. The chi-squared value (with three degrees of freedom) is 10.52 ($P < .015$).

They also conclude that

> the sports of basketball, hockey, and football have remarkably similar percentages of late-game reversals, ranging from 18.8% to 22.6%. The sport that is an anomaly is baseball. Only 6.5% of baseball games resulted in late reversals. . . . [The chi-squared test] is statistically significant due almost entirely to baseball.

In this particular analysis, the authors are comparing four populations (games from each of the four sports) on the basis of a categorical variable with two categories (late-game leader wins and late-game leader loses). The appropriate null hypothesis is then

H_0:   The true proportion in each category (leader wins, leader loses) is the same for all four sports.

Based on the reported value of the chi-squared statistics and the associated *P*-value, this null hypothesis is rejected, leading to the conclusion that the category proportions are not the same for all four sports.

The validity of the chi-squared test requires that the sample sizes be large enough so that no expected counts are less than 5. Is this reasonable here? The accompanying MINITAB output shows the expected cell counts and the computation of the X^2 statistic.

Chi-Square Test

Expected counts are printed below observed counts

	leader w	leader l	Total
1	150	39	189
	155.28	33.72	
2	86	6	92
	75.59	16.41	
3	65	15	80
	65.73	14.27	
4	72	21	93
	76.41	16.59	
Total	373	81	454

Chi-Sq = 0.180 + 0.827 +
         1.435 + 6.607 +
         0.008 + 0.037 +
         0.254 + 1.171 = 10.518
DF = 3, P-Value = 0.015

The smallest expected count is 14.27, so the sample sizes are large enough to justify the use of the X^2 test. Note also that the two cells in the table that correspond to baseball contribute a total of $1.435 + 6.607 = 8.042$ to the value of the X^2 statistic of 10.518. This is due to the large discrepancies between the observed and expected counts for these two cells. There is reasonable agreement between the observed and the expected counts in the other cells. This is probably the basis for the authors' conclusion that baseball is the only anomaly and that the other sports were similar.

Summary of Key Concepts and Formulas

Term or Formula	Comment
One-way frequency table	A compact way of summarizing data on a categorical variable; it gives the number of times each of the possible categories in the data set occurs (the frequencies).
Goodness-of-fit statistic, $$X^2 = \sum_{\text{all cells}} \frac{\left(\begin{array}{c}\text{observed} \quad\ \text{expected} \\ \text{cell count} - \text{cell count}\end{array}\right)^2}{\text{expected cell count}}$$	A statistic used to provide a comparison between observed counts and those expected when a given hypothesis is true. When none of the expected counts are too small, X^2 has approximately a chi-squared distribution.
X^2 test in a one-way frequency table: H_0: π_1 = hypothesized proportion for category 1 $\vdots$ π_k = hypothesized proportion for category k	A hypothesis test performed to determine whether the true category proportions are different from those specified by the given null hypothesis.
Two-way frequency table (contingency table)	A rectangular table used to summarize a bivariate categorical data set; two-way tables are used to compare several populations on the basis of a categorical variable or to identify whether an association exists between two categorical variables.
X^2 test for comparing two or more populations: H_0: The true category proportions are the same for all the populations.	The hypothesis test performed to determine whether the true category proportions are the same for all of the populations to be compared.
X^2 test for independence: H_0: The two variables defining the table are independent.	The hypothesis test performed to determine whether an association exists between two categorical variables.

Supplementary Exercises 12.34 – 12.42

12.34   In a study of 2989 cancer deaths, the location of death (home, acute-care hospital, or chronic-care facility) and age at death were recorded, resulting in the accompanying two-way frequency table ("Where Cancer Patients Die," *Public Health Reports* (1983): 173). Using a .01 significance level, test the null hypothesis that age at death and location of death are independent.

	Location		
Age	**Home**	**Acute-care**	**Chronic-care**
15–54	94	418	23
55–64	116	524	34
65–74	156	581	109
Over 74	138	558	238

12.35 Each driver in a sample of size 1024 was classified according to both seat-belt usage and gender to obtain the accompanying 2 × 2 frequency table ("What Kinds of People Do Not Use Seat Belts," *Amer. J. of Public Health* (1977): 1043–1049). Does the data strongly suggest an association between gender and seat-belt usage? Use a .05 significance level.

Seat-belt Usage

Gender	Don't Use	Use
Male	192	272
Female	284	276

12.36 The *Los Angeles Times* (July 29, 1983) conducted a survey to find out why some Californians don't register to vote. Random samples of 100 Latinos, 100 Anglos, and 100 African-Americans who were not registered to vote were selected. The resulting data is summarized in the accompanying table. Does the data suggest that the true proportion falling into each response category is not the same for Latinos, Anglos, and African-Americans? Use a .05 significance level.

Race

Reason For Not Registering	Latino	Anglo	African-American
Not a Citizen	45	8	0
Not Interested	19	33	19
Can't Meet Residency Requirements	9	35	23
Distrust of Politics	5	10	8
Too Difficult to Register	10	10	27
Other Reason	12	4	23

12.37 The relative importance attached to work and home life by high school students was examined in "Work Role Salience as a Determinant of Career Maturity in High School Students" (*J. Vocational Behavior* (1984): 30–44). Does the data summarized in the accompanying two-way frequency table suggest that gender and relative importance assigned to work and home are not independent? Test using a .05 level of significance.

Relative Importance

Gender	Work > Home	Work = Home	Work < Home
Female	68	26	94
Male	75	19	57

12.38 The accompanying 2 × 2 frequency table is the result of classifying random samples of 112 librarians and 108 faculty members of the California State University system with respect to gender ("Job Satisfaction Among

Faculty and Librarians: A Study of Gender, Autonomy, and Decision Making Opportunities," *J. of Library Admin.* (1984): 43–56). Does the data strongly suggest that librarians and faculty members differ with respect to the proportion of males and females? Test using a .05 level of significance.

	Faculty	Librarians
Male	56	59
Female	52	53

12.39 Is there any relationship between the age of an investor and the rate of return that the investor expects from an investment? A random sample of 972 common stock investors was selected, and each was placed in one of four age categories and in one of four categories according to rate believed attainable ("Patterns of Investment Strategy and Behavior Among Individual Investors," *J. of Business* (1977): 296–333). The resulting data is given in the accompanying table. Does there appear to be an association between age and rate believed attainable? Test the appropriate hypotheses using a .01 significance level.

Rate Believed Attainable

Investor Age	0–5%	6–10%	11–15%	Over 15%
Under 45	15	51	51	29
45–54	31	133	70	48
55–64	59	139	35	20
65 and Over	84	157	32	18

12.40 The article "Participation of Senior Citizens in the Swine Flu Inoculation Program" (*J. of Gerontology* (1979): 201–208) described a study of the factors thought to influence a person's decision to obtain a flu vaccination. Each member of a sample of 122 senior citizens was classified according to belief about the likelihood of getting the flu and vaccine status to obtain the two-way frequency table shown here. Using a .05 significance level, test to determine whether there is an association between belief and vaccine status.

Vaccine Status

Belief	Received Vaccine	Didn't Receive Vaccine
Very Unlikely	25	24
Unlikely	30	11
Likely	6	8
Don't Know	5	13

12.41 The article "An Instant Shot of 'Ah': Cocaine Use Among Methadone Clients" (*J. of Psychoactive Drugs*

(1984): 217–227) reported the accompanying data on frequency of cocaine use for individuals in three different treatment groups. Does the data suggest that the true proportion of individuals in each of the different cocaine-use categories differs for the three treatments? Carry out an appropriate test at level .05.

Treatment

Cocaine Use	A	B	C
None	149	75	8
1–2 Times	26	27	15
3–6 Times	6	20	11
At Least 7 Times	4	10	10

References

Agresti, Alan, and B. Finlay. *Statistical Methods for the Social Sciences,* 3rd ed. New Jersey: Prentice-Hall, 1997. (This book includes a good discussion of measures of association for two-way frequency tables.)

Everitt, B. S. *The Analysis of Contingency Tables.* New York: Halsted Press, 1977. (A compact but informative survey of methods for analyzing categorical data.)

Mosteller, Frederick, and Robert Rourke. *Sturdy Statistics.* Reading, Mass.: Addison-Wesley, 1973. (Contains several very readable chapters on the varied uses of the chi-squared statistic.)

12.42   The article "Identification of Cola Beverages" (*J. of Appl. Psych.* (1962): 356–360) reported on an experiment in which each of 79 subjects was presented with glasses of cola in pairs and asked to identify which glass contained a specific brand of cola. The accompanying data appeared in the article. Does this data suggest that individuals' abilities to make correct identification differ for the different brands of cola?

Number of Correct Identifications

Cola	0	1	2	3 or 4
Coca-Cola	13	23	24	19
Pepsi Cola	12	20	26	21
Royal Crown	18	28	19	14

GRAPHING CALCULATOR EXPLORATIONS

12.1 The Goodness-of-Fit Test

As computers have seemingly taken over the world, many strange new words have been added to our collective vocabulary. One of these "words" is WYSIWYG, an acronym for "What you see is what you get." Goodness-of-fit tests on your calculator are an example of this idea, in that the table of observed and expected counts corresponds very closely to the Lists on your calculator. Because of this match, the calculator commands will also appear very similar to the chi-squared formula in a goodness-of-fit test, making the calculator procedure very easy to remember.

We will use Example 12.2, the patient admission data. The goodness-of-fit procedure, as it develops, will have two parts: First, we will calculate the value of X^2, then we will evaluate its statistical significance. We begin by entering the observed counts for the categories in List1. For your reference, observed counts follow.

Category	Within 7 Days	Between 8 and 30 Days	Between 31 and 90 Days	More Than 90 Days
Observed count	11	24	69	96

Now we have to compute the expected counts. It seems natural to calculate the expected counts as was done in the text, and then enter them in List2. However, an alternate approach is often easier. Since expected counts usually follow from the hypothesized category proportions, you can enter the proportions in List2 and use them to calculate the expected values. Putting the hypothesized proportions in List2, we have

$$15/365 \rightarrow \text{List2}(1)$$
$$46/365 \rightarrow \text{List2}(2)$$
$$120/365 \rightarrow \text{List2}(3)$$
$$184/365 \rightarrow \text{List2}(4)$$

At this point, your data edit screen should look something like the accompanying figure.

The expected values are calculated by multiplying the proportions in List2 by the total number of observations. You can find the total number of observations by using the statistics capabilities of the calculator to find the sum of the numbers in List1. You can get lots of statistics, such as the mean, sum, and standard deviation, by pressing the "stat" button. On some calculators, it is possible to get the sum alone by keying in something like "sum(List1)." If your calculator has a "sum" list function, key in

Sum(List1) * List2 $\rightarrow$ List3.

If not, you can find the sum (200 in this example) and then key in

200 * List2 $\rightarrow$ List3.

You should now see three lists, as shown, on your calculator screen.

Note that, except for differences due to rounding, the numbers in List3 are the expected values from the original problem in the text. Now we are in a position to actually calculate X^2. The formula for X^2 is a sum of quantities of the form

$$\frac{(\text{observed cell count} - \text{expected cell count})^2}{(\text{expected cell count})}$$

On your calculator, these quantities can be calculated all at once and then stored in List4:

(List1 − List3)^2/List3 → List4.

The value of the X^2 statistic is then found by finding the sum of the numbers in List4. The calculator we are using has a list function called sum, so we enter "sum(List4)" and the calculator returns: 1.389411232. The X^2 from the calculator is slightly different from the calculations in the text; this is due to rounding.

Now that you have calculated the value of the X^2 statistic, you can proceed to assess its significance. Since the goodness-of-fit test is a one-tailed test, we must find the area to the right of $X^2 = 1.389411232$ under the chi-squared curve with 3 degrees of freedom. At this point it is possible to be misled by the menu system of your calculator. There may be a "Chi square test" on your calculator, but it will *not* be the test for goodness-of-fit. (The "Chi square test" on the calculator is the test for association or homogeneity in a two-way table and will be the topic of the next Graphing Calculator Exploration.)

To find the *P*-value on your calculator, you must navigate your menu system to a function that returns cumulative probabilities for X^2 values — most likely called a "Chisqcdf" or "χ^2 cdf." The two parameters necessary to find the *P*-value are the X^2 statistic value and the number of degrees of freedom. Some calculators, as we saw with the normal distribution calculator exploration, use the "cdf" loosely; they actually find the probability of a statistic being *between* two values. In the case of the chi-squared distribution, this is not too much of a problem because the X^2 statistic cannot be less than zero. If your calculator is one that uses the "cdf" loosely, you should be able to find the cumulative area by keying in (using the current problem numbers) something like

χ^2 cdf(0, 1.389411232, 3)

The calculator should return .291981336. The actual *P*-value, remember, will be the area to the right of X^2, 1 − .291981336 = .708018664. You can then round off this value to some reasonable number of decimal places.

In the case of the chi-squared calculations, your calculator may differ in its syntax from what we have written here. Your calculator manual will once again come to the rescue if there are differences.

12.2 Homogeneity, Independence, and the Chi-Squared Test

In Graphing Calculator Exploration 12.1, we discussed the use of the chi-squared statistic to assess the "goodness of fit" of data to a model. Using the chi-squared statistic for the assessment of homogeneity of proportions and the independence of two categorical variables requires a bit more discussion. The calculator procedures are more complicated than the goodness-of-fit procedure because there is more than one row of data to consider. However, once the data is entered into the calculator, the calculation of the expected values is an automatic feature built into the calculator — a great time saver!

There are three steps in performing a chi-squared test for independence or homogeneity:

1. Prepare to enter the data.
2. Enter the data.
3. Perform the chi-squared calulations.

Preparing to enter the data sounds fairly easy, possibly something like (a) pull up a chair, (b) sharpen your pencil, (c) adjust the lighting. Actually, we're not referring to your work environment, but preparation of the *calculator* for data entry for chi squared! In past Explorations, you have entered data at the cursor into individual places on the screen, and sets of data into lists. The chi-squared data entry will be a little different, because you will enter data into a "matrix." Be careful at this point! Your calculator may have another possibility for data entry, that being "tables." Look in your manual under *matrix*. Matrices (plural of matrix) are different from both lists and tables.

The details for working with matrices on calculators will differ from machine to machine, but generally you must be able to (a) create or redefine a matrix so that it has the same number of rows and columns as the table you are analyzing, and (b) enter numbers into that matrix. Your calculator will already have some predefined matrices, and these are usually labeled "A," "B," etc. Some calculators may have a "matrix" or "matr" key, or you may have to navigate your menu system to get to a menu item with these terms. If you are following along with your calculator, your goal is to navigate to a matrix screen and set up the matrices for data entry. Two common methods for indicating what matrices are already defined are shown in the accompanying figures.

```
Matrix
Mat A  :  2x 3
Mat B  :  2x 2
Mat C  :  None
```

```
Names  Math  Edit
1: [A]   2x3
2: [B]   2x2
3: [C]
4: [D]
5: [E]
6: [F]
7: [G]
```

These screens show that two matrices are defined. The first one, matrix "A," has two rows and three columns; the second, matrix "B," has two rows and two columns. To change, or edit, a matrix, you must move your cursor to that matrix and press Enter. Editing matrices is another place where calculators will differ, and we refer you to your manual for edification. On the calculator we are using, we move the cursor over to "Edit" and press Enter. The calculator then shows this screen:

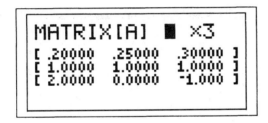

```
MATRIX[A] ■ x3
[ .20000   .25000   .30000 ]
[ 1.0000   1.0000   1.0000 ]
[ 2.0000   0.0000  ‾1.000  ]
```

We will use Example 12.7, data from a survey on contraceptive use among teens. That data, you may notice, uses a 3×2 table. Our plan is to use matrix A for the observed values, but our existing matrix A is a 3×3. We must change this 3×3 matrix into a 3×2 and enter the data from Example 12.7. The matrix then appears as

```
MATRIX[A] 3 x2
[ 210.00   350.00          ]
[ 190.00   320.00          ]
[ 400.00   530.00          ]
```

We also need a matrix for the expected values for the chi-squared procedure. (Some calculators may have a "dedicated" matrix for this purpose. If so, it will be named "Expected" or "MatAns" or something similar.) The calculator we are using requires us to specify a matrix to hold the soon-to-be calculated expected values, so we will resize matrix "B" to a 3×2. After resizing the necessary matrices and entering the data from Example 12.7, you are now ready to proceed with the actual chi-squared calculations. The chi-squared test is a statistical hypothesis test, so your navigation through your calculator's menus should be a familiar one. When you get to the menu item, "χ^2-test," press Enter or Execute as appropriate. You will be asked to indicate which matrix contains the observed values, matrix A in our case. If your calculator does not have a dedicated matrix for the expected values, you must specify that matrix also, in our case matrix B. Then move your cursor to "Calculate" or "Execute" as appropriate.

Your calculator should return a screen that looks something like the accompanying display. The value of the chi-squared statistic, the P-value, and the number of degrees of freedom are all reported. Once again, be sure to compare your numbers with the "correct" answer. If your results differ significantly, enough that round-off error seems unlikely, review your methods and see where you might have made an error. Especially remember to check the original data — incorrect data entry is a common source of errors when working with calculators.

χ^2 Test
$\chi^2 = 6.5721853$
p= .0373997
df= 2.0000000

13

Simple Linear Regression and Correlation: Inferential Methods

Regression and correlation were introduced in Chapter 5 as techniques for describing and summarizing bivariate data consisting of (x, y) pairs. For example, consider a regression of the dependent variable y = average SAT score on the independent variable x = spending per pupil (thousands of dollars) based on a sample of n = 44 New Jersey school districts ("Cost-Effectiveness in Public Education." *Chance* (1995): 38–41). A scatter plot of the data shows a linear pattern; the sample correlation coefficient is r = .4; and the equation of the least squares line is $\hat{y} = 766 + 15.0x$. When x = 10 is substituted into this equation, $\hat{y}$ is calculated to be 916. This number can be interpreted either as a *point estimate* of the average score for all districts that spend $10,000 per pupil or as a *point prediction* for the score in a single district that spends this amount of money per pupil. In this chapter, we develop inferential methods for this type of data, including a confidence interval (interval estimate) for a mean y value, a prediction interval for a single y value, and a test of hypotheses regarding the extent of correlation in the entire population of (x, y) pairs.

13.1   The Simple Linear Regression Model

A *deterministic relationship* is one in which the value of y is completely determined by the value of an independent variable x. Such a relationship can be described using traditional mathematical notation such as y = f(x), where f(x) is a specified function of x. For example, we might have

$$y = f(x) = 10 + 2x$$

or

$$y = f(x) = 4 - (10)^{2x}$$

However, in most situations, the variables of interest are not deterministically related. For example, the value of $y =$ first-year-college grade point average is certainly not determined solely by $x =$ high-school grade point average, and $y =$ crop yield is determined partly by factors other than $x =$ amount of fertilizer used.

A description of the relation between two variables x and y that are not deterministically related can be given by specifying a **probabilistic model.** The general form of an **additive probabilistic model** allows y to be larger or smaller than $f(x)$ by a random amount, e. The **model equation** is of the form

$$y = \text{deterministic function of } x + \text{random deviation}$$
$$= f(x) + e$$

Let x^* denote some particular value of x, and suppose that an observation on y is made when $x = x^*$. Then

$$y > f(x^*) \quad \text{if } e > 0$$
$$y < f(x^*) \quad \text{if } e < 0$$
$$y = f(x^*) \quad \text{if } e = 0$$

Thinking geometrically, if $e > 0$, the observed point (x^*, y) will lie above the graph of $y = f(x)$; $e < 0$ implies that this point will fall below the graph. This is illustrated in Figure 13.1. If $f(x)$ is a function used in a probabilistic model relating y to x, and observations on y are made for various values of x, the resulting (x, y) points will be distributed about the graph of $f(x)$, some falling above it and others below it.

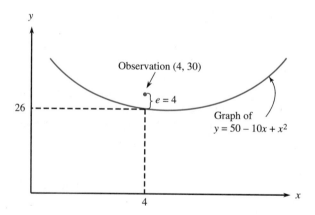

FIGURE 13.1 A deviation from the deterministic part of a probabilistic model

Simple Linear Regression

The simple linear regression model is a special case of the general probabilistic model in which the deterministic function $f(x)$ is linear (so its graph is a straight line).

> ### Definition
>
> The **simple linear regression model** assumes that there is a line with vertical or y intercept α and slope β, called the **true** or **population regression line.** When a value of the independent variable x is fixed and an observation on the dependent variable y is made,
>
> $$y = \alpha + \beta x + e$$
>
> Without the random deviation e, all observed (x, y) points would fall exactly on the population regression line. The inclusion of e in the model equation allows points to deviate from the line by random amounts.

Figure 13.2 shows several observations in relation to the population regression line.

FIGURE 13.2 Several observations resulting from the simple linear regression model

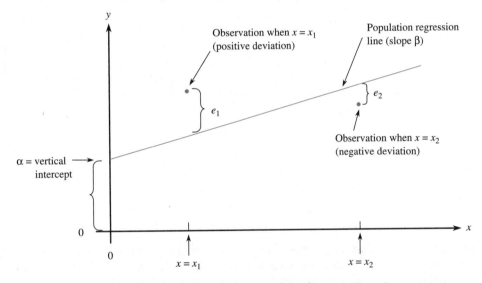

Before we make an observation on y for any particular value of x, we are uncertain about the value of e. It could be negative, positive, or even zero. Also, it might be quite large in magnitude (a point far from the population regression line) or quite small (a point very close to the line). In this chapter, we will make some assumptions about the distribution of e in repeated sampling at any particular x value.

Basic Assumptions of the Simple Linear Regression Model

1. The distribution of e at any particular x value has mean value 0 ($\mu_e = 0$).
2. The standard deviation of e (which describes the spread of its distribution) is the same for any particular value of x. This standard deviation is denoted by σ.

(continued)

3. The distribution of e at any particular x value is normal.

4. The random deviations $e_1, e_2, \ldots, e_n$ associated with different observations are independent of one another.

Randomness in e implies that y itself is subject to uncertainty. The foregoing assumptions about the distribution of e imply that the distribution of y values in repeated sampling satisfies certain properties. Consider y when x has some fixed value x^*, so that

$$y = \alpha + \beta x^* + e$$

Because α and β are fixed numbers, so is $\alpha + \beta x^*$. The sum of a fixed number and a normally distributed variable is again a normally distributed variable (the bell-shaped curve is simply relocated), so y itself has a normal distribution. Furthermore, $\mu_e = 0$ implies that the mean value of y is just $\alpha + \beta x^*$, the height of the population regression line above the value x^*. Lastly, because there is no variability in the fixed number $\alpha + \beta x^*$, the standard deviation of y is the same as that of e. These properties are summarized in the following box.

For any fixed x value, y itself has a normal distribution, with

$$\left(\begin{array}{c} \text{mean } y \text{ value} \\ \text{for fixed } x \end{array} \right) = \left(\begin{array}{c} \text{height of the population} \\ \text{regression line above } x \end{array} \right) = \alpha + \beta x$$

and

$$(\text{standard deviation of } y \text{ for fixed } x) = \sigma$$

The slope β of the population regression line is the *average* change in y associated with a 1-unit increase in x. The vertical intercept α is the height of the population line when $x = 0$. The value of σ determines the extent to which (x, y) observations deviate from the population line; when σ is small, most observations will be quite close to the line, but with large σ, there are likely to be some substantial deviations.

The key features of the model are illustrated in Figures 13.3 and 13.4. Notice that the three normal curves in Figure 13.3 have identical spreads. This is a consequence of $\sigma_e = \sigma$, which does not depend on x.

EXAMPLE 13.1 The authors of the article "On Weight Loss by Wrestlers Who Have Been Standing on Their Heads" (Sixth International Conference on Statistics, Combinatorics, and Related Areas, Forum for Interdisciplinary Mathematics, 1999) state

FIGURE 13.3 Illustration of the simple linear regression model

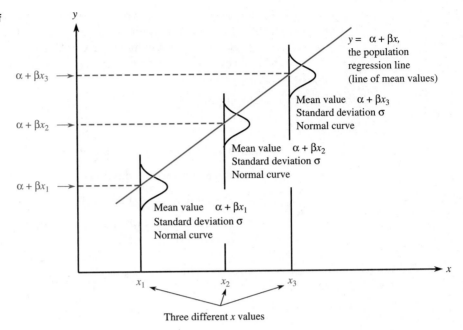

Three different x values

FIGURE 13.4 Data from the simple linear regression model: (a) σ small; (b) σ large

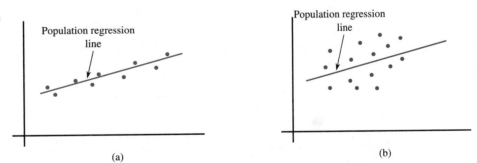

Amateur wrestlers who are overweight near the end of the weight certification period, but just barely so, have been known to stand on their heads for a minute or two, get on their feet, step back on the scale, and establish that they are in the desired weight class. Using a headstand as the method of last resort has become a fairly common practice in amateur wrestling.

Does this really work? Data was collected in an experiment where weight loss was recorded for each wrestler after exercising for 15 minutes and then doing a headstand for 1 minute and 45 seconds. Based on this data, the authors of the article concluded that there was in fact a demonstrable weight loss that was greater than that for a control group that exercised for 15 minutes but did not do the headstand.

(The authors give a plausible explanation for why this might be the case based on the way blood and other body fluids collect in the head during the headstand and the effect of weighing while these fluids are draining immediately after standing.) The authors also concluded that a simple linear regression model was a reasonable way to relate the variable

y = weight loss (in pounds)

and

x = body weight prior to exercise and headstand (in pounds)

Suppose that the actual model equation has

$$\alpha = 0 \qquad \beta = .001 \qquad \sigma = .09$$

(these values are consistent with the findings in the article). The population regression line is shown in Figure 13.5.

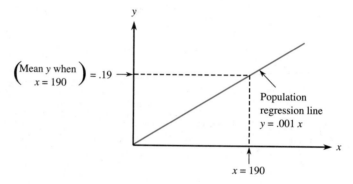

FIGURE 13.5 The population regression line for Example 13.1

If the distribution of the random errors at any fixed weight (x value) is normal, then the variable y = weight loss is normally distributed with

$$\mu_y = 0 + .001x \quad \text{and} \quad \sigma_y = .09$$

For example, when $x = 190$ (corresponding to a 190-lb wrestler), weight loss has mean value

$$\mu_y = 0 + .001(190) = .19$$

Since the standard deviation of y is $\sigma = .09$, the interval $.19 \pm 2(.09) = (.01, .37)$ includes y values that are within 2 standard deviations of the mean value for y when $x = 190$. Roughly 95% of the weight loss observations made for 190-lb wrestlers will be in this range.

The slope $\beta = .001$ is the change in average weight associated with each additional pound of body weight.

More insight into model properties can be gained by thinking of the population of all (x, y) pairs as consisting of many smaller populations. Each one of these smaller populations contains pairs for which x has a fixed value. Suppose, for example, that the variables

$$x = \text{grade point average in major courses}$$

and

$$y = \text{starting salary after graduation}$$

are related according to the simple linear regression model. Then there is the population of all pairs with $x = 3.20$, the population of all pairs having $x = 2.75$, and so on. The model assumes that for each such population, y is normally distributed with the same standard deviation, and the *mean y value* (rather than y itself) is linearly related to x.

In practice, the judgment as to whether the simple linear regression model is appropriate must be based on how the data was collected and on a scatter plot of the data. The sample observations should be independent of one another. In addition, the scatter plot should show a linear rather than a curved pattern, and the vertical spread of points should be relatively homogeneous throughout the range of x values. Figure 13.6 shows plots with three different patterns; the first is consistent with the model assumptions.

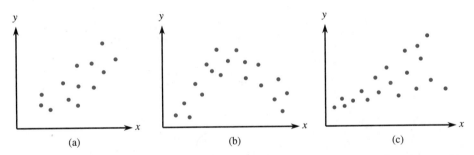

FIGURE 13.6 Some commonly encountered patterns in scatter plots: (a) Consistent with the simple linear regression model; (b) Suggests a nonlinear probabilistic model; (c) Suggests that variability in *y* changes with *x*

Estimating the Population Regression Line

The values of α and β (vertical intercept and slope of the population regression line) will almost never be known to an investigator. Instead, these values must first be estimated from the sample data $(x_1, y_1), \ldots, (x_n, y_n)$. We now assume that these n (x, y) pairs were obtained independently of one another, and that each observed y is related to the corresponding x via the model equation for simple linear regression.

Let a and b denote point estimates of α and β, respectively. These estimates come from applying the method of least squares introduced in Chapter 5; the least squares line has a smaller sum of squared vertical deviations from points in the scatter plot than does any other line.

The point estimates of β, the slope, and α, the y intercept of the population regression line, are the slope and y intercept, respectively, of the least squares line. That is,

$$b = \text{point estimate of } \beta = \frac{S_{xy}}{S_{xx}}$$

$$a = \text{point estimate of } \alpha = \bar{y} - b\bar{x}$$

where

$$S_{xy} = \Sigma xy - \frac{(\Sigma x)(\Sigma y)}{n} \quad \text{and} \quad S_{xx} = \Sigma x^2 - \frac{(\Sigma x)^2}{n}$$

The estimated regression line is then just the least squares line

$$\hat{y} = a + bx$$

Let x^* denote a specified value of the predictor variable x. Then $a + bx^*$ has two different interpretations:

1. It is a point estimate of the mean y value when $x = x^*$.
2. It is a point prediction of an individual y value to be observed when $x = x^*$.

EXAMPLE 13.2 Medical researchers have noted that adolescent females are much more likely to deliver low birth weight babies than are adult females. Because low birth weight babies have higher mortality rates, there have been a number of studies examining the relationship between birth weight and mother's age for babies born to young mothers. One such study is described in the article "The Risk of Teen Mothers Having Low Birth Weight Babies: Implications of Recent Medical Research for School Health Personnel" (*J. of School Health* (1998): 271–274). The accompanying data on

x = maternal age (years)

and

y = birth weight of baby (grams)

is consistent with summary values given in the referenced article and also with data published by the National Center for Health Statistics.

					Observation					
	1	**2**	**3**	**4**	**5**	**6**	**7**	**8**	**9**	**10**
x	15	17	18	15	16	19	17	16	18	19
y	2289	3393	3271	2648	2897	3327	2970	2535	3138	3573

A scatter plot (Figure 13.7) supports the appropriateness of the simple linear regression model.

The summary statistics are

$$n = 10 \qquad \Sigma x = 170 \qquad \Sigma y = 30{,}041$$
$$\Sigma x^2 = 2910 \qquad \Sigma xy = 515{,}600 \qquad \Sigma y^2 = 91{,}785{,}351$$

FIGURE 13.7 Scatter plot of
the data from Example 13.2

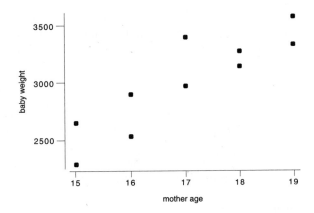

from which

$$S_{xy} = \Sigma\, xy - \frac{(\Sigma\, x)(\Sigma\, y)}{n} = 515{,}600 - \frac{(170)(30{,}041)}{10} = 4903.0$$

$$S_{xx} = \Sigma\, x^2 - \frac{(\Sigma\, x)^2}{n} = 2910 - \frac{(170)^2}{10} = 20.0$$

$$\bar{x} = \frac{170}{10} = 17.0 \qquad \bar{y} = \frac{30041}{10} = 3004.1$$

This gives

$$b = \frac{S_{xy}}{S_{xx}} = \frac{4903.0}{20.0} = 245.15$$

$$a = \bar{y} - b\bar{x} = 3004.1 - (245.15)(17.0) = -1163.45$$

The equation of the estimated regression line is then

$$\hat{y} = a + bx = -1163.45 + 245.15x$$

A point estimate of the average birth weight of babies born to 18-year-old mothers results from substituting $x = 18$ into the estimated equation:

$$\text{(estimated average } y \text{ value when } x = 18) = a + bx$$
$$= -1163.45 + 245.15(18)$$
$$= 3249.25 \text{ grams}$$

Similarly, we would predict the birth weight of a baby to be born to a particular 18-year-old mother to be

$$\text{(predicted } y \text{ value when } x = 18) = a + b(18) = 3249.25$$

The point estimate and the point prediction are identical, because the same x value was used in each calculation. However, the interpretation of each is different. One represents our prediction of the weight of a single baby whose mother is

18, whereas the other represents our estimate of the average weight of *all* babies born to 18-year-old mothers. This distinction will become important in Section 13.4 when we consider interval estimates and predictions.

In Example 13.2, the x values in the sample ranged from 15 to 19. An estimate or prediction should not be calculated for any x value much outside this range. Without sample data for such values, there is no hard evidence that the estimated linear relationship can be extrapolated very far. Statisticians refer to this potential pitfall as the **danger of extrapolation.**

Estimating σ^2 and σ

The value of σ determines the extent to which observed points (x, y) tend to fall close to or far away from the population regression line. A point estimate of σ is based on

$$\text{SSResid} = \Sigma (y - \hat{y})^2$$

where $\hat{y}_1 = a + bx_1, \ldots, \hat{y}_n = a + bx_n$ are the fitted or predicted y values and the residuals are $y_1 - \hat{y}_1, \ldots, y_n - \hat{y}_n$. SSResid is a measure of the extent to which the sample data spreads out about the estimated regression line.

Definition

The statistic for estimating the variance σ^2 is

$$s_e^2 = \frac{\text{SSResid}}{n - 2}$$

where

$$\text{SSResid} = \Sigma (y - \hat{y})^2 = \Sigma y^2 - a \Sigma y - b \Sigma xy$$

The subscript e in s_e^2 reminds us that we are estimating the variance of the "errors" or residuals.

The estimate of σ is the **estimated standard deviation**

$$s_e = \sqrt{s_e^2}$$

The number of degrees of freedom associated with estimating σ^2 or σ in simple linear regression is $n - 2$.

The estimates and number of degrees of freedom here have analogues in our previous work involving a single sample $x_1, x_2, \ldots, x_n$. The sample variance s^2 had numerator $\Sigma (x - \bar{x})^2$, a sum of squared deviations (residuals), and denominator $n - 1$, the number of degrees of freedom associated with s^2 and s. The use of $\bar{x}$ as an estimate of μ in the formula for s^2 reduces the number of degrees of freedom by 1, from n to $n - 1$. In simple linear regression, estimation of α and β results in a loss of 2 degrees of freedom, leaving $n - 2$ as df for SSResid, s_e^2, and s_e.

The coefficient of determination was defined previously (see Chapter 5) as

$$r^2 = 1 - \frac{\text{SSResid}}{\text{SSTo}}$$

where

$$\text{SSTo} = \Sigma (y - \bar{y})^2 = \Sigma y^2 - \frac{(\Sigma y)^2}{n} = S_{yy}$$

The value of r^2 can now be interpreted as the proportion of observed y variation that can be explained by (or attributed to) the model relationship. The estimate s_e also gives another assessment of model performance. Roughly speaking, the value of σ represents the magnitude of a typical deviation of a point (x, y) in the population from the population regression line. Similarly, in a rough sense, s_e is the magnitude of a typical sample deviation (residual) from the least squares line. The smaller the value of s_e, the closer the points in the sample fall to the line, and the better the line does in predicting y from x.

EXAMPLE 13.3   Forest managers are increasingly concerned about the damage done to animal populations when forests are clear-cut. Woodpeckers are a valuable forest asset, both because they provide nest and roost holes for other animals and birds and because they prey on many forest insect pests. The article "Artificial Trees as a Cavity Substrate for Woodpeckers" (*J. of Wildlife Mgmt.* (1983): 790–798) reported on a study of how woodpeckers behaved when provided with polystyrene cylinders as an alternative roost and nest cavity substrate. We give selected values of $x =$ ambient temperature (°C) and $y =$ cavity depth (cm); these values were read from a scatter plot that appeared in the article.

Obs.	Temp.	Depth	Pred. *y* value	Residual
1	−6	21.1	22.195	−1.095
2	−3	26.0	21.160	4.840
3	−2	18.0	20.815	−2.815
4	1	19.2	19.780	−.580
5	6	16.9	18.055	−1.155
6	10	18.1	16.675	1.425
7	11	16.8	16.330	.470
8	19	11.8	13.569	−1.769
9	21	11.0	12.879	−1.879
10	23	12.1	12.189	−.089
11	25	14.8	11.499	3.301
12	26	10.5	11.154	−.654

Our plot (Figure 13.8, page 648), as well as the original plot, gives evidence of a strong negative linear relationship between x and y.

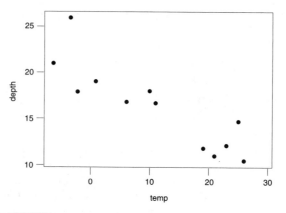

FIGURE 13.8 MINITAB scatter plot for Example 13.3

The summary statistics are

$$n = 12 \qquad \Sigma x = 131 \qquad \Sigma y = 196.3$$
$$\Sigma x^2 = 2939 \qquad \Sigma xy = 1622.3 \qquad \Sigma y^2 = 3445.25$$

from which we calculate

$$b = -.345043 \qquad a = 20.125053$$
$$\text{SSResid} = 54.4655 \qquad \text{SSTo} = 234.1093$$

Thus,

$$r^2 = 1 - \frac{\text{SSResid}}{\text{SSTo}} = 1 - \frac{54.4655}{234.1093} = 1 - .233 = .767$$
$$s_e^2 = \frac{\text{SSResid}}{n - 2} = \frac{54.4655}{10} = 5.447$$

and

$$s_e = \sqrt{5.447} = 2.33$$

Approximately 76.7% of the observed variation in cavity depth y can be attributed to the probabilistic linear relationship with ambient temperature. The magnitude of a typical sample deviation from the least squares line is about 2.3, which is reasonably small in comparison to the y values themselves. The model appears to be useful for estimation and prediction; in the next section, we show how a model utility test can be used to judge whether this is indeed the case.

A key assumption of the simple linear regression model is that the random deviation e in the model equation is normally distributed. In Section 13.3, we will indicate how the residuals can be used to determine whether this is plausible.

Exercises 13.1 – 13.11

13.1 Let x be the size of a house (sq ft) and y be the amount of natural gas used (therms) during a specified period. Suppose that for a particular community, x and y are related according to the simple linear regression model with

$$\beta = \text{slope of population regression line} = .017$$

$$\alpha = y \text{ intercept of population regression line} = -5.0$$

a. What is the equation of the population regression line?
b. Graph the population regression line by first finding the point on the line corresponding to $x = 1000$ and then the point corresponding to $x = 2000$, and drawing a line through these points.
c. What is the mean value of gas usage for houses with 2100 sq ft of space?
d. What is the average change in usage associated with a 1-sq-ft increase in size?
e. What is the average change in usage associated with a 100-sq-ft increase in size?
f. Would you use the model to predict mean usage for a 500-sq-ft house? Why or why not? (*Note:* There are no small houses in the community in which this model is valid.)

13.2 The flow rate y (m^3/min) in a device used for air-quality measurement depends on the pressure drop x (in of water) across the device's filter. Suppose that for x values between 5 and 20, these two variables are related according to the simple linear regression model with true regression line $y = -.12 + .095x$.

a. What is the true average flow rate for a pressure drop of 10 in? A drop of 15 in?
b. What is the true average change in flow rate associated with a 1-in increase in pressure drop? Explain.
c. What is the average change in flow rate when pressure drop decreases by 5 in?

13.3 Data presented in the article "Manganese Intake and Serum Manganese Concentration of Human Milk-Fed and Formula-Fed Infants" (*Amer. J. of Clinical Nutrition* (1984): 872–878) suggests that a simple linear regression model is reasonable for describing the relationship between y = serum manganese (Mn) and x = Mn intake (μg/kg/day). Suppose that the true regression line is $y = -2 + 1.4x$ and that $\sigma = 1.2$. Then for a fixed x value, y has a normal distribution with mean $-2 + 1.4x$ and standard deviation 1.2.

a. What is the mean value of serum Mn when Mn intake is 4.0? When Mn intake is 4.5?
b. What is the probability that an infant whose Mn intake is 4.0 will have serum Mn greater than 5?
c. Approximately what proportion of infants whose Mn intake is 5 will have a serum Mn greater than 5? Less than 3.8?

13.4 A sample of small cars was selected, and the values of x = horsepower and y = fuel efficiency (mpg) were determined for each car. Fitting the simple linear regression model gave the estimated regression equation $\hat{y} = 44.0 - .150x$.

a. How would you interpret $b = -.150$?
b. Substituting $x = 100$ gives $\hat{y} = 29.0$. Give two different interpretations of this number.
c. What happens if you predict efficiency for a car with a 300 horsepower engine? Why do you think this has occurred?
d. Interpret $r^2 = .680$ in the context of this problem.
e. Interpret $s_e = 3.0$ in the context of this problem.

13.5 Suppose that a simple linear regression model is appropriate for describing the relationship between y = house price and x = house size (sq ft) for houses in a large city. The true regression line is $y = 23,000 + 47x$ and $\sigma = 5000$.

a. What is the average change in price associated with one extra sq ft of space? With an additional 100 sq ft of space?
b. What proportion of 1800-sq-ft homes would be priced over $110,000? Under $100,000?

13.6 **a.** Explain the difference between the line $y = \alpha + \beta x$ and the line $\hat{y} = a + bx$.
b. Explain the difference between β and b.
c. Let x^* denote a particular value of the independent variable. Explain the difference between $\alpha + \beta x^*$ and $a + bx^*$.
d. Explain the difference between σ and s_e.

13.7 Legumes, such as peas and beans, are important crops whose production is greatly affected by pests. The article "Influence of Wind Speed on Residence Time of *Uroleucon ambrosiae Alatae* on Bean Plants" (*Envir. Entom.* (1991): 1375–1380) reported on a study in which aphids were placed on a bean plant, and the elapsed time until half of the aphids had departed was observed. Data on x = wind speed (m/sec) and y = residence half-time was given and used to produce the following information.

$a = .0119$ $b = 3.4307$ $n = 13$

$SSTo = 73.937$ $SSResid = 27.890$

a. What percentage of observed variation in residence half-time can be attributed to the simple linear regression model?

b. Give a point estimate of σ and interpret the estimate.

c. Estimate the mean change in residence half-time associated with a 1-m/sec increase in wind speed.

d. Calculate a point estimate of true average residence half-time when wind speed is 1 m/sec.

13.8 The accompanying data on x = treadmill run time to exhaustion (min) and y = 20-km ski time (min) was taken from the article "Physiological Characteristics and Performance of Top U.S. Biathletes" (*Medicine and Science in Sports and Exercise* (1995): 1302–1310):

x	7.7	8.4	8.7	9.0	9.6	9.6
y	71.0	71.4	65.0	68.7	64.4	69.4

x	10.0	10.2	10.4	11.0	11.7
y	63.0	64.6	66.9	62.6	61.7

$\Sigma x = 106.3$ $\Sigma y = 1040.95$ $\Sigma x^2 = 728.70$

$\Sigma xy = 7009.91$ $\Sigma y^2 = 48,390.79$

a. Does a scatter plot suggest that the simple linear regression model is appropriate?

b. Determine the equation of the estimated regression line, and draw the line on your scatter plot.

c. What is your estimate of the average change in ski time associated with a 1-minute increase in treadmill time?

d. What would you predict ski time to be for an individual whose treadmill time is 10 min?

e. Should the model be used as a basis for predicting ski time when treadmill time is 15 min? Explain.

f. Calculate and interpret the value of r^2.

g. Calculate and interpret the value of s_e.

13.9 The accompanying summary quantities resulted from a study in which x was the number of photocopy machines serviced during a routine service call and y was the total service time (min):

$n = 16$ $\Sigma (y - \bar{y})^2 = 22,398.05$ $\Sigma (y - \hat{y})^2 = 2620.57$

a. What proportion of observed variation in total service time can be explained by a linear probabilistic relationship between total service time and the number of machines serviced?

b. Calculate the value of the estimated standard deviation s_e. What is the number of degrees of freedom associated with this estimate?

13.10 Exercise 5.45 described a regression situation in which y = hardness of molded plastic and x = amount of time elapsed since termination of the molding process. Summary quantities included $n = 15$, $SSResid = 1235.470$, and $SSTo = 25,321.368$.

a. Calculate a point estimate of σ. On how many degrees of freedom is the estimate based?

b. What percentage of observed variation in hardness can be explained by the simple linear regression model relationship between hardness and elapsed time?

13.11 The accompanying data on x = advertising share and y = market share for a particular brand of cigarettes during ten randomly selected years is from the article "Testing Alternative Econometric Models on the Existence of Advertising Threshold Effect" (*J. of Marketing Research* (1984): 298–308).

x	.103	.072	.071	.077	.086
y	.135	.125	.120	.086	.079

x	.047	.060	.050	.070	.052
y	.076	.065	.059	.051	.039

a. Construct a scatter plot for this data. Do you think the simple linear regression model would be appropriate for describing the relationship between x and y?

b. Calculate the equation of the estimated regression line and use it to obtain the predicted market share when the advertising share is .09.

c. Compute r^2. How would you interpret this value?

d. Calculate a point estimate of σ. On how many degrees of freedom is your estimate based?

13.2 Inferences Concerning the Slope of the Population Regression Line

The slope coefficient β in the simple linear regression model is the average or expected change in the dependent variable y associated with a 1-unit increase in the value of the independent variable x. For example, consider x = the size of a house (sq ft) and y = selling price of the house. If we assume that the simple linear regression model is appropriate for the population of houses in a particular city, β would be the average increase in selling price associated with a 1-sq-ft increase in size. As another example, if x = amount of time per week a computer system is used and y = the resulting annual maintenance expense, then β would be the expected change in expense associated with using the computer system one additional hour per week.

Since the value of β is almost always unknown, it will have to be estimated from sample data: n independently selected observations $(x_1, y_1), \ldots, (x_n, y_n)$. The slope of the least squares line gives a point estimate. As with any point estimate, though, it is desirable to have some indication of how accurately b estimates β. In some situations, the value of the statistic b may vary greatly from sample to sample, so b computed from a single sample may well be rather different from the true slope β. In other situations, almost all possible samples yield b values quite close to β, so the error of estimation is almost sure to be small. To proceed further, we need some facts about the sampling distribution of b: information concerning the shape of the sampling distribution curve, where the curve is centered relative to β, and how much the curve spreads out about its center.

Properties of the Sampling Distribution of b

When the four basic assumptions of the simple linear regression model are satisfied, the following conditions are met:

1. The mean value of b is β. That is, $\mu_b = \beta$, so the sampling distribution of b is always centered at the value of β. Thus, b is an unbiased statistic for estimating β.

2. The standard deviation of the statistic b is

$$\sigma_b = \frac{\sigma}{\sqrt{S_{xx}}}$$

3. The statistic b has a normal distribution (a consequence of the model assumption that the random deviation e is normally distributed).

The fact that b is unbiased means only that the sampling distribution is centered at the right place; it gives no information about dispersion. If σ_b is large, the normal sampling distribution curve will be quite spread out around β, and an estimate far from β may well result. For σ_b to be small, the numerator σ should be small (little variability about the population line) and/or the denominator $\sqrt{S_{xx}}$, or equivalently $S_{xx} = \Sigma (x - \bar{x})^2$ itself, should be large. Because $\Sigma (x - \bar{x})^2$ is a measure of how much the observed x values spread out, β will tend to be more precisely estimated when the x values in our sample are spread out than when they are close together.

The normality of b implies that the standardized variable

$$z = \frac{b - \beta}{\sigma_b}$$

has a standard normal distribution. However, inferential methods cannot be based on this variable, because the value of σ_b is not available (since the unknown σ appears in the numerator of σ_b). The obvious way out of this dilemma is to estimate σ with s_e, yielding an estimated standard deviation.

The **estimated standard deviation of the statistic b** is

$$s_b = \frac{s_e}{\sqrt{S_{xx}}}$$

When the four basic assumptions of the simple linear regression model are satisfied, the probability distribution of the standardized variable

$$t = \frac{b - \beta}{s_b}$$

is the t distribution with df $= (n - 2)$.

In the same way that $t = (\bar{x} - \mu)/(s/\sqrt{n})$ was used in Chapter 9 to develop a confidence interval for μ, the t variable in the preceding box can be employed to give a confidence interval (interval estimate) for β.

Confidence Interval for β

When the four basic assumptions of the simple linear regression model are satisfied, a **confidence interval for β,** the slope of the population regression line, has the form

$$b \pm (t \text{ critical value}) \cdot s_b$$

where the t critical value is based on df $= n - 2$. Appendix Table III gives critical values corresponding to the most frequently used confidence levels.

The interval estimate of β is centered at b and extends out from the center by an amount that depends on the sampling variability of b. When s_b is small, the interval will be narrow, implying that the investigator has relatively precise knowledge of β.

EXAMPLE 13.4   Is cardiovascular fitness (as measured by time to exhaustion running on a treadmill) related to an athlete's performance in a 20-km ski race? The accompanying data on

x = treadmill run time to exhaustion (min)

and

y = 20-km ski time (min)

was taken from the article "Physiological Characteristics and Performance of Top U.S. Biathletes" (*Medicine and Science in Sports and Exercise* (1995): 1302–1310):

x	7.7	8.4	8.7	9.0	9.6	9.6	10.0	10.2	10.4	11.0	11.7
y	71.0	71.4	65.0	68.7	64.4	69.4	63.0	64.6	66.9	62.6	61.7

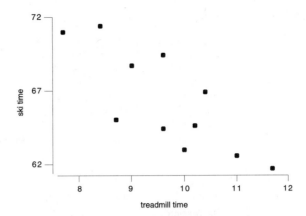

A scatter plot of the data appears in the preceding figure. The plot shows a linear pattern and the vertical spread of points does not appear to be changing over the range of x values in the sample. If we assume that the distribution of errors at any given x value is approximately normal, the simple linear regression model seems appropriate.

The slope β in this context is the average change in ski time associated with a 1-minute increase in treadmill time. The scatter plot shows a negative linear relationship, so the point estimate of β will be negative.

Straightforward calculation gives

$$n = 11 \qquad \Sigma x = 106.3 \qquad \Sigma y = 728.70$$
$$\Sigma x^2 = 1040.95 \qquad \Sigma xy = 7009.91 \qquad \Sigma y = 48{,}390.79$$

from which

$$b = -2.3335 \qquad a = 88.796$$

$$\text{SSResid} = 43.097 \qquad \text{SSSTo} = 117.727$$

$r^2 = .634$ (63.4% of the observed variation in ski time can be explained by the simple linear regression model)

$$s_e^2 = 4.789 \qquad s_e = 2.188$$

$$s_b = \frac{s_e}{\sqrt{S_{xx}}} = \frac{2.188}{3.702} = .591$$

Calculation of the 95% confidence interval for β requires a t critical value based on df $= n - 2 = 11 - 2 = 9$, which (from Appendix Table III) is 2.26. The resulting interval is then

$$b \pm (t \text{ critical value}) \cdot s_b = -2.3335 \pm (2.26)(.591)$$
$$= -2.335 \pm 1.336$$
$$= (-3.671, -.999)$$

We interpret this interval as follows: Based on the sample data, we are 95% confident that the true average decrease in ski time associated with a one-minute increase in treadmill time is between 1 and 3.7 minutes.

FIGURE 13.9 Partial MINITAB output for the data of Example 13.4

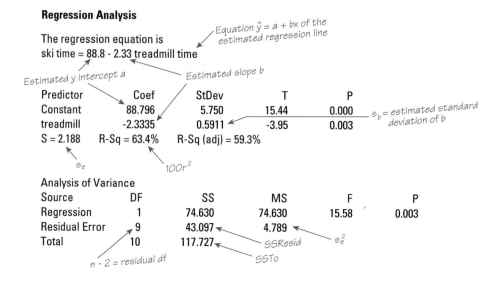

Regression Analysis

The regression equation is
ski time = 88.8 - 2.33 treadmill time

Equation $\hat{y} = a + bx$ of the estimated regression line

Estimated y intercept a   *Estimated slope b*

Predictor	Coef	StDev	T	P
Constant	88.796	5.750	15.44	0.000
treadmill	-2.3335	0.5911	-3.95	0.003

s_b = estimated standard deviation of b

S = 2.188   R-Sq = 63.4%   R-Sq (adj) = 59.3%

s_e   *$100r^2$*

Analysis of Variance

Source	DF	SS	MS	F	P
Regression	1	74.630	74.630	15.58	0.003
Residual Error	9	43.097	4.789		
Total	10	117.727			

$n - 2$ = residual df   *SSResid*   *s_e^2*   *SSTo*

Output from any of the standard statistical computer packages routinely includes the computed values of a, b, SSResid, s_e, SSTo, r^2, and s_b. Figure 13.9 displays partial MINITAB output for the data of Example 13.4. The format from other packages is very similar. Rounding will occasionally lead to small discrepancies between hand-calculated and computer-calculated values, but there are no such discrepancies for this example.

Hypothesis Tests Concerning β

Hypotheses about β can be tested using a t test very similar to the t tests discussed in previous chapters. The null hypothesis states that β has a specified hypothesized value. The t statistic results from standardizing b, the point estimate of β, under the assumption that H_0 is true. When H_0 is true, the sampling distribution of this statistic is the t distribution with df $= n - 2$.

Summary of Hypothesis Tests Concerning β

Null hypothesis: H_0: β = hypothesized value

Test statistic: $t = \dfrac{b - \text{hypothesized value}}{s_b}$

The test is based on df $= n - 2$.

Alternative hypothesis:	*P-value*:
H_a: $\beta >$ hypothesized value	Area to the right of the computed t under the appropriate t curve
H_a: $\beta <$ hypothesized value	Area to the left of the computed t under the appropriate t curve
H_a: $\beta \neq$ hypothesized value	(i) $2 \cdot$ (area to the right of t) if t is positive
	or
	(ii) $2 \cdot$ (area to the left of the t) if t is negative

(continued)

Assumptions: For this test to be appropriate, the four basic assumptions of the simple linear regression model must be met:
1. The distribution of e at any particular x value has mean value 0 ($\mu_e = 0$).
2. The standard deviation of e is σ, which does not depend on x.
3. The distribution of e at any particular x value is normal.
4. The random deviations $e_1, e_2, \ldots, e_n$ associated with different observations are independent of one another.

Very frequently, the null hypothesis of interest is that $\beta = 0$. When this is the case, the population regression line is a horizontal line, and the value of y in the simple linear regression model does not depend on x. That is,

$$y = \alpha + 0 \cdot x + e$$

or equivalently,

$$y = \alpha + e$$

In this situation, knowledge of x is of no use in predicting y. On the other hand, if $\beta \neq 0$, there is a useful linear relationship between x and y, and knowledge of x is useful for predicting y. This is illustrated by the scatter plots of Figure 13.10.

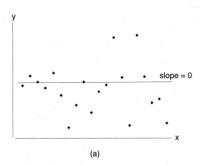
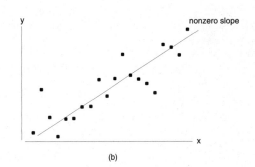

FIGURE 13.10 (a) $\beta = 0$; (b) $\beta \neq 0$

The test of H_0: $\beta = 0$ versus H_a: $\beta \neq 0$ is called the *model utility test for simple linear regression.*

The Model Utility Test for Simple Linear Regression

The **model utility test for simple linear regression** is the test of

H_0: $\beta = 0$

versus

H_a: $\beta \neq 0$

(continued)

The null hypothesis specifies that there is *no* useful linear relationship between x and y, whereas the alternative hypothesis specifies that there *is* a useful linear relationship between x and y. If H_0 is rejected, we conclude that the simple linear regression model is useful for predicting y. The test procedure in the previous box (with hypothesized value $= 0$) is used to carry out the model utility test; in particular, the test statistic is the **t ratio** $\dfrac{b}{s_b}$.

If a scatter plot and the value of r^2 do not provide convincing evidence for a useful linear relationship, we recommend that the model utility test be carried out before the estimated model is used to make other inferences.

EXAMPLE 13.5   Computers seem to have invaded almost every aspect of our lives. In recent years, this has led to an explosion in the number of investigations whose objective has been to study and document the varied effects of the computer revolution. Many of these have dealt with the work environment of people who spend a substantial amount of time staring at computer monitors. Image quality of monitors is an important characteristic, affecting, among other things, extent of eye strain and work efficiency. The article "Image Quality Determines Differences in Reading Performance and Perceived Image Quality with CRT and Hard Copy Displays" (*Human Factors* (1991): 459–469) reported on an experiment in which image quality (x) and average time for a group of subjects to read certain passages (y, in seconds) were determined. The accompanying data was read from a graph in the article:

x	4.30	4.55	5.55	5.65	5.95	6.30	6.45	6.45
y	8.0	8.3	7.8	7.25	7.7	7.5	7.6	7.2

The summary statistics necessary for a simple linear regression analysis are as follows:

$$n = 8 \qquad \Sigma x = 45.20 \qquad \Sigma x^2 = 260.215$$
$$\Sigma y = 61.35 \qquad \Sigma y^2 = 471.4325 \qquad \Sigma xy = 344.9425$$

from which

$$b = -.348501 \qquad a = 9.637778 \qquad \text{SSResid} = .367626$$
$$s_e = .247530 \qquad r^2 = .615 \qquad S_{xx} = 4.835$$

Since $r^2 = .615$, about 61.5% of observed variation in reading time can be explained by the simple linear regression model. It appears from this that there is a useful linear relation between the two variables, but a confirmation requires a formal model utility test. We will use a significance level of .05 to carry out this test.

1. $\beta =$ the true average change in reading time associated with a 1-unit increase in image quality
2. $H_0: \beta = 0$
3. $H_a: \beta \neq 0$

4. $\alpha = .05$

5. *Test statistic*: $t = \dfrac{b - \text{hypothesized value}}{s_b} = \dfrac{b - 0}{s_b} = \dfrac{b}{s_b}$

6. *Assumptions*:   The accompanying scatter plot of the data shows a linear pattern and the variability of points does not appear to be changing with x. Assuming that the distribution of errors at any given x value is approximately normal, the assumptions of the simple linear regression model are appropriate.

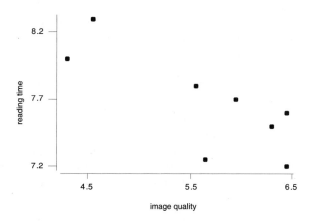

7. *Calculation*:   The calculation of t requires

$$s_b = \frac{s_e}{\sqrt{S_{xx}}} = \frac{.247530}{2.1989} = .11257$$

yielding

$$t = \frac{-.348501 - 0}{.11257} = -3.10$$

8. *P-value*:   Appendix Table IV shows that for a t test based on 6 df, $P(t > 3.10) = P(t < -3.10) = .011$. The inequality in H_a requires a two-tailed test, so P-value $= 2(.011) = .022$.

9. *Conclusion*:   Since .022 is smaller than the significance level .05, H_0 is rejected. We conclude, as did the investigators, that there is a useful linear relationship between image quality and reading time.

Figure 13.11 (page 658) shows partial MINITAB output from a simple linear regression analysis. The **Coef** column gives $b = -.3485$; $s_b = .1125$ is in the **Stdev** column; the **T** column (for t ratio) contains the value of the test statistic for testing $H_0: \beta = 0$, $t = (b - 0)/s_b = b/s_b$; and the P-value for the model utility test is given in the last column as .021 (slightly different from the one given in step 8 because of rounding). Other commonly used statistical packages also include this information in their output.

FIGURE 13.11 MINITAB output for the data of Example 13.5

Regression Analysis

The regression equation is
reading time = 9.64 - 0.349 image quality

Predictor	Coef	StDev	T	P
Constant	9.6378	0.6419	15.01	0.000
image qu	-0.3485	0.1125	-3.10	0.021

S = 0.2475 R-Sq = 61.5% R-Sq(adj) = 55.1%

Analysis of Variance

Source	DF	SS	MS	F	P
Regression	1	0.58722	0.58722	9.59	0.021
Residual Error	6	0.36746	0.06124		
Total	7	0.95469			

When $H_0: \beta = 0$ cannot be rejected by the model utility test at a reasonably small significance level, the search for a useful model must continue. One possibility is to relate y to x using a nonlinear model, an appropriate strategy if the scatter plot shows curvature. Alternatively, a multiple regression model using more than one predictor variable may be employed. The next chapter introduces such models.

Exercises 13.12 – 13.26

13.12 a. What is the difference between σ and σ_b?
b. What is the difference between σ_b and s_b?

13.13 a. Suppose that a single y observation is made at each of the x values 5, 10, 15, 20, and 25. If $\sigma = 4$, what is σ_b, the standard deviation of the statistic b?
b. Now suppose that a second observation is made at every x value listed in part **a** (for a total of 10 observations). Is the resulting value of σ_b half of what it was in part **a**?
c. How many observations at each x value in part **a** are required to yield a σ_b value that is half the value calculated in part **a**? Verify your conjecture.

13.14 Refer back to Example 13.3 in which the simple linear regression model was fit to data on x = ambient temperature and y = cavity depth. For the purpose of estimating β as accurately as possible, would it have been preferable to make observations at the x values $x_1 = -6$, $x_2 = -5$, $x_3 = -4$, $x_4 = -3$, $x_5 = -2$, $x_6 = -1$, $x_7 = 21$, $x_8 = 22$, $x_9 = 23$, $x_{10} = 24$, $x_{11} = 25$, and $x_{12} = 26$? Explain your reasoning.

13.15 Exercise 13.10 presented information from a study in which y was the hardness of molded plastic and x was the time elapsed since termination of the molding process. Summary quantities included

$$n = 15 \quad b = 2.50 \quad \text{SSResid} = 1235.470$$
$$\Sigma (x - \bar{x})^2 = 4024.20$$

a. Calculate s_b, the estimated standard deviation of the statistic b.
b. Obtain a 95% confidence interval for β, the slope of the true regression line.
c. Does the interval in part **b** suggest that β has been precisely estimated? Explain.

13.16 A study was carried out to relate sales revenue y (in thousands of dollars) to advertising expenditure x (also in thousands of dollars) for fast-food outlets during a 3-month period. A sample of 15 outlets yielded the accompanying summary quantities.

$$\Sigma x = 14.10 \quad \Sigma y = 1438.50 \quad \Sigma x^2 = 13.92$$
$$\Sigma y^2 = 140,354 \quad \Sigma xy = 1387.20$$
$$\Sigma (y - \bar{y})^2 = 2401.85 \quad \Sigma (y - \hat{y})^2 = 561.46$$

a. What proportion of observed variation in sales revenue can be attributed to the linear relationship between revenue and advertising expenditure?

b. Calculate s_e and s_b.

c. Obtain a 90% confidence interval for β, the average change in revenue associated with a $1000 (that is, 1-unit) increase in advertising expenditure.

13.17 An experiment to study the relationship between x = time spent exercising (min) and y = amount of oxygen consumed during the exercise period resulted in the following summary statistics.

$$n = 20 \quad \Sigma x = 50 \quad \Sigma y = 16{,}705 \quad \Sigma x^2 = 150$$

$$\Sigma y^2 = 14{,}194{,}231 \quad \Sigma xy = 44{,}194$$

a. Estimate the slope and y intercept of the population regression line.

b. One sample observation on oxygen usage was 757 for a 2-min exercise period. What amount of oxygen consumption would you predict for this exercise period, and what is the corresponding residual?

c. Compute a 99% confidence interval for the true average change in oxygen consumption associated with a 1-min increase in exercise time.

13.18 Exercise 5.66 presented data on x = average hourly wage and y = quit rate for a sample of industries. See the accompanying MINITAB output.

The regression equation is
quit rate = 4.86 - 0.347 wage

Predictor	Coef	Stdev	t-ratio	p
Constant	4.8615	0.5201	9.35	0.000
wage	-0.34655	0.05866	-5.91	0.000

s = 0.4862 R-sq = 72.9% R-sq(adj) = 70.8%

Analysis of Variance

SOURCE	DF	SS	MS	F	p
Regression	1	8.2507	8.2507	34.90	0.000
Error	13	3.0733	0.2364		
Total	14	11.3240			

a. Based on the given P-value, does there appear to be a useful linear relationship between average wage and quit rate? Explain your reasoning.

b. Calculate an estimate of the average change in quit rate associated with a $1 increase in average hourly wage, and do so in a way that conveys information about the precision and reliability of the estimate.

13.19 The article "Cost-Effectiveness in Public Education" cited in the chapter introduction reported that, for a sample of $n = 44$ school districts, a regression of y = average SAT score on x = expenditure per pupil (thousands of dollars) gave $b = 15.0$ and $s_b = 5.3$.

a. Does the simple linear regression model specify a useful relationship between x and y?

b. Calculate and interpret a confidence interval for β based on a 95% confidence level.

13.20 Exercise 5.40 described a study reported in the 1991 volume of the *Amer. J. of Phys. Anthropology* in which the objective was to predict age (y) from percentage of a tooth's root with transparent dentine. The accompanying data is for anterior teeth.

x	15	19	31	39	41	44	47	48	55	65
y	23	52	65	55	32	60	78	59	61	60

Use the accompanying MINITAB output to decide whether the simple linear regression model is useful.

The regression equation is
age = 32.1 + 0.555 percent

Predictor	Coef	Stdev	t-ratio	p
Constant	32.08	13.32	2.41	0.043
percent	0.5549	0.3101	1.79	0.111

s = 14.30 R-sq = 28.6% R-sq(adj) = 19.7%

Analysis of Variance

SOURCE	DF	SS	MS	F	p
Regression	1	654.8	654.8	3.20	0.111
Error	8	1635.7	204.5		
Total	9	2290.5			

13.21 The accompanying data was read from a plot (and is a subset of the complete data set) given in the article "Cognitive Slowing in Closed-Head Injury" (*Brain and Cognition* (1996): 429–440). The data represents the mean response times for a group of individuals with closed-head injury (CHI) and a matched control group without head injury on 10 different tasks. Each observation was based on a different study, and used different subjects, so it is reasonable to assume that the observations are independent.

	Mean Response Time	
Study	**Control**	**CHI**
1	250	303
2	360	491
3	475	659
4	525	683
5	610	922
6	740	1044
7	880	1421
8	920	1329
9	1010	1481
10	1200	1815

a. Fit a linear regression model that would allow you to predict the mean response time for those suffering a closed-head injury from the mean response time on the same task for individuals with no head injury.

b. Does the sample data support the hypothesis that there is a useful linear relationship between the mean response time for individuals with no head injury and the mean response time for individuals with CHI? Test the appropriate hypotheses using $\alpha = .05$.

c. It is also possible to test hypotheses about the y intercept in a linear regression model. For this data, the null hypothesis $H_0: \alpha = 0$ cannot be rejected at the .05 significance level, suggesting that a model with a y intercept of 0 might be an appropriate model. Fitting such a model results in an estimated regression equation of

CHI = 1.48 Control

Interpret the estimated slope of 1.48.

13.22 The article "Effects of Enhanced UV-B Radiation on Ribulose-1,5-Biphosphate, Carboxylase in Pea and Soybean" (*Environ. and Exper. Botany* (1984): 131–143) included the accompanying data on pea plants, with $y = $ sunburn index and $x = $ distance (cm) from an ultraviolet light source.

x	18	21	25	26	30	32	36	40
y	4.0	3.7	3.0	2.9	2.6	2.5	2.2	2.0

x	40	50	51	54	61	62	63
y	2.1	1.5	1.5	1.5	1.3	1.2	1.1

$$\Sigma x = 609 \qquad \Sigma y = 33.1 \qquad \Sigma x^2 = 28{,}037$$
$$\Sigma xy = 1156.8 \qquad \Sigma y^2 = 84.45$$

Estimate the mean change in the sunburn index associated with an increase of 1 cm in distance in a way that includes information about the precision of estimation.

13.23 Exercise 13.16 described a regression analysis in which $y = $ sales revenue and $x = $ advertising expenditure. Summary quantities given there yield

$$n = 15 \qquad b = 52.57 \qquad s_b = 8.05$$

a. Test the hypothesis $H_0: \beta = 0$ versus $H_a: \beta \neq 0$ using a significance level of .05. What does your conclusion say about the nature of the relationship between x and y?

b. Consider the hypothesis $H_0: \beta = 40$ versus $H_a: \beta > 40$. The null hypothesis states that the average change in sales revenue associated with a 1-unit increase in advertising expenditure is (at most) \$40,000. Carry out a test using significance level .01.

13.24 The article "Technology, Productivity and Industry Structure" (*Tech. Forecasting and Social Change* (1983): 1–13) included the accompanying data on $x = $ research

and development expenditure and $y = $ growth rate for eight different industries.

x	2024	5038	905	3572	1157	327	378	191
y	1.90	3.96	2.44	.88	.37	−.90	.49	1.01

a. Would a simple linear regression model provide useful information for predicting growth rate from research and development expenditure? Use a .05 level of significance.

b. Use a 90% confidence interval to estimate the average change in growth rate associated with a 1-unit increase in expenditure. Interpret the resulting interval.

13.25 The article "Effect of Temperature on the pH of Skim Milk" (*J. of Dairy Research* (1988): 277–280) reported on a study involving $x = $ temperature (°C) under specified experimental conditions and $y = $ milk pH. The accompanying data (read from a graph) is a representative subset of that which appeared in the article:

x	4	4	24	24	25	38	38	40
y	6.85	6.79	6.63	6.65	6.72	6.62	6.57	6.52

x	45	50	55	56	60	67	70	78
y	6.50	6.48	6.42	6.41	6.38	6.34	6.32	6.34

$$\Sigma x = 678 \qquad \Sigma y = 104.54 \qquad \Sigma x^2 = 36{,}056$$
$$\Sigma xy = 4376.36 \qquad \Sigma y^2 = 683.4470$$

Does this data strongly suggest that there is a negative (inverse) linear relationship between temperature and pH? State and test the relevant hypotheses using a significance level of .01.

13.26 In anthropological studies, an important characteristic of fossils is cranial capacity. Frequently skulls are at least partially decomposed, so it is necessary to use other characteristics to obtain information about capacity. One such measure that has been used is the length of the lambda-opisthion chord. An article that appeared in the 1971 *Amer. Journal of Physical Anthropology* entitled "Vertesszollos and the Presapiens Theory" reported the accompanying data for $n = 7$ *Homo erectus* fossils.

x (chord length in mm)	78	75	78	81
y (capacity in cm^3)	850	775	750	975

x (chord length in mm)	84	86	87
y (capacity in cm^3)	915	1015	1030

Suppose that from previous evidence, anthropologists had believed that for each 1-mm increase in chord length, cranial capacity would be expected to increase by 20 cm^3. Does this new experimental data strongly contradict prior belief?

13.3   Checking Model Adequacy

The simple linear regression model equation is

$$y = \alpha + \beta x + e$$

where e represents the random deviation of an observed y value from the population regression line $\alpha + \beta x$. The inferential methods presented in the previous section require some assumptions about e. Key assumptions follow:

1. At any particular x value, the distribution of e is a normal distribution.
2. At any particular x value, the standard deviation of e is σ, which is constant over all values of x.

Inferences based on the simple linear regression model continue to be reliable when model assumptions are slightly violated (for example, mild nonnormality of the random deviation distribution). However, using an estimated model in the face of grossly violated assumptions can result in very misleading conclusions. Therefore, it is desirable to have easily applied methods available for identifying such serious violations and for suggesting how a satisfactory model can be obtained.

Residual Analysis

If the deviations $e_1, e_2, \ldots, e_n$ from the population line were available, they could be examined for any inconsistencies with model assumptions. For example, a normal probability plot would suggest whether the normality assumption was tenable. But, since

$$e_1 = y_1 - (\alpha + \beta x_1)$$
$$\vdots$$
$$e_n = y_n - (\alpha + \beta x_n)$$

these deviations can be calculated only if the equation of the population line is known. In practice, this will never be the case. Instead, diagnostic checks must be based on the residuals

$$y_1 - \hat{y}_1 = y_1 - (a + bx_1)$$
$$\vdots$$
$$y_n - \hat{y}_n = y_n - (a + bx_n)$$

which are the deviations from the *estimated* line.

The values of the residuals will vary from sample to sample. When all model assumptions are met, the mean value of the residuals at any particular x value is zero. Any observation that gives a very large positive or negative residual should be examined carefully for any anomalous circumstances, such as a recording error or exceptional experimental conditions. Identifying residuals with unusually large magnitudes is made easier by inspecting **standardized residuals.**

Recall that a quantity is standardized by subtracting its mean value (zero in this case) and dividing by its true or estimated standard deviation. Thus,

$$\text{standardized residual} = \frac{\text{residual}}{\text{estimated standard deviation of residual}}$$

The value of a standardized residual tells how many standard deviations the corresponding residual lies from its expected value, zero.

Because residuals at different x values have different standard deviations (depending on the value of x for that observation),* computing the standardized residuals can be tedious. Fortunately, many computer regression programs provide standardized residuals as part of the output.

In Chapter 7, a normal probability plot was introduced as a technique for deciding whether the n observations in a random sample could plausibly have come from a normal population distribution. To assess whether the assumption that $e_1, e_2, \ldots,$ e_n all come from the same normal distribution is reasonable, we recommend a normal probability plot of the standardized residuals.

EXAMPLE 13.6   Landslides are common events in tree-growing regions of the Pacific Northwest, so their effect on timber growth is of special concern to foresters. The article "Effects of Landslide Erosion on Subsequent Douglas Fir Growth and Stocking Levels in the Western Cascades, Oregon" (*Soil Science Soc. of Amer. J.* (1984): 667–671) reported on the results of a study in which growth in a landslide area was compared with growth in a previously clear-cut area. We present data on clear-cut growth, with x = tree age (years) and y = 5-year height growth (cm). The scatter plot in Figure 13.12 is consistent with the assumptions of the simple linear regression model.

FIGURE 13.12 MINITAB scatter plot for the data of Example 13.6

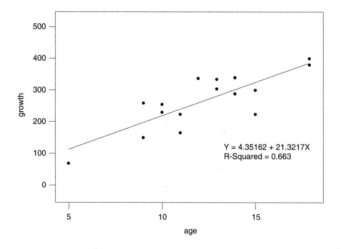

The residuals, their standard deviations, and the standardized residuals are given in Table 13.1. Except for $x = 5$ and $x = 18$, the two most extreme values, the residuals have roughly equal standard deviations. The residual with the largest magnitude, -99.2, initially seems quite extreme, but the corresponding standardized residual is only -2.04. That is, the residual is approximately 2 standard deviations below its expected value of zero, which is not terribly unusual in a sample of this size.

---

*The estimated standard deviation of the ith residual, $y_i - \hat{y}_i$, is $s_e\sqrt{1 - \dfrac{1}{n} - \dfrac{(x - \bar{x})^2}{S_{xx}}}$.

Table 13.1 *Data, residuals, and standardized residuals for Example 13.6*

Obs.	x	y	$\hat{y}$	Residual	Estimated Standard Deviation of Residual	Standardized Residual
1	5	70	111.0	−41.0	40.9	−1.00
2	9	150	196.2	−46.2	48.1	−.96
3	9	260	196.2	63.8	48.1	1.33
4	10	230	217.6	12.4	49.0	.25
5	10	255	217.6	37.4	49.0	.76
6	11	165	238.9	−73.9	49.5	−1.49
7	11	225	238.9	−13.9	49.5	−.28
8	12	340	260.2	79.8	49.8	1.60
9	13	305	281.5	23.5	49.7	.47
10	13	335	281.5	53.5	49.7	1.08
11	14	290	302.9	−12.9	49.4	−.26
12	14	340	302.9	37.1	49.4	.75
13	15	225	324.2	−99.2	48.7	−2.04
14	15	300	324.2	−24.2	48.7	−.50
15	18	380	388.1	−8.1	44.6	−.18
16	18	400	388.1	11.9	44.6	.27

On the standardized scale, no residual here is surprisingly large. Before standardization, there are some large residuals simply because there appears to be a substantial amount of variability about the true regression line ($s_e = 51.43$, $r^2 = .683$).

Figure 13.13 displays a normal probability plot of the standardized residuals and also one of the residuals. Notice that in this case the plots are nearly identical; it is usually the case that the two plots are similar. Although it is preferable to work with the standardized residuals, if you do not have access to a computer package or

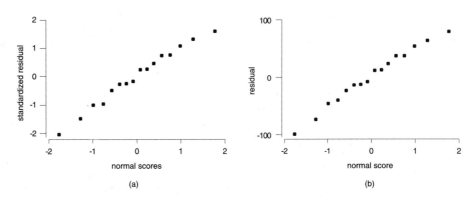

(a)   (b)

FIGURE 13.13 Normal probability plots for Example 13.6 (from MINITAB): (a) Standardized residuals; (b) Residuals

calculator that will produce standardized residuals, a plot of the unstandardized residuals should suffice. Few plots are straighter than these! The plots would not cause us to question the assumption of normality.

Plotting the Residuals

A plot of the (x, residual) pairs is called a **residual plot** and a plot of the (x, standardized residual) pairs is a **standardized residual plot.** Residual and standardized residual plots typically exhibit the same general shapes. If you are using a computer package or graphing calculator that calculates standardized residuals, we recommend using the standardized residual plot. If not, it is acceptable to use the residual plot instead.

A standardized residual or residual plot is often helpful in identifying unusual or highly influential observations and in checking for violations of model assumptions. A desirable plot is one that exhibits no particular pattern (such as curvature or a much greater spread in one part of the plot than in another), and one that has no point that is far removed from all the others. A point falling far above or below the horizontal line at height zero corresponds to a large standardized residual, which may indicate some kind of unusual behavior, such as a recording error, a nonstandard experimental condition, or an atypical experimental subject. A point that has an x value that differs greatly from others in the data set could have exerted excessive influence in determining the fitted line.

A standardized residual plot such as the one pictured in Figure 13.14(a) is desirable, since no point lies much outside the horizontal band between −2 and 2 (so there is no unusually large residual corresponding to an outlying observation); there is no point far to the left or right of the others (thus no observation that might greatly influence the fit), and there is no pattern to indicate that the model should somehow be modified. When the plot has the appearance of Figure 13.14(b), the fitted model should be changed to incorporate curvature (a nonlinear model).

The increasing spread from left to right in Figure 13.14(c) suggests that the variance of y is not the same at each x value but rather increases with x. A straight-line model may still be appropriate, but the best-fit line should be selected by using *weighted least squares* rather than ordinary least squares. This involves giving more weight to observations in the region exhibiting low variability and less weight to observations in the region exhibiting high variability. A specialized regression analysis text or a knowledgeable statistician should be consulted for details.

The standardized residual plots of Figure 13.14(d)–(e) show an extreme outlier and a potentially influential observation, respectively. Consider deleting the observation corresponding to such a point from the data set and refitting the same model. Substantial changes in estimates and various other quantities warn of instability in the data. The investigator should certainly carry out a more careful analysis and perhaps collect more data before drawing any firm conclusions. Improved computing power has allowed statisticians to develop and implement a variety of diagnostics for identifying unusual observations in a regression data set.

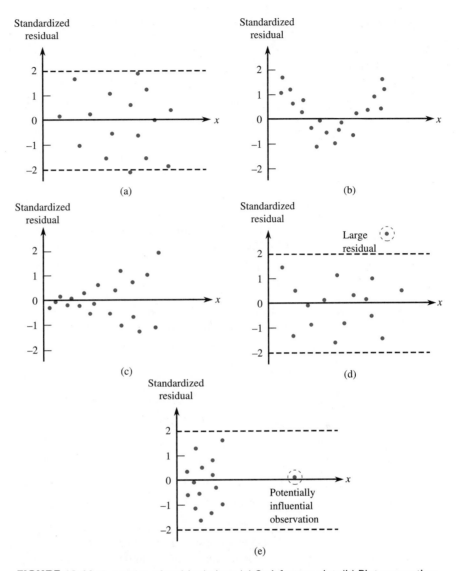

FIGURE 13.14 Examples of residual plots: (a) Satisfactory plot; (b) Plot suggesting that a curvilinear regression model is needed; (c) Plot indicating nonconstant variance; (d) Plot showing a large residual; (e) Plot showing a potentially influential observation

EXAMPLE 13.7   Figures 13.15(a) and (b) on page 666 display a standardized residual plot and a residual plot, respectively, for the data of Example 13.6 regarding tree age and 5-year growth. The first observation was at $x_1 = 5$ and the corresponding standardized residual was -1.00, so the first plotted point in the standardized residual plot is $(5, -1.00)$. Other points are similarly obtained and plotted. The plot shows no unusual behavior that might call for model modifications or further analysis.

FIGURE 13.15 Plots for the data of Example 13.6 (from MINITAB): (a) Standardized residual plot; (b) Residual plot

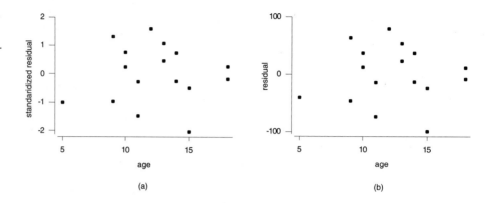

(a)   (b)

Note that the general pattern in the residual plot is similar to that of the standardized residual plot.

EXAMPLE 13.8

The article "Snow Cover and Temperature Relationships in North America and Eurasia" (*J. of Climate and Appl. Meteorology* (1983): 460–469) explored the relationship between October–November continental snow cover (x, in millions of km^2) and December–February temperature (y, in °C). The following data refers to Eurasia during the $n = 13$ time periods 1969–1970, 1970–1971, . . . , 1981–1982:

x	13.00	12.75	16.70	18.85	16.60	15.35	13.90
y	−13.5	−15.7	−15.5	−14.7	−16.1	−14.6	−13.4
Std. resid.	−.11	−2.19	−.36	1.23	−.91	−.12	.34

x	22.40	16.20	16.70	13.65	13.90	14.75
y	−18.9	−14.8	−13.6	−14.0	−12.0	−13.5
Std. resid.	−1.54	.04	1.25	−.28	1.54	.58

A simple linear regression analysis done by the authors yielded $r^2 = .52$, $r = −.72$, suggesting a significant linear relationship. This is confirmed by a model utility test. The scatter plot and standardized residual plot are displayed in Figure 13.16. There are no unusual patterns, although one standardized residual, −2.19, is a bit on the large side. The most interesting feature is the observation (22.40, −18.9), corresponding to a point far to the right of the others in these plots. This observation may have had a substantial influence on all aspects of the fit. The estimated slope when all 13 observations are included is $b = −.459$, and $s_b = .133$. When the potentially influential observation is deleted, the estimate of β based on the remaining 12 observations is $b = −.228$. Thus,

$$\text{change in slope} = \text{original } b − \text{new } b$$
$$= −.459 − (−.288)$$
$$= −.231$$

FIGURE 13.16 Plots for the data of Example 13.8 (from MINITAB): (a) Scatter plot; (b) Standardized residual plot

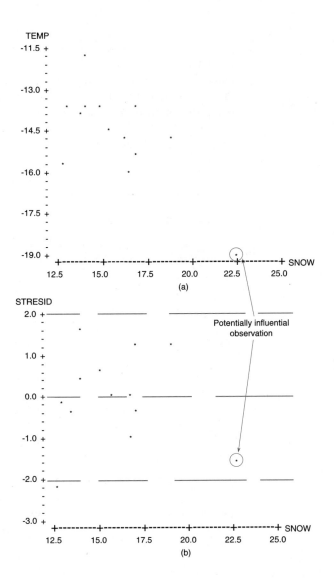

The change expressed in standard deviations is $-.231/.133 = -1.74$. Because b has changed by substantially more than 1 standard deviation, the observation under consideration appears to be highly influential.

Additionally, r^2 based just on the 12 observations is only .13, and the t ratio for testing $\beta = 0$ is not significant. Evidence for a linear relationship is much less conclusive in light of this analysis. The investigators should seek a climatological explanation for the influential observation and collect more data, which can be used to find an effective relationship.

EXAMPLE 13.9

Example 13.4 presented data on x = treadmill time and y = ski time. A simple linear regression model was fit to the data prior to constructing a confidence interval for β, the average change in ski time associated with a 1-minute increase in treadmill time. The validity of the t confidence interval depends on the assumptions that the distribution of the residuals from the population regression line at any fixed x is approximately normal and that the variance of this distribution does not depend on x. Constructing a normal probability plot of the standardized residuals and a standardized residual plot will provide insight as to whether these assumptions are in fact reasonable.

MINITAB was used to fit the simple linear regression model and compute the standardized residuals, resulting in the values shown in Table 13.2.

Table 13.2 *Data, residuals, and standardized residuals for Example 13.9*

Obs.	Treadmill	Ski Time	Residual	Std. Residual
1	7.7	71.000	.172	.10
2	8.4	71.400	2.206	1.13
3	8.7	65.000	−3.494	−1.74
4	9.0	68.700	.906	.44
5	9.6	64.400	−1.994	−.96
6	9.6	69.400	3.006	1.44
7	10.0	63.000	−2.461	−1.18
8	10.2	64.600	−.394	−.19
9	10.4	66.900	2.373	1.16
10	11.0	62.600	−.527	−.27
11	11.7	61.700	.206	.12

Figure 13.17 shows a normal probability plot of the standardized residuals and a standardized residual plot. The normal probability plot is quite straight and the standardized residual plot does not show evidence of any patterns or of increasing spread. These observations support the use of the t confidence interval in Example 13.4.

FIGURE 13.17 Normal probability plot and standardized residual plot for Example 13.9

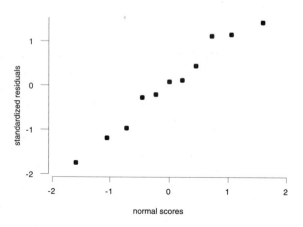

(*continued*)

FIGURE 13.17 (Continued)

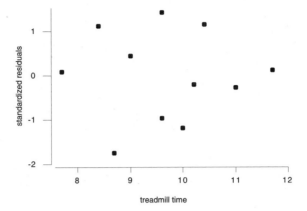

EXAMPLE 13.10

(Example 13.5 continued)   Use of the model utility test in Example 13.5 led to the conclusion that there was a useful linear relationship between

  y = average reading time

and

  x = image quality

The validity of this test requires that the random error distribution be normal and that the variability of the random error does not change with x. Let's construct a normal probability plot of the standardized residuals and a standardized residual plot to see whether these assumptions are reasonable. Table 13.3 gives the residuals and standardized residuals that are needed to construct the plots in Figure 13.18 (page 670).

Table 13.3 *Data, residuals, and standardized residuals for Example 13.10*

Obs	Image Quality	Reading Time	Residual	Std. Residual
1	4.30	8.0000	−.1392	−.80
2	4.55	8.3000	.2479	1.27
3	5.55	7.8000	.0964	.42
4	5.65	7.2500	−.4188	−1.81
5	5.95	7.7000	.1358	.59
6	6.30	7.5000	.0578	.26
7	6.45	7.6000	.2101	.98
8	6.45	7.2000	−.1899	−.89

FIGURE 13.18 Normal probability plot and standardized residual plot for Example 13.10

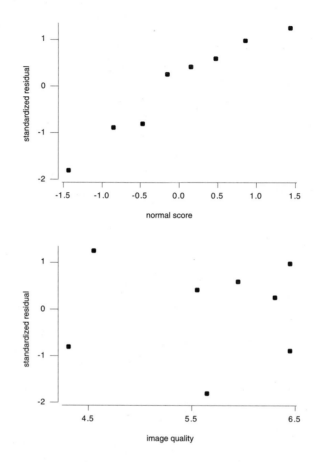

The normal probability plot is quite straight, and the standardized residual plot does not show evidence of any pattern or increasing spread. This supports the use of the model utility test in Example 13.5.

Occasionally you will see a residual plot or a standardized residual plot with $\hat{y}$ plotted on the horizontal axis rather than x. Since $\hat{y}$ is just a linear function of x, using $\hat{y}$ rather than x will change the scale of the horizontal axis but will not change the pattern of the points in the plot. As a consequence, plots that use $\hat{y}$ on the horizontal axis can be interpreted in the same manner as plots that use x.

When the distribution of the random deviation e has heavier tails than does the normal distribution, observations with large standardized residuals are not that unusual. Such observations can have great effects on the estimated regression line when the least squares approach is used. In recent years, statisticians have proposed a number of alternative methods — called **robust,** or **resistant,** methods — for fitting a line. Such methods give less weight to outlying observations than does the method of least squares, without deleting them from the data set. The most widely used robust procedures require a substantial amount of computation, so a good computer program is necessary. Associated confidence-interval and hypothesis-testing procedures are still in the developmental stage.

Exercises 13.27 – 13.33

13.27 Exercise 13.8 gave data on x = treadmill run time to exhaustion and y = 20-km ski time. The x values and corresponding standardized residuals from a simple linear regression are as follows.

x	7.7	8.4	8.7	9.0	9.6	9.6
St. resid.	.10	1.13	−1.74	.44	−.96	1.44

x	10.0	10.2	10.4	11.0	11.7
St. resid.	−1.18	−.19	1.16	−.27	.12

Construct a standardized residual plot. Does the plot exhibit any unusual features?

13.28 The article "Vital Dimensions in Volume Perception: Can the Eye Fool the Stomach?" (*Journal of Marketing Research* (1999): 313–326) gave the accompanying data on the dimensions of 27 representative food products (Gerber baby food, Cheez Whiz, Skippy Peanut Butter, and Ahmed's tandoori paste, to name a few).

Row	Max. Width	Min. Width
1	2.50	1.80
2	2.90	2.70
3	2.15	2.00
4	2.90	2.60
5	3.20	3.15
6	2.00	1.80
7	1.60	1.50
8	4.80	3.80
9	5.90	5.00
10	5.80	4.75
11	2.90	2.80
12	2.45	2.10
13	2.60	2.20
14	2.60	2.60
15	2.70	2.60
16	3.10	2.90
17	5.10	5.10
18	10.20	10.20
19	3.50	3.50
20	2.70	1.20
21	3.00	1.70

(continued)

Row	Max. Width	Min. Width
22	2.70	1.75
23	2.50	1.70
24	2.40	1.20
25	4.40	1.20
26	7.50	7.50
27	4.25	4.25

a. Fit the simple linear regression model that would allow prediction of the maximum width (in cm) of a food container based on its minimum width (in cm).
b. Calculate the standardized residuals (or just the residuals if you don't have access to a computer program that gives standardized residuals), and plot them to determine whether there are any outliers. Would eliminating these outliers increase the accuracy of predictions using this regression? If so, eliminate the outlying data point (a 1-liter coke bottle) and refit the regression.
c. Interpret the estimated slope and, if appropriate, the intercept.
d. Do you think that the assumptions of the simple linear regression model are reasonable? Give statistical evidence for your answer.

13.29 The authors of the article "Age, Spacing and Growth Rate of *Tamarix* as an Indication of Lake Boundary Fluctuations at Sebkhet Kelbia, Tunisia" (*J. of Arid Environ.* (1982): 43–51) used a simple linear regression model to describe the relationship between y = vigor (average width in centimeters of the last two annual rings) and x = stem density (stems/m^2). The estimated model was based on the following data. Also given are the standardized residuals.

x	4	5	6	9	14
y	.75	1.20	.55	.60	.65
St. resid.	−.28	1.92	−.90	−.28	.54

x	15	15	19	21	22
y	.55	0	.35	.45	.40
St. resid.	.24	−2.05	−.12	.60	.52

a. What assumptions are required for the simple linear regression model to be appropriate?

b. Construct a normal probability plot of the standardized residuals. Does the assumption that the random deviation distribution is normal appear to be reasonable? Explain.

c. Construct a standardized residual plot. Are there any unusually large residuals?

d. Is there anything about the standardized residual plot that would cause you to question the use of the simple linear regression model to describe the relationship between x and y?

13.30 The article "Effects of Gamma Radiation on Juvenile and Mature Cuttings of Quaking Aspen" (*Forest Science* (1967): 240–245) reported the following data on x = exposure time to radiation (kR/16 hours) and y = dry weight of roots (mg):

x	0	2	4	6	8
y	110	123	119	86	62

a. Construct a scatter plot for this data. Does the plot suggest that the simple linear regression model might be appropriate?

b. The estimated regression line for this data is $\hat{y} = 127 - 6.65x$ and the standardized residuals are as given.

x	0	2	4	6	8
St. resid.	−1.55	.68	1.25	−.05	−1.06

Construct a standardized residual plot. What does the plot suggest about the adequacy of the simple linear regression model?

13.31 Carbon aerosols have been identified as a contributing factor in a number of air quality problems. In a chemical analysis of diesel engine exhaust, x = mass ($\mu g/cm^2$) and y = elemental carbon ($\mu g/cm^2$) were recorded ("Comparison of Solvent Extraction and Thermal Optical Carbon Analysis Methods: Application to Diesel Vehicle Exhaust Aerosol" *Environ. Science Tech.* (1984): 231–234). The estimated regression line for this data set is $\hat{y} = 31 + .737x$. The accompanying table gives the observed x and y values and the corresponding standardized residuals.

x	164.2	156.9	109.8	111.4	87.0
y	181	156	115	132	96
St. resid.	2.52	.82	.27	1.64	.08

(continued)

x	161.8	230.9	106.5	97.6	79.7
y	170	193	110	94	77
St. resid.	1.72	−.73	.05	−.77	−1.11

x	118.7	248.8	102.4	64.2	89.4
y	106	204	98	76	89
St. resid.	−1.07	−.95	−.73	−.20	−.68

x	108.1	89.4	76.4	131.7	100.8
y	102	91	97	128	88
St. resid.	−.75	−.51	.85	0.00	−1.49

x	78.9	387.8	135.0	82.9	117.9
y	86	310	141	90	130
St. resid.	−.27	−.89	.91	−.18	1.05

a. Construct a standardized residual plot. Are there any unusually large residuals? Do you think that there are any influential observations?

b. Is there any pattern in the standardized residual plot that would indicate that the simple linear regression model is not appropriate?

c. Based on your plot in part **a**, do you think that it is reasonable to assume that the variance of y is the same at each x value? Explain.

13.32 An investigation of the relationship between traffic flow x (thousands of cars per 24 hr) and lead content y of bark on trees near the highway ($\mu g/g$ dry weight) yielded the accompanying data. A simple linear regression model was fit, and the resulting estimated regression line was $\hat{y} = 28.7 + 33.3x$. Both residuals and standardized residuals are also given.

x	8.3	8.3	12.1	12.1	17.0
y	227	312	362	521	640
Residual	−78.1	6.9	−69.6	89.4	45.3
St. resid.	−.99	.09	−.81	1.04	.51

x	17.0	17.0	24.3	24.3	24.3
y	539	728	945	738	759
Residual	−55.7	133.3	107.2	−99.8	−78.8
St. resid.	−.63	1.51	1.35	−1.25	−.99

a. Plot the (x, residual) pairs. Does the resulting plot suggest that a simple linear regression model is an appropriate choice? Explain your reasoning.

b. Construct a standardized residual plot. Does the plot differ significantly in general appearance from the plot in part **a**?

13.33 The accompanying data on x = U.S. population (millions) and y = crime index (millions) appeared in the article "The Normal Distribution of Crime" (*J. of Police Science and Admin.* (1975): 312–318). The author comments that "The simple linear regression analysis remains one of the most useful tools for crime prediction." When observations are made sequentially in time, the residuals or standardized residuals should be plotted in time order (that is, first the one for time $t = 1$ (1963 here), then the

one for time $t = 2$, and so on). Notice that here x increases with time, so an equivalent plot is of residuals or standardized residuals versus x. Using $\hat{y} = -47.26 + .260x$, calculate the residuals and plot the (x, residual) pairs. Does the plot exhibit a pattern that casts doubt on the appropriateness of the simple linear regression model? Explain.

Year	1963	1964	1965	1966	1967	1968	1969	1970	1971	1972	1973
x	188.5	191.3	193.8	195.9	197.9	199.9	201.9	203.2	206.3	208.2	209.9
y	2.26	2.60	2.78	3.24	3.80	4.47	4.99	5.57	6.00	5.89	8.64

13.4 Inferences Based on the Estimated Regression Line (Optional)

The number obtained by substituting a particular x value x^* into the equation of the estimated regression line has two different interpretations. It is a point estimate of the average y value when $x = x^*$, and it is also a point prediction of a single y value to be observed when $x = x^*$. How precise is this estimate or prediction? That is, how close might $a + bx^*$ be to the actual mean value, $\alpha + \beta x^*$, or to a particular y observation? Because both a and b vary in value from sample to sample (each one is a statistic), the statistic $a + bx^*$ also has different values for different samples. The way in which this statistic varies in value with different samples is summarized by its sampling distribution. Properties of the sampling distribution are used to obtain both a confidence interval formula for $\alpha + \beta x^*$ and a prediction interval formula for a particular y observation. The width of the corresponding interval conveys information about the precision of the estimate or prediction.

Properties of the Sampling Distribution of $a + bx$ for a Fixed x Value

Let x^* denote a particular value of the independent variable x. When the four basic assumptions of the simple linear regression model are satisfied, the sampling distribution of the statistic $a + bx^*$ has the following properties:

1. The mean value of $a + bx^*$ is $\alpha + \beta x^*$, so $a + bx^*$ is an unbiased statistic for estimating the average y value when $x = x^*$.

2. The standard deviation of the statistic $a + bx^*$, denoted by σ_{a+bx^*}, is given by

$$\sigma_{a+bx^*} = \sigma\sqrt{\frac{1}{n} + \frac{(x^* - \bar{x})^2}{S_{xx}}}$$

3. The distribution of $a + bx^*$ is normal.

As you can see from the formula for σ_{a+bx^*}, the standard deviation of $a + bx^*$ is larger when $(x^* - \bar{x})^2$ is large than when it is small; that is, $a + bx^*$ tends to be a more precise estimate of $\alpha + \beta x^*$ when x^* is close to the center of the x values at which observations were made than when x^* is far from the center.

The standard deviation σ_{a+bx^*} cannot be calculated from the sample data, because the value of σ is unknown. However, it can be estimated by using s_e in place of σ. Using its mean and estimated standard deviation to standardize $a + bx^*$ gives a variable with a t distribution.

The **estimated standard deviation of the statistic $a + bx^*$,** denoted by s_{a+bx^*}, is given by

$$s_{a+bx^*} = s_e\sqrt{\frac{1}{n} + \frac{(x^* - \bar{x})^2}{S_{xx}}}$$

When the four basic assumptions of the simple linear regression model are satisfied, the probability distribution of the standardized variable

$$t = \frac{a + bx^* - (\alpha + \beta x^*)}{s_{a+bx^*}}$$

is the t distribution with df $= n - 2$.

Inferences About the Mean y Value $\alpha + \beta x^*$

In previous chapters, standardized variables were manipulated algebraically to give confidence intervals of the form

(point estimate) $\pm$ (critical value) $\cdot$ (estimated standard deviation)

A parallel argument leads to the following interval.

Confidence Interval for a Mean y Value

When the basic assumptions of the simple linear regression model are met, a **confidence interval for $\alpha + \beta x^*$,** the average y value when x has value x^*, is

$a + bx^* \pm (t$ critical value$) \cdot s_{a+bx^*}$

where the t critical value is based on df $= n - 2$. Appendix Table III gives critical values corresponding to the most frequently used confidence levels.

Because s_{a+bx^*} is larger the farther x^* is from $\bar{x}$, the confidence interval widens as x^* moves away from the center of the data.

EXAMPLE 13.11   Physical characteristics of sharks are of interest to surfers and scuba divers, as well as marine researchers. The accompanying data on $x =$ length (ft) and $y =$ jaw width (in) for 44 sharks was found in articles appearing in the magazines *Skin Diver* and *Scuba News*:

x	18.7	12.3	18.6	16.4	15.7	18.3	14.6	15.8	14.9	17.6	12.1
y	17.5	12.3	21.8	17.2	16.2	19.9	13.9	14.7	15.1	18.5	12.0

x	16.4	16.7	17.8	16.2	12.6	17.8	13.8	12.2	15.2	14.7	12.4
y	13.8	15.2	18.2	16.7	11.6	17.4	14.2	14.8	15.9	15.3	11.9

x	13.2	15.8	14.3	16.6	9.4	18.2	13.2	13.6	15.3	16.1	13.5
y	11.6	14.3	13.3	15.8	10.2	19.0	16.8	14.2	16.9	16.0	15.9

x	19.1	16.2	22.8	16.8	13.6	13.2	15.7	19.7	18.7	13.2	16.8
y	17.9	15.7	21.2	16.3	13.0	13.3	14.3	21.3	20.8	12.2	16.9

Because it is difficult to measure jaw width in living sharks, researchers would like to determine whether it is possible to estimate jaw width from body length, which is more easily measured. A scatter plot of the data (Figure 13.19) shows a linear pattern and is consistent with use of the simple linear regression model.

FIGURE 13.19 A scatter plot for the data of Example 13.11

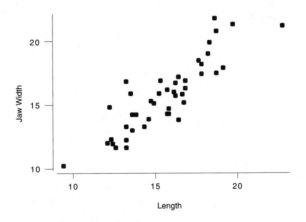

From the accompanying MINITAB output, it is easily verified that

$$a = .688 \qquad b = .96345 \qquad \text{SSResid} = 79.49$$
$$\text{SSTo} = 339.02 \qquad s_e = 1.376 \qquad r^2 = .766$$

Regression Analysis

The regression equation is
Jaw Width = 0.69 + 0.963 Length

Predictor	Coef	StDev	T	P
Constant	0.688	1.299	0.53	0.599
Length	0.96345	0.08228	11.71	0.000

S = 1.376   R-Sq = 76.6%   R-Sq(adj) = 76.0%

Analysis of Variance

Source	DF	SS	MS	F	P
Regression	1	259.53	259.53	137.12	0.000
Residual Error	42	79.49	1.89		
Total	43	339.02			

From the data, we can also compute $S_{xx} = 279.8718$. Since $r^2 = .766$, the simple linear regression model explains 76.6% of the variability in jaw width. The model utility test also confirms the usefulness of this model (P-value $= .000$).

Let's use the data to compute a 90% confidence interval for the mean jaw width for 15-ft-long sharks. The mean jaw width when length is 15 is $\alpha + \beta(15)$. The point estimate is

$$a + b(15) = .688 + .96345(15) = 15.140 \text{ in}$$

Since

$$\bar{x} = \frac{\Sigma x}{n} = \frac{685.80}{44} = 15.586$$

the estimated standard deviation of $a + b(15)$ is

$$s_{a+b(15)} = s_e\sqrt{\frac{1}{n} + \frac{(15 - \bar{x})^2}{S_{xx}}}$$

$$= (1.376)\sqrt{\frac{1}{44} + \frac{(15 - 15.586)^2}{279.8718}}$$

$$= .213$$

The t critical value for df $= 42$ is 1.68 (using the tabled value for df $= 40$). We now have all of the relevant quantities needed to compute a 90% confidence interval:

$$a + b(15) \pm (t \text{ critical value}) \cdot s_{a+b(15)} = 15.140 \pm (1.68)(.213)$$
$$= 15.140 \pm .358$$
$$= (14.782, 15.498)$$

Based on this sample data, we can be 90% confident that the mean jaw width for sharks whose length is 15 ft is between 14.782 and 15.498 in. As with all confidence intervals, the 90% confidence level means that we have used a *method* to construct this interval estimate that has a 10% error rate.

We have just considered estimation of the mean y value at a fixed $x = x^*$. When the basic assumptions of the simple linear regression model are met, the true value of this mean is $\alpha + \beta x^*$. The reason that our point estimate, $a + bx^*$ is not exactly equal to $\alpha + \beta x^*$ is that the values of α and β are not known, so they have been estimated from sample data. As a result, the estimate $a + bx^*$ is subject to sampling variability and the extent to which the estimated line might differ from the true population line is reflected in the width of the confidence interval.

A Prediction Interval for a Single y

We now turn our attention to the problem of predicting a single y value at a particular $x = x^*$ (rather than estimating the mean y value when $x = x^*$). This is equivalent to trying to predict the y value of an individual point in a scatter plot of the population. If we use the estimated regression line to obtain a point prediction,

$a + bx^*$, there are two reasons why this prediction will probably not be exactly equal to the true y value. First, as was the case when estimating a mean y value, the estimated line is not going to be exactly equal to the true population regression line. But, in the case of predicting a single y value, there is an additional source of error — even if we knew the true population line, individual points would not fall exactly on the population line. This implies that there is more uncertainty associated with predicting a single y value at a particular x^* than with estimating the mean y value at x^*. This extra uncertainty is reflected in the width of the corresponding intervals.

An interval for a single y value, y^*, is called a **prediction interval** (to distinguish it from the confidence interval for a mean y value). The interpretation of a prediction interval is very similar to the interpretation of a confidence interval. A 95% prediction interval for y^* is constructed using a method for which 95% of all possible samples would yield interval limits capturing y^*; only 5% of all samples would give an interval that did not include y^*.

Manipulation of a standardized variable similar to the one from which a confidence interval was obtained gives the following prediction interval.

Prediction Interval for a Single y Value

When the four basic assumptions of the simple linear regression model are met, a **prediction interval for y^*,** a single y observation made when $x = x^*$, has the form

$$a + bx^* \pm (t \text{ critical value}) \cdot \sqrt{s_e^2 + s_{a+bx^*}^2}$$

The prediction interval and confidence interval are centered at exactly the same place, $a + bx^*$. The addition of s_e^2 under the square root symbol makes the prediction interval wider — often substantially so — than the confidence interval.

EXAMPLE 13.12   (Example 13.11 continued)   In Example 13.11, we computed a 90% confidence interval for the mean jaw width of sharks whose length is 15 ft. Suppose that we are interested in predicting the jaw width of a single shark of length 15 ft. The required calculations for a 90% prediction interval for y^* are

$$a + b(15) = .688 + .96245(15) = 15.140$$
$$s_e^2 = (1.376)^2 = 1.8934$$
$$s_{a+b(15)}^2 = (.213)^2 = .0454$$

The t critical value for df $= 42$ and a 90% prediction level is 1.68 (using the tabled value for df $= 40$). Substitution into the prediction interval formula then gives

$$a + b(15) \pm (t \text{ critical value})\sqrt{s_e^2 + s_{a+b(15)}^2} = 15.140 \pm (1.68)\sqrt{1.9388}$$
$$= 15.140 \pm 2.339$$
$$= (12.801, 17.479)$$

We can be 90% confident that an individual shark with length 15 ft will have a jaw width between 12.801 and 17.479 in. Notice that, as expected, this 90% prediction

FIGURE 13.20 MINITAB output for the data of Example 13.12

Regression Analysis

The regression equation is
Jaw Width = 0.69 + 0.963 Length

Predictor	Coef	StDev	T	P
Constant	0.688	1.299	0.53	0.599
Length	0.96345	0.08228	11.71	0.000

S = 1.376   R-Sq = 76.6%   R-Sq(adj) = 76.0%

Analysis of Variance

Source	DF	SS	MS	F	P
Regression	1	259.53	259.53	137.12	0.000
Residual Error	42	79.49	1.89		
Total	43	339.02			

Predicted Values

	Fit	StDev Fit	95.0% CI	95.0% PI
$x^* = 15 \rightarrow$	15.140	0.213	(14.710, 15.569)	(12.330, 17.949)
$x^* = 20 \rightarrow$	19.957	0.418	(19.113, 20.801)	(17.055, 22.859)

interval is much wider than the 90% confidence interval when $x^* = 15$ from Example 13.11.

Figure 13.20 gives MINITAB output that includes a 95% confidence interval and a 95% prediction interval both when $x^* = 15$ and $x^* = 20$. The intervals for $x^* = 20$ are both wider than the corresponding intervals for $x^* = 15$ because 20 is farther from $\bar{x}$ (the center of the sample x values) than is 15. Each prediction interval is wider than the corresponding confidence interval. Figure 13.21 is a MINITAB plot

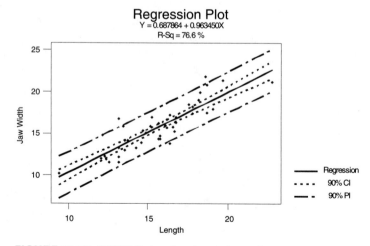

FIGURE 13.21 MINITAB plot showing estimated regression line and 90% confidence and prediction limits for the data of Example 13.12

that shows the estimated regression line as well as 90% confidence limits and prediction limits.

Exercises 13.34 – 13.48

13.34 Explain the difference between a confidence interval and a prediction interval. How can a prediction level of 95% be interpreted?

13.35 Suppose that a regression data set is given and you are asked to obtain a confidence interval. How would you tell from the phrasing of the request whether the interval is for β or for $\alpha + \beta x^*$?

13.36 In Exercise 13.17, we considered a regression of y = oxygen consumption on x = time spent exercising. Summary quantities given there yield

$$n = 20 \qquad \bar{x} = 2.50 \qquad S_{xx} = 25$$
$$b = 97.26 \qquad a = 592.10 \qquad s_e = 16.486$$

a. Calculate $s_{a+b(2.0)}$, the estimated standard deviation of the statistic $a + b(2.0)$.
b. Without any further calculation, what is $s_{a+b(3.0)}$, and what reasoning did you use to obtain it?
c. Calculate the estimated standard deviation of the statistic $a + b(2.8)$.
d. For what value x^* is the estimated standard deviation of $a + bx^*$ smallest, and why?

13.37 Example 13.3 gave data on x = ambient temperature (°C) and y = cavity depth (cm).
a. Calculate a confidence interval using a 95% confidence level for the true average change in depth associated with a 1° increase in temperature.
b. Calculate a confidence interval using a 95% confidence level for the true average depth when temperature is 1°C.

13.38 The data of Exercise 13.25, in which x = milk temperature and y = milk pH, yields

$$n = 16 \qquad \bar{x} = 43.375 \qquad S_{xx} = 7325.75$$
$$b = -.00730608 \qquad a = 6.843345 \qquad s_e = .0356$$

a. Obtain a 95% confidence interval for $\alpha + \beta(40)$, the true average milk pH when the milk temperature is 40°C.
b. Calculate a 99% confidence interval for the true average milk pH when the milk temperature is 35°C.
c. Would you recommend using the data to calculate a

95% confidence interval for the true average pH when the temperature is 90°C? Why or why not?

13.39 Return to the regression of y = milk pH on x = milk temperature described in Exercise 13.38.
a. Obtain a 95% prediction interval for a single pH observation to be made when milk temperature = 40°C.
b. Calculate a 99% prediction interval for a single pH observation when milk temperature = 35°C.
c. When the milk temperature is 60°C, would a 99% prediction interval be wider than the intervals of parts **a** and **b**? Answer without calculating the interval.

13.40 An experiment was carried out by geologists to see how the time necessary to drill a distance of 5 ft in rock (y, in min) depended on the depth at which the drilling began (x, in ft, between 0 and 400). We show part of the MINITAB output obtained from fitting the simple linear regression model ("Mining Information," *Am. Stat.* (1991): 4–9).

The regression equation is
time = 4.79 + 0.0144 depth

Predictor	Coef	Stdev	t-ratio	p
Constant	4.7896	0.6663	7.19	0.000
depth	0.014388	0.002847	5.05	0.000

s = 1.432   R-sq = 63.0%   R-sq(adj) = 60.5%

Analysis of Variance

SOURCE	DF	SS	MS	F	p
Regression	1	52.378	52.378	25.54	0.000
Error	15	30.768	2.051		
Total	16	83.146			

a. What proportion of observed variation in time can be explained by the simple linear regression model?
b. Does the simple linear regression model appear to be useful?
c. MINITAB reported that $s_{a+b(200)}$ = .347. Calculate a confidence interval at the 95% confidence level for the true average time when depth = 200 ft.
d. A single observation on time is to be made when drilling starts at a depth of 200 ft. Use a 95% prediction interval to predict the resulting value of time.

e. MINITAB gave (8.147, 10.065) as a 95% confidence interval for true average time when depth = 300. Calculate a 99% confidence interval for this average.

13.41 According to "Reproductive Biology of the Aquatic Salamander *Amphiuma tridactylum* in Louisiana" (*Journal of Herpetology*) 1999: 100–105, the size of a female salamander's snout is correlated with the number of eggs in her clutch. The following data is consistent with summary quantities reported in the article. MINITAB output is also included.

Snout-Vent Length	32	53	53	53	54
Clutch Size	45	215	160	170	190

Snout-Vent Length	57	57	58	58	59
Clutch Size	200	270	175	245	215

Snout-Vent Length	63	63	64	67
Clutch Size	170	240	245	280

The regression equation is
y = - 133 + 5.92 x

Predictor	Coef	StDev	T	P
Constant	-133.02	64.30	-2.07	0.061
x	5.919	1.127	5.25	0.000

S = 33.90   R-Sq = 69.7%   R-Sq(adj) = 67.2%

Additional summary statistics are

$n = 14$ $\bar{x} = 56.5$ $\bar{y} = 201.4$

$\Sigma x^2 = 45,958$ $\Sigma y^2 = 613,550$ $\Sigma xy = 164,969$

a. What is the equation of the regression line for predicting clutch size based on snout-vent length?
b. Calculate the standard deviation of b.
c. Test the hypothesis that the slope of the population line is positive.
d. Predict the clutch size for a salamander with a snout-vent length of 65 using a 95% interval.
e. Predict the clutch size for a salamander with snout-vent length of 105.

13.42 The article first introduced in Exercise 13.28 of Section 13.3 gave data on the dimensions of 27 representative food products.
a. Use the information given there to test the hypothesis that there is a positive linear relationship between the minimum width and the maximum width of an object.
b. Calculate and interpret s_e.
c. Calculate a 95% confidence interval for the mean maximum width of products with a minimum width of 6 cm.
d. Calculate a 95% prediction interval for the maximum width of a food package with a minimum width of 6 cm.

13.43 The shelf life of packaged food depends on many factors. Dry cereal is considered to be a moisture-sensitive product (no one likes soggy cereal!), with the shelf life determined primarily by moisture content. In a study of the shelf life of one particular brand of cereal, x = time on shelf (stored at 73°F and 50% relative humidity) and y = moisture content were recorded. The resulting data is from "Computer Simulation Speeds Shelf Life Assessments" (*Package Engr.* (1983): 72–73).

x	0	3	6	8	10	13	16
y	2.8	3.0	3.1	3.2	3.4	3.4	3.5

x	20	24	27	30	34	37	41
y	3.1	3.8	4.0	4.1	4.3	4.4	4.9

a. Summary quantities are

$\Sigma x = 269$ $\Sigma y = 51$ $\Sigma xy = 1081.5$

$\Sigma y^2 = 7745$ $\Sigma x^2 = 190.78$

Find the equation of the estimated regression line for predicting moisture content from time on the shelf.
b. Does the simple linear regression model provide useful information for predicting moisture content from knowledge of shelf time?
c. Find a 95% interval for the moisture content of an individual box of cereal that has been on the shelf 30 days.
d. According to the article, taste tests indicate that this brand of cereal is unacceptably soggy when the moisture content exceeds 4.1. Based on your interval in part **c**, do you think that a box of cereal that has been on the shelf 30 days will be acceptable? Explain.

13.44 For the cereal data of Exercise 13.43, the average x value is 19.21. Would a 95% confidence interval with $x^* = 20$ or $x^* = 17$ be wider? Explain. Answer the same question for a prediction interval.

13.45 High blood-lead levels are associated with a number of different health problems. The article "A Study of the Relationship between Blood Lead Levels and Occupational Lead Levels" (*Am. Stat.* (1983): 471) gave data on x = air-lead level ($\mu g/m^3$) and y = blood-lead level ($\mu g/dL$). Summary quantities (based on a subset of the data given in a plot in the article) are

$n = 15$ $\Sigma x = 1350$ $\Sigma y = 600$

$\Sigma x^2 = 155,400$ $\Sigma y^2 = 24,869.33$ $\Sigma xy = 57,760$

a. Find the equation of the estimated regression line.
b. Estimate the mean blood-lead level for people who work where the air-lead level is 100 $\mu g/m^3$ using a 90% interval.
c. Construct a 90% prediction interval for the blood-

lead level of a particular person who works where the air-lead level is 100 $\mu g/m^3$.

d. Explain the difference in interpretation of the intervals computed in parts **b** and **c**.

13.46 A regression of y = sunburn index for a pea plant on x = distance from an ultraviolet light source was considered in Exercise 13.22. The data and summary statistics presented there give

$$n = 15 \qquad \bar{x} = 40.60 \qquad \Sigma (x - \bar{x})^2 = 3311.60$$
$$b = -.0565 \qquad a = 4.500 \qquad \text{SSResid} = .8430$$

a. Calculate a 95% confidence interval for the true average sunburn index when the distance from the light source is 35 cm.

b. When two 95% confidence intervals are computed, it can be shown that the simultaneous confidence level is at least $[100 - 2(5)]\% = 90\%$. That is, if both intervals are computed for a first sample, for a second sample, yet again for a third, and so on, in the long run at least 90% of the samples will result in intervals both of which capture the values of the corresponding population characteristics. Calculate confidence intervals for the true mean sunburn index when the distance is 35 cm and when the distance is 45 cm in such a way that the simultaneous confidence level is at least 90%.

c. If two 99% intervals were computed, what do you think could be said about the simultaneous confidence level?

d. If a 95% confidence interval were computed for the true mean index when $x = 35$, another 95% confidence interval were computed when $x = 40$, and yet another one when $x = 45$, what do you think would be the simultaneous confidence level for the three resulting intervals?

e. Return to part **d** and answer the question posed there if the individual confidence level for each interval were 99%.

13.47 By analogy with the discussion in Exercise 13.46, when two different prediction intervals are computed, each at the 95% prediction level, the *simultaneous prediction level* is at least $[100 - 2(5)]\% = 90\%$.

a. Return to Exercise 13.46 and obtain prediction intervals for sunburn index both when distance is 35 cm and when distance is 45 cm, so that the simultaneous prediction level is at least 90%.

b. If three different 99% prediction intervals are calculated for distances (cm) of 35, 40, and 45, respectively, what can be said about the simultaneous prediction level?

13.48 The article "Performance Test Conducted for a Gas Air-Conditioning System" (*Am. Soc. of Heating, Refrigerating, and Air Cond. Engr.* (1969): 54) reported the following data on maximum outdoor temperature (x) and hours of chiller operation per day (y) for a 3-ton residential gas air-conditioning system:

x	72	78	80	86	88	92
y	4.8	7.2	9.5	14.5	15.7	17.9

Suppose that the system is actually a prototype model, and the manufacturer does not wish to produce this model unless the data strongly indicates that when maximum outdoor temperature is 82°F, the true average number of hours of chiller operation is less than 12. The appropriate hypothesis is then

$$H_0: \quad \alpha + \beta(82) = 12 \qquad \text{versus} \qquad H_a: \quad \alpha + \beta(82) < 12$$

Use the statistic

$$t = \frac{a + b(82) - 12}{s_{a+b(82)}}$$

which has a t distribution based on $(n - 2)$ df when H_0 is true, to test the hypotheses at significance level .01.

13.5 Inferences About the Population Correlation Coefficient (Optional)

The sample correlation coefficient r, defined in Chapter 5, measures how strongly the x and y values in a *sample* of pairs are linearly related to one another. There is an analogous measure of how strongly x and y are related in the entire *population* of pairs from which the sample $(x_1, y_1), \ldots, (x_n, y_n)$ was obtained. It is called the **population correlation coefficient** and is denoted by ρ. As with r, ρ must be between

−1 and 1, and it assesses the extent of any *linear* association in the population. To have $\rho = 1$ or -1, all (x, y) pairs in the population must lie exactly on a straight line. The value of ρ is a population characteristic and is generally unknown. The sample correlation coefficient, r, can be used as the basis for making inferences about ρ.

A Test for Independence ($\rho = 0$)

Investigators are often interested in detecting not just linear association but association of *any* kind. When there is no association of any type between x and y values, statisticians say that the two variables are *independent*. In general, $\rho = 0$ is not equivalent to the independence of x and y. However, there is one special, yet frequently occurring, situation in which the two conditions ($\rho = 0$ and independence) are identical. This is when the pairs in the population have what is called a **bivariate normal distribution.** The essential feature of such a distribution is that for *any* fixed x value, the distribution of associated y values is normal, *and* for any fixed y value, the distribution of x values is normal. As an example, suppose that height x and weight y have a bivariate normal distribution in the American adult male population. (There is good empirical evidence for this.) Then when $x = 68$ in, weight y has a normal distribution; when $x = 72$ in, weight is normally distributed; when $y = 160$ lb, height x has a normal distribution; when $y = 175$ lb, height has a normal distribution; and so on. In this example, of course, x and y are not independent, since a large height value tends to be associated with a large weight value and a small height value with a small weight value.

There is no easy way to check the assumption of bivariate normality, especially when the sample size n is small. A partial check can be based on the following property: If (x, y) has a bivariate normal distribution, then x alone has a normal distribution and so does y. This suggests constructing a normal probability plot of $x_1, x_2, \ldots, x_n$ and a separate normal probability plot of $y_1, \ldots, y_n$. If either plot shows a substantial departure from a straight line, bivariate normality is a questionable assumption. If both plots are reasonably straight, bivariate normality is plausible, although no guarantee can be given.

The test of independence (zero correlation) is a t test. The formula for the test statistic essentially involves standardizing the estimate r under the assumption that the null hypothesis $H_0: \rho = 0$ is true.

A Test for Independence in a Bivariate Normal Population

Null hypothesis: $H_0: \rho = 0$

Test statistic: $t = \dfrac{r}{\sqrt{\dfrac{1 - r^2}{n - 2}}}$

The t critical value is based on df $= n - 2$.

Alternative hypothesis: *P-value*:

$H_a: \rho > 0$ (positive dependence)   Area under the appropriate t curve to the right of the computed t

(*continued*)

Alternative hypothesis: *P-value*:

$H_a: \rho < 0$ (negative dependence)   Area under the appropriate t curve to the left of the
                                            computed t

$H_a: \rho \neq 0$ (dependence)   (i) $2 \cdot$ (area to the right of t) if t is positive

                                            or

                                            (ii) $2 \cdot$ (area to the left of t) if t is negative

Assumption: r is the correlation coefficient for a *random sample* from a *bivariate normal population*.

EXAMPLE 13.13   Is there any relationship between the age of an individual and the age at which that individual plans to retire? The authors of the article "Predictors of Planned Retirement Age" (*Psychology and Aging* (1995): 76–83) hypothesized that younger people would plan to retire earlier than would older people. They selected a random sample of $n = 303$ workers from a wide variety of job positions and recorded $x =$ current age and $y =$ planned retirement age for each individual in the sample. The sample correlation coefficient was $r = .35$. Does this data support the stated hypothesis? Let's carry out a test using a significance level of .01.

1. $\rho =$ the correlation between age and planned retirement age in the population from which the 303 workers were selected

2. $H_0: \rho = 0$

3. $H_a: \rho > 0$ (large values of $y =$ planned retirement age tend to be associated with large values of $x =$ age)

4. $\alpha = .01$

5. *Test statistic*: $t = \dfrac{r}{\sqrt{\dfrac{1 - r^2}{n - 2}}}$

6. *Assumptions*:   Without the actual data values, it is not possible to assess whether the assumption of bivariate normality is reasonable. For purposes of this example, we will assume that it is reasonable and proceed with the test. Had the data been available, we would have looked at normal probability plots of the x values and of the y values.

7. *Calculation*: $t = \dfrac{.35}{\sqrt{\dfrac{1 - (.35)^2}{301}}} = \dfrac{.35}{\sqrt{.0029153}} = 6.48$

8. *P-value*:   The t curve with 301 df is essentially indistinguishable from the z curve, and Appendix Table IV shows that the area under this curve and to the right of 6.48 is approximately zero. That is, *P-value* ≈ 0.

9. *Conclusion*:   The *P*-value is smaller than any reasonable significance level; in particular, *P*-value $\leq .01$. We reject H_0, as did the authors of the article, and confirm their conjecture. Notice, though, that according to the guidelines

described in Chapter 5, $r = .35$ suggests only a weak linear relationship. Since $r^2 = .1225$, fitting the simple linear regression model to the data would result in only about 12% of observed variation in planned retirement age being explained.

In the context of regression analysis, the hypothesis of no linear relationship ($H_0: \beta = 0$) was tested using the t ratio $t = b/s_b$. Some algebraic manipulation shows that

$$\frac{r}{\sqrt{\dfrac{1 - r^2}{n - 2}}} = \frac{b}{s_b}$$

so the model utility test can also be used to test for independence in a bivariate normal population. The reason for using the formula for t that involves r is that when we are interested only in correlation, the extra effort involved in computing the regression quantities b, a, SSResid, s_e, and s_b need not be expended.

Other inferential procedures for drawing conclusions about ρ — a confidence interval or a test of hypothesis with nonzero hypothesized value — are somewhat complicated. One of the chapter references can be consulted for details.

Exercises 13.49 – 13.57

13.49 Discuss the difference between r and ρ.

13.50 **a.** If the sample correlation coefficient is equal to 1, is it necessarily true that $\rho = 1$?
b. If $\rho = 1$, is it necessarily true that $r = 1$?

13.51 A sample of $n = 353$ college faculty members was obtained, and the values of $x =$ teaching evaluation index and $y =$ annual raise were determined ("Determination of Faculty Pay: An Agency Theory Perspective," *Academy of Management J.* (1992): 921–955). The resulting value of r was .11. Does there appear to be a linear association between these variables in the population from which the sample was selected? Carry out a test of hypothesis using a significance level of .05. Does the conclusion surprise you? Explain.

13.52 It seems plausible that higher rent for retail space could be justified only by a higher level of sales. A random sample of $n = 53$ specialty stores in a chain was selected, and the values of $x =$ annual dollar rent per square foot and $y =$ annual dollar sales per square foot were determined, resulting in $r = .37$ ("Association of Shopping Center Anchors with Performance of a Nonanchor Specialty

Chain Store," *J. of Retailing* (1985): 61–74). Carry out a test at significance level .05 to see whether there is in fact a positive linear association between x and y in the population of all such stores.

13.53 Television is regarded by many as a prime culprit for the difficulty many students have in performing well in school. The article "The Impact of Athletics, Part-Time Employment, and Other Activities on Academic Achievement" (*J. of College Student Development* (1992): 447–453) reported that for a random sample of $n = 528$ college students, the sample correlation coefficient between time spent watching television (x) and grade point average (y) was $r = -.26$.
a. Does this suggest that there is a negative correlation between these two variables in the population from which the 528 students were selected? Use a test with significance level .01.
b. If y were regressed on x, would the regression explain a substantial percentage of the observed variation in grade point average? Explain your reasoning.

13.54 The accompanying summary quantities for $x =$ particulate pollution ($\mu g/m^3$) and $y =$ luminance (.01 cd/m²)

were calculated from a representative sample of data that appeared in the article "Luminance and Polarization of the Sky Light at Seville (Spain) Measured in White Light" (*Atmos. Environ.* (1988): 595–599).

$$n = 15 \qquad \Sigma x = 860 \qquad \Sigma y = 348$$
$$\Sigma x^2 = 56{,}700 \qquad \Sigma y^2 = 8954 \qquad \Sigma xy = 22{,}265$$

a. Test to see whether there is a positive correlation between particulate pollution and luminance in the population from which the data was selected.
b. What proportion of observed variation in luminance can be attributed to the approximate linear relationship between luminance and particulate pollution?

13.55 In a study of bacterial concentration in surface and subsurface water ("Pb and Bacteria in a Surface Microlayer" *J. of Marine Research* (1982): 1200–1206), the accompanying data was obtained.

		Concentration (× 10^6/mL)			
Surface	48.6	24.3	15.9	8.29	5.75
Subsurface	5.46	6.89	3.38	3.72	3.12

(continued)

	Concentration (× 10^6/mL)			
Surface	10.8	4.71	8.26	9.41
Subsurface	3.39	4.17	4.06	5.16

Summary quantities are

$$\Sigma x = 136.02 \qquad \Sigma y = 39.35$$
$$\Sigma x^2 = 3602.65 \qquad \Sigma y^2 = 184.27 \qquad \Sigma xy = 673.65$$

Using a significance level of .05, determine whether the data supports the hypothesis of a linear relationship between surface and subsurface concentration.

13.56 A sample of $n = 500$ (x, y) pairs was collected and a test of H_0: $\rho = 0$ versus H_a: $\rho \neq 0$ was carried out. The resulting P-value was computed to be .00032.
a. What conclusion would be appropriate at level of significance .001?
b. Does this small P-value indicate that there is a very strong linear relationship between x and y (a value of ρ that differs considerably from zero)? Explain.

13.57 A sample of $n = 10{,}000$ (x, y) pairs resulted in $r = .022$. Test H_0: $\rho = 0$ versus H_a: $\rho \neq 0$ at level .05. Is the result statistically significant? Comment on the practical significance of your analysis.

13.6 Interpreting the Results of Statistical Analyses

Although regression analysis can be used as a tool for summarizing bivariate data, more often than not it is also used to enable researchers to make inferences about the way in which two variables are related.

What to Look For in Published Data

Here are some things to consider when you evaluate research that involves fitting a simple linear regression model.

- Which variable is the dependent variable? Is it a numerical (rather than qualitative) variable?

- If sample data has been used to estimate the coefficients in a simple linear regression model, is it reasonable to think that the basic assumptions required for inference are met?

- Does the model appear to be useful? Are the results of a model utility test reported? What is the P-value associated with the test?

- Has the model been used in an appropriate way? Has the regression equation been used to predict y for values of the independent variable that are outside the range of the data?

■ If a correlation coefficient is reported, is it accompanied by a test of significance? Are the results of the test interpreted properly?

The effects of caffeine were examined in the article "Withdrawal Syndrome After the Double-Blind Cessation of Caffeine Consumption" (*New England J. of Medicine* (1992): 1109–1113). The authors found that the dose of caffeine was significantly correlated with a measure of insomnia and also with latency on a test of reaction time. They reported $r = .26$ (*P*-value $= .042$) for insomnia and $r = .31$ (*P*-value $= .014$) for latency. Since both *P*-values are small, the authors concluded that the population correlation coefficients differ from zero. We should also note, however, that the reported correlation coefficients are not particularly large and do not indicate a strong linear relationship.

Regression analysis was used by the authors of the article "On the Generality of the 'Sit and Reach' Test: An Analysis of Flexibility Data for an Aging Population" (*Research Quarterly for Exercise and Sport* (1990): 326–330). A linear regression model was used to describe the relationship between $y =$ head rotation to the left (in degrees) and $x =$ age (in years). The authors report that "regression analysis showed an age effect for leftward head rotation (significant beta weight, $t_{78} = 3.10$, $P < .003$)." The term *beta weight* refers to the β term in the linear regression model. A model utility test was performed, and the value of the test statistic was 3.10. The test was based on 78 degrees of freedom (this is the meaning of the subscript on t_{78}). The associated *P*-value leads to the rejection of $H_0: \beta = 0$ and to the authors's stated conclusion of an "age effect." The reported regression equation is

$$\hat{y} = 87.5 - .31x$$

The coefficient of age in this equation ($-.31$) indicates that, on average, leftward head rotation decreases by about .31 degrees with each additional year of age. (It is interesting to note that age did not turn out to be an important predictor of rightward motion!)

The article does not provide enough information to determine whether the basic assumptions required for the validity of the test reported are reasonable, and the authors do not address this issue directly. If raw data were available, we could check on this by looking at such things as a standardized residual plot and various normal probability plots.

Summary of Key Concepts and Formulas

Term or Formula	Comment
Simple linear regression model, $y = \alpha + \beta x + e$	This model assumes that there is a line with slope β and y intercept α, called the population (true) regression line, such that an observation deviates from the line by a random amount e. The random deviation is assumed to have a normal distribution with mean zero and standard deviation σ, and random deviation for different observations are assumed to be independent of one another.
Estimated regression line, $\hat{y} = a + bx$	The least squares line introduced in Chapter 5.

$$s_e = \sqrt{\frac{\text{SSResid}}{n-2}}$$

The point estimate of the standard deviation σ, with associated degrees of freedom $n - 2$.

$$s_b = \frac{s_e}{\sqrt{S_{xx}}}$$

The estimated standard deviation of the statistic b.

$$b \pm (t \text{ critical value})s_b$$

A confidence interval for the slope β of the population regression line, where the t critical value is based on $(n - 2)$ df.

$$t = \frac{b - \text{hyp. val.}}{s_b}$$

The test statistic for testing hypotheses about β. The test is based on $(n - 2)$ df.

Model utility test, with test statistic $t = \dfrac{b}{s_b}$

A test of $H_0: \beta = 0$, which asserts that there is no useful linear relationship between x and y, versus $H_a: \beta \neq 0$, the claim that there is a useful linear relationship.

Residual analysis

Methods based on the residuals or standardized residuals for checking the assumptions of a regression model.

Standardized residual

A residual divided by its standard deviation.

Standardized residual plot

A plot of the $(x, \text{standard residual})$ pairs. A pattern in this plot suggests a problem with the simple linear regression model.

$$s_{a+bx^*} = s_e\sqrt{\frac{1}{n} + \frac{(x^* - \bar{x})^2}{S_{xx}}}$$

The estimated standard deviation of the statistic $a + bx^*$, where x^* denotes a particular value of x.

$$a + bx^* \pm (t \text{ crit. val.})s_{a+bx^*}$$

A confidence interval for $\alpha + \beta x^*$, the average value of y when $x = x^*$.

$$a + bx^* \pm (t \text{ crit. val.})\sqrt{s_e^2 + s_{a+bx^*}^2}$$

A prediction interval for a single y value to be observed when $x = x^*$.

Population correlation coefficient ρ

A measure of the extent to which the x and y values in an entire population are linearly related.

$$t = \frac{r}{\sqrt{\dfrac{1 - r^2}{n-2}}}$$

The test statistic for testing $H_0: \rho = 0$, according to which (assuming a bivariate normal population distribution) x and y are independent of one another.

Supplementary Exercises 13.58 – 13.78

13.58 The effects of grazing animals on grasslands have been the focus of numerous investigations by ecologists. One such study, reported in "The Ecology of Plants, Large Mammalian Herbivores, and Drought in Yellowstone National Park" (*Ecology* (1992): 2043–2058), proposed using the simple linear regression model to relate y = green biomass concentration (g/cm^3) to x = elapsed time since snowmelt (days).

a. The estimated regression equation was given as $\hat{y} = 106.3 - .640x$. What is the estimate of average change in biomass concentration associated with a 1-day increase in elapsed time?

b. What value of biomass concentration would you predict when elapsed time is 40 days?

c. The sample size was $n = 58$, and the reported value of the coefficient of determination was .470. Does this suggest that there is a useful linear relationship between the two variables? Carry out an appropriate test.

13.59 A random sample of $n = 347$ students was selected, and each one was asked to complete several questionnaires, from which a Coping Humor Scale value x and a Depression Scale value y were determined ("Depression and Sense of Humor" (*Psychological Reports* (1994):

1473–1474). The resulting value of the sample correlation coefficient was $-.18$.

a. The investigators reported that P-value $< .05$. Do you agree?

b. Is the sign of r consistent with your intuition? Explain. (Higher scale values correspond to more developed sense of humor and greater extent of depression.)

c. Would the simple linear regression model give accurate predictions? Why or why not?

13.60 Exercise 5.62 gave data on x = bound rubber content (%) and y = tensile modulus (MPa), a measure of coupling effectiveness. A scatter plot shows a substantial linear pattern. Summary quantities are

$$n = 11 \qquad \Sigma x = 292.9 \qquad \Sigma x^2 = 8141.75$$

$$\Sigma y = 69.03 \qquad \Sigma y^2 = 442.1903 \qquad \Sigma xy = 1890.200$$

a. Calculate point estimates for the slope and vertical intercept of the population regression line.

b. Calculate a point estimate for the standard deviation σ.

c. Using a .01 significance level, does the data suggest the existence of a useful linear relationship between x and y?

13.61 Return to the summary quantities of Exercise 13.60.

a. Use a 95% confidence interval to estimate the true mean tensile modulus when the bound rubber content is 20%.

b. Use a 95% prediction interval to predict the value of the tensile modulus resulting from a single observation made when the rubber content is 20%.

c. What aspect of the problem description in part **a** tells you that the requested confidence interval is not for β?

13.62 Data on x = depth of flooding and y = flood damage was given in Exercise 5.63. Summary quantities are

$$n = 13 \qquad \Sigma x = 91 \qquad \Sigma x^2 = 819$$

$$\Sigma y = 470 \qquad \Sigma y^2 = 19{,}118 \qquad \Sigma xy = 3867$$

a. Does the data suggest the existence of a positive linear relationship (one in which an increase in y tends to be associated with an increase in x)? Test using a .05 significance level.

b. Predict flood damage resulting from a claim made when depth of flooding is 3.5 ft, and do so in a way that conveys information about the precision of the prediction.

13.63 Exercise 13.8 gave data on x = treadmill run time to exhaustion and y = 20-km ski time for a sample of 11 biathletes. Use the accompanying MINITAB output to answer the following questions.

The regression equation is
ski = 88.8 - 2.33 tread

Predictor	Coef	Stdev	t-ratio	p
Constant	88.796	5.750	15.44	0.000
tread	-2.3335	0.5911	-3.95	0.003

s = 2.188   R-sq = 63.4%   R-sq(adj) = 59.3%

Analysis of Variance

SOURCE	DF	SS	MS	F	p
Regression	1	74.630	74.630	15.58	0.003
Error	9	43.097	4.789		
Total	10	117.727			

a. Carry out a test at significance level .01 to decide whether the simple linear regression model is useful.

b. Estimate the average change in ski time associated with a 1-minute increase in treadmill time, and do so in a way that conveys information about the precision of estimation.

c. MINITAB reported that $s_{a+b(10)} = .689$. Predict ski time for a single biathlete whose treadmill time is 10 min, and do so in a way that conveys information about the precision of prediction.

d. MINITAB also reported that $s_{a+b(11)} = 1.029$. Why is this larger than $s_{a+b(10)}$?

13.64 Exercise 5.42 presented data on x = squawfish length and y = maximum size of salmonid consumed, both in mm. Use the accompanying MINITAB output along with the values $\bar{x} = 343.27$ and $S_{xx} = 69{,}112.18$ to answer the following questions.

The regression equation is
size = -89.1 + 0.729 length

Predictor	Coef	Stdev	t-ratio	p
Constant	-89.09	16.83	-5.29	0.000
length	0.72907	0.04778	15.26	0.000

s = 12.56   R-sq = 96.3%   R-sq(adj) = 95.9%

Analysis of Variance

SOURCE	DF	SS	MS	F	p
Regression	1	36736	36736	232.87	0.000
Error	9	1420	158		
Total	10	38156			

a. Does there appear to be a useful linear relationship between length and size?

b. Does it appear that the average change in maximum size associated with a 1-mm increase in length is less than .8 mm? State and test the appropriate hypotheses.

c. Estimate average maximum size when length is 325 mm in a way that conveys information about the precision of estimation.

d. How would the estimate when length is 250 mm compare to the estimate of part **c**? Answer without actually calculating the new estimate.

13.65 A sample of $n = 61$ penguin burrows was selected, and values of both $y =$ trail length (m) and $x =$ soil hardness (force required to penetrate the substrate to a depth of 12 cm with a certain gauge, in kg) were determined for each one ("Effects of Substrate on the Distribution of Magellanic Penguin Burrows," *The Auk* (1991): 923–933). The equation of the least squares line was $\hat{y} = 11.607 - 1.4187x$, and $r^2 = .386$.
  a. Does the relationship between soil hardness and trail length appear to be linear, with shorter trails associated with harder soil (as the article asserted)? Carry out an appropriate test of hypotheses.
  b. Using $s_e = 2.35$, $\bar{x} = 4.5$, and $\Sigma (x - \bar{x})^2 = 250$, predict trail length when soil hardness is 6.0 in a way that conveys information about the reliability and precision of the prediction.
  c. Would you use the simple linear regression model to predict trail length when hardness is 10.0? Explain your reasoning.

13.66 The article "Photocharge Effects in Dye Sensitized Ag[Br,I] Emulsions at Millisecond Range Exposures" (*Photographic Science and Engr.* (1981): 138–144) gave the accompanying data on $x =$ % light absorption and $y =$ peak photovoltage.

| x | 4.0 | 8.7 | 12.7 | 19.1 | 21.4 | 24.6 | 28.9 | 29.8 | 30.5 |
| y | .12 | .28 | .55 | .68 | .85 | 1.02 | 1.15 | 1.34 | 1.29 |

$$\Sigma x = 179.7 \qquad \Sigma x^2 = 4334.41$$
$$\Sigma y = 7.28 \qquad \Sigma y^2 = 7.4028 \qquad \Sigma xy = 178.683$$

  a. Construct a scatter plot of the data. What does it suggest?
  b. Assuming that the simple linear regression model is appropriate, obtain the equation of the estimated regression line.
  c. How much of the observed variation in peak photovoltage can be explained by the model relationship?
  d. Predict peak photovoltage when percent absorption is 19.1, and compute the value of the corresponding residual.
  e. The authors claimed that there is a useful linear relationship between the two variables. Do you agree? Carry out a formal test.
  f. Give an estimate of the average change in peak photovoltage associated with a 1% increase in light

absorption. Your estimate should convey information about the precision of estimation.
  g. Give an estimate of true average peak photovoltage when % light absorption is 20, and do so in a way that conveys information about precision.

13.67 Reduced visual performance with increasing age has been a much-studied phenomenon in recent years. This decline is due partly to changes in optical properties of the eye itself and partly to neural degeneration throughout the visual system. As one aspect of this problem, the article "Morphometry of Nerve Fiber Bundle Pores in the Optic Nerve Head of the Human" (*Exp. Eye Research* (1988): 559–568) presented the accompanying data on $x =$ age and $y =$ percentage of the cribriform area of the lamina scleralis occupied by pores.

| x | 22 | 25 | 27 | 39 | 42 | 43 | 44 | 46 | 46 |
| y | 75 | 62 | 50 | 49 | 54 | 49 | 59 | 47 | 54 |

| x | 48 | 50 | 57 | 58 | 63 | 63 | 74 | 74 |
| y | 52 | 58 | 49 | 52 | 49 | 31 | 42 | 41 |

  a. Suppose that the researchers had believed a priori that the average decrease in percentage area associated with a 1-year age increase was .5%. Does the data contradict this prior belief? State and test the appropriate hypotheses using a .10 significance level.
  b. Estimate true average percentage area covered by pores for all 50-year-olds in the population in a way that conveys information about the precision of estimation.

13.68 Occasionally an investigator may wish to compute a confidence interval for α, the y intercept of the true regression line, or test hypotheses about α. The estimated y intercept is simply the height of the estimated line when $x = 0$, since $a + b(0) = a$. This implies that s_a, the estimated standard deviation of the statistic a, results from substituting $x^* = 0$ in the formula for s_{a+bx^*}. The desired confidence interval is then

$$a \pm (t \text{ critical value}) s_a$$

and a test statistic is

$$t = \frac{a - \text{hypothesized value}}{s_a}$$

  a. The article "Comparison of Winter-Nocturnal Geostationary Satellite Infrared-Surface Temperature with Shelter-Height Temperature in Florida" (*Remote Sensing of the Environ.* (1983): 313–327) used the simple linear regression model to relate surface temperature as measured by a satellite (y) to actual air temperature (x) as determined from a thermocouple placed on a traversing vehicle. Selected data is given (read from a scatter plot in the article).

x	-2	-1	0	1	2	3	4	5	6	7
y	-3.9	-2.1	-2.0	-1.2	0	1.9	.6	2.1	1.2	3.0

Estimate the true regression line.

b. Compute the estimated standard deviation s_a. Carry out a test at level of significance .05 to see whether the y intercept of the true regression line differs from zero.

c. Compute a 95% confidence interval for α. Does the result indicate that $\alpha = 0$ is plausible? Explain.

13.69 Example 13.3 gave data on x = ambient temperature and y = cavity depth. Use the accompanying MINITAB output to estimate the true average depth when temperature = 0°C with a 95% confidence interval. (*Hint:* Refer to the previous exercise.)

The regression equation is
depth = 20.1 - 0.345 temp

Predictor	Coef	Stdev
Constant	20.1251	0.9402
temp	-0.34504	0.06008

S = 2.334   R-SQ = 76.7%

13.70 In some studies, an investigator has n (x, y) pairs sampled from one population and m (x, y) pairs from a second population. Let β and β' denote the slopes of the first and second population lines, respectively, and let b and b' denote the estimated slopes calculated from the first and second samples, respectively. The investigator may then wish to test the null hypothesis $H_0: \beta - \beta' = 0$ (that is, $\beta = \beta'$) against an appropriate alternative hypothesis. Suppose that σ^2, the variance about the population line, is the same for both populations. Then this common variance can be estimated by

$$s^2 = \frac{\text{SSResid} + \text{SSResid}'}{(n-2) + (m-2)}$$

where SSResid and SSResid' are the residual sums of squares for the first and second samples, respectively. With S_{xx} and S'_{xx} denoting the quantity $\Sigma (x - \bar{x})^2$ for the first and second samples, respectively, the test statistic is

$$t = \frac{b - b'}{\sqrt{\dfrac{s^2}{S_{xx}} + \dfrac{s^2}{S'_{xx}}}}$$

When H_0 is true, this statistic has a t distribution based on $(n + m - 4)$ df.

The given data is a subset of the data in the article "Diet and Foraging Model of *Bufa marinus* and *Lepto-*

dactylus ocellatus" (*J. of Herpetology* (1984): 138–146). The independent variable x is body length (cm) and the dependent variable y is mouth width (cm), with $n = 9$ observations for one type of nocturnal frog and $m = 8$ observations for a second type. Test at level .05 to see whether the slopes of the true regression lines for the two different frog populations are identical. (Summary statistics are given in the third table.)

Leptodactylus ocellatus

x	3.8	4.0	4.9	7.1	8.1	8.5	8.9	9.1	9.8
y	1.0	1.2	1.7	2.0	2.7	2.5	2.4	2.9	3.2

Bufa marinus

x	3.8	4.3	6.2	6.3	7.8	8.5	9.0	10.0
y	1.6	1.7	2.3	2.5	3.2	3.0	3.5	3.8

	Leptodactylus	Bufa
Sample size:	9	8
Σx	64.2	55.9
Σx^2	500.78	425.15
Σy	19.6	21.6
Σy^2	47.28	62.92
Σxy	153.36	163.36

13.71 Consider the following four (x, y) data sets: the first three have the same x values, so these values are listed only once (from Frank Anscombe, "Graphs in Statistical Analysis" *Amer. Statistician* (1973): 17–21).

Data Set	1–3	1	2	3	4	4
Variable	x	y	y	y	x	y
	10.0	8.04	9.14	7.46	8.0	6.58
	8.0	6.95	8.14	6.77	8.0	5.76
	13.0	7.58	8.74	12.74	8.0	7.71
	9.0	8.81	8.77	7.11	8.0	8.84
	11.0	8.33	9.26	7.81	8.0	8.47
	14.0	9.96	8.10	8.84	8.0	7.04
	6.0	7.24	6.13	6.08	8.0	5.25
	4.0	4.26	3.10	5.39	19.0	12.50
	12.0	10.84	9.13	8.15	8.0	5.56
	7.0	4.82	7.26	6.42	8.0	7.91
	5.0	5.68	4.74	5.73	8.0	6.89

For each of these data sets, the values of the summary quantities, $\bar{x}, \bar{y}, \Sigma (x - \bar{x})^2$, and $\Sigma (x - \bar{x})(y - \bar{y})$, are identical, so all quantities computed from these will be identical for the four sets: the estimated regression line, SSResid, s_e^2, r^2, and so on. The summary quantities provide no way of distinguishing among the four data sets. Based on a scatter plot for each set, comment on the appropriateness or inappropriateness of fitting the simple linear regression model in each case.

13.72 The accompanying scatter plot, based on 34 sediment samples with x = sediment depth (cm) and y = oil and grease content (mg/kg), appeared in the article "Mined Land Reclamation Using Polluted Urban Navigable Waterway Sediments" (*J. of Environ. Quality* (1984): 415–422). Discuss the effect that the observation (20, 33,000) will have on the estimated regression line. If this point were omitted, what can you say about the slope of the estimated regression line? What do you think will happen to the slope if this observation is included in the computations?

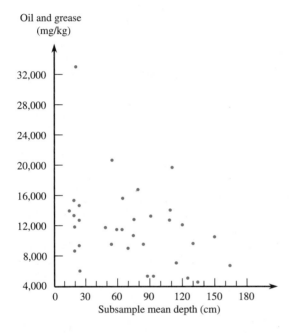

Oil and grease (mg/kg)

Subsample mean depth (cm)

13.73 The article "Improving Fermentation Productivity with Reverse Osmosis" (*Food Tech.* (1984): 92–96) gave the following data (read from a scatter plot) on y = glucose concentration (g/L) and x = fermentation time (days) for a blend of malt liquor.

x	1	2	3	4	5	6	7	8
y	74	54	52	51	52	53	58	71

a. Use the data to calculate the estimated regression line.
b. Does the data indicate a linear relationship between y and x? Test using a .10 significance level.
c. Using the estimated regression line of part **a,** compute the residuals and construct a plot of the residuals versus x (that is, of the (x, residual) pairs).
d. Based on the plot in part **c,** do you think that a lin-

ear model is appropriate for describing the relationship between y and x? Explain.

13.74 The employee relations manager of a large company was concerned that raises given to employees during a recent period might not have been based strictly on objective performance criteria. A sample of n = 20 employees was selected, and the values of x, a quantitative measure of productivity, and y, the percentage salary increase, were determined for each one. A computer package was used to fit the simple linear regression model, and the resulting output gave the P-value = .0076 for the model utility test. Does the percentage raise appear to be linearly related to productivity? Explain.

13.75 The article "Statistical Comparison of Heavy Metal Concentrations in Various Louisiana Sediments" (*Environ. Monitoring and Assessment* (1984): 163–170) gave the accompanying data on depth (m), zinc concentration (ppm), and iron concentration (%) for 17 core samples.

Core	Depth	Zinc	Iron
1	.2	86	3.4
2	2.0	77	2.9
3	5.8	91	3.1
4	6.5	86	3.4
5	7.6	81	3.2
6	12.2	87	2.9
7	16.4	94	3.2
8	20.8	92	3.4
9	22.5	90	3.1
10	29.0	108	4.0
11	31.7	112	3.4
12	38.0	101	3.6
13	41.5	88	3.7
14	60.0	99	3.5
15	61.5	90	3.4
16	72.0	98	3.5
17	104.0	70	4.8

a. Using a .05 significance level, test appropriate hypotheses to determine whether a correlation exists between depth and zinc concentration.
b. Using a .05 significance level, does the data strongly suggest a correlation between depth and iron concentration?
c. Calculate the slope and intercept of the estimated regression line relating y = iron concentration and x = depth.
d. Use the estimated regression equation to construct a 95% prediction interval for the iron concentration of a single core sample taken at a depth of 50 m.

e. Compute and interpret a 95% interval estimate for the true average iron concentration of core samples taken at 70 m.

13.76 The accompanying figure is from the article "Root and Shoot Competition Intensity Along a Soil Depth Gradient" (*Ecology* (1995): 673–682). It shows the relationship between above-ground biomass and soil depth within the experimental plots. The relationship is described by the linear equation: biomass $= -9.85 + 25.29$(soil depth) ($r^2 = .65$; $P < 0.001$; $n = 55$). Do you think the simple linear regression model is appropriate here? Explain. What would you expect to see in a plot of the standardized residuals versus x?

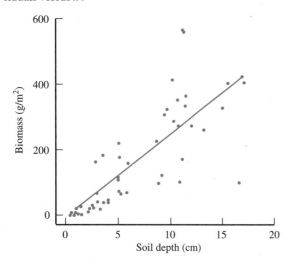

13.77 Give a brief answer, comment, or explanation for each of the following.

a. What is the difference between $e_1, e_2, \ldots, e_n$ and the n residuals?

b. The simple linear regression model states that $y = \alpha + \beta x$.

c. Does it make sense to test hypotheses about b?

d. SSResid is always positive.

e. A student reported that a data set consisting of $n = 6$ observations yielded residuals 2, 0, 5, 3, 0, and 1 from the least squares line.

f. A research report included the following summary quantities obtained from a simple linear regression analysis:

$$\Sigma\,(y - \bar{y})^2 = 615 \qquad \Sigma\,(y - \hat{y})^2 = 731$$

13.78 Some straightforward but slightly tedious algebra shows that

$$\text{SSResid} = (1 - r^2)\,\Sigma\,(y - \bar{y})^2$$

from which it follows that

$$s_e = \sqrt{\frac{n-1}{n-2}}\left(\sqrt{1 - r^2}\right)s_y$$

Unless n is quite small, $(n-1)/(n-2) \approx 1$, so

$$s_e \approx \left(\sqrt{1 - r^2}\right)s_y$$

a. For what value of r is s_e as large as s_y? What is the least squares line in this case?

b. For what values of r will s_e be much smaller than s_y?

References

See the references at the end of Chapter 5.

GRAPHING CALCULATOR EXPLORATION

13.1 Inference for the Regression Slope

By this time you should be an old hand at performing hypothesis tests. You can navigate your calculator menu with aplomb, and punch keys with the best of them! With this background, inferences concerning the slope of a population regression line using your calculator will hold few surprises.

Enter the data from Example 13.4, the treadmill data, into your calculator. Enter the treadmill run time in List1 and the ski time in List2. Navigating the menu system on the calculator should lead you to a screen similar to the one in the accompanying figure.

```
LinRegTTest
Xlist: L1
Ylist: L2
Freq: 1
β & ρ: ≠0 <0 >0:
RegEq:
Calculate
```

Here, L1 and L2 are the default values chosen by the calculator; if you entered the data in a different list, you will have to change these lists. The line, β & ρ: $\neq 0$ <0 >0, is where the alternative hypothesis is selected. For this example, we have selected the alternative hypothesis $\beta \neq 0$.

Placing the calculator's cursor on Calculate and pressing Enter produces the following screen: (This is another one of those larger-than-life presentations.)

```
LinRegTTest
Y=a+bx
β ≠ 0 and ρ ≠0
t=-3.9478
p=.0034
df=9.0000
a=88.7956
b=-2.3335
s=2.1883
r² = .6339
r=-.7962
```

Comparing the screen with quantities calculated in the text, you will find all the information needed to complete the hypothesis test. The notation is standard, except for the equation, $s = 2.1883$. This s, after checking with the quantities in the discussion of Example 13.4, turns out to be s_e, the estimated standard deviation of the random deviations. It is NOT the estimated standard deviation of the sample slope, s_b. This is unfortunate, since to construct a confidence interval for b we need to calculate $b \pm (t$ critical value$) \cdot s_b$.

Don't worry! You can construct the confidence interval for b with the aid of algebra and the statistics capabilities of your calculator. The standard error for b can be calculated from the information given and the sample variance of the treadmill times:

$$s_b = \frac{s_e}{\sqrt{S_{xx}}} = \frac{s_e}{\sqrt{\Sigma (x - \bar{x})^2}} = \frac{s_e}{\sqrt{(n - 1)s_x^2}} = \frac{s_e}{s_x \sqrt{n - 1}}$$

There are 11 data points; $s_x = 1.1707030$ (from the statistics menu for List1). Substituting gives

$$s_b = \frac{s_e}{s_x \sqrt{n - 1}} = \frac{2.188286324}{1.1707030\sqrt{10}} = .5910952$$

With the information now available, you can calculate the 95% confidence interval for the population slope:

$$b \pm (t \text{ critical value}) \cdot s_b = -2.3335 \pm (2.26) \cdot .5910952$$
$$= (-3.6694, -.9976)$$

The computations necessary to carry out the model utility test are also provided by the calculator. For the data of Example 13.4, the value of the test statistic for testing $H_0: \beta = 0$ versus $H_a: \beta \neq 0$ is given as $t = 3.9478$, with associated P-value $= .0034$. This information can be used to complete the steps required to reach a conclusion in the model utility test.

14

Multiple Regression Analysis

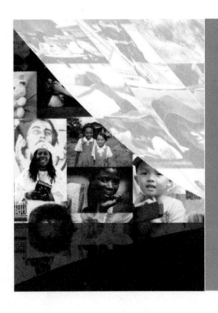

INTRODUCTION

The general objective of regression analysis is to establish a useful relationship between a dependent variable y and one or more independent (i.e., predictor or explanatory) variables. The simple linear regression model $y = \alpha + \beta x + e$, discussed in Chapter 13, has been used successfully by many investigators in a wide variety of disciplines to relate y to a single predictor variable x. In many situations, however, there will not be a strong relationship between y and any single predictor variable, but knowing the values of several independent variables may considerably reduce uncertainty concerning the associated y value. For example, some variation in house prices in a large city can certainly be attributed to house size, but knowledge of size by itself would not usually enable a bank appraiser to accurately assess (predict) a home's value. Price is also determined to some extent by other variables, such as age, lot size, number of bedrooms and bathrooms, distance from schools, and so on. As another example, let y denote a quantitative measure of wrinkle resistance in cotton cellulose fabric. Then the value of y may depend on such variable quantities as curing temperature, curing time, concentration of formaldehyde, and concentration of the catalyst sulfur dioxide used in the production process.

In this chapter, we extend the regression methodology developed in the previous chapter to *multiple regression models*, which include at least two predictor variables. Fortunately, many of the concepts developed in the context of simple linear regression carry over to multiple regression with little or no modification. The calculations required to fit such a model and make further inferences are *much* more tedious than those for simple linear regression, so a computer is an indispensable tool for multiple regression analysis. The computer's ability to perform a huge number of computations in a very short time has spurred the development of new methods for analyzing large data sets with many predictor variables. These include techniques for fitting numerous alternative models and choosing between them, tools for identifying influential observations, and both algebraic and graphical diagnostics designed to reveal potential violations of model assumptions. A single chapter can do little more than scratch the surface of this important subject area.

14.1 Multiple Regression Models

The relationship between a dependent or response variable y and two or more independent or predictor variables is deterministic if the value of y is completely determined, with no uncertainty, once values of the independent variables have been specified. Consider, for example, a school district in which teachers with no prior teaching experience and no college credits beyond a bachelor's degree start at an annual salary of $28,000. Suppose that for each year of teaching experience up to 20 years, a teacher receives an additional $800 per year and that each unit of post-college coursework up to a limit of 75 results in an extra $60 per year. Define three variables:

y = salary of a teacher who has at most 20 years teaching experience and at most 75 postcollege units

x_1 = number of years teaching experience

x_2 = number of postcollege units

Previously, x_1 and x_2 denoted the first two observations on the single variable x. In the usual notation for multiple regression, however, x_1 and x_2 represent two different variables.

The value of y is entirely determined by values of x_1 and x_2 through the equation

$$y = 28{,}000 + 800x_1 + 60x_2$$

Thus, if $x_1 = 10$ and $x_2 = 30$,

$$
\begin{aligned}
y &= 28{,}000 + (800)(10) + (60)(30) \\
&= 28{,}000 + 8000 + 1800 \\
&= 37{,}800
\end{aligned}
$$

If two different teachers both have the same x_1 values and the same x_2 values, they will also have identical y values.

Only rarely is y deterministically related to predictors $x_1, \ldots, x_k$. A probabilistic model results from adding a random deviation e to a deterministic function of the x_i's.

Definition

A **general additive multiple regression model,** which relates a dependent variable y to k predictor variables $x_1, x_2, \ldots, x_k$, is given by the model equation

$$y = \alpha + \beta_1 x_1 + \beta_2 x_2 + \cdots + \beta_k x_k + e$$

The random deviation e is assumed to be normally distributed with mean value 0 and variance σ^2 for any particular values of $x_1, \ldots, x_k$. This implies that for fixed $x_1, x_2, \ldots, x_k$ values, y has a normal distribution with variance σ^2 and

(mean y value for fixed $x_1, \ldots, x_k$ values) $= \alpha + \beta_1 x_1 + \beta_2 x_2 + \cdots + \beta_k x_k$

The β_i's are called **population regression coefficients**; each β_i can be interpreted as the true average change in y when the predictor x_i increases by 1 unit *and* the values of all the other predictors remain fixed.

The deterministic portion $\alpha + \beta_1 x_1 + \cdots + \beta_k x_k$ is called the **population regression function.**

As in simple linear regression, if σ^2 is quite close to 0, any particular observed y will tend to be quite near its mean value. When σ^2 is large, many of the y observations may deviate substantially from their mean y values.

EXAMPLE 14.1   Cardiorespiratory fitness is widely recognized as a major component of overall physical well-being. Direct measurement of maximal oxygen uptake (VO_2max) is the single best measure of such fitness, but direct measurement is time-consuming and expensive. It is therefore desirable to have a prediction equation for VO_2max in terms of easily obtained quantities. Consider the variables

$y = VO_2max$ (L/min)

$x_1 = $ weight (kg)

$x_2 = $ age (yr)

$x_3 = $ time necessary to walk a mile (min)

$x_4 = $ heart rate at the end of the walk (beats/min)

One possible population model for male students that is consistent with information given in the article "Validation of the Rockport Fitness Walking Test in College Males and Females" (*Research Quarterly for Exercise and Sport* (1994): 152–158) is

$$y = 5.0 + .01x_1 - .05x_2 - .13x_3 - .01x_4 + e \qquad \sigma = .4$$

The population regression function is

(mean y value for fixed $x_1, \ldots, x_4$) $= 5.0 + .01x_1 - .05x_2 - .13x_3 - .01x_4$

For individuals whose weight is 76 kg, age is 20 yr, walk time is 12 min, and heart rate is 140 beats/min,

(mean value of VO_2max) $= 5.0 + .01(76) - .05(20) - .13(12) - .01(140) = 1.80$

With $2\sigma = .80$, it is quite likely that an actual y value will be within .80 of the mean value (i.e., in the interval from 1.00 to 2.60).

The true average change in VO_2max when weight increases by 1 kg and the other predictors remain fixed is $\beta_1 = .01$. Similarly, since $\beta_2 = -.05$, when age increases by 1 year and the other three predictors remain constant, we expect VO_2max to decrease by .05 L/min.

A Special Case: Polynomial Regression

Consider again the case of a single independent variable x, and suppose that a scatter plot of the n sample (x, y) pairs has the appearance of Figure 14.1. The simple linear regression model is clearly not appropriate, but it does look as though a parabola (quadratic function) with equation $y = \alpha + \beta_1 x + \beta_2 x^2$ would provide a very good fit to the data for appropriately chosen values of α, β_1, and β_2. Just as the inclusion of the random deviation e in simple linear regression allowed an observation to deviate from the population regression line by a random amount, adding e to this quadratic function yields a probabilistic model in which an observation is allowed to fall above or below the parabola. The model equation is

$$y = \alpha + \beta_1 x + \beta_2 x^2 + e$$

FIGURE 14.1 A scatter plot that suggests the appropriateness of a quadratic probabilistic model

Let's rewrite the model equation by using x_1 to denote x and x_2 to denote x^2. The model equation then becomes

$$y = \alpha + \beta_1 x_1 + \beta_2 x_2 + e$$

This is a special case of the general multiple regression model with $k = 2$. You may wonder about the legitimacy of allowing one predictor variable to be a mathematical function of another predictor — here, $x_2 = (x_1)^2$. However, there is absolutely nothing in the general multiple regression model that prevents this. *In the model* $y = \alpha + \beta_1 x_1 + \cdots + \beta_k x_k + e$, *it is permissible to have several predictors that are mathematical functions of other predictors.* For example, starting with the two independent variables x_1 and x_2, we could create a model with $k = 4$ predictors in which x_1 and x_2 themselves are the first two predictor variables and $x_3 = (x_1)^2$, $x_4 = x_1 x_2$. (We will soon discuss the consequences of using a predictor such as x_4.) In particular, the general polynomial regression model begins with a single independent variable x and creates predictors $x_1 = x$, $x_2 = x^2$, $x_3 = x^3$, ..., $x_k = x^k$ for some specified value of k.

Definition

The **kth-degree polynomial regression model**

$$y = \alpha + \beta_1 x + \beta_2 x^2 + \cdots + \beta_k x^k + e$$

is a special case of the general multiple regression model with $x_1 = x$, $x_2 = x^2, \ldots, x_k = x^k$.

The **population regression function** (mean value of y for fixed values of the predictors) is

$$\alpha + \beta_1 x + \cdots + \beta_k x^k$$

The most important special case other than simple linear regression ($k = 1$) is the **quadratic regression model**

$$y = \alpha + \beta_1 x + \beta_2 x^2 + e$$

This model replaces the line of mean values $\alpha + \beta x$ in simple linear regression with a parabolic curve of mean values $\alpha + \beta_1 x + \beta_2 x^2$. If $\beta_2 > 0$, the curve opens upward, whereas if $\beta_2 < 0$, the curve opens downward (Figure 14.2). A less frequently encountered special case is that of cubic regression, in which $k = 3$.

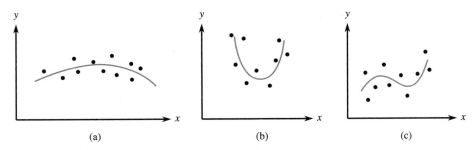

FIGURE 14.2 Polynomial regression: (a) Quadratic regression model with $\beta_2 < 0$; (b) Quadratic regression model with $\beta_2 > 0$; (c) Cubic regression model with $\beta_3 > 0$

EXAMPLE 14.2

Researchers have examined a variety of climatic variables in an attempt to gain an understanding of the mechanisms that govern rainfall runoff. The article "The Applicability of Morton's and Penman's Evapotranspiration Estimates in Rainfall Runoff Modeling" (*Water Resources Bull.* (1991): 611–620) reported on a study in which data on

x = cloud cover

and

y = daily sunshine (hr)

was gathered from a number of different locations. The authors used a cubic regression model to relate these variables (see Figure 14.2(c)). Suppose that the actual model equation for a particular location is

$$y = 11 - .400x - .250x^2 + .005x^3 + e$$

Then the population regression function is

(mean daily sunshine for given cloud cover x) = $11 - .400x - .250x^2 + .005x^3$

For example,

(mean daily sunshine when cloud cover is 4) = $11 - (.400)(4) - (.250)(4)^2 + (.005)(4)^3$
$$= 5.72$$

If $\sigma = 1$, it is quite likely that an observation on daily sunshine made when $x = 4$ would be between 3.72 and 7.72 hr.

———

The interpretation of β_i previously given for the general multiple regression model cannot be applied in polynomial regression. This is because all predictors are functions of the single variable x, so $x_i = x^i$ cannot be increased by 1 unit without changing the values of all the other predictor variables as well. *In general, the interpretation of regression coefficients requires extra care when some predictor variables are mathematical functions of other variables.*

Interaction Between Variables

Suppose that an industrial chemist is interested in the relationship between product yield (y) from a certain chemical reaction and two independent variables, $x_1 = $ reaction temperature and $x_2 = $ pressure at which the reaction is carried out. The chemist initially suggests that for temperature values between 80 and 110 in combination with pressure values ranging from 50 to 70, the relationship can be well described by the probabilistic model

$$y = 1200 + 15x_1 - 35x_2 + e$$

The regression function, which gives the mean y value for any specified values of x_1 and x_2, is then $1200 + 15x_1 - 35x_2$. Consider this mean y value for three different particular temperature values:

$x_1 = 90$:   mean y value = $1200 + (15)(90) - 35x_2 = 2550 - 35x_2$

$x_1 = 95$:   mean y value = $2625 - 35x_2$

$x_1 = 100$:   mean y value = $2700 - 35x_2$

Graphs of these three mean value functions (each a function only of pressure x_2, since the temperature value has been specified) are shown in Figure 14.3(a). Each graph is a straight line, and the three lines are parallel, each one having slope -35. Because of this, the average change in yield when pressure x_2 is increased by 1 unit is -35, irrespective of the fixed temperature value.

Since chemical theory suggests that the decline in average yield when pressure x_2 increases should be more rapid for a high temperature than for a low temperature, the chemist now has reason to doubt the appropriateness of the proposed model. Rather than the lines being parallel, the line for a temperature of 100 should be steeper than the line for a temperature of 95, and that line in turn should be steeper than the one for $x_1 = 90$. A model that has this property includes, in addi-

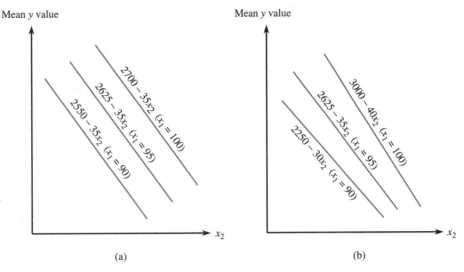

FIGURE 14.3 Graphs of mean y value for two different models: (a) $1200 + 15x_1 - 35x_2$; (b) $-4500 + 75x_1 + 60x_2 - x_1x_2$

tion to predictors x_1 and x_2 separately, a third predictor variable $x_3 = x_1x_2$. One such model is

$$y = -4500 + 75x_1 + 60x_2 - x_1x_2 + e$$

which has regression function $-4500 + 75x_1 + 60x_2 - x_1x_2$. Then

$$\text{(mean } y \text{ value when } x_1 = 100) = -4500 + (75)(100) + 60x_2 - 100x_2$$
$$= 3000 - 40x_2$$

whereas

$$\text{(mean } y \text{ value when } x_1 = 95) = 2625 - 35x_2$$
$$\text{(mean } y \text{ value when } x_1 = 90) = 2250 - 30x_2$$

These are graphed in Figure 14.3(b), where it is clear that the three slopes are different. In fact, each different value of x_1 yields a different slope, so the average change in yield associated with a 1-unit increase in x_2 depends on the value of x_1. When this is the case, the two variables are said to *interact*.

Definition

If the change in the mean y value associated with a 1-unit increase in one independent variable depends on the value of a second independent variable, there is **interaction** between these two variables. When the variables are denoted by x_1 and x_2, such interaction can be modeled by including x_1x_2, the product of the variables that interact, as a predictor variable.

The general equation for a multiple regression model based on two independent variables x_1 and x_2 that also includes an interaction predictor is

$$y = \alpha + \beta_1 x_1 + \beta_2 x_2 + \beta_3 x_1 x_2 + e$$

When x_1 and x_2 do interact, this model will usually give a much better fit to resulting sample data — and thus explain more variation in y— than would the no-interaction model. Failure to consider a model with interaction often leads an investigator to conclude incorrectly that there is no strong relationship between y and a set of independent variables.

More than one interaction predictor can be included in the model when more than two independent variables are available. If, for example, there are three independent variables x_1, x_2, and x_3, one possible model is

$$y = \alpha + \beta_1 x_1 + \beta_2 x_2 + \beta_3 x_3 + \beta_4 x_4 + \beta_5 x_5 + \beta_6 x_6 + e$$

where

$$x_4 = x_1 x_2 \qquad x_5 = x_1 x_3 \qquad x_6 = x_2 x_3$$

One could even include a three-way interaction predictor $x_7 = x_1 x_2 x_3$ (the product of all three independent variables), although in practice this is rarely done.

In applied work, quadratic terms, such as x_1^2 and x_2^2, are often included to model a curved relationship between y and several independent variables. For example, a frequently used model involving just two independent variables x_1 and x_2 but $k = 5$ predictors is the *full quadratic* or **complete second-order model**

$$y = \alpha + \beta_1 x_1 + \beta_2 x_2 + \beta_3 x_1 x_2 + \beta_4 x_1^2 + \beta_5 x_2^2 + e$$

This model replaces the straight lines of Figure 14.3 with parabolas (each one is the graph of the regression function for different values of x_2 when x_1 has a fixed value). With four independent variables, one could examine a model containing four quadratic predictors and six two-way interaction predictor variables. Clearly, with just a few independent variables, one could examine a great many different multiple regression models. In the last section of this chapter, we briefly discuss methods for selecting one model from a number of competing models.

Qualitative Predictor Variables

Up to this point, we have explicitly considered the inclusion only of quantitative (numerical) predictor variables in a multiple regression model. Using a simple numerical coding, qualitative (categorical) variables can also be incorporated into a model. Let's focus first on a dichotomous variable, one with just two possible categories: male or female, U.S. or foreign manufacture, a house with or without a view, and so on. With any such variable, we associate a numerical variable x whose possible values are 0 and 1, where 0 is identified with one category (for example, married) and $x = 1$ with the other possible category (not married). This 0–1 variable is often called a **dummy variable** or **indicator variable.**

EXAMPLE 14.3   The article "Grade Level and Gender Differences in Writing Self-Beliefs of Middle School Students" (*Contemporary Educational Psych.* (1999): 390–405) considered relating writing competence score to a number of predictor variables, including perceived value of writing and gender. Writing competence and perceived value of writing were both represented by a numerically scaled variable, but gender is a qualitative predictor.

Let

$$y = \text{writing competence score}$$

$$x_1 = \begin{cases} 0 & \text{if male} \\ 1 & \text{if female} \end{cases}$$

$$x_2 = \text{perceived value of writing}$$

One possible multiple regression model is

$$y = \alpha + \beta_1 x_1 + \beta_2 x_2 + e$$

Considering the mean y value first when $x_1 = 0$ and then when $x_1 = 1$ yields

$$\text{average score} = \alpha + \beta_2 x_2 \qquad \text{when } x_1 = 0 \,(\text{males})$$
$$\text{average score} = \alpha + \beta_1 + \beta_2 x_2 \qquad \text{when } x_1 = 1 \,(\text{females})$$

The coefficient β_1 is the difference in average writing competence score between males and females when perceived value of writing is held fixed.

A second possibility is a model with an interaction term:

$$y = \alpha + \beta_1 x_1 + \beta_2 x_2 + \beta_3 x_1 x_2 + e$$

The regression function for this model is $\alpha + \beta_1 x_1 + \beta_2 x_2 + \beta_3 x_3$, where $x_3 = x_1 x_2$. Now the two cases $x_1 = 0$ and $x_1 = 1$ give

$$\text{average score} = \alpha + \beta_2 x_2 \qquad \text{when } x_1 = 0$$
$$\text{average score} = \alpha + \beta_1 + (\beta_2 + \beta_3) x_2 \qquad \text{when } x_1 = 1$$

For each model, the graph of the average writing competence score, when regarded as a function of perceived value of writing is a line for either gender (Figure 14.4).

FIGURE 14.4 Regression functions for models with one qualitative variable (x_1) and one quantitative variable (x_2): (a) No interaction; (b) Interaction

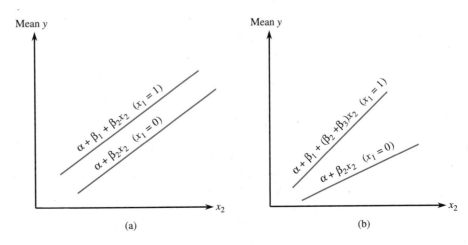

In the no-interaction model, the coefficient on x_2 is β_2 both when $x_1 = 0$ and when $x_1 = 1$, so the two lines are parallel, although their intercepts are different (unless $\beta_1 = 0$). With interaction, the lines not only have different intercepts but also different slopes (unless $\beta_3 = 0$). For this latter model, the change in average writing competence score when perceived value of writing increases by 1 unit depends on gender — the two variables *perceived value* and *gender* interact.

You might think that the way to handle a three-category situation is to define a single numerical variable with coded values such as 0, 1, and 2 corresponding to the three categories. This is incorrect, because it imposes an ordering on the categories that is not necessarily implied by the problem. The correct approach to modeling a categorical variable with three categories is to define *two* different 0–1 variables.

EXAMPLE 14.4   The article "The Effect of Ownership on the Organization Structure in Small Firms" (*Admin. Sci. Quarterly* (1984): 232–237) considers three different management status categories — owner-successors, professional managers, and owner-founders — in relating degree of horizontal differentiation (y) to firm size. Let

$$x_1 = \begin{cases} 1 & \text{if the firm is managed by owner-successors} \\ 0 & \text{otherwise} \end{cases}$$

$$x_2 = \begin{cases} 1 & \text{if the firm is managed by professional managers} \\ 0 & \text{otherwise} \end{cases}$$

$x_3 =$ firm size

Thus, $x_1 = 1$, $x_2 = 0$ indicates owner-successors; $x_1 = 0$, $x_2 = 1$, professional managers; and $x_1 = x_2 = 0$, owner-founders. ($x_1 = x_2 = 1$ is not possible.) The author suggests an interaction model of the form

$$y = \alpha + \beta_1 x_1 + \beta_2 x_2 + \beta_3 x_3 + \beta_4 x_1 x_3 + \beta_5 x_2 x_3 + e$$

This model allows the change in mean differentiation when size increases by 1 to be different for all three management status categories.

In general, incorporating a categorical variable with c possible categories into a regression model requires the use of $c - 1$ indicator variables. Thus, even one such categorical variable can add many predictors to a model.

Nonlinear Multiple Regression Models

Many nonlinear relationships can be put in the form $y = \alpha + \beta_1 x_1 + \cdots + \beta_k x_k + e$ by transforming one or more of the variables. An appropriate transformation could be suggested by theory or by various plots of the data (for example, residual plots after fitting a particular model). There are also relationships that cannot be linearized by transformations, in which case more complicated methods of analysis must be used. A general discussion of nonlinear regression is beyond the scope of this text; you can learn more by consulting one of the sources listed in the Chapter 5 references.

Exercises 14.1 – 14.14

14.1 Explain the difference between a deterministic and a probabilistic model. Give an example of a dependent variable y and two or more independent variables that might be related to y deterministically. Give an example of a dependent variable y and two or more independent variables that might be related to y in a probabilistic fashion.

14.2 A number of recent investigations have focused on the problem of assessing loads that can be manually handled in a safe manner. The article "Anthropometric, Muscle Strength, and Spinal Mobility Characteristics as Predictors in the Rating of Acceptable Loads in Parcel Sorting" (*Ergonomics* (1992): 1033–1044) proposed using a regression model to relate the dependent variable

y = individual's rating of acceptable load (kg)

to $k = 3$ independent (predictor) variables:

x_1 = extent of left lateral bending (cm)

x_2 = dynamic hand grip endurance (sec)

x_3 = trunk extension ratio (N/kg)

Suppose that the model equation is

$y = 30 + .90x_1 + .08x_2 - 4.50x_3 + e$

and that $\sigma = 5$.

a. What is the population regression function?
b. What are the values of the population regression coefficients?
c. Interpret the value of β_1.
d. Interpret the value of β_3.
e. What is the mean value of rating of acceptable load when extent of left lateral bending is 25 cm, dynamic hand grip endurance is 200 sec, and trunk extension ratio is 10 N/kg?
f. If repeated observations on rating are made on different individuals, all of whom have the values of x_1, x_2, and x_3 specified in part **e**, in the long run approximately what percentage of ratings will be between 13.5 kg and 33.5 kg?

14.3 The following statement appeared in the article "Dimensions of Adjustment Among College Women" (*J. College Student Development* (1998): 364):

> Regression analyses indicated that academic adjustment and race made independent contributions

to academic achievement, as measured by current GPA.

Suppose

y = current GPA

x_1 = academic adjustment score

x_2 = race (with white = 0, other = 1)

What multiple regression model is suggested by the statement? Did you include an interaction term in the model? Why or why not?

14.4 According to "Assessing the Validity of the Postmaterialism Index" (*American Political Science Review* (1999): 649–664), one may be able to predict an individual's level of support for ecology based on demographic and ideological characteristics. The multiple regression model proposed by the authors was

$$\hat{y} = 3.60 - .01x_1 + .01x_2 - .07x_3 + .12x_4 \\ + .02x_5 - .04x_6 - .01x_7 - .04x_8 - .02x_9 + e$$

where the variables are defined as follows

y = ecology score (higher values indicate a greater concern for ecology)

x_1 = age times 10

x_2 = income (in thousands of dollars)

x_3 = gender (1 = male, 0 = female)

x_4 = race (1 = white, 0 = nonwhite)

x_5 = education (in years)

x_6 = ideology (4 = conservative, 3 = right of center, 2 = middle of the road, 1 = left of center, and 0 = liberal)

x_7 = social class (4 = upper, 3 = upper middle, 2 = middle, 1 = lower middle, 0 = lower)

x_8 = postmaterialist (1 if postmaterialist, 0 otherwise)

x_9 = materialist (1 if materialist, 0 otherwise)

a. Suppose you knew a person with the following characteristics: a 25-year-old, white female with a college degree (16 years of education), who has a $32,000-per-year job, is from the upper middle class and considers herself left of center, but who is neither a materialist nor a postmaterialist. Predict her ecology score.

b. If the woman described in part **a** were Hispanic rather than white, how would the prediction change?

c. Given that the other variables are the same, what is the estimated mean difference in ecology score for men and women?

d. How would your interpret the coefficient of x_2?

e. Comment on the numerical coding of the ideology and social class variables. Can you suggest a better way of incorporating these two variables into the model?

14.5 The article "The Influence of Temperature and Sunshine on the Alpha-Acid Contents of Hops" (*Ag. Meteorology* (1974): 375–382) used a multiple regression model to relate y = yield of hops to x_1 = mean temperature (°C) between date of coming into hop and date of picking and x_2 = mean percentage of sunshine during the same period. The model equation proposed is

$$y = 415.11 - 6.60x_1 - 4.50x_2 + e$$

a. Suppose that this equation does indeed describe the true relationship. What mean yield corresponds to a temperature of 20 and a sunshine percentage of 40?

b. What is the mean yield when the mean temperature and percentage of sunshine are 18.9 and 43, respectively?

c. Interpret the values of the population regression coefficients.

14.6 The article "Readability of Liquid Crystal Displays: A Response Surface" (*Human Factors* (1983): 185–190) used a multiple regression model with four independent variables, where

y = error percentage for subjects reading a four-digit liquid crystal display

x_1 = level of backlight (from 0 to 122 cd/m²)

x_2 = character subtense (from .025° to 1.34°)

x_3 = viewing angle (from 0° to 60°)

x_4 = level of ambient light (from 20 to 1500 lx)

The model equation suggested in the article is

$$y = 1.52 + .02x_1 - 1.40x_2 + .02x_3 - .0006x_4 + e$$

a. Assume that this is the correct equation. What is the mean value of y when $x_1 = 10$, $x_2 = .5$, $x_3 = 50$, and $x_4 = 100$?

b. What mean error percentage is associated with a backlight level of 20, character subtense of .5, viewing angle of 10, and ambient light level of 30?

c. Interpret the values of β_2 and β_3.

14.7 The article "Pulp Brightness Reversion: Influence of Residual Lignin on the Brightness Reversion of Bleached Sulfite and Kraft Pulps" (*TAPPI* (1964): 653–662) proposed a quadratic regression model to describe the relationship between x = degree of delignification during the processing of wood pulp for paper and y = total chlorine content. Suppose that the actual model is

$$y = 220 + 75x - 4x^2 + e$$

a. Graph the regression function $220 + 75x - 4x^2$ over x values between 2 and 12. (Substitute $x = 2, 4, 6, 8, 10$, and 12 to find points on the graph, and connect them with a smooth curve.)

b. Would mean chlorine content be higher for a degree of delignification value of 8 or 10?

c. What is the change in mean chlorine content when the degree of delignification increases from 8 to 9? From 9 to 10?

14.8 The relationship between yield of maize, date of planting, and planting density was investigated in the article "Development of a Model for Use in Maize Replant Decisions" (*Agronomy J.* (1980): 459–464). Let

y = percent maize yield

x_1 = planting date (days after April 20)

x_2 = planting density (plants/ha)

The regression model with both quadratic terms ($y = \alpha + \beta_1 x_1 + \beta_2 x_2 + \beta_3 x_3 + \beta_4 x_4 + e$, where $x_3 = x_1^2$ and $x_4 = x_2^2$) provides a good description of the relationship between y and the independent variables.

a. If $\alpha = 21.09$, $\beta_1 = .653$, $\beta_2 = .0022$, $\beta_3 = -.0206$, and $\beta_4 = .00004$, what is the population regression function?

b. Use the regression function in part **a** to determine the mean yield for a plot planted on May 6 with a density of 41,180 plants/ha.

c. Would the mean yield be higher for a planting date of May 6 or May 22 (for the same density)?

d. Is it legitimate to interpret $\beta_1 = .653$ as the true average change in yield when planting date increases by one day and the values of the other three predictors are held fixed? Why or why not?

14.9 Suppose that the variables y, x_1, and x_2 are related by the regression model

$$y = 1.8 + .1x_1 + .8x_2 + e$$

a. Construct a graph (similar to that of Figure 14.4) showing the relationship between mean y and x_2 for fixed values 10, 20, and 30 of x_1.

b. Construct a graph depicting the relationship between mean y and x_1 for fixed values 50, 55, and 60 of x_2.

c. What aspect of the graphs in parts **a** and **b** can be attributed to the lack of an interaction between x_1 and x_2?

d. Suppose the interaction term $.03x_3$, where $x_3 = x_1x_2$, is added to the regression model equation. Using this new model, construct the graphs described in parts **a** and **b**. How do they differ from those obtained in parts **a** and **b**?

14.10 A manufacturer of wood stoves collected data on y = particulate matter concentration and x_1 = flue temperature for three different air intake settings (low, medium, high).

a. Write a model equation that includes dummy variables to incorporate intake setting, and interpret all the β coefficients.

b. What additional predictors would be needed to incorporate interaction between temperature and intake setting?

14.11 Consider a regression analysis with three independent variables x_1, x_2, and x_3. Give the equation for the following regression models.

a. The model that includes as predictors all independent variables but no quadratic or interaction terms.

b. The model that includes as predictors all independent variables and all quadratic terms

c. All models that include as predictors all independent variables, no quadratic terms, and exactly one interaction term

d. The model that includes as predictors all independent variables, all quadratic terms, and all interaction terms (the full quadratic model)

14.12 The article "The Value and the Limitations of High-Speed Turbo-Exhausters for the Removal of Tar-Fog from Carburetted Water-Gas" (*Soc. Chem. Industry*

J. (1946): 166–168) presented data on y = tar content (grains/100 ft^3) of a gas stream as a function of x_1 = rotor speed (rev/min) and x_2 = gas inlet temperature (°F). A regression model using x_1, x_2, $x_3 = x_2^2$, and $x_4 = x_1x_2$ was suggested:

mean y value = $86.8 - .123x_1 + 5.09x_2 - .0709x_3 + .001x_4$

a. According to this model, what is the mean y value if $x_1 = 3200$ and $x_2 = 57$?

b. For this particular model, does it make sense to interpret the value of any individual β_i (β_1, β_2, β_3, or β_4) in the way we have previously suggested? Explain.

14.13 Consider the dependent variable y = fuel efficiency of a car (mpg).

a. Suppose that you want to incorporate size class of car, with four categories (subcompact, compact, midsize, and large), into a regression model that also includes x_1 = age of car and x_2 = engine size. Define the necessary dummy variables, and write out the complete model equation.

b. Suppose that you want to incorporate interaction between age and size class. What additional predictors would be needed to accomplish this?

14.14 If we knew the width and height of cylindrical tin cans of food, could we predict the volume of these cans with precision and accuracy?

a. Give the equation that would allow us to make such predictions.

b. Is the relationship between volume and its predictors, height and width, a linear one?

c. Should we use an additive multiple regression model to predict a volume of a can from its height and width? Explain.

d. If you were to take logarithms of each side of your equation in part **a**, would the relationship be linear?

14.2   Fitting a Model and Assessing Its Utility

In Section 14.1, we discussed and contrasted multiple regression models containing several different types of predictors. Let's now suppose that a particular set of k predictor variables $x_1, x_2, \ldots, x_k$ has been selected for inclusion in the model

$$y = \alpha + \beta_1x_1 + \beta_2x_2 + \cdots + \beta_kx_k + e$$

It is then necessary to estimate the model coefficients $\alpha, \beta_1, \ldots, \beta_k$ and the regression function $\alpha + \beta_1x_1 + \cdots + \beta_kx_k$ (mean y value for specified values of the predictors), assess the model's utility, and perhaps use the estimated model to make further inferences. All this, of course, requires sample data. As before, n denotes the number of observations in the sample. With just one predictor variable, the sample consisted of n (x, y) pairs. Now each observation consists of $k + 1$ numbers: a value of x_1, a value of $x_2, \ldots$, a value of x_k, and the associated value of y. The n observations are assumed to have been selected independently of one another.

EXAMPLE 14.5

Soil and sediment adsorption, the extent to which chemicals collect in a condensed form on the surface, is an important characteristic because it influences the effectiveness of pesticides and various agricultural chemicals. The article "Adsorption of Phosphates, Arsenate, Methanearsenate, and Calcodylate by Lake and Stream Sediments: Comparisons with Soils" (*J. of Environ. Qual.* (1984): 499–504) presented the accompanying data consisting of $n = 13$ (x_1, x_2, y) triples and proposed the model

$$y = \alpha + \beta_1 x_1 + \beta_2 x_2 + e$$

for relating

y = phosphate adsorption index

x_1 = amount of extractable iron

x_2 = amount of extractable aluminum

Observation	x_1 extractable iron	x_2 extractable aluminum	y adsorption index
1	61	13	4
2	175	21	18
3	111	24	14
4	124	23	18
5	130	64	26
6	173	38	26
7	169	33	21
8	169	61	30
9	160	39	28
10	244	71	36
11	257	112	65
12	333	88	62
13	199	54	40

As in simple linear regression, the principle of least squares is used to estimate the coefficients $\alpha, \beta_1, \ldots, \beta_k$. For specified estimates $a, b_1, \ldots, b_k$,

$$y - (a + b_1 x_1 + b_2 x_2 + \cdots + b_k x_k)$$

is the deviation between the observed y value for a particular observation and the predicted value using the estimated regression function $a + b_1x_1 + \cdots + b_kx_k$. For example, the first observation in the data set of Example 14.5 is $(x_1, x_2, y) = (61, 13, 4)$. The resulting deviation between observed and predicted y values is

$$4 - [a + b_1(61) + b_2(13)]$$

Deviations corresponding to other observations are expressed in a similar manner. The principle of least squares then says to use as estimates of α, β_1, and β_2 the values of a, b_1, and b_2 that minimize the sum of these squared deviations.

Definition

According to the principle of least squares, the fit of a particluar estimated regression function $a + b_1x_1 + \cdots + b_kx_k$ to the observed data is measured by the sum of squared deviations between the observed y values and the y values predicted by the estimated function:

$$\Sigma\,[y - (a + b_1x_1 + \cdots + b_kx_k)]^2$$

The **least squares estimates** of α, $\beta_1, \ldots, \beta_k$ are those values of a, $b_1, \ldots, b_k$ that make this sum of squared deviations as small as possible.

The least squares estimates for a given data set are obtained by solving a system of $k + 1$ equations in the $k + 1$ unknowns a, $b_1, \ldots, b_k$ (called the *normal equations*). In the case $k = 1$, simple linear regression, there are only two equations, and we gave their general solution — the expressions for b and a — in Chapter 5. For $k \geq 2$, it is not as easy to write general expressions for the estimates without using advanced mathematical notation. Fortunately, the computer saves us! Formulas for the estimates have been programmed into all the commonly used statistical software packages.

EXAMPLE 14.6   Figure 14.5 displays MINITAB output from a regression command requesting that the model $y = \alpha + \beta_1x_1 + \beta_2x_2 + e$ be fit to the phosphate adsorption data of

FIGURE 14.5 MINITAB output for the regression analysis of Example 14.6

THE REGRESSION EQUATION IS
HPO = -7.35 + 0.113 FE + 0.349 AL

Predictor	Coef	Stdev	t-ratio	p
Constant	-7.351	3.485	-2.11	0.061
FE	0.11273	0.02969	3.80	0.004
AL	0.34900	0.07131	4.89	0.000

s = 4.379   R-sq = 94.8%   R-sq(adj) = 93.8%

Coefficient of multiple determination = .948

Analysis of Variance

SOURCE	DF	SS	MS	F	p
Regression	2	3529.9	1765.0	92.03	0.000
Error	10	191.8	19.2		
Total	12	3721.7			

P-value for model utility test

Example 14.5. (We named the dependent variable HPO, an abbreviation for the chlorate H_2PO_4 (dihydrogen phosphate), and the two predictor variables x_1 and x_2 were named FE and AL, the standard abbreviations for iron and aluminum, respectively.) Focus on the column labeled **Coef** (for coefficient) in the table near the top of the figure. The three numbers in this column are the estimated model coefficients:

$$a = -7.351 \quad \text{(the estimate of the constant term } \alpha)$$
$$b_1 = .11273 \quad \text{(the estimate of the coefficient } \beta_1)$$
$$b_2 = .34900 \quad \text{(the estimate of the coefficient } \beta_2)$$

Thus, we estimate that the average change in HPO associated with a 1-unit increase in FE while AL remains fixed is .11273. A similar interpretation applies to b_2. The estimated regression function is

$$(\text{estimated mean value of } y \text{ for specified } x_1 \text{ and } x_2 \text{ values}) = -7.351 + .11273x_1 + .34900x_2$$

Substituting $x_1 = 150$ and $x_2 = 60$ gives

$$-7.351 + .11273(150) + .34900(60) = 30.5$$

which can be interpreted either as a point estimate for the mean value of HPO or as a point prediction for a single HPO value.

Is the Model Useful?

The utility of an estimated model can be assessed by examining the extent to which predicted y values based on the estimated regression function are close to the y values actually observed.

Definition

The first predicted value $\hat{y}_1$ is obtained by taking the values of the predictor variables $x_1, x_2, \ldots, x_k$ for the first sample observation and substituting these values into the extimated regression function. Doing this successively for the remaining observations yields the **predicted values** $\hat{y}_2, \ldots, \hat{y}_n$. The **residuals** are then the differences $y_1 - \hat{y}_1, y_2 - \hat{y}_2, \ldots, y_n - \hat{y}_n$, between the observed and predicted y values.

The predicted values and residuals are defined here exactly as they were in simple linear regression, but computation of the $\hat{y}$ values is more tedious because there is more than one predictor. Fortunately, the $\hat{y}$'s and $y - \hat{y}$'s are automatically computed and displayed in the output of all good statistical software packages. Consider again the phosphate adsorption data discussed in Examples 14.5 and 14.6. Since the first y observation, $y_1 = 4$, was made with $x_1 = 61$ and $x_2 = 13$, the first predicted value is

$$\hat{y}_1 = -7.351 + .11273(61) + .34900(13) \approx 4.06$$

The first residual is then

$$y_1 - \hat{y}_1 = 4 - 4.06 = -.06$$

The other predicted values and residuals are computed in a similar fashion. The sum of residuals from a least squares fit should, except for rounding effects, be zero.

As in simple linear regression, the sum of squared residuals is the basis for several important summary quantities that are indicative of a model's utility.

Definition

The **residual (or error) sum of squares, SSResid,** and **total sum of squares, SSTo,** are given by

$$\text{SSResid} = \Sigma (y - \hat{y})^2 \qquad \text{SSTo} = \Sigma (y - \bar{y})^2$$

where $\bar{y}$ is the mean of the y observations in the sample.

The number of degrees of freedom associated with SSResid is $n - (k + 1)$, because $k + 1$ df are lost in estimating the $k + 1$ coefficients $\alpha, \beta_1, \ldots, \beta_k$.

An estimate of the random deviation variance σ^2 is given by

$$s_e^2 = \frac{\text{SSResid}}{n - (k + 1)}$$

and $s_e = \sqrt{s_e^2}$ is the estimate of σ.

The **coefficient of multiple determination, R^2,** interpreted as the proportion of variation in observed y values that is explained by the fitted model, is

$$R^2 = 1 - \frac{\text{SSResid}}{\text{SSTo}}$$

EXAMPLE 14.7

Looking again at Example 14.6, which contains MINITAB output for the adsorption data fit by a two-predictor model, residual sum of squares is found in the **Error** row and **SS** column of the table headed **Analysis of Variance**: SSResid = 191.8. The associated number of degrees of freedom is $n - (k + 1) = 13 - (2 + 1) = 10$, which appears in the **DF** column just to the left of SSResid. The sample average y value is $\bar{y} = 29.85$, and SSTo = $\Sigma (y - 29.85)^2 = 3721.7$ appears in the **Total** row and **SS** column of the **Analysis of Variance** table just under the value of SSResid. The values of s_e^2, s_e, and R^2 are then

$$s_e^2 = \frac{\text{SSResid}}{n - (k + 1)} = \frac{191.8}{10} = 19.18 \approx 19.2$$

(in the **MS** column of the MINITAB output)

$$s_e = \sqrt{s_e^2} = \sqrt{19.18} = 4.379$$

(which appears just above the **Analysis of Variance** table)

$$R^2 = 1 - \frac{\text{SSResid}}{\text{SSTo}} = 1 - \frac{191.8}{3721.7} = 1 - .052 = .948$$

Thus, the percentage of variation explained is $100R^2 = 94.8\%$, which appears on the output as **R-sq** = 94.8%. The values of R^2 and s_e suggest that the chosen model has been very successful in relating y to the predictors.

Generally speaking, a desirable model is one that results in both a large R^2 value and a small s_e value. However, there is a catch. These two conditions can be achieved by fitting a model that contains a large number of predictors. Such a model may be successful in explaining y variation, but it almost always specifies a relationship that is unrealistic and difficult to interpret. What we really want is a simple model — one with relatively few predictors whose roles are easily interpreted and which also does a good job of explaining variation in y.

All statistical software packages include R^2 and s_e in their output, and most give SSResid also. In addition, some packages compute the quantity called **adjusted R^2**:

$$\text{adjusted } R^2 = 1 - \left[\frac{n-1}{n-(k+1)}\right] \cdot \frac{\text{SSResid}}{\text{SSTo}}$$

Since the quantity in square brackets exceeds 1, the number subtracted from 1 is larger than SSResid/SSTo, so adjusted R^2 is smaller than R^2. The value of R^2 must be between 0 and 1, but adjusted R^2 can be negative. If a large R^2 has been achieved through using just a few model predictors, adjusted R^2 will differ little from R^2. However, the adjustment can be substantial when a great many predictors (relative to the number of observations) have been used or when R^2 itself is small to moderate (which could happen even when there is no relationship between y and the predictors). In Example 14.6, adjusted $R^2 = .938$, not much less than R^2, since the model included only two predictor variables.

F Distributions

The model utility test in simple linear regression was based on the fact that when $H_0: \beta = 0$ is true, the test statistic $t = (b - 0)/s_b$ has a t distribution. The model utility test for multiple regression uses a type of probability distribution called an F distribution. We digress briefly to describe some general properties of F distributions.

An F distribution always arises in connection with a ratio in which the numerator involves one sum of squares and the denominator involves a second sum of squares. Each sum of squares has associated with it a specified number of degrees of freedom, so a particular F distribution is determined by fixing values of $\text{df}_1 = $ numerator df and $\text{df}_2 = $ denominator df. There is a different F distribution for each different df_1, df_2 combination. For example, there is an F distribution based on 4 numerator df and 12 denominator df, another F distribution based on 3 numerator df and 20 denominator df, and so on. A typical F curve for fixed numerator and denominator df appears in Figure 14.6. All F tests presented in this book are upper-tailed. Recall that for an upper-tailed t test, the P-value is the area under the associated t curve to the right of the calculated t. Similarly, the P-value for an upper-tailed F test is the area under the associated F curve to the right of the calculated F. Figure 14.6 illustrates this for a test based on $\text{df}_1 = 4$ and $\text{df}_2 = 6$.

FIGURE 14.6 A *P*-value for an upper-tailed *F* test

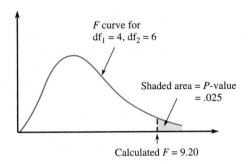

Unfortunately, tabulation of these upper-tail areas is much more cumbersome than for *t* distributions, because here two df's are involved. For each of a number of different *F* distributions, our *F* table (Appendix Table VII) tabulates only four numbers: the values that capture tail areas .10, .05, .01, and .001. Different columns correspond to different values of df_1 (numerator df), and each different group of rows is for a different value of df_2. Figure 14.7 shows how this is used to obtain *P*-value information.

FIGURE 14.7 Obtaining *P*-value information from the *F* table

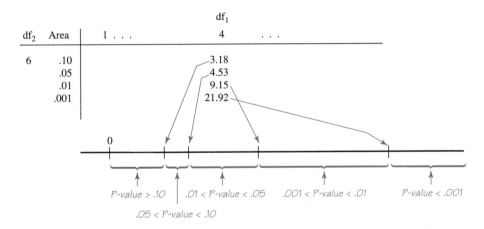

For example, for a test with $df_1 = 4$ and $df_2 = 6$,

calculated $F = 5.70 \Rightarrow .01 < P\text{-value} < .05$

calculated $F = 2.16 \Rightarrow P\text{-value} > .10$

calculated $F = 25.03 \Rightarrow P\text{-value} < .001$

Only if calculated *F* equals a tabulated value do we obtain an exact *P*-value (e.g., if calculated $F = 4.53$, then $P\text{-value} = .05$). If $.01 < P\text{-value} < .05$, we should reject H_0 at a significance level of .05 but not at a level of .01. When $P\text{-value} < .001$, H_0 would be rejected at any reasonable significance level.

The *F* Test for Model Utility

In the simple linear model with regression function $\alpha + \beta x$, if $\beta = 0$, there is no useful linear relationship between *y* and the single predictor variable *x*. Similarly, if all k coefficients $\beta_1, \beta_2, \ldots, \beta_k$, are zero in the general k-predictor multiple regression model, there is no useful linear relationship between y and *any* of the predictor variables $x_1, x_2, \ldots, x_k$ included in the model. Before using an estimated model to make further inferences (for example, predictions and estimates of mean values), it is desirable to confirm the model's utility through a formal test procedure.

Recall that SSTo is a measure of total variation in the observed y values and that SSResid measures the amount of total variation that has not been explained by the fitted model. The difference between total and error sums of squares is itself a sum of squares, called **regression sum of squares** and denoted by SSRegr:

SSRegr $=$ SSTo $-$ SSResid

SSRegr is interpreted as the amount of total variation that *has* been explained by the model. Intuitively, the model should be judged useful if SSRegr is large relative to SSResid, and this is achieved by using a small number of predictors relative to the sample size. The number of degrees of freedom associated with SSRegr is k, the number of model predictors, and df for SSResid is $n - (k + 1)$. The model utility F test is based on the following distributional result.

When all k β_i's are zero in the model $y = \alpha + \beta_1 x_1 + \cdots + \beta_k x_k + e$ and when the distribution of e is normal with mean 0 and variance σ^2 for any particular values of $x_1, x_2, \ldots, x_k$, the statistic

$$F = \frac{\dfrac{\text{SSRegr}}{k}}{\dfrac{\text{SSResid}}{n - (k + 1)}}$$

has an F probability distribution based on k numerator df and $n - (k + 1)$ denominator df.

The value of F tends to be larger when at least one β_i is not zero than when all the β_i's are zero, since more variation is typically explained by the model in the former case than in the latter. An F statistic value far out in the upper tail of the associated F distribution can be more plausibly attributed to at least one nonzero β_i than to something very unusual having occurred when all β_i's are zero. This is why the F test for model utility is upper-tailed.

The F Test for Utility of the Model $y = \alpha + \beta_1 x_1 + \cdots + \beta_k x_k + e$

Null hypothesis: $H_0: \beta_1 = \beta_2 = \cdots = \beta_k = 0$
(There is no useful linear relationship between y and *any* of the predictors.)

Alternative hypothesis: H_a: At least one among $\beta_1, \ldots, \beta_k$ is not zero.
(There is a useful linear relationship between y and *at least one* of the predictors.)

(continued)

Test statistic: $\quad F = \dfrac{\dfrac{SSRegr}{k}}{\dfrac{SSResid}{n-(k+1)}}$

where $SSRegr = SSTo - SSResid$.

An equivalent formula is

$$F = \frac{\dfrac{R^2}{k}}{\dfrac{1-R^2}{n-(k+1)}}$$

The test is upper-tailed, and the information in Appendix Table VII is used to obtain a bound or bounds on the *P*-value using numerator df $= k$ and denominator df $= n - (k+1)$.

Assumptions: For any particular combination of predictor variable values, the distribution of *e,* the random deviation, is *normal* with mean 0 and *constant variance.*

The null hypothesis is the claim that the model is not useful. Unless H_0 can be rejected at a small level of significance, the model has not demonstrated its utility, in which case the investigator must search further for a model that can be judged useful. The alternative formula for F allows the test to be carried out when only R^2, k, and n are available, as is frequently the case in published articles.

EXAMPLE 14.8   The model fit to the phosphate adsorption data introduced in Example 14.5 involved $k = 2$ predictors. The MINITAB output in Figure 14.5 contains the relevant information for carrying out the model utility test.

1. The model is $y = \alpha + \beta_1 x_1 + \beta_2 x_2 + e$, where $y =$ phosphate adsorption, $x_1 =$ extractable Fe, and $x_2 =$ extractable Al.

2. $H_0: \beta_1 = \beta_2 = 0$

3. $H_a:$ At least one of the two β_i's is not zero.

4. *Significance level:* $\alpha = .05$

5. *Test statistic:* $\quad F = \dfrac{\dfrac{SSRegr}{k}}{\dfrac{SSResid}{n-(k+1)}}$

6. *Assumptions:*   The accompanying table gives the residuals and standardized residuals (from MINITAB) for the model under consideration.

AL	FE	HPO	Residual	St Resid
13	61	4.00	-0.06	-0.02
21	175	18.00	-1.71	-0.46
24	111	14.00	0.46	0.11
23	124	18.00	3.34	0.83
64	130	26.00	-3.64	-1.04
38	173	26.00	0.59	0.14
33	169	21.00	-2.22	-0.54
61	169	30.00	-2.99	-0.73
39	160	28.00	3.70	0.89
71	244	36.00	-8.94	-2.22
112	257	65.00	4.29	1.44
88	333	62.00	1.10	0.37
54	199	40.00	6.07	1.45

A normal probability plot of the standardized residuals is shown here. The plot is quite straight, indicating that the assumption of normality of the random deviation distribution is reasonable.

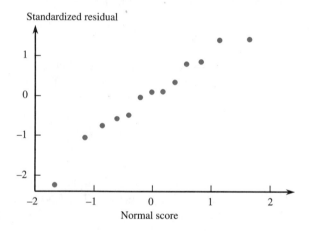

7. Directly from the **Analysis of Variance** table in Figure 14.5, the **SS** column gives SSRegr = 3529.9 and SSResid = 191.8. Thus,

$$F = \frac{\dfrac{3529.9}{2}}{\dfrac{191.8}{10}} = \frac{1764.95}{19.18} = 92.02 \quad (92.03 \text{ in Figure 14.5})$$

8. Appendix Table VII shows that for a test based on $df_1 = k = 2$ and $df_2 = n - (k + 1) = 13 - (2 + 1) = 10$, the value 14.91 captures upper-tail F curve area .001. Since calculated $F = 92.02 > 14.91$, it follows that P-value $< .001$. In fact, Figure 14.5 shows that to three decimal places, P-value $= 0$.

9. Because P-value $< .001 \le .05 = \alpha$, H_0 should be rejected. The conclusion would be the same using $\alpha = .01$ or $\alpha = .001$. The utility of the model is resoundingly confirmed.

EXAMPLE 14.9

A multiple regression analysis presented in the article "The Politics of Bureaucratic Discretion: Educational Access as an Urban Service" (*Amer. J. Political Sci.* (1991): 155–177) considered a model in which the dependent variable was

y = percentage of school board members in a school district who are black

and the predictors were

x_1 = black-to-white income ratio in the district

x_2 = percentage of whites in the district below the poverty line

x_3 = indicator for whether district was in the South

x_4 = percentage of blacks in the district with a high-school education

x_5 = black population percentage in the district

Summary quantities included $n = 140$ and $R^2 = .749$.

1. The fitted model was $y = \alpha + \beta_1 x_1 + \cdots + \beta_5 x_5 + e$.
2. $H_0: \beta_1 = \beta_2 = \beta_3 = \beta_4 = \beta_5 = 0$
3. H_a: At least one among $\beta_1, \dots, \beta_5$ is not zero.
4. *Significance level*: $\alpha = .01$
5. *Test statistic*: $F = \dfrac{\dfrac{R^2}{k}}{\dfrac{1 - R^2}{n - (k + 1)}}$
6. *Assumptions*:   The raw data was not given in this article, so we are unable to compute standardized residuals or construct a normal probability plot. For this test to be valid, we must be willing to assume that the random deviation distribution is normal.
7. $F = \dfrac{\dfrac{.749}{5}}{\dfrac{.251}{140 - (5 + 1)}} = \dfrac{.1498}{.001873} = 80.0$
8. The test is based on $df_1 = k = 5$ and $df_2 = n - (k + 1) = 134$. This latter df is not included in the F table. However, the .001 cutoff value for $df_2 = 120$ is 4.42, and for $df_2 = 240$ it is 4.25; so for $df_2 = 134$, the cutoff value is roughly 4.4. Clearly, 80.0 greatly exceeds this value, implying that P-value $< .001$.
9. Since P-value $< .001 \le .01 = \alpha$, H_0 is rejected at significance level .01. There appears to be a useful linear relationship between y and at least one of the five predictors.

In the next section, we presume that a model has been judged useful by the F test and then show how the estimated coefficients and regression function can be used to draw further conclusions. However, you should realize that in many applications, there will be more than one model whose utility could be confirmed by the F test. Suppose, for example, that data has been collected on six independent variables x_1, $x_2, \ldots, x_6$. Then a model with all six of these as predictors might be judged useful, another useful model might have only x_1, x_3, x_5, and x_6 as predictors, and yet another useful model might incorporate the six predictors x_1, x_4, x_6, $x_7 = x_1 x_4$, $x_8 = x_1^2$, and $x_9 = x_6^2$. In the last section of this chapter, we briefly consider strategies for selecting a model.

Exercises 14.15 – 14.35

14.15 The article cited in Example 14.1 recommended the following estimated regression equation for relating $y = VO_2max$ (L/min) to the predictors $x_1 = $ gender (female $= 0$, male $= 1$), $x_2 = $ weight (lb), $x_3 = $ 1-mile walk time (min), and $x_4 = $ heart rate at the end of the walk (beats/min):

$$\hat{y} = 3.5959 + .6566x_1 + .0096x_2 - .0996x_3 - .0080x_4$$

a. How would you interpret the value of $b_3 = -.0996$?
b. How would you interpret the value of $b_1 = .6566$?
c. Suppose that an observation made on a male whose weight was 80 kg, walk time was 11 min, and heart rate was 140 beats/min resulted in $VO_2max = 3.15$. What would you have predicted for VO_2max, and what is the corresponding residual?
d. It SSResid = 30.1033 and SSTo = 102.3922, what proportion of observed variation in VO_2max can be attributed to the model relationship?
e. The sample size was $n = 196$. Calculate an estimate of σ.

14.16 When coastal power stations take in large quantities of cooling water, it is inevitable that a number of fish are drawn in with the water. Various methods have been designed to screen out the fish. The article "Multiple Regression Analysis for Forecasting Critical Fish Influxes at Power Station Intakes" (*J. Appl. Ecol.* (1983): 33–42) examined intake fish catch at an English power plant and several other variables thought to affect fish intake:

   $y = $ fish intake (number of fish)

   $x_1 = $ water temperature (°C)

   $x_2 = $ number of pumps running

   $x_3 = $ sea state (values 0, 1, 2, or 3)

   $x_4 = $ speed (knots)

Part of the data given in the article was used to obtain the estimated regression equation

$$\hat{y} = 92 - 2.18x_1 - 19.20x_2 - 9.38x_3 + 2.32x_4$$

(based on $n = 26$). SSRegr = 1486.9 and SSResid = 2230.2 were also calculated.
a. Interpret the values of b_1 and b_4.
b. What proportion of observed variation in fish intake can be explained by the model relationship?
c. Estimate the value of σ.
d. Calculate adjusted R^2. How does it compare to R^2 itself?

14.17 Obtain as much information as you can about the P-value for an upper-tailed F test in each of the following situations.
a. $df_1 = 3$, $df_2 = 15$, calculated $F = 4.23$
b. $df_1 = 4$, $df_2 = 18$, calculated $F = 1.95$
c. $df_1 = 5$, $df_2 = 20$, calculated $F = 4.10$
d. $df_1 = 4$, $df_2 = 35$, calculated $F = 4.58$

14.18 Obtain as much information as you can about the P-value for the F test for model utility in each of the following situations.
a. $k = 2$, $n = 21$, calculated $F = 2.47$
b. $k = 8$, $n = 25$, calculated $F = 5.98$
c. $k = 5$, $n = 26$, calculated $F = 3.00$
d. The full quadratic model based on x_1, x_2 is fit, $n = 20$, and calculated $F = 8.25$.
e. $k = 5$, $n = 100$, calculated $F = 2.33$

14.19 The ability of ecologists to identify regions of greatest species richness could have an impact on the preservation of genetic diversity, a major objective of the World Conservation Strategy. The article "Prediction of Rarities from Habitat Variables: Coastal Plain Plants on Nova Scotian Lakeshores" (*Ecology* (1992): 1852–1859) used a sample of $n = 37$ lakes to obtain the estimated regression equation

$$\hat{y} = 3.89 + .033x_1 + .024x_2 + .023x_3 - .0080x_4$$
$$- .13x_5 - .72x_6$$

where $y = $ species richness, $x_1 = $ watershed area, $x_2 = $ shore width, $x_3 = $ drainage (%), $x_4 = $ water color (total color units), x_5 = sand (%), and x_6 = alkalinity. The coefficient of multiple determination was reported as R^2 = .83. Use a test with significance level .01 to decide whether the chosen model is useful.

14.20   The article "Impacts of On-Campus and Off-Campus Work on First-Year Cognitive Outcomes" (*J. of College Student Devel.* (1994): 364–370) reported on a study in which y = spring 1992 math comprehension score was regressed against x_1 = fall 1991 test score, x_2 = fall 1991 academic motivation, x_3 = age, x_4 = number of credit hours, x_5 = residence (1 if on campus, 0 otherwise), x_6 = hours worked on campus, and x_7 = hours worked off campus. The sample size was n = 210, and R^2 = .543. Test to see whether there is a useful linear relationship between y and at least one of the predictors.

14.21   Return to the situation described in Exercise 14.15, and carry out the model utility test to decide whether there is a useful linear relationship between y and at least one of the four predictors.

14.22   Is the model fit in Exercise 14.16 useful? Carry out a test using a significance level of .10.

14.23   The accompanying MINITAB output results from fitting the model described in Exercise 14.12 to data.

Predictor	Coef	Stdev	t-ratio
Constant	86.85	85.39	1.02
X1	-0.12297	0.03276	-3.75
X2	5.090	1.969	2.58
X3	-0.07092	0.01799	-3.94
X4	0.0015380	0.0005560	2.77

s = 4.784   R-sq = 90.8%   R-sq(adj) = 89.4%

ANALYSIS OF VARIANCE

	DF	SS	MS
Regression	4	5896.6	1474.2
Error	26	595.1	22.9
Total	30	6491.7	

a. What is the estimated regression equation?
b. Using a .01 significance level, perform the model utility test.
c. Interpret the values of R^2 and s_e given in the output.

14.24   For the multiple regression model in Exercise 14.4, the value of R^2 was .06 and the adjusted R^2 was .06. The model was based on a data set with 1136 observations. Perform a model utility test for this regression.

14.25   *This exercise requires the use of a computer package.* The article "Movement and Habitat Use by Lake

Whitefish During Spawning in a Boreal Lake: Integrating Acoustic Telemetry and Geographic Information Systems" (*Transactions of the American Fisheries Society*) 1999: 939–952 included the accompanying data on 17 fish caught in 1995 and 1996.

Year	Fish Number	Weight (g)	Length (mm)	Age (years)
1995	1	776	410	9
	2	580	368	11
	3	539	357	15
	4	648	373	12
	5	538	361	9
	6	891	385	9
	7	673	380	10
	8	783	400	12
1996	9	571	407	12
	10	627	410	13
	11	727	421	12
	12	867	446	19
	13	1042	478	19
	14	804	441	18
	15	832	454	12
	16	764	440	12
	17	727	427	12

a. Fit a multiple regression model to describe the relationship between weight and the predictors *length* and *age.*
b. Carry out the model utility test to determine whether the predictors *length* and *age,* together, are useful for predicting weight.

14.26   *This exercise requires the use of a computer package.* The authors of the article "Absolute Versus per Unit Body Length Speed of Prey as an Estimator of Vulnerability to Predation" (*Animal Behaviour* (1999): 347–352) found that the speed of a prey (twips/s) and the length of a prey (twips × 100) are good predictors of the time (s) required to catch the prey. (A twip is a measure of distance used by programmers.) Data was collected in an experiment where subjects were asked to "catch" an animal of prey moving across his or her computer screen by clicking on it with the mouse. The investigators varied the length of the prey and the speed with which the prey moved across the screen. The following data is consistent with summary values and a graph given in the article. Each value represents the average catch time over all subjects. The order of the various speed–length combinations was randomized for each subject.

Subject	Prey Length	Prey Speed	Catch Time
1	7	20	1.10
2	6	20	1.20
3	5	20	1.23
4	4	20	1.40
5	3	20	1.50
6	3	40	1.40
7	4	40	1.36
8	6	40	1.30
9	7	40	1.28
10	7	80	1.40
11	6	60	1.38
12	5	80	1.40
13	7	100	1.43
14	6	100	1.43
15	7	120	1.70
16	5	80	1.50
17	3	80	1.40
18	6	100	1.50
19	3	120	1.90

a. Fit a multiple regression model for predicting catch time using prey length and speed as predictors.

b. Predict the catch time for an animal of prey whose length is 6 and whose speed is 50.

c. Is the multiple regression model useful for predicting catch time? Test the relevant hypotheses using $\alpha = .05$.

d. The authors of the article suggest that a simple linear regression model with the single predictor

$$x = \frac{\text{length}}{\text{speed}}$$

might be a better model for predicting catch time. Calculate the x values and use them to fit this linear regression model.

e. Which of the two models considered (the multiple regression model from part **a** or the simple linear regression model from part **d**) would you recommend for predicting catch time? Justify your choice.

14.27   *This exercise requires the use of a computer package.* The article "Vital Dimensions in Volume Perception: Can the Eye Fool the Stomach?" (*Journal of Marketing*

Data for Exercise 14.27

Row	Material	Height	Maxwidth	Minwidth	Elongation	Volume
1	glass	7.7	2.50	1.80	1.50	125
2	glass	6.2	2.90	2.70	1.07	135
3	glass	8.5	2.15	2.00	1.98	175
4	glass	10.4	2.90	2.60	1.79	285
5	plastic	8.0	3.20	3.15	1.25	330
6	glass	8.7	2.00	1.80	2.17	90
7	glass	10.2	1.60	1.50	3.19	120
8	plastic	10.5	4.80	3.80	1.09	520
9	plastic	3.4	5.90	5.00	.29	330
10	plastic	6.9	5.80	4.75	.59	570
11	tin	10.9	2.90	2.80	1.88	340
12	plastic	9.7	2.45	2.10	1.98	175
13	glass	10.1	2.60	2.20	1.94	240
14	glass	13.0	2.60	2.60	2.50	240
15	glass	13.0	2.70	2.60	2.41	360
16	glass	11.0	3.10	2.90	1.77	310
17	cardboard	8.7	5.10	5.10	.85	635
18	cardboard	17.1	10.20	10.20	.84	1250
19	glass	16.5	3.50	3.50	2.36	650
20	glass	16.5	2.70	1.20	3.06	305
21	glass	9.7	3.00	1.70	1.62	315
22	glass	17.8	2.70	1.75	3.30	305
23	glass	14.0	2.50	1.70	2.80	245
24	glass	13.6	2.40	1.20	2.83	200
25	plastic	27.9	4.40	1.20	3.17	1205
26	tin	19.5	7.50	7.50	1.30	2330
27	tin	13.8	4.25	4.25	1.62	730

Research (1999): 313–326) gave the accompanying data on dimensions of 27 representative food products.

a. Fit a multiple regression model for predicting the volume (in ml) of a package based on its minimum width, maximum width, and elongation score.

b. Why should we consider adjusted R^2 instead of R^2 when attempting to determine the quality of fit of the data to our model?

c. Perform a model utility test.

14.28 The article "The Caseload Controversy and the Study of Criminal Courts" (*J. Criminal Law and Criminology* (1979): 89–101) used a multiple regression analysis to help assess the impact of judicial caseload on the processing of criminal court cases. Data was collected in the Chicago criminal courts on the following variables:

y = number of indictments

x_1 = number of cases on the docket

x_2 = number of cases pending in criminal court trial system

The estimated regression equation (based on $n = 367$ observations) was

$$\hat{y} = 28 - .05x_1 - .003x_2 + .00002x_3$$

where $x_3 = x_1x_2$.

a. The reported value of R^2 was .16. Conduct the model utility test. Use a .05 significance level.

b. Given the results of the test in part **a**, does it surprise you that the R^2 value is so low? Can you think of a possible explanation for this?

c. How does adjusted R^2 compare to R^2?

14.29 The article "The Undrained Strength of Some Thawed Permafrost Soils" (*Canad. Geotech. J.* (1979): 420–427) contained the accompanying data on y = shear strength of sandy soil (kPa), x_1 = depth (m), and x_2 = water content (%). The predicted values and residuals were computed using the estimated regression equation $\hat{y} = -151.36 - 16.22x_1 + 13.48x_2 + .094x_3 - .253x_4 + .492x_5$, where $x_3 = x_1^2$, $x_4 = x_2^2$, and $x_5 = x_1x_2$.

y	x_1	x_2	Predicted y	Residual
14.7	8.9	31.5	23.35	−8.65
48.0	36.6	27.0	46.38	1.62
25.6	36.8	25.9	27.13	−1.53
10.0	6.1	39.1	10.99	−.99
16.0	6.9	39.2	14.10	1.90
16.8	6.9	38.3	16.54	.26
20.7	7.3	33.9	23.34	−2.64
38.8	8.4	33.8	25.43	13.37

(continued)

y	x_1	x_2	Predicted y	Residual
16.9	6.5	27.9	15.63	1.27
27.0	8.0	33.1	24.29	2.71
16.0	4.5	26.3	15.36	.64
24.9	9.9	37.8	29.61	−4.71
7.3	2.9	34.6	15.38	−8.08
12.8	2.0	36.4	7.96	4.84

a. Use the given information to compute SSResid, SSTo, and SSRegr.

b. Calculate R^2 for this regression model. How would you interpret this value?

c. Use the value of R^2 from part **b** and a .05 level of significance to conduct the appropriate model utility test.

14.30 The article "Readability of Liquid Crystal Displays: A Response Surface" (*Human Factors* (1983): 185–190) used the estimated regression equation

$$\hat{y} = 1.52 + .02x_1 - 1.40x_2 + .02x_3 - .0006x_4$$

to describe the relationship between y = error percentage for subjects reading a four-digit liquid crystal display and the independent variables x_1 = level of backlight, x_2 = character subtense, x_3 = viewing angle, and x_4 = level of ambient light. From a table given in the article, SSRegr = 19.2, SSResid = 20.0, and $n = 30$.

a. Does the estimated regression equation specify a useful relationship between y and the independent variables? Use the model utility test with a .05 significance level.

b. Calculate R^2 and s_e for this model. Interpret these values.

c. Do you think that the estimated regression equation would provide reasonably accurate predictions of error rate? Explain.

14.31 The article "Effect of Manual Defoliation on Pole Bean Yield" (*J. Econ. Ent.* (1984): 1019–1023) used a quadratic regression model to describe the relationship between y = yield (kg/plot) and x = defoliation level (a proportion between 0 and 1). The estimated regression equation based on $n = 24$ was $\hat{y} = 12.39 + 6.67x_1 - 15.25x_2$, where $x_1 = x$ and $x_2 = x^2$. The article also reported that R^2 for this model was .902. Does the quadratic model specify a useful relationship between y and x? Carry out the appropriate test using a .01 level of significance.

14.32 Suppose that a multiple regression data set consists of $n = 15$ observations. For what values of k, the number of model predictors, would the corresponding model with $R^2 = .90$ be judged useful at significance level .05? Does such a large R^2 value necessarily imply a useful model? Explain.

14.33 *This exercise requires the use of a computer package.* Use the data given in Exercise 14.29 to verify that the true regression function

mean y value = $\alpha + \beta_1 x_1 + \beta_2 x_2 + \beta_3 x_3 + \beta_4 x_4 + \beta_5 x_5$

is estimated by

$$\hat{y} = -151.36 - 16.22 x_1 + 13.48 x_2 + .094 x_3 - .253 x_4 + .492 x_5$$

14.34 *This exercise requires the use of a computer package.* The accompanying data resulted from a study of the relationship between y = brightness of finished paper and the independent variables x_1 = hydrogen peroxide (% by weight), x_2 = sodium hydroxide (% by weight), x_3 = silicate (% by weight), and x_4 = process temperature ("Advantages of CE-HDP Bleaching for High Brightness Kraft Pulp Production," *TAPPI* (1964): 107A–173A).

x_1	x_2	x_3	x_4	y
.2	.2	1.5	145	83.9
.4	.2	1.5	145	84.9
.2	.4	1.5	145	83.4
.4	.4	1.5	145	84.2
.2	.2	3.5	145	83.8
.4	.2	3.5	145	84.7
.2	.4	3.5	145	84.0
.4	.4	3.5	145	84.8
.2	.2	1.5	175	84.5
.4	.2	1.5	175	86.0
.2	.4	1.5	175	82.6
.4	.4	1.5	175	85.1
.2	.2	3.5	175	84.5
.4	.2	3.5	175	86.0
.2	.4	3.5	175	84.0
.4	.4	3.5	175	85.4
.1	.3	2.5	160	82.9
.5	.3	2.5	160	85.5
.3	.1	2.5	160	85.2
.3	.5	2.5	160	84.5
.3	.3	.5	160	84.7
.3	.3	4.5	160	85.0
.3	.3	2.5	130	84.9
.3	.3	2.5	190	84.0
.3	.3	2.5	160	84.5
.3	.3	2.5	160	84.7
.3	.3	2.5	160	84.6

(continued)

x_1	x_2	x_3	x_4	y
.3	.3	2.5	160	84.9
.3	.3	2.5	160	84.9
.3	.3	2.5	160	84.5
.3	.3	2.5	160	84.6

a. Find the estimated regression equation for the model that includes all independent variables, all quadratic terms, and all interaction terms.

b. Using a .05 significance level, perform the model utility test.

c. Interpret the values of the following quantities: SSResid, R^2, s_e.

14.35 *This exercise requires the use of a computer package.* The cotton aphid poses a threat to cotton crops in Iraq. The accompanying data on

y = infestation rate (aphids/100 leaves)

x_1 = mean temperature (°C)

x_2 = mean relative humidity

appeared in the article "Estimation of the Economic Threshold of Infestation for Cotton Aphid" (*Mesopotamia J. Ag.* (1982): 71–75). Use the data to find the estimated regression equation and assess the utility of the multiple regression model

$$y = \alpha + \beta_1 x_1 + \beta_2 x_2 + e$$

y	x_1	x_2	y	x_1	x_2
61	21.0	57.0	77	24.8	48.0
87	28.3	41.5	93	26.0	56.0
98	27.5	58.0	100	27.1	31.0
104	26.8	36.5	118	29.0	41.0
102	28.3	40.0	74	34.0	25.0
63	30.5	34.0	43	28.3	13.0
27	30.8	37.0	19	31.0	19.0
14	33.6	20.0	23	31.8	17.0
30	31.3	21.0	25	33.5	18.5
67	33.0	24.5	40	34.5	16.0
6	34.3	6.0	21	34.3	26.0
18	33.0	21.0	23	26.5	26.0
42	32.0	28.0	56	27.3	24.5
60	27.8	39.0	59	25.8	29.0
82	25.0	41.0	89	18.5	53.5
77	26.0	51.0	102	19.0	48.0
108	18.0	70.0	97	16.3	79.5

14.3   Inferences Based on an Estimated Model

In the previous section, we discussed estimating the coefficients $\alpha, \beta_1, \ldots, \beta_k$ in the model $y = \alpha + \beta_1 x_1 + \cdots + \beta_k x_k + e$ (using the principle of least squares) and then showed how the usefulness of the model could be confirmed by application of the F test for model utility. If $H_0: \beta_1 = \cdots = \beta_k = 0$ cannot be rejected at a reasonably small level of significance, it must be concluded that the model does not specify a useful relationship between y and any of the predictor variables $x_1, \ldots, x_k$. The investigator must then search further for a model that does describe a useful relationship, perhaps by introducing different predictors or making variable transformations. Only if H_0 can be rejected is it appropriate to proceed further with the chosen model and make inferences based on the estimated coefficients $a, b_1, \ldots, b_k$ and the estimated regression function $y = a + b_1 x_1 + \cdots + b_k x_k$. In this section, we will consider two different types of inferential problems. One type involves drawing a conclusion about an individual regression coefficient β_i— either computing a confidence interval for β_i or testing a hypothesis concerning β_i. The second type of problem involves first fixing values of $x_1, \ldots, x_k$ and then computing either a point estimate or a confidence interval for the corresponding mean y value or predicting a future y value with a point prediction or a prediction interval.

Inferences About a Regression Coefficient

A confidence interval for and hypothesis test concerning the slope coefficient β in simple linear regression were based on facts about the sampling distribution of the statistic b used to obtain a point estimate of β. Similarly, in multiple regression, procedures for making inferences about β_i are derived from properties of the sampling distribution of b_i. Unfortunately, formulas for b_i and its standard deviation σ_{b_i} are quite complicated and cannot be stated concisely except by using some advanced mathematical notation.

The Sampling Distribution of b_i

Let b_i denote the statistic for estimating (via the principle of least squares) the coefficient β_i in the model $y = \alpha + \beta_1 x_1 + \cdots + \beta_k x_k + e$. Assumptions about this model given in Section 14.1 imply the following:

1. b_i has a normal distribution.

2. $\mu_{b_i} = \beta_i$, so b_i is an unbiased statistic for estimating β_i.

3. The standard deviation of b_i, σ_{b_i}, involves σ^2 and a complicated function of all the values of $x_1, x_2, \ldots, x_k$ in the sample. The estimated standard deviation s_{b_i} results from replacing σ^2 with s_e^2 in the formula for σ_{b_i}.

A consequence of the properties of the sampling distribution of b_i is that the standardized variable

$$t = \frac{b_i - \beta_i}{s_{b_i}}$$

has a t distribution with df $= n - (k + 1)$. This leads to a confidence interval for β_i.

A Confidence Interval for β_i

A confidence interval for β_i is

$$b_i \pm (t\text{ critical value}) \cdot s_{b_i}$$

The t critical value is based on df $= n - (k + 1)$.

Any good statistical computer package will provide both the estimated coefficients $a, b_1, \ldots, b_k$ and their estimated standard deviations $s_a, s_{b_1}, \ldots, s_{b_k}$.

EXAMPLE 14.10

Reconsider the regression of phosphate adsorption (y) against extractable iron (x_1) and extractable aluminum (x_2), with the sample data given in Example 14.5 and MINITAB output displayed in Figure 14.5. Suppose that the investigator had required a 95% confidence interval for β_1, the average change in phosphate adsorption when extractable iron increases by 1 unit and extractable aluminum is held fixed. The necessary quantities are

$b_1 = .11273 \qquad s_{b_1} = .02969 \qquad$ (from the **Coef** and **Stdev** columns in the output)

df $= n - (k + 1) = 13 - (2 + 1) = 10$

t critical value $= 2.23$

The resulting confidence interval is

$$b_i \pm (t\text{ critical value}) \cdot s_{b_i} = .11273 \pm (2.23)(.02969) = .11273 \pm .06621 = (.047, .179)$$

The standardized variable $t = (b_i - \beta_i)/s_{b_i}$ is also the basis for a test statistic for testing hypotheses about β_i.

Testing Hypotheses About β_i

Null hypothesis: $\quad H_0: \beta_i =$ hypothesized value

Test statistic: $\quad t = \dfrac{b_i - \text{hypothesized value}}{s_{b_i}}$

The test is based on df $= n - (k + 1)$.

(continued)

Alternative hypothesis: *P-value*:

H_a: β_i > hypothesized value   Area under the appropriate *t* curve to the right of the computed *t*

H_a: β_i < hypothesized value   Area under the appropriate *t* curve to the left of the computed *t*

H_a: β_i ≠ hypothesized value   (i) 2 · (area to the right of *t*) if *t* is positive

                                        or

                                (ii) 2 · (area to the left of *t*) if *t* is negative

Assumptions: For any particular combination of predictor variable values, the distribution of *e*, the random deviation from the model, is *normal* with mean 0 and *constant variance*.

The most frequently tested null hypothesis in this situation is H_0: β_i = 0. The interpretation of H_0 is that *as long as the other predictors* $x_1, \ldots, x_{i-1}, x_{i+1}, \ldots, x_k$ *remain in the model,* the predictor x_i provides no useful information about *y*, and so it can be eliminated. The test statistic for H_0: β_i = 0 simplifies to the *t* **ratio**

$$t = \frac{b_i}{s_{b_i}}$$

EXAMPLE 14.11

Economists are always interested in studying the factors affecting the price paid for a product. The article "Testing for Imperfect Competition at the Fulton Fish Market" (*Rand J. of Econ.* (1995): 75–92) considered a regression of *y* = price paid for whiting on eight different predictors. (The Fulton Fish Market, located in New York City, is the largest such market in the United States; the many dealers make it highly competitive, and whiting is one of the most frequently purchased types of fish.) There were *n* = 54 observations, and the coefficient of multiple determination was $R^2 = .992$, indicating a highly useful model (model utility *F* = 697.5, *P*-value = .000). One of the predictors used was x_3, an indicator variable for whether a purchaser was Asian. The values $b_3 = -.0463$ and $s_{b_3} = .0191$ were given. Provided that all seven other predictors remain in the model, does x_3 appear to provide useful information about *y*?

1. β_3 is the average difference in price paid by Asian and non-Asian purchasers when all other predictors are held fixed.

2. H_0: $\beta_3 = 0$

3. H_a: $\beta_3 \neq 0$

4. A significance level of $\alpha = .05$ was suggested in the article.

5. *Test statistic*: $t = \dfrac{b_3}{s_{b_3}}$

6. *Assumptions*:   Without raw data, we cannot assess the reasonableness of the assumptions. If we had the data, a normal probability plot or boxplot of the standardized residuals would be a good place to start. For purposes of this example, we will assume that it is reasonable to proceed with the test.

7. The test statistic value is

$$t = \frac{b_3}{s_{b_3}} = \frac{-.0463}{.0191} = -2.42$$

8. The test is based on $n - (k + 1) = 54 - (8 + 1) = 45$ df. An examination of the 40 and 60 df columns of Appendix Table IV shows that the P-value is roughly $2 \cdot (.010) = .020$.

9. Because P-value $= .020 \leq .05 = \alpha$, H_0 should be rejected. The predictor that indicates whether a purchaser is Asian does appear to provide useful information about price, over and above the information contained in the other predictors. The author of the article indicated that this result is inconsistent with the model of perfect competition.

EXAMPLE 14.12

Our analysis of the phosphate adsorption data introduced in Example 14.5 has so far focused on the model $y = \alpha + \beta_1 x_1 + \beta_2 x_2 + e$ in which x_1 (extractable iron) and x_2 (extractable aluminum) affect the response separately. Suppose that the researcher wishes to investigate the possibility of interaction between x_1 and x_2 through fitting the model with predictors x_1, x_2, and $x_3 = x_1 x_2$. We list a few of the sample values of y, x_1, and x_2, along with the corresponding values of x_3.

Observation	y	x_1	x_2	$x_3 = x_1 x_2$
1	4	61	13	793
2	18	175	21	3675
3	14	111	24	2664
⋮	⋮	⋮	⋮	⋮
13	40	199	54	10,746

In practice, a statistical software package would automatically compute x_3 values upon request once x_1 and x_2 values had been entered, so hand computations would not be necessary. Figure 14.8 displays partial MINITAB output resulting from a request to fit this model. Let's use the output to see whether inclusion of the interaction predictor is justified.

FIGURE 14.8 MINITAB output for model with interaction fit to the phosphate adsorption data ($x_1 = $ FE, $x_2 = $ AL, $x_3 = x_1 x_2 = $ FEAL)

Regression Analysis
The regression equation is
HPO = -2.37 + 0.0828 FE + 0.246 AL +0.000528 FEAL

Predictor	Coef	StDev	T	P
Constant	-2.368	7.179	-0.33	0.749
FE	0.08279	0.04818	1.72	0.120
AL	0.2460	0.1481	1.66	0.131
FEAL	0.0005278	0.0006610	0.80	0.445

S = 4.461   R-Sq = 95.2%   R-Sq(adj) = 93.6%

1. The model is $y = \alpha + \beta_1 x_1 + \beta_2 x_2 + \beta_3 x_3 + e$, where $x_3 = x_1 x_2$.

2. $H_0: \beta_3 = 0$

3. $H_a: \beta_3 \neq 0$

4. *Test statistic*:   The t ratio for β_3 is

$$t = \frac{b_3}{s_{b_3}}$$

5. *Significance level*: $\alpha = .05$

6. *Assumptions*:   The residuals and standardized residuals (from MINITAB) for this model are as follows:

Obs	FE	HPO	Residual	St Resid
1	61	4.00	-2.30	-0.94
2	175	18.00	-1.23	-0.33
3	111	14.00	-0.13	-0.03
4	124	18.00	2.94	0.72
5	130	26.00	-2.53	-0.77
6	173	26.00	1.23	0.30
7	169	21.00	-1.68	-0.41
8	169	30.00	-2.07	-0.52
9	160	28.00	4.24	1.01
10	244	36.00	-8.44	-2.08
11	257	65.00	3.35	1.19
12	333	62.00	-0.31	-0.13
13	199	40.00	6.94	1.68

A normal probability plot of the standardized residuals is shown here. The plot is quite straight, indicating that normality of the random error distribution is plausible.

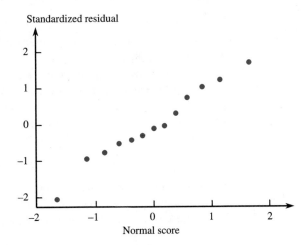

Standardized residual

Normal score

7. Directly from the t ratio column in Figure 14.8, we find $t = .80$.

8. Figure 14.8 shows that the P-value is .445.

9. Because .445 > .05, H_0 cannot be rejected. The interaction predictor does not appear to be useful and can be removed from the model.

─────────

An interesting aspect of the computer output for the interaction model in the foregoing example is that the t ratios for β_1, for β_2, and for β_3 (1.72, 1.66, and .80) are all relatively small and the corresponding P-values are large, yet $R^2 = .952$ is quite large. The high R^2 value suggests a useful model (this can be confirmed by the model utility test), yet the size of each t ratio and P-value might tempt us to conclude that all three β_i's are zero. This sounds like a contradiction, but it involves a misinterpretation of the t ratios. For example, the t ratio for β_3, $t = b_3/s_{b_3}$, tests the null hypothesis $H_0: \beta_3 = 0$ *when x_1 and x_2 are included in the model.* The smallness of a given t ratio suggests that the associated predictor can be dropped from the model *as long as the other predictors are retained.* The fact that all t ratios are small in this example does not, therefore, allow us simultaneously to delete all predictors. The data does suggest deleting x_3, because the model that includes x_1 and x_2 has already been found to give a very good fit to the data and is simple to interpret. In Section 14.4, we comment further on why it might happen that all t ratios are small even when the model seems very useful.

The model utility test amounts to testing a simultaneous claim about the values of all β_i's—that they are all zero. There is also an F test for testing a hypothesis involving a specified subset consisting of at least two β_i's. For example, we might fit the model

$$y = \alpha + \beta_1 x_1 + \beta_2 x_2 + \beta_3 x_1 x_2 + \beta_4 x_1^2 + \beta_5 x_2^2 + e$$

and then wish to test $H_0: \beta_3 = \beta_4 = \beta_5 = 0$ (which says that the second-order predictors contribute nothing to the model). Please see one of the chapter references for Chapter 5 for further details.

Inferences Based on the Estimated Regression Function

The estimated regression line $\hat{y} = a + bx$ in simple linear regression was used both to estimate the mean y value when x had a specified value and to predict the associated y value for a single observation made at a particular x value. The estimated regression function for the model $y = \alpha + \beta_1 x_1 + \cdots + \beta_k x_k + e$ can be used in the same two ways. When fixed values of the predictor variables $x_1, x_2, \ldots, x_k$ are substituted into the estimated regression function

$$\hat{y} = a + b_1 x_1 + b_2 x_2 + \cdots + b_k x_k$$

the result can be used either as a point estimate of the corresponding mean y value or as a point prediction of the y value that will result from a single observation when the x_i's have the specified values.

EXAMPLE 14.13   Precise information concerning bus transit times is important when making transportation planning decisions. The article "Factors Affecting Running Time on Transit Routes" (*Transportation Research* (1983): 107–113) reported on an empirical study based on data gathered in Cincinnati, Ohio. The variables of interest were

y = running time per mile during the morning peak period (in seconds)

x_1 = number of passenger boardings per mile

x_2 = number of passenger alightings per mile

x_3 = number of signalized intersections per mile

x_4 = proportion of a route on which parking is allowed

The values of x_1 and x_2 were not necessarily equal, because observations were made over partial segments of routes (so not all passengers entered or exited on the segments). The estimated regression function was

$$\hat{y} = 169.50 + 5.07x_1 + 4.53x_2 + 6.61x_3 + 67.70x_4$$

Consider the predictor variable values $x_1 = 4.5, x_2 = 5.5, x_3 = 5$, and $x_4 = .1$. Then a point estimate for true average running time per mile when the x_i's have these values is

$$\hat{y} = 169.50 + (5.07)(4.5) + (4.53)(5.5) + (6.61)(5) + (67.70)(.1) = 257.05$$

This value 257.05 is also the predicted running time per mile for a single trip when x_1, x_2, x_3, and x_4 have the given values.

Remember that before the sample observations $y_1, y_2, \ldots, y_n$ are obtained, a and all b_i's are statistics (because they all involve the y_i's). This implies that for fixed values of $x_1, x_2, \ldots, x_k$, the estimated regression function $a + b_1x_1 + \cdots + b_kx_k$ is a statistic (its value varies from sample to sample). To obtain a confidence interval for the mean y value for specified $x_1, \ldots, x_k$ values, we need some facts about the sampling distribution of this statistic.

Properties of the Sampling Distribution of $\hat{y} = a + b_1x_1 + \cdots + b_kx_k$

Assumptions about the model $y = \alpha + \beta_1x_1 + \cdots + \beta_kx_k + e$ given in Section 14.1 imply that for fixed values of the predictors $x_1, x_2, \ldots, x_k$, the statistic $\hat{y} = a + b_1x_1 + \cdots + b_kx_k$ satisfies the following properties:

1. It has a normal distribution.
2. Its mean value is $\alpha + \beta_1x_1 + \cdots + \beta_kx_k$. That is, the statistic is unbiased for estimating the mean y value when $x_1, \ldots, x_k$ are fixed.
3. The standard deviation of $\hat{y}$ involves σ^2 and a complicated function of all the sample predictor variable values. The estimated standard deviation of $\hat{y}$, denoted by $s_{\hat{y}}$, results from replacing σ^2 with s_e^2 in this function.

The standardized variable

$$t = \frac{\hat{y} - (\alpha + \beta_1x_1 + \cdots + \beta_kx_k)}{s_{\hat{y}}}$$

then has a t distribution with $n - (k + 1)$ df.

The formula for $s_{\hat{y}}$ has been programmed into the most widely used statistical computer packages, and its value for specified $x_1, \ldots, x_k$ is available upon request. Manipulation of the t variable as before gives a confidence interval formula. A prediction interval formula is based on a similar standardized variable.

For fixed values of $x_1, x_2, \ldots, x_k$, a **confidence interval for the mean y value** — that is, for $\alpha + \beta_1 x_1 + \cdots + \beta_k x_k$ — is

$$\hat{y} \pm (t \text{ critical value}) \cdot s_{\hat{y}}$$

A prediction interval for a single y value is

$$\hat{y} \pm (t \text{ critical value}) \cdot \sqrt{s_e^2 + s_{\hat{y}}^2}$$

The t critical value in both intervals is based on $n - (k + 1)$ df.

EXAMPLE 14.14

(Example 14.10 continued)   Figure 14.5 shows MINITAB output from fitting the model $y = \alpha + \beta_1 x_1 + \beta_2 x_2 + e$ to the phosphate data introduced previously. A request for estimation and prediction information when $x_1 = 150$ and $x_2 = 40$ resulted in the following additional output:

Fit	StDev Fit	95.0% CI	95.0% PI
23.52	1.31	(20.60, 26.44)	(13.33, 33.70)

The t critical value for a 95% confidence level when df $= 10$ is 2.23, and the corresponding interval is

$$23.52 \pm (2.23)(1.31) = 23.52 \pm 2.92$$
$$= (20.60, 26.44)$$

Using $s_e^2 = 4.379$ along with the given values of $\hat{y}$, $s_{\hat{y}}$, and t critical value in the prediction interval formula results in (13.33, 33.70). Both intervals are centered at $\hat{y}$, but the prediction interval is much wider than the confidence interval.

The danger of extrapolation in simple linear regression is that if the x value of interest is much outside the interval of x values in the sample, the postulated model might no longer be valid, and even if it were, $\alpha + \beta x$ could still be quite poorly estimated (s_{a+bx} could be large). There is a similar danger in multiple regression, but it is not always obvious when the x_i values of interest involve an extrapolation from sample data. As an example, suppose that a single y observation is made for each of the following 13 (x_1, x_2) pairs:

Obs.	1	2	3	4	5	6	7	8	9	10	11	12	13
x_1	0	0	0	0	0	-5	5	5	5	10	-5	-5	-10
x_2	10	5	0	-5	-10	-5	0	5	-5	0	0	5	0

The x_1 values range between -10 and 10, as do the x_2 values. After fitting the model $y = \alpha + \beta_1 x_1 + \beta_2 x_2 + e$, we might then want a confidence interval for the mean y

FIGURE 14.9 The danger of extrapolation: • denotes an (x_1, x_2) pair for which the sample contains a y observation; ○ denotes an (x_1, x_2) pair well outside the sample region, although the individual values $x_1 = 10$ and $x_2 = 10$ are within x_1 and x_2 sample ranges separately

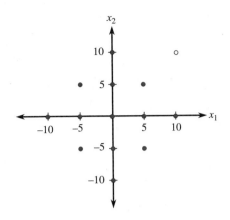

value when $x_1 = 10$ and $x_2 = 10$. However, whereas each of these values separately is within the range of sample x_i values, Figure 14.9 shows that the point $(x_1, x_2) = (10, 10)$ is actually far from (x_1, x_2) pairs in the sample. Thus, drawing a conclusion about y when $x_1 = 10$ and $x_2 = 10$ involves a substantial extrapolation. In particular, the estimated standard deviation of $a + b_1(10) + b_2(10)$ would probably be quite large even if the model were valid near this point.

When more than two predictor variables are included in the model, we cannot tell from a plot like that of Figure 14.9 whether the $x_1, x_2, \ldots, x_k$ values of interest involve an extrapolation. It is then best to compare $s_{\hat{y}}$'s based on $a + b_1 x_1 + \cdots + b_k x_k$ for the values of interest with the $s_{\hat{y}}$'s corresponding to $x_1, x_2, \ldots, x_k$ values in the sample (MINITAB gives these on request). Extrapolation is indicated by a value of the former standard deviation (at the x_i values of interest) that is much larger than the standard deviations for sampled values.

Exercises 14.36 – 14.50

14.36 Explain why it is preferable to perform a model utility test before using an estimated regression model to make predictions or to estimate the mean y value for specified values of the independent variables.

14.37 The article "Zoning and Industrial Land Values: The Case of Philadelphia" (*AREUEA J.* (1991): 154–159) considered a regression model to relate the value of a vacant lot in Philadelphia to a number of different predictor variables.
a. One of the predictors was x_3 = the distance from the city's major east-west thoroughfare, for which $b_3 = -.489$ and $s_{b_3} = .1044$. The model contained seven predictors, and the coefficients were estimated from a sample of 100 vacant lots. Calculate and interpret a confidence interval for β_3 using a confidence level of 95%.
b. Another predictor was x_1, an indicator variable for whether the lot was zoned for residential use, for which $b_1 = -.183$ and $s_{b_1} = .3055$. Carry out a test of $H_0: \beta_1 = 0$ versus $H_a: \beta_1 \neq 0$, and interpret the conclusion in the context of the problem.

14.38 Twenty-six observations from the article "Multiple Regression Analysis for Forecasting Critical Fish Influxes at Power Station Intakes" (see Exercise 14.16) were used to fit a multiple regression model relating y = number of fish at intake to the independent variables x_1 = water temperature (°C), x_2 = number of pumps running, x_3 = sea state (taking values 0, 1, 2, or 3), and x_4 = speed (knots). Partial MINITAB output follows.

THE REGRESSION EQUATION is

$Y = 92.0 - 2.18X1 - 19.2X2 - 9.38X3 + 2.32X4$

Predictor	Coef	Stdev	t-ratio
Constant	91.98	42.07	2.19
X1	-2.179	1.087	-2.00
X2	-19.189	9.215	-2.08
X3	-9.378	4.356	-2.15
X4	2.3205	0.7686	3.02

$s = 10.53$ R-sq = 39.0%   R-sq(adj) = 27.3%

a. Construct a 95% confidence interval for β_3, the coefficient of x_3 = sea state. Interpret the resulting interval.

b. Construct a 90% confidence interval for the mean change in y associated with a 1° increase in temperature when number of pumps, sea state, and speed remain fixed.

14.39 A study reported in the article "Leakage of Intracellular Substances from Alfalfa Roots at Various Subfreezing Temperatures" (*Crop Sci.* (1991): 1575–1578) considered a quadratic regression model to relate y = MDH activity (a measure of the extent to which cellular membranes suffer extensive damage from freezing) to x = electrical conductivity (which describes the damage in the early stages of freezing). The estimated regression function was

$$\hat{y} = -.1838 + .0272x + .0446x^2$$

with $R^2 = .860$.

a. Supposing that $n = 50$, does the quadratic model appear to be useful? Test the appropriate hypotheses.

b. Using $s_{b_2} = .0103$, carry out a test at significance level .01 to decide whether the quadratic predictor x^2 is important.

c. If the standard deviation of the statistic $\hat{y} = a + b_1(40) + b_2(1600)$ is $s_{\hat{y}} = .120$, calculate a confidence interval with confidence level 90% for true average MDH activity when conductivity is 40.

14.40 The article first introduced in Exercise 14.27 of Section 14.2 gave data on the dimensions of 27 representative food products. Use the multiple regression model fit in Exercise 14.27.

a. Could any of these variables be eliminated from a regression with the purpose of predicting volume?

b. Predict the volume of a package with a minimum width of 2.5 cm, a maximum width of 3.0 cm, and an elongation of 1.55.

c. Calculate a 95% prediction interval for a package with a minimum width of 2.5 cm, a maximum width of 3.0 cm, and an elongation of 1.55.

14.41 Data from a random sample of 107 students taking a managerial accounting course was used to obtain the estimated regression equation $2.178 + .469x_1 + 3.369x_2 + 3.054x_3$, where y = student's exam score, x_1 = student's expected score on the exam, x_2 = time spent studying (hr/week), and x_3 = student's grade point average or GPA ("Effort, Expectation and Academic Performance in Managerial Cost Accounting," *J. Acctg. Educ.* (1989): 57–68). The value of R^2 was .686, and the estimated standard deviations of the statistics b_1, b_2, and b_3 were .090, .456, and 1.457, respectively.

a. How would you interpret the estimated coefficient .469?

b. Does there appear to be a useful linear relationship between exam score and at least one of the three predictors?

c. Calculate a confidence interval for the mean change in exam score associated with a 1-hour increase in study time when expected score and GPA remain fixed.

d. Obtain a point prediction of the exam score for a student who expects a 75, has studied 8 hours per week, and has a GPA of 2.8.

e. If the standard deviation of the statistic on which the prediction of part **d** is based is 1.2 and SSTo = 10,200, calculate a 95% prediction interval for the score of the student described in part **d**.

14.42 Benevolence payments are monies collected by a church to fund activities and ministries beyond those provided by the church to its own members. The article "Optimal Church Size: The Bigger the Better" (*J. for the Scientific Study of Religion* (1993): 231–241) considered a regression of y = benevolence payments on x_1 = number of church members, $x_2 = x_1^2$, and x_3 = an indicator variable for urban versus nonurban churches.

a. The sample size was $n = 300$, and adjusted $R^2 = .774$. Does at least one of the three predictors provide useful information about y?

b. The article reported that $b_3 = 101.1$ and $s_{b_3} = 625.8$. Should the indicator variable be retained in the model? Test the relevant hypotheses.

14.43 The estimated regression equation

$$\hat{y} = 28 - .05x_1 - .003x_2 + .00002x_3$$

where

y = number of indictments disposed of in a given month

x_1 = number of cases on judge's docket

x_2 = number of cases pending in the criminal trial court

$x_3 = x_1 x_2$

appeared in the article "The Caseload Controversy and the Study of Criminal Courts" (*J. Criminal Law and Criminology* (1979): 89–101). This equation was based on $n = 367$ observations. The b_i's and their associated standard deviations are given in the accompanying table.

		Estimated Standard
i	b_i	**Deviation of b_i**
1	−.05	.03
2	−.003	.0024
3	.00002	.000009

Is inclusion of the interaction predictor important? Test $H_0: \beta_3 = 0$ using a .05 level of significance.

14.44 The accompanying data was obtained from a study of a certain method for preparing pure alcohol from refinery streams ("Direct Hydration of Olefins" *Indus. and Engr. Chem.* (1961): 209–211). The independent variable x is volume, and the dependent variable y is amount of isobutylene converted.

x	1	1	2	4	4	4	6
y	23.0	24.5	28.0	30.9	32.0	33.6	20.0

MINITAB output — the result of fitting a quadratic regression model where $x_1 = x$ and $x_2 = x^2$ — is given. Would a linear regression have sufficed? That is, is the quadratic predictor important? Use a level .05 test.

THE REGRESSION EQUATION is
Y = 13.6 + 11.4X1 - 1.72X2

Predictor	Coef	Stdev	t-ratio
Constant	13.636	1.896	7.19
X1	11.406	1.356	8.41
X2	-1.7155	0.2036	-8.42

s = 1.428   R-sq = 94.7%   R-sq(adj) = 92.1%

14.45 The article "Bank Full Discharge of Rivers" (*Water Resources J.* (1978): 1141–1154) reported data on $y =$ discharge amount (m^2/s), $x_1 =$ flow area (m^2), and $x_2 =$ slope of the water surface (m/m) obtained at $n = 10$ floodplain stations. A multiple regression model using x_1, x_2, and $x_3 = x_1 x_2$ was fit to this data, and partial MINITAB output appears here.

THE REGRESSION EQUATION is
Y = -3.14 + 1.70X1 + 96.1X2 + 8.38X3

Predictor	Coef	Stdev	t-ratio
Constant	-3.14	14.54	-0.22
X1	1.697	1.431	1.19
X2	96.1	702.7	0.14
X3	8.38	199.0	0.04

s = 17.58   R-sq = 73.2%   R-sq(adj) = 59.9%

a. Perform the model utility test using a significance level of .05.

b. Is the interaction term important? Test using a .05 significance level.

c. Does it bother you that the model utility test indicates a useful model, but all values in the **t-ratio** column of the output are small? Explain.

14.46 In the article "An Ultracentrifuge Flour Absorption Method" (*Cereal Chem.* (1978): 96–101), the authors discussed the relationship between water absorption for wheat flour and various characteristics of the flour. The model $y = \alpha + \beta_1 x_1 + \beta_2 x_2 + e$ was used to relate $y =$ absorption (%) to $x_1 =$ flour protein (%) and $x_2 =$ starch damage (Farrand units). MINITAB output based on $n = 28$ observations is given. Use a significance level of .05 for all tests requested.

THE REGRESSION EQUATION is
Y = 19.4 + 1.44X1 + .336X2

Predictor	Coef	Stdev	t-ratio	p
Constant	19.440	2.188	8.88	.000
X1	1.4423	0.2076	6.95	.000
X2	0.33563	0.01814	18.51	.000

s = 1.094   R-sq = 96.4%   R-sq(adj) = 96.2%

Analysis of Variance

SOURCE	DF	SS	MS	F	p
Regression	2	812.380	406.190	339.3	.000
Error	25	29.928	1.197		
Total	27	842.307			

a. Does the model appear to be useful? Test the relevant hypotheses.

b. Calculate and interpret a 95% confidence interval for β_2.

c. Conduct tests for each of the following pairs of hypotheses.

   i. $H_0: \beta_1 = 0$ versus $H_a: \beta_1 \neq 0$
   ii. $H_0: \beta_2 = 0$ versus $H_a: \beta_2 \neq 0$

d. Based on the results of part **c**, would you conclude that both independent variables are important? Explain.

e. An estimate of the mean water absorption for wheat with 11.7% protein and a starch damage of 57 is desired. Compute a 90% confidence interval for $\alpha + \beta_1(11.7) + \beta_2(57)$ if the estimated standard deviation of $a + b_1(11.7) + b_2(57)$ is .522. Interpret the resulting interval.

f. A single shipment of wheat is received. For this particular shipment, $x_1 = 11.7$ and $x_2 = 57$. Predict the water absorption for this shipment (a single y value) using a 90% interval.

14.47 Exercise 14.25 gave data on fish weight, length, age, and year caught. A multiple regression model was fit to describe the relationship between weight and the predictors length and age.

a. Could either length or age be eliminated from the model without significantly affecting your ability to make accurate predictions? Why or why not? Give statistical evidence to support your conclusion.

b. Create a dummy (indicator) variable for year caught. Fit a multiple regression model that includes year, length, and age as predictors of weight. Is there evidence that year is a useful predictor given that length and age are included in the model? Test the relevant hypotheses using $\alpha = .05$.

14.48 The article "Predicting Marathon Time from Anaerobic Threshold Measurements" (*The Physician and Sportsmed.* (1984): 95–98) gave data on $y = $ maximum heart rate (beats/min), $x_1 = $ age, and $x_2 = $ weight (kg) for $n = 18$ marathon runners. The estimated regression equation for the model $y = \alpha + \beta_1 x_1 + \beta_2 x_2 + e$ was $\hat{y} = 179 - .8x_1 + .5x_2$ and SSRegr $= 649.75$, SSResid $= 538.03$.

a. Is the model useful for predicting maximum heart rate? Use a significance level of .10.

b. Using $s_{b_1} = .280$, calculate and interpret a 95% confidence interval for β_1.

c. Predict the maximum heart rate of a particular runner who is 43 years old and weighs 65 kg, using a 99% interval. The estimated standard deviation of the statistic $a + b_1(43) + b_2(65)$ is 3.52.

d. Use a 90% interval to estimate the average maximum heart rate for all marathon runners who are 30 years old and weigh 77.2 kg. The estimated standard deviation of $a + b_1(30) + b_2(77.2)$ is 2.97.

e. Would a 90% prediction interval for a single 30-year-old runner weighing 77.2 kg be wider or narrower than the interval computed in part **d**? Explain. (You need not compute the interval.)

14.49 The effect of manganese (Mn) on wheat growth is examined in the article "Manganese Deficiency and Toxicity Effects on Growth, Development and Nutrient Composition in Wheat" (*Agronomy J.* (1984): 213–217). A quadratic regression model was used to relate $y = $ plant height (cm) to $x = $ log (added Mn), with μM as the units for added Mn. The accompanying data was read from a scatter plot in the article. Also given is MINITAB output, where $x_1 = x$ and $x_2 = x^2$. Use a .05 significance level for any hypothesis tests needed to answer the questions that follow.

x	−1	−.4	0	.2	1	2	2.8	3.2	3.4	4
y	32	37	44	45	46	42	42	40	37	30

THE REGRESSION EQUATION is
Y = 41.7 + 6.58X1 - 2.36X2

Predictor	Coef	Stdev	t-ratio
Constant	41.7422	0.8522	48.98
X1	6.581	1.002	6.57
X2	-2.3621	0.3073	-7.69

s = 1.963   R-sq = 89.8%   R-sq(adj) = 86.9%

Analysis of Variance

SOURCE	DF	SS	MS
Regression	2	237.520	118.760
Error	7	26.980	3.854
Total	9	264.500	

a. Is the quadratic model useful for describing the relationship between y and x?

b. Are both the linear and quadratic predictors important? Could either one be eliminated from the model? Explain.

c. Give a 95% confidence interval for the mean y value when $x = 2$. The estimated standard deviation of $a + b_1(2) + b_2(4)$ is 1.037. Interpret the resulting interval.

d. Estimate the mean height for wheat treated with 10 μM of Mn using a 90% interval. (*Note:* The estimated standard deviation of $a + b_1(1) + b_2(1)$ is 1.031 and $\log(10) = 1$.)

14.50 *This exercise requires the use of a computer package.* A study on the effect of applying fertilizer in bands is described in the article "Fertilizer Placement Effects on Growth, Yield, and Chemical Composition of Burley Tobacco" (*Agronomy J.* (1984): 183–188). The accompanying data was taken from a scatter plot appearing in the article, with $y = $ plant Mn (μg/g dry weight) and $x = $ distance from the fertilizer band (cm). The authors suggest a quadratic regression model.

x	0	10	20	30	40
y	110	90	76	72	70

a. Use a suitable computer package to find the estimated quadratic regression equation.

b. Perform the model utility test.

c. Interpret the values of R^2 and s_e.

d. Are both the linear and quadratic predictors important? Carry out the necessary hypothesis tests and interpret the results.

e. Find a 90% confidence interval for the mean plant Mn for plants that are 30 cm from the fertilizer band.

14.4 Other Issues in Multiple Regression

Primary objectives in multiple regression include estimating a mean y value, predicting an individual y value, and gaining insight into how changes in predictor variable values affect y. Often an investigator has data on a number of predictor variables that might be incorporated into a model to be used for such purposes. Some of these may actually be unrelated or only very weakly related to y or they may contain information that duplicates information provided by other predictors. If all these predictors are included in the model, many model coefficients have to be estimated. This reduces the number of degrees of freedom associated with SSResid, leading to a deterioration in the degree of precision associated with other inferences (for example, wide confidence and prediction intervals). A model with many predictors can also be cumbersome to use and difficult to interpret.

In this section, we first introduce some guidelines and procedures for selecting a set of useful predictors. In choosing a model, the analyst should examine the data carefully for evidence of unusual observations or potentially troublesome patterns. It is important to identify unusually deviant or influential observations and to look for possible inconsistencies with model assumptions. Our discussion of multiple regression closes with a brief mention of some diagnostic methods designed for these purposes.

Variable Selection

Suppose that an investigator has data on p predictors $x_1, x_2, \ldots, x_p$, which are candidates for use in building a model. Some of these predictors might be specified functions of others, for example, $x_3 = x_1 x_2$, $x_4 = x_1^2$, and so on. The objective is then to select a set of these predictors that in some sense specifies a best model (of the general additive form considered in the two previous sections). Fitting a model that uses a specified k predictors requires that $k + 1$ model coefficients (α and the k corresponding β's) be estimated. Generally speaking, the number of observations, n, should be at least twice the number of predictors in the largest model under consideration to ensure reasonably accurate coefficient estimates and a sufficient number of degrees of freedom associated with SSResid.

If p is not too large, a good statistical computer package can quickly fit a model based on each different subset of the p predictors. Consider the case $p = 4$. There are two possibilities for each predictor — it could be included or not included in a model — so the number of possible models in this case is $2 \cdot 2 \cdot 2 \cdot 2 = 2^4 = 16$ (including the model with all four predictors and the model with only the constant term and none of the four predictors). These 16 possibilities are displayed in the accompanying table.

Predictors included	None	x_1	x_2	x_3	x_4	x_1, x_2	x_1, x_3	x_1, x_4	x_2, x_3	x_2, x_4
Number of predictors in model	0	1	1	1	1	2	2	2	2	2

Predictors included	x_3, x_4	x_1, x_2, x_3	x_1, x_2, x_4	x_1, x_3, x_4	x_2, x_3, x_4	x_1, x_2, x_3, x_4
Number of predictors in model	2	3	3	3	3	4

More generally, when there are p candidate predictors, the number of possible models is 2^p. The number of possible models is, therefore, substantial if p is even moderately large — for example, 1024 possible models when $p = 10$ and 32,768 possibilities when $p = 15$.

Model selection methods can be divided into two types. There are those based on fitting every possible model, computing one or more summary quantities from each fit, and comparing these quantities to identify the most satisfactory models. Several of the most powerful statistical computer packages have an **all-subsets** option, which will give limited output from fitting each possible model. Methods of the second type are appropriate when p is so large that it is not feasible to examine all subsets. These methods are often referred to as **automatic selection,** or **stepwise, procedures.** The general idea is either to begin with the p predictor model and delete predictors one by one until all remaining predictors are judged important, or to begin with no predictors and add predictors until no predictor not in the model seems important. With present-day computing power, the value of p for which examination of all subsets is feasible is surprisingly large, so automatic selection procedures are not as important as they once were.

Suppose then that p is small enough that all subsets can be fit. What characteristic(s) of the estimated models should be examined in the search for a best model? An obvious and appealing candidate is the coefficient of multiple determination, R^2, which measures the proportion of observed y variation explained by the model. Certainly a model with a large R^2 value is preferable to another model that contains the same number of predictors but has a much smaller R^2 value. Thus, if the model with predictors x_1 and x_2 has $R^2 = .765$ and the model with predictors x_1 and x_3 has $R^2 = .626$, the latter model would almost surely be eliminated from further consideration.

However, using R^2 to choose between models containing different numbers of predictors is not so straightforward, because adding a predictor to a model can never decrease the value of R^2. Let

$$R^2_{(1)} = \text{largest } R^2 \text{ for any one-predictor model}$$
$$R^2_{(2)} = \text{largest } R^2 \text{ for any two-predictor model}$$
$$\vdots$$

Then $R^2_{(1)} \le R^2_{(2)} \le \cdots \le R^2_{(p-1)} \le R^2_{(p)}$. When statisticians base model selection on R^2, the objective is not simply to find the model with the largest R^2 value; the model with p predictors does that. Instead, we should look for a model that contains relatively few predictors but has a large R^2 value and is such that no other model containing more predictors gives much of an improvement in R^2. Suppose, for example, that $p = 5$ and that

$$R^2_{(1)} = .427 \qquad R^2_{(2)} = .733 \qquad R^2_{(3)} = .885 \qquad R^2_{(4)} = .898 \qquad R^2_{(5)} = .901$$

Then the best three-predictor model appears to be a good choice, since it substantially improves on the best one- and two-predictor models, whereas very little is gained by using the best four-predictor model or all five predictors.

A small increase in R^2 resulting from the addition of a predictor to a model may be offset by the increased complexity of the new model and the reduction in degrees of freedom associated with SSResid. This has led statisticians to consider an adjusted R^2, which can either decrease or increase when a predictor is added to a model. It follows that adjusted R^2 for the best k-predictor model (the one with coefficient of multiple determination $R^2_{(k)}$) may be larger than adjusted R^2 for the best model based on $k + 1$ predictors. Adjusted R^2 formalizes the notion of diminishing returns as more predictors are added: Small increases in R^2 are outweighed by corresponding decreases in degrees of freedom associated with SSResid. A reasonable strategy in model selection is to identify the model with the largest value of adjusted R^2 (the corresponding number of predictors k is often much smaller than p) and then consider only that model and any others whose adjusted R^2 values are nearly as large.

EXAMPLE 14.15

The article "Anatomical Factors Influencing Wood Specific Gravity of Slash Pines and the Implications for the Development of a High-Quality Pulpwood" (*TAPPI* (1964): 401–404) reported the results of an experiment in which 20 specimens of slash pine wood were analyzed. A primary objective was to relate wood specific gravity (y) to various other wood characteristics. The independent variables on which observations were made were

x_1 = number of fibers/mm^2 in springwood

x_2 = number of fibers/mm^2 in summerwood

x_3 = % springwood

x_4 = % springwood light absorption

x_5 = % summerwood light absorption

The data is displayed in the accompanying table.

x_1	x_2	x_3	x_4	x_5	y
573	1059	46.5	53.8	84.1	.534
651	1356	52.7	54.5	88.7	.535
606	1273	49.4	52.1	92.0	.570
630	1151	48.9	50.3	87.9	.528
547	1135	53.1	51.9	91.5	.548
557	1236	54.9	55.2	91.4	.555
489	1231	56.2	45.5	82.4	.481
685	1564	56.6	44.3	91.3	.516
536	1182	59.2	46.4	85.4	.475
685	1564	63.1	56.4	91.4	.486
664	1588	50.6	48.1	86.7	.554
703	1335	51.9	48.4	81.2	.519
653	1395	62.5	51.9	89.2	.492
586	1114	50.5	56.5	88.9	.517

(*continued*)

x_1	x_2	x_3	x_4	x_5	y
534	1143	52.1	57.0	88.9	.502
523	1320	50.5	61.2	91.9	.508
580	1249	54.6	60.8	95.4	.520
448	1028	52.2	53.4	91.8	.506
476	1057	42.9	53.2	92.9	.595
528	1057	42.4	56.6	90.0	.568

Consider $x_1, x_2, \ldots, x_5$ as the set of potential predictors ($p = 5$, with no derived predictors such as squares or interaction terms as candidates for inclusion). Then there are $2^5 = 32$ possible models, among which 5 consist of a single predictor, 10 involve two predictors, 10 others involve three predictors, and 5 include four predictor variables. We used a statistical computer package to fit each possible model and extracted both R^2 and adjusted R^2 from the output. To save space, results appear for all one- and four-predictor models but only for the five best two- and three-predictor models.

It is immediately clear that the best three-predictor models offer considerable improvement with respect to both R^2 and adjusted R^2 over any model containing one or two predictors and that the model with all five predictors is inferior to the best four-predictor model. The best four-predictor model has the largest value of adjusted R^2 (.709), but the best three-predictor model is not far behind. The five models that seem most appealing to us are shaded in the display of results. The selection of a single model would have to be based on a more detailed comparison of these five models, with special attention given to residuals and other diagnostics (to be mentioned shortly) that bear on model adequacy. In particular, it is dangerous to embrace the model with the largest adjusted R^2 value automatically and reject all other models out of hand.

Models with One Predictor

Predictor	x_3	x_5	x_2	x_4	x_1
R^2	$R^2_{(1)} = .564$	.106	.053	.020	.008
Adjusted R^2	.539	.057	.001	$-.034$	$-.047$

Models with Two Predictors

Predictors	x_3, x_5	x_2, x_3	x_1, x_3	x_3, x_4	x_2, x_5
R^2	$R^2_{(2)} = .655$	.621	.603	.564	.158
Adjusted R^2	.614	.576	.556	.513	.059

Models with Three Predictors

Predictors	x_1, x_3, x_5	x_3, x_4, x_5	x_2, x_3, x_5	x_1, x_2, x_3	x_2, x_3, x_4
R^2	$R^2_{(3)} = .723$	.712	.711	.622	.611
Adjusted R^2	.671	.659	.657	.551	.550

Models with Four Predictors

Predictors	x_1, x_3, x_4, x_5	x_2, x_3, x_4, x_5	x_1, x_2, x_3, x_5	x_1, x_2, x_3, x_4	x_1, x_2, x_4, x_5
R^2	$R_{(4)}^2 = .770$	.748	.727	.622	.239
Adjusted R^2	.709	.681	.654	.522	.036

The model with all five predictors included has $R_{(5)}^2 = .770$, adjusted $R^2 = .689$.

Various other criteria have been proposed and used for model selection after fitting all subsets. A chapter reference can be consulted for more details.

When using particular criteria as a basis for model selection, many of the 2^p possible subset models are not serious candidates because of poor criteria values. For example, if $p = 15$, there are 5005 different models consisting of six predictor variables, many of which typically have small R^2 and adjusted R^2 values. An investigator usually wishes to consider only a few of the best models of each different size (a model whose criteria value is close to the best one may be easier to interpret than the best model or may include a predictor that the investigator thinks should be in the selected model). In recent years, statisticians have developed computer programs to achieve this without actually fitting all possible models. One version of such a program has been implemented in MINITAB and can be used as long as $p \leq 20$. (There are roughly one million possible models when $p = 20$, so fitting them all would be out of the question.) The user specifies a number between 1 and 10 as the number of models of each given size for which output will be provided. In addition to R^2 and adjusted R^2, values of another criterion, called *Mallow's C_p*, are included. A good model according to this criterion is one that has small C_p (for accurate predictions) and $C_p \approx k + 1$ (for unbiasedness in estimating model coefficients). After the choice of models is narrowed, the analyst can request more detail on each finalist.

EXAMPLE 14.16

The accompanying data was taken from the article "Applying Stepwise Multiple Regression Analysis to the Reaction of Formaldehyde with Cotton Cellulose" (*Textile Research J.* (1984): 157–165). The dependent variable

y = durable press rating

is a quantitative measure of wrinkle resistance. The four independent variables used in the model building process are

x_1 = HCHO (formaldehyde) concentration

x_2 = catalyst ratio

x_3 = curing temperature

x_4 = curing time

In addition to these variables, the investigators considered as potential predictors $x_1^2, x_2^2, x_3^2, x_4^2$, and all six interactions $x_1 x_2, \ldots, x_3 x_4$, a total of $p = 14$ candidates.

Figure 14.10 displays output for the best three subset models of each size from $k = 4$ to $k = 9$ predictor variables and also for the model with all 14 predictors.

Observation	x_1	x_2	x_3	x_4	y	Observation	x_1	x_2	x_3	x_4	y
1	8	4	100	1	1.4	16	4	10	160	5	4.6
2	2	4	180	7	2.2	17	4	13	100	7	4.3
3	7	4	180	1	4.6	18	10	10	120	7	4.9
4	10	7	120	5	4.9	19	5	4	100	1	1.7
5	7	4	180	5	4.6	20	8	13	140	1	4.6
6	7	7	180	1	4.7	21	10	1	180	1	2.6
7	7	13	140	1	4.6	22	2	13	140	1	3.1
8	5	4	160	7	4.5	23	6	13	180	7	4.7
9	4	7	140	3	4.8	24	7	1	120	7	2.5
10	5	1	100	7	1.4	25	5	13	140	1	4.5
11	8	10	140	3	4.7	26	8	1	160	7	2.1
12	2	4	100	3	1.6	27	4	1	180	7	1.8
13	4	10	180	3	4.5	28	6	1	160	1	1.5
14	6	7	120	7	4.7	29	4	1	100	1	1.3
15	10	13	180	3	4.8	30	7	10	100	7	4.6

FIGURE 14.10 MINITAB output for the data of Example 14.16

```
                                                  c   c   c   c   t
                                          t   t   c   o   o   a   a   e
                                  c   c   e   o   n   n   t   t   m
                                  o   a   m   m   n   *   *   *   *   *
                          t   t   n   t   p   e   *   t   t   t   t   t
                  c   c   e   i   s   s   s   c   e   i   e   i   i
             Adj. o   a   m   m   q   q   q   q   a   m   m   m   m   m
Vars R-sq R-sq C-p    s    n   t   p   e   d   d   d   d   t   p   e   p   e   e
  4  82.2 79.4 13.9 0.63607    X       X           X                   X
  4  82.1 79.2 14.1 0.63843    X                   X       X           X
  4  81.7 78.8 14.8 0.64455    X                   X                   X                   X
  5  86.7 83.9  7.4 0.56196    X           X   X                       X   X
  5  85.5 82.5  9.6 0.58571    X       X   X   X                       X
  5  85.5 82.5  9.6 0.58614    X               X   X           X       X
  6  88.5 85.5  5.9 0.53337    X               X   X                   X   X   X
  6  88.2 85.1  6.5 0.54094    X       X   X   X                       X                   X
  6  88.0 84.9  6.8 0.54386    X               X   X                   X   X               X
  7  89.9 86.7  5.2 0.51045    X   X   X   X   X   X                    X
  7  89.9 86.7  5.3 0.51140    X   X           X   X   X   X            X
  7  89.7 86.4  5.7 0.51683    X   X           X   X   X                X   X
  8  91.5 88.3  4.2 0.47933    X   X   X   X   X   X                    X                   X
  8  90.6 87.0  6.0 0.50535    X   X           X   X   X   X            X                   X
  8  90.5 86.8  6.1 0.50793    X   X           X   X   X                X   X   X
  9  91.7 87.9  5.9 0.48650  X X   X   X   X   X   X                    X                   X
  9  91.6 87.8  6.0 0.48853    X   X   X   X   X   X                    X       X           X
  9  91.6 87.8  6.0 0.48860    X   X   X   X   X   X                    X   X               X
 14  92.1 84.8 15.0 0.54639  X X   X   X   X   X   X   X   X   X   X   X   X   X
```

The choice of a best model here, as often happens, is not clear-cut. We certainly don't see the benefit of including more than $k = 8$ predictor variables (after that, adjusted R^2 begins to decrease), nor would we suggest a model with fewer than five predictors (adjusted R^2 is still increasing and C_p is large). Based only on this output, the best six-predictor model is a reasonable choice. The corresponding estimated regression function is

$$-1.218 + .9599x_2 - .0373x_1^2 - .0389x_2^2 + .0037x_1x_3 + .019x_1x_4 - .0013x_2x_3$$

Another good candidate is the best seven-predictor model. Although it includes one more predictor than the model just suggested, only one of the seven predictors is an interaction term (x_1x_3), so model interpretation is somewhat easier. (Notice, though, that none of the best three models with seven predictors results simply from adding a single predictor to the best six-predictor model.) Since every good model includes x_1, x_2, x_3, and x_4 in some predictor, it appears that HCHO concentration, catalyst ratio, curing time, and curing temperature are all important determinants of durable press rating.

The most easily understood and implemented automatic selection procedure is referred to as **backward elimination.** It involves starting with the model that contains all p potential predictors and then deleting them one by one until all remaining predictors seem important. The first step is to specify the value of a positive constant t_{out}, which is used to decide whether deletion should be continued. After fitting the p predictor model, the t ratios $b_1/s_{b_1}, b_2/s_{b_2}, \ldots, b_p/s_{b_p}$ are examined. The predictor variable whose t ratio is closest to zero, whether positive or negative, is the obvious candidate for deletion. The corresponding predictor is eliminated from the model if this t ratio satisfies the inequalities $-t_{out} \leq t$ ratio $\leq t_{out}$. Suppose that this is the case. The model with the remaining $p - 1$ predictors is then fit, and again the predictor with the t ratio closest to zero is eliminated, provided that it satisfies $-t_{out} \leq t$ ratio $\leq t_{out}$. The procedure continues until, at some stage, no t ratio satisfies $-t_{out} \leq t$ ratio $\leq t_{out}$ (all are either greater than t_{out} or less than $-t_{out}$). The chosen model is then the last one fit (though some analysts recommend examining other models of the same size). It is customary to use $t_{out} = 2$, since for many different values of df, this corresponds to a two-tailed test with approximate significance level .05.*

EXAMPLE 14.17

Figure 14.11 (page 742) shows MINITAB output resulting from the application of the backward elimination procedure with $t_{out} = 2$ to the specific gravity data introduced in Example 14.15. The t ratio closest to zero when the model with all five predictors was fit was .12, so the corresponding predictor x_2 was deleted from the model. When the model with the four remaining predictors was fit, the t ratio closest to

---

*Some computer packages base the procedure on the squares of the t ratios, which are F ratios, and continue to delete as long as F ratio $\leq F_{out}$ for at least one predictor. The predictor with the smallest F ratio is eliminated. $F_{out} = 4$ corresponds to $t_{out} = 2$.

FIGURE 14.11 MINITAB
output for Example 14.17

STEP	1	2	3	4
CONSTANT	0.4421	0.4384	0.4381	0.5179
X1	0.00011	0.00011	0.00012	
T-RATIO	1.17	1.95	1.98	
X2	0.00001			
T-RATIO	0.12			
X3	-0.00531	-0.00526	-0.00498	-0.00438
T-RATIO	-5.70	-6.56	-5.96	-5.20
X4	-0.0018	-0.0019		
T-RATIO	-1.63	-1.76		
X5	0.0044	0.0044	0.0031	0.0027
T-RATIO	3.01	3.31	2.63	2.12
S	0.0180	0.0174	0.0185	0.0200
R-SQ	77.05	77.03	72.27	65.50

zero was -1.76, which satisfied $-2 \leq -1.76 \leq 2$. Thus, the corresponding predictor, x_4, was eliminated, leaving x_1, x_3, and x_5 still in the model. The next predictor to be dropped was x_1, because its t ratio, 1.98, was the one closest to zero and (barely) satisfied $-2 \leq 1.98 \leq 2$. When the model with predictors x_3 and x_5 was fit, neither t ratio satisfied $-2 \leq t$ ratio ≤ 2, so the procedure terminated. Looking back to Example 14.15, this is actually the best model containing only two predictors. However, x_1 just barely met the elimination criterion, so the model with predictors x_1, x_3, and x_5 should also be given serious consideration.

———

Unfortunately, the backward elimination method does not always terminate with a model that is the best of its size, and this is also true of other automatic selection procedures. For example, the authors of the article mentioned in Example 14.16 used an automatic procedure to obtain a six-predictor model with $R^2 = .77$, whereas all of the 10 best six-predictor models have R^2 values of at least .87. Because of this, we recommend using a statistical software package that will identify best subsets of different sizes whenever possible.

Checks on Model Adequacy

In Chapter 13, we discussed some informal techniques for checking the adequacy of the simple linear regression model. Most of these were based on plots involving the standardized residuals. The formula for standardizing residuals in multiple regression is quite complicated, but it has been programmed into many statistical computer packages. Once the standardized residuals resulting from the fit of a particular model have been computed, plots similar to those discussed previously are useful in diagnosing model defects. A normal probability plot of the standardized residuals that departs too much from a straight line casts doubt on the assumption that the random deviation e has a normal distribution. Plots of the standardized residuals against each predictor variable in the model — that is, a plot of $(x_1$, standardized residual) pairs, another of $(x_2$, standardized residual) pairs, and so on — are analogous to the standardized residual versus x plot discussed and illustrated in the previous chapter. The appearance of any discernible pattern in these plots (for example, curvature or increasing spread from left to right) points to the need for model modification. If observations have been made over time, a periodic pattern when the standardized residuals are plotted in time order suggests that successive observations were not independent. Models that incorporate dependence of successive observations are substantially more complicated than those we have presented here. They are especially important in econometrics, which involves using statistical methodology to model economic data. Please consult one of the references in Chapter 5 for more information.

One other aspect of model adequacy that has received much attention from statisticians in recent years is the identification of any observations in the data set that may have been highly influential in estimating model coefficients. Recall that in simple linear regression, an observation with potentially high influence is one whose x value places it far to the right or left of the other points in the scatter plot or standardized residual plot. If a multiple regression model involves only two predictor variables x_1 and x_2, an observation with potentially large influence can be revealed by examining a plot of (x_1, x_2) pairs. Any point in this plot that is far away from the others corresponds to an observation that, if deleted from the sample, may cause coefficient estimates and other quantities to change considerably. The detection of influential observations when the model contains at least three predictors is more difficult. Recent research has yielded several helpful diagnostic quantities. One of these has been implemented in MINITAB, and a large value of this quantity automatically results in the corresponding observation being identified as one that might have great influence. Deleting the observation and refitting the model will reveal the extent of actual influence (which depends on how consistent the corresponding y observation is with the rest of the data).

Multicollinearity

When the simple linear regression model is fit using sample data in which the x values are all close to one another, small changes in observed y values can cause the values of the estimated coefficients a and b to change considerably. This may well result in standard deviations σ_b and σ_a that are quite large, so that estimates of β and α from any given sample are likely to differ greatly from the true values. There is an analogous condition that leads to this same type of behavior in multiple regression: a configuration of predictor variable values that is likely to result in poorly estimated model coefficients.

When the model to be fit includes k predictors $x_1, x_2, \ldots, x_k$, there is said to be **multicollinearity** if there is a strong linear relationship between values of the predictors. Severe multicollinearity leads to instability of estimated coefficients as well as various other problems. Such a relationship is difficult to visualize when $k > 2$, so statisticians have developed various quantitative indicators to measure the extent of multicollinearity in a data set. The most straightforward approach involves computing R^2 values for regressions in which the dependent variable is taken to be one of the k x's and the predictors are the remaining $(k-1)$ x's. For example, when $k = 3$, there are three relevant regressions:

1. Dependent variable = x_1, predictor variables = x_2 and x_3
2. Dependent variable = x_2, predictor variables = x_1 and x_3
3. Dependent variable = x_3, predictor variables = x_1 and x_2

Each regression yields an R^2 value. In general, there are k such regressions and, therefore, k resulting R^2 values. If one or more of these R^2 values is large (close to 1), multicollinearity is present. MINITAB will print a message saying that the predictors are highly correlated when at least one of these R^2 values exceeds .99 and will refuse to include a predictor in the model if the corresponding R^2 value is larger than .9999. Other analysts are more conservative and would judge multicollinearity to be a potential problem if any of these R^2 values exceeded .9.

When the values of predictor variables are under the control of the investigator, which often happens in scientific experimentation, a careful choice of values will preclude multicollinearity from arising. Multicollinearity does frequently occur in social science and business applications of regression analysis, where data results simply from observation rather than from intervention by an experimenter. Statisticians have proposed various remedies for the problems associated with multicollinearity in such situations, but a discussion would take us beyond the scope of this book. (After all, we want to leave something for your next statistics course!)

Exercises 14.51 – 14.62

14.51 The article "The Caseload Controversy and the Study of Criminal Courts" (*J. Criminal Law and Criminology* (1979): 89–101) used multiple regression to analyze a data set consisting of observations on the variables

y = length of sentence in trial case (months)

x_1 = seriousness of first offense

x_2 = dummy variable indicating type of trial (bench or jury)

x_3 = number of legal motions

x_4 = measure of delay for confined defendants (0 for those not confined)

x_5 = measure of judge's caseload

The estimated regression equation proposed by the authors is

$$\hat{y} = 12.6 + .59x_1 - 70.8x_2 - 33.6x_3 - 15.5x_5 + .0007x_6 + 3x_7 - 41.5x_8$$

where $x_6 = x_4^2$, $x_7 = x_1x_2$, and $x_8 = x_3x_5$. How do you think the authors might have arrived at this particular model?

14.52 The article "Histologic Estimation of Age at Death Using the Anterior Cortex of the Femur" (*Amer. J. Phys. Anthro.* (1991): 171–179) developed multiple regression models to relate y = age at death to a number of different histologic predictor variables. Stepwise regression (an automatic selection procedure favored over the backward elimination method by many statisticians) was used to identify models shown in the accompanying table, for the case of females.

Predictors	R^2	Adjusted R^2
x_6	.60	.60
x_3, x_6	.66	.66
x_3, x_5	.68	.68
x_2, x_3, x_5	.69	.68
x_2, x_3, x_5, x_6	.70	.69
x_2, x_3, x_4, x_5, x_6	.71	.70
$x_1, x_2, x_3, x_4, x_5, x_6$	.71	.70
$x_1, x_2, x_3, x_4, x_5, x_6, x_7$	.71	.70
$x_1, x_2, x_3, x_5, x_6, x_7, x_8$	.72	.71
$x_1, x_2, x_3, x_4, x_5, x_6, x_7, x_8$	.72	.70

Based on the given information, which of these models would you recommend using, and why?

14.53 The article "The Analysis and Selection of Variables in Linear Regression" (*Biometrics* (1976): 1–49) reports on an analysis of data taken from issues of *Motor Trend* magazine. The dependent variable y was gas mileage, there were $n = 32$ observations, and the independent variables were x_1 = engine type (1 = straight, 0 = V), x_2 = number of cylinders, x_3 = transmission type (1 = manual, 0 = automatic), x_4 = number of transmission speeds, x_5 = engine size, x_6 = horsepower, x_7 = number of carburetor barrels, x_8 = final drive ratio, x_9 = weight, and x_{10} = quarter-mile time. The R^2 and adjusted R^2 values are given in the accompanying table for the best model using k predictors for $k = 1, \ldots, 10$. Which model would you select? Explain your choice and the criteria used to reach your decision.

k	Variable Included	R^2	Adjusted R^2
1	x_9	.756	.748
2	x_2, x_9	.833	.821
3	x_3, x_9, x_{10}	.852	.836
4	x_3, x_6, x_9, x_{10}	.860	.839
5	$x_3, x_5, x_6, x_9, x_{10}$	.866	.840
6	$x_3, x_5, x_6, x_8, x_9, x_{10}$	.869	.837
7	$x_3, x_4, x_5, x_6, x_8, x_9, x_{10}$	.870	.832
8	$x_3, x_4, x_5, x_6, x_7, x_8, x_9, x_{10}$	.871	.826
9	$x_1, x_3, x_4, x_5, x_6, x_7, x_8, x_9, x_{10}$	.871	.818
10	All independent variables	.871	.809

14.54 The article "Estimation of the Economic Threshold of Infestation for Cotton Aphid" (*Mesopotamia J. Ag.* (1982): 71–75) gave $n = 34$ observations on

y = infestation rate (number of aphids per 100 leaves)
x_1 = mean temperature (°C)
x_2 = mean relative humidity

Partial SAS computer output resulting from fitting the model $y = \alpha + \beta_1 x_1 + \beta_2 x_2 + e$ to the data given in the article is shown.

		R-SQUARE	0.5008
		ADJ R-SQ	0.4533

VARIABLE	DF	PARAMETER ESTIMATE	STANDARD ERROR	T FOR H0: PARAMETER = 0
INTERCEP	1	15.667014	84.157420	0.186
TEMP	1	-0.360928	2.330835	-0.155
RH	1	1.959715	0.576946	2.877

If the method of backward elimination is to be employed, which variable would be the first candidate for elimina-

tion? Using the criterion of eliminating a variable if its t ratio satisfies $-2 \le t$ ratio ≤ 2, can it be eliminated from the model?

14.55 The following statement is from the article "Blood Cadmium Levels in Nonexposed Male Subjects Living in the Rome Area: Relationship to Selected Cardiovascular Risk Factors" (*Microchemical J.* (1998): 173–179):

A multiple regression analyzed ln(blood cadmium level) as a dependent variable, with a model including age, alcohol consumption, daily cigarette consumption, driving habits, body mass index, skinfold thickness, high density lipoprotein cholesterol, non-high density lipoprotein cholesterol, and ln(triglyceride level) as possible predictors. This model explained 30.92% of the total variance in ln(blood cadmium level).

a. A total of $n = 1856$ men participated in this study. Perform a model utility test to determine whether the predictor variables, taken as a group are useful in predicting ln(blood cadmium level).
b. The article went on to state:

Smoking habits, the most important predictor of ln(blood cadmium level), explained 30.26% of the total variation, and alcohol consumption was the only other variable influencing ln(blood cadmium level).

These statements were based on the result of a forward stepwise regression. In forward stepwise regression, variables are added to the model one at a time. At each step, the variable that results in the largest improvement is added to the model. This continues until there is no variable whose addition will result in a significant improvement. Describe what you think happened at each step of the stepwise procedure.

14.56 For the multiple regression model in Exercise 14.4, the estimated standard deviations for each of the coefficients (excluding the constant) are (in order) .01, .01, .03, .06, .01, .01, .03, .04, and .05.
a. If a backward elimination variable selection process were to be used, which of the nine predictors would be the first candidate for elimination? Would it be eliminated?
b. Based on the information given, can you tell what the second candidate for elimination would be? Explain.

14.57 The following table showing the results of two multiple regressions appears in "Does the Composition of the Compensation Committee Influence CEO Compensation Practices?" (*Financial Management* (1999): 41–53):

Independent Variable: Log of Compensation

Dependent Variables	Log of 1992 Compensation	
	b	T
Intercept	−2.508	4.80
Stock ownership	−.002	−.58
Log of sales	.372	6.56
CEO tenure	.005	.56
Stock returns	.003	2.12
Return on equity	.005	2.79
Insider	.135	1.35

Dependent Variables	Log of 1991 Compensation	
	b	T
Intercept	−2.098	−3.92
Stock ownership	−.005	−1.73
Log of sales	.286	4.92
CEO tenure	.020	2.16
Stock returns	.006	3.88
Return on equity	.001	.33
Insider	.061	.57

a. Do both regressions show a similar pattern?
b. Calculate the standard error for the coefficient corresponding to the log of sales for the 1992 regression.
c. For the 1991 regression, which variable would be considered first in a backward elimination variable selection procedure? Would it be eliminated?
d. Would the same variable identified in part **c** be the first to be considered in a backward elimination process for the 1992 regression?
e. Assuming the number of observations is 160, determine the P-value for a test of the hypothesis that β_1 in the 1991 model is negative.

14.58 Suppose you were considering a multiple regression analysis with y = house price, x_1 = number of bedrooms, x_2 = number of bathrooms, and x_3 = total number of rooms. Do you think that multicollinearity might be a problem? Explain.

14.59 *This exercise requires use of a computer package.* Using a statistical computer package, compare the best one-, two-, three-, and four-predictor models for the data given in Exercise 14.58. Does this variable-selection procedure lead you to the same choice of model as the backward elimination used in Exercise 14.58?

14.60 The accompanying data is from the article "Breeding Success of the Common Puffin on Different Habitats at Great Island, Newfoundland" (*Ecol. Monographs* (1972): 246–252). The variables considered are y = nesting frequency (burrows per 9 m^2), x_1 = grass cover (%), x_2 = mean soil depth (cm), x_3 = angle of slope (degrees), and x_4 = distance from cliff edge (m).

y	x_1	x_2	x_3	x_4
16	45	39.2	38	3
15	65	47.0	36	12
10	40	24.3	14	18
7	20	30.0	16	21
11	40	47.6	6	27
7	80	47.6	9	36
4	80	45.6	7	39
0	15	27.8	8	45
0	0	41.9	8	54
0	20	36.8	5	60
15	40	34.9	31	3
21	60	45.2	37	12
12	95	32.9	24	18
8	50	26.6	11	24
9	80	32.7	10	30
6	80	38.1	5	36
16	60	37.1	35	6
25	60	47.1	35	12
13	85	34.0	23	18
13	90	43.6	12	21
11	20	30.8	9	27
3	85	34.6	6	33
0	30	37.7	8	42
0	75	45.5	5	48
0	15	51.4	8	54
18	40	32.1	36	6
19	40	35.4	37	9
8	90	30.2	11	18
12	80	33.9	9	24
10	80	40.2	11	30
3	75	33.5	7	36
0	65	40.3	10	42
0	60	31.4	5	39
0	70	32.7	2	48
0	35	38.1	8	51
0	80	40.3	12	45
0	50	43.1	13	51
0	50	42.0	3	57

MINITAB output resulting from application of the backward elimination procedure is also given. Explain what action was taken at each step and why.

STEP	1	2	3
CONSTANT	12.29	11.45	13.96
X_1	-0.019		
T-RATIO	-0.70		
X_2	-0.043	-0.043	
T-RATIO	-1.42	-1.42	
X_3	0.224	0.225	0.176
T-RATIO	2.91	2.94	2.54
X_4	-0.182	-0.203	-0.293
T-RATIO	-2.12	-2.55	-6.01
S	2.83	2.81	2.85
R-SQ	86.20	85.99	85.16

14.61 *This exercise requires use of a computer package.* The accompanying $n = 25$ observations on y = catch at intake (number of fish), x_1 = water temperature (°C), x_2 = minimum tide height (m), x_3 = number of pumps running, x_4 = speed (knots), and x_5 = wind-range of direction (degrees) constitute a subset of the data that appeared in the article "Multiple Regression Analysis for Forecasting Critical Fish Influxes at Power Station Intakes" (*J. Applied Ecol.* (1983): 33–42). Use the variable-selection procedures discussed in this section to formulate a model.

y	x_1	x_2	x_3	x_4	x_5
17	6.7	.5	4	10	50
42	7.8	1.0	4	24	30
1	9.9	1.2	4	17	120

(continued)

y	x_1	x_2	x_3	x_4	x_5
11	10.1	.5	4	23	30
8	10.0	.9	4	18	20
30	8.7	.8	4	9	160
2	10.3	1.5	4	13	40
6	10.5	.3	4	10	150
11	11.0	1.2	3	9	50
14	11.2	.6	3	7	100
53	12.9	1.8	3	10	90
9	13.2	.2	3	12	50
4	16.2	.7	3	6	80
3	15.8	1.6	3	7	120
7	16.2	.4	3	10	50
9	15.8	1.2	3	9	60
10	16.0	.8	3	12	90
7	16.2	1.2	3	5	160
12	17.1	.7	3	10	90
12	17.5	.8	3	12	110
26	17.5	1.2	3	18	130
14	17.4	.8	3	9	60
18	17.4	1.1	3	13	30
14	17.8	.5	3	8	160
5	18.0	1.6	3	10	40

14.62 Suppose that $R^2 = .723$ for the model containing predictors x_1, x_4, x_5, and x_8 and $R^2 = .689$ for the model with predictors x_1, x_3, x_5, and x_6.
a. What can you say about R^2 for the model containing predictors x_1, x_3, x_4, x_5, x_6, and x_8? Explain.
b. What can you say about R^2 for the model containing predictors x_1 and x_4? Explain.

Summary of Key Concepts and Formulas

Term or Formula	Comment
Additive multiple regression model, $y = \alpha + \beta_1 x_1 + \cdots + \beta_k x_k + e$	This equation specifies a general probabilistic relationship between y and k predictor variables $x_1, x_2, \ldots, x_k$, where $\beta_1, \ldots, \beta_k$ are population regression coefficients and $\alpha + \beta_1 x_1 + \cdots + \beta_k x_k$ is the population regression function (the mean value of y for fixed values of $x_1, \ldots, x_k$).
Estimated regression function, $\hat{y} = a + b_1 x_1 + \cdots + b_k x_k$	The estimates $a, b_1, \ldots, b_k$ of $\alpha, \beta_1, \ldots, \beta_k$ result from applying the principle of least squares.
Coefficient of multiple determination, $R^2 = 1 - \dfrac{\text{SSResid}}{\text{SSTo}}$	The proportion of observed y variation that can be explained by the model relationship, where SSResid is defined as it is in simple linear regression but is now based on $n - (k + 1)$ degrees of freedom.

Adjusted R^2

A downward adjustment of R^2 that depends on the number of predictors k relative to the sample size n.

F distribution

A type of probability distribution used in many different inferential procedures. A particular F distribution results from specifying both numerator df and denominator df.

$$F = \frac{\dfrac{R^2}{k}}{\dfrac{(1 - R^2)}{n - (k + 1)}}$$

OR

$$F = \frac{\dfrac{\text{SSRegr}}{k}}{\dfrac{\text{SSResid}}{n - (k + 1)}}$$

The test statistic for testing $H_0: \beta_1 = \beta_2 = \cdots = \beta_k = 0$, which asserts that there is no useful linear relationship between y and any of the model predictors. The F test is upper-tailed and is based on numerator df $= k$ and denominator df $= n - (k + 1)$.

$b_i \pm (t \text{ critical values})s_{b_i}$

A confidence interval for the regression coefficient β_i (based on $n - (k + 1)$ df).

$$t = \frac{b_i - \text{hyp. val.}}{s_{b_i}}$$

The test statistic for testing hypotheses about β_i (based on $n - (k + 1)$ df). The most important case is $H_0: \beta_i = 0$, according to which the predictor x_i is not useful as long as all other predictors remain in the model.

$\hat{y} \pm (t \text{ crit. val.})s_{\hat{y}}$

$\hat{y} \pm (t \text{ crit. val.})\sqrt{s_e^2 + s_{\hat{y}}^2}$

A confidence interval for a mean y value and a prediction interval for a single y value when $x_1, \ldots, x_k$ have specified values, where $\hat{y} = a + b_1 x_1 + \cdots + b_k x_k$ and $s_{\hat{y}}$ denotes the estimated standard deviation of $\hat{y}$.

Model selection

An investigator may have data on many predictors that could be included in a model. There are two different approaches to choosing a model: (1) Use an *all-subsets procedure* to identify the best models of each different size (one-predictor models, two-predictor models, etc.) and then compare these according to criteria such as R^2 and adjusted R^2; (2) employ an *automatic selection procedure,* which either successively eliminates predictors until a stopping point is reached (backward elimination) or starts with no predictors and adds them successively until the inclusion of additional predictors cannot be justified.

Supplementary Exercises 14.63 – 14.73

14.63 What factors affect the quality of education in our schools? The article "How Much Is Enough? Applying Regression to a School Finance Case" (*Statistics and the Law,* New York: Wiley, 1986: 257–287) used data from $n = 88$ school districts to carry out a regression analysis based on the following variables:

$y = $ average language score for fourth-grade students

Background variables:

$x_1 = $ occupational index (% managerial and professional workers in the community)

$x_2 = $ median income

Peer group variables:

$x_3 = $ % Title I enrollment

$x_4 = $ Hispanic enrollment ($= 1$ if $> 5\%$ and $= 0$ otherwise)

$x_5 = $ logarithm of fourth-grade enrollment

$x_6 = $ prior test score (from 4 years previous to year under study)

School variables:

$x_7 = $ administrator–teacher ratio

$x_8 = $ pupil–teacher ratio

x_9 = certificated staff–pupil ratio

x_{10}, x_{11} = indicator variables for average teaching experience ($x_{10} = 1$ if less than 3 years and 0 otherwise; $x_{11} = 1$ if more than 6 years and 0 otherwise)

The accompanying table gives estimated coefficients, values of t ratios (b_1/s_{b_1}, b_2/s_{b_2}, and so on), and other relevant information.

Variable	Estimated Coefficient	t ratio
Constant	70.67	6.08
x_1	.458	3.08
x_2	−.000141	−.59
x_3	−58.47	−4.02
x_4	−3.668	−1.73
x_5	−2.942	−2.34
x_6	.200	3.60
x_7	17.93	.42
x_8	.689	.42
x_9	−.403	−.27
x_{10}	−.614	−1.30
x_{11}	−.382	−.58

$s_e = 5.57$ $R^2 = .64$

a. Is there a useful linear relationship between y and at least one of the predictors? Test the relevant hypotheses using a .01 significance level.

b. What is the value of adjusted R^2?

c. Calculate and interpret a 95% confidence interval for β_1.

d. Does it appear that one or more predictors could be eliminated from the model without discarding useful information? If so, which one would you eliminate first? Explain your reasoning.

e. Suppose you wanted to decide whether any of the peer group variables provided useful information about y (provided that all other predictors remained in the model). What hypotheses would you test (a single H_0 and H_a)? Could one of the test procedures presented in this chapter be used? Explain.

14.64 The article "Innovation, R&D Productivity, and Global Market Share in the Pharmaceutical Industry" (*Review of Indus. Org.* (1995): 197–207) presented the results of a multiple regression analysis in which the dependent variable was y = the number of patented R&D compounds per R&D employee. The number of observations was $n = 78$, the model contained six predictors, and SSResid = 1.02, $R^2 = .50$.

a. Is the chosen model useful? Test the relevant hypotheses.

b. What is the value of adjusted R^2?

c. The predictor x_6 was an indicator variable with value 1 if a firm was headquartered outside the United States. The article reported that $b_6 = .001$ and the corresponding t ratio was −.92. Provided that the other five predictors remain in the model, is x_6 useful? Test the relevant hypotheses.

14.65 The accompanying data on y = glucose concentration (g/L) and x = fermentation time (days) for a particular blend of malt liquor was read from a scatter plot in the article "Improving Fermentation Productivity with Reverse Osmosis" (*Food Tech.* (1984): 92–96):

x	1	2	3	4	5	6	7	8
y	74	54	52	51	52	53	58	71

a. Construct a scatter plot for this data. Based on the scatter plot, what type of model would you suggest?

b. MINITAB output resulting from fitting a multiple regression model with $x_1 = x$ and $x_2 = x^2$ is shown here. Does this quadratic model specify a useful relationship between y and x?

THE REGRESSION EQUATION is
Y = 84.5 - 15.9X1 + 1.77X2

Predictor	Coef	Stdev	t-ratio
Constant	84.482	4.904	17.23
X1	-15.875	2.500	-6.35
X2	1.7679	0.2712	6.52

s = 3.515   R-sq = 89.5%   R-sq(adj) = 85.3%

c. Could the quadratic term have been eliminated? That is, would a simple linear model have sufficed? Test using a .05 significance level.

14.66 Much interest in management circles has recently focused on how employee compensation is related to various company characteristics. The article "Determinants of R and D Compensation Strategies" (*Personnel Psych.* (1984): 635–650) proposed a quantitative scale for y = base salary for employees of high-tech companies. The following estimated multiple regression equation was then presented:

$$\hat{y} = 2.60 + .125x_1 + .893x_2 + .057x_3 - .014x_4$$

where x_1 = sales volume (in millions of dollars), x_2 = stage in product life cycles (1 = growth, 0 = mature), x_3 = profitability (%), and x_4 = attrition rate (%).

a. There were $n = 33$ firms in the sample and $R^2 = .69$. Is the fitted model useful?

b. Predict base compensation for a growth stage firm with sales volume $50 million, profitability 8%, and attrition rate 12%.

c. The estimated standard deviations for the coefficient estimates were .064, .141, .014, and .005 for b_1, b_2, b_3, and b_4, respectively. Should any of the predictors be deleted from the model? Explain.

d. β_2 is the difference between average base compensation for growth stage and mature stage firms when all other predictors are held fixed. Use the information in part **c** to calculate a 95% confidence interval for β_2.

14.67 If $n = 21$ and $k = 10$, for what values of R^2 would adjusted R^2 be negative?

14.68 A study of total body electrical conductivity was described in the article "Measurement of Total Body Electrical Conductivity: A New Method for Estimation of Body Composition" (*Amer. J. Clinical Nutr.* (1983): 735–739). Nineteen observations were given for the variables y = total body electrical conductivity, x_1 = age (years), x_2 = sex (0 = male, 1 = female), x_3 = body mass (kg/m^2), x_4 = body fat (kg), and x_5 = lean body mass (kg).

a. The backward elimination method of variable selection was employed, and MINITAB output is given here. Explain what occurred at each step in the process.

```
STEPWISE REGRESSION OF Y ON 5
PREDICTORS, WITH N = 19
```

STEP	1	2	3
CONSTANT	-6.193	-13.285	-15.175
X1	0.31	0.36	0.38
T-RATIO	1.90	2.36	2.96
X2	-7.9	-7.5	-7.0
T-RATIO	-1.93	-1.89	-2.04
X3	-0.43		
T-RATIO	-0.72		
X4	0.22	0.03	
T-RATIO	0.78	0.29	
X5	0.365	0.339	0.378
T-RATIO	2.16	2.09	4.18
S	5.16	5.07	4.91
R-SQ	79.53	78.70	78.57

b. MINITAB output for the multiple regression model relating y to x_1, x_2, and x_5 is shown here. Interpret the values of R^2 and s_e.

```
THE REGRESSION EQUATION is
Y = -15.2 + 0.377X1 - 6.99X2 + 0.378X5
```

Predictor	Coef	Stdev	t-ratio
Constant	-15.175	9.620	-1.58
X1	0.3771	0.1273	2.96
X2	-6.988	3.425	-2.04
X5	0.37779	0.09047	4.18

s = 4.914   R-sq = 78.6%   R-sq(adj) = 74.3%

c. Interpret the value of b_2 in the estimated regression equation.

d. If the estimated standard deviation of the statistic $a + b_1(31) + b_2(1) + b_5(52.7)$ is 1.42, give a 95% confidence interval for the true mean total body electrical conductivity of all 31-year-old females whose lean body mass is 52.7 kg.

14.69 The article "Creep and Fatigue Characteristics of Ferrocement Slabs" (*J. Ferrocement* (1984): 309–322) reported data on y = tensile strength (MPa), x_1 = slab thickness (cm), x_2 = load (kg), x_3 = age at loading (days), and x_4 = time under test (days) resulting from stress tests of $n = 9$ reinforced concrete slabs. The backward elimination method of variable selection was applied. Partial MINITAB output follows. Explain what action was taken at each step in the process. MINITAB output for the selected model is also given. Use the estimated regression equation to predict tensile strength for a slab that is 25 cm thick, 150 days old, and is subjected to a load of 200 kg for 50 days.

STEP	1	2	3
CONSTANT	8.496	12.670	12.989
X1	-0.29	-0.42	-0.49
T-RATIO	-1.33	-2.89	-3.14
X2	0.0104	0.0110	0.0116
T-RATIO	6.30	7.40	7.33
X3	0.0059		
T-RATIO	0.83		
X4	-0.023	-0.023	
T-RATIO	-1.48	-1.53	
S	0.533	0.516	0.570
R-SQ	95.81	95.10	92.82

```
THE REGRESSION EQUATION is
Y = 13.0 - .487X1 + 0.116X2
```

Predictor	Coef	Stdev	T-ratio
Constant	12.989	3.640	3.57
X1	-0.4867	0.1549	-3.14
X2	0.011569	0.001579	7.33

s = 0.5698   R-sq = 92.8%   R-sq(adj) = 90.4%

14.70   A study of pregnant grey seals involved $n = 25$ observations on the variables y = fetus progesterone level (μg), x_1 = fetus sex (0 = male, 1 = female), x_2 = fetus length (cm), and x_3 = fetus weight (g). MINITAB output for the model using all three independent variables is given ("Gonadotrophin and Progesterone Concentration in Placenta of Grey Seals," *J. Repro. and Fertility* (1984): 521–528).

THE REGRESSION EQUATION is
Y = -1.98 - 1.87X1 + .234X2 + .0001X3

Predictor	Coef	Stdev	t-ratio
Constant	-1.982	4.290	-0.46
X1	-1.871	1.709	-1.09
X2	0.2340	0.1906	1.23
X3	0.000060	0.002020	0.03

s = 4.189   R-sq = 55.2%   R-sq(adj) = 48.8%

Analysis of Variance

SOURCE	DF	SS	MS	F	p
Regression	3	454.63	151.54	8.63	.001
Error	21	368.51	17.55		
Total	24	823.15			

a. Use information from the MINITAB output to test the hypothesis $H_0: \beta_1 = \beta_2 = \beta_3 = 0$.
b. Using an elimination criterion of $-2 \le t$ ratio ≤ 2, should any variable be eliminated? If so, which one?
c. MINITAB output for the regression using only x_1 = sex and x_2 = length is given. Would you recommend keeping both x_1 and x_2 in the model? Explain.

THE REGRESSION EQUATION is
Y = -2.09 - 1.87X1 + .240X2

Predictor	Coef	Stdev	t-ratio
Constant	-2.090	2.212	-0.94
X1	-1.865	1.661	-1.12
X2	0.23952	0.04604	5.20

s = 4.093   R-sq = 55.2%   R-sq(adj) = 51.2%

d. After elimination of both x_3 and x_1, the estimated regression equation is $\hat{y} = -2.61 + .231x_2$. The corresponding values of R^2 and s_e are .527 and 4.116, respectively. Interpret these values.
e. Referring to part **d,** how would you interpret the value of $b_2 = .231$? Does it make sense to interpret the value of a as the estimate of average progesterone level when length is zero? Explain.

14.71   The authors of the article "Influence of Temperature and Salinity on Development of White Perch Eggs"

(*Trans. Amer. Fisheries Soc.* (1982): 396–398) used a quadratic regression model to describe the relationship between y = percentage hatch of white perch eggs and x = water temperature (°C). The estimated regression equation was given as $\hat{y} = -41.0 + 14.6x - .55x^2$.
a. Graph the curve corresponding to this equation by plotting the (x, y) pairs using x values of 10, 12, 14, 16, 18, 20, 22, and 24 (these are the temperatures used in the article). Draw a smooth curve through the points in your plot.
b. The authors make the claim that the optimal temperature for hatch is 14°C. Based on your graph in part **a,** does this statement seem reasonable? Explain.

14.72   A sample of $n = 20$ companies was selected, and the values of y = stock price and $k = 15$ predictor variables (such as quarterly dividend, previous year's earnings, debt ratio, and so on) were determined. When the multiple regression model with these 15 predictors was fit to the data, $R^2 = .90$ resulted.
a. Does the model appear to specify a useful relationship between y and the predictor variables? Carry out a test at a significance level of .05. (The value 5.86 captures upper-tail area .05 under the F curve with 15 numerator df and 4 denominator df.)
b. Based on the result of part **a,** does a high R^2 value by itself imply that a model is useful? Under what circumstances might you be suspicious of a model with a high R^2 value?
c. With n and k as given here, how large would R^2 have to be for the model to be judged useful at the .05 level of significance?

14.73   *This exercise requires the use of a computer package.* The article "Entry and Profitability in a Rate-Free Savings and Loan Market" (*Quarterly Review of Econ. and Business* (1978): 87–95) gave the accompanying data on y = profit margin of savings and loan companies in a given year, x_1 = net revenues in that year, and x_2 = number of savings and loan branch offices.

x_1	x_2	y	x_1	x_2	y
3.92	7298	.75	3.78	6672	.84
3.61	6855	.71	3.82	6890	.79
3.32	6636	.66	3.97	7115	.70
3.07	6506	.61	4.07	7327	.68
3.06	6450	.70	4.25	7546	.72
3.11	6402	.72	4.41	7931	.55
3.21	6368	.77	4.49	8097	.63

(*continued*)

x_1	x_2	y	x_1	x_2	y
3.26	6340	.74	4.70	8468	.56
3.42	6349	.90	4.58	8717	.41
3.42	6352	.82	4.69	8991	.51
3.45	6361	.75	4.71	9179	.47
3.58	6369	.77	4.78	9318	.32
3.66	6546	.78			

a. Fit a multiple regression model using both independent variables.

b. Use the F test to determine whether the model provides useful information for predicting profit margin.

c. Interpret the values of R^2 and s_e.

d. Would a regression model using a single independent variable (x_1 alone or x_2 alone) have sufficed? Explain.

e. Plot the (x_1, x_2) pairs. Does the plot indicate any sample observation that may have been highly influential in estimating the model coefficients? Explain. Do you see any evidence of multicollinearity? Explain.

References

Consult the references listed at the end of Chapter 5.

Kerlinger, Fred N. and Elazar J. Pedhazar, *Multiple Regression in Behavioral Research.* Austin, Texas: Holt, Rinehart & Winston, Inc., 1973. (A readable introduction to multiple regression).

15

The Analysis of Variance

Methods for testing $H_0: \mu_1 - \mu_2 = 0$ (that is, $\mu_1 = \mu_2$), where μ_1 and μ_2 are the means of two different populations or the true mean responses when two different treatments are applied, were discussed in Chapter 11. Many investigations involve a comparison of more than two population or treatment means. For example, an investigation carried out to study purchasers of luxury automobiles ("Measuring Values Can Sharpen Segmentation in the Luxury Car Market," *J. of Advertising Research* (1995): 9–22) reported data on a number of different attributes that might affect purchase decisions, including comfort, safety, styling, durability, and reliability. Here is summary information on the level of importance of speed, rated on a seven-point scale.

Type of Car	American	German	Japanese
Sample size	58	38	59
Sample mean rating	3.05	2.87	2.67

Let μ_1, μ_2, and μ_3 denote the true average (i.e., population mean) rating on this attribute for owners of American, German, and Japanese luxury cars, respectively. Does the data support the claim that $\mu_1 = \mu_2 = \mu_3$, or does it appear that at least two of the μ's are different from one another? This is an example of a **single-factor analysis of variance (ANOVA)** problem, in which the objective is to decide whether the means for more than two populations or treatments are identical. The first several sections of this chapter discuss various aspects of single-factor ANOVA. In Sections 15.3 and 15.4, we consider more complex ANOVA situations and methods.

15.1 Single-Factor ANOVA and the *F* Test

When two or more populations or treatments are being compared, the characteristic that distinguishes the populations or treatments from one another is called the **factor** under investigation. For example, an experiment might be carried out to compare three different methods for teaching reading (three different treatments), in which case the factor of interest would be *teaching method.* This is a qualitative factor. If the growth of fish raised in waters having different salinity levels — 0%, 10%, 20%, and 30% — is of interest, the factor *salinity level* is quantitative.

A **single-factor analysis of variance** (**ANOVA**) problem involves a comparison of k population or treatment means $\mu_1, \mu_2, \ldots, \mu_k$. The objective is to test

$$H_0: \quad \mu_1 = \mu_2 = \cdots = \mu_k$$

against

$$H_a: \quad \text{At least two of the } \mu\text{'s are different.}$$

The analysis is based on k independently selected samples, one from each population or for each treatment. In the case of populations, a random sample from each population is selected independently of that from any other population. When comparing treatments, the experimental units (subjects or objects) that receive any particular treatment are chosen at random from those available for the experiment. A comparison of treatments based on independently selected experimental units is often referred to as a **completely randomized design.**

The decision as to whether the null hypothesis of single-factor ANOVA should be rejected depends on how substantially the samples from the different populations or treatments differ from one another. Figure 15.1 displays two possible data

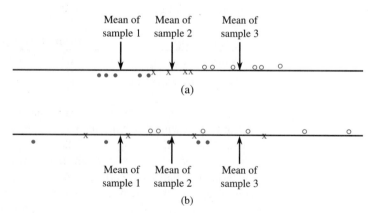

FIGURE 15.1 Two possible ANOVA data sets when three populations are under investigation:
- • = observation from population 1
- x = observation from population 2
- ○ = observation from population 3

sets that might result when observations are selected from each of three populations under study. Each data set consists of five observations from the first population, four from the second, and six from the third. For both data sets, the three sample means are located by vertical line segments. The means of the two samples from population 1 are identical, and a similar statement holds for the two samples from population 2 and those from population 3.

Almost anyone, after looking at the data set in Figure 15.1(a), would readily agree that the claim $\mu_1 = \mu_2 = \mu_3$ appears to be false. Not only are the three sample means different, but there is very clear separation between the samples. In other words, differences between the three sample means are quite large relative to variability within each sample. (If all data sets gave such obvious messages, statisticians would not be in such great demand.) The situation pictured in Figure 15.1(b) is much less clear-cut. The sample means are as different as they were in the first data set, but there is now considerable overlap among the three samples. The separation between sample means here can plausibly be attributed to substantial variability in the populations (and therefore the samples) rather than to differences between μ_1, μ_2, and μ_3. The phrase *analysis of variance* comes from the idea of analyzing variability in the data to see how much can be attributed to differences in the μ's and how much is due to variability in the individual populations. In Figure 15.1(a), there is little within-sample variability relative to the amount of between-sample variability, whereas in Figure 15.1(b), a great deal more of the total variability is due to variation within each sample. If differences between the sample means can be explained by within-sample variability, there is no compelling reason for rejecting H_0.

Notation and Assumptions

Notation in single-factor ANOVA is a natural extension of the notation used in Chapter 11 for comparing two population or treatment means.

ANOVA Notation

k = number of populations or treatments being compared

Population or treatment	1	2	. . .	k
Population or treatment mean	μ_1	μ_2	. . .	μ_k
Population or treatment variance	σ_1^2	σ_2^2	. . .	σ_k^2
Sample size	n_1	n_2	. . .	n_k
Sample mean	$\bar{x}_1$	$\bar{x}_2$	. . .	$\bar{x}_k$
Sample variance	s_1^2	s_2^2	. . .	s_k^2

$N = n_1 + n_2 + \cdots + n_k$ (the total number of observations in the data set)

T = grand total = sum of all N observations = $n_1\bar{x}_1 + n_2\bar{x}_2 + \cdots + n_k\bar{x}_k$

$\bar{\bar{x}}$ = grand mean = $\dfrac{T}{N}$

A decision between H_0 and H_a will be based on examining the $\bar{x}$ values to see whether observed discrepancies are small enough to be attributable simply to sampling variability or whether an alternative explanation for the differences is necessary.

EXAMPLE 15.1

The article "Compression of Single-Wall Corrugated Shipping Containers Using Fixed and Floating Text Platens" (*J. Testing and Evaluation* (1992): 318–320) describes an experiment in which several different types of boxes were compared with respect to compression strength (lb). Table 15.1 presents the results of a single-factor experiment involving $k = 4$ types of boxes (the sample means and standard deviations are in close agreement with values given in the paper).

Table 15.1 *Data and summary quantities for Example 15.1*

Type of Box	Compression Strength (lb)						Sample Mean	Sample SD
1	655.5	788.3	734.3	721.4	679.1	699.4	713.00	46.55
2	789.2	772.5	786.9	686.1	732.1	774.8	756.93	40.34
3	737.1	639.0	696.3	671.7	717.2	727.1	698.07	37.20
4	535.1	628.7	542.4	559.0	586.9	520.0	562.02	39.87
							$\bar{\bar{x}} = 682.50$	

With μ_i denoting the true average compression strength for boxes of type i ($i = 1, 2, 3, 4$), the null hypothesis is $H_0: \mu_1 = \mu_2 = \mu_3 = \mu_4$. Figure 15.2(a) shows a comparative boxplot for the four samples. There is a substantial amount of overlap among observations on the first three types of boxes, but compression strengths for the fourth type appear considerably smaller than for the other types. This suggests that H_0 is not true. If the comparative boxplot looked instead like that of Figure 15.2(b), it would not be so obvious whether H_0 were true or false. In situations such as this, we need a formal test procedure.

FIGURE 15.2 Boxplots for Example 15.1: (a) Original data

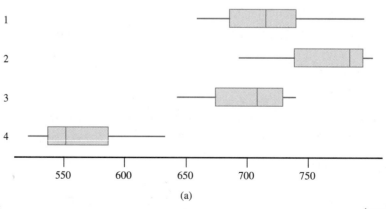

(a)

(continued)

FIGURE 15.2 (Continued)
(b) Altered data

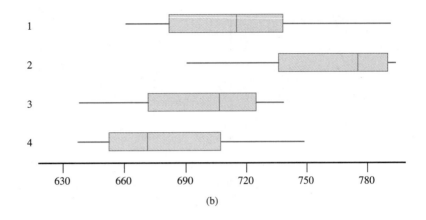

(b)

As with the inferential methods of previous chapters, the validity of the ANOVA test for H_0: $\mu_1 = \mu_2 = \cdots = \mu_k$ requires some assumptions.

Assumptions for ANOVA

1. Each of the k population or treatment response distributions is normal.
2. $\sigma_1 = \sigma_2 = \cdots = \sigma_k$ (The k normal distributions have identical standard deviations.)
3. The observations in the sample from any particular one of the k populations or treatments are independent of one another.
4. When comparing population means, k random samples are selected independently of one another. When comparing treatment means, treatments are assigned at random to subjects or objects (or vice versa).

In practice, the test based on these assumptions will work well as long as the assumptions are not too badly violated. If the sample sizes are reasonably large, normal probability plots or boxplots of the data in each sample are helpful in checking the assumption of normality. Often, however, sample sizes are so small that a separate normal probability plot for each sample is of little value in checking normality. A single combined plot results from first subtracting $\bar{x}_1$ from each observation in the first sample, $\bar{x}_2$ from each value in the second sample, and so on, and then constructing a normal probability plot of all N deviations from their respective means. The plot should be reasonably straight. Figure 15.3 (page 758) shows such a plot for the data of Example 15.1.

There is a formal procedure for testing the equality of population standard deviations. Unfortunately, it is quite sensitive to even a small departure from the normality assumption, so we do not recommend its use. Instead, we suggest that the F test (to be described subsequently) can safely be used if the largest of the sample standard deviations is at most twice the smallest one. The largest standard deviation

FIGURE 15.3 A normal probability plot based on the data of Example 15.1

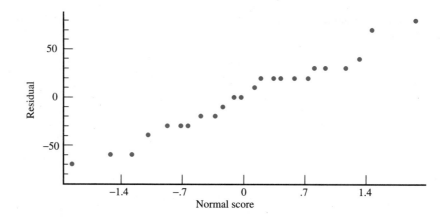

in Example 15.1 is $s_1 = 46.55$, which is only about 1.25 times the smallest one ($s_3 = 37.20$). The book *Beyond ANOVA: The Basics of Applied Statistics* (see the chapter references) is a good source for alternative methods of analysis if there appears to be a violation of assumptions.

The test procedure is based on the following measures of variation in the data.

Definition

A measure of disparity among the sample means is the **treatment sum of squares,** denoted by **SSTr** and given by

$$\text{SSTr} = n_1(\bar{x}_1 - \bar{\bar{x}})^2 + n_2(\bar{x}_2 - \bar{\bar{x}})^2 + \cdots + n_k(\bar{x}_k - \bar{\bar{x}})^2$$

A measure of variation within the k samples, called **error sum of squares** and denoted by **SSE,** is

$$\text{SSE} = (n_1 - 1)s_1^2 + (n_2 - 1)s_2^2 + \cdots + (n_k - 1)s_k^2$$

Each sum of squares has an associated df:

treatment df $= k - 1$ error df $= N - k$

A **mean square** is a sum of squares divided by its df. In particular,

$$\text{mean square for treatments} = \text{MSTr} = \frac{\text{SSTr}}{k - 1}$$

$$\text{mean square for error} = \text{MSE} = \frac{\text{SSE}}{N - k}$$

The error df comes from adding the df's associated with each of the sample variances:

$$(n_1 - 1) + (n_2 - 1) + \cdots + (n_k - 1) = n_1 + n_2 + \cdots + n_k - 1 - 1 - \cdots - 1 = N - k$$

EXAMPLE 15.2   Let's return to the compression strength data of Example 15.1. With $\bar{x}_1 = 713.00$, $\bar{x}_2 = 756.93$, $\bar{x}_3 = 698.07$, $\bar{x}_4 = 562.02$, $\bar{\bar{x}} = 682.50$, and $n_1 = n_2 = n_3 = n_4 = 6$,

$$SSTr = n_1(\bar{x}_1 - \bar{\bar{x}})^2 + n_2(\bar{x}_2 - \bar{\bar{x}})^2 + \cdots + n_k(\bar{x}_k - \bar{\bar{x}})^2$$
$$= 6(713.00 - 682.50)^2 + 6(756.93 - 682.50)^2 + 6(698.07 - 682.50)^2$$
$$+ 6(562.02 - 682.50)^2$$
$$= 5581.50 + 33{,}238.95 + 1454.55 + 87{,}092.58$$
$$= 127{,}367.58$$

Since $s_1 = 46.55$, $s_2 = 40.34$, $s_3 = 37.20$, and $s_4 = 39.87$,

$$SSE = (n_1 - 1)s_1^2 + (n_2 - 1)s_2^2 + \cdots + (n_k - 1)s_k^2$$
$$= (6 - 1)(46.55)^2 + (6 - 1)(40.34)^2 + (6 - 1)(37.20)^2 + (6 - 1)(39.87)^2$$
$$= 33{,}838.375$$

Degrees of freedom are

$$\text{treatment df} = k - 1 = 3 \qquad \text{error df} = N - k = 6 + 6 + 6 + 6 - 4 = 20$$

from which

$$MSTr = \frac{SSTr}{k - 1} = \frac{127{,}367.58}{3} = 42{,}455.86$$

$$MSE = \frac{SSE}{N - k} = \frac{33{,}838.375}{20} = 1691.92$$

Both MSTr and MSE are quantities whose values can be calculated once sample data is available; that is, they are statistics. Each of these statistics will vary in value from data set to data set. For example, repeating the experiment of Example 15.1 with the same sample sizes might yield

$$\bar{x}_1 = 703.24 \qquad \bar{x}_2 = 769.15 \qquad \bar{x}_3 = 678.83 \qquad \bar{x}_4 = 575.50$$
$$s_1 = 42.69 \qquad s_2 = 41.35 \qquad s_3 = 35.02 \qquad s_4 = 41.20$$

from which

$$MSTr = 38{,}796.30 \quad \text{(previously 42,455.86)}$$
$$MSE = 1614.02 \quad \text{(previously 1691.92)}$$

Both MSTr and MSE have sampling distributions and, in particular, mean values. The following box describes the key relationship between μ_{MSTr} and μ_{MSE}, the mean values of the two statistics.

When H_0 is true ($\mu_1 = \mu_2 = \cdots = \mu_k$),

$$\mu_{MSTr} = \mu_{MSE}$$

However, when H_0 is false,

$$\mu_{MSTr} > \mu_{MSE}$$

and the greater the differences among the μ's, the larger μ_{MSTr} will be relative to μ_{MSE}.

According to this result, when H_0 is true, we expect the two mean squares to be close to one another, whereas we expect MSTr to substantially exceed MSE when some μ's differ greatly from others. Thus, a calculated MSTr that is much larger than MSE casts doubt on H_0. In Example 15.2, MSTr = 42,455.86 and MSE = 1691.92, so MSTr is about 25 times as large as MSE. Can this be attributed solely to sampling variability, or is the ratio MSTr/MSE of sufficient magnitude to suggest that H_0 is false? Before we can describe a formal test procedure, it is necessary to revisit F distributions first introduced in multiple regression analysis (Chapter 14).

Many ANOVA test procedures are based on probability distributions called F distributions. As explained in Chapter 14, an F distribution always arises in connection with a ratio. A particular F distribution is obtained by specifying both numerator degrees of freedom (df_1) and denominator degrees of freedom (df_2). Figure 15.4 shows an F curve for a particular choice of df_1 and df_2. All F tests in this book are upper-tailed, so the P-value is the area under the F curve to the right of the calculated F.

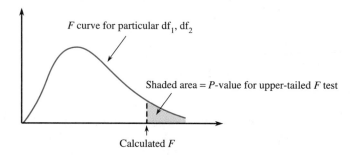

FIGURE 15.4 An F curve and P-value for an upper-tailed test

Tabulation of these upper-tail areas is very cumbersome, because there are two df's rather than just one (as in the case of t distributions). For selected df_1, df_2 pairs, our F table (Appendix Table VII) gives only the four numbers that capture tail areas .10, .05, .01, and .001, respectively. Here are the four numbers for $df_1 = 4$, $df_2 = 10$ along with the statements that can be made about the P-value:

Tail area	.10	.05	.01	.001
Value	2.61	3.48	5.99	11.28
	↑	↑	↑	↑	↑
	a	b	c	d	e

a. $F < 2.61$ → tail area = P-value > .10

b. $2.61 < F < 3.48$ → $.05 < P$-value < .10

c. $3.48 < F < 5.99$ → $.01 < P$-value < .05

d. $5.99 < F < 11.28$ → $.001 < P$-value < .01

e. $F > 11.28$ → P-value < .001

Thus, if $F = 7.12$, $.001 < P\text{-value} < .01$. If a test with $\alpha = .05$ is used, H_0 should be rejected, since $P\text{-value} \le \alpha$. The most frequently used statistical computer packages will provide exact P-values for F tests.

The Single-Factor ANOVA *F* Test

Null hypothesis: $H_0\colon \mu_1 = \mu_2 = \cdots = \mu_k$

Test statistic: $F = \dfrac{\text{MSTr}}{\text{MSE}}$

When H_0 is true and the ANOVA assumptions are reasonable, F has an F distribution with $\text{df}_1 = k - 1$ and $\text{df}_2 = N - k$.

Values of F more contradictory to H_0 than what was calculated are values even farther out in the upper tail, so the P-value is the area captured in the upper tail of the corresponding F curve.

EXAMPLE 15.3

The two mean squares for the compression strength data given in Example 15.1 were calculated in Example 15.2 as

$$\text{MSTr} = 42{,}455.86 \qquad \text{MSE} = 1691.92$$

The value of the F statistic is then

$$F = \frac{\text{MSTr}}{\text{MSE}} = \frac{42{,}455.86}{1691.92} = 25.09$$

With $\text{df}_1 = k - 1 = 3$ and $\text{df}_2 = N - k = 24 - 4 = 20$, Appendix Table VII shows that 8.10 captures tail area .001. Since $25.09 > 8.10$, it follows that $P\text{-value} = $ captured tail area $< .001$. The P-value is smaller than any reasonable α, so there is compelling evidence for rejecting $H_0\colon \mu_1 = \mu_2 = \cdots = \mu_k$. True average compression strength appears to depend in some way on the type of box used.

EXAMPLE 15.4

The article "The Soundtrack of Recklessness: Musical Preferences and Reckless Behavior Among Adolescents (*J. Adolescent Research* (1992): 313–331) described a study whose purpose was to determine whether adolescents who preferred certain types of music reported higher rates of reckless behaviors, such as speeding, drug use, shoplifting, and unprotected sex.

Independently chosen random samples were selected from each of four groups of students with different musical preferences at a large high school: (1) acoustic/pop, (2) mainstream rock, (3) hard rock, and (4) heavy metal. Each student in these samples was asked how many times they had engaged in various reckless activities during the last year. The table on page 762 lists data on driving over 80 mph that is consistent with summary quantities given in the article. (The sample sizes in the article were much larger, but for purposes of this example, we use $n_1 = n_2 = n_3 = n_4 = 20$.)

	Musical Preference			
	Acoustic/ Pop	**Mainstream Rock**	**Hard Rock**	**Heavy Metal**
	2	3	3	4
	3	2	4	3
	4	1	3	4
	1	2	1	3
	3	3	2	3
	3	4	1	3
	3	3	4	3
	3	2	2	3
	2	4	2	2
	2	4	2	4
	1	4	3	4
	3	4	3	5
	2	2	4	4
	2	3	3	5
	2	2	3	3
	3	2	2	4
	2	2	3	5
	2	3	4	4
	3	1	2	2
	4	3	4	3
n	20	20	20	20
$\bar{x}$	2.50	2.70	2.75	3.55
s	.827	.979	.967	.887
s^2	.6839	.9584	.9351	.7868

Also, $N = 80$, grand total $= 230.0$, and $\bar{\bar{x}} = 230.0/80 = 2.875$.

Let's carry out an F test to see whether true mean number of times driving over 80 mph varies with musical preference.

1. Let μ_1, μ_2, μ_3, and μ_4 denote the true mean number of times driving over 80 mph for individuals preferring acoustic/pop, mainstream rock, hard rock, and heavy metal, respectively.

2. $H_0: \mu_1 = \mu_2 = \mu_3 = \mu_4$

3. H_a: At least two among μ_1, μ_2, μ_3, and μ_4 are different.

4. *Significance level*: $\alpha = .01$

5. *Test statistic*: $F = \dfrac{MSTr}{MSE}$

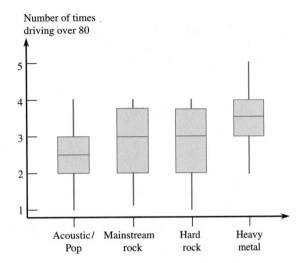

FIGURE 15.5 Boxplots for data of Example 15.4

6. *Assumptions*: Figure 15.5 shows boxplots for the four samples. The boxplots are roughly symmetric, and there are no outliers. The largest standard deviation ($s_2 = .979$) is not more than twice as big as the smallest ($s_1 = .827$). The data consists of independently chosen random samples. The assumptions of ANOVA are reasonable.

7. *Computation*:

$$\text{SSTr} = n_1(\bar{x}_1 - \bar{\bar{x}})^2 + n_2(\bar{x}_2 - \bar{\bar{x}})^2 + \cdots + n_k(\bar{x}_k - \bar{\bar{x}})^2$$
$$= 20(2.50 - 2.875)^2 + 20(2.70 - 2.875)^2$$
$$\quad + 20(2.75 - 2.875)^2 + 20(3.55 - 2.875)^2$$
$$= 12.850$$

treatment df $= k - 1 = 3$

$$\text{SSE} = (n_1 - 1)s_1^2 + (n_2 - 1)s_2^2 + \cdots + (n_k - 1)s_k^2$$
$$= 19(.6839) + 19(.9584) + 19(.9351) + 19(.7868)$$
$$= 63.9020$$

error df $= N - k = 80 - 4 = 76$

Thus,

$$F = \frac{\text{MSTr}}{\text{MSE}} = \frac{\text{SSTr/df}}{\text{SSE/df}} = \frac{12.850/3}{63.9020/76} = \frac{4.283}{.841} = 5.09$$

8. *P-value*: Appendix Table VII shows that for $df_1 = 3$ and $df_2 = 90$ (the closest tabled df to df $= 76$), the value 4.01 captures upper-tail area .01 and 5.91 captures upper-tail area .001. Because $4.01 < F = 5.09 < 5.91$, it follows that $.001 < P\text{-value} < .01$.

9. *Conclusion*: Since $P\text{-value} \leq \alpha$, we reject H_0. The mean number of times driving over 80 mph does not appear to be the same for all four musical preference groups.

Summarizing an ANOVA

Analysis of variance calculations are often summarized in a tabular format called an ANOVA table. To understand such a table, we must define one more sum of squares.

Total sum of squares, denoted by **SSTo,** is given by

$$SSTo = \sum_{\text{all } N \text{ obs.}} (x - \bar{\bar{x}})^2$$

with associated df $N - 1$.

The relationship between the three sums of squares is

$$SSTo = SSTr + SSE$$

which is often called the *fundamental identity for single-factor ANOVA.*

The quantity SSTo, the sum of squared deviations about the grand mean, is a measure of total variability in the data set consisting of all k samples. The quantity SSE results from measuring variability separately within each sample and then combining as indicated in the formula for SSE. Such within-sample variability is present regardless of whether H_0 is true. The magnitude of SSTr, on the other hand, has much to do with the status of H_0 (whether it is true or false). The more the μ's differ from one another, the larger SSTr will tend to be. Thus, SSTr represents variation that can (at least to some extent) be explained by any differences between means. An informal paraphrase of the fundamental identity is

total variation = explained variation + unexplained variation

Once any two of the SS's have been calculated, the remaining one is easily obtained from the fundamental identity. Often SSTo and SSTr are calculated first (using computational formulas given in the chapter appendix), and then SSE is obtained by subtraction: SSE = SSTo − SSTr. All df's, SS's, and MS's are entered in an ANOVA table as displayed in Table 15.2. The *P*-value will usually appear to the right of *F* when the analysis is done by a statistical software package.

Table 15.2 *General format for a single-factor ANOVA table*

Source of Variation	df	Sum of Squares	Mean Square	F
Treatments	$k - 1$	SSTr	$MSTr = \dfrac{SSTr}{k - 1}$	$F = \dfrac{MSTr}{MSE}$
Error	$N - k$	SSE	$MSE = \dfrac{SSE}{N - k}$	
Total	$N - 1$	SSTo		

An ANOVA table from MINITAB for the reckless driving data of Example 15.4 is shown in Table 15.3. The *P*-value is .003, consistent with our previous conclusion that $.001 < P\text{-value} < .01$.

Table 15.3 *An ANOVA table from MINITAB for the data of Example 15.4*

One-way Analysis of Variance

Analysis of Variance

Source	DF	SS	MS	F	P
Factor	3	12.850	4.283	5.09	0.003
Error	76	63.900	0.841		
Total	79	76.750			

Exercises 15.1 – 15.18

15.1 Give as much information as you can about the *P*-value for an upper-tailed *F* test in each of the following situations.

    a. $df_1 = 4$, $df_2 = 15$, $F = 5.37$
    b. $df_1 = 4$, $df_2 = 15$, $F = 1.90$
    c. $df_1 = 4$, $df_2 = 15$, $F = 4.89$
    d. $df_1 = 3$, $df_2 = 20$, $F = 14.48$
    e. $df_1 = 3$, $df_2 = 20$, $F = 2.69$
    f. $df_1 = 4$, $df_2 = 50$, $F = 3.24$

15.2 Give as much information as you can about the *P*-value of the single-factor ANOVA *F* test in each of the following situations.

    a. $k = 5, n_1 = n_2 = n_3 = n_4 = n_5 = 4, F = 5.37$
    b. $k = 5, n_1 = n_2 = n_3 = 5, n_4 = n_5 = 4, F = 2.83$
    c. $k = 3, n_1 = 4, n_2 = 5, n_3 = 6, F = 5.02$
    d. $k = 3, n_1 = n_2 = 4, n_3 = 6, F = 15.90$
    e. $k = 4, n_1 = n_2 = 15, n_3 = 12, n_4 = 10, F = 1.75$

15.3 Employees of a certain state university system can choose from among four different health plans. Each plan differs somewhat from the others in terms of hospitalization coverage. Four samples of recently hospitalized individuals were selected, each sample consisting of people covered by a different health plan. The length of the hospital stay (number of days) was determined for each individual selected.

    a. What hypotheses would you test to decide whether average length of stay was related to health plan? (*Note:* Carefully define the population characteristics of interest.)

    b. If each sample consisted of eight individuals and the value of the ANOVA *F* statistic was $F = 4.37$, what conclusion would be appropriate for a test with $\alpha = .01$?

    c. Answer the question posed in part **b** if the *F* value given there resulted from sample sizes $n_1 = 9, n_2 = 8, n_3 = 7$, and $n_4 = 8$.

15.4 The article "Heavy Drinking and Problems Among Wine Drinkers" (*Journal of Studies on Alcohol* (1999): 467–471) analyzed drinking problems among Canadians. For each of several different groups of drinkers, the mean and standard deviation of "highest number of drinks consumed" were calculated:

	$\bar{x}$	s	n
Beer only	7.52	6.41	1256
Wine only	2.69	2.66	1107
Spirits only	5.51	6.44	759
Beer and wine	5.39	4.07	1334
Beer and spirits	9.16	7.38	1039
Wine and spirits	4.03	3.03	1057
Beer, wine, and spirits	6.75	5.49	2151

Assume that each of the seven samples studied can be viewed as a random sample for the respective group. Is there sufficient evidence to conclude that the mean value of highest number of drinks consumed is not the same for all seven groups?

15.5 Suppose that a random sample of size $n = 5$ was selected from the vineyard properties for sale in Sonoma County, California, in each of 3 years. The following data is consistent with summary information on price per acre (in dollars, rounded to the nearest thousand) for disease-resistant grape vineyards in Sonoma County (*Wines and Vines* (November 1999)).

1996	30,000	34,000	36,000	38,000	40,000
1997	30,000	35,000	37,000	38,000	40,000
1998	40,000	41,000	43,000	44,000	50,000

a. Construct boxplots for each of the 3 years on a common axis, and label each by year. Comment on the similarities and differences.

b. Carry out an ANOVA to determine whether there is evidence to support the claim that the mean price per acre for vineyard land in Sonoma County was not the same for the 3 years considered. Use a significance level of .05 for your test.

15.6 The article "An Analysis of Job Sharing, Full-Time, and Part-Time Arrangements" (*Amer. Business Review* (1989): 34–40) reported on a study of hospital employees from three different groups. Each employee reported a level of satisfaction with his or her work schedule (1 = very dissatisfied to 7 = very satisfied), as shown in the accompanying table.

Group	Sample Size	Sample Mean
1. Job sharers	24	6.60
2. Full-time employees	24	5.37
3. Part-time employees	20	5.20

a. What are numerator and denominator df's for the *F* test?

b. The test statistic value is $F = 6.62$. Use a test with significance level .05 to decide whether it is plausible that true average satisfaction levels are identical for the three groups.

c. What is the value of MSE?

15.7 High productivity and carbohydrate storage ability of the Jerusalem artichoke make it a promising agricultural crop. The article "Leaf Gas Exchange and Tuber Yield in Jerusalem Artichoke Cultivars" (*Field Crops Research* (1991): 241–252) reported on various plant characteristics. Consider the accompanying data on chlorophyll concentration (gm/m^2) for four varieties of Jerusalem artichoke:

Variety	BI	RO	WA	TO
Sample mean	.30	.24	.41	.33

Suppose that the sample sizes were 5, 5, 4, and 6, respectively, and also that MSE = .0130. Does the data suggest that true average chlorophyll concentration depends on the variety? State and test the appropriate hypotheses using a significance level of .05.

15.8 It has been reported that varying work schedules can lead to a variety of health problems for workers. The article "Nutrient Intake in Day Workers and Shift Workers" (*Work and Stress* (1994): 332–342) reported on blood glucose levels (mmol/L) for day-shift workers and workers on two different types of rotating shifts. The sample sizes were $n_1 = 37$ for the day shift, $n_2 = 34$ for the second shift, and $n_3 = 25$ for the third shift. A single-factor ANOVA resulted in $F = 3.834$. At a significance level of .05, does true average blood glucose level appear to depend on the type of shift?

15.9 The chapter introduction discussed a study in which luxury car owners were asked to rate the importance of various attributes on a quantitative scale. Summary information for the importance of speed is as follows:

Type of Car	American	German	Japanese
Sample size	58	38	59
Sample mean	3.05	2.87	2.67

In addition, SSE = 459.04. Carry out a test of hypothesis to decide whether true average importance rating of speed is the same for owners of the three types of cars.

15.10 The article "Utilizing Feedback and Goal Setting to Increase Performance Skills of Managers" (*Academy of Mgmt. J.* (1979): 516–526) reported the results of an experiment to compare three different interviewing techniques for employee evaluations. One method allowed the employee being evaluated to discuss previous evaluations, the second involved setting goals for the employee, and the third did not allow either feedback or goal setting. After the interviews were concluded, the evaluated employee was asked to indicate how satisfied he or she was with the interview. (A numerical scale was used to quantify level of satisfaction.) The authors used ANOVA to compare the three interview techniques. An *F* statistic value of 4.12 was reported.

a. Suppose that a total of 33 subjects were used, with each technique applied to 11 of them. Use this information to conduct a level .05 test of the null hypothesis of no difference in mean satisfaction level for the three interview techniques.

b. The actual number of subjects on which each technique was used was 45. After studying the *F* table, explain why the conclusion in part **a** still holds.

15.11 An investigation carried out to study the toxic effects of mercury was described in the article "Comparative Responses of the Action of Different Mercury Compounds on Barley" (*Intl. J. of Environ. Studies* (1983): 323–327). Ten different concentrations of mercury (0, 1, 5, 10, 50, 100, 200, 300, 400, and 500 mg/L) were compared with respect to their effects on average dry weight (per 100 seven-day-old seedlings). The basic experiment was replicated four times for a total of 40 dry-weight observations (four for each treatment level). The article reported an ANOVA F statistic value of 1.895. Using a significance level of .05, test the null hypothesis that the true mean dry weight is the same for all ten concentration levels.

15.12 The accompanying data on calcium content of wheat is consistent with summary quantities that appeared in the article "Mineral Contents of Cereal Grains as Affected by Storage and Insect Infestation" (*J. Stored Prod. Res.* (1992): 147–151). Four different storage times were considered. Partial output from the SAS computer package is shown at the bottom of the page.

Storage Period	Observations					
0 months	58.75	57.94	58.91	56.85	55.21	57.30
1 month	58.87	56.43	56.51	57.67	59.75	58.48
2 months	59.13	60.38	58.01	59.95	59.51	60.34
4 months	62.32	58.76	60.03	59.36	59.61	61.95

a. Verify that the sums of squares and df's are as given in the ANOVA table.
b. Is there sufficient evidence to conclude that the mean calcium content is not the same for the four different storage times? Use the value of F from the ANOVA table to test the appropriate hypotheses at significance level .05.

15.13 In an experiment to investigate the performance of four different brands of spark plugs intended for use on a 125-cc motorcycle, five plugs of each brand were tested, and the number of miles (at a constant speed) until failure was observed. A partially completed ANOVA table is given. Fill in the missing entries, and test the relevant hypotheses using a .05 level of significance.

Source of Variation	df	Sum of Squares	Mean Square	F
Brands				
Error		235,419.04		
Total		310,500.76		

15.14 The partially completed ANOVA table given in this problem is taken from the article "Perception of Spatial Incongruity" (*J. of Nervous and Mental Disease* (1961): 222), in which the abilities of three different groups to identify a perceptual incongruity were assessed and compared. All individuals in the experiment had been hospitalized to undergo psychiatric treatment. There were 21 individuals in the depressive group, 32 individuals in the functional "other" group, and 21 individuals in the brain-damaged group. Complete the ANOVA table. Carry out the appropriate test of hypothesis (use $\alpha = .01$), and interpret your results.

Source of Variation	df	Sum of Squares	Mean Square	F
Treatments			76.09	
Error				
Total		1123.14		

15.15 Research carried out to investigate the relationship between smoking status of workers and short-term absenteeism rate (hr/mo) yielded the accompanying summary information ("Work-Related Consequences of Smoking Cessation," *Academy of Mgmt. J.* (1989): 606–621). In addition, $F = 2.56$. Construct an ANOVA table, and then state and test the appropriate hypotheses using a .01 significance level.

Status	Sample Size	Sample Mean
Continuous smoker	96	2.15
Recent ex-smoker	34	2.21
Long-term ex-smoker	86	1.47
Never smoked	206	1.69

15.16 Parents are frequently concerned when their child seems slow to begin walking (although when the child finally walks, the resulting havoc sometimes has the parents wishing they could turn back the clock!). The article "Walking in the Newborn" (*Science*, 176 (1972): 314–315) reported on an experiment in which the effects of several different treatments on the age at which a child first walks were compared. Children in the first group were given

Output for Exercise 15.12   Dependent Variable: CALCIUM

Source	DF	Sum of Squares	Mean Square	F Value	Pr > F
Model	3	32.13815000	10.71271667	6.51	0.0030
Error	20	32.90103333	1.64505167		
Corrected Total	23	65.03918333			

R-Square	C.V.	Root MSE	CALCIUM Mean
0.494135	2.180018	1.282596	58.8341667

special walking exercises for 12 min per day beginning at age 1 week and lasting 7 weeks. The second group of children received daily exercises but not the walking exercises administered to the first group. The third and fourth groups were control groups: They received no special treatment and differed only in that the third group's progress was checked weekly, whereas the fourth group's progress was checked just once at the end of the study. Observations on age (in months) when the children first walked are shown in the accompanying table. Also given is the ANOVA table, obtained from the SPSS computer package.

	Age			n	Total
Treatment 1	9.00	9.50	9.75	6	60.75
	10.00	13.00	9.50		
Treatment 2	11.00	10.00	10.00	6	68.25
	11.75	10.50	15.00		
Treatment 3	11.50	12.00	9.00	6	70.25
	11.50	13.25	13.00		
Treatment 4	13.25	11.50	12.00	5	61.75
	13.50	11.50			

Analysis of Variance

Source	df	Sum of sq.	Mean Sq.	F Ratio	F Prob
Between Groups	3	14.778	4.926	2.142	.129
Within Groups	19	43.690	2.299		
Total	22	58.467			

a. Verify the entries in the ANOVA table.
b. State and test the relevant hypotheses using a significance level of .05.

15.17 The fog index is a measure of reading difficulty based on the average number of words per sentence and the percentage of words with three or more syllables.

High values of the fog index are associated with difficult reading levels. Independent random samples of six advertisements were taken from three different magazines, and fog indices were computed to obtain the data given in the accompanying table ("Readability Levels of Magazine Advertisements," *J. of Ad. Research* (1981): 45–50). Construct an ANOVA table, and then use a significance level of .01 to test the null hypothesis of no difference between the mean fog index levels for advertisements appearing in the three magazines.

Scientific

American	15.75	11.55	11.16	9.92	9.23	8.20
Fortune	12.63	11.46	10.77	9.93	9.87	9.42
New Yorker	9.27	8.28	8.15	6.37	6.37	5.66

15.18 Some investigators think that the concentration (μg/mL) of a particular antigen in supernatant fluids could be related to onset of meningitis in infants. The accompanying data is typical of that given in plots in the article "Type-Specific Capsular Antigen Is Associated with Virulence in Late-Onset Group B Streptococcal Type III Disease" (*Infection and Immunity* (1984): 124–129). Construct an ANOVA table, and use it to test the null hypothesis of no difference in mean antigen concentrations for the three groups.

Asymptomatic infants	1.56	1.06	.87	1.39	.71	.87
Infants with late-onset sepsis	1.51	1.78	1.45	1.13	1.87	1.89
	1.07	1.72				
Infants with late-onset meningitis	1.21	1.34	1.95	2.27	.88	1.67
	2.57					

15.2 Multiple Comparisons

When $H_0: \mu_1 = \mu_2 = \cdots = \mu_k$ is rejected by the F test, we believe that there are differences among the k population or treatment means. A natural question to ask at this point is, Which means differ? For example, with $k = 4$, it might be the case that $\mu_1 = \mu_2 = \mu_4$, with μ_3 different from the other three. Another possibility is that $\mu_1 = \mu_4$ and $\mu_2 = \mu_3$. Still another possibility is that all four means are different from one another. A **multiple comparison procedure** is a method for identifying differences among the μ's once the hypothesis of overall equality has been rejected. We will present one such method, the **Tukey–Kramer** (T–K) multiple comparison procedure.

The T–K procedure is based on computing confidence intervals for the difference between each possible pair of μ's. In the case $k = 3$, there are three differences to consider:

$$\mu_1 - \mu_2 \qquad \mu_1 - \mu_3 \qquad \mu_2 - \mu_3$$

(The difference $\mu_2 - \mu_1$ is not considered, since the interval for $\mu_1 - \mu_2$ provides the same information. Similarly, intervals for $\mu_3 - \mu_1$ and $\mu_3 - \mu_2$ are not necessary.) Once all confidence intervals have been computed, each is examined to determine whether the interval includes zero. If a particular interval does not include zero, the two means are declared to be "significantly different" from one another. An interval that does include zero supports the conclusion that there is no significant difference between the means involved.

Suppose, for example, that $k = 3$ and the three confidence intervals are:

Difference	T–K Interval
$\mu_1 - \mu_2$	$(-.9, 3.5)$
$\mu_1 - \mu_3$	$(2.6, 7.0)$
$\mu_2 - \mu_3$	$(1.2, 5.7)$

Since the interval for $\mu_1 - \mu_2$ includes zero, we judge that μ_1 and μ_2 do not differ significantly. The other two intervals do not include zero, so we conclude that $\mu_1 \neq \mu_3$ and $\mu_2 \neq \mu_3$.

The T–K intervals are based on critical values for a probability distribution called the *Studentized range distribution*. These critical values appear in Appendix Table VIII. Enter the table at the column corresponding to the number of populations or treatments being compared, move down to the rows corresponding to error df, and select either the value for a 95% confidence level or the one for a 99% level.

The Tukey–Kramer Multiple Comparison Procedure

When there are k populations or treatments being compared, $k(k - 1)/2$ confidence intervals must be computed. If we denote the relevant Studentized range critical value by q, the intervals are as follows:

$$\text{For } \mu_i - \mu_j: \quad (\bar{x}_i - \bar{x}_j) \pm q\sqrt{\frac{\text{MSE}}{2}\left(\frac{1}{n_i} + \frac{1}{n_j}\right)}$$

Two means are judged to differ significantly if the corresponding interval does not include zero.

If the sample sizes are all the same, let n denote the common value of $n_1, \ldots, n_k$. In this case, the $\pm$ factor for each interval is the same quantity

$$q\sqrt{\frac{\text{MSE}}{n}}$$

EXAMPLE 15.5

The degree of success at mastering a skill often depends on the method used to learn the skill. The article "Effects of Occluded Vision and Imagery on Putting Golf Balls" (*Perceptual and Motor Skills* (1995): 179–186) reported on a study involving the following four learning methods: (1) visual contact and imagery, (2) nonvisual contact and imagery, (3) visual contact, and (4) control. There were 20 subjects randomly assigned to each method. The accompanying summary data on putting performance was reported.

Method	1	2	3	4
$\bar{x}$	16.30	15.25	12.05	9.30
s	2.03	3.23	2.91	2.85

It is easily verified that MSTr = 202.283, MSE = 7.7861, and F = 25.98. With df_1 = 3 and df_2 = 76, Appendix Table VII shows that *P*-value < .001. The authors concluded that the mean performance is not the same for the four methods.

Appendix Table VIII gives the 95% Studentized range critical value q = 3.74 (using error df = 60, the closest tabled value to df = 76). The first two T–K intervals are

$$\mu_1 - \mu_2: \quad (16.30 - 15.25) \pm 3.74\sqrt{7.7861\left(\frac{1}{20} + \frac{1}{20}\right)}$$

$$= 1.05 \pm 2.334$$
$$= (-1.284, 3.384) \quad \leftarrow \text{Includes 0.}$$

$$\mu_1 - \mu_3: \quad (16.30 - 12.05) \pm 3.74\sqrt{7.7861\left(\frac{1}{20} + \frac{1}{20}\right)}$$

$$= 4.25 \pm 2.334$$
$$= (1.916, 6.584) \quad \leftarrow \text{Does not include 0.}$$

The remaining intervals are

$\mu_1 - \mu_4$	$7.00 \pm 2.334 = (4.666, 9.334)$	← Does not include 0.
$\mu_2 - \mu_3$	$3.20 \pm 2.334 = (.866, 5.534)$	← Does not include 0.
$\mu_2 - \mu_4$	$5.95 \pm 2.334 = (3.616, 8.284)$	← Does not include 0.
$\mu_3 - \mu_4$	$2.75 \pm 2.334 = (.416, 5.084)$	← Does not include 0.

We would conclude that μ_1 is not significantly different from μ_2, but that all other means differ from one another. That is, there is no significant difference between the visual contact and imagery and the nonvisual contact and imagery treatments. However, the visual contact and the control treatments differ from each other and from treatments 1 and 2.

MINITAB can be used to construct T–K intervals if raw data is available. Typical output (based on Example 15.5) is shown in Figure 15.6. From the output, we see that the confidence interval for $\mu_1 - \mu_2$ is $(-1.284, 3.384)$, that for $\mu_2 - \mu_4$ is $(3.616, 8.284)$, and so on.

FIGURE 15.6 The T–K
intervals for Example 15.5
(from MINITAB)

Tukey's pairwise comparisons
Family error rate = 0.0500
Individual error rate = 0.0111
Critical value = 3.74
Intervals for (column level mean) - (row level mean)

	1	2	3
	1	2	3
2	-1.284		
	3.384		
3	1.916	.866	
	6.584	5.534	
4	4.666	3.616	.416
	9.334	8.284	5.084

Why calculate the Tukey–Kramer intervals rather than use the t confidence interval for a difference between μ's (from Chapter 11)? The answer is that the T–K intervals control the **simultaneous confidence level** at approximately 95% (or 99%). That is, if the procedure is used repeatedly on many different data sets, in the long run only about 5% (or 1%) of the time would at least one of the intervals not include the value of what it is estimating. Consider using separate 95% t intervals, each one having a 5% error rate. Then the chance that at least one interval would make an incorrect statement about a difference in μ's increases dramatically with the number of intervals calculated. The MINITAB output of Figure 15.6 shows that to achieve a simultaneous confidence level of about 95% (experimentwise or "family" error rate of 5%) when $k = 4$ and error df = 76, the individual confidence level must be 98.89% (individual error rate 1.11%).

An effective display for summarizing the results of any multiple comparison procedure involves listing the $\bar{x}$'s and underscoring pairs judged to be not significantly different. The process for constructing such a display is described in the accompanying box.

Summarizing the Results of the Tukey–Kramer Procedure

1. List the sample means in increasing order, identifying the corresponding population just above the value of each $\bar{x}$.

2. Use the T–K intervals to determine the group of means that do not differ significantly from the first in the list. Draw a horizontal line extending from the smallest mean to the last mean in the group identified. For example, if there are five means, arranged in order,

Population	3	2	1	4	5
Sample mean	$\bar{x}_3$	$\bar{x}_2$	$\bar{x}_1$	$\bar{x}_4$	$\bar{x}_5$

and μ_3 is judged to be not significantly different from μ_2 or μ_1 but is judged to be significantly different from μ_4 and μ_5, draw the following line:

Population	3	2	1	4	5
Sample mean	$\bar{x}_3$	$\bar{x}_2$	$\bar{x}_1$	$\bar{x}_4$	$\bar{x}_5$

(*continued*)

3. Use the T–K intervals to determine the group of means that are not significantly different from the second smallest. (You need consider only means that appear to the right of the mean under consideration.) If there is already a line connecting the second smallest mean with all means in the new group identified, no new line need be drawn. If this entire group of means is not underscored with a single line, draw a line extending from the second smallest to the last mean in the new group. Continuing with our example, if μ_2 is not significantly different from μ_1 but is significantly different from μ_4 and μ_5, no new line need be drawn. However, if μ_2 is not significantly different from either μ_1 or μ_4 but is judged to be different from μ_5, a second line is drawn as shown:

Population	3	2	1	4	5
Sample mean	$\bar{x}_3$	$\bar{x}_2$	$\bar{x}_1$	$\bar{x}_4$	$\bar{x}_5$

4. Continue considering the means in the order listed, adding new lines as needed.

To illustrate this summary procedure, suppose that four samples with $\bar{x}_1 = 19$, $\bar{x}_2 = 27$, $\bar{x}_3 = 24$, and $\bar{x}_4 = 10$ are used to test $H_0: \mu_1 = \mu_2 = \mu_3 = \mu_4$, and that this hypothesis is rejected. Suppose the T–K confidence intervals indicate that μ_2 is significantly different from both μ_1 and μ_4, and that there are no other significant differences. The resulting summary display would then be

Population	4	1	3	2
Sample mean	10	19	24	27

EXAMPLE 15.6

A biologist wished to study the effects of ethanol on sleep time. A sample of 20 rats, matched for age and other characteristics, was selected, and each rat was given an oral injection having a particular concentration of ethanol per body weight. The rapid eye movement (REM) sleep time for each rat was then recorded for a 24-hour period, with the results shown in the accompanying table.

Treatment						$\bar{x}$
1. 0 (control)	88.6	73.2	91.4	68.0	75.2	79.28
2. 1 g/kg	63.0	53.9	69.2	50.1	71.5	61.54
3. 2 g/kg	44.9	59.5	40.2	56.3	38.7	47.92
4. 4 g/kg	31.0	39.6	45.3	25.2	22.7	32.76

Table 15.4 (an ANOVA table from SAS) leads to the conclusion that true average REM sleep time depends on the treatment used; the P-value for the F test is .0001.

Table 15.4 *SAS ANOVA table*

Analysis of Variance Procedure

Dependent Variable: TIME

Source	DF	Sum of Squares	Mean Square	F Value	Pr > F
Model	3	5882.35750	1960.78583	21.09	0.0001
Error	16	1487.40000	92.96250		
Corrected Total	19	7369.75750			

The T–K intervals are

Difference	Interval	Includes 0?
$\mu_1 - \mu_2$	17.74 ± 17.446	no
$\mu_1 - \mu_3$	31.36 ± 17.446	no
$\mu_1 - \mu_4$	46.24 ± 17.446	no
$\mu_2 - \mu_3$	13.08 ± 17.446	yes
$\mu_2 - \mu_4$	28.78 ± 17.446	no
$\mu_3 - \mu_4$	15.16 ± 17.446	yes

The only T–K intervals that include zero are those for $\mu_2 - \mu_3$ and $\mu_3 - \mu_4$. The corresponding underscoring pattern is

$\overline{x}_4$	$\overline{x}_3$	$\overline{x}_2$	$\overline{x}_1$
32.76	47.92	61.54	79.28

Figure 15.7 displays SAS output that agrees with our underscoring; letters are used to indicate groupings in place of the underscoring.

FIGURE 15.7 SAS output
for Example 15.6

Alpha = 0.05   df = 16   MSE = 92.9625
Critical Value of Studentized Range = 4.046
Minimum Significant Difference = 17.446
Means with the same letter are not significantly different.

Tukey Grouping		Mean	N	TREATMENT
	A	79.280	5	0 (control)
	B	61.540	5	1 gm/kg
C	B	47.920	5	2 gm/kg
C		32.760	5	4 gm/kg

EXAMPLE 15.7

How satisfied are college students with dormitory roommates? The article "Roommate Satisfaction and Ethnic Identity in Mixed-Race and White University Roommate Dyads" (*J. College Student Development* (1998): 194–199) investigated differences among randomly assigned African-American/White, Asian/White, Hispanic/White, and White/White roommate pairs. The researchers used a one-way ANOVA to analyze scores on the Roommate Relationship Inventory to see whether a difference in mean score existed for the four types of roommate pairs. They reported

> significant differences among the means ($P < .01$). Follow-up Tukey [intervals]...indicated differences between White dyads ($M = 77.49$) and African-American/White dyads ($M = 71.27$). ... No other significant differences were found.

Although the mean satisfaction score for the Asian/White and Hispanic/White groups were not given, they must have been between 77.49, the mean for the White/White pairs, and 71.27, the mean for the African-American/White pairs. (If they had been larger than 77.49, they would have been significantly different from the African-American/White pairs mean, and if they had been smaller than 71.27, they would have been significantly different from the White/White pairs mean.) An underscoring consistent with the reported information is

$$
\begin{array}{c c c}
 & \text{Hispanic/White} & \\
 & \text{and} & \\
\text{White/White} & \text{Asian/White} & \text{African-American/White} \\
\underline{\hphantom{White/White}} & \underline{\hphantom{Asian/White}} & \\
 & \underline{\hphantom{Asian/White African-American/White}} &
\end{array}
$$

Exercises 15.19 – 15.29

15.19 Leaf surface area is an important variable in plant gas-exchange rates. The article "Fluidized Bed Coating of Conifer Needles with Glass Beads for Determination of Leaf Surface Area" (*Forest Sci.* (1980): 29–32) included an analysis of dry matter per unit surface area (mg/cm²) for trees raised under three different growing conditions. Let μ_1, μ_2, and μ_3 represent the true mean dry matter per unit surface area for the growing conditions 1, 2, and 3, respectively. The given 95% simultaneous confidence intervals are based on summary quantities that appear in the article:

Difference	$\mu_1 - \mu_2$	$\mu_1 - \mu_3$	$\mu_2 - \mu_3$
Interval	$(-3.11, -1.11)$	$(-4.06, -2.06)$	$(-1.95, .05)$

Which of the following four statements do you think describes the relationship between μ_1, μ_2, and μ_3? Explain your choice.

  i. $\mu_1 = \mu_2$, and μ_3 differs from μ_1 and μ_2.
  ii. $\mu_1 = \mu_3$, and μ_2 differs from μ_1 and μ_3.
  iii. $\mu_2 = \mu_3$, and μ_1 differs from μ_2 and μ_3.
  iv. All three μ's are different from one another.

15.20 The article referenced in Example 15.5 reported on a study involving the following four learning methods: (1) visual contact and imagery, (2) nonvisual contact and imagery, (3) visual contact, and (4) control. There were 20 subjects assigned to each method. The accompanying summary data on putting performance was reported.

Method	1	2	3	4
$\bar{x}$	16.30	15.25	12.05	9.30
s	2.03	3.23	2.91	2.85

These in turn give MSE = 7.7861 and $F = 25.98$, from which P-value $< .001$, so H_0 can be rejected at significance level .01. Calculate the 99% T–K intervals, indicate which methods differ significantly from one another, and summarize the results by underscoring. How do the 99% T–K intervals compare to the 95% T–K intervals of Example 15.5?

15.21 The accompanying data resulted from a flammability study in which specimens of five different fabrics were tested to determine burn times.

	1	17.8	16.2	15.9	15.5	
	2	13.2	10.4	11.3		
Fabric	3	11.8	11.0	9.2	10.0	
	4	16.5	15.3	14.1	15.0	13.9
	5	13.9	10.8	12.8	11.7	

$$MSTr = 23.67$$
$$MSE = 1.39$$
$$F = 17.08$$
$$P\text{-value} = .000$$

The accompanying output gives the T–K intervals as calculated by MINITAB. Identify significant differences and give the underscoring pattern.

Individual error rate = 0.00750
Critical value = 4.37
Intervals for (column level mean) - (row level mean)

	1	2	3	4
	1.938			
2	7.495			
	3.278	-1.645		
3	8.422	3.912		
	-1.050	-5.983	-6.900	
4	3.830	-0.670	-2.020	
	1.478	-3.445	-4.372	0.220
5	6.622	2.112	0.772	5.100

15.22 Exercise 15.6 presented the accompanying summary information on satisfaction levels for employees on three different work schedules:

$$n_1 = 24 \qquad n_2 = 24 \qquad n_3 = 20$$
$$\bar{x}_1 = 6.60 \qquad \bar{x}_2 = 5.37 \qquad \bar{x}_3 = 5.20$$

MSE was computed to be 2.028. Calculate the 95% T–K intervals, and identify any significant differences.

15.23 Sample mean chlorophyll concentrations for the four Jerusalem artichoke varieties introduced in Exercise 15.7 were .30, .24, .41, and .33, with corresponding sample sizes of 5, 5, 4, and 6, respectively. In addition, MSE = .0130. Calculate the 95% T–K intervals, and then use the underscoring procedure described in this section to identify significant differences among the varieties.

15.24 Do lizards play a role in spreading plant seeds? Some research carried out in South Africa would suggest so ("Dispersal of Namaqua Fig (*Ficus cordata cordata*) Seeds by the Augrabies Flat Lizard (*Platysaurus broadleyi*)," *Journal of Herpetology* (1999): 328–330). The re-searchers collected 400 seeds of this particular type of fig, 100 of which were from each treatment: lizard dung, bird dung, rock hyrax dung, and uneaten figs. They planted these seeds in batches of 5, and for each group of 5 they recorded how many of the seeds germinated. This resulted in 20 observations for each treatment. The treatment means and standard deviations are given in the accompanying table.

Treatment	n	$\bar{x}$	s
Uneaten figs	20	2.40	.30
Lizard dung	20	2.35	.33
Bird dung	20	1.70	.34
Hyrax dung	20	1.45	.28

a. Construct the appropriate ANOVA table, and test the hypothesis that there is no difference between mean number of seeds germinating for the four treatments.
b. Is there evidence that seeds eaten and then excreted by lizards germinate at a higher rate than those eaten and then excreted by birds? Give statistical evidence to support your answer.

15.25 The article "Growth Response in Radish to Sequential and Simultaneous Exposures of NO_2 and SO_2" (*Environ. Pollution* (1984): 303–325) compared a control group (no exposure), a sequential exposure group (plants exposed to one pollutant followed by exposure to the second four weeks later), and a simultaneous-exposure group (plants exposed to both pollutants at the same time). The article states, "Sequential exposure to the two pollutants had no effect on growth compared to the control. Simultaneous exposure to the gases significantly reduced plant growth." Let $\bar{x}_1$, $\bar{x}_2$, and $\bar{x}_3$ represent the sample means for the control, sequential, and simultaneous groups, respectively. Suppose that $\bar{x}_1 > \bar{x}_2 > \bar{x}_3$. Use the given information to construct a table where the sample means are listed in increasing order, with those that are judged not to be significantly different underscored.

15.26 The nutritional quality of shrubs commonly used for feed by rabbits was the focus of a study summarized in the article "Estimation of Browse by Size Classes for Snowshoe Hare" (*J. Wildlife Mgmt.* (1980): 34–40). The energy content (cal/g) of three sizes (4 mm or less, 5–7 mm, and 8–10 mm) of serviceberries was studied. Let μ_1, μ_2, and μ_3 denote the true energy content for the three size classes. Suppose that 95% simultaneous confidence intervals for $\mu_1 - \mu_2$, $\mu_1 - \mu_3$, and $\mu_2 - \mu_3$ are $(-10, 290)$, $(150, 450)$, and $(10, 310)$, respectively. How would you interpret these intervals?

15.27 The accompanying underscoring pattern appeared in the article "Effect of SO_2 on Transpiration, Chlorophyll

Content, Growth and Injury in Young Seedlings of Woody Angiosperms" (*Canadian J. Forest Research* (1980): 78–81). Water loss of plants (*Acer saccharinum*) exposed to 0, 2, 4, 8, and 16 hours of fumigation was recorded, and a multiple comparison procedure was used to detect differences among the mean water losses for the different fumigation durations. How would you interpret this pattern?

Duration of fumigation	16	0	8	2	4
Sample mean water loss	27.57	28.23	30.21	31.16	36.21

15.28 Samples of six different brands of diet or imitation margarine were analyzed to determine the level of physiologically active polyunsaturated fatty acids (PAPUFA, in percent), resulting in the data shown in the accompanying table. (The data is fictitious, but the sample means agree with data reported in *Consumer Reports*.)

Imperial	14.1	13.6	14.4	14.3	
Parkay	12.8	12.5	13.4	13.0	12.3
Blue Bonnet	13.5	13.4	14.1	14.3	
Chiffon	13.2	12.7	12.6	13.9	
Mazola	16.8	17.2	16.4	17.3	18.0
Fleischmann's	18.1	17.2	18.7	18.4	

a. Test for differences among the true average PAPUFA percentages for the different brands. Use $\alpha = .05$.

b. Use the T–K procedure to compute 95% simultaneous confidence intervals for all differences between means and interpret the resulting intervals.

15.29 Consider the accompanying data on plant growth after the application of different types of growth hormone.

	1	13	17	7	14
	2	21	13	20	17
Hormone	**3**	18	14	17	21
	4	7	11	18	10
	5	6	11	15	8

a. Carry out the *F* test at level $\alpha = .05$.

b. What happens when the T–K procedure is applied? (*Note*: This "contradiction" can occur when H_0 is "barely" rejected. It happens because the test and the multiple comparison method are based on different distributions. Consult your friendly neighborhood statistician for more information.)

15.3   The *F* Test for a Randomized Block Experiment

We saw in Chapter 11 that when two treatments are to be compared, a paired experiment is often more effective than one involving two independent samples. This is because pairing can considerably reduce the extraneous variation in subjects or experimental units. A similar result can be achieved when more than two treatments are to be compared. Suppose that four different pesticides (the treatments) are being considered for application to a particular crop. There are 20 plots of land available for planting. If 5 of these plots are randomly selected to receive pesticide 1, 5 of the remaining 15 randomly selected for pesticide 2, and so on, the result is a *completely randomized experiment,* and the data should be analyzed using single-factor ANOVA. The disadvantage of this experiment is that if there are any substantial differences in characteristics of the plots that could affect yield, a separate assessment of any differences between treatments will not be possible (plot effects are confounded with treatment effects).

Here is an alternative experiment. Consider separating the 20 plots into five groups, each consisting of 4 plots. Within each group, the plots are as much alike as possible with respect to characteristics affecting yield. Then within each group, one plot is randomly selected for pesticide 1, a second plot is randomly chosen to receive pesticide 2, and so on. The homogeneous groups are called *blocks,* and the random allocation of treatments within each block as described results in a *randomized block experiment.*

Definition

Suppose that experimental units (individuals or objects to which the treatments are applied) are first separated into groups consisting of k units in such a way that the units within each group are as similar as possible. Within any particular group, the treatments are then randomly allocated so that each unit in a group receives a different treatment. The groups are often called **blocks,** and the experimental design is referred to as a **randomized block design.**

EXAMPLE 15.8

High energy costs have made consumers and home builders increasingly aware of whether household appliances are energy efficient. A large developer carried out a study to compare electricity usage for four different residential air-conditioning systems being considered for tract homes. Each system was installed in five homes, and the resulting electricity usage (in KWh) was monitored for a 1-month period. Because the developer realized that many characteristics of a home could affect usage (for example, floor space, type of insulation, directional orientation, and type of roof and exterior), care was taken to ensure that extraneous variation in such characteristics did not influence the conclusions. Homes selected for the experiment were grouped into five blocks consisting of four homes each in such a way that the four homes within any given block were as similar as possible. Resulting data is displayed in Table 15.5, in which rows correspond to the different treatments (air-conditioning systems) and columns correspond to the different blocks.

Table 15.5 *Data from the randomized block experiment of Example 15.8*

		Block					
	Treatment	1	2	3	4	5	Treatment Average
	1	116	118	97	101	115	109.40
	2	171	131	105	107	129	128.60
	3	138	131	115	93	110	117.40
	4	144	141	115	93	99	118.40
Block average		142.25	130.25	108.00	98.50	113.25	**Grand mean** 118.45

Later in this section we will analyze this data to see whether electricity usage depends on which system is used.

The hypotheses of interest and assumptions underlying the analysis of a randomized block design are similar to those for a completely randomized design.

Assumptions and Hypotheses

The single observation made on any particular treatment in a given block is assumed to be selected from a normal distribution. The variance of this distribution is σ^2, the same for each block–treatment combination. However, the mean value may depend separately both on the treatment applied and on the block. The hypotheses of interest are as follows:

H_0:   The mean value does not depend on which treatment is applied.

H_a: The mean value does depend on which treatment is applied.

The μ notation used previously is no longer appropriate for stating hypotheses, since an observation's mean value may depend on both the treatment applied and the block utilized.

The F Test

The key to analyzing data from a randomized block experiment is to represent SSTo, which measures total variation, as a sum of three pieces: SSTr, SSE, and a block sum of squares SSBl. The latter SS incorporates any variation due to differences between the blocks, which will be substantial if there was great heterogeneity in experimental units (for example, plots) prior to creating the blocks. Once the four sums of squares have been computed, the test statistic is an F ratio MSTr/MSE, but error df is no longer $N - k$ as in single-factor ANOVA.

Summary of the Randomized Block F Test

Notation: Let

  k = number of treatments

  l = number of blocks

  $\bar{x}_i$ = average of all observations for treatment i

  $\bar{b}_i$ = average of all observations in block i

  $\bar{\bar{x}}$ = average of all kl observations in the experiment (the grand mean)

Sums of squares and associated df's are as follows.

Sum of Squares	Symbol	df	Formula
Treatments	SSTr	$k - 1$	$l[(\bar{x}_1 - \bar{\bar{x}})^2 + (\bar{x}_2 - \bar{\bar{x}})^2 + \cdots + (\bar{x}_k - \bar{\bar{x}})^2]$
Blocks	SSBl	$l - 1$	$k[(\bar{b}_1 - \bar{\bar{x}})^2 + (\bar{b}_2 - \bar{\bar{x}})^2 + \cdots + (\bar{b}_k - \bar{\bar{x}})^2]$
Error	SSE	$(k - 1)(l - 1)$	by subtraction
Total	SSTo	$kl - 1$	$\displaystyle\sum_{\substack{\text{all } kl \\ \text{obs.}}} (x - \bar{\bar{x}})^2$

(*continued*)

SSE is obtained by subtraction through the use of the fundamental identity

$$\text{SSTo} = \text{SSTr} + \text{SSBl} + \text{SSE}$$

Test statistic: $\quad F = \dfrac{\text{MSTr}}{\text{MSE}}$

where

$$\text{MSTr} = \frac{\text{SSTr}}{k - 1} \quad \text{and} \quad \text{MSE} = \frac{\text{SSE}}{(k - 1)(l - 1)}$$

The test is based on $\text{df}_1 = k - 1$ and $\text{df}_2 = (k - 1)(l - 1)$. *P*-value information comes from Appendix Table VII.

Alternative formulas for the SS's appropriate for efficient hand computation appear in the chapter appendix.

Calculations for this *F* test are usually summarized in an ANOVA table. The table is similar to the one for single-factor ANOVA except that blocks are an extra source of variation, so four rows are included rather than just three.

Table 15.6 shows a mean square for blocks as well as for treatments and error. Sometimes the *F* ratio MSBl/MSE is also computed. A large value of this ratio suggests that blocking was effective in filtering out extraneous variation.

Table 15.6 *The ANOVA table for a randomized block experiment*

Source of Variation	df	Sum of Squares	Mean Square	F
Treatments	$k - 1$	SSTr	$\text{MSTr} = \dfrac{\text{SSTr}}{k - 1}$	$F = \dfrac{\text{MSTr}}{\text{MSE}}$
Blocks	$l - 1$	SSBl	$\text{MSBl} = \dfrac{\text{SSBl}}{l - 1}$	
Error	$(k - 1)(l - 1)$	SSE	$\text{MSE} = \dfrac{\text{SSE}}{(k - 1)(l - 1)}$	
Total	$kl - 1$	SSTo		

EXAMPLE 15.9   Reconsider the electricity usage data given in Example 15.8.

H_0: The mean electricity usage does not depend on which air conditioning system is used.

H_a: The mean electricity usage does depend on which system is used.

Test statistic: $\quad F = \dfrac{\text{MSTr}}{\text{MSE}}$

From Example 15.8,

$$\bar{\bar{x}} = 118.45$$

$$\bar{x}_1 = 109.40 \qquad \bar{x}_2 = 128.60 \qquad \bar{x}_3 = 117.40 \qquad \bar{x}_4 = 118.40$$
(these are the four row averages)

$$\bar{b}_1 = 142.25 \qquad \bar{b}_2 = 130.25 \qquad \bar{b}_3 = 108.00 \qquad \bar{b}_4 = 98.50 \qquad \bar{b}_5 = 113.25$$
(the five column averages)

Using the individual observations given previously,

$$\text{SSTo} = \sum_{\substack{\text{all } kl \\ \text{obs.}}} (x - \bar{\bar{x}})^2$$

$$= (116 - 118.45)^2 + (118 - 118.45)^2 + \cdots + (99 - 118.45)^2 = 7594.95$$

The other sums of squares are

$$\text{SSTr} = l[(\bar{x}_1 - \bar{\bar{x}})^2 + (\bar{x}_2 - \bar{\bar{x}})^2 + \cdots + (\bar{x}_k - \bar{\bar{x}})^2]$$
$$= 5[(109.4 - 118.45)^2 + (128.6 - 118.45)^2$$
$$+ (117.4 - 118.45)^2 + (118.4 - 118.45)^2]$$
$$= 930.15$$

$$\text{SSBl} = k[(\bar{b}_1 - \bar{\bar{x}})^2 + (\bar{b}_2 - \bar{\bar{x}})^2 + \cdots + (\bar{b}_k - \bar{\bar{x}})^2]$$
$$= 4[(142.254 - 118.45)^2 + (130.25 - 118.45)^2 + \cdots + (113.25 - 118.45)^2]$$
$$= 4959.70$$

$$\text{SSE} = \text{SSTo} - \text{SSTr} - \text{SSBl}$$
$$= 7594.95 - 930.15 - 4959.70$$
$$= 1705.10$$

The remaining calculations are displayed in the accompanying ANOVA table.

Source of Variation	df	Sum of Squares	Mean Square	F
Treatments	3	930.15	310.05	$\dfrac{310.05}{142.09} = 2.18$
Blocks	4	4959.70	1239.93	
Error	12	1705.10	142.09	
Total	19	7594.95		

In Appendix Table VII with $df_1 = 3$ and $df_2 = 12$, the value 2.61 corresponds to a P-value of .10. Since $2.18 < 2.61$, P-value $> .10$ and H_0 cannot be rejected. Mean electricity usage does not seem to depend on which of the four air-conditioning systems is used.

In many studies, all k of the treatments can be applied to the same experimental unit, so there is no need to group different experimental units to form blocks. For example, an experiment to compare effects of four different gasoline additives on automobile engine efficiency could be carried out by selecting just 5 engines and using all four treatments on each one rather than using 20 engines and blocking them. Each engine by itself then constitutes a block. As another example, a manufacturing company might wish to compare outputs for three different packaging machines. Because output could be affected by which operator is using the machine, a design that controls for the effects of operator variation is desirable. One possibility is to use 15 operators grouped into homogeneous blocks of 5 operators each, but such homogeneity within each block may be difficult to achieve. An alternative approach is to use only 5 operators and have each one operate all three machines. There are then three observations in each block, all three with the same operator.

EXAMPLE 15.10

In the article "The Effects of a Pneumatic Stool and a One-Legged Stool on Lower Limb Joint Load and Muscular Activity During Sitting and Rising" (*Ergonomics* (1993): 519–535), the following data is given on the effort (measured on the Borg Scale) required by a subject to rise from a sitting position for each of four different stools. Because it was suspected that different people could exhibit large differences in effort, even for the same type of stool, a sample of nine people was selected and each tested on all four stools, with the following results:

		Subject								
		1	**2**	**3**	**4**	**5**	**6**	**7**	**8**	**9**
	A	12	10	7	7	8	9	8	7	9
Type of	**B**	15	14	14	11	11	11	12	11	13
stool	**C**	12	13	13	10	8	11	12	8	10
	D	10	12	9	9	7	10	11	7	8

For each person, the order in which the stools were tested was randomized. This is a randomized block experiment, with subjects playing the role of blocks. The test consists of these hypotheses:

H_0: Mean effort does not depend on type of stool.

H_a: Mean effort does depend on type of stool.

Test statistic: $F = \dfrac{\text{MSTr}}{\text{MSE}}$

Computations are summarized in Table 15.7, an ANOVA table from MINITAB.

Table 15.7 *ANOVA Table for Example 15.10*

Two-way Analysis of Variance
Analysis of Variance for Effort

Source	DF	SS	MS	F	P
Stool	3	81.19	27.06	22.36	0.000
Block	8	66.50	8.31	6.87	0.000
Error	24	29.06	1.21		
Total	35	176.75			

The test statistic value is 22.36, with P-value = .000. Since P-value $< \alpha$, we reject H_0. There is sufficient evidence to conclude that the mean effort required is not the same for all four stool types.

Experiments such as the one described in Example 15.10, in which repeated observations are made on the same experimental unit, are sometimes called *repeated measures designs*. Such designs should not be used when application of the first several treatments somehow affects responses to later treatments. This would be the case if treatments were different methods for learning the same skill, so that if all treatments were given to the same subject, the response to the treatment given last would presumably be much better than the response to the treatment initially applied.

Multiple Comparisons

As in single-factor ANOVA, once H_0 has been rejected, further analysis of the data is appropriate to identify significant differences among the treatments. The Tukey–Kramer method is easily adapted for this purpose.

Declare that treatments i and j differ significantly if the interval

$$\bar{x}_i - \bar{x}_j \pm q\sqrt{\frac{MSE}{l}}$$

does not include zero, where q is based on a comparison of k treatments and error df $= (k - 1)(l - 1)$.

EXAMPLE 15.11   In Example 15.10, we had $k = 4$ and error df $= 24$, from which $q = 4.91$ for a 99% simultaneous confidence level. The $\pm$ factor for each interval is

$$q\sqrt{\frac{MSE}{l}} = 4.91\sqrt{\frac{1.21}{9}} = 1.80$$

The four treatment means arranged in order are

Treatment	A	D	C	B
Sample mean	8.556	9.222	10.778	12.444

It is easily verified that corresponding underscoring pattern is as follows:

Treatment	A	D	C	B
Sample mean	8.556	9.222	10.778	12.444

Exercises 15.30 – 15.38

15.30 A particular county employs three assessors, who are responsible for determining the value of residential property in the county. To see whether these assessors differ systematically in their appraisals, 5 houses are selected, and each assessor is asked to determine the market value of each house. Explain why a randomized block experiment (with blocks corresponding to the 5 houses) was used rather than a completely randomized experiment involving a total of 15 houses with each assessor asked to appraise 5 different houses (a different group of 5 for each assessor).

15.31 The accompanying display is a partially completed ANOVA table for the experiment described in Exercise 15.30 (with houses representing blocks and assessors representing treatments).

Source of Variation	df	Sum of Squares	Mean Square	F
Treatments		11.7		
Blocks		113.5		
Error				
Total		250.8		

a. Fill in the missing entries in the ANOVA table.
b. Use the ANOVA *F* statistic and a .05 level of significance to test the null hypothesis of no difference between assessors.

15.32 Land-treatment wastewater-processing systems work by removing nutrients and thereby discharging water of better quality. The land used is often planted with a crop such as corn, because plant uptake removes nitrogen from the water and sale of the crop helps reduce the costs of wastewater treatment. The concentration of nitrogen in treated water was observed from 1975 to 1979 under wastewater application rates of none, .05 m/week, and .1 m/week. A randomized block ANOVA was performed, with the 5 years serving as blocks. The accompanying display is a partially completed ANOVA table from the article "Quality of Percolate Water After Treatment of a Municipal Wastewater Effluent by a Crop Irrigation System" (*J. Environ. Quality* (1984): 256–264).

Source of Variation	df	Sum of Squares	Mean Square	F
Treatments		1835.2		
Blocks				
Error		206.1		
Total	14	2134.1		

a. Complete the ANOVA table.
b. Is there sufficient evidence to reject the null hypothesis of no difference between the true mean nitrogen concentrations for the three application rates? Use $\alpha = .05$.

15.33 In a comparison of the energy efficiency of three types of ovens (conventional, biradiant, and convection), the energy used in cooking was measured for eight different foods (one-layer cake, two-layer cake, biscuits, bread, frozen pie, baked potatoes, lasagna, and meat loaf). Since a comparison between the three types of ovens is desired, a randomized block ANOVA (with the eight foods serving as blocks) will be used. Suppose calculations result in the quantities SSTo = 4.57, SSTr = 3.97, and SSBl = .2503. (A similar study is described in the article "Optimizing Oven Radiant Energy Use," *Home Ec. Res. J.* (1980): 242–251.) Construct an ANOVA table and test the null hypothesis of no difference in mean energy use for the three types of ovens. Use a .01 significance level.

15.34 The article "Rate of Stuttering Adaptation Under Two Electro-Shock Conditions" (*Behavior Res. Therapy* (1967): 49–54) gave adaptation scores for three different treatments: no shock (treatment 1), shock following each stuttered word (treatment 2), and shock during each moment of stuttering (treatment 3). These treatments were used on each of 18 stutterers. The 18 subjects were viewed as blocks, and the data were analyzed using a randomized block ANOVA. Summary quantities are SSTr = 28.78, SSBl = 2977.67, and SSE = 469.55. Construct the ANOVA table, and test at significance level .05 to see whether true average adaptation score depends on the treatment given.

15.35 The accompanying table shows average height of cotton plants during 1978–1980 under three different effluent application rates (350, 440, and 515 mm) ("Drip Irrigation of Cotton with Treated Municipal Effluents: Yield Response," *J. Environ. Quality* (1984): 231–238).

	Application Rate		
Year	**350**	**440**	**515**
1978	166	176	177
1979	109	126	136
1980	140	155	156

a. This data was analyzed using a randomized block ANOVA, with years serving as blocks. Explain why this would be better for comparing treatments than a completely randomized ANOVA.

b. With treatments 1, 2, and 3 denoting the application rates 350, 440, and 515, respectively, summary quantities are

$$\Sigma(x - \bar{\bar{x}})^2 = 4266.0 \qquad \bar{\bar{x}} = 149.00$$

$$\bar{x}_1 = 138.33 \qquad \bar{x}_2 = 152.33 \qquad \bar{x}_3 = 156.33$$

$$\bar{b}_1 = 173.00 \qquad \bar{b}_2 = 123.67 \qquad \bar{b}_3 = 150.33$$

Construct an ANOVA table. Using $\alpha = .05$, determine whether the true mean height differs for the three effluent rates.

c. Identify significant differences among the rates. (*Hint*: $q = 5.04$)

15.36 The article "Responsiveness of Food Sales to Shelf Space Changes in Supermarkets" (*J. Mktg. Research* (1964): 63–67) described an experiment to assess the effect of allotted shelf space on product sales. Two of the products studied were baking soda (a staple product) and Tang (considered to be an impulse product). Six stores (blocks) were used in the experiment, and six different shelf-space allotments were tried for 1 week each. Space allotments of 2, 4, 6, 8, 10, and 12 ft were used for baking soda, and 6, 9, 12, 15, 18, and 21 ft were used for Tang. The author speculated that sales of staple goods would not be sensitive to changes in shelf space, whereas sales of impulse products would be affected by changes in shelf space. Data on number of boxes of baking soda and on number of containers of Tang sold during a 1-week period is given in the accompanying tables. Construct ANOVA tables and test the appropriate hypotheses. Use a significance level of .05. Was the author correct in his speculation that sales of the staple product would not be affected by shelf space allocation, whereas sales of the impulse product would be affected? Explain.

Baking Soda

		Shelf Space					
		2	4	6	8	10	12
	1	36	42	36	40	30	22
	2	74	61	65	67	83	84
Store	3	40	58	42	73	69	63
	4	43	65	65	41	43	47
	5	27	33	35	17	40	26
	6	23	31	36	38	42	37

Tang

		Shelf Space					
		6	9	12	15	18	21
	1	30	35	25	25	38	31
	2	47	59	43	62	65	48
Store	3	47	55	48	54	36	54
	4	29	19	41	27	33	39
	5	17	11	25	23	24	26
	6	22	9	19	18	25	22

15.37 The article "Measuring Treatment Effects Through Comparisons Along Plot Boundaries" (*Forest Sci.* (1980): 704–709) reported the results of a randomized block experiment. Five different sources of pine seed were used in each of four blocks. The accompanying table gives data on plant height (m). Does the data provide sufficient evidence to conclude that the true mean height is not the same for all five seed sources? Use a .05 level of significance.

		Block			
		I	II	III	IV
	1	7.1	5.8	7.2	6.9
	2	6.2	5.3	7.7	4.7
Source	3	7.9	5.4	8.6	6.2
	4	9.0	5.9	5.7	7.3
	5	7.0	6.3	4.4	6.1

15.38 The article cited in Exercise 15.37 also gave the accompanying data on survival rate for five different seed sources.

		Block			
		I	II	III	IV
	1	62.5	87.5	50.0	70.3
	2	50.0	50.0	54.7	59.4
Source	3	93.8	92.2	87.5	87.5
	4	96.9	76.6	70.3	65.6
	5	56.3	50.0	45.3	56.3

a. Construct an ANOVA table and test the null hypothesis of no difference in true mean survival rates for the five seed sources. Use $\alpha = .05$.

b. Use multiple comparisons to identify significant differences among the seed sources. (*Hint*: $q = 5.04$)

15.4   Two-Factor ANOVA

An investigator will often be interested in assessing the effects of two different factors on a response variable. Consider the following examples.

1. A physical education researcher wishes to know how body density of football players varies with position played (a categorical factor with categories defensive back, offensive back, defensive lineman, and offensive lineman) and level of play (a second categorical factor with categories professional, college Division I, college Division II, and college Division III).

2. An agricultural scientist is interested in seeing how yield of tomatoes is affected by choice of variety planted (a categorical factor, with each category corresponding to a different variety) and planting density (a quantitative factor, with a level corresponding to each planting density being considered).

3. An applied chemist might wish to investigate how shear strength of a particular adhesive varies with application temperature (a quantitative factor with levels 250°F, 260°F, and 270°F) and application pressure (a quantitative factor with levels 110 lb/in^2, 120 lb/in^2, 130 lb/in^2, and 140 lb/in^2).

Let's label the two factors under study factor A and factor B. Even when a factor is categorical, it simplifies terminology to refer to the categories as **levels.** Thus, in the first example, the categorical factor *position played* has four levels. The number of levels of factor A is denoted by k, and l denotes the number of levels of factor B, as shown in Table 15.8. This rectangular table contains a row corresponding to each level of factor A and a column corresponding to each level of factor B. Each cell in the table corresponds to a particular level of factor A in combination with a particular level of factor B. Because there are l cells in each row and k rows, there are kl cells in the table. The kl different combinations of factor A and factor B levels are often referred to as **treatments.** For example, if there are three tomato varieties and four different planting densities under consideration, the number of treatments is 12.

Table 15.8 *A table of factor combinations for a two-way ANOVA experiment*

Suppose that an experiment is carried out, resulting in a data set that contains some number of observations for each of the kl treatments. In general, there could be more observations for some treatments than for others, and there may even be a few treatments for which no observations are available. An experimenter may set out to make the same number of observations on each treatment, but occasionally forces beyond the experimenter's control — the death of an experimental subject, malfunctioning equipment, and so on — result in different sample sizes for some treatments. However, such imbalance in sample sizes makes analysis of the data rather difficult. **We will restrict consideration to data sets containing the same number of observations on each treatment and let m denote this number.**

Notation

k = number of levels of factor A

l = number of levels of factor B

kl = number of treatments   (each one a combination of a factor A level and a factor B level)

m = number of observations on each treatment

EXAMPLE 15.12

An experiment was carried out to assess the effects of tomato variety (factor A, with $k = 3$ levels) and planting density (factor B, with l levels — 10, 20, 30, and 40 thousand plants per hectare) on yield. Each of the $kl = 12$ treatments was used on $m = 3$ plots, resulting in the data set consisting of $klm = 36$ observations shown in Table 15.9 (adapted from "Effects of Plant Density on Tomato Yields in Western Nigeria," *Exper. Ag.* (1976): 43–47).

Table 15.9 *Data from the two-factor experiment of Example 15.12*

		Density (B)			
		1	**2**	**3**	**4**
	1	7.9, 9.2, 10.5	11.2, 12.8, 13.3	12.1, 12.6, 14.0	9.1, 10.8, 12.5
Variety (A)	**2**	8.1, 8.6, 10.1	11.5, 12.7, 13.7	13.7, 14.4, 15.4	11.3, 12.5, 14.5
	3	15.3, 16.1, 17.5	16.6, 18.5, 19.2	18.0, 20.8, 21.0	17.2, 18.4, 18.9

Sample average yields for each treatment, each level of factor A, and each level of factor B are important summary quantities. These can be displayed in a rectangular table (see Table 15.10).

A plot of these sample averages is also quite informative. First, construct horizontal and vertical axes, and scale the vertical axis in units of the response variable (yield). Then mark a point on the horizontal axis for each level of one of the factors (either A or B can be chosen). Now above each such mark, plot a point for the sample average response for each level of the other factor. Finally, connect all points corresponding to the same level of the other factor using straight line segments.

Table 15.10 *Sample means for the 12 treatments*

		Factor B (Planting Density)				Sample Average Yield for Each Level of Factor A
		1	2	3	4	
Factor A (Variety)	1	9.20	12.43	12.90	10.80	11.33
	2	8.93	12.63	14.50	12.77	12.21
	3	16.30	18.10	19.93	18.17	18.13
Sample Average Yield for Each Level of Factor B		11.48	14.39	15.78	13.91	Grand mean = 13.89

Figure 15.8 displays one such plot for which factor A levels mark the horizontal axis and another plot for which factor B levels mark this axis; usually only one of the two plots is constructed.

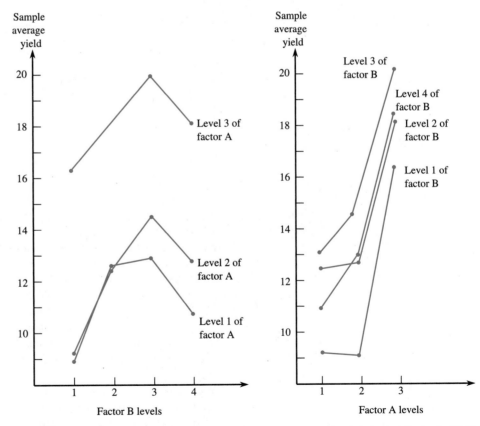

FIGURE 15.8 Graphs of treatment sample average responses for the data of Example 15.12

Interaction

An important aspect of two-factor studies involves assessing how simultaneous changes in the levels of both factors affect the response. As a simple example, suppose that an automobile manufacturer is studying engine efficiency (miles per gallon) for two different engine sizes (factor A, with $k = 2$ levels) in combination with two different carburetor designs (factor B with $l = 2$ levels). Consider the two possible sets of true average responses displayed in Figure 15.9. In Figure 15.9(a), when factor A changes from level 1 to level 2 and factor B remains at level 1 (the change within the first column), the true average response increases by 2. Similarly, when factor B changes from level 1 to level 2 and factor A is fixed at level 1 (the change within the first row), the true average response increases by 3. And when the levels of both factors are changed from 1 to 2, the true average response increases by 5, which is the sum of the two "one-at-a-time" increases. This is because the change in true average response when the level of either factor changes from 1 to 2 is the same for each level of the other factor: The change within either row is 3, and the change within either column is 2. In this case, changes in levels of the two factors affect the true average response separately, or in an additive manner.

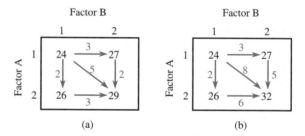

FIGURE 15.9 Two possible sets of true average responses when $k = 2$ and $l = 2$

The changes in true average responses in the first row and in the first column of Figure 15.9(b) are 3 and 2, respectively, exactly as in Figure 15.9(a). However, the change in true average response when the levels of both factors change simultaneously from 1 to 2 is 8, which is much larger than the separate changes suggest. In this case, there is interaction between the two factors, so that the effect of simultaneous changes cannot be determined from the individual effects of separate changes. This is because in Figure 15.9(b), the change in going from the first to the second column is different for the two rows, and the change in going from the first to the second row is different for the two columns. That is, *the change in true average response when the level of one factor changes depends on the level of the other factor.* This is not true in Figure 15.9(a).

When there are more than two levels of either factor, a graph of true average responses, similar to that for sample average responses in Figure 15.8, provides insight into how changes in the level of one factor depend on the level of the other factor. Figure 15.10 shows several possible such graphs when $k = 4$ and $l = 3$. The most general situation is pictured in Figure 15.10(a). There, the change in true av-

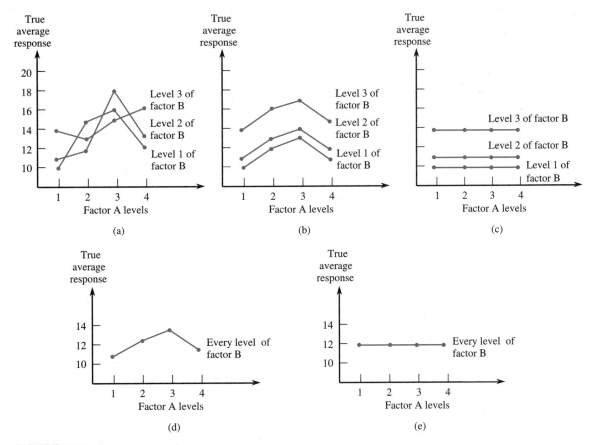

FIGURE 15.10 Some graphs of true average responses

erage response when the level of B is changed (a vertical distance) depends on the level of A. An analogous property would hold if the picture were redrawn so that levels of B were marked on the horizontal axis. This is a prototypical picture suggesting **interaction** between the factors — the change in true average response when the level of one factor changes depends on the level of the other factor.

There is no interaction between the factors when the connected line segments are parallel, as in Figure 15.10(b). Then the change in true average response when the level of one factor changes is the same for each level of the other factor (the vertical distances are the same for each level of factor A). Figure 15.10(c) illustrates an even more restrictive situation — there is no interaction between factors, and, in addition, the true average response does not depend on the level of factor A. Only when the graph looks like this can it be said that factor A has no effect on the responses. Similarly, the graph in Figure 15.10(d) indicates no interaction and no dependence on the level of factor B. A final case, pictured in Figure 15.10(e), shows a single set of four points connected by horizontal line segments, which indicates that the true average response is identical for every level of either factor.

If the graphs of true average responses are connected line segments that are parallel, there is no interaction between the factors. In this case, the change in true average response when the level of one factor is changed is the same for each level of the other factor. Special cases of no interaction are as follows:

1. The true average response is the same for each level of factor A (no factor A main effects).
2. The true average response is the same for each level of factor B (no factor B main effects).

The graphs of Figure 15.10 depict true average responses, quantities whose values are fixed but unknown to an investigator. Figure 15.8 contains graphs of the sample average responses based on data resulting from an experiment. These sample averages are, of course, subject to variability because there is sampling variation in the individual observations. If the experiment discussed in Example 15.12 were repeated, the resulting graphs of sample averages would probably look somewhat different from the graph in Figure 15.8 — perhaps a great deal different if there were substantial underlying variability in responses. Even when there is no interaction among factors, the connected line segments in the sample mean picture will not typically be exactly parallel, and they may deviate quite a bit from parallelism in the presence of substantial underlying variability. Similarly, there might actually be no factor A effects (Figure 15.10(c)), yet the sample graphs would not usually be exactly horizontal. The sample pictures give us insight, but formal inferential procedures are necessary to draw sound conclusions about the nature of the true average responses for different factor levels.

Hypotheses and F Tests

Basic Assumptions for Two-Factor ANOVA

The observations on any particular treatment are independently selected from a normal distribution with variance σ^2 (the same variance for each treatment), and samples from different treatments are independent of one another.

Because of the normality assumption, tests based on F statistics and distributions are appropriate. The necessary sums of squares result from breaking up $SSTo = \Sigma (x - \bar{\bar{x}})^2$ into four parts, which reflect random variation and variation attributable to various factor effects:

$$SSTo = SSA + SSB + SSAB + SSE$$

where

1. SSTo is total sum of squares, with associated df $= klm - 1$.
2. SSA is the factor A main effect sum of squares, with associated df $= k - 1$.

3. SSB is the factor B main effect sum of squares, with associated df $= l - 1$.

4. SSAB is the interaction sum of squares, with associated df $= (k - 1)(l - 1)$.

5. SSE is error sum of squares, with associated df $= kl(m - 1)$.

The formulas for these sums of squares are similar to those given in previous sections, but we will not give them here. The standard statistical computer packages will calculate all sums of squares and other necessary quantities. The magnitude of SSE is related entirely to the amount of underlying variability (as specified by σ^2) in the distributions being sampled. It has nothing to do with values of the various true average responses. SSAB reflects in part underlying variability, but its value is also affected by whether there is interaction between the factors. Generally speaking, the more extensive the amount of interaction (that is, the further the graphs of true average responses are from being parallel), the larger the value of SSAB tends to be. The test statistic for testing the null hypothesis that there is no interaction between factors is the ratio MSAB/MSE. A large value of this statistic suggests that interaction effects are present.

The absence of factor A effects and of factor B effects are both special cases of no-interaction situations. If the data suggests that interaction is present, it does not make sense to investigate effects of one factor without reference to the other factor. Our recommendation is that hypotheses concerning the presence or absence of separate factor effects be tested only if the hypothesis of no interaction is not rejected. Then the factor A main effect sum of squares SSA will reflect random variation and, in addition, any differences between true average responses for different levels of factor A. The same applies to SSB.

Two-Factor ANOVA Hypotheses and Tests

1. H_0: There is no interaction between factors.

   H_a: There is interaction between factors.

   Test statistic: $F_{AB} = \dfrac{\text{MSAB}}{\text{MSE}}$, based on $df_1 = (k - 1)(l - 1)$ and $df_2 = kl(m - 1)$.

   The following two hypotheses should be tested only if the hypothesis of no interaction is not rejected.

2. H_0: There are no factor A main effects (true average response is the same for each level of factor A).

   H_a: H_0 is not true.

   Test statistic: $F_A = \dfrac{\text{MSA}}{\text{MSE}}$, based on $df_1 = k - 1$ and $df_2 = kl(m - 1)$.

3. H_0: There are no factor B main effects.

   H_a: H_0 is not true.

   Test statistic: $F_B = \dfrac{\text{MSB}}{\text{MSE}}$, based on $df_1 = l - 1$ and $df_2 = kl(m - 1)$.

Computations are typically summarized in an ANOVA table, as shown in Table 15.11.

Table 15.11

Source of Variation	df	Sum of Squares	Mean Square	F
A main effects	$k - 1$	SSA	$MSA = \dfrac{SSA}{k-1}$	$F = \dfrac{MSA}{MSE}$
B main effects	$l - 1$	SSB	$MSB = \dfrac{SSB}{l-1}$	$F = \dfrac{MSB}{MSE}$
AB interaction	$(k-1)(l-1)$	SSAB	$MSAB = \dfrac{SSAB}{(k-1)(l-1)}$	$F = \dfrac{MSAB}{MSE}$
Error	$kl(m-1)$	SSE	$MSE = \dfrac{SSE}{kl(m-1)}$	
Total	$klm - 1$	SSTo		

EXAMPLE 15.13   An ANOVA table for the tomato-yield data of Example 15.12 is given in Table 15.12.

Table 15.12 *Analysis of variance on yield*

Source	df	Sum of Squares	Mean Square	F
Variety	2	327.60	163.80	103.70
Density	3	86.69	28.90	18.30
Interaction	6	8.03	1.34	.85
Error	24	38.04	1.58	
Total	35	460.36		

1. Test of H_0: no interaction between variety and density:

   calculated $F_{AB} = .85$, based on $df_1 = 6$, $df_2 = 24$

   From Appendix Table VII, the smallest value for these df's is 2.04, so P-value $>$ .10. There is no evidence of interaction, so it is appropriate to carry out further tests concerning the presence of main effects.

2. Test of H_0: factor A (variety) main effects are absent:

   $F_A = 103.7$, based on $df_1 = 2$, $df_2 = 24$

   Appendix Table VII shows that P-value $< .001$. We therefore reject H_0 and conclude that true average yield does depend on variety.

3. Test of H_0: factor B (density) main effects are absent:

   $F_B = 18.3$, based on $df_1 = 3$, $df_2 = 24$

Again, P-value $< .001$, so we reject H_0 and conclude that true average yield does depend on planting density.

After the null hypothesis of no factor A main effects has been rejected, significant differences in factor A levels can be identified by using a multiple comparisons procedure. In particular, the Tukey–Kramer method described previously can be applied. The quantities $\bar{x}_1, \bar{x}_2, \ldots, \bar{x}_k$ are now the sample average responses for levels $1, \ldots, k$ of factor A, and error df is $kl(m - 1)$. A similar comment applies to factor B main effects and significant differences in factor B levels.

The Case $m = 1$

There is a problem with the foregoing analysis when $m = 1$ (one observation on each treatment). Although we did not give the formula, MSE is an estimate of σ^2 obtained by computing a separate sample variance s^2 for the m observations on each treatment and then averaging these kl sample variances. With only one observation on each treatment, there is no way to estimate σ^2 separately from each of the treatments.

One way out of this dilemma is to assume a priori that there is no interaction between factors. This should, of course, be done only when the investigator has sound reasons, based on a thorough understanding of the problem, for believing that the factors contribute separately to the response. Having made this assumption, what would otherwise be interaction sum of squares can then be used for SSE. The fundamental identity becomes

$$\text{SSTo} = \text{SSA} + \text{SSB} + \text{SSE}$$

with the four associated df $kl - 1$, $k - 1$, $l - 1$, and $(k - 1)(l - 1)$.

Table 15.13 gives the corresponding ANOVA table. F_A is the test statistic for testing the null hypothesis that true average responses are identical for all factor A levels. F_B plays a similar role for factor B main effects. The analysis of data from a randomized block experiment in fact assumed no interaction between treatments

Table 15.13 *ANOVA table for two-factor experiment with $m = 1$*

Source of Variation	df	Sum of Squares	Mean Square	F
Factor A	$k - 1$	SSA	$\text{MSA} = \dfrac{\text{SSA}}{k - 1}$	$F = \dfrac{\text{MSA}}{\text{MSE}}$
Factor B	$l - 1$	SSB	$\text{MSB} = \dfrac{\text{SSB}}{l - 1}$	$F = \dfrac{\text{SSB}}{\text{MSE}}$
Error	$(k - 1)(l - 1)$	SSE	$\text{MSE} = \dfrac{\text{SSE}}{(k - 1)(l - 1)}$	
Total	$kl - 1$	SSTo		

and blocks. If SSTr is relabeled SSA and SSBl is relabeled SSB, the formulas for all sums of squares given in Section 15.3 are valid here ($\bar{x}_1, \ldots, \bar{x}_k$ and $\bar{b}_1, \ldots, \bar{b}_l$ are now the sample average responses for factor A levels and factor B levels, respectively).

EXAMPLE 15.14   When metal pipe is buried in soil, it is desirable to apply a coating to retard corrosion. Four different coatings are under consideration for use with pipe that will ultimately be buried in three types of soil. An experiment to investigate the effects of these coatings and soils was carried out by first selecting 12 pipe segments and applying each coating to three segments. The segments were then buried in soil for a specified period in such a way that each soil type received one piece with each coating. The resulting data (depth of corrosion) and ANOVA table are given in Table 15.14. Assuming that there is no interaction between coating type and soil type, let's test at level .05 for the presence of separate factor A (coating) and factor B (soil) effects.

Table 15.14 *Data and ANOVA table for Example 15.14*

		Factor B (Soil)			Sample Average
		1	**2**	**3**	
Factor A (Coating)	1	64	49	50	54.33
	2	53	51	48	50.67
	3	47	45	50	47.33
	4	51	43	52	48.67
Sample average		53.75	47.00	50.00	$\bar{\bar{x}} = 50.25$

Source of Variation	df	Sum of Squares	Mean Square	F
Factor A	3	83.5	27.8	$F_A = \dfrac{27.8}{20.6} = 1.3$
Factor B	2	91.5	45.8	$F_B = \dfrac{45.8}{20.6} = 2.2$
Error	6	123.3	20.6	
Total	11	298.3		

Appendix Table VII shows that *P*-value > .10 for both tests. It appears that the true average response (amount of corrosion) depends neither on the coating used nor on the type of soil in which pipe is buried.

Exercises 15.39 – 15.51

15.39 The behavior of undergraduate students when exposed to various odors was examined by the authors of the article "Effects of Environmental Odor and Coping Style on Negative Affect, Anger, Arousal and Escape" (*J. Applied Social Psychology* (1999): 245–260). The following table was constructed using data on reported discomfort level (measured on a scale from 1 to 5). There were 24 students in each odor–gender combination.

Type of Odor	Male $\bar{x}$	Female $\bar{x}$
No odor	1.36	1.68
Rotten egg	1.83	2.33
Skunk	2.42	2.69
Cigarette ash	2.74	3.16

a. Construct a graph (similar to those of Figure 15.8) that shows the mean discomfort level on the vertical axis. Mark the four odor categories on the horizontal axis. Then plot the four means for the males and connect them with line segments. Plot the four means for females and connect them with line segments.
b. Interpret the interaction plot. Do you think that there is an interaction between gender and type of odor?

15.40 Many students report that test anxiety affects their performance on exams. A study of the effect of anxiety and instructional mode (lecture versus independent study) on test performance was described in the article "Interactive Effects of Achievement Anxiety, Academic Achievement, and Instructional Mode on Performance and Course Attitudes" (*Home Ec. Research J.* (1980): 216–227). Students classified as belonging to either high- or low-achievement anxiety groups (factor A) were assigned to one of the two instructional modes (factor B). Mean test scores for the four treatments (factor level combinations) are given in the accompanying table. Use these means to construct a graph (similar to those of Figure 15.8) of the treatment sample averages. Does the picture suggest the existence of an interaction between factors? Explain.

Anxiety Group	Instructional Mode	
	Lecture	Independent Study
High	145.8	144.3
Low	142.9	144.8

15.41 The accompanying plot appeared in the article "Group Process-Work Outcome Relationships: A Note

on the Moderating Impact of Self-Esteem" (*Academy of Mgmt. J.* (1982): 575–585). The response variable was tension, a measure of job strain. The two factors of interest were peer-group interaction (with two levels — high and low) and self-esteem (also with two levels — high and low).

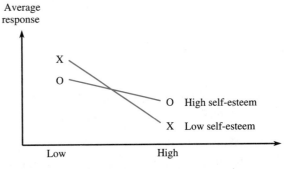

a. Does this plot suggest an interaction between peer-group interaction and self-esteem? Explain.
b. The authors of the article state, "Peer group interaction had a stronger effect on individuals with lower self-esteem than on those with higher self-esteem." Do you agree with this statement? Explain.

15.42 Explain why the individual effects of factor A or factor B cannot be interpreted when an AB interaction is present.

15.43 The article "Experimental Analysis of Prey Selection by Largemouth Bass" (*Trans. Amer. Fisheries Soc.* (1991): 500–508) gave an ANOVA summary in which the response variable was a certain preference index, there were three sizes of bass, and there were two different species of prey. Three observations were made for each size–species combination. Sums of squares for size, species, and interaction were reported as .088, .048, and .048, respectively, and SSTo = .316. Test the relevant hypotheses using a significance level of .01.

15.44 The following partially completed ANOVA table approximately matches summary statistics given in the article "From Here to Equity: The Influence of Status on Student Access to and Understanding of Science" (*Science Education* (1999): 577–602). The study described in this article attempted to quantify the effect of socioeconomic status on learning science. The response variable was "Rate of Talk" (the number of on-task talk speech acts per minute) during group work. Data was also collected on the variable *socioeconomic status* (low, middle, high) and the variable *gender* (female, male). The author allowed for an interaction between gender and status. Assume that there were 12 students at each status–gender level combination, for a total of 72 subjects.

Source	df	SS	MS	F
Status	2			14.49
Gender	1			.15
Interaction	2			.95
Error	66		.0120	
Total	71			

a. Fill in the missing numbers in the ANOVA table.
b. Is there a significant interaction between status and gender?
c. Is there a difference between the mean "Rate of Talk" scores for girls and boys?
d. Is there a difference between the mean "Rate of Talk" scores across the three different status groups?

15.45 Is self-esteem related to year in college or to fraternity membership? The authors of the article "Self-Esteem Among College Men as a Function of Greek Affiliation and Year in College" (*J. College Student Development* (1998): 611–613) interviewed 75 college men who were members of fraternities and 57 who were not. In addition to Greek status (in a fraternity, not in a fraternity), year in college (freshman, sophomore, junior, senior) was also noted. Subjects completed a 10-item, self-esteem inventory, which was used to compute a self-esteem score. The article reported the following *F* statistic values from a two-way ANOVA:

Source	F
Greek status	20.53
Year	2.59
Interaction	.70

a. Is there evidence of a significant interaction between year and Greek status?
b. Is the main effect for Greek status significant?
c. Is the main effect for year significant?

15.46 Three ultrasonic devices (factor A, with levels 20, 30, and 40 kHz) were tested for effectiveness under two test conditions (factor B, with levels plentiful food supply and restricted food supply). Daily food consumption was recorded for three rats under each factor–level combination for a total of 18 observations. Data compatible with summary values given in the article "Variables Affecting Ultrasound Repellency in Philippine Rats" (*J. Wildlife Mgmt.* (1982): 148–155) was used to obtain the sums of squares given in the accompanying ANOVA table. Complete the table, and use it to test the relevant hypotheses.

Source of Variation	df	Sum of Squares	Mean Square	F
A main effects		4206		
B main effects		1782		
AB interaction				
Error		2911		
Total		10,846		

15.47 The article "Learning, Opportunity to Cheat, and Amount of Reward" (*J. Exper. Educ.* (1977): 30–40) described a study to determine the effects of expectations concerning cheating during a test and perceived payoff on test performance. Subjects, students at UCLA, were randomly assigned to a particular factor–level combination. Factor A was expectation of opportunity to cheat, with levels high, medium, and low. Those in the high group were asked to study and then recall a list of words. For the first four lists, they were left alone in a room with the door closed, so they could look at the original list of words if they wanted to. The medium group was asked to study and recall the list while left alone but with the door open. For the low group, the experimenter remained in the room. For study and recall of a fifth list, the experimenter stayed in the room for all three groups, thus precluding any cheating on the fifth list. Score on the fifth test was the response variable. The second factor (B) under study was the perceived payoff, with a high and a low level. The high payoff group was told that if they scored above average on the test, they would receive 2 hours of credit rather than just 1 hour (subjects were fulfilling a course requirement by participating in experiments). The low group was not given any extra incentive for scoring above the average. The article gave the following statistics: $F_A = 4.99$, $F_B = 4.81$, $F_{AB} < 1$, error df ≈ 120. Test the null hypothesis of no interaction between the factors. If appropriate, test the null hypotheses of no factor A and no factor B effects. Use $\alpha = .05$.

15.48 The accompanying (slightly modified) ANOVA table appeared in the article "An Experimental Test of Mate Defense in an Iguanid Lizard" (*Ecology* (1991): 1218–1224). The response variable was territory size.

Source of Variation	df	SS
Age	1	.614
Sex	1	1.754
Interaction	1	.146
Error	80	5.624

a. How many age classes were there?
b. How many observations were made for each age–sex combination?

c. What conclusions can be drawn about how the factors affect the response variable?

15.49 Identification of gender in human skeletons is an important part of many anthropological studies. An experiment conducted to determine whether measurements of the sacrum could be used to determine gender was described in the article "Univariate and Multivariate Methods for Sexing the Sacrum" (*Am. J. Phys. Anthro.* (1978): 103–110). Sacra from skeletons of individuals of known race (factor A, with two levels — Caucasian and black) and gender (factor B, with two levels — male and female) were measured and the lengths recorded. Data compatible with summary quantities given in the article was used to compute the following: SSA = 857, SSB = 291, SSAB = 32, SSE = 5541, and error df = 36.

   a. Use a significance level of .01 to test the null hypothesis of no interaction between race and gender.

   b. Using a .01 significance level, test to determine whether the true average length differs for the two races.

   c. Using a .01 significance level, test to determine whether the true average length differs for males and females.

15.50 The article "Food Consumption and Energy Requirements of Captive Bald Eagles" (*J. Wildlife Mgmt.* (1982): 646–654) investigated mean gross daily energy intake (the response variable) for different diet types (factor A, with three levels — fish, rabbit, and mallard) and temperatures (factor B, with three levels). Summary quantities given in the article were used to generate data, resulting in SSA = 18,138, SSB = 5182, SSAB = 1737, SSE = 11,291, and error df = 36. Construct an ANOVA table and test the relevant hypotheses.

15.51 The effect of three different soil types and three phosphate application rates on total phosphorus uptake (mg) of white clover was examined in the article "A Glasshouse Comparison of Six Phosphate Fertilisers" (*N. Zeal. J. Exp. Ag.* (1984): 131–140). Only one observation was obtained for each factor–level combination. Assuming there is no interaction between soil type and application rate, use the accompanying data to construct an ANOVA table and to test the null hypotheses of no main effects due to soil type and of no main effects due to application rate. Also identify significant differences among soil types.

Application Rate (kgP/ha)	Soil Type		
	Ramiha	Konini	Wainui
0	1.29	10.42	17.10
75	11.73	21.08	23.69
150	17.63	31.37	32.88

15.5 Interpreting the Results of Statistical Analyses

The analysis of variance procedures introduced in this chapter are used to compare more than two population or treatment means. When a single-factor analysis of variance has been used to test the null hypothesis of equal population or treatment means, the value of the *F* statistic and the associated *P*-value usually are reported. It is also fairly common to see the supporting calculations summarized in an ANOVA table, although this is not always the case.

What to Look For in Published Data

Here are some questions to ask when reading an article that includes a description of a single-factor analysis of variance.

   ■ Are the assumptions required for the validity of the ANOVA procedure reasonable? Specifically, are the samples independently chosen, or is there random assignment of treatments? Is it reasonable to think that the population or treatment response distributions are normal in shape? Are the reported sample standard deviations consistent with the assumption of equal population or treatment variances?

   ■ What is the *P*-value associated with the test? Does the *P*-value lead to rejection of the null hypothesis?

- If the ANOVA F test led to rejection of the null hypothesis, was a multiple comparisons procedure used to identify differences in the means? Are the results of the multiple comparisons procedure interpreted properly?
- Are the conclusions drawn consistent with the results of the hypothesis test and the multiple comparisons procedure? If H_0 was rejected, does this indicate practical significance or only statistical significance?

As an example, consider an analysis of variance performed by the authors of "Perceived Age and Attractiveness of Models in Cigarette Advertisements" (*J. of Marketing* (1992): 22–37). The authors hypothesized that certain cigarette brands aimed their advertising at the young and that the average age of readers of magazines in which the cigarette ads appeared was not the same for the 12 brands of cigarettes studied. This hypothesis was tested using a single-factor analysis of variance. The accompanying ANOVA table is from the article.

Source	df	SS	MS	F
Treatments (brands)	11	592.66	53.88	4.13
Error	175	2282.98	13.05	
Total	186	2875.64		

It was reported that the hypothesis of equal mean ages for the 12 brands considered in the study was rejected ($F = 4.13$, P-value $< .001$). This was interpreted as support for the researchers' hypothesis that some brands targeted the young. (Does this conclusion necessarily follow from the result of the ANOVA F test?)

To clarify these results, a multiple comparisons procedure could be used to identify significant differences among brands, although this was not done in the reported research. The article did include sample sizes and sample means for the 12 brands, and this information is reproduced in the accompanying table.

Brand	Sample Size	Average Audience Age for Magazines in Which Ad Appears (Years)
Lucky Strike Lights	3	28.5
Newport Lights	5	31.1
Camel Lights	21	31.3
Kool Milds	24	31.3
Newport	12	31.3
Winston Lights	15	31.3
Winston	8	31.6
Virginia Slims Lights	32	33.9
Marlboro	26	34.3
More	17	34.8
Benson Hedges Light	23	31.9
Carlton	1	41.0

Notice that the sample sizes vary greatly. We might also worry about the validity of the assumptions required for the ANOVA F test. It would be of interest to examine the sample standard deviations and, if the original data was available, to look at various plots of the data.

Summary of Key Concepts and Formulas

Term or Formula	Comment
Single-factor analysis of variance (ANOVA)	A test procedure for determining whether there are significant differences among k population or treatment means. The hypotheses tested are $H_0: \mu_1 = \cdots = \mu_k$ versus H_a: at least two μ's differ.
Treatment sum of squares: $\text{SSTr} = n_1(\bar{x}_1 - \bar{\bar{x}})^2 + \cdots + n_k(\bar{x}_k - \bar{\bar{x}})^2$	A measure of how different the k sample means $\bar{x}_1, \bar{x}_2, \ldots, \bar{x}_k$ are from one another; associated df $= k - 1$.
Error sum of squares: $\text{SSE} = (n_1 - 1)s_1^2 + \cdots + (n_k - 1)s_k^2$	A measure of the amount of variability within the individual samples; associated df $= N - k$, where $N = n_1 + \cdots + n_k$.
Mean square	A sum of squares divided by its df. For single-factor ANOVA, $\text{MSTr} = \text{SSTR}/(k - 1)$ and $\text{MSE} = \text{SSE}/(N - k)$.
$F = \dfrac{\text{MSTr}}{\text{MSE}}$	The test statistic for testing $H_0: \mu_1 = \cdots = \mu_k$ in a single-factor ANOVA. When H_0 is true, F has an F distribution with numerator df $= k - 1$ and denominator df $= N - k$.
$\text{SSTo} = \text{SSTr} + \text{SSE}$	The fundamental identity in single-factor ANOVA, where SSTo = total sum of squares $= \Sigma (x - \bar{\bar{x}})^2$.
Tukey–Kramer multiple comparison procedure	A procedure for identifying significant differences among the μ's once the hypothesis $H_0: \mu_1 = \cdots = \mu_k$ has been rejected by the ANOVA F test.
Randomized block design	An experimental design that controls for extraneous variation when comparing treatments. The experimental units are grouped into homogeneous *blocks* so that within each block, the units are as similar as possible. Then each treatment is used on exactly one experimental unit in every block (each treatment appears once in every block).
Randomized block F test	The four sums of squares for a randomized block design — SSTo, SSTr, SSBl, and SSE (with df $kl - 1$, $k - 1$, $l - 1$, and $(k - 1)(l - 1)$, respectively) — are related by $\text{SSTo} = \text{SSTr} + \text{SSBl} + \text{SSE}$. Usually SSE is obtained by subtraction once the other three have been calculated using computing formulas. The null hypothesis is that the true average response does not depend on which treatment is applied. With mean squares $\text{MSTr} = \text{SSTr}/(k - 1)$ and $\text{MSE} = \text{SSE}/(k - 1)(l - 1)$, the test statistic is $F = \text{MSTr}/\text{MSE}$, based on $\text{df}_1 = k - 1$ and $\text{df}_2 = (k - 1)(l - 1)$.
Interaction between factors	Two factors are said to interact if the average change in response associated with changing the level of one factor depends on the level of the other factor.
Two-factor ANOVA	When there are k levels of factor A and l levels of factor B, and m (>1) observations made for each combination of A–B levels, total sum of squares SSTo can be decomposed into SSA (sum of squares for A main effects), SSB, SSAB

(interaction sum of squares), and SSE. Associated df are $klm - 1, k - 1,$ $l - 1, (k - 1)(l - 1),$ and $kl(m - 1),$ respectively. The null hypothesis of no interaction between the two factors is tested using $F_{AB} = MSAB/MSE,$ where $MSAB = SSAB/(k - 1)(l - 1)$ and $MSE = SSE/kl(m - 1).$ If this null hypothesis cannot be rejected, tests for A and B main effects are based on $F_A = MSA/MSE$ and $F_B = MSB/MSE,$ respectively.

APPENDIX

ANOVA Computations
Single-Factor ANOVA

Let T_1 denote the sum of the n_1 observations in the sample from the first population or treatment, and let $T_2, \ldots, T_k$ denote the other sample totals. Also let T represent the sum of all N observations — the **grand total** — and

$$CF = \text{correction factor} = \frac{T^2}{N}$$

Then

$$SSTo = \sum_{\substack{\text{all } N \\ \text{obs.}}} x^2 - CF$$

$$SSTr = \frac{T_1^2}{n_1} + \frac{T_2^2}{n_2} + \cdots + \frac{T_k^2}{n_k} - CF$$

$$SSE = SSTo - SSTr$$

EXAMPLE

	1	4.2	3.7	5.0	4.8	$T_1 = 17.7$	$n_1 = 4$
Treatment	**2**	5.7	6.2	6.4		$T_2 = 18.3$	$n_2 = 3$
	3	4.6	3.2	3.5	3.9	$T_3 = 15.2$	$n_3 = 4$
						$T = 51.2$	$N = 11$

$$CF = \text{correction factor} = \frac{T^2}{N} = \frac{(51.2)^2}{11} = 238.31$$

$$SSTr = \frac{T_1^2}{n_1} + \frac{T_2^2}{n_2} + \cdots + \frac{T_k^2}{n_k} - CF$$

$$= \frac{(17.7)^2}{4} + \frac{(18.3)^2}{3} + \frac{(15.2)^2}{4} - 238.31 = 9.40$$

$$SSTo = \sum_{\substack{\text{all } N \\ \text{obs.}}} x^2 - CF = (4.2)^2 + (3.7)^2 + \cdots + (3.9)^2 - 238.31 = 11.81$$

$$SSE = SSTo - SSTr = 11.81 - 9.40 = 2.41$$

Randomized Block Experiment

Let $T_1, T_2, \ldots, T_k$ denote the treatment totals and $B_1, B_2, \ldots, B_l$ represent the block totals. Also, let T be the grand total of all kl observations and

$$\text{CF} = \text{correction factor} = \frac{T^2}{kl}$$

Then

$$\text{SSTo} = \sum_{\substack{\text{all } kl \\ \text{obs.}}} x^2 - \text{CF}$$

$$\text{SSTr} = \frac{1}{l}[T_1^2 + T_2^2 + \cdots + T_k^2] - \text{CF}$$

$$\text{SSBl} = \frac{1}{k}[B_1^2 + B_2^2 + \cdots + B_l^2] - \text{CF}$$

$$\text{SSE} = \text{SSTo} - \text{SSTr} - \text{SSBl}$$

EXAMPLE

		Block				
		1	**2**	**3**	**4**	
	1	4.2	3.7	5.0	4.8	$T_1 = 17.7$
Treatment **2**		5.2	4.5	6.7	5.4	$T_2 = 21.8$
	3	3.4	3.2	5.1	3.9	$T_3 = 15.6$
		$B_1 = 12.8$	$B_2 = 11.4$	$B_3 = 16.8$	$B_4 = 14.1$	$T = 55.1$

$$\text{CF} = \text{correction factor} = \frac{T^2}{kl} = \frac{(55.1)^2}{(3)(4)} = \frac{3036.01}{12} = 253.00$$

$$\text{SSTr} = \frac{1}{l}[T_1^2 + T_2^2 + \cdots + T_k^2] - \text{CF}$$

$$= \frac{1}{4}[(17.7)^2 + (21.8)^2 + (15.6)^2] - 253.00 = 4.97$$

$$\text{SSBl} = \frac{1}{k}[B_1^2 + B_2^2 + \cdots + B_l^2] - \text{CF}$$

$$= \frac{1}{3}[(12.8)^2 + (11.4)^2 + (16.8)^2 + (14.1)^2] - 253.00 = 5.28$$

$$\text{SSTo} = \sum_{\substack{\text{all } kl \\ \text{obs.}}} x^2 - \text{CF} = (4.2)^2 + \cdots + (3.9)^2 - 253.00 = 10.73$$

$$\text{SSE} = \text{SSTo} - \text{SSTr} - \text{SSBl} = 10.73 - 4.97 - 5.28 = .48$$

Supplementary Exercises 15.52 – 15.65

15.52 A study was carried out to compare aptitudes and achievements of three different groups of college students ("A Comparison of Three Groups of Academically At-Risk College Students," *J. of College Student Development* (1995): 270–279):

1. Students diagnosed as learning disabled
2. Students who identified themselves as learning disabled
3. Students who were low achievers

The Scholastic Abilities Test for Adults was given to each student. Consider the following summary data on writing composition score:

$$n_1 = 30 \qquad n_2 = 30 \qquad n_3 = 30$$
$$\bar{x}_1 = 9.40 \qquad \bar{x}_2 = 11.63 \qquad \bar{x}_3 = 11.00$$
$$SSE = 749.85$$

Does it appear that population mean score is not the same for the three types of students? Carry out a test of hypothesis. Does your conclusion depend on whether a significance level of .05 or .01 is used?

15.53 Tobacco companies have denied engaging in any youth-directed marketing efforts, yet critics have observed that adolescents are attracted by the images portrayed in many advertisements. The article "Perceived Age and Attractiveness of Models in Cigarette Advertisements" (*J. of Marketing* (1992): 22–37) reported on the results of a study that focused on certain aspects of this issue. Consider the accompanying data on average median audience ages for readers of magazines in which a sample of cigarette advertisements appeared (obtained from marketing data):

Brand	Sample Size	Average Audience Age
Lucky Strike Lights	3	28.5
Newport Lights	5	31.1
Camel Lights	21	31.3
Kool Milds	24	31.3
Newport	12	31.3
Winston Lights	15	31.3
Winston	8	31.6
Virginia Slims Lights	32	33.9
Marlboro	26	34.3
More	17	34.8
Benson Hedges Lights	23	35.1
Carlton	1	41.0
	187	$\bar{\bar{x}} = 32.96$

Let $\mu_1, \mu_2, \ldots, \mu_{12}$ denote the average audience ages for magazines in which advertisements for the various brands appeared. Test $H_0: \mu_1 = \mu_2 = \cdots = \mu_{12}$ versus H_a: At least two μ's are different, by first filling in the missing entries in the accompanying ANOVA table.

Source	df	SS	MS	F
Treatments (brands)		592.66		
Error				
Total		2875.64		

15.54 Eye inflammation can be induced by the endotoxin lipopolysaccharide (LPS). A random sample of 35 rats was randomly divided into five groups of 7 rats each (a completely randomized design). Rats within each group received the same dose of LPS. Data on vascular permeability was read from plots that appeared in the article "Endotoxin-Induced Uveitis in the Rat: Observations on Altered Vascular Permeability, Clinical Findings, and Histology" (*Exper. Eye Research* (1984): 665–676). Use the accompanying SAS output to test the null hypothesis of no difference between the five treatment (dose) means.

Dependent Variable: VASPERM

Source	DF	Sum of Squares	Mean Square	F Value	Pr > F
Model	4	291.0428571	72.7607143	8.04	0.0002
Error	30	270.5000000	9.0166667		
Corrected Total	34	561.5428571			

15.55 Most large companies have established grievance procedures for their employees. One question of interest to employers is why certain groups within a company have higher grievance rates than others. The study described in the article "Grievance Rates and Technology" (*Academy of Mgmt. J.* (1979): 810–815) distinguished four types of jobs. These types were labeled apathetic, erratic, strategic, and conservative. Suppose that a total of 52 work groups were selected (13 of each type), and a measure of grievance rate was determined for each one. Here are the resulting $\bar{x}$ values:

Group	Apathetic	Erratic	Strategic	Conservative
Sample mean	2.96	5.05	8.74	4.91

In addition, SSTo = 682.10 and SSE = 506.19. Test at significance level .05 to see whether true average grievance rate depends on the type of job. If appropriate, use the T–K method to identify significant differences among job types.

15.56 The results of a study on the effectiveness of line drying on the smoothness and stiffness of fabric was summarized in the article "Line-Dried vs. Machine-Dried Fabrics: Comparison of Appearance, Hand, and Consumer Acceptance" (*Home Ec. Res. J.* (1984): 27–35). Smoothness scores were given for nine different types of fabric and five different drying methods — (1) machine dry, (2) line dry, (3) line dry followed by 15-min tumble, (4) line dry with softener, and (5) line dry with air movement — as given in the accompanying table. Using a .05 significance level, test to see whether there is a difference in the true mean smoothness scores for the drying methods. (There may be "extraneous" variation due to different fabrics.)

Fabric	Drying Method				
	1	**2**	**3**	**4**	**5**
Crepe	3.3	2.5	2.8	2.5	1.9
Doubleknit	3.6	2.0	3.6	2.4	2.3
Twill	4.2	3.4	3.8	3.1	3.1
Twill mix	3.4	2.4	2.9	1.6	1.7
Terry	3.8	1.3	2.8	2.0	1.6
Broadcloth	2.2	1.5	2.7	1.5	1.9
Sheeting	3.5	2.1	2.8	2.1	2.2
Corduroy	3.6	1.3	2.8	1.7	1.8
Denim	2.6	1.4	2.4	1.3	1.6

15.57 In many countries, grains and cereals are the primary food source. The authors of the article "Mineral Contents of Cereal Grains as Affected by Storage and Insect Infestation" (*J. Stored Prod. Res.* (1992): 147–151) investigated the effects of storage period on the mineral content of maize. Four storage periods were considered: 0 months (no storage), 1 month, 2 months, and 4 months. Twenty-four containers of maize were randomly divided into four groups of six each. The iron content (mg/100 g dry weight) of the first group of six was measured immediately (0 months of storage), the second six were measured after 1 month in storage, etc. The following summary quantities are consistent with information in the article:

Storage Period	$\bar{x}$	s^2
0	4.923	.000107
1	4.923	.000067
2	4.917	.000147
4	4.902	.000057

a. Use a test with $\alpha = .05$ to decide whether true average iron content is the same for all four storage periods.
b. If appropriate, carry out a multiple comparison analysis.

15.58 Controlling a filling operation with multiple fillers requires adjustment of the individual units. Data resulting from a sample of size 5 from each pocket of a 12-pocket filler was given in the article "Evaluating Variability of Filling Operations" (*Food Technology* (1984): 51–55). Data for the first five pockets is given in the accompanying table.

Pocket	Fill (oz)				
1	10.2	10.0	9.8	10.4	10.0
2	9.9	10.0	9.9	10.1	10.0
3	10.1	9.9	9.8	9.9	9.7
4	10.0	9.7	9.9	9.7	9.6
5	10.2	9.8	9.9	9.7	9.8

Use the ANOVA *F* test to determine whether the null hypothesis of no difference in the mean fill weight of the five pockets can be rejected. If so, use an appropriate technique to determine where the differences lie.

15.59 *This problem requires the use of a computer package.* The effect of oxygen concentration on fermentation end products was examined in the article "Effects of Oxygen on Pyruvate Formate Lyase in Situ and Sugar Metabolism of *Streptococcus mutans* and *Streptococcus samguis*" (*Infection and Immunity* (1985): 129–134). Four oxygen concentrations (0, 46, 92, and 138 μM) and two types of sugar (galactose and glucose) were used. The accompanying table gives two observations on amount of ethanol (μmol/mg) for each sugar–oxygen concentration combination. Construct an ANOVA table and test the relevant hypotheses.

Oxygen Concentration	Galactose	Glucose
0	.59, .30	.25, .03
46	.44, .18	.13, .02
92	.22, .23	.07, .00
138	.12, .13	.00, .01

15.60 The ANOVA table on page 804 is from the article "Bacteriological and Chemical Variations and Their Inter-Relationships in a Slightly Polluted Water-Body" (*Inter. J. Environ. Studies* (1984): 121–129). A water specimen was taken every month for a year at each of 15 designated locations on the Lago di Piediluco in Italy. The ammonia-nitrogen concentration was determined for each specimen and the resulting data analyzed using a two-way ANOVA. The researchers were willing to assume that there was no interaction between the two factors *location* and *month*. Complete the given ANOVA table, and use it to perform the tests required to determine whether the true mean concentration differs by location or by month of year. Use a .05 significance level for both tests.

Source	df	Sum of Squares	Mean Square	F
Location		.6		
Month	11	2.3		
Error				
Total	179	6.4		

15.61 The water absorption of two types of mortar used to repair damaged cement was discussed in the article "Polymer Mortar Composite Matrices for Maintenance-Free, Highly Durable Ferrocement" (*J. Ferrocement* (1984): 337–345). Specimens of ordinary cement mortar (OCM) and polymer cement mortar (PCM) were submerged for varying lengths of time (5, 9, 24, or 48 hr) and water absorption (% by weight) was recorded. With mortar type as factor A (with two levels) and submersion period as factor B (with four levels), three observations were made for each factor–level combination. Data included in the article was used to compute the following sums of squares: SSA = 322.667, SSB = 35.623, SSAB = 8.557, SSTo = 372.113. Use this information to construct an ANOVA table, and then use a .05 significance level to test the appropriate hypotheses.

15.62 Suppose that each observation in a single-factor ANOVA data set is multiplied by a constant c (a change in units; for example, $c = 2.54$ changes observations from inches to centimeters). How does this affect MSTr, MSE, and the test statistic F? Is this reasonable? Explain.

15.63 Three different brands of automobile batteries, each one having a 42-month warranty, were included in a study of battery lifetime. A random sample of batteries of each brand was selected and lifetime (in months) was determined, resulting in the following data:

Brand 1	45	38	52	47	45	42	43
Brand 2	39	44	50	54	48	46	40
Brand 3	50	46	43	48	57	44	48

State and test the appropriate hypotheses using a significance level of .05. Be sure to summarize your calculations in an ANOVA table.

15.64 Let $c_1, c_2, \ldots, c_k$ denote k specified numbers, and consider the quantity θ defined by

$$\theta = c_1\mu_1 + c_2\mu_2 + \cdots + c_k\mu_k$$

A confidence interval for θ is then

$$c_1\bar{x}_1 + \cdots + c_k\bar{x}_k \pm (t \text{ critical val.})\sqrt{\text{MSE}\left(\frac{c_1^2}{n_1} + \cdots + \frac{c_k^2}{n_k}\right)}$$

where the t critical value is based on an error df of $N - k$.

For example, in a study carried out to compare pain relievers with respect to true average time to relief, suppose that brands 1, 2, and 3 are nationally available, whereas brands 4 and 5 are sold only by two large chains of drug stores. An investigator might then wish to consider

$$\theta = \frac{1}{3}\mu_1 + \frac{1}{3}\mu_2 + \frac{1}{3}\mu_3 - \frac{1}{2}\mu_4 - \frac{1}{2}\mu_5$$

which, in essence, compares the average time to relief of the national brands to the average for the house brands.

Refer to Exercise 15.63 and suppose that brand 1 is a store brand and brands 2 and 3 are national brands. Obtain a 95% confidence interval for $\theta = \mu_1 - \frac{1}{2}\mu_2 - \frac{1}{2}\mu_3$.

15.65 One of the assumptions that underlies the validity of the ANOVA F test is that the population or treatment response variances $\sigma_1^2, \sigma_2^2, \ldots, \sigma_k^2$ should be identical regardless of whether H_0 is true: the assumption of constant variance across populations or treatments. In some situations, the x values themselves may not satisfy this assumption, yet a transformation using some specified mathematical function (for example, taking the logarithm or the square root) will give observations that have (approximately) constant variance. The ANOVA F test can then be applied to the transformed data. When observations are made on a counting variable (x = number of something), statisticians have found that taking the square root will frequently "stabilize the variance." In an experiment to compare the quality of four different brands of videotape, cassettes of a specified length were selected, and the number of flaws in each was determined.

Brand 1	10	14	5	12	8
Brand 2	17	14	8	9	12
Brand 3	13	18	15	18	10
Brand 4	14	22	12	16	17

Make a square-root transformation, and analyze the resulting data by using the ANOVA F test at significance level .01.

References

Miller, Rupert. *Beyond ANOVA: The Basics of Applied Statistics*. New York: Wiley, 1986. (This book contains a wealth of information concerning violations of basic assumptions and alternative methods of analysis.)

Neter, John, William Wasserman, and Michael Kutner. *Applied Linear Statistical Models.* New York: McGraw-Hill, 1996. (The latter half of the book contains a readable survey of ANOVA and experimental design.)

GRAPHING CALCULATOR EXPLORATION

15.1 Single-Factor ANOVA

Single-factor ANOVA is at the limit of the capability of a graphing calculator unless you have a programmable calculator and *reliable* programs. Nevertheless, for small ANOVA problems, the calculator can be used effectively when access to a computer is limited.

By now you should be familiar with the capabilities of your calculator, so we don't need to go into a great deal of detail here. Begin by entering the data, one treatment per List. To illustrate, we will use the data from Example 15.1 on box compression strength. Put the data from box type 1 into List1, data from box type 2 into List2, etc. Boxplots and normal probability plots (discussed in previous Graphing Calculator Explorations) can be used to assess the plausibility of ANOVA assumptions.

After entering the data, navigate the menus to the ANOVA screen. Screens will differ from one calculator to another, but two common versions are shown in the accompanying figures.

```
ANOVA
How Many: 4
List1      :List1
List2      :List2
List3      :List3
List4      :List4
Execute
```

```
ANOVA(List1, List2, List3, List4)
```

Press either "Execute" or the Enter key to initiate the ANOVA computations on the calculator. The result should look like one of the two "larger-than-life" displays shown below.

```
ANOVA
   F       =25.0943
   p       =5.5254499E-7
   xpσn−1=41.1333
        Fdf=3
        SS=127374.755
        MS=1691.94875
        Edf=20
        SS=33838.9750
        MS=1691.94875
```

```
One-way ANOVA
   F=25.0942894
   p=5.5254499E-7
Factor
   df=3.0000000
   SS=127374.755
   MS=42458.2515
Error
   df=20.0000000
   SS=33838.9750
   MS=1691.94875
   Sxp=41.1333046
```

From the information provided by the calculator, you can reproduce the single-factor ANOVA table. Check to make sure you understand the correspondence between the calculator output and the ANOVA table. Notice that the total sum of squares does not appear on the calculator screen and must be calculated by summing the treatment and error sums of squares. The quantity labeled Sxp is $\sqrt{MSE}$, which is an estimate of the common population standard deviation σ.

Source of Variation	df	Sum of Squares	Mean Square	F
Treatments	3	127374.755	42458.2515	25.094
Error	20	33838.9750	1691.94875	
Total	23	???		

The Tukey–Kramer multiple comparison procedure could be undertaken by combining the information from statistics calculated from the individual lists and the calculator output, but these procedures are not usually built-in functions of the calculator. If you anticipate performing many single-factor ANOVAs on your calculator, you might consider downloading a good program from the Web. However, these programs vary in quality, so be sure to check such a program against a known correct solution!

Appendix: Statistical Tables

Table I

Random numbers

Row																			
1	4	5	1	8	5	0	3	3	7	1	2	8	4	5	1	1	0	9	5	7
2	4	2	5	5	8	0	4	5	7	0	7	0	3	6	6	1	3	1	3	1
3	8	9	9	3	4	3	5	0	6	3	9	1	1	8	2	6	9	2	0	9
4	8	9	0	7	2	9	9	0	4	7	6	7	4	7	1	3	4	3	5	3
5	5	7	3	1	0	3	7	4	7	8	5	2	0	1	3	7	7	6	3	6
6	0	9	3	8	7	6	7	9	9	5	6	2	5	6	5	8	4	2	6	4
7	4	1	0	1	0	2	2	0	4	7	5	1	1	9	4	7	9	7	5	1
8	6	4	7	3	6	3	4	5	1	2	3	1	1	8	0	0	4	8	2	0
9	8	0	2	8	7	9	3	8	4	0	4	2	0	8	9	1	2	3	3	2
10	9	4	6	0	6	9	7	8	8	2	5	2	9	6	0	1	4	6	0	5
11	6	6	9	5	7	4	4	6	3	2	0	6	0	8	9	1	3	6	1	8
12	0	7	1	7	7	7	2	9	7	8	7	5	8	8	6	9	8	4	1	0
13	6	1	3	0	9	7	3	3	6	6	0	4	1	8	3	2	6	7	6	8
14	2	2	3	6	2	1	3	0	2	2	6	6	9	7	0	2	1	2	5	8
15	0	7	1	7	4	2	0	0	0	1	3	1	2	0	4	7	8	4	1	0
16	6	6	5	1	6	1	8	1	5	5	2	6	2	0	1	1	5	2	3	6
17	9	9	6	2	5	3	5	9	8	3	7	5	0	1	3	9	3	8	0	8
18	9	9	9	6	1	2	9	3	4	6	5	6	4	6	5	8	2	7	4	0
19	2	5	6	3	1	9	8	1	1	0	3	5	6	7	9	1	4	5	2	0
20	5	1	1	9	8	1	2	1	1	6	9	8	1	8	1	9	9	1	2	0
21	1	9	8	0	7	4	6	8	4	0	3	0	8	1	1	0	6	2	3	2
22	9	7	0	9	6	3	8	9	9	7	0	6	5	4	3	6	5	0	3	2
23	1	7	6	4	8	2	0	3	9	6	3	6	2	1	0	7	7	3	1	7
24	6	2	5	8	2	0	7	8	6	4	6	6	8	9	2	0	6	9	0	4
25	1	5	7	1	1	1	9	5	1	4	5	2	8	3	4	3	0	7	3	5
26	1	4	6	6	5	6	0	1	9	4	0	5	2	7	6	4	3	6	8	8
27	1	8	5	0	2	1	6	8	0	7	7	2	6	2	6	7	5	4	8	7
28	7	8	7	4	6	5	4	3	7	9	3	9	2	7	9	5	4	2	3	1
29	1	6	3	2	8	3	7	3	0	7	2	4	8	0	9	9	9	4	7	0
30	2	8	9	0	8	1	6	8	1	7	3	1	3	0	9	7	2	5	7	9
31	0	7	8	8	6	5	7	5	5	4	0	0	3	4	1	2	7	3	7	9
32	8	4	0	1	4	5	1	9	1	1	2	1	5	3	2	8	5	5	7	5
33	7	3	5	9	7	0	4	9	1	2	1	3	2	5	1	9	3	3	8	3
34	4	7	2	6	7	6	9	9	2	7	8	7	5	5	5	2	4	4	3	4
35	9	3	3	7	0	7	0	5	7	5	6	9	5	4	3	1	4	6	6	8
36	0	2	4	9	7	8	1	6	3	8	7	8	0	5	6	7	2	7	5	0
37	7	1	0	1	8	4	7	1	2	9	3	8	0	0	8	7	9	2	8	6
38	9	7	9	4	4	5	3	1	9	3	4	5	0	6	3	5	9	6	9	8
39	0	4	2	5	0	0	9	9	6	4	0	6	9	0	3	8	3	5	7	2
40	0	7	1	2	3	6	1	7	9	3	9	5	4	6	8	4	8	8	0	6
41	3	5	6	6	2	4	4	5	6	3	7	8	7	6	5	2	0	4	3	2
42	6	6	8	5	5	2	9	7	9	3	3	1	6	9	5	9	7	1	1	2
43	9	5	0	4	3	1	1	7	3	9	2	7	7	4	7	0	3	1	2	8
44	5	1	7	8	9	4	7	2	9	2	8	9	9	8	0	6	3	7	2	1
45	1	6	3	9	4	1	3	2	1	1	8	5	6	3	4	1	9	3	1	7
46	4	4	8	6	4	0	3	8	3	8	3	5	9	5	9	4	8	3	9	4
47	7	7	6	6	4	5	4	4	8	4	4	0	3	9	8	5	2	0	2	3
48	2	5	6	6	3	7	0	6	5	6	9	0	1	9	5	2	6	9	1	2
49	9	4	0	4	7	5	3	2	8	7	2	7	4	9	3	9	6	5	5	6
50	7	3	1	5	6	6	5	0	3	5	3	7	2	8	6	2	4	1	8	7

Table I   Random Numbers   809

Table I
Random numbers (continued)

Row																			
51	7	5	8	2	8	8	8	7	6	4	1	1	0	2	3	1	9	3	6	0
52	3	3	6	0	9	1	1	0	3	2	7	8	2	0	5	3	4	8	9	8
53	0	2	9	6	9	8	9	3	8	1	5	3	9	9	7	0	7	7	1	6
54	8	5	9	6	2	9	6	8	2	1	2	4	7	0	6	8	3	4	6	1
55	5	4	7	6	1	0	0	1	0	4	6	1	4	1	5	0	9	6	5	5
56	5	0	3	6	4	1	9	8	4	4	1	2	0	2	5	1	8	1	2	1
57	0	2	6	3	7	5	1	1	6	6	0	5	8	1	2	3	3	6	1	3
58	3	8	1	6	3	8	1	4	5	2	9	4	2	5	7	3	2	3	1	8
59	9	1	5	6	0	6	5	6	6	3	6	2	3	0	0	0	1	8	5	9
60	5	3	5	6	3	9	5	4	7	3	6	6	7	5	0	1	5	6	7	3
61	9	6	6	4	5	7	7	6	1	5	4	4	8	0	6	5	7	6	3	0
62	6	3	0	6	7	9	5	5	4	6	2	2	8	4	4	0	0	9	9	8
63	8	5	8	3	5	2	0	6	6	0	0	6	0	6	3	0	1	7	0	5
64	3	8	2	4	9	0	9	2	6	2	9	5	1	9	1	9	0	8	3	3
65	1	4	4	1	1	7	4	6	3	6	5	6	5	5	7	7	0	3	5	8
66	5	9	9	5	3	7	2	5	1	7	1	1	0	7	1	0	9	2	8	8
67	8	7	1	7	5	2	5	6	8	7	9	9	1	3	9	6	4	9	3	0
68	6	7	2	3	1	4	9	2	1	7	0	8	6	7	8	9	9	4	7	4
69	2	3	2	8	7	0	9	7	1	1	1	2	8	2	9	1	0	6	7	7
70	2	9	5	7	8	4	7	9	0	3	6	9	2	0	6	0	6	2	6	8
71	4	8	9	8	3	2	7	6	9	1	9	8	6	9	5	2	4	9	9	9
72	1	5	6	5	7	7	5	4	3	4	3	8	1	8	9	9	4	4	1	1
73	1	8	1	1	7	2	8	5	5	8	9	9	9	6	2	0	1	6	6	7
74	5	7	7	0	9	5	5	6	8	6	8	2	2	6	0	5	5	1	8	7
75	1	8	6	0	5	4	8	3	4	5	3	5	8	7	7	7	8	5	7	0
76	2	6	6	7	9	4	2	2	8	7	4	3	4	9	6	1	9	4	3	9
77	3	6	6	4	5	7	8	3	0	2	8	4	6	7	2	1	4	5	2	3
78	0	7	8	0	1	2	1	1	3	4	2	1	6	9	3	3	5	4	0	4
79	8	3	6	0	5	7	7	9	1	5	8	8	4	9	5	7	2	2	7	6
80	5	3	6	9	0	6	3	8	7	5	9	5	9	7	4	2	5	6	2	9
81	0	9	3	7	7	2	8	6	4	3	2	9	4	8	2	9	9	6	9	9
82	9	4	7	4	0	0	0	3	5	4	6	6	2	6	2	3	6	1	1	4
83	5	5	4	1	7	8	6	4	2	3	2	9	8	4	6	3	8	3	0	5
84	5	3	0	0	5	4	8	0	7	4	7	6	2	1	1	2	1	2	6	9
85	3	3	0	9	3	2	9	4	0	5	5	4	8	7	5	7	5	3	8	8
86	3	0	5	7	1	9	5	8	0	0	4	5	3	0	3	0	2	7	6	7
87	5	0	8	6	0	8	1	6	2	0	8	6	5	4	0	7	2	9	1	0
88	3	6	4	7	8	2	3	5	7	9	8	5	2	7	6	9	0	2	4	9
89	9	0	4	4	9	1	6	8	5	2	8	9	0	7	5	7	2	5	1	8
90	9	5	2	6	9	3	9	6	5	1	8	8	7	8	2	0	4	4	7	9
91	9	4	5	7	0	3	4	6	4	2	5	4	8	6	1	1	9	1	8	8
92	8	1	1	8	0	5	4	2	8	5	3	3	3	0	1	1	4	4	8	3
93	6	9	4	7	8	3	3	9	1	2	5	0	1	2	3	0	1	1	2	5
94	0	0	6	8	8	7	2	4	4	7	6	6	0	3	4	7	5	6	8	2
95	5	3	3	9	3	8	4	9	1	9	1	7	8	4	5	2	2	5	4	4
96	2	5	6	2	7	6	0	3	8	1	4	4	2	6	8	3	6	3	2	8
97	7	4	3	7	9	6	8	6	2	8	3	8	4	2	2	0	7	0	5	3
98	1	9	0	8	8	0	1	2	2	2	7	5	6	5	5	7	8	7	2	6
99	2	4	8	0	2	5	2	7	0	5	9	6	6	1	5	8	7	9	7	5
100	4	1	7	8	6	7	1	1	5	8	9	4	8	9	8	3	0	9	0	7

Table II

Standard normal probabilities (cumulative z curve areas)

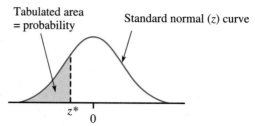

Tabulated area = probability

Standard normal (z) curve

z^*	.00	.01	.02	.03	.04	.05	.06	.07	.08	.09
−3.8	.0001	.0001	.0001	.0001	.0001	.0001	.0001	.0001	.0001	.0000
−3.7	.0001	.0001	.0001	.0001	.0001	.0001	.0001	.0001	.0001	.0001
−3.6	.0002	.0002	.0001	.0001	.0001	.0001	.0001	.0001	.0001	.0001
−3.5	.0002	.0002	.0002	.0002	.0002	.0002	.0002	.0002	.0002	.0002
−3.4	.0003	.0003	.0003	.0003	.0003	.0003	.0003	.0003	.0003	.0002
−3.3	.0005	.0005	.0005	.0004	.0004	.0004	.0004	.0004	.0004	.0003
−3.2	.0007	.0007	.0006	.0006	.0006	.0006	.0006	.0005	.0005	.0005
−3.1	.0010	.0009	.0009	.0009	.0008	.0008	.0008	.0008	.0007	.0007
−3.0	.0013	.0013	.0013	.0012	.0012	.0011	.0011	.0011	.0010	.0010
−2.9	.0019	.0018	.0018	.0017	.0016	.0016	.0015	.0015	.0014	.0014
−2.8	.0026	.0025	.0024	.0023	.0023	.0022	.0021	.0021	.0020	.0019
−2.7	.0035	.0034	.0033	.0032	.0031	.0030	.0029	.0028	.0027	.0026
−2.6	.0047	.0045	.0044	.0043	.0041	.0040	.0039	.0038	.0037	.0036
−2.5	.0062	.0060	.0059	.0057	.0055	.0054	.0052	.0051	.0049	.0048
−2.4	.0082	.0080	.0078	.0075	.0073	.0071	.0069	.0068	.0066	.0064
−2.3	.0107	.0104	.0102	.0099	.0096	.0094	.0091	.0089	.0087	.0084
−2.2	.0139	.0136	.0132	.0129	.0125	.0122	.0119	.0116	.0113	.0110
−2.1	.0179	.0174	.0170	.0166	.0162	.0158	.0154	.0150	.0146	.0143
−2.0	.0228	.0222	.0217	.0212	.0207	.0202	.0197	.0192	.0188	.0183
−1.9	.0287	.0281	.0274	.0268	.0262	.0256	.0250	.0244	.0239	.0233
−1.8	.0359	.0351	.0344	.0336	.0329	.0322	.0314	.0307	.0301	.0294
−1.7	.0446	.0436	.0427	.0418	.0409	.0401	.0392	.0384	.0375	.0367
−1.6	.0548	.0537	.0526	.0516	.0505	.0495	.0485	.0475	.0465	.0455
−1.5	.0668	.0655	.0643	.0630	.0618	.0606	.0594	.0582	.0571	.0559
−1.4	.0808	.0793	.0778	.0764	.0749	.0735	.0721	.0708	.0694	.0681
−1.3	.0968	.0951	.0934	.0918	.0901	.0885	.0869	.0853	.0838	.0823
−1.2	.1151	.1131	.1112	.1093	.1075	.1056	.1038	.1020	.1003	.0985
−1.1	.1357	.1335	.1314	.1292	.1271	.1251	.1230	.1210	.1190	.1170
−1.0	.1587	.1562	.1539	.1515	.1492	.1469	.1446	.1423	.1401	.1379
−0.9	.1841	.1814	.1788	.1762	.1736	.1711	.1685	.1660	.1635	.1611
−0.8	.2119	.2090	.2061	.2033	.2005	.1977	.1949	.1922	.1894	.1867
−0.7	.2420	.2389	.2358	.2327	.2296	.2266	.2236	.2206	.2177	.2148
−0.6	.2743	.2709	.2676	.2643	.2611	.2578	.2546	.2514	.2483	.2451
−0.5	.3085	.3050	.3015	.2981	.2946	.2912	.2877	.2843	.2810	.2776
−0.4	.3446	.3409	.3372	.3336	.3300	.3264	.3228	.3192	.3156	.3121
−0.3	.3821	.3783	.3745	.3707	.3669	.3632	.3594	.3557	.3520	.3483
−0.2	.4207	.4168	.4129	.4090	.4052	.4013	.3974	.3936	.3897	.3859
−0.1	.4602	.4562	.4522	.4483	.4443	.4404	.4364	.4325	.4286	.4247
−0.0	.5000	.4960	.4920	.4880	.4840	.4801	.4761	.4721	.4681	.4641

Table II Standard Normal Probabilities   811

Table II
Standard normal
probabilities (continued)

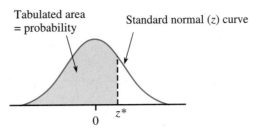

Tabulated area = probability

Standard normal (z) curve

0   z*

z^*	.00	.01	.02	.03	.04	.05	.06	.07	.08	.09
0.0	.5000	.5040	.5080	.5120	.5160	.5199	.5239	.5279	.5319	.5359
0.1	.5398	.5438	.5478	.5517	.5557	.5596	.5636	.5675	.5714	.5753
0.2	.5793	.5832	.5871	.5910	.5948	.5987	.6026	.6064	.6103	.6141
0.3	.6179	.6217	.6255	.6293	.6331	.6368	.6406	.6443	.6480	.6517
0.4	.6554	.6591	.6628	.6664	.6700	.6736	.6772	.6808	.6844	.6879
0.5	.6915	.6950	.6985	.7019	.7054	.7088	.7123	.7157	.7190	.7224
0.6	.7257	.7291	.7324	.7357	.7389	.7422	.7454	.7486	.7517	.7549
0.7	.7580	.7611	.7642	.7673	.7704	.7734	.7764	.7794	.7823	.7852
0.8	.7881	.7910	.7939	.7967	.7995	.8023	.8051	.8078	.8106	.8133
0.9	.8159	.8186	.8212	.8238	.8264	.8289	.8315	.8340	.8365	.8389
1.0	.8413	.8438	.8461	.8485	.8508	.8531	.8554	.8577	.8599	.8621
1.1	.8643	.8665	.8686	.8708	.8729	.8749	.8770	.8790	.8810	.8830
1.2	.8849	.8869	.8888	.8907	.8925	.8944	.8962	.8980	.8997	.9015
1.3	.9032	.9049	.9066	.9082	.9099	.9115	.9131	.9147	.9162	.9177
1.4	.9192	.9207	.9222	.9236	.9251	.9265	.9279	.9292	.9306	.9319
1.5	.9332	.9345	.9357	.9370	.9382	.9394	.9406	.9418	.9429	.9441
1.6	.9452	.9463	.9474	.9484	.9495	.9505	.9515	.9525	.9535	.9545
1.7	.9554	.9564	.9573	.9582	.9591	.9599	.9608	.9616	.9625	.9633
1.8	.9641	.9649	.9656	.9664	.9671	.9678	.9686	.9693	.9699	.9706
1.9	.9713	.9719	.9726	.9732	.9738	.9744	.9750	.9756	.9761	.9767
2.0	.9772	.9778	.9783	.9788	.9793	.9798	.9803	.9808	.9812	.9817
2.1	.9821	.9826	.9830	.9834	.9838	.9842	.9846	.9850	.9854	.9857
2.2	.9861	.9864	.9868	.9871	.9875	.9878	.9881	.9884	.9887	.9890
2.3	.9893	.9896	.9898	.9901	.9904	.9906	.9909	.9911	.9913	.9916
2.4	.9918	.9920	.9922	.9925	.9927	.9929	.9931	.9932	.9934	.9936
2.5	.9938	.9940	.9941	.9943	.9945	.9946	.9948	.9949	.9951	.9952
2.6	.9953	.9955	.9956	.9957	.9959	.9960	.9961	.9962	.9963	.9964
2.7	.9965	.9966	.9967	.9968	.9969	.9970	.9971	.9972	.9973	.9974
2.8	.9974	.9975	.9976	.9977	.9977	.9978	.9979	.9979	.9980	.9981
2.9	.9981	.9982	.9982	.9983	.9984	.9984	.9985	.9985	.9986	.9986
3.0	.9987	.9987	.9987	.9988	.9988	.9989	.9989	.9989	.9990	.9990
3.1	.9990	.9991	.9991	.9991	.9992	.9992	.9992	.9992	.9993	.9993
3.2	.9993	.9993	.9994	.9994	.9994	.9994	.9994	.9995	.9995	.9995
3.3	.9995	.9995	.9995	.9996	.9996	.9996	.9996	.9996	.9996	.9997
3.4	.9997	.9997	.9997	.9997	.9997	.9997	.9997	.9997	.9997	.9998
3.5	.9998	.9998	.9998	.9998	.9998	.9998	.9998	.9998	.9998	.9998
3.6	.9998	.9998	.9999	.9999	.9999	.9999	.9999	.9999	.9999	.9999
3.7	.9999	.9999	.9999	.9999	.9999	.9999	.9999	.9999	.9999	.9999
3.8	.9999	.9999	.9999	.9999	.9999	.9999	.9999	.9999	.9999	1.0000

Table III

t critical values

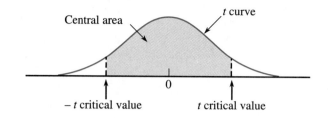

Central area

t curve

− *t* critical value   0   *t* critical value

Central area captured:		.80	.90	.95	.98	.99	.998	.999
Confidence level:		80%	90%	95%	98%	99%	99.8%	99.9%
	1	3.08	6.31	12.71	31.82	63.66	318.31	636.62
	2	1.89	2.92	4.30	6.97	9.93	23.33	31.60
	3	1.64	2.35	3.18	4.54	5.84	10.21	12.92
	4	1.53	2.13	2.78	3.75	4.60	7.17	8.61
	5	1.48	2.02	2.57	3.37	4.03	5.89	6.86
	6	1.44	1.94	2.45	3.14	3.71	5.21	5.96
	7	1.42	1.90	2.37	3.00	3.50	4.79	5.41
	8	1.40	1.86	2.31	2.90	3.36	4.50	5.04
	9	1.38	1.83	2.26	2.82	3.25	4.30	4.78
	10	1.37	1.81	2.23	2.76	3.17	4.14	4.59
	11	1.36	1.80	2.20	2.72	3.11	4.03	4.44
	12	1.36	1.78	2.18	2.68	3.06	3.93	4.32
	13	1.35	1.77	2.16	2.65	3.01	3.85	4.22
	14	1.35	1.76	2.15	2.62	2.98	3.79	4.14
	15	1.34	1.75	2.13	2.60	2.95	3.73	4.07
	16	1.34	1.75	2.12	2.58	2.92	3.69	4.02
Degrees of	17	1.33	1.74	2.11	2.57	2.90	3.65	3.97
freedom	18	1.33	1.73	2.10	2.55	2.88	3.61	3.92
	19	1.33	1.73	2.09	2.54	2.86	3.58	3.88
	20	1.33	1.73	2.09	2.53	2.85	3.55	3.85
	21	1.32	1.72	2.08	2.52	2.83	3.53	3.82
	22	1.32	1.72	2.07	2.51	2.82	3.51	3.79
	23	1.32	1.71	2.07	2.50	2.81	3.49	3.77
	24	1.32	1.71	2.06	2.49	2.80	3.47	3.75
	25	1.32	1.71	2.06	2.49	2.79	3.45	3.73
	26	1.32	1.71	2.06	2.48	2.78	3.44	3.71
	27	1.31	1.70	2.05	2.47	2.77	3.42	3.69
	28	1.31	1.70	2.05	2.47	2.76	3.41	3.67
	29	1.31	1.70	2.05	2.46	2.76	3.40	3.66
	30	1.31	1.70	2.04	2.46	2.75	3.39	3.65
	40	1.30	1.68	2.02	2.42	2.70	3.31	3.55
	60	1.30	1.67	2.00	2.39	2.66	3.23	3.46
	120	1.29	1.66	1.98	2.36	2.62	3.16	3.37
z critical values	∞	1.28	1.645	1.96	2.33	2.58	3.09	3.29

Table IV Tail Areas for *t* Curves   813

Table IV

Tail areas for t curves

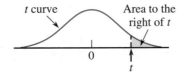

df	1	2	3	4	5	6	7	8	9	10	11	12
t												
0.0	.500	.500	.500	.500	.500	.500	.500	.500	.500	.500	.500	.500
0.1	.468	.465	.463	.463	.462	.462	.462	.461	.461	.461	.461	.461
0.2	.437	.430	.427	.426	.425	.424	.424	.423	.423	.423	.423	.422
0.3	.407	.396	.392	.390	.388	.387	.386	.386	.386	.385	.385	.385
0.4	.379	.364	.358	.355	.353	.352	.351	.350	.349	.349	.348	.348
0.5	.352	.333	.326	.322	.319	.317	.316	.315	.315	.314	.313	.313
0.6	.328	.305	.295	.290	.287	.285	.284	.283	.282	.281	.280	.280
0.7	.306	.278	.267	.261	.258	.255	.253	.252	.251	.250	.249	.249
0.8	.285	.254	.241	.234	.230	.227	.225	.223	.222	.221	.220	.220
0.9	.267	.232	.217	.210	.205	.201	.199	.197	.196	.195	.194	.193
1.0	.250	.211	.196	.187	.182	.178	.175	.173	.172	.170	.169	.169
1.1	.235	.193	.176	.167	.162	.157	.154	.152	.150	.149	.147	.146
1.2	.221	.177	.158	.148	.142	.138	.135	.132	.130	.129	.128	.127
1.3	.209	.162	.142	.132	.125	.121	.117	.115	.113	.111	.110	.109
1.4	.197	.148	.128	.117	.110	.106	.102	.100	.098	.096	.095	.093
1.5	.187	.136	.115	.104	.097	.092	.089	.086	.084	.082	.081	.080
1.6	.178	.125	.104	.092	.085	.080	.077	.074	.072	.070	.069	.068
1.7	.169	.116	.094	.082	.075	.070	.066	.064	.062	.060	.059	.057
1.8	.161	.107	.085	.073	.066	.061	.057	.055	.053	.051	.050	.049
1.9	.154	.099	.077	.065	.058	.053	.050	.047	.045	.043	.042	.041
2.0	.148	.092	.070	.058	.051	.046	.043	.040	.038	.037	.035	.034
2.1	.141	.085	.063	.052	.045	.040	.037	.034	.033	.031	.030	.029
2.2	.136	.079	.058	.046	.040	.035	.032	.029	.028	.026	.025	.024
2.3	.131	.074	.052	.041	.035	.031	.027	.025	.023	.022	.021	.020
2.4	.126	.069	.048	.037	.031	.027	.024	.022	.020	.019	.018	.017
2.5	.121	.065	.044	.033	.027	.023	.020	.018	.017	.016	.015	.014
2.6	.117	.061	.040	.030	.024	.020	.018	.016	.014	.013	.012	.012
2.7	.113	.057	.037	.027	.021	.018	.015	.014	.012	.011	.010	.010
2.8	.109	.054	.034	.024	.019	.016	.013	.012	.010	.009	.009	.008
2.9	.106	.051	.031	.022	.017	.014	.011	.010	.009	.008	.007	.007
3.0	.102	.048	.029	.020	.015	.012	.010	.009	.007	.007	.006	.006
3.1	.099	.045	.027	.018	.013	.011	.009	.007	.006	.006	.005	.005
3.2	.096	.043	.025	.016	.012	.009	.008	.006	.005	.005	.004	.004
3.3	.094	.040	.023	.015	.011	.008	.007	.005	.005	.004	.004	.003
3.4	.091	.038	.021	.014	.010	.007	.006	.005	.004	.003	.003	.003
3.5	.089	.036	.020	.012	.009	.006	.005	.004	.003	.003	.002	.002
3.6	.086	.035	.018	.011	.008	.006	.004	.004	.003	.002	.002	.002
3.7	.084	.033	.017	.010	.007	.005	.004	.003	.002	.002	.002	.002
3.8	.082	.031	.016	.010	.006	.004	.003	.003	.002	.002	.001	.001
3.9	.080	.030	.015	.009	.006	.004	.003	.002	.002	.001	.001	.001
4.0	.078	.029	.014	.008	.005	.004	.003	.002	.002	.001	.001	.001

(continued)

Table IV

Tail areas for t curves
(continued)

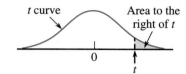

df t	13	14	15	16	17	18	19	20	21	22	23	24
0.0	.500	.500	.500	.500	.500	.500	.500	.500	.500	.500	.500	.500
0.1	.461	.461	.461	.461	.461	.461	.461	.461	.461	.461	.461	.461
0.2	.422	.422	.422	.422	.422	.422	.422	.422	.422	.422	.422	.422
0.3	.384	.384	.384	.384	.384	.384	.384	.384	.384	.383	.383	.383
0.4	.348	.347	.347	.347	.347	.347	.347	.347	.347	.347	.346	.346
0.5	.313	.312	.312	.312	.312	.312	.311	.311	.311	.311	.311	.311
0.6	.279	.279	.279	.278	.278	.278	.278	.278	.278	.277	.277	.277
0.7	.248	.247	.247	.247	.247	.246	.246	.246	.246	.246	.245	.245
0.8	.219	.218	.218	.218	.217	.217	.217	.217	.216	.216	.216	.216
0.9	.192	.191	.191	.191	.190	.190	.190	.189	.189	.189	.189	.189
1.0	.168	.167	.167	.166	.166	.165	.165	.165	.164	.164	.164	.164
1.1	.146	.144	.144	.144	.143	.143	.143	.142	.142	.142	.141	.141
1.2	.126	.124	.124	.124	.123	.123	.122	.122	.122	.121	.121	.121
1.3	.108	.107	.107	.106	.105	.105	.105	.104	.104	.104	.103	.103
1.4	.092	.091	.091	.090	.090	.089	.089	.089	.088	.088	.087	.087
1.5	.079	.077	.077	.077	.076	.075	.075	.075	.074	.074	.074	.073
1.6	.067	.065	.065	.065	.064	.064	.063	.063	.062	.062	.062	.061
1.7	.056	.055	.055	.054	.054	.053	.053	.052	.052	.052	.051	.051
1.8	.048	.046	.046	.045	.045	.044	.044	.043	.043	.043	.042	.042
1.9	.040	.038	.038	.038	.037	.037	.036	.036	.036	.035	.035	.035
2.0	.033	.032	.032	.031	.031	.030	.030	.030	.029	.029	.029	.028
2.1	.028	.027	.027	.026	.025	.025	.025	.024	.024	.024	.023	.023
2.2	.023	.022	.022	.021	.021	.021	.020	.020	.020	.019	.019	.019
2.3	.019	.018	.018	.018	.017	.017	.016	.016	.016	.016	.015	.015
2.4	.016	.015	.015	.014	.014	.014	.013	.013	.013	.013	.012	.012
2.5	.013	.012	.012	.012	.011	.011	.011	.011	.010	.010	.010	.010
2.6	.011	.010	.010	.010	.009	.009	.009	.009	.008	.008	.008	.008
2.7	.009	.008	.008	.008	.008	.007	.007	.007	.007	.007	.006	.006
2.8	.008	.007	.007	.006	.006	.006	.006	.006	.005	.005	.005	.005
2.9	.006	.005	.005	.005	.005	.005	.005	.004	.004	.004	.004	.004
3.0	.005	.004	.004	.004	.004	.004	.004	.004	.003	.003	.003	.003
3.1	.004	.004	.004	.003	.003	.003	.003	.003	.003	.003	.003	.002
3.2	.003	.003	.003	.003	.003	.002	.002	.002	.002	.002	.002	.002
3.3	.003	.002	.002	.002	.002	.002	.002	.002	.002	.002	.002	.001
3.4	.002	.002	.002	.002	.002	.002	.002	.001	.001	.001	.001	.001
3.5	.002	.002	.002	.001	.001	.001	.001	.001	.001	.001	.001	.001
3.6	.002	.001	.001	.001	.001	.001	.001	.001	.001	.001	.001	.001
3.7	.001	.001	.001	.001	.001	.001	.001	.001	.001	.001	.001	.001
3.8	.001	.001	.001	.001	.001	.001	.001	.001	.001	.000	.000	.000
3.9	.001	.001	.001	.001	.001	.001	.000	.000	.000	.000	.000	.000
4.0	.001	.001	.001	.001	.000	.000	.000	.000	.000	.000	.000	.000

Table IV Tail Areas for *t* Curves   815

Table IV

*Tail areas for t curves
(continued)*

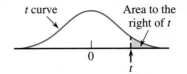

t curve

Area to the right of *t*

0

t

df *t*	25	26	27	28	29	30	35	40	60	120	∞(=z)
0.0	.500	.500	.500	.500	.500	.500	.500	.500	.500	.500	.500
0.1	.461	.461	.461	.461	.461	.461	.460	.460	.460	.460	.460
0.2	.422	.422	.421	.421	.421	.421	.421	.421	.421	.421	.421
0.3	.383	.383	.383	.383	.383	.383	.383	.383	.383	.382	.382
0.4	.346	.346	.346	.346	.346	.346	.346	.346	.345	.345	.345
0.5	.311	.311	.311	.310	.310	.310	.310	.310	.309	.309	.309
0.6	.277	.277	.277	.277	.277	.277	.276	.276	.275	.275	.274
0.7	.245	.245	.245	.245	.245	.245	.244	.244	.243	.243	.242
0.8	.216	.215	.215	.215	.215	.215	.215	.214	.213	.213	.212
0.9	.188	.188	.188	.188	.188	.188	.187	.187	.186	.185	.184
1.0	.163	.163	.163	.163	.163	.163	.162	.162	.161	.160	.159
1.1	.141	.141	.141	.140	.140	.140	.139	.139	.138	.137	.136
1.2	.121	.120	.120	.120	.120	.120	.119	.119	.117	.116	.115
1.3	.103	.103	.102	.102	.102	.102	.101	.101	.099	.098	.097
1.4	.087	.087	.086	.086	.086	.086	.085	.085	.083	.082	.081
1.5	.073	.073	.073	.072	.072	.072	.071	.071	.069	.068	.067
1.6	.061	.061	.061	.060	.060	.060	.059	.059	.057	.056	.055
1.7	.051	.051	.050	.050	.050	.050	.049	.048	.047	.046	.045
1.8	.042	.042	.042	.041	.041	.041	.040	.040	.038	.037	.036
1.9	.035	.034	.034	.034	.034	.034	.033	.032	.031	.030	.029
2.0	.028	.028	.028	.028	.027	.027	.027	.026	.025	.024	.023
2.1	.023	.023	.023	.022	.022	.022	.022	.021	.020	.019	.018
2.2	.019	.018	.018	.018	.018	.018	.017	.017	.016	.015	.014
2.3	.015	.015	.015	.015	.014	.014	.014	.013	.012	.012	.011
2.4	.012	.012	.012	.012	.012	.011	.011	.011	.010	.009	.008
2.5	.010	.010	.009	.009	.009	.009	.009	.008	.008	.007	.006
2.6	.008	.008	.007	.007	.007	.007	.007	.007	.006	.005	.005
2.7	.006	.006	.006	.006	.006	.006	.005	.005	.004	.004	.003
2.8	.005	.005	.005	.005	.005	.004	.004	.004	.003	.003	.003
2.9	.004	.004	.004	.004	.004	.003	.003	.003	.003	.002	.002
3.0	.003	.003	.003	.003	.003	.003	.002	.002	.002	.002	.001
3.1	.002	.002	.002	.002	.002	.002	.002	.002	.001	.001	.001
3.2	.002	.002	.002	.002	.002	.002	.001	.001	.001	.001	.001
3.3	.001	.001	.001	.001	.001	.001	.001	.001	.001	.001	.000
3.4	.001	.001	.001	.001	.001	.001	.001	.001	.001	.000	.000
3.5	.001	.001	.001	.001	.001	.001	.001	.001	.000	.000	.000
3.6	.001	.001	.001	.001	.001	.001	.000	.000	.000	.000	.000
3.7	.001	.001	.000	.000	.000	.000	.000	.000	.000	.000	.000
3.8	.000	.000	.000	.000	.000	.000	.000	.000	.000	.000	.000
3.9	.000	.000	.000	.000	.000	.000	.000	.000	.000	.000	.000
4.0	.000	.000	.000	.000	.000	.000	.000	.000	.000	.000	.000

Table V *Curves of β = P(type II error) for t tests*

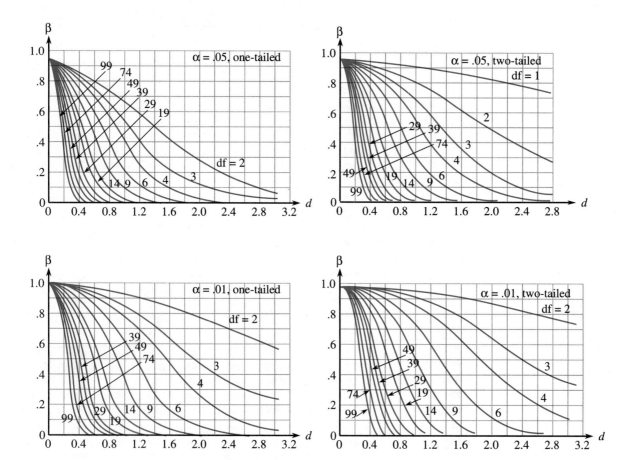

Table VI P-value Information for the Rank Sum Test 817

Table VI *P-value information for the rank-sum test*

		Upper-tailed test		Lower-tailed test		Two-tailed test	
n_1	n_2	P-value < .05 if rank-sum is greater than or equal to	P-value < .01 if rank-sum is greater than or equal to	P-value < .05 if rank-sum is less than or equal to	P-value < .01 if rank-sum is less than or equal to	P-value < .05 if rank-sum is not between*	P-value < .01 if rank-sum is not between
3	3	15	—	6	—	—	—
3	4	17	—	7	—	18,6	—
3	5	20	21	7	6	21,6	—
3	6	22	24	8	6	23,7	—
3	7	24	27	9	6	26,7	27,6
3	8	27	29	9	7	28,8	30,6
4	3	21	—	11	—	—	—
4	4	24	26	12	10	25,11	—
4	5	27	30	13	10	29,11	30,10
4	6	30	33	14	11	32,12	34,10
4	7	33	36	15	12	35,13	37,11
4	8	36	40	16	12	38,14	41,11
5	3	29	30	16	15	30,15	—
5	4	32	35	18	15	34,16	35,15
5	5	36	39	19	16	37,18	39,16
5	6	40	43	20	17	41,19	44,16
5	7	43	47	22	18	45,20	48,17
5	8	47	51	23	19	49,21	52,18
6	3	37	39	23	21	38,22	—
6	4	41	44	25	22	43,23	45,21
6	5	46	49	26	23	47,25	50,22
6	6	50	54	28	24	52,26	55,23
6	7	54	58	30	26	56,28	60,24
6	8	58	63	32	27	61,29	65,25
7	3	46	49	31	28	48,29	49,28
7	4	51	54	33	30	53,31	55,29
7	5	56	60	35	31	58,33	61,30
7	6	61	65	37	33	63,35	67,31
7	7	66	71	39	34	68,37	72,33
7	8	71	76	41	36	73,39	78,34
8	3	57	59	39	37	58,38	60,36
8	4	62	66	42	38	64,40	67,37
8	5	68	72	44	40	70,42	73,39
8	6	73	78	47	42	76,44	80,40
8	7	79	84	49	44	81,47	86,42
8	8	84	90	52	46	87,49	92,44

*Including endpoints. For example, when $n_1 = 3$ and $n_2 = 4$, P-value < .05 if $6 \leq$ rank-sum ≤ 18.

Table VII *Values that capture specified upper-tail F curve areas*

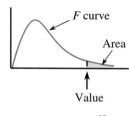

df_1

df_2	Area	1	2	3	4	5	6	7	8	9	10
1	.10	39.86	49.50	53.59	55.83	57.24	58.20	58.91	59.44	59.86	60.19
	.05	161.4	199.5	215.7	224.6	230.2	234.0	236.8	238.9	240.5	241.9
	.01	4052	5000	5403	5625	5764	5859	5928	5981	6022	6056
2	.10	8.53	9.00	9.16	9.24	9.29	9.33	9.35	9.37	9.38	9.39
	.05	18.51	19.00	19.16	19.25	19.30	19.33	19.35	19.37	19.38	19.40
	.01	98.50	99.00	99.17	99.25	99.30	99.33	99.36	99.37	99.39	99.40
	.001	998.5	999.0	999.2	999.2	999.3	999.3	999.4	999.4	999.4	999.4
3	.10	5.54	5.46	5.39	5.34	5.31	5.28	5.27	5.25	5.24	5.23
	.05	10.13	9.55	9.28	9.12	9.01	8.94	8.89	8.85	8.81	8.79
	.01	34.12	30.82	29.46	28.71	28.24	27.91	27.67	27.49	27.35	27.23
	.001	167.0	148.5	141.1	137.1	134.6	132.8	131.6	130.6	129.9	129.2
4	.10	4.54	4.32	4.19	4.11	4.05	4.01	3.98	3.95	3.94	3.92
	.05	7.71	6.94	6.59	6.39	6.26	6.16	6.09	6.04	6.00	5.96
	.01	21.20	18.00	16.69	15.98	15.52	15.21	14.98	14.80	14.66	14.55
	.001	74.14	61.25	56.18	53.44	51.71	50.53	49.66	49.00	48.47	48.05
5	.10	4.06	3.78	3.62	3.52	3.45	3.40	3.37	3.34	3.32	3.30
	.05	6.61	5.79	5.41	5.19	5.05	4.95	4.88	4.82	4.77	4.74
	.01	16.26	13.27	12.06	11.39	10.97	10.67	10.46	10.29	10.16	10.05
	.001	47.18	37.12	33.20	31.09	29.75	28.83	28.16	27.65	27.24	26.92
6	.10	3.78	3.46	3.29	3.18	3.11	3.05	3.01	2.98	2.96	2.94
	.05	5.99	5.14	4.76	4.53	4.39	4.28	4.21	4.15	4.10	4.06
	.01	13.75	10.92	9.78	9.15	8.75	8.47	8.26	8.10	7.98	7.87
	.001	35.51	27.00	23.70	21.92	20.80	20.03	19.46	19.03	18.69	18.41
7	.10	3.59	3.26	3.07	2.96	2.88	2.83	2.78	2.75	2.72	2.70
	.05	5.59	4.74	4.35	4.12	3.97	3.87	3.79	3.73	3.68	3.64
	.01	12.25	9.55	8.45	7.85	7.46	7.19	6.99	6.84	6.72	6.62
	.001	29.25	21.69	18.77	17.20	16.21	15.52	15.02	14.63	14.33	14.08
8	.10	3.46	3.11	2.92	2.81	2.73	2.67	2.62	2.59	2.56	2.54
	.05	5.32	4.46	4.07	3.84	3.69	3.58	3.50	3.44	3.39	3.35
	.01	11.26	8.65	7.59	7.01	6.63	6.37	6.18	6.03	5.91	5.81
	.001	25.41	18.49	15.83	14.39	13.48	12.86	12.40	12.05	11.77	11.54
9	.10	3.36	3.01	2.81	2.69	2.61	2.55	2.51	2.47	2.44	2.42
	.05	5.12	4.26	3.86	3.63	3.48	3.37	3.29	3.23	3.18	3.14
	.01	10.56	8.02	6.99	6.42	6.06	5.80	5.61	5.47	5.35	5.26
	.001	22.86	16.39	13.90	12.56	11.71	11.13	10.70	10.37	10.11	9.89

Table VII Values That Capture Specified Upper-tail *F* Curve Areas   819

Table VII *Values that capture specified upper-tail F curve areas (continued)*

						df$_1$					
df$_2$	Area	1	2	3	4	5	6	7	8	9	10
10	.10	3.29	2.92	2.73	2.61	2.52	2.46	2.41	2.38	2.35	2.32
	.05	4.96	4.10	3.71	3.48	3.33	3.22	3.14	3.07	3.02	2.98
	.01	10.04	7.56	6.55	5.99	5.64	5.39	5.20	5.06	4.94	4.85
	.001	21.04	14.91	12.55	11.28	10.48	9.93	9.52	9.20	8.96	8.75
11	.10	3.23	2.86	2.66	2.54	2.45	2.39	2.34	2.30	2.27	2.25
	.05	4.84	3.98	3.59	3.36	3.20	3.09	3.01	2.95	2.90	2.85
	.01	9.65	7.21	6.22	5.67	5.32	5.07	4.89	4.74	4.63	4.54
	.001	19.69	13.81	11.56	10.35	9.58	9.05	8.66	8.35	8.12	7.92
12	.10	3.18	2.81	2.61	2.48	2.39	2.33	2.28	2.24	2.21	2.19
	.05	4.75	3.89	3.49	3.26	3.11	3.00	2.91	2.85	2.80	2.75
	.01	9.33	6.93	5.95	5.41	5.06	4.82	4.64	4.50	4.39	4.30
	.001	18.64	12.97	10.80	9.63	8.89	8.38	8.00	7.71	7.48	7.29
13	.10	3.14	2.76	2.56	2.43	2.35	2.28	2.23	2.20	2.16	2.14
	.05	4.67	3.81	3.41	3.18	3.03	2.92	2.83	2.77	2.71	2.67
	.01	9.07	6.70	5.74	5.21	4.86	4.62	4.44	4.30	4.19	4.10
	.001	17.82	12.31	10.21	9.07	8.35	7.86	7.49	7.21	6.98	6.80
14	.10	3.10	2.73	2.52	2.39	2.31	2.24	2.19	2.15	2.12	2.10
	.05	4.60	3.74	3.34	3.11	2.96	2.85	2.76	2.70	2.65	2.60
	.01	8.86	6.51	5.56	5.04	4.69	4.46	4.28	4.14	4.03	3.94
	.001	17.14	11.78	9.73	8.62	7.92	7.44	7.08	6.80	6.58	6.40
15	.10	3.07	2.70	2.49	2.36	2.27	2.21	2.16	2.12	2.09	2.06
	.05	4.54	3.68	3.29	3.06	2.90	2.79	2.71	2.64	2.59	2.54
	.01	8.68	6.36	5.42	4.89	4.56	4.32	4.14	4.00	3.89	3.80
	.001	16.59	11.34	9.34	8.25	7.57	7.09	6.74	6.47	6.26	6.08
16	.10	3.05	2.67	2.46	2.33	2.24	2.18	2.13	2.09	2.06	2.03
	.05	4.49	3.63	3.24	3.01	2.85	2.74	2.66	2.59	2.54	2.49
	.01	8.53	6.23	5.29	4.77	4.44	4.20	4.03	3.89	3.78	3.69
	.001	16.12	10.97	9.01	7.94	7.27	6.80	6.46	6.19	5.98	5.81
17	.10	3.03	2.64	2.44	2.31	2.22	2.15	2.10	2.06	2.03	2.00
	.05	4.45	3.59	3.20	2.96	2.81	2.70	2.61	2.55	2.49	2.45
	.01	8.40	6.11	5.18	4.67	4.34	4.10	3.93	3.79	3.68	3.59
	.001	15.72	10.66	8.73	7.68	7.02	6.56	6.22	5.96	5.75	5.58
18	.10	3.01	2.62	2.42	2.29	2.20	2.13	2.08	2.04	2.00	1.98
	.05	4.41	3.55	3.16	2.93	2.77	2.66	2.58	2.51	2.46	2.41
	.01	8.29	6.01	5.09	4.58	4.25	4.01	3.84	3.71	3.60	3.51
	.001	15.38	10.39	8.49	7.46	6.81	6.35	6.02	5.76	5.56	5.39
19	.10	2.99	2.61	2.40	2.27	2.18	2.11	2.06	2.02	1.98	1.96
	.05	4.38	3.52	3.13	2.90	2.74	2.63	2.54	2.48	2.42	2.38
	.01	8.18	5.93	5.01	4.50	4.17	3.94	3.77	3.63	3.52	3.43
	.001	15.08	10.16	8.28	7.27	6.62	6.18	5.85	5.59	5.39	5.22

(*continued*)

Table VII *Values that capture specified upper-tail F curve areas (continued)*

df_2	Area	df_1									
		1	2	3	4	5	6	7	8	9	10
20	.10	2.97	2.59	2.38	2.25	2.16	2.09	2.04	2.00	1.96	1.94
	.05	4.35	3.49	3.10	2.87	2.71	2.60	2.51	2.45	2.39	2.35
	.01	8.10	5.85	4.94	4.43	4.10	3.87	3.70	3.56	3.46	3.37
	.001	14.82	9.95	8.10	7.10	6.46	6.02	5.69	5.44	5.24	5.08
21	.10	2.96	2.57	2.36	2.23	2.14	2.08	2.02	1.98	1.95	1.92
	.05	4.32	3.47	3.07	2.84	2.68	2.57	2.49	2.42	2.37	2.32
	.01	8.02	5.78	4.87	4.37	4.04	3.81	3.64	3.51	3.40	3.31
	.001	14.59	9.77	7.94	6.95	6.32	5.88	5.56	5.31	5.11	4.95
22	.10	2.95	2.56	2.35	2.22	2.13	2.06	2.01	1.97	1.93	1.90
	.05	4.30	3.44	3.05	2.82	2.66	2.55	2.46	2.40	2.34	2.30
	.01	7.95	5.72	4.82	4.31	3.99	3.76	3.59	3.45	3.35	3.26
	.001	14.38	9.61	7.80	6.81	6.19	5.76	5.44	5.19	4.99	4.83
23	.10	2.94	2.55	2.34	2.21	2.11	2.05	1.99	1.95	1.92	1.89
	.05	4.28	3.42	3.03	2.80	2.64	2.53	2.44	2.37	2.32	2.27
	.01	7.88	5.66	4.76	4.26	3.94	3.71	3.54	3.41	3.30	3.21
	.001	14.20	9.47	7.67	6.70	6.08	5.65	5.33	5.09	4.89	4.73
24	.10	2.93	2.54	2.33	2.19	2.10	2.04	1.98	1.94	1.91	1.88
	.05	4.26	3.40	3.01	2.78	2.62	2.51	2.42	2.36	2.30	2.25
	.01	7.82	5.61	4.72	4.22	3.90	3.67	3.50	3.36	3.26	3.17
	.001	14.03	9.34	7.55	6.59	5.98	5.55	5.23	4.99	4.80	4.64
25	.10	2.92	2.53	2.32	2.18	2.09	2.02	1.97	1.93	1.89	1.87
	.05	4.24	3.39	2.99	2.76	2.60	2.49	2.40	2.34	2.28	2.24
	.01	7.77	5.57	4.68	4.18	3.85	3.63	3.46	3.32	3.22	3.13
	.001	13.88	9.22	7.45	6.49	5.89	5.46	5.15	4.91	4.71	4.56
26	.10	2.91	2.52	2.31	2.17	2.08	2.01	1.96	1.92	1.88	1.86
	.05	4.23	3.37	2.98	2.74	2.59	2.47	2.39	2.32	2.27	2.22
	.01	7.72	5.53	4.64	4.14	3.82	3.59	3.42	3.29	3.18	3.09
	.001	13.74	9.12	7.36	6.41	5.80	5.38	5.07	4.83	4.64	4.48
27	.10	2.90	2.51	2.30	2.17	2.07	2.00	1.95	1.91	1.87	1.85
	.05	4.21	3.35	2.96	2.73	2.57	2.46	2.37	2.31	2.25	2.20
	.01	7.68	5.49	4.60	4.11	3.78	3.56	3.39	3.26	3.15	3.06
	.001	13.61	9.02	7.27	6.33	5.73	5.31	5.00	4.76	4.57	4.41
28	.10	2.89	2.50	2.29	2.16	2.06	2.00	1.94	1.90	1.87	1.84
	.05	4.20	3.34	2.95	2.71	2.56	2.45	2.36	2.29	2.24	2.19
	.01	7.64	5.45	4.57	4.07	3.75	3.53	3.36	3.23	3.12	3.03
	.001	13.50	8.93	7.19	6.25	5.66	5.24	4.93	4.69	4.50	4.35
29	.10	2.89	2.50	2.28	2.15	2.06	1.99	1.93	1.89	1.86	1.83
	.05	4.18	3.33	2.93	2.70	2.55	2.43	2.35	2.28	2.22	2.18
	.01	7.60	5.42	4.54	4.04	3.73	3.50	3.33	3.20	3.09	3.00
	.001	13.39	8.85	7.12	6.19	5.59	5.18	4.87	4.64	4.45	4.29

Table VII Values That Capture Specified Upper-tail *F* Curve Areas   821

Table VII *Values that capture specified upper-tail F curve areas (continued)*

df$_2$	Area	df$_1$ 1	2	3	4	5	6	7	8	9	10
30	.10	2.88	2.49	2.28	2.14	2.05	1.98	1.93	1.88	1.85	1.82
	.05	4.17	3.32	2.92	2.69	2.53	2.42	2.33	2.27	2.21	2.16
	.01	7.56	5.39	4.51	4.02	3.70	3.47	3.30	3.17	3.07	2.98
	.001	13.29	8.77	7.05	6.12	5.53	5.12	4.82	4.58	4.39	4.24
40	.10	2.84	2.44	2.23	2.09	2.00	1.93	1.87	1.83	1.79	1.76
	.05	4.08	3.23	2.84	2.61	2.45	2.34	2.25	2.18	2.12	2.08
	.01	7.31	5.18	4.31	3.83	3.51	3.29	3.12	2.99	2.89	2.80
	.001	12.61	8.25	6.59	5.70	5.13	4.73	4.44	4.21	4.02	3.87
60	.10	2.79	2.39	2.18	2.04	1.95	1.87	1.82	1.77	1.74	1.71
	.05	4.00	3.15	2.76	2.53	2.37	2.25	2.17	2.10	2.04	1.99
	.01	7.08	4.98	4.13	3.65	3.34	3.12	2.95	2.82	2.72	2.63
	.001	11.97	7.77	6.17	5.31	4.76	4.37	4.09	3.86	3.69	3.54
90	.10	2.76	2.36	2.15	2.01	1.91	1.84	1.78	1.74	1.70	1.67
	.05	3.95	3.10	2.71	2.47	2.32	2.20	2.11	2.04	1.99	1.94
	.01	6.93	4.85	4.01	3.53	3.23	3.01	2.84	2.72	2.61	2.52
	.001	11.57	7.47	5.91	5.06	4.53	4.15	3.87	3.65	3.48	3.34
120	.10	2.75	2.35	2.13	1.99	1.90	1.82	1.77	1.72	1.68	1.65
	.05	3.92	3.07	2.68	2.45	2.29	2.18	2.09	2.02	1.96	1.91
	.01	6.85	4.79	3.95	3.48	3.17	2.96	2.79	2.66	2.56	2.47
	.001	11.38	7.32	5.78	4.95	4.42	4.04	3.77	3.55	3.38	3.24
240	.10	2.73	2.32	2.10	1.97	1.87	1.80	1.74	1.70	1.65	1.63
	.05	3.88	3.03	2.64	2.41	2.25	2.14	2.04	1.98	1.92	1.87
	.01	6.74	4.69	3.86	3.40	3.09	2.88	2.71	2.59	2.48	2.40
	.001	11.10	7.11	5.60	4.78	4.25	3.89	3.62	3.41	3.24	3.09
∞	.10	2.71	2.30	2.08	1.94	1.85	1.77	1.72	1.67	1.63	1.60
	.05	3.84	3.00	2.60	2.37	2.21	2.10	2.01	1.94	1.88	1.83
	.01	6.63	4.61	3.78	3.32	3.02	2.80	2.64	2.51	2.41	2.32
	.001	10.83	6.91	5.42	4.62	4.10	3.74	3.47	3.27	3.10	2.96

Table VIII

Critical values q for the Studentized range distribution

Error df	Confidence level	\multicolumn{8}{c}{Number of populations, treatments, or levels being compared}							
		3	4	5	6	7	8	9	10
5	95%	4.60	5.22	5.67	6.03	6.33	6.58	6.80	6.99
	99%	6.98	7.80	8.42	8.91	9.32	9.67	9.97	10.24
6	95%	4.34	4.90	5.30	5.63	5.90	6.12	6.32	6.49
	99%	6.33	7.03	7.56	7.97	8.32	8.61	8.87	9.10
7	95%	4.16	4.68	5.06	5.36	5.61	5.82	6.00	6.16
	99%	5.92	6.54	7.01	7.37	7.68	7.94	8.17	8.37
8	95%	4.04	4.53	4.89	5.17	5.40	5.60	5.77	5.92
	99%	5.64	6.20	6.62	6.96	7.24	7.47	7.68	7.86
9	95%	3.95	4.41	4.76	5.02	5.24	5.43	5.59	5.74
	99%	5.43	5.96	6.35	6.66	6.91	7.13	7.33	7.49
10	95%	3.88	4.33	4.65	4.91	5.12	5.30	5.46	5.60
	99%	5.27	5.77	6.14	6.43	6.67	6.87	7.05	7.21
11	95%	3.82	4.26	4.57	4.82	5.03	5.20	5.35	5.49
	99%	5.15	5.62	5.97	6.25	6.48	6.67	6.84	6.99
12	95%	3.77	4.20	4.51	4.75	4.95	5.12	5.27	5.39
	99%	5.05	5.50	5.84	6.10	6.32	6.51	6.67	6.81
13	95%	3.73	4.15	4.45	4.69	4.88	5.05	5.19	5.32
	99%	4.96	5.40	5.73	5.98	6.19	6.37	6.53	6.67
14	95%	3.70	4.11	4.41	4.64	4.83	4.99	5.13	5.25
	99%	4.89	5.32	5.63	5.88	6.08	6.26	6.41	6.54
15	95%	3.67	4.08	4.37	4.59	4.78	4.94	5.08	5.20
	99%	4.84	5.25	5.56	5.80	5.99	6.16	6.31	6.44
16	95%	3.65	4.05	4.33	4.56	4.74	4.90	5.03	5.15
	99%	4.79	5.19	5.49	5.72	5.92	6.08	6.22	6.35
17	95%	3.63	4.02	4.30	4.52	4.70	4.86	4.99	5.11
	99%	4.74	5.14	5.43	5.66	5.85	6.01	6.15	6.27
18	95%	3.61	4.00	4.28	4.49	4.67	4.82	4.96	5.07
	99%	4.70	5.09	5.38	5.60	5.79	5.94	6.08	6.20
19	95%	3.59	3.98	4.25	4.47	4.65	4.79	4.92	5.04
	99%	4.67	5.05	5.33	5.55	5.73	5.89	6.02	6.14
20	95%	3.58	3.96	4.23	4.45	4.62	4.77	4.90	5.01
	99%	4.64	5.02	5.29	5.51	5.69	5.84	5.97	6.09
24	95%	3.53	3.90	4.17	4.37	4.54	4.68	4.81	4.92
	99%	4.55	4.91	5.17	5.37	5.54	5.69	5.81	5.92
30	95%	3.49	3.85	4.10	4.30	4.46	4.60	4.72	4.82
	99%	4.45	4.80	5.05	5.24	5.40	5.54	5.65	5.76
40	95%	3.44	3.79	4.04	4.23	4.39	4.52	4.63	4.73
	99%	4.37	4.70	4.93	5.11	5.26	5.39	5.50	5.60
60	95%	3.40	3.74	3.98	4.16	4.31	4.44	4.55	4.65
	99%	4.28	4.59	4.82	4.99	5.13	5.25	5.36	5.45
120	95%	3.36	3.68	3.92	4.10	4.24	4.36	4.47	4.56
	99%	4.20	4.50	4.71	4.87	5.01	5.12	5.21	5.30
∞	95%	3.31	3.63	3.86	4.03	4.17	4.29	4.39	4.47
	99%	4.12	4.40	4.60	4.76	4.88	4.99	5.08	5.16

Table IX Upper-tail Areas for Chi-squared Distributions 823

Table IX

Upper-tail areas for chi-squared distributions

Right-tail area	df = 1	df = 2	df = 3	df = 4	df = 5
> .100	< 2.70	< 4.60	< 6.25	< 7.77	< 9.23
0.100	2.70	4.60	6.25	7.77	9.23
0.095	2.78	4.70	6.36	7.90	9.37
0.090	2.87	4.81	6.49	8.04	9.52
0.085	2.96	4.93	6.62	8.18	9.67
0.080	3.06	5.05	6.75	8.33	9.83
0.075	3.17	5.18	6.90	8.49	10.00
0.070	3.28	5.31	7.06	8.66	10.19
0.065	3.40	5.46	7.22	8.84	10.38
0.060	3.53	5.62	7.40	9.04	10.59
0.055	3.68	5.80	7.60	9.25	10.82
0.050	3.84	5.99	7.81	9.48	11.07
0.045	4.01	6.20	8.04	9.74	11.34
0.040	4.21	6.43	8.31	10.02	11.64
0.035	4.44	6.70	8.60	10.34	11.98
0.030	4.70	7.01	8.94	10.71	12.37
0.025	5.02	7.37	9.34	11.14	12.83
0.020	5.41	7.82	9.83	11.66	13.38
0.015	5.91	8.39	10.46	12.33	14.09
0.010	6.63	9.21	11.34	13.27	15.08
0.005	7.87	10.59	12.83	14.86	16.74
0.001	10.82	13.81	16.26	18.46	20.51
<0.001	>10.82	>13.81	>16.26	>18.46	>20.51

Right-tail area	df = 6	df = 7	df = 8	df = 9	df = 10
> .100	<10.64	<12.01	<13.36	<14.68	<15.98
0.100	10.64	12.01	13.36	14.68	15.98
0.095	10.79	12.17	13.52	14.85	16.16
0.090	10.94	12.33	13.69	15.03	16.35
0.085	11.11	12.50	13.87	15.22	16.54
0.080	11.28	12.69	14.06	15.42	16.75
0.075	11.46	12.88	14.26	15.63	16.97
0.070	11.65	13.08	14.48	15.85	17.20
0.065	11.86	13.30	14.71	16.09	17.44
0.060	12.08	13.53	14.95	16.34	17.71
0.055	12.33	13.79	15.22	16.62	17.99
0.050	12.59	14.06	15.50	16.91	18.30
0.045	12.87	14.36	15.82	17.24	18.64
0.040	13.19	14.70	16.17	17.60	19.02
0.035	13.55	15.07	16.56	18.01	19.44
0.030	13.96	15.50	17.01	18.47	19.92
0.025	14.44	16.01	17.53	19.02	20.48
0.020	15.03	16.62	18.16	19.67	21.16
0.015	15.77	17.39	18.97	20.51	22.02
0.010	16.81	18.47	20.09	21.66	23.20
0.005	18.54	20.27	21.95	23.58	25.18
0.001	22.45	24.32	26.12	27.87	29.58
<0.001	>22.45	>24.32	>26.12	>27.87	>29.58

(continued)

Table IX

Upper-tail areas for chi-squared distributions (continued)

Right-tail area	df = 11	df = 12	df = 13	df = 14	df = 15
> .100	<17.27	<18.54	<19.81	<21.06	<22.30
0.100	17.27	18.54	19.81	21.06	22.30
0.095	17.45	18.74	20.00	21.26	22.51
0.090	17.65	18.93	20.21	21.47	22.73
0.085	17.85	19.14	20.42	21.69	22.95
0.080	18.06	19.36	20.65	21.93	23.19
0.075	18.29	19.60	20.89	22.17	23.45
0.070	18.53	19.84	21.15	22.44	23.72
0.065	18.78	20.11	21.42	22.71	24.00
0.060	19.06	20.39	21.71	23.01	24.31
0.055	19.35	20.69	22.02	23.33	24.63
0.050	19.67	21.02	22.36	23.68	24.99
0.045	20.02	21.38	22.73	24.06	25.38
0.040	20.41	21.78	23.14	24.48	25.81
0.035	20.84	22.23	23.60	24.95	26.29
0.030	21.34	22.74	24.12	25.49	26.84
0.025	21.92	23.33	24.73	26.11	27.48
0.020	22.61	24.05	25.47	26.87	28.25
0.015	23.50	24.96	26.40	27.82	29.23
0.010	24.72	26.21	27.68	29.14	30.57
0.005	26.75	28.29	29.81	31.31	32.80
0.001	31.26	32.90	34.52	36.12	37.69
<0.001	>31.26	>32.90	>34.52	>36.12	>37.69

Right-tail area	df = 16	df = 17	df = 18	df = 19	df = 20
> .100	<23.54	<24.77	<25.98	<27.20	<28.41
0.100	23.54	24.76	25.98	27.20	28.41
0.095	23.75	24.98	26.21	27.43	28.64
0.090	23.97	25.21	26.44	27.66	28.88
0.085	24.21	25.45	26.68	27.91	29.14
0.080	24.45	25.70	26.94	28.18	29.40
0.075	24.71	25.97	27.21	28.45	29.69
0.070	24.99	26.25	27.50	28.75	29.99
0.065	25.28	26.55	27.81	29.06	30.30
0.060	25.59	26.87	28.13	29.39	30.64
0.055	25.93	27.21	28.48	29.75	31.01
0.050	26.29	27.58	28.86	30.14	31.41
0.045	26.69	27.99	29.28	30.56	31.84
0.040	27.13	28.44	29.74	31.03	32.32
0.035	27.62	28.94	30.25	31.56	32.85
0.030	28.19	29.52	30.84	32.15	33.46
0.025	28.84	30.19	31.52	32.85	34.16
0.020	29.63	30.99	32.34	33.68	35.01
0.015	30.62	32.01	33.38	34.74	36.09
0.010	32.00	33.40	34.80	36.19	37.56
0.005	34.26	35.71	37.15	38.58	39.99
0.001	39.25	40.78	42.31	43.81	45.31
<0.001	>39.25	>40.78	>42.31	>43.81	>45.31

Table X Binomial Probabilities 825

Table X *Binomial probabilities*

n = 5

							π						
x	0.05	0.1	0.2	0.25	0.3	0.4	0.5	0.6	0.7	0.75	0.8	0.9	0.95
0	.774	.590	.328	.237	.168	.078	.031	.010	.002	.001	.000	.000	.000
1	.203	.329	.409	.396	.360	.259	.157	.077	.029	.015	.007	.000	.000
2	.022	.072	.205	.263	.309	.346	.312	.230	.132	.088	.051	.009	.001
3	.001	.009	.051	.088	.132	.230	.312	.346	.309	.263	.205	.072	.022
4	.000	.000	.007	.015	.029	.077	.157	.259	.360	.396	.409	.329	.203
5	.000	.000	.000	.001	.002	.010	.031	.078	.168	.237	.328	.590	.774

n = 10

							π						
x	0.05	0.1	0.2	0.25	0.3	0.4	0.5	0.6	0.7	0.75	0.8	0.9	0.95
0	.599	.349	.107	.056	.028	.006	.001	.000	.000	.000	.000	.000	.000
1	.315	.387	.268	.188	.121	.040	.010	.002	.000	.000	.000	.000	.000
2	.075	.194	.302	.282	.233	.121	.044	.011	.001	.000	.000	.000	.000
3	.010	.057	.201	.250	.267	.215	.117	.042	.009	.003	.001	.000	.000
4	.001	.011	.088	.146	.200	.251	.205	.111	.037	.016	.006	.000	.000
5	.000	.001	.026	.058	.103	.201	.246	.201	.103	.058	.026	.001	.000
6	.000	.000	.006	.016	.037	.111	.205	.251	.200	.146	.088	.011	.001
7	.000	.000	.001	.003	.009	.042	.117	.215	.267	.250	.201	.057	.010
8	.000	.000	.000	.000	.001	.011	.044	.121	.233	.282	.302	.194	.075
9	.000	.000	.000	.000	.000	.002	.010	.040	.121	.188	.268	.387	.315
10	.000	.000	.000	.000	.000	.000	.001	.006	.028	.056	.107	.349	.599

(*continued*)

Table X *Binomial probabilities (continued)*

$n = 15$

x	0.05	0.1	0.2	0.25	0.3	0.4	0.5	0.6	0.7	0.75	0.8	0.9	0.95
0	.463	.206	.035	.013	.005	.000	.000	.000	.000	.000	.000	.000	.000
1	.366	.343	.132	.067	.030	.005	.000	.000	.000	.000	.000	.000	.000
2	.135	.267	.231	.156	.092	.022	.004	.000	.000	.000	.000	.000	.000
3	.031	.128	.250	.225	.170	.064	.014	.002	.000	.000	.000	.000	.000
4	.004	.043	.188	.225	.218	.126	.041	.007	.001	.000	.000	.000	.000
5	.001	.011	.103	.166	.207	.196	.092	.025	.003	.001	.000	.000	.000
6	.000	.002	.043	.091	.147	.207	.153	.061	.011	.003	.001	.000	.000
7	.000	.000	.014	.040	.081	.177	.196	.118	.035	.013	.003	.000	.000
8	.000	.000	.003	.013	.035	.118	.196	.177	.081	.040	.014	.000	.000
9	.000	.000	.001	.003	.011	.061	.153	.207	.147	.091	.043	.002	.000
10	.000	.000	.000	.001	.003	.025	.092	.196	.207	.166	.103	.011	.001
11	.000	.000	.000	.000	.001	.007	.041	.126	.218	.225	.188	.043	.004
12	.000	.000	.000	.000	.000	.002	.014	.064	.170	.225	.250	.128	.031
13	.000	.000	.000	.000	.000	.000	.004	.022	.092	.156	.231	.267	.135
14	.000	.000	.000	.000	.000	.000	.000	.005	.030	.067	.132	.343	.366
15	.000	.000	.000	.000	.000	.000	.000	.000	.005	.013	.035	.206	.463

Table X Binomial Probabilities   827

Table X *Binomial probabilities (continued)*

$n = 20$

x	0.05	0.1	0.2	0.25	0.3	0.4	π 0.5	0.6	0.7	0.75	0.8	0.9	0.95
0	.358	.122	.012	.003	.001	.000	.000	.000	.000	.000	.000	.000	.000
1	.377	.270	.058	.021	.007	.000	.000	.000	.000	.000	.000	.000	.000
2	.189	.285	.137	.067	.028	.003	.000	.000	.000	.000	.000	.000	.000
3	.060	.190	.205	.134	.072	.012	.001	.000	.000	.000	.000	.000	.000
4	.013	.090	.218	.190	.130	.035	.005	.000	.000	.000	.000	.000	.000
5	.002	.032	.175	.202	.179	.075	.015	.001	.000	.000	.000	.000	.000
6	.000	.009	.109	.169	.192	.124	.037	.005	.000	.000	.000	.000	.000
7	.000	.002	.055	.112	.164	.166	.074	.015	.001	.000	.000	.000	.000
8	.000	.000	.022	.061	.114	.180	.120	.035	.004	.001	.000	.000	.000
9	.000	.000	.007	.027	.065	.160	.160	.071	.012	.003	.000	.000	.000
10	.000	.000	.002	.010	.031	.117	.176	.117	.031	.010	.002	.000	.000
11	.000	.000	.000	.003	.012	.071	.160	.160	.065	.027	.007	.000	.000
12	.000	.000	.000	.001	.004	.035	.120	.180	.114	.061	.022	.000	.000
13	.000	.000	.000	.000	.001	.015	.074	.166	.164	.112	.055	.002	.000
14	.000	.000	.000	.000	.000	.005	.037	.124	.192	.169	.109	.009	.000
15	.000	.000	.000	.000	.000	.001	.015	.075	.179	.202	.175	.032	.002
16	.000	.000	.000	.000	.000	.000	.005	.035	.130	.190	.218	.090	.013
17	.000	.000	.000	.000	.000	.000	.001	.012	.072	.134	.205	.190	.060
18	.000	.000	.000	.000	.000	.000	.000	.003	.028	.067	.137	.285	.189
19	.000	.000	.000	.000	.000	.000	.000	.000	.007	.021	.058	.270	.377
20	.000	.000	.000	.000	.000	.000	.000	.000	.001	.003	.012	.122	.358

(continued)

Table X *Binomial probabilities (continued)*

n = 25

x	0.05	0.1	0.2	0.25	0.3	0.4	0.5	0.6	0.7	0.75	0.8	0.9	0.95
0	.277	.072	.004	.001	.000	.000	.000	.000	.000	.000	.000	.000	.000
1	.365	.199	.023	.006	.002	.000	.000	.000	.000	.000	.000	.000	.000
2	.231	.266	.071	.025	.007	.000	.000	.000	.000	.000	.000	.000	.000
3	.093	.227	.136	.064	.024	.002	.000	.000	.000	.000	.000	.000	.000
4	.027	.138	.187	.118	.057	.007	.000	.000	.000	.000	.000	.000	.000
5	.006	.065	.196	.164	.103	.020	.002	.000	.000	.000	.000	.000	.000
6	.001	.024	.163	.183	.148	.045	.005	.000	.000	.000	.000	.000	.000
7	.000	.007	.111	.166	.171	.080	.015	.001	.000	.000	.000	.000	.000
8	.000	.002	.062	.124	.165	.120	.032	.003	.000	.000	.000	.000	.000
9	.000	.000	.030	.078	.134	.151	.061	.009	.000	.000	.000	.000	.000
10	.000	.000	.011	.042	.091	.161	.097	.021	.002	.000	.000	.000	.000
11	.000	.000	.004	.019	.054	.146	.133	.044	.004	.001	.000	.000	.000
12	.000	.000	.002	.007	.027	.114	.155	.076	.011	.002	.000	.000	.000
13	.000	.000	.000	.002	.011	.076	.155	.114	.027	.007	.002	.000	.000
14	.000	.000	.000	.001	.004	.044	.133	.146	.054	.019	.004	.000	.000
15	.000	.000	.000	.000	.002	.021	.097	.161	.091	.042	.011	.000	.000
16	.000	.000	.000	.000	.000	.009	.061	.151	.134	.078	.030	.000	.000
17	.000	.000	.000	.000	.000	.003	.032	.120	.165	.124	.062	.002	.000
18	.000	.000	.000	.000	.000	.001	.015	.080	.171	.166	.111	.007	.000
19	.000	.000	.000	.000	.000	.000	.005	.045	.148	.183	.163	.024	.001
20	.000	.000	.000	.000	.000	.000	.002	.020	.103	.164	.196	.065	.006
21	.000	.000	.000	.000	.000	.000	.000	.007	.057	.118	.187	.138	.027
22	.000	.000	.000	.000	.000	.000	.000	.002	.024	.064	.136	.227	.093
23	.000	.000	.000	.000	.000	.000	.000	.000	.007	.025	.071	.266	.231
24	.000	.000	.000	.000	.000	.000	.000	.000	.002	.006	.023	.199	.365
25	.000	.000	.000	.000	.000	.000	.000	.000	.000	.001	.004	.072	.277

Answers to Selected Odd-Numbered Exercises

CHAPTER 1

1.1 Descriptive statistics is made up of those methods whose purpose is to organize and summarize the values in a data set. Inferential statistics refers to those procedures whose purpose is to allow us to draw a conclusion about a population based on information from a sample.   **1.3** The population is the entire student body (the 15,000) students. The sample consists of the 200 students interviewed.   **1.5** The population consists of all single-family homes in Northridge. The sample consists of the 100 homes selected for inspection.   **1.7** The population consists of all 5000 bricks in the lot. The sample consists of the 100 bricks selected for inspection.

2.1 a. numerical (discrete)   **b.** categorical   **c.** numerical (continuous)   **d.** numerical (continuous)   **e.** categorical   **2.3 a.** discrete   **b.** continuous   **c.** discrete   **d.** discrete   **e.** continuous   **f.** continuous   **g.** continuous   **h.** discrete   **2.5** observational study   **2.7** Cause-and-effect conclusions can't be drawn from this study. Seat belt usage is a possible confounding factor.   **2.9** No, it is an observational study. Cause-and-effect conclusions cannot be made on the basis of an observational study.   **2.13 a.** Stratifying would be worthwhile because cost of books may vary by class standing.   **b.** Stratifying would be worthwhile because cost of books may vary by field of study.   **c.** Stratifying would not be worthwhile because it is unlikely that the alphabetic groups would be any more homogeneous than the population as a whole.   **2.15** Researcher B's plan would result in a simple random sample of trees and would avoid possible confounding factors related to location in the field.   **2.17 a.** The population consists of all 16–24-year-olds who live in New York state, but not in New York City. **2.19** The sample consisted of volunteers, included only women, included only students from a private university, and included only students attending school in Texas.   **2.21** selection bias **2.23 a.** strength of binding   **b.** type of glue   **c.** Number of pages in the book and whether the book is paperback or hardback. Other factors might include type of paper (rough or glossy), proportion of paper made with recycled paper, and whether there are any tear-out pages in the book.   **2.25 a.** Age, weight, lean body mass, and capacity to lift weights were all dealt with by direct control.   **b.** Yes. Knowing whether they were receiving creatine might have influenced the effort a subject put into his workout.   **c.** Yes, especially if the trainer is the one taking the measurements. If the trainer knows which men are receiving creatine, it might influence the way in which he works with the subjects.   **2.29** Gender is a useful blocking variable if men and women of the same status tend to differ with respect to "rate of talk." Blocking on gender would make it easier to assess the effect of status on rate of talk.   **2.31** It is unlikely that a person who is described as a moderate drinker would be unable to tell whether they were drinking vodka tonics or virgin tonics.   **2.35** Stratifying by room type would probably be a good strategy.

3.1 b. A bar chart is better in this case because there are 7 categories, which is on the high side for a pie chart.   **3.3** The percent with medical career aspirations has been rising for women, but for men it peaked in 1980 and then declined in 1985 and 1990.   **3.5** The bar chart shows

the percent spent on health care increasing over time at a fairly steady rate.   **3.9 b.** .4759
3.15 a. 2.2   **b.** In the row with the stem of 8. The leaf of 9 would be placed to the right of the
other leaves.   **c.** A large number of flow rates are between 6.0 and 8.0; a typical flow rate
would be 6.9 or 7.0.   **d.** There appears to be a lot of variability in the flow rates.   **e.** The
distribution is not symmetric.   **f.** 18.9 appears to be an outlier.

3.17

1H	8
2L	024
2H	98756978585587787 9667585
3L	14
3H	6
4L	
4H	79
5L	4

Stems: Ones
Leaves: Tenths

3.19

64	64, 70, 35, 33
65	26, 83, 06, 27
66	14, 05, 94
67	70, 70, 90, 00, 98, 45, 13
68	50, 73, 70, 90
69	36, 27, 00, 04
70	05, 40, 22, 11, 51, 50
71	31, 69, 68, 05, 65, 13
72	09, 80

Stems: Hundreds
Leaves: Ones

3.21 a. Relative frequencies: .4167, .1792, .1500, .0708, .1000, .0375, .0458   **b.** .7459   **c.** .2541
d. .1833   **e.** The frequencies (and relative frequencies) tend to decrease as the number of
impairments increases.   **3.23 b.** The histogram is positively skewed.   **c.** .016   **d.** between
5 and 10 (inclusive): .636; strictly between 5 and 10: .459   **e.** .033   **3.25 b.** .34, .54   **c.** .60
d. No. There are 8 observations in the interval 40–<50, but we don't know how many of these
are equal to 40.   **3.27 d.** .8652   **e.** .5506   **f.** about 20.46 months   **g.** about 29.21 months
3.29 b. .0603   **c.** .2603   **3.31** Almost all of the differences are positive, indicating that the
runners slow down. The histogram is positively skewed. A typical difference is about 150. About
2% of the runners ran the late distance more quickly than the early distance.   **3.33 a.** .362,
.638   **b.** .894, .830   **3.35 b.** estimated proportion = .34, actual proportion = .3378
c. from density histogram, approximately .09; from frequency distribution, approximately
.1036   **3.37 I:** symmetric;   **II:** positively skewed;   **III:** bimodal and approximately symmetric;   **IV:** bimodal;   **V:** negatively skewed   **3.41** The histogram is positively skewed.
3.43 b. .0908   **c.** .2382   **3.45 c.** Both histograms are positively skewed.

3.49 a and c.

Class Interval	Frequency	Relative Frequency	Cum. Rel. Freq.
0–<.5	5	.1064	.1064
.5–<1.0	9	.1915	.2979
1.0–<1.5	10	.2128	.5107
1.5–<2.0	9	.1915	.7022
2.0–<2.5	3	.0638	.7660
2.5–<3.0	5	.1064	.8724
3.0–<3.5	1	.0213	.8937
3.5–<4.0	3	.0638	.9575
4.0–<4.5	1	.0213	.9788
4.5–<5.0	1	.0213	1.0001

b. The histogram is positively skewed.   **d. i.** .2979   **ii.** .4043   **iii.** .2978
3.51 b. .3648, .9177

CHAPTER 4

4.1 $\bar{x}$ = 11.61, median = 10.05; The mean is larger because of the two large values in the data
set.   **4.3** 122.57

4.5 a.

32	55
33	49
34	
35	6699
36	34469
37	03345
38	9
39	2347
40	23
41	
42	4

Since the distribution is approximately symmetric, the mean and median should be similar in value.   **b.** $\bar{x} = 370.69$, median $= 369.50$ **c.** The largest value could be increased without limit without affecting the value of the median. It could be decreased to 370 without affecting the value of the median.   **4.7 a.** Because whether total personal income is judged as high or low depends on how many people are in the area.   **b.** There are not as many people in the area.   **c.** Approximately 1,620,500   **4.9** The statement could be correct if there were a small group of residents with very high wages.   **4.11** Average for diet 1 is larger than that for diet 2.   **4.13** median $= 4286$, $\bar{x} = 3968.67$ **4.15** median $= 68$; 20% trimmed mean $= 66.17$ **4.17** median $= 11$; the median is not influenced by the unusually large value of 33.   **4.19** $s^2 = 1.70$, $s = 1.3038$ **4.23** The variance and standard deviation are not affected by adding or subtracting a number from each observation.   **4.25** $s = 20.86$ **4.27** iqr/1.35 is less than the value of s. This suggests that a histogram for this data would be heavy-tailed compared to a normal curve.   **4.29 a.** There are no mild or extreme outliers.   **4.31 a.** lower quartile $= 7.82$, upper quartile $= 8.01$, iqr $= .19$ **b.** The boxplot indicates that the distribution is not symmetric, with a longer tail on the upper end.   **4.35 a.** 40 is one standard deviation above the mean and 30 is one standard deviation below the mean. 45 is two standard deviations above the mean and 25 is two standard deviations below the mean.   **b.** at least 75%   **c.** at most 11%   **d.** approximately 95% between 25 and 45; approximately .3% less than 20 or greater than 50; approximately .15% less than 20   **4.37 a.** at least 75%   **b.** (2.90, 70.94)   **c.** If the distribution of NO_2 values was normal, approximately 2.5% of the values would be less than -9.64 (two standard deviations below the mean). Since NO_2 concentration can't be negative, the distribution shape must not resemble a normal curve.   **4.39** The z score is larger for the second exam, so performance was better on the second exam.   **4.41 a.** approximately 68%   **b.** approximately 5%   **c.** approximately 13.5%   **d.** because the histogram is well approximated by a normal curve   **4.43** The proportion is at most .16.   **4.45** Because the number of answers changed from right to wrong cannot be negative and one standard deviation below the mean is a negative number, the distribution can't be well described by a normal curve. From Chebyshev's Rule, the proportion who changed at least 6 answers from right to wrong is at most .106.   **4.47 b. i.** 13.158   **ii.** 16.25   **iii.** 6.923   **iv.** 21   **v.** 11.842   **4.49 b.** For the cancer group: $\bar{x} = 22.81$, median $= 16$; for the no-cancer group: $\bar{x} = 19.15$, median $= 12$.   **c.** For the cancer group: $s = 31.6553$; for the no-cancer group: $s = 16.99$. There is more variability in the cancer sample values than in the no-cancer sample values.   **4.51** It implies that the median salary is less than $1 million and that the histogram would be positively skewed or that there are outliers with very large values.   **4.53 b.** The mean and median are not the same because the distribution is not symmetric.   **c.** In the interval (0, 45.71). Note: 0 is used for the lower endpoint because score cannot be negative.   **4.55** Various graphical and numerical methods can be used for comparison. In general, they suggest that milk volume for nonsmoking mothers is greater than that for smoking mothers.   **4.59** Since the mean is larger than the median, this suggests that the distribution of values is positively skewed or has some outliers with very large values.   **4.61 a.** $\bar{x} = 22.15$, $s = 11.366$ **b.** 10% trimmed mean $= 19.4375$ **c.** lower quartile $= 18$, upper quartile $= 20.5$, iqr $= 2.5$ **d.** 25 and 28 are mild outliers, and 69 is an extreme outlier **4.63 a.** If one observation is deleted from each end, the trimming percent is 6.67%. If two observations are deleted from each end, the trimming percent is 13.33%. **b.** 6.67% trimmed mean = 10.66; 13.33% trimmed mean = 10.58 **c.** Averaging the two trimmed means from part **b** gives 10.62. This is a reasonable approach since 10% is halfway between 6.67% and 13.33%.

CHAPTER 5

5.1 a. There is a tendency for y to increase as x does (a positive relationship). **b.** There are several observations that have identical x values but different y values. Thus, the value of y is not determined solely by x. **5.3 a.** Both boxplots have two outliers on the high side. Whereas the minimum and lower quartiles are similar for x and y, the median and upper quartile are higher for x. **b.** The plot shows a linear pattern. The two points that stand out in the plot look like they conform to the linear pattern seen in the rest of the data. **5.7 a.** The graft weight ratio tends to decrease as recipient body weight increases, but the relationship does not look linear. **5.9 a.** positive **b.** negative **c.** positive **d.** no correlation **e.** positive correlation **f.** positive **g.** negative **h.** no correlation **5.13 a.** There appears to be a strong positive correlation. **b.** $r = .9366$ indicates a strong positive linear relationship. **c.** If x and y had a correlation coefficient of 1, all of the points would lie on a straight line, but the line would not necessarily have slope 1 and intercept 0. **5.15 a.** $r = .39$ **5.17** No. An r value of $-.085$ indicates at best a weak relationship between support for environmental spending and degree of belief. **5.19** In this case, r cannot be negative. **5.21** $r_s = .039$, which is very close to zero. **5.23** $r_s = .9504$ **5.27 a.** $\hat{y} = .365 + .967x$ **b.** $\hat{y} = .8485$ **5.29 a.** slope = 244.9, intercept = -275.1 **b.** 244.9 **c.** 214.7 **d.** No, the predicted value for height = 1 is negative, which is not a possible value of breaking strength. **5.31 a.** $\hat{y} = 31.040 - 5.792x$ **b.** -5.792 (a decrease in percent crown dieback) **c.** 7.872 **d.** No, because 5.6 is outside the range of the data. **5.33** The relationship observed over the range of the data may not continue to hold outside this range. **5.37 a.** 76.64% of the variation in clutch size can be explained by the linear relationship with snout-vent length. **b.** $s_e = 29.25$ **5.39 a.** $\hat{y} = 914.5$, residual = -21.5 **5.41 a.** Predicted values: 40.4048, 42.6245, 49.2837, 53.7231, 54.8330, 56.4978, 58.1626, 58.7175, 62.6020, 68.1513; Residuals: -17.4048, 9.3755, 15.7163, 1.2769, -22.8330, 3.5022, 19.8374, .2825, -1.6020, -8.1513 **b.** SSResid = 1635.6833, $r^2 = .2859$ **5.43 a.** $\hat{y} = 94.33 - 15.388x$ **b.** SSTo = 5534.91667, SSResid = 285.82261 **c.** $r^2 = .94836$ **d.** $s_e = 5.34624$ **e.** Since the slope is negative, $r = -\sqrt{r^2} = -.97384$ **5.45** $r^2 = .9512$. This means that 95.12% of the variability in hardness can be explained by the linear relationship with elapsed time. **5.47** There is a curved pattern in the plot, suggesting that the linear model may not be appropriate. **5.51 a.** When $r = 0$, $s_e = s_y$. The least squares line is then a horizontal line with slope = 0 and intercept $\bar{y}$. **b.** When r is close to 1, s_e will be much smaller than s_y. **c.** $s_e = 1.5$ **5.53 b.** The plot of $\log(y)$ versus $\log(x)$ looks the straightest. The line for predicting $y' = \log(y)$ from $x' = \log(x)$ is $\hat{y}' = 1.61867 - .31646x'$. When $x = 25$, the predicted y is 15.0064. **5.55 a.** $r = -.717$ **b.** $r = -.835$ for the transformed data **5.57 a.** The plot does show a positive relationship. **b.** The plot appears straighter. **c.** The plot also appears straighter, and the points conform more closely to a line. **d.** The plot shows curvature in the opposite direction as the plot in part **a**. **5.59** The plot suggests fitting a quadratic curve to the data. **5.61 a.** The plot does not show a strong linear pattern. **b.** $\hat{y} = 32.254$ **d.** With the outlier deleted, $\hat{y} = 36.4175 - .1978x$, $r^2 = .027$. There is little evidence of a linear relationship. **5.63 a.** $\hat{y} = 13.9617 + 3.1703x$ **b.** The scatter plot shows some curvature, suggesting it might be possible to do better with a nonlinear model. **c.** 34.5687 **d.** No, because $x = 18$ is outside the range of the data. **15.65 a.** $r = 0$ **b.** For example, $y = 1$ (Any y value greater than .973 will work.) **c.** For example, $y = -1$ (Any y value less than $-.973$ will work.) **15.67 b.** $a = 5.550$, $b = 2.1975$ **c.** $r = .5815$ **d.** $r_s = .625$ **15.69 a.** $r = -.981$. This value of r suggests a strong negative linear relationship. **b.** There is a slight curvature to the plot.

CHAPTER 6

6.3 a. Sample space = {*AA, AM, MA, MM*}   **c.** *B* = {*MA, AM, AA*}; *C* = {*AM, MA*}; *D* = {*MM*}; *D* is a simple event   **d.** *B and C* = {*AM, MA*}; *B or C* = {*MA, AM, AA*}
6.5 b. A^c = {(2, 0), (2, 1), (2, 2), (2, 3), (2, 4)}; $A \cup B$ = {(0, 0), (0, 1), (0, 2), (0, 3), (0, 4), (1, 0), (1, 1), (1, 2), (1, 3), (1, 4), (2, 0), (2, 1)}; $A \cap B$ = {(0, 0), (0, 1), (1, 0), (1, 1)}   **c.** *A* and *C* are not disjoint. *B* and *C* are disjoint.   **6.7 b.** *A* = {3, 4, 5}   **c.** *C* = {125, 15, 215, 25, 5}
6.9 a. *A* = {*NN, DNN, NDN*}   **b.** B = {*DDNN, NDDN, DNDN*}   **c.** There are an infinite number of possible outcomes.   **6.13 a.** Sample space = {0, 1, 2, . . . , 999, 1000}   **b.** *E* = {0, 1, 2, 3, 4, 5}; *G* = {6, 7, 8, 9, 10}; *F* = {11, 12, 13, . . . , 19, 20}   **c.** Good: more than 5 but fewer than 11 errors; Fair: more than 10 but fewer than 21 errors.   **6.15 a.** .45   **b.** .70   **c.** .75
6.17 a. .0119   **b.** .00000238   **c.** .012   **6.19 a.** .07   **b.** .30   **c.** .57   **6.21** The statement is not true because the two events are not disjoint.   **6.23 b.** .25   **c.** .3333   **d.** 0   **e.** .2917   **6.25 a.** The ten simple events are (*B, C*), (*B, M*), (*B, P*), (*B, S*), (*C, M*), (*C, P*), (*C, S*), (*M, P*), (*M, S*), (*P, S*)   **b.** .1   **c.** .4   **d.** .3   **6.27 a.** .1   **b.** .3   **c.** .7   **d.** .6
6.29 a. .9   **b.** .771   **6.33 a.** .77, .31   **b.** .871   **c.** .169, the conditional probability is smaller   **6.35 a. i.** .7418   **ii.** .8594   **6.37** They are dependent.   **6.39** They are dependent.   **6.41 a.** .001   **b.** .999   **c.** .009   **6.43 a.** .7   **b.** .5   **c.** .714286   **d.** no   **e.** They are independent.   **6.45 a. i.** .51   **ii.** .56   **iii.** .45   **iv.** .36   **v.** .72   **b.** No, because $P(F) \neq P(F \mid C)$.   **c.** No, because $P(F) \neq P(F \mid O)$.   **6.47 a.** .336   **b.** .56   **c.** .06   **d.** .180

6.49

.12	.20	.08
.18	.30	.12

6.51 a. *CC, CN, NC, NN*   **b.** .00391   **c.** .00383   **6.53 a.** 6/10   **b.** 5/9   **c.** 1/3
6.55 a. .60   **b.** .15   **6.57 a.** No, because $P(A \cap B)$ is not known and the events are not disjoint.   **b.** No, $P(A \cap B)$ cannot be larger than $P(A)$.   **c.** .6, .4   **6.59** .08
6.61 a. .081   **b.** .08325   **c.** .97297   **d.** .19   **6.63** .286   **6.65** .818   **6.67 b.** .025   **c.** .069   **6.69 a.** .018   **b.** .444, .278, .278   **6.71 b.** .27   **c.** .43   **d.** not independent   **e.** .5263   **f.** .1395   **6.73 a.** .4247   **b.** .3139   **c.** .4375   **d.** .0052   **e.** .2014   **f.** .1962   **g.** .3350   **h.** .1774   **i.** .4876   **j.** .2954   **k.** .5659   **6.75 a.** Results will vary from one simulation to another, but values should be around .26.   **6.77 a.** Results will vary from one simulation to another, but values should be around .85.   **6.79 a.** .00243   **b.** .00275   **6.83 a.** .4   **b.** .75   **c.** .25   **6.85** .49   **6.87** R_1 and R_2 are not independent, since $P(R_2 \mid R_1) = 1139/2516 = .4527$, which is not equal to $P(R_2) = .4529$. However, these two probabilities do not differ much, and from a practical point of view they can be regarded as independent.   **6.89 a.** $P(A_1) = .2, P(A_2 \mid A_1) = .143$, and $P(A_1 \cap A_2) = .0286$ **b.** .324
6.91 a. .8   **b.** .7917   **c.** .7826   **d.** .4957   **6.93 a.** .48   **b.** .244   **c.** .024   **d.** .144

CHAPTER 7

7.1 a. discrete   **b.** continuous   **c.** discrete   **d.** discrete   **e.** continuous   **7.3** Possible *y* values are the positive integers.   **7.5** *y* is a continuous variable with possible values in the interval 0 to 100.   **7.7 a.** 3, 4, 5, 6, 7   **b.** −3, −2, −1, 1, 2, 3   **c.** 0, 1, 2   **d.** 0, 1
7.9 a. .01   **b.** About 20% of the cartons have exactly one broken egg.   **c.** .95   **d.** .85, $P(x < 2)$ is smaller than $P(x \leq 2)$ because *x* is discrete and the latter probability includes $P(x = 2)$.   **e.** .10   **f.** .95   **7.11 a.** .82   **b.** .18   **c.** .65, .27

7.15 a.

x	0	1	2	3	4
p(*x*)	.4096	.4096	.1536	.0256	.0016

b. 0 and 1 both have probability .4096   **c.** .1808   **7.17 a.** The smallest *y* value is 1, which results from the outcome *S*. The second smallest *y* value is 2, which results from the outcome *FS*.

b. the positive integers: 1, 2, 3, . . .

c.

y	1	2	3	4	5
$p(y)$	.7	$(.3)(.7)$	$(.3)^2(.7)$	$(.3)^3(.7)$	$(.3)^4(.7)$

In general, $p(y) = (.30)^{y-1}(.7)$ for $y = 1, 2, 3, \ldots$

7.19

y	0	1	2	3
$p(y)$	.16	.33	.32	.19

7.21 a. .5   **b.** .2   **c.** $\mu = 5$ **7.23 b.** .08   **c.** .36   **d.** .40, .40, they are equal because the associated areas are the same   **7.25 a.** .5, .25   **b.** .25   **c.** 18   **7.27 a.** .375, .21875, .34375   **b.** .4012   **7.29 a.** $\mu_y = .56$ **b.** .65   **c.** Because the probability distribution indicates that there are more 0's and 1's than 2's, 3's, or 4's   **7.31** $\sigma_x^2 = 2.52, \sigma_x = 1.5875$ **7.33 a.** 4.12   **b.** 1.9456, 1.3948   **c.** .72   **d.** .07   **7.35** Under the royalty plan, the mean amount received would be $10,550. Since this exceeds $10,000, a case could be made for choosing the royalty plan. On the other hand, with the royalty plan the probability that the amount received exceeds $10,000 is only .25, whereas the probability that the amount received is less than $10,000 is .35, so a case could also be made for taking the flat payment. The choice would depend on how confident the author was.   **7.37 a.** discrete; there are only 6 possible values   **b.** .4, .86   **c.** 126.76, 13.45   **7.39** 88.5, 20.25   **7.41 a.** 2.8, 1.66   **b.** .7, .61   **c.** 8.4, 14.94   **d.** 7, 61   **e.** 3.5, 2.27   **f.** 15.4, 75.94   **7.43 a.** .0486   **b.** .6561   **c.** .3439   **7.45 a.** .302   **b.** .322   **c.** .966

7.47

x	0	1	2	3	4	5
$p(x)$	.03125	.15625	.3125	.3125	.15625	.03125

7.49 If the graphologist is guessing, $P(x \geq 6) = .377$.   **7.51 a.** .722   **b.** .484   **c.** 17.5, 2.2913   **d.** .618   **7.53** Since the sampling is done without replacement and the sample size is more than 5% of the population size, the number of invalid signatures cannot be considered to have a binomial distribution.   **7.55 a.** .017   **b.** .811, .425   **c.** The error probability in part **a** decreases, whereas the error probability in part **b** increases.   **7.57 a.** geometric   **b.** .09   **c.** .729   **7.59 a.** .9599   **b.** .2483   **c.** .1151   **d.** .9976   **e.** .6887   **f.** .6826   **g.** 1   **7.61 a.** .9909   **b.** .9909   **c.** .1093   **d.** .1267   **e.** .0706   **f.** .0228   **g.** .9996   **h.** 1   **7.63 a.** -1.96 **b.** -2.33 **c.** -1.645 **d.** 2.05   **e.** 2.33   **f.** 1.28   **7.65 a.** 1.96   **b.** 1.645   **c.** 2.33   **d.** 1.75   **7.67 a.** .5   **b.** .9772   **c.** .9772   **d.** .8185   **e.** .9938   **f.** 1   **7.69** .3174   **7.71 a.** .0228   **b.** .8400   **c.** .25   **7.73** .0730 or 7.3%   **7.75 a.** .1587   **b.** 51.4 minutes   **c.** 41.65 minutes   **7.77** There is curvature in the plot, indicating that the cadmium concentration distribution is not normal.   **7.79** The plot exhibits a linear pattern, so it is reasonable to think that the bearing load life distribution is approximately normal.   **7.81** The plot exhibits a linear pattern, so it is reasonable to think that the disk diameter distribution is approximately normal.   **7.83 b.** The histogram is positively skewed.   **c.** A transformation could result in a more symmetric distribution.   **7.85 a.** The histogram is positively skewed.   **b.** The histogram of the square-root transformed data is still positively skewed.   **7.87** The transformation of the Mn data has been successful, resulting in a much more symmetric distribution. The transformation was not as successful with the Cu and the Zn data.   **7.89 a.** .0016   **b.** .9946   **c.** .7925   **7.91 a.** .1114   **b.** .0409   **c.** .0968   **d.** .9429   **e.** .9001   **7.93 a.** .7960   **b.** .7016   **c.** .0143   **d.** .0500   **7.95 a.** Since $n\pi < 10$, it would not be appropriate to use the normal approximation to the binomial distribution in this case.   **b.** .8708   **7.97 a.** Since more than 5% of the population is sampled and sampling is without replacement, the binomial distribution should not be used.   **b.** 27.5, 4.4651   **c.** No, it increases by a factor of $\sqrt{2}$.   **7.99 a.** .70   **b.** .45   **c.** .55   **d.** .71   **e.** .65   **f.** .45   **7.101 a.** .64   **b.** .256   **c.** The fifth battery must be the second acceptable battery selected; $p(5) = .02048$ **d.** $p(y) = (y - 1)(.2)^{y-2}(.8)^2$ for $y = 2, 3, 4, \ldots$

7.103 a. .5934   **b.** yes, probability ≈ 0 **c.** .000107   **d.** $c = 28.026$ **7.105 a.** .9876
b. .98884   **c.** $c = 6.608$ **7.107 a.** .6826   **b.** .5   **c.** .0013   **d.** 5.28   **e.** .3174
7.109 a.

y	0	1	2	3	4
$p(y)$	1/16	7/16	5/16	2/16	1/16

$\mu_y = 1.6875$

b.

y	0	1	2	3	4
$p(y)$	.0256	.3264	.3456	.1728	.1296

$\mu_y = 2.0544$

c.

z	1	2	3	4
$p(z)$	2/16	4/16	8/16	2/16

$\mu_z = 2.375$

7.111

x	1	2	3
$p(x)$	1/2	1/3	1/6

7.113

y	0	1	2	3
$p(y)$	.1552	.2688	.29952	.27648

7.115 yes, your score was in the top 5.82%   **7.117** Bulbs should be replaced after about
658 hours.   **7.119 a.** .8247   **b.** .0516   **c.** .6826   **d.** .0030   **e.** $P(x < 261) = .3783$

CHAPTER 8

8.1 A statistic is computed from the observations in a sample, whereas a population characteristic is a quantity that describes the whole population.   **8.3 a.** population characteristic
b. statistic   **c.** population characteristic   **d.** statistic   **e.** statistic
8.7 a.

t value	6	11	15	21	25	30
Probability	1/6	1/6	1/6	1/6	1/6	1/6

b. $\mu = 9, \mu_t = 18$
8.9

Statistic #1:	Value	2.67	3.00	3.33	3.67
	Probability	.1	.4	.3	.2
Statistic #2:	Value	3	4		
	Probability	.7	.3		
Statistic #3:	Value	2.5	3.0	3.5	
	Probability	.1	.5	.4	

8.11 a. 11.8   **8.13** Histograms for samples of size 10 would be centered in the same place
but would be less spread out than for samples of size 5.   **8.15** for $n = 36, 50, 100,$ and 400
8.17 a. $\mu_{\bar{x}} = 40, \sigma_{\bar{x}} = .625$, approximately normal   **b.** .5762   **c.** .2628   **8.19 a.** $\mu_{\bar{x}} = 2$,
$\sigma_{\bar{x}} = .267$ **b.** for $n = 20, \mu_{\bar{x}} = 2, \sigma_{\bar{x}} = .179$ **c.** for $n = 100, \mu_{\bar{x}} = 2, \sigma_{\bar{x}} = .08$
8.21 a. .8185, .0013   **b.** .9772, 0   **8.23** .0026   **8.25** approximately 0   **8.27 a.** .65,

.15083   **b.** .65, .10665   **c.** .65, .08708   **d.** .65, .06745   **e.** .65, .04770   **f.** .65, .03373
8.29 a. $\mu_p = .005$, $\sigma_p = .007$ **b.** Since $n\pi < 10$, the sampling distribution of p is not well approximated by a normal curve.   **c.** n would have to be at least 2000   **8.31 a.** for $\pi = .5$: $\mu_p = .5$, $\sigma_p = .0333$; for $\pi = .6$: $\mu_p = .6$, $\sigma_p = .0327$; in both cases the sampling distribution of p is approximately normal.   **b.** .0013 (for $\pi = .5$), .5000 (for $\pi = .6$)   **c.** The probability when $\pi = .5$ would decrease and the probability when $\pi = .6$ would remain the same.   **8.33 a.** .9744
b. approximately 0   **8.35 a.** The sampling distribution of $\bar{x}$ will be approximately normal with mean 50 and standard deviation .1.   **b.** .9876   **c.** .5   **8.37 a.** $\mu_{\bar{x}} = 52$, $\sigma_{\bar{x}} = .33$
b. approximately 1, approximately 0   **8.39 a.** .0207   **b.** The probability of observing such a sample is approximately 0. It is very unlikely that $\pi = .4$.

CHAPTER 9

9.1 Statistic II would be preferred because it is unbiased and has a smaller standard deviation.
9.3 .252   **9.5** .262   **9.7 a.** $s = 1.886$ **b.** sample median $= 11.35$ **c.** Trimming one observation from each side gives 8.3% trimmed mean $= 11.73$ **d.** 14.381   **9.9 a.** 120.6
b. 1,206,000   **c.** .8   **d.** sample median $= 120$ **9.11 a.** 1.96   **b.** 1.645   **c.** 2.58
d. 1.28   **e.** 1.44   **9.13 a.** As the confidence level increases, the width of the interval increases.   **b.** As the sample size increases, the width of the interval decreases.   **c.** As the value of p gets farther from .5, the width of the interval decreases.   **9.15** (.382, .695)
9.19 (.586, .714), yes   **9.21 a.** (.669, .773)   **b.** (.240, .318)   **c.** narrower
9.23 a. (.4140, .4656)   **b.** The entire interval is below .5. It is not plausible that a majority of U.S. adults feel that it makes no difference which party is in control.   **9.25** $n = 2401$
9.27 $n = 385$ **9.29 a.** 90%   **b.** 95%   **c.** 95%   **d.** 99%   **e.** 1%   **f.** .5%
g. 5%   **9.31 a.** $\bar{x} = 115$ **b.** (114.4, 115.6) is the 90% interval and (114.1, 115.9) is the 99% interval   **9.33 a.** narrower, since the z critical value would have been smaller   **b.** The statement is incorrect. The 95% refers to the percentage of all possible samples that would result in an interval that contains μ, not to the chance that any specific interval contains μ.   **c.** Although we would expect 95% of such intervals to include μ *in the long run,* we would not expect to see *exactly* 95 of any specific 100 intervals.   **9.35 a.** (.409, .591)   **b.** No. It does imply that at least some students lie to their mothers, but the mean could be positive even when there are some 0's in the population.   **9.37 a.** LBO firms: $(-.1866, 1.4294)$; non-LBO firms: $(.5435, 1.3573)$
b. the sample size is larger   **9.39** (1.79, 2.61)   **9.41** (19.51, 24.29)   **9.43** (5.77, 28.23)
9.45 No. The sample size is small and there are two outliers in the data set, indicating that the assumption of a normal population distribution is not reasonable.   **9.47** $n = 246$
9.51 (9.976, 11.824)   **9.53 a.** (199.407, 216.193)   **b.** Since the sample size is small, the population distribution must be normal.   **9.55** (.241, .319)

CHAPTER 10

10.1 not a legitimate hypothesis since $\bar{x}$ is not a *population* characteristic   **10.3** Choose the first pair of hypotheses, since this places the burden of proof on the welding contractor to show that the welds meet specifications.   **10.5** $H_0: \pi = .6$, $H_a: \pi > .6$ **10.7** $H_0: \mu = 170$, $H_a: \mu < 170$ **10.9** $H_0: \mu = 40$, $H_a: \mu \neq 40$ **10.11 a.** They failed to reject the null hypothesis because their conclusion of no evidence of increased risk is consistent with H_0.   **b.** Type II, since a type I error results only when a true null hypothesis is rejected.   **c.** Since the null hypothesis is initially assumed to be true and is rejected only if there is strong evidence against it, if we fail to reject the null hypothesis we have not "proven" it to be true. We can say only that there is not strong evidence against it.   **10.13 a.** Pizza Hut's decision is consistent with rejecting the null hypothesis.   **b.** Rejecting H_0 when it is true is a type I error, so if they incorrectly reject H_0, they are making a type I error.   **10.15 a.** A type I error is returning to the supplier a shipment that is not of inferior quality. A type II error is accepting a shipment of inferior quality.
b. The calculator manufacturer would most likely consider a type II error more serious, because

they would then end up producing defective calculators.   **c.** From the supplier's point of view, a type I error is more serious, because the supplier would end up having a good shipment returned. **10.17 a.** The manufacturer claims that the percentage defective is 10%. Since we would probably not object if the proportion defective is less than 10%, we would want to determine whether the proportion defective exceeds the value claimed by the manufacturer.   **b.** A type I error is concluding that the proportion defective exceeds 10% when, in fact, it does not. The consequence of this decision would be the filing of false advertising charges against a manufacturer who is not guilty of false advertising. A type II error is concluding that there is no evidence that the proportion defective is greater than 10% when, in fact, it is. The consequence of this type of error is that the manufacturer would be permitted to continue with a false advertisement. **10.19 a.** $H_0: \pi = .02$, $H_a: \pi < .02$ **b.** A type I error is changing to robots when in fact they are not superior to humans. A type II error is not changing to robots when in fact they are superior to humans.   **c.** Since a type I error would result in a substantial monetary loss to the company and also the loss of jobs for former employees, a small α should be used.   **10.21** .001, .021, and .047   **10.23 a.** .0808   **b.** .1762   **c.** .0250   **d.** .0071   **e.** .5675   **10.25** $z = 4.78$, P-value ≈ 0, reject H_0 **10.27** $z = -8.24$, P-value ≈ 0, reject H_0 **10.29 a.** $z = -13.91$, P-value ≈ 0, reject H_0 **b.** A type I error is concluding that there are not enough valid signatures, when in fact there were. If a type I error were to occur, the recall would be tossed out when it should have been placed on the ballot.   **c.** A type II error would be concluding that there were enough valid signatures when in fact there were not. If a type II error were to occur, the recall would be placed on the ballot when it should have been tossed out.   **d.** Type I   **e.** Yes, the test indicates that there is convincing evidence that the number of valid signatures is too small to qualify for the ballot.   **10.31** $z = 1.78$, P-value $= .0375$, fail to reject H_0 **10.33** $z = .75$, P-value $= .2266$, fail to reject H_0 **10.35** $z = 3.20$, P-value $= .0007$, reject H_0 **10.37** $z = -9.35$, P-value ≈ 0, reject H_0 **10.39 a.** $z = 3.00$ **b.** .0014   **c.** reject H_0 **10.41 a.** .040   **b.** .003   **c.** .019   **d.** 0   **e.** .13   **f.** .13   **g.** 0   **10.43 a.** reject H_0 **b.** reject H_0 **c.** fail to reject H_0 **10.45 a.** P-value $= .136$, fail to reject H_0 **b.** P-value $= .136$, fail to reject H_0 **c.** P-value $= .016$, fail to reject H_0 **d.** P-value $= .002$, reject H_0 for any $\alpha > .002$ **10.47** $t = -3.89$, P-value $\approx .000$, reject H_0 **10.49** $t \approx -14.75$, P-value ≈ 0, reject H_0 **10.51** $t = -5.05$, P-value ≈ 0, reject H_0 **10.53 a.** Since the boxplot is nearly symmetric with no outliers and the normal probability plot is reasonably straight, the t test is appropriate.   **c.** $t = 1.21$, P-value $\approx .256$, fail to reject H_0 **10.55 a.** $t = 1.875$, P-value $\approx .041$, reject H_0 **b.** yes   **10.57 a.** The sample is randomly selected, the sample size is large, and σ is known.   **b.** A type I error is concluding that the water discharged is too hot when in fact it is not. A type II error is concluding that the water is not too hot, when in fact it is too hot.   **c.** .0359   **e.** .3745   **f.** approximately 0   **g.** fail to reject H_0, type II error   **10.59 a.** $z = 1.71$, P-value $= .0436$, reject H_0, yes   **b.** .3520   **10.61 a. i.** $\beta \approx .85$ **ii.** $\beta \approx .55$ **iii.** $\beta \approx .10$ **iv.** $\beta \approx 0$ **b.** β increases   **10.63 a.** *   **b.** †   **c.** none   **d.** ‡   **10.65** $z = 2.25$, P-value $= .0122$, reject H_0 at $\alpha = .05$, but not at $\alpha = .01$ **10.67 a.** If the distribution were normal, the Empirical Rule implies that about 16% of the values would be less than the value that is one standard deviation below the mean. Since the standard deviation is larger than the mean, this value is negative. Caffeine consumption can't be negative, so it is not plausible that the distribution is normal. However, since the sample size is large ($n = 47$), the t test can still be used.   **b.** $t = .44$, P-value $\approx .346$, fail to reject H_0 **10.69** $z = 4.19$, P-value ≈ 0, reject H_0 **10.71** $t = 2.99$, P-value $\approx .002$, reject H_0 **10.73** $t = -2.87$, P-value $\approx .005$, reject H_0

CHAPTER 11   **11.1** approximately normal with mean 5 and standard deviation .529   **11.3 a.** $H_0: \mu_1 - \mu_2 = 10$; $H_a: \mu_1 - \mu_2 > 10$ **b.** $H_0: \mu_1 - \mu_2 = -10$; $H_a: \mu_1 - \mu_2 < -10$ **11.5** for small prey: $t = 1.27$, df $= 7$, P-value $\approx .23$, fail to reject H_0; for medium prey: $t = 1.83$, df $= 10$, P-value $\approx .102$, fail to reject H_0; for large prey: $t = -18.08$, df $= 12$, P-value ≈ 0, reject H_0 **11.7 a.** yes   **b.** yes   **c.** no, would need to know the sample means and standard deviations   **11.9 a.** $t = -6.56$, df $= 201$, P-value ≈ 0, reject H_0 **b.** $t = 6.25$, df $= 201$, P-value ≈ 0, reject H_0 **c.** $t = .07$, df $= 201$, P-value $\approx .92$, fail to reject H_0 **11.11 a.** $t = 1.77$, df $= 37$, P-value $\approx .08$, fail to reject H_0 **b.** $t = -1.28$, df $= 29$, P-value $\approx .204$, fail to reject H_0 **c.** $t = -7.1$, df $= 332$, P-value ≈ 0, reject H_0 **11.13 a.** $H_0: \mu = 100,000$ would be rejected in favor of $H_a: \mu > 100,000$ only for $\alpha > .212$ **b.** $t = 2.72$, df $= 473$, P-value $\approx .003$, reject H_0 **11.15** $t = 3.1$, df $= 113$, P-value $\approx .001$, reject H_0 **11.17 a.** $t = 2.93$, df $= 54$, P-value $\approx .006$, reject H_0 **b.** $t = .15$, df $= 68$, P-value $\approx .842$, fail to reject H_0 **c.** df $= 65$, $(.242, 1.458)$ **11.19** $t = 113.2$, df $= 45$, P-value ≈ 0, reject H_0 **11.21** $t = 3.30$, df $= 98$, P-value $\approx .001$, reject H_0 **11.23** $t = 2.97$, df $= 10$, P-value $= .007$, reject H_0 **11.25** $t = 2.94$, df $= 26$, P-value $\approx .004$, reject H_0 **11.27** $t = 2.43$, df $= 32$, P-value $\approx .011$, fail to reject H_0 **11.29** df $= 23$, $(-100.22, 668.22)$ **11.31 a.** $s_p^2 = .00004$, $t = 9.86$, df $= 24$, P-value ≈ 0, reject H_0 **b.** $t = 9.98$, df $= 22$, P-value ≈ 0, conclusion is the same as for the test in part **a** **11.35 b.** $(-1.03, -.63)$ **11.37** $t = 9.91$, P-value ≈ 0, reject H_0 **11.39** $t = -1.34$, P-value $\approx .206$, fail to reject H_0 **11.41 a.** $t = 3.95$, P-value ≈ 0, reject H_0 **11.43** $t = 0.047$, P-value $\approx .5$, fail to reject H_0 **11.45** $z = -1.67$, P-value $= .048$, reject H_0 **11.47** $z = -4.12$, P-value ≈ 0, reject H_0 **11.49** $z = -4.09$, P-value ≈ 0, reject H_0 **11.51** $z = -1.84$, P-value $= .066$, fail to reject H_0 **11.53** $z = 6.21$, P-value ≈ 0, reject H_0 **11.55** $(-.2053, .0777)$ **11.57** $(-.2743, -.0823)$, the interval does not include 0. The proportion with decayed teeth is lower for children drinking fluoridated water by somewhere between .0823 and .2743. **11.59** rank sum $= 83$, P-value $> .05$, fail to reject H_0 **11.61** rank sum $= 48$, P-value $> .05$, fail to reject H_0 **11.63** rank sum $= 53$, P-value $> .05$, fail to reject H_0 **11.67** $t = 5.03$, df $= 100$, P-value ≈ 0, reject H_0 **11.69 a.** $t = -1.44$, df $= 108$, P-value $\approx .082$, fail to reject H_0 **b.** $t = -2.85$, df $= 112$, P-value $\approx .002$, reject H_0 **c.** $t = .145$, df $= 96$, P-value $\approx .920$, fail to reject H_0 **d.** $t = -1.72$, df $= 106$, P-value $\approx .092$, fail to reject H_0 **11.71** $t = 2.55$, df $= 77$, P-value $\approx .012$, reject H_0 at $\alpha = .05$, but not at $\alpha = .01$ **11.73** $t = .504$, df $= 288$, P-value $\approx .618$, fail to reject H_0 **11.75 a.** $t = -3.86$, df $= 3$, P-value $\approx .032$, reject H_0 **b.** Since the sample sizes are small, the t test requires that the distribution of number of flashes for both single-peak and multiple-peak storms be normal. **11.77** $t = 2.13$, df $= 21$, P-value $\approx .024$, reject H_0 **11.79** $t = 3.13$, df $= 35$, P-value $\approx .002$, reject H_0 **11.81 a.** $t = 2.02$, df $= 17$, P-value $\approx .062$, fail to reject H_0 **b.** $t = 1.10$, df $= 22$, P-value $\approx .284$, fail to reject H_0 **c.** $(1161.83, 1838.17)$ **d.** $(1027.19, 1886.81)$ **11.83** No, because the large-sample z test for comparing two proportions requires that the samples be independent. **11.85 a.** $(-.186, .111)$ **b.** It must be reasonable to assume that the distribution of differences is normal. **11.87** df $= 7$, $(-47.21, 75.01)$; because 0 is in the interval, there may be no difference between the true mean age at death for male and female SIDS victims. **11.89** $t = 3.59$, df $= 6$, P-value $\approx .006$, reject H_0

CHAPTER 12

12.1 a. $.020 < P\text{-value} < .025$ **b.** $.040 < P\text{-value} < .045$ **c.** $.035 < P\text{-value} < .040$ **d.** $P\text{-value} < .001$ **e.** $P\text{-value} > .10$ **12.3 a.** $X^2 = 19.0$, P-value $< .001$, reject H_0 **b.** If $n = 40$, the chi-squared test should not be used, since one of the expected counts would be less than 5. **12.5** $X^2 = 11.242$, $.010 < P\text{-value} < .015$, fail to reject H_0 at $\alpha = .01$ **12.7 a.** $X^2 = 8216.5$, P-value $< .001$, reject H_0 **c.** $X^2 = 10.748$, P-value $> .100$, fail to reject H_0 **12.9** $X^2 = 253.482$, P-value $< .001$, reject H_0 **12.11** $X^2 = .931$, P-value $> .100$, fail to reject H_0 **12.13** $X^2 = 1.47$, P-value $> .100$, fail to reject H_0 **12.15 a.** df $= 10$ **b.** df $= 12$ **c.** df $= 15$ **12.17** $X^2 = 8.6837$, $.065 < P\text{-value} < .070$, fail to reject H_0 **12.19 a.** $X^2 = 6.852$, P-value $> .100$, fail to reject H_0 **b.** yes **12.21** $X^2 = 73.789$, P-value $< .001$, reject H_0 **12.23** $X^2 = 6.621$, P-value $\approx .085$ **12.25** $X^2 = 10.976$, P-value $< .001$, reject H_0 **12.27** $X^2 = 22.855$, P-value $< .001$, reject H_0 **12.29** $X^2 = 1.978$, P-value $> .100$, fail to reject H_0 **12.31** $X^2 = 22.373$, P-value $< .001$, reject H_0 **12.33** $X^2 = 9.844$, $.005 < P\text{-value} < .010$, reject H_0 **12.35** $X^2 = 8.89$, $.001 < P\text{-value} < .005$, reject H_0 **12.37** $X^2 = 6.54$, $.035 < P\text{-value} < .040$, reject H_0 **12.39** $X^2 = 103.87$, P-value $< .001$, reject H_0 **12.41** $X^2 = 75.35$, P-value $< .001$, reject H_0; note that there are two cells with expected counts less than 5. If the last two rows of the table are combined, $X^2 = 73.03$, P-value $< .001$, reject H_0

CHAPTER 13

13.1 a. $y = -5.0 + .017x$ **c.** 30.7   **d.** .017   **e.** 1.7   **f.** No, the model should not be used to predict outside the range of the data.   **13.3 a.** 3.6, 4.3   **b.** .1210   **c.** .5, .1587   **13.5 a.** 47, 4700   **b.** .3156, .0643   **13.7 a.** $r^2 = .6228$ **b.** $s_e = 1.5923$ **c.** $b = 3.4307$ **d.** 3.4426   **13.9 a.** $r^2 = .883$ **b.** $s_e = 13.682$, df $= 14$ **13.11 b.** $\hat{y} = -.002274 + 1.2467x$, predicted value when $x = .09$ is .1099   **c.** $r^2 = .436$. This means that 43.6% of the total variability in market share can be explained by the linear relationship with advertising share.   **d.** $s_e = .0263$, df $= 8$ **13.13 a.** .253   **b.** .179, no   **c.** It would require 4 observations at each x value.   **13.15 a.** $s_e = 9.7486$, $s_b = .1537$ **b.** (2.17, 2.83)   **c.** Yes, because the interval is narrow   **13.17 a.** $a = 592.1$, $b = 97.26$ **b.** $\hat{y} = 786.62$, residual $= -29.62$ **c.** (87.76, 106.76)   **13.19 a.** $t = 2.8302$, P-value $= .008$, reject H_0 and conclude that the model is useful   **b.** (4.294, 25.706); based on this interval, we estimate that change in mean SAT score associated with an increase on \$1000 in expenditure per child is between 4.294 and 25.706   **13.21 a.** $\hat{y} = -96.67 + 1.595x$ **b.** $t = 27.17$, P-value ≈ 0, reject H_0 and conclude that the model is useful   **c.** The mean response time for individuals with closed-head injury is 1.48 times as great as for those with no head injury.   **13.23 a.** $t = 6.53$, P-value ≈ 0, reject H_0 and conclude that the model is useful   **b.** $t = 1.56$, P-value $= .079$, fail to reject H_0 **13.25** $t = -17.57$, P-value ≈ 0, reject H_0 **13.27** The standardized residual plot does not exhibit any unusual features.   **13.29 b.** The plot is reasonably straight. The assumption of normality is plausible.   **c.** There are two residuals that are unusually large.   **13.31 a.** There are several large residuals and potentially influential observations.   **13.33** The residuals are positive in 1963 and 1964, and then they are negative from 1965 through 1972. This pattern in the residual plot casts doubt on the appropriateness of the simple linear regression model.   **13.35** If the request is for a confidence interval for β, the wording will be something like "estimate the *change* in the average y value," whereas if the request is for a confidence interval for $\alpha + \beta x$, the wording would be something like "estimate the *average y value*."   **13.37 a.** $(-.4788, -.2113)$ **b.** (17.78, 21.78)   **13.39 a.** (6.4722, 6.6300)   **b.** (6.4779, 6.6973)   **c.** Yes, because $x^* = 60$ is farther from $\bar{x}$.   **13.41 a.** $\hat{y} = -133.02 + 5.919x$ **b.** $s_b = 1.127$ **c.** $t = 5.25$, P-value ≈ 0, reject H_0 **d.** 251.74   **e.** Shouldn't use regression equation to predict for $x = 105$ because it is outside the range of the data.   **13.43 a.** $\hat{y} = 2.78551 + .04462x$ **b.** $t = 10.85$, P-value ≈ 0, reject H_0 and conclude that the model is useful   **c.** (3.671, 4.577)   **d.** Since 4.1 is in the interval, it is plausible that a box that has been on the shelf 30 days will not be acceptable.   **13.45 a.** $\hat{y} = 30.019 + .1109x$ **b.** (38.354, 43.864)   **c.** (30.311, 51.907)   **d.** The interval in part **b** is for the mean blood lead level for all people who work where the air lead level is 100. The interval in part **c** is for a particular individual who works in a place where the air lead level is 100.   **13.47 a.** prediction interval when $x = 35$: (1.9519, 3.0931); prediction interval when $x = 45$: (1.3879, 2.5271)   **b.** The simultaneous confidence level would be at least 97%.   **13.51** $t = 2.07$, P-value $\approx .036$, reject H_0 **13.53 a.** $t = -6.175$, P-value ≈ 0, reject H_0 **b.** no, $r^2 = .0676$ **13.55** $r = .574$, $t = 1.85$, P-value $\approx .1$, fail to reject H_0 **13.57** $t = 2.2$, P-value $= .028$, reject H_0 **13.61 a.** (4.903, 5.632)   **b.** (4.4078, 6.1272)   **13.63 a.** From the MINITAB output, $t = -3.95$, P-value $= .003$, reject H_0 and conclude that the model is useful   **b.** $(-3.6694, -.9976)$ **c.** (60.276, 70.646)   **d.** because $x = 11$ is farther from $\bar{x}$ than is $x = 10$ **13.65 a.** $t = -6.09$, P-value ≈ 0, reject H_0 **c.** The predicted value when $x = 10$ is negative, and y cannot be negative. $x = 10$ is probably outside of the range of the data, and the model should not be used to predict for $x = 10$.   **13.67 a.** $t = .49$, P-value $= .624$, fail to reject H_0 **b.** A 95% confidence interval is (47.083, 54.009).   **13.69** (18.0284, 22.2218)   **13.71** A linear model is not appropriate for data sets 2, 3, and 4.   **13.73 a.** $\hat{y} = 57.964 + .0357x$ **b.** $t = 0.023$, P-value ≈ 1, fail to reject H_0 **d.** The residual plot has a distinct curved pattern. The simple linear regression model is not appropriate.   **13.75 a.** $t = -0.06$, P-value $= .922$, fail to reject H_0 **b.** $t = 4.35$, P-value ≈ 0, reject H_0 **c.** $a = 3.0781$, $b = .01161$ **d.** (2.9778, 4.3392)   **e.** (3.6197, 4.1619)

CHAPTER 14

14.1 A deterministic model does not have the random error component, e, whereas a probabilistic model does contain such a component.   **14.3** $y = \alpha + \beta_1 x_1 + \beta_2 x_2 + e$. An interaction

term is not included because the statement indicates that x_1 and x_2 make independent contributions. **14.5 a.** 103.11 **b.** 96.87 **c.** $\beta_1 = -6.6$; this is the expected change in yield associated with a 1-unit increase in mean temperature when mean percentage sunshine remains fixed. $\beta_2 = -4.5$; this is the expected change in yield associated with a 1-unit increase in mean percentage sunshine when mean temperature remains fixed. **14.7 b.** The mean chlorine content is higher for $x = 10$. **c.** When degree of delignification increases from 8 to 9, the mean chlorine content increases by 7. Mean chlorine content decreases by 1 when degree of delignification increases from 9 to 10. **14.9 c.** The parallel lines in each graph are attributable to the lack of interaction between the two independent variables. **d.** Because there is an interaction term, the lines are not parallel. **14.11 a.** $y = \alpha + \beta_1 x_1 + \beta_2 x_2 + \beta_3 x_3 + e$
b. $y = \alpha + \beta_1 x_1 + \beta_2 x_2 + \beta_3 x_3 + \beta_4 x_1^2 + \beta_5 x_2^2 + \beta_6 x_3^2 + e$
c. $y = \alpha + \beta_1 x_1 + \beta_2 x_2 + \beta_3 x_3 + \beta_4 x_1 x_2 + e$; $y = \alpha + \beta_1 x_1 + \beta_2 x_2 + \beta_3 x_3 + \beta_4 x_1 x_3 + e$; $y = \alpha + \beta_1 x_1 + \beta_2 x_2 + \beta_3 x_3 + \beta_4 x_2 x_3 + e$
d. $y = \alpha + \beta_1 x_1 + \beta_2 x_2 + \beta_3 x_3 + \beta_4 x_1^2 + \beta_5 x_2^2 + \beta_6 x_3^2 + \beta_7 x_1 x_2 + \beta_8 x_1 x_3 + \beta_9 x_2 x_3 + e$
14.13 a. Three dummy variables would be needed to incorporate a nonnumerical variable with four categories. **14.15 a.** $b_3 = -.0996$ is the estimated change in mean VO_2 max associated with a 1-unit increase in 1 mile walk time when the values of the other predictor variables are fixed. **b.** $b_1 = .6566$ is the estimated difference in mean VO_2 max for males and females when the values of the other predictors are fixed. **c.** $\hat{y} = 2.8049$, residual $= .3451$ **d.** $R^2 = .706$ **e.** $s_e = .397$ **14.17 a.** $.01 < P$-value $< .05$ **b.** P-value $> .10$ **c.** P-value $= .01$ **d.** $.001 < P$-value $< .01$ **14.19** $F = 24.41$, P-value $< .001$, reject H_0 and conclude that the model is useful **14.21** P-value $< .001$, reject H_0 and conclude that the model is useful **14.23 a.** $\hat{y} = 86.65 - .12297 x_1 + 5.090 x_2 - .07092 x_3 + .001538 x_4$ **b.** $F = 64.4$, P-value $< .001$, reject H_0 and conclude that the model is useful **c.** $R^2 = .908$. This means that 90.8% of the variability in the observed tar content values can be explained by the fitted model. $s_e = 4.784$. This means that the typical distance of an observation from the corresponding mean value is 4.784. **14.25 a.** $\hat{y} = -510.9 + 3.0633 length - 1.113 age$ **b.** $F = 10.21$, $.001 < P$-value $< .01$, reject H_0 and conclude that the model is useful **14.27 a.** $\hat{y} = -859.2 + 23.72 minwidth + 225.81 maxwidth + 225.24 elongation$ **b.** Because adjusted R^2 takes the number of predictors in the model into account **c.** $F = 16.03$, P-value $< .001$, reject H_0 and conclude that the model is useful **14.29 a.** SSResid $= 390.4347$, SSTo $= 1618.2093$, SSRegr $= 1227.7746$ **b.** $R^2 = .759$. This means that 75.9% of the variability in the observed shear strength values can be explained by the fitted model. **c.** $F = 5.039$, $.01 < P$-value $< .05$, reject H_0 and conclude that the model is useful **14.31** $F = 96.64$, P-value $< .001$, reject H_0 and conclude that the model is useful **14.35** $\hat{y} = 35.8 - .68 x_1 + 1.28 x_2$, $F = 18.95$, P-value $< .001$, reject H_0 and conclude that the model is useful **14.37 a.** $(-.697, -.281)$; With 95% confidence, we can say that the change in the mean value of a vacant lot associated with a 1-unit increase in distance from the city's major east–west thoroughfare is a decrease of somewhere between .281 and .697, assuming all other predictors remain constant. **b.** $t = -.599$, P-value $\approx .550$, fail to reject H_0 **14.39 a.** $F = 144.36$, P-value $< .001$, reject H_0 and conclude that the model is useful **b.** $t = 4.33$, P-value ≈ 0, reject H_0 and conclude that the quadratic term is important **c.** $(72.0626, 72.4658)$ **14.41 a.** .469 is an estimate of the expected change in mean score associated with a 1-unit increase in the student's expected score, assuming time spent studying and GPA are fixed. **b.** $F = 75.01$, P-value $< .001$, reject H_0 and conclude that the model is useful **c.** $(2.466, 4.272)$ **d.** 72.856 **e.** $(61.562, 84.150)$ **14.43** $t = 2.22$, P-value $= .028$, reject H_0 and conclude that the interaction term is important **14.45 a.** $F = 5.47$, $.01 < P$-value $< .05$, reject H_0 and conclude that the model is useful **b.** $t = .04$, P-value $\approx .924$, fail to reject H_0 and conclude that the interaction term is not needed in the model **c.** No. The model utility test looks at the predictors as a group. The t test looks at the predictors one at a time and tests to see whether an individual predictor is useful *given* that the other predictors are included in the model. **14.49 a.** $F = 30.81$, P-value $< .001$, reject H_0 and conclude that the model is useful **b.** for β_1: $t = 6.57$, P-value $= 0$, reject H_0; for β_2: $t = -7.69$, P-value $= 0$, reject H_0 **c.** $(43.0, 47.92)$ **d.** $(44.0, 47.92)$ **14.53** The model using the three variables x_3, x_9, and x_{10} is a reasonable choice. **14.55 a.** $F \approx 86$, P-value $< .001$, reject H_0 and conclude that the model is useful **b.** The predictor added at the first step was smoking habits. At step 2, alcohol

consumption was added. No other variables were added to the model. **14.57 b.** .0567
c. return on equity, yes **d.** no, CEO tenure would be considered first **e.** P-value $\approx$ .04
14.59 Reasonable choices would be the model with predictors x_2 and x_4 or possibly using x_2, x_3,
and x_4. In either case, the set of predictors is not the same as in Exercise 14.58. **14.63 a.** $F =$
12.28, P-value $<$ .001, reject H_0 and conclude that the model is useful **b.** .5879 **c.** (.1606,
.7554) **d.** x_9 would be considered first, and it would be eliminated. **e.** H_0: $\beta_3 = \beta_4 = \beta_5 =$
$\beta_6 = 0$. None of the procedures in this chapter can be used to test hypotheses of this form.
14.65 b. $F = 21.25$, .001 $<$ P-value $<$ .01, reject H_0 and conclude that the model is useful
c. $t = 6.52$, P-value $\approx$ 0, reject H_0, the quadratic term should not be eliminated **14.67** Adjusted R^2 will be negative for values of $R^2 <$ .5. **14.69** The model with all four predictors was
fit. The variable *age at loading* was eliminated at step 2, because it had the t ratio closest to 0 and
it was between -2 and 2. The variable *time* was deleted at step 3 because it had the t ratio closest
to 0 and it was between -2 and 2. No other variables could be eliminated from the model. The
final model uses predictors *slab thickness* and *load.* **14.71 b.** The claim is reasonable because
14 is close to where the curve has its highest value. **14.73 a.** $\hat{y} = 1.56 + .0237x_1 - .000249x_2$
b. $F = 70.66$, P-value $<$ .001, reject H_0 and conclude that the model is useful **c.** $R^2 = .865$.
This means that 86.5% of the variability in observed profit margin values can be explained by the
fitted model. $s_e = .0533$. This a typical prediction error (distance of a predicted value from the
true mean value). **d.** No. Neither x_1 nor x_2 can be eliminated from the model.

CHAPTER 15

15.1 a. .001 $<$ P-value $<$ .01 **b.** P-value $>$ .10 **c.** P-value $=$ .01 **d.** P-value $<$ .001
e. .05 $<$ P-value $<$ .10 **f.** .01 $<$ P-value $<$ .05 (Using $df_1 = 4$ and $df_2 = 60$ in F table)
15.3 a. H_0: $\mu_1 = \mu_2 = \mu_3 = \mu_4$, H_a: at least two of the μ's are different **b.** .01 $<$ P-value $<$ .05,
fail to reject H_0 **c.** .01 $<$ P-value $<$ .05; conclusion would be the same **15.5 a.** 1996 and
1997 boxplots are similar, but the 1998 boxplot lies entirely to the right of those for 1996 and
1997, indicating higher prices in 1998. The boxplot for 1998 also has a longer tail on the upper
end than the other two boxplots. **b.** $F = 6.83$, P-value $\approx$ .01, reject H_0 **15.7** $F = 1.71$,
P-value $>$ .10, fail to reject H_0 **15.9** $F = 0.70$, P-value $>$ .10, fail to reject H_0 **15.11** $F = 1.895$,
.05 $<$ P-value $<$ .10, fail to reject H_0 **15.13** $F = 1.70$, P-value $>$ .10, fail to reject H_0
15.15 $F = 2.56$, .05 $<$ P-value $<$ .10, fail to reject H_0 **15.17** $F = 6.97$, .001 $<$ P-value $<$ .01,
reject H_0 **15.19** Since the intervals for $\mu_1 - \mu_2$ and $\mu_1 - \mu_3$ do not contain zero, μ_1 is judged
to be different from μ_2 and μ_3. Since the interval for $\mu_2 - \mu_3$ contains zero, μ_2 and μ_3 are not
significantly different. Statement (**iii**) is the correct choice.

15.21

Fabric	3	2	5	4	1
Mean	10.5	11.63	12.3	14.96	16.35

15.23 No differences are detected. This is in agreement with the results of Exercise 15.7.

15.25

Group	Simultaneous	Sequential	Control
Mean	$\bar{x}_3$	$\bar{x}_2$	$\bar{x}_1$

15.27 Mean water loss for 4 hours of fumigation is significantly different from the other means.
The mean water loss for 2 hours of fumigation is different from that for 16 and 0 hours but not
from 8 hours. The mean water losses for duration 16, 0, and 8 hours are not significantly different
from one another. **15.29 a.** $F = 3.30$, .01 $<$ P-value $<$ .05, reject H_0 **b.** No significant
differences are identified by the T–K method.

15.31 a.

Source	df	SS	MS	F
Treatments	2	11.7	5.85	.37
Blocks	4	113.5	28.375	
Error	8	125.6	15.7	
Total	14	250.8		

b. $F = .37$, P-value $> .10$, fail to reject H_0 **15.33** $F = 79.4$, P-value $< .001$, reject H_0
15.35 a. Other environmental factors (amount of rainfall, number of days of cloudy weather, average daily temperature, etc.) vary from year to year. Using a randomized block design helps control for variation in these other factors. **b.** $F = 15.03$, $.01 < P$-value $< .05$, reject H_0
c.

Application rate	1	2	3
Mean	138.33	152.33	156.33

15.37 $F = .868$, P-value $> .10$, fail to reject H_0 **15.39 b.** The lines are very close to parallel. The plot does not show evidence of an interaction between gender and type of odor.
15.41 a. The plot does suggest an interaction between peer group and self-esteem. **b.** The change in average response is greater for the low-self-esteem group than it is for the high-self-esteem group when changing from low to high peer group. The authors are correct in their statement. **15.43** test for interaction: $F = 2.18$, P-value $> .10$, fail to reject H_0; test for main effect of size: $F = 4.00$, $.01 < P$-value $< .05$, fail to reject H_0; test for main effect of species: $F = 4.36$, $.05 < P$-value $< .10$, fail to reject H_0 **15.45 a.** $F = 0.70$, P-value $> .10$, fail to reject H_0 **b.** $F = 20.53$, P-value $< .001$, reject H_0 **c.** $F = 2.59$, $.05 < P$-value $< .10$, fail to reject H_0 **15.47** test for interaction: $F < 1$, P-value $> .10$, fail to reject H_0; test for main effect of A: $F = 4.99$, $.01 < P$-value $< .05$, reject H_0; test for main effect of B: $F = 4.81$, $.01 < P$-value $< .05$, reject H_0 **15.49 a.** test for interaction: $F = 0.21$, P-value $> .10$, fail to reject H_0 **b.** test for main effect of race: $F = 5.57$, $.01 < P$-value $< .05$, fail to reject H_0 **c.** test for main effect of gender: $F = 1.89$, P-value $> .10$, fail to reject H_0 **15.51** test for main effect of rate: $F = 76.14$, P-value $< .001$, reject H_0; test for main effect of soil type: $F = 54.14$, P-value $< .001$, reject H_0.

Soil type	Ramina	Konini	Wainui
Mean	10.217	20.957	24.557

15.53 $F = 4.13$, P-value $< .001$, reject H_0

15.55 $F = 5.56$, $.001 < P$-value $< .01$, reject H_0

Group	Apathetic	Conservative	Erratic	Strategic
Mean	2.96	4.91	5.05	8.74

15.57 a. $F = 6.24$, $.001 < P$-value $< .01$, reject H_0
b.

Storage period	4	5	2	1
Mean	4.902	4.917	4.923	4.923

15.59 test for interaction: $F = 0.44$, P-value $> .10$, fail to reject H_0; test for main effect of oxygen: $F = 2.76$, P-value $> .10$, fail to reject H_0; test for main effect of sugar: $F = 13.29$, $.001 < P$-value $< .01$, reject H_0 **15.61** test for interaction: $F = 8.67$, $.001 < P$-value $< .01$, reject H_0; significant interaction, so no tests for main effects should be performed **15.63** $F = 0.907$, P-value $> .10$, fail to reject H_0 **15.65** $F = 3.23$, $.01 < P$-value $< .05$, fail to reject H_0

Index